IFIP Advances in Information and Communication Technology 413

IFIP – The International Federation for Information Processing

IFIP was founded in 1960 under the auspices of UNESCO, following the First World Computer Congress held in Paris the previous year. An umbrella organization for societies working in information processing, IFIP's aim is two-fold: to support information processing within its member countries and to encourage technology transfer to developing nations. As its mission statement clearly states,

> IFIP's mission is to be the leading, truly international, apolitical organization which encourages and assists in the development, exploitation and application of information technology for the benefit of all people.

IFIP is a non-profitmaking organization, run almost solely by 2500 volunteers. It operates through a number of technical committees, which organize events and publications. IFIP's events range from an international congress to local seminars, but the most important are:

- The IFIP World Computer Congress, held every second year;
- Open conferences;
- Working conferences.

The flagship event is the IFIP World Computer Congress, at which both invited and contributed papers are presented. Contributed papers are rigorously refereed and the rejection rate is high.

As with the Congress, participation in the open conferences is open to all and papers may be invited or submitted. Again, submitted papers are stringently refereed.

The working conferences are structured differently. They are usually run by a working group and attendance is small and by invitation only. Their purpose is to create an atmosphere conducive to innovation and development. Refereeing is also rigorous and papers are subjected to extensive group discussion.

Publications arising from IFIP events vary. The papers presented at the IFIP World Computer Congress and at open conferences are published as conference proceedings, while the results of the working conferences are often published as collections of selected and edited papers.

Any national society whose primary activity is about information processing may apply to become a full member of IFIP, although full membership is restricted to one society per country. Full members are entitled to vote at the annual General Assembly, National societies preferring a less committed involvement may apply for associate or corresponding membership. Associate members enjoy the same benefits as full members, but without voting rights. Corresponding members are not represented in IFIP bodies. Affiliated membership is open to non-national societies, and individual and honorary membership schemes are also offered.

Jiří Hřebíček Gerald Schimak
Miroslav Kubásek Andrea E. Rizzoli (Eds.)

Environmental Software Systems

Fostering Information Sharing

10th IFIP WG 5.11 International Symposium, ISESS 2013
Neusiedl am See, Austria, October 9-11, 2013
Proceedings

Springer

Volume Editors

Jiří Hřebíček
Miroslav Kubásek
Masaryk University
Institute of Biostatistics and Analyses
Kamenice 126/3, 62500 Brno, Czech Republic
E-mail: {hrebicek, kubasek}@iba.muni.cz

Gerald Schimak
Austrian Institute of Technology (AIT)
Bundesstrasse 60, 2444 Seibersdorf, Austria
E-mail: gerald.schimak@ait.ac.at

Andrea E. Rizzoli
Scuola Universitaria Professionale della Svizzera Italiana
Istituto Dalle Molle di Studi sull'Intelligenza Artificiale (IDSIA)
Galleria 2, 6928 Manno, Switzerland
E-mail: andrea@idsia.ch

ISSN 1868-4238 e-ISSN 1868-422X
ISBN 978-3-662-51443-6 e-ISBN 978-3-642-41151-9
DOI 10.1007/978-3-642-41151-9
Springer Heidelberg New York Dordrecht London

CR Subject Classification (1998): I.2, H.2, H.4, H.3, C.2, H.5, D.2

Typesetting: Camera-ready by author, data conversion by Scientific Publishing Services, Chennai, India

Printed on acid-free paper

Springer is part of Springer Science+Business Media (www.springer.com)

Preface

1 ISESS – The Conference

1.1 IFIP WG 5.11 and ISESS

The International Symposium on Environmental Software Systems (ISESS) is one of several overlapping forums discussing issues of Environmental Information Systems, Environmental Decision Support Systems, Environmental Software Systems, Environmental Informatics, Eco-Informatics or Enviromatics – whichever of the terms you prefer, at a certain time of day.

ISESS was founded by Gerald Schimak and Ralf Denzer in 1995, with great support from the German Informatics Society Working Group 4.6 "Computer Science for Environmental Protection" and the International Federation for Information Processing (IFIP) Working Group 5.11 "Computers and Environment".

As with many things, ISESS started over a couple of drinks (how else, for a German), when a visiting friend, David Russell from Pennsylvania State University issued an invitation to organize some sort of workshop or conference at his campus. Ralf Denzer stated: "When we prepared for the first ISESS at Penn State, it was at best a shot in the dark, although I am prepared to swear every oath that neither the Pink Panther nor Inspector Clouseau had any part of the plot."

The symposium has been held in a number of countries since: the United States, Canada, Austria, New Zealand, Switzerland, Portugal, the Czech Republic, Spain, Italy; in some years, joint sessions were held in conjunction with the biannual meeting of iEMSs (International Environmental Modelling and Software Society). ISESS is an official IFIP event which is directly supported by WG 5.11.

The WG itself was led by Giorgio Guariso (1991-1999), Ralf Denzer (1999-2005) and Dave Swayne (2005-2011). Since 2011 it has been in the hands of Gerald Schimak. Several individuals have served as vice-chairs respectively as secretaries of the WG and many members of the WG have been active supporters for a long time without holding an official position and whom we would like to express our thanks herewith.

1.2 The Why

As we can all feel in our personal and daily life the state of our environment is changing rapidly. Whatever we are going to invent, produce or market has effects or impacts on our environment or our physical health, positively or negatively. Sometimes it's like a vicious circle.

For more than two decades ISESS has brought together researchers dealing with environmental challenges and tried to provide solutions using forward-looking and leading-edge IT technology. Climate change has been in every mouth and the impacts can be seen again and most recently after the huge and unexpected flooding in Europe this summer. A flooding counted even more disastrous compared to the 100-years flooding in 2002 in Europe. This example stands as only one of the drivers to improve our knowledge on the content side as well as on the technology side to allow a better preparedness and mitigation of events.

As the challenges and problems are known ever since, the dynamics and impacts on each individual, the communication and information tools and the computational paradigms have changed more drastically than ever expected.

ISESS 2013 clues together researchers and users to overcome this manifold of technology changes and dynamics by using their best knowledge to improve for the society:

- the usage and understanding of environmental data and information;
- the openness of data: e.g. to open up data for third parties (starting up open data initiatives);
- the usability of new ICT technology: i.e. examine the usability of new technology for a broader communication of environmental, health, emergency or crises information (e.g. APPs for smartphones);
- monitoring leading edge technology for future internet applications
- obeying the outreach into other closely related domains like emergency or crises and disaster management;
- collaboration with decision makers in policy, legislation, authorities, emergency organizations.

Thus, the need for a conference like ISESS has been unbroken since 1995. We are looking forward to many follow-up workshops and conferences.

1.3 Fostering Sharing Environmental Information in EU - SEIS

The EU's Shared Environmental Information System (SEIS) sets out seven principles for better management of environmental data and information, which already guide the actions of the EU Member States, the European Commission (EC) and the European Environment Agency (EEA). The principles establish that shared environmental information should be:

- managed as close as possible to its source;
- collected once and shared with others for many purposes;
- readily available to easily fulfil reporting obligations;
- readily accessible to end-users at all levels for the design of new policies;
- accessible to enable comparisons of the environment at the appropriate geographic scale;
- fully available to the general public, to enable citizen participation;
- supported through common, free and open software standards.

The Digital Agenda for Europe (DAE)[1] initiative supports the SEIS principle aiming to generate information sharing through common, free, open source soft-

[1] http://ec.europa.eu/digital-agenda/

ware. For example, an EU legal framework for licensing open source software now exists (ISA EUPL[2]) and governments are increasingly keen to develop its use. Actions under the DAE are creating a safer and better-performing digital environment, improving access conditions and interoperability and stimulating the growth of cross-border eGovernment services. The open source and open data movement offers significant opportunities for further developing the SEIS but more open-source empowered applications ensuring compatibility across EU member states will be needed for managing and sharing environmental information.

The EC outlines six priority areas (SWD(2013) 18 final) to bring the most value added to SEIS in the short to medium term:

1. Assessing the current capacities within the EU member states to meet the necessary information and data quality criteria required at the EU and international level with the full involvement of the EEA;
2. Streamlining EU reporting requirements towards fully on-line reporting;
3. Piloting the development of Structured Implementation and Information Frameworks (SIIFs)[3] for selected thematic areas, linking these location-based information systems to EU level eReporting systems in order to make reported information usable in various contexts and at different geographic scales;
4. Improving public access to environment information while contributing to – and benefitting from – the EU Open Data strategy (COM (2011)882). The recently adopted Commission report (COM (2012)774) on the implementation of the Public Access to Environmental Information Directive 2003/4/EC outlines the improvements and the challenges related to the public access to information;
5. Improving participation of the public in the collection and dissemination of environment data and information, through further improvements under the DAE and the EU eGovernment Action Plan (COM (2010)743) of eEnvironment services and flexible on-line communication and information systems;
6. Assessing, with the assistance of the EEA, the implementation of the Infrastructure for Spatial Information in Europe Directive (INSPIRE) provides the framework for a distributed network of databases, linked by common standards and protocols to ensure compatibility and inter-operability. It already sets out much of the core architecture for SEIS, i.e. by ensuring that electronic data content and services at national and regional levels are implemented in accordance with common standards, are easily accessible, and can be combined across administrative borders.

[2] http://joinup.ec.europa.eu/software/page/eupl/introduction-eupl-licence

[3] A SIIF would aim, together with the range of SEIS initiatives, to help Member States set up transparent information systems that make this information accessible online. For example, one would be able to identify on a map abstraction points, source protection zones, treatment plants and distribution networks and have links to related information such as leakage reduction programmes.

From a longer-term point of view, we could foresee SEIS being the conduit towards a system of reporting / monitoring, (partly) driven by the reduction of administrative burden that modern ICT facilitates. It could be the structure around which the "knowledge-base" underpinning the environment is constructed. In addition, SEIS could finally provide a platform open to those beyond the policy domain – e.g. scientists, academicians, economists, researchers, and citizens – providing a neutral interface between environment policy-making and the fundamental information that should underpin it and vice versa.

1.4 What the Future Might Behold?

ISESS also wants to be a privileged outpost to spot future trends in environmental software systems and to discuss such developments among the researchers and the institutions who could put the new developments at the service of the citizens and the community. In the past few years we have witnessed the amazing growth of social networks, as of today more than one seventh of the world population has some sort of social profile on the Internet. There is a growing interest in exploring how this information infrastructure can be exploited to provide useful outcomes, in the so called "social computation" research area. Examples range from digitising ancient textbooks to collaborative translation of texts, but the lowest common denominator are the distribution of complex tasks in smaller, much easier chunks, to a huge number of people, and rewarding their efforts thanks to different strategies, from simple monetary compensation to social rewards (reputation), to entertainment (the so-called "gamification" of social computation tasks).

As researchers in environmental software systems our antennae are raised and we are looking forward to new applications of social computing to the environmental domain. The most obvious applications are in the area of "crowdsensing" (people as sensors), where qualitative and quantitative environmental data can be collected at low cost by a large number of human sensors, and also in the area of "crowdsourcing", where collective human behavior is driven by appropriate information channeled through the social networks.

We hope that ISESS will be a fertile ground for the development of innovative and effective software systems that will help us to close the loop: information from the people, and information to the people, for a better environment.

2 ISESS Framework

2.1 Predecessors of ISESS 2013

The first and second proceedings of the International Symposium on Environmental Software Systems (ISESS 1995 and ISESS 1997) were published and printed under official IFIP Chapman&Hall:

- R. Denzer, D. Russell, G. Schimak (eds.), Environmental Software Systems, Chapman & Hall, 1996, ISBN 0 412 73730 2.

- R. Denzer, D.A. Swayne, G. Schimak (eds.), Environmental Software Systems Vol. 2, Chapman & Hall, 1998, ISBN 0 412 81740 3.
- Then IFIP changed publisher and the third proceedings of ISESS 1999 was published and printed under official IFIP publisher Kluwer:
- R. Denzer, D.A. Swayne, M. Purvis, G. Schimak (eds.), Environmental Software Systems Vol. 3 - Environmental Information and Environmental Decision Support, Kluwer Academic Publishers, 2000, ISBN 0 7923 7832 6.
- The fourth and fifth proceedings of ISESS 2001 and ISESS 2003 were published and printed by organizers of symposium themselves under IFIP ISBN:
- D.A. Swayne, R. Denzer, G. Schimak (eds.), Environmental Software Systems Vol. 4 - Environmental Information and Indicators, International Federation for Information Processing, 2001, ISBN 3 901882 14 6.
- G. Schimak, D.A. Swayne, N.T. Quinn, R. Denzer (eds.), Environmental Software Systems Vol. 5 - Environmental Knowledge and Information Systems, International Federation for Information Processing, 2003, ISBN 3 901882 16 2.

In a similar way the sixth and seventh proceedings of ISESS 2005 and ISESS 2007 were published but this time on CD and USB-Stick under IFIP ISBN:

- D.A. Swayne, T. Jakeman (eds.), Environmental Software Systems, Vol. 6 - Environmental Risk Assessment Systems, International Federation for Information Processing, 2005, IFIP ISBN 3-901882-21-9.
- D.A. Swayne, J. Hřebíček (eds.), Environmental Software Systems, Vol. 7 - Dimensions of Environmental Informatics, International Federation for Information Processing, 2007, IFIP ISBN: 978-3-901882-22-7.

The eighth proceedings of ISESS 2009 were published again on USB-Stick, but not under IFIP ISBN:

- D.A. Swayne, R. Soncini-Sessa (eds.), Environmental Software Systems, Vol. 8, 2009, C.R.L.E University of Guelph ISBN: 978-3-901882-364.

Since 2011 we are following the IFIP Advances in Information and Communication Technology (AICT) publishing procedure which is done in conjunction with Springer. The proceedings of ISESS 2011 have been published:

- J. Hřebíček, G. Schimak, R. Denzer (eds.), Environmental Software Systems. Vol. 9 – Frameworks of eEnvrionment. 2011; ISBN 978-3-642-22284-9; ISSN 1868-4238; IFIP AICT 359.

2.2 Collaboration with IFIP and Springer

Since 2011, the editors of these proceedings have decided to use the IFIP publishing channel which is Springer and to publish ISESS 2013 proceedings electronically (as a CD) in the IFIP Advances in Information and Communication Technology (AICT) series.

The IFIP AICT series publishes state-of-the-art results in science and of information and communication technologies. The scope of the series includes: foundations of computer science; software theory and practice; education; computer applications in technology; communication systems; systems modeling and optimization; information systems; computers and society; computer systems technology; security; artificial intelligence; and human-computer interaction.

The principal aim of the IFIP AICT series is to encourag education and the dissemination and exchange of information about all aspects of computing.

Merging knowledge and interests of members of WG 5.11 "Computers and Environment" and the ISESS 2013 conference dedicated to information exchange among scientists and businesses involved in the development and use of environmental informatics for the delivery of future oriented solutions and applied research results in Europe and worldwide.

Therefore the conference focuses specifically on the following topics:

- Environmental Application in the scope of the Future Internet
- Smart and mobile devices used for Environmental Applications
- Linked Open Data – fostering Environmental Services
- Information Tools for Global Environmental Assessment
- Environmental Applications in Risk & Crises Management
- SEIS as a part of the 7th Environment Action Programme of EU
- Human Interaction and Human factors driving future EIS/EDSS developments

The ISESS 2013 is the meeting place for experts on leading edge technologies, and it aims to foster the standardization and integration of environmental data and information flows, which are essential pre-requirements for managing our natural resources in a framework of sustainable development.

The ISESS 2013 is open to governmental institutions, international and intergovernmental organisations, environmental agencies and networks, scientists, academicians, politicians, businesses, public administration and decision makers in the field of environmental information, experts from ICT industry, specialists of theoretical and applied informatics, consultants, students, and the concerned public.

The editors hope that you experience new knowledge and ideas when reading the IFIP AICT proceedings of ISESS 2013 and its web, where you can meet reviewed contributions of authors and experts on leading edge ICT technologies for environment of the above mentioned topics.

3 Word of IFIP WG 5.11 Chairman

ISESS 2013, which we all hope will be as fruitful as its predecessors, is organised under the umbrella of IFIP and represents itself the IFIP WG 5.11 "Computers and Environment".

The vision and mission of the WG 5.11 was predominantly to foster the improved application of Information Technology in environmental research, monitoring, assessment, management, and policy. This is still valid but over the last years the picture has slightly changed and new challenges have popped up.

Challenges, stemming from requirements realizing a Single Information Space for Environment (SISE) and Shared Environmental Information System (SEIS) or when setting up the Strategies for the Digital Agenda for Europe, are asking for new services, new collaboration strategies in decision making and information exchange cross-border. This is not only valid for Europe but also applicable worldwide, if we consider the ICT needs in Copernicus (earlier GMES) or GEOSS in ICT and Future Internet development.

Beside that, within the last two years we recognized a tremendous technology jump in the usage of mobile devices in our domain. Mobile devices, sensitive in many directions are on the market; e.g. devices suitable to record or sense personal exposure information valuable for health protection, as well as to report environmental issues or deliver support in emergency or crises situations. The mass on information is increasing faster than ever.

Thus my personal vision and hopefully also the one of IFIP WG 5.11 for the future is, in order to achieve our mission mentioned above:

- To provide and broaden our platform / forum amongst and between ICT and Environmental professionals.
- Perform knowledge-centric conferences worldwide to exchange information about state-of-the-art technology and prepare the ground for the future.
- Provide expert advice to government, multinational organizations and industry.
- Provide strategy and policy makers with intuitive ICT concepts and solutions.

A permanent review and update of the ISESS membership list is a quality indication for allowing high level expertise in our domain and up-to-date knowledge provision to our audience.

3.1 Acknowledgements

The editors are grateful to all authors for their valuable contributions to the conference and the members of Program Committee for reviewing about one hundred submitted abstracts and selected papers. Special acknowledgements belong to representatives of International Federation for Information Processing, European Environment Agency, DG Environment and DG CONNECT of the European Commission as well as the Joint Research Centre for their kind support and useful advice with respect to the collaboration and organization of ISESS 2013.

We would like to thank the following institutions and partners:

- the AIT Austrian Institute of Technology, Department Safety & Security, which has organized this conference in Austria,
- the Masaryk University, its Institute of Biostatistics and Analyses,
- Austrian Environment Agency (Umweltbundesamt),

- the company Czech Software First and Mrs. Radka Novotná for their administrative support and Miroslav Kubásek, Jiří Kalina and Jakub Gregor - the conference web administration and editorial team for their enormous work in providing the ISESS web page.

Last but not the least we thank all partners involved in the preparation of this conference as well as all sponsors for their valuable cooperation and patience.

October 2013

Jiří Hřebíček
Gerald Schimak
Miroslav Kubásek
Andrea Rizzoli

Organization

Program Committee Chairs

Gerald Schimak (Chair)	Austrian Institute of Technology, Austria
Jiří Hřebíček (Vice-chair)	Masaryk University, Czech Republic
Arne J. Berre	SINTEF, Norway
Ladislav Dušek	Masaryk University, Czech Republic
Steven Frysinger	James Madison University, USA
Denis Havlik	Austrian Institute of Technology, Austria
Jose Lorenzo	Atos Spain, Spain

Publication Chair

Miroslav Kubásek	Masaryk University, Czech Republic

Program Committee Members

Robert Argent	University of Melbourne, Australia
Alessandro Annoni	European Commission, DG Joint Research Centre, Italy
Ioannis N. Athanasiadis	Democritus University of Thrace, Greece
Guiseppe Avelino	Elsag Datamat, Italy
Florian Bauer	Renewable Energy and Energy Efficiency Partnership, UK
Arne J. Berre	SINTEF, Norway
Ralf Denzer	Saarland University of Applied Sciences, Germany
Ladislav Dušek	Masaryk University, Czech Republic
Giorgio Guariso	University of Milan, Italy
Werner Geiger	Research Center Karlsruhe, Germany
Lars Gidhagen	Swedish Meteorological and Hydrological Institute, Sweden
Eric Gontier	Flemish Institute for Technological Research (VITO), Belgium
Reiner Güttler	Saarland State University for Applied Scinces, Germany
Hugo de Groof	European Commission, DG Environment, Belgium
Steven Frysinger	James Madison University, USA
Denis Havlik	Austrian Institute of Technology

Jiří Hradec	Czech Environmental Information Agency
Jiří Hřebíček	Masaryk University, Czech Republic
Karel Charvát	Czech Centre for Science and Society
Anthony Jakeman	Australian National University
Stefan Jensen	European Environment Agency, Denmark
Martin Kaltenböck	Semantic Web Company & OGD Austria
Kostas Karatzas	Aristotle University, Greece
Jana Klánová	Masaryk University, Czech Republic
Milan Konečný	Masaryk University, Czech Republic
Horst Kremers	CODATA - Germany
Rudolf Legat	Umweltbundesamt, Austria
Jose Lorenzo	Atos Spain
Thomas Maurer	Federal Institute of Hydrology, Germany
Johannes Mayer	Umweltbundesamt, Austria
Jadranka Pečar-Ilić	Rudjer Boskovic Institute, Croatia
Werner Pillmann	International Society for Environmental Protection, Austria
Tomáš Pitner	Masaryk University, Czech Republic
Jaroslav Pokorný	Charles University, Czech Republic
Nigel Quinn	Berkeley Research Lab and University of Berkeley, USA
Iva Ritschelová	Czech Statistical Office
Andrea E. Rizzoli	Scuola universitaria professionale della Svizzera italiana, Switzerland
Gerald Schimak	Austrian Institute of Technology
Paul Smith	European Commission, DG Joint Research Centre, Italy
Chris Steenmans	European Environment Agency, Denmark
Alberto Susini	Geneva Labour inspectorate (OCIRT), Switzerland
David Swayne	University of Guelph, Canada
Thomas Usländer	Fraunhofer IOSB, Germany

Table of Contents

Environmental Application in the Scope of the Future Internet

The HABITATS Approach to Build the INSPIRE Infrastructure 1
Karel Charvát, Otakar Čerba, Štěpán Kafka, Tomáš Mildorf, and Přemysl Vohnout

Dynamic Data Driven Ensemble for Wildfire Behaviour Assessment: A Case Study 11
Margherita Di Leo, Daniele de Rigo, Dario Rodriguez-Aseretto, Claudio Bosco, Thomas Petroliagkis, Andrea Camia, and Jesús San-Miguel-Ayanz

Automated Semantic Validation of Crowdsourced Local Information – The Case of the Web Application "Climate Twins" 23
Alexander Kaufmann, Jan Peters-Anders, Sinan Yurtsever, and Luca Petronzio

Information System Efficiency as an Attribute in Environmental Information Systems 31
Miloš Koch and Zuzana Chvátalová

PROMETHEE-GAIA Method as a Support of the Decision-Making Process in Evaluating Technical Facilities 44
Alena Kocmanová, Marie Dočekalová, and Jiří Luňáček

Incorporate Sensor Data and Dynamic Modeling for Real Time Control of Sewer Systems 54
Erik de Rooij and Klaas-Jan van Heeringen

Data Fusion in the Environmental Domain 62
Hylke van der Schaaf, Mike Kobernus, Markus Falgenhauer, Jasmin Pielorz, and Kym Watson

Approaching Cross-Domain Search in Environmental Applications – Towards Linked Data 71
Gerald Schimak, Luca Petronzio, and Tomas Pariente Lobo

ENVIROFI - Bringing Biodiversity to the Future Internet 85
Katharina Schleidt, Nina Laurenne, Andrea Giacomelli, and Denis Havlik

Acquisition and Representation of Knowledge for Atmospheric New Particle Formation ... 98
Markus Stocker, Elham Baranizadeh, Amar Hamed, Mauno Rönkkö, Annele Virtanen, Ari Laaksonen, Harri Portin, Mika Komppula, and Mikko Kolehmainen

The Future Internet Enablement of the Environment Information Space ... 109
Thomas Usländer, Arne J. Berre, Carlos Granell, Denis Havlik, José Lorenzo, Zoheir Sabeur, and Stefano Modafferi

Smart and Mobile Devices Used for Environmental Applications

A Model Integration Framework for Assessing Integrated Landscape Management Strategies ... 121
Jared M. Abodeely, David J. Muth, Joshua B. Koch, and Kenneth M. Bryden

E-SMART: Environmental Sensing for Monitoring and Advising in Real-Time ... 129
Matthew Aitkenhead, David Donnelly, Malcolm Coull, and Helaina Black

Identifying Smart Solutions for Fighting Illegal Logging and Timber Trade ... 143
Ioannis N. Athanasiadis, Despina Anastasiadou, Kostas Koulinas, and Fotis Kiourtsis

Using Smartphones to Profile Mobility Patterns in a Living Lab for the Transition to E-mobility ... 154
Francesca Cellina, Anna Förster, Davide Rivola, Luca Pampuri, Roman Rudel, and Andrea Emilio Rizzoli

Robust and Trusted Crowd-Sourcing and Crowd-Tasking in the Future Internet ... 164
Denis Havlik, Maria Egly, Hermann Huber, Peter Kutschera, Markus Falgenhauer, and Markus Cizek

Mapping of Illegal Dumps in the Czech Republic – Using a Crowd-Sourcing Approach ... 177
Miroslav Kubásek

A Mobile Application for Reporting Disease Incidents ... 188
Hylke van der Schaaf, Ente Rood, and Kym Watson

"My Environment" – A Dashboard for Environmental Information on Mobile Devices 196
Thorsten Schlachter, Clemens Düpmeier, Rainer Weidemann, Wolfgang Schillinger, and Nina Bayer

Information Tools for Global Environmental Assessment

Estimating Impacts of Environmental Interventions in Monitoring Programs Requires Conceptual Data Models and Robust Statistical Processing (Position Paper) 204
Ladislav Dušek, Jana Klánová, Jiří Jarkovský, Jakub Gregor, Richard Hůlek, Ivan Holoubek, and Jiří Hřebíček

Tools for Collection, Analysis and Visualization of Data from the Stockholm Convention Global Monitoring Plan on Persistent Organic Pollutants 222
Jakub Gregor, Richard Hůlek, Jiří Jarkovský, Jana Borůvková, Jiří Kalina, Kateřina Šebková, Daniel Schwarz, Jana Klánová, and Ladislav Dušek

GENASIS System Architecture: On the Way from Environmental Data Repository towards a Research Infrastructure 230
Richard Hůlek, Jiří Jarkovský, Miroslav Kubásek, Jakub Gregor, Jiří Hřebíček, Ladislav Dušek, Jana Klánová, Kateřina Šebková, Jana Borůvková, and Ivan Holoubek

Integration of R Statistical Environment into ICT Infrastructure of GMP and GENASIS 240
Richard Hůlek, Jiří Kalina, Ladislav Dušek, and Jiří Jarkovský

GENASIS: On-Line Data Browser for Environmental Monitoring and Associated Information Systems 253
Jiří Jarkovský, Ladislav Dušek, Miroslav Kubásek, Richard Hůlek, Jakub Gregor, Jiří Hřebíček, Jana Klánová, Kateřina Šebková, Jana Borůvková, and Ivan Holoubek

Development of a Catchment Management Tool to Assess Environmental Risk from Nutrient Loadings Using Open Source GIS ... 262
Ian Packham, Eva Mockler, and Michael Bruen

Continental-Scale Living Forest Biomass and Carbon Stock: A Robust Fuzzy Ensemble of IPCC Tier 1 Maps for Europe 271
Daniele de Rigo, José I. Barredo, Lorenzo Busetto, Giovanni Caudullo, and Jesús San-Miguel-Ayanz

Environmental Applications in Risk and Crises Management

Geo-analysis of Landscape Level Degradation and Natural Risk Formation under Uncertainty: A Case Study of Selected Czech Urban Watercourses ... 285
Radka Báčová, Petr Kubíček, Jiří Jakubínský, Eva Svobodová, and Vladimír Herber

The Use of Simulations for the Need of Environment Protection ... 294
Jiří Barta

Protection of Information and Communication Systems ... 302
Jiří Barta, Veronika Sadovská, Albert Srník, and Jiří F. Urbánek

Crime Open Data Aggregation and Management for the Design of Safer Spaces in Urban Environments ... 311
Antonios Bonatsos, Lee Middleton, Panos Melas, and Zoheir Sabeur

Multi-scale Robust Modelling of Landslide Susceptibility: Regional Rapid Assessment and Catchment Robust Fuzzy Ensemble ... 321
Claudio Bosco, Daniele de Rigo, Tom Dijkstra, Graham Sander, and Janusz Wasowski

RainPortal – A Web Portal for Providing Climate Change Related Precipitation Data Using SUDPLAN Services ... 336
Peter Kutschera, Jonas Olsson, and Lars Gidhagen

Crowdtasking – A New Concept for Volunteer Management in Disaster Relief ... 345
Georg Neubauer, Andrea Nowak, Bettina Jager, Christian Kloyber, Christian Flachberger, Gerry Foitik, and Gerald Schimak

GIS-Based Route Planning for HAZMAT Transportation ... 357
Alexander Preda, Mauno Rönkkö, Stefan Pickl, and Mikko Kolehmainen

An Architecture for Adaptive Robust Modelling of Wildfire Behaviour under Deep Uncertainty ... 367
Daniele de Rigo, Dario Rodriguez-Aseretto, Claudio Bosco, Margherita Di Leo, and Jesús San-Miguel-Ayanz

TiTiMaKe: A Distributed Service Architecture for Security Applications ... 381
Mauno Rönkkö, Markus Stocker, Markku Huttunen, Ville Kotovirta, and Mikko Kolehmainen

Modelling of Environmental Risk Management under Information Asymmetry ... 391
Petr Šauer, Petr Fiala, and Antonín Dvořák

SEIS as a Part of the 7th Environment Action Programme of EU

An Information System Supporting WFD Reporting 403
Thorsten Hell, Eckhard Kohlhas, Sascha Schlobinski, Ralf Denzer, and Reiner Güttler

Web 3D Visualization of Noise Mapping for Extended INSPIRE Buildings Model 414
Lukáš Herman and Tomáš Řezník

National INSPIRE Geoportal of the Czech Republic 425
Jan Prášek, Jiří Valta, and Jiří Hřebíček

INSPIREd Air Quality Reporting: European Air Quality e-Reporting Based on INSPIRE 439
Katharina Schleidt

Human Interaction and Human Factors Driving Future EIS/EDSS Developments

Using the HOS Method for Evaluating the Efficiency of Environmental Information Systems 451
Zuzana Chvátalová and Miloš Koch

How to Build Integrated Climate Change Enabled EDSS 464
Ralf Denzer, Sascha Schlobinski, Lars Gidhagen, and Thorsten Hell

Qualified Safety Management Ensures Sustainable Human Life Quality 472
Dana Procházková

Novel Approaches for Web-Based Access to Climate Change Adaptation Information – MEDIATION Adaptation Platform and ci:grasp-2 489
Markus Wrobel, Alexander Bisaro, Dominik Reusser, and Jürgen P. Kropp

Environmental Management/-Accounting and -Statistics

Integration of Macro and Micro Sustainability Issues: The Need for 'Engineering Accounting' 500
Joanne L. Tingey-Holyoak, Roger L. Burritt, and John D. Pisaniello

Benefits of Environmental Management Systems in the Czech Republic 508
Jaroslava Hyršlová, Miroslav Hájek, and Gabriela Rajdlová

The Comparison of Selected Risk Management Methods for Project Management ... 517
Vladěna Obrová and Lenka Smolíková

A Framework for Monitoring and Evaluation of Learning Processes ... 525
Lucie Pekárková, Patrícia Eibenová, and Tomáš Pitner

Classifying Environmental Monitoring Systems ... 533
Mauno Rönkkö, Ville Kotovirta, and Mikko Kolehmainen

Waste Management Statistics in the Czech Republic ... 543
Egor Sidorov, Iva Ritschelová, and Jiří Hřebíček

Methodology and Information System for Evaluating Environmental Protection Expenditure Efficiency at the Local Level ... 560
Jana Soukopová and Eduard Bakoš

Modelling and Simulation Support of EMS Processes ... 571
Jan Ministr

Information Systems and Applications

Green Computing Practices as a Part of the Way to the Sustainable Development ... 579
Zuzana Bezáková

The Plan4business Approach to Transfer Open Data into Real Estate Businesses ... 588
Jan Ježek, Tomáš Mildorf, Karel Charvát Jr., and Karel Charvát

SVOD – System for Visualizing of Oncological Data and Their Semantic Enhancement ... 597
Miroslav Kubásek, Jiří Hřebíček, Ladislav Dušek, Jan Mužík, and Jiří Kalina

Intelligent Facility Management for Sustainability and Risk Management ... 608
Adam Kučera and Tomáš Pitner

Visual Analytics in Environmental Research: A Survey on Challenges, Methods and Available Tools ... 618
Martin Komenda and Daniel Schwarz

Information System for Global Sustainability Reporting ... 630
Ondřej Popelka, Michal Hodinka, Jiří Hřebíček, and Oldřich Trenz

The Model of Territory Unit Evaluation for Allocation of Resources on Flood Protection ... 641
Jana Soukopová, Eduard Bakoš, Jan Šelešovský, and Lenka Furová

Decision Support of Waste Management Expenditures Efficiency Assessment 651
Jana Soukopová, Ivan Malý, Jiří Hřebíček, and Michal Struk

GeneDBase – Genetic Database of Selected Species of Mammals in the Czech Republic 661
Luděk Skočovský, Tomáš Šikula, Martin Ernst, and David Kouřil

Vattenwebb: A Transparent Service to Support Decision Makers in Achieving Improved Water Status 669
Lena Strömbäck, Niclas Hjerdt, Lena Eriksson Bram, and Per Lewau

Pedestrian Localization in Closed Environments: Android Prototype 679
Jonáš Ševčík

Workshop ENVIP 2013, Environmental Information Systems and Services - Infrastructures and Platforms 2013

Environmental Infrastructures and Platforms with Citizens Observatories and Linked Open Data 688
Arne J. Berre, Sven Schade, and Dumitru Roman

Author Index 697

The HABITATS Approach to Build the INSPIRE Infrastructure

Karel Charvát, Otakar Čerba, Štěpán Kafka, Tomáš Mildorf, and Přemysl Vohnout

Help Service - Remote Sensing, Benešov, Czech Republic
kch@hsrs.cz, ota.cerba@gmail.com, kafka@email.cz,
{mildorf,vohnout}@ccss.cz

Abstract. Different initiatives focused on spatial data in Europe should not be isolated but closely connected. The portfolio of such initiatives is very extensive. On the one hand, it covers all European activities such as INSPIRE or GMES, on the other hand, there are products of modern approaches based on neogeography and Volunteered Geographic Information (e.g. OpenStreetMap). Data are published by various regional or local authorities, non-governmental organisations, public bodies, research projects as well as by different commercial subjects. The focus of the HABITATS project was to build an environment that enables to share and combine data in order to reach new data, information and knowledge. On the basis of different pilots, HABITATS defined and tested harmonisation rules for spatial environmental data and designed the concept of Reference Laboratory as a tool for testing the interoperability and supporting unification of outputs cross different pilots.

Keywords: HABITATS, reference laboratory, INSPIRE, SDI, environment, social validation, volunteered geographic information.

1 Introduction

Advances in information and communication technologies contributed to increase the awareness about environmental protection all over the world over the last decades. More attention and is paid to the environment and changes towards the environment protection are made in policies, organisational structures, data collection and data management.

The Europe's nature conservation policy including the Birds and Habitats Directives was complemented by the Directive 2007/2/EC of the European Parliament and of the Council of 14 March 2007 establishing an Infrastructure for Spatial Information in the European Community (INSPIRE). INSPIRE supports environmental policies and policies or activities which may have impact on the environment. INSPIRE, which is based on spatial data collected by Member States (national spatial data infrastructures, SDI), paves the way for seamless data sharing and data reuse in Europe. Next to INSPIRE, there are other initiatives that support data sharing in different areas, such as Copernicus (previously known as Global Monitoring for Environment and Security) or the Shared Environmental Information System (SEIS).

J. Hřebíček et al. (Eds.): ISESS 2013, IFIP AICT 413, pp. 1–10, 2013.

This paper tackles the issues of spatial data management and SDI building based on the developments and results of the HABITATS project. HABITATS was an EU funded project focusing on adoption of the INSPIRE standards through a participatory process to design and validate environmental geospatial data, metadata and service specifications. HABITATS built an environment that enables combining data from various sources to gain new data, information and knowledge.

Spatial data in the context of the above mentioned initiatives help us to protect our lands and properties and the environment we live in and to shape our future environment. The attention is given to participation of citizens in decision making processes so they can influence where a factory or a new road will be built. Popular is collection of data using smart phones and other devices by volunteers. Citizens can for example report potholes in their street or fly tipping (dumping waste illegally) in their neighbourhood or map real world objects and phenomena.

The data sources for HABITATS were supported by the above mentioned initiatives as well as voluntary based projects such as OpenStreetMap. On the basis of different pilots, HABITATS defined and tested harmonisation rules for spatial environmental data and designed the concept of Reference Laboratory as a tool for testing the interoperability and supporting unification of outputs cross different pilots. The pilot applications tested ontology concepts that are described later in this paper.

2 HABITATS Networking Services and Architecture

The intention of the HABITATS Networking Services is to provide shift from the classic INSPIRE, GEOSS and GMESS architectures towards a solution which will support local and regional SDI building and their interaction with INSPIRE and which will move standard SDI model towards neogeography and the spider web paradigm of distributed systems. The way to test and provide it is Reference Laboratory as key tool of the HABITATS Networking Architecture.

The HABITATS Networking Architecture defines a platform-neutral SDI with a basic set of networking services in compliance with the INSPIRE Directive for sharing environmental data, especially that related to the 4 INSPIRE spatial data themes

- Sea-Regions (SR);
- Bio-geographical Regions (BR);
- HABITATS and Biotopes (HB);
- Species Distribution (SD).

This will resulted in a European metadata profile for these four data themes, which are an extension of the INSPIRE metadata profile. Our intention was not only to follow the INSPIRE profile for discovery services, but also to reflect the extension the profiles for using data; a link to data modelling activities is therefore necessary. This profile is further open to extension by single countries or user groups, but the aim is that it is respected as a minimum set.

The set of HABITATS Networking Services was implemented on the HABITATS Reference Laboratory (RL) geoportal platform[1] introduced in the next section. This acts as a client of the 7 HABITATS pilots that provide a rich set of cross-pilots and inter-regional services.

In order to validate the HABITATS networking services architecture, a series of specific service applets were deployed and tested for data sharing using the HABITATS RL geoportal platform.

Comparing network services and spatial data services, the INSPIRE Forum[2] definitions are:

- Spatial data services are all services (Discovery, View, Download, Transformation, Invoke, Other) regarding spatial data;
- Network services (Discovery, View, Download, Transformation, Invoke):
  - non-compliant network services are INSPIRE compliant services with respect to functionality;
  - compliant network services are compliant services with respect to functionality and quality of service.

The HABITATS networking architecture supports INSPIRE Network services, but needs to go behind this concept. INSPIRE networking services are in principle limited only to the management of existing data and metadata. The HABITATS Networking Services also support such functionality with data and metadata management, data and metadata collection, working with non-spatial data, etc.

The HABITATS service applets re-use existing applications where possible and are themselves designed for re-use. The selection of the specific services to deploy is primarily a user-driven process, as defined in the user scenarios and requirements. The prototype set of Network Service Applets to be installed in validation pilot platforms were drafted as:

- A series of specific networking service applets deployed and tested for data sharing within the project using the Network Service Architecture:
  - Interoperability Services;
  - Enabling Services:
    - Visualisation of information layers;
    - Overlay of information from different sources;
    - Spatial and Temporal Analysis;
  - "quick" and "light" on-demand applets to meet validation pilot expectations and user needs;
  - Usability, simplicity and openness to rapid prototyping mash-ups;

- A set of specific service applets that allow users to identify, access, use and reuse habitats-related data, designed and deployed on-demand to meet user needs;

[1] At www.habitats.cz
[2] http://inspire-forum.jrc.ec.europa.eu/mod/groups/topicposts.php?topic=11135&group_guid=8651

- Mobile Apps allowing to use advantage of the HABITATS RL;
- "Quick Prototyping" service applets respecting the HABITATS service architecture and developed on-demand.

These are based on the outcomes from the earlier tasks and work with the HABITATS RL, and will now lead into the interface tools and toolkit.

The HABITATS Networking Architecture aims to extend the principles of the INSPIRE networking architecture, because INSPIRE doesn't cover important aspects such as data management and data collection. So all components of the INSPIRE networking architecture will be included in the HABITATS architecture, but this concept will be extended by other functionalities. From this point of view principles of GEOSS and GMES and also principles of Shared Environmental Information System (SEIS) and Single Information Space in Europe for the Environment (SISE) have influenced the HABITATS architecture and its networked services.

The development of the network service architecture process was initiated through a state of the art analysis of existing SDI, to find out more about already existing infrastructures and to examine how data should be shared and what services are required to enable sharing. When designing the networking architecture, a set of specific networking service applets was deployed and tested for data sharing within the project. Also the potential for re-use of existing application was taken into account.

3 HABITATS Reference Laboratory

The HABITATS Reference Laboratory (RL) is the central hub of the HABITATS Networking Architecture. It consists of several layers, which include [2]:

- Data layers – management data and files on storage, eventually guarantee access to external sensors;
- Server (engine) layer – defines tools, which guarantee basic services on the server side – supplying service;
- Client layer – is client side of Web services, which guarantee access of users to services;
- Application layer - is some form of wrapping elementary client services into application or into such form, which could be used by other web tools;
- Presentation layer - contain such web tools, which allows to combine and to publish single objects from the application level as part of Web presentation.

Figure 1 shows the different layers of the HABITATS Networking Architecture.

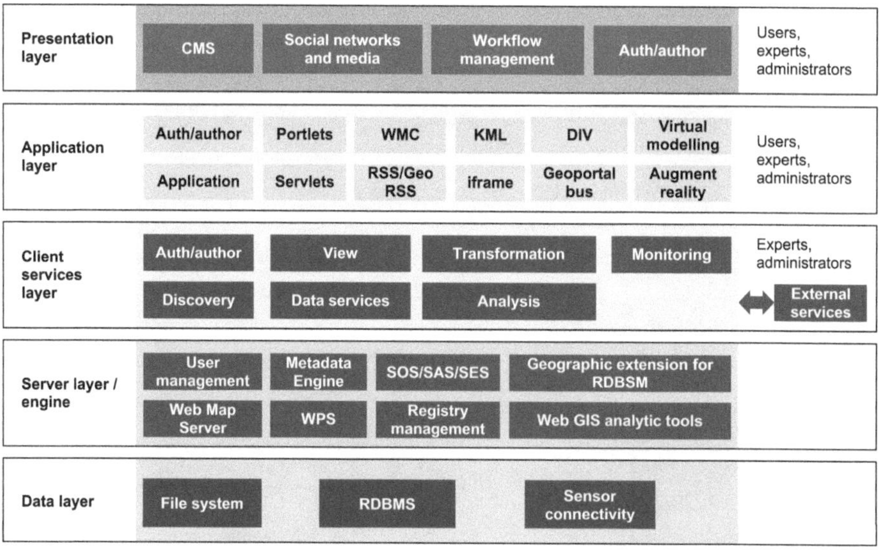

Fig. 1. HABITATS Networking Architecture [3]

The final implementation of the HABITATS services anticipates that selected concrete services will be deployed for every pilot, and that there will be one central platform (i.e. the Reference Laboratory).

The HABITATS RL provides the Networking Architecture supporting both Networking Services and Spatial Data services, that support the SDI network services to enable trans-European sharing of habitats-related spatial data between public authorities and other stakeholders in the Community, enabling the creation of value added services. The HABITATS RL is focused on implementation of:

- Cross pilot scenarios based on sharing of data among more pilots;
- Validation platform for testing of conformity of implemented pilot services;
- Services supporting global discovery, view and downloading;
- Repository for common metadata;
- Repository for pan-European datasets such as Natura 2000, Corine Land Cover, Urban Atlas and Open Street Maps;
- Interlink with social networks.

The HABITATS RL provides the Networking Architecture, that supports the SDI network services to enable trans-European sharing of habitats-related spatial data between public authorities and other stakeholders in the Community enabling the creation of value added services.

The HABITATS RL enables deployment of specific service applets, including interoperability and enabling services, on-demand from user communities and the pilots for initial implementation and validation. It is being developed further to include an invoking service toolkit integrating the service applets with the goal of facilitating the development of end-user services accessing habitats-related spatial data over time.

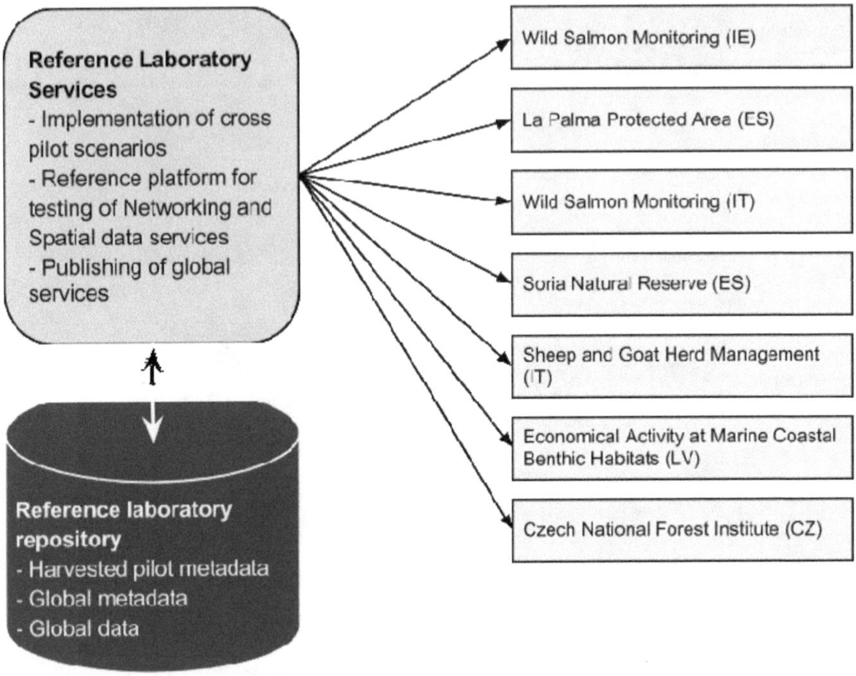

Fig. 2. Reference Laboratory and its pilot applications [3]

However, applications are the key objectives and final goal of using the HABITATS RL. As the HABITATS RL is just a geoportal tool to help to build applications that address the Pilots' typical use cases.

The HABITATS RL is:

- an interface that enables interactive search, portrayal, evaluation, sharing, analyse and reuse of spatial and non-spatial data;
- a solution based on interoperable standards (OGC, W3C, OASIS, ISO). It is interconnected to other resources through the Internet. It helps to create a distributed structure of information and knowledge with spatial position.

However, the RL is not a central data storage nor a closed web application with maps. It is a geoportal with:

- Independent components;
- Composition according to user requirements;
- Based on SOA;
- Possibility to integrate with other resources;
- Maximum openness;
- Open Source;
- Open Standards;

- Extension to non-GIS community;
- Open Search;
- Administration of other (non-spatial) data sources.

The HABITATS RL allows deployment of the current state of the art of technological solutions, which will be tested and adopted by the HABITATS partners and user stakeholders. It allows testing of current existing technology and generation of further research tasks driven by users. The HABITATS RL also collects information coming from other projects, which is an important input for the HABITATS analysis and public discussion. The methods of social assessment will be an increasingly important part of the HABITATS RL.

Thus the HAITATS RL's networking services aim to help HABITATS to extend user-centric, co-design approaches into the arena of standards design and adoption processes, considering standards initiatives such as INSPIRE, OGC, UNSDI to be significant social, economic and institutional innovations. The elements of the approach are maintained, applying the model at all levels from the global scale to the local and regional policies that frame many HABITATS validation pilots. Community building activities follow a Web 2.0 approach to capture the knowledge in active user communities with a strong interest in contributing to the standards development process. By inviting a broad multi-sectoral and inter-disciplinary range of concerned stakeholders to participate into the HABITATS network, a viral motivation spiral is set off. A peer-to-peer approach to opening up information sources and providing access to content ensures a rapid extension of the critical mass of environmental data established by project partners.

4 Ontology Concepts Applied by the HABITATS

4.1 Domain Ontology

Domain ontology describes the fundamental concepts and relationships which are important in a particular domain. It represents a domain vocabulary as well as a formal and formalized specification of the meaning of terms. According to the Protégé web site (http://protege.stanford.edu/) ontologies range "from taxonomies and classifications, database schemas, to fully axiomatised theories". Ontologies could be an essential part of many applications of information technologies (including geoinformation technologies) such as portals, information management and integration systems or semantic web services.

The principles of ontologies could be summarized to the following three points:

1. Ontologies serve as a tools used for a description of real or abstract objects or processes;
2. Ontologies as a kind of data source have to use formal recommended to use and distinction between two words connected with ontologies – formal (application of formal terms) and formalised (formal way of coding) ways of data structures;

3. It is not possible to develop a universal ontology, since its properties depend on various factors such as way of use.

Common domain ontologies using the standard data format OWL (Web Ontology Language, see below) can contain five crucial elements or groups of elements [4]:

1. Classes represent and describe concepts (particular parts of domain). They are also able to interconnect these parts. They form together with individuals a hierarchy of ontology. The hierarchy is based on is-a (is a part) relation. There are two types of classes in ontologies – primitive classes (described by necessary conditions) and defined classes (described by necessary and sufficient conditions). To build ontological classes the formulation of descriptive logic (such as union, intersection or quantifiers) are used;
2. Individuals are very similar to classes. The indivisibility of individuals is the main difference between these two components. While class can contain other class or classes, the same construction cannot be implemented to individuals (individuals cannot be divided);
3. Properties or relations are designated for definition, interconnection, description and detail specification of classes, individuals and their connections. There are two basic types of relations in ontologies: DataType Properties (connecting classes or individuals with concrete data types) and Object Properties (describing relations between classes or individuals);
4. Axioms express explicitly rules managing relation between elements of ontology. They enable to describe rules such as equivalence, subsumption, disjunction etc. Ontologies use two specific axiomatic constructions – closure axiom (combination of existing and universal quantifier) and coverage axiom (definition of closed set of sub-classes);
5. Annotations represent an analogy to metadata records. They are based on Dublin Core standard (http://dublincore.org/) and enable not only to describe the whole ontology but also its elements by properties such as label or language.

The methodologies constitute very important but underestimated part of ontologies – designing of ontologies as well as development and managing phase. There are many various methods such as On-To-Knowledge, METHONTOLOGY (the most popular methodology; standard IEEE Std 1074-1995), Cyc (probably the oldest methodology), Gruninger and Fox, DILIGENT, KACTUS, SENSUS, Noy and Macguinness or Uschold & King. They are used to construct and maintain ontologies. It is necessary to mention that most of the methodologies are based on detail preparatory phase which is very often underestimated in practice. This phase can be composed of an identification of the ontology (title, purpose, scope, coding, granularity), planning of conceptualisation and an integration of existing ontologies and other sources.

4.2 General Example Using Ontology in a Pilot Application

Ontologies were used in selected pilot activities in the HABITATS project. They are not designed for spatial data harmonisation, but they should be a part of the harmonisation process. They can serve as a description of various heterogeneities and mapping of various types of concepts. The real harmonisation is performed by harmonisation tool (GIS, ETL etc.) and using information and knowledge formalised and stored in ontology.

As an example, the ontological system transforming various types of taxonomies (classification systems) of land cover and land use data was used. Land cover and land use, widely used by the INSPIRE initiative, form an essential part of the HABITATS pilots (forestry, sea regions, tourism, etc.).

The ontology is composed of three main components:

1. File Parameter.owl that is imported into ontologies containing various classification systems, describes the properties of all classes of these nomenclatures. The file contains not only all the parameters (the current version consists of nine parameters), but also their values and relations (object properties) that assign values to the parameters of each class of classification systems. Current version covers nine properties with values arranged in multi-level hierarchy;
2. Files containing all elements of the particular classification systems (CORINE Land Cover, PELCOM, GlobCorine, LUCAS, UMD, NLCD etc.). All classes of ontologies are described through the properties using the parameters and their values;
3. Ontology is used to transform each other classification systems (file Trasformation.owl) – this ontology imports files describing particular classification systems. After their evaluation (reasoning) is obtained multi-hierarchical structure providing information about the interactions of described taxonomies.

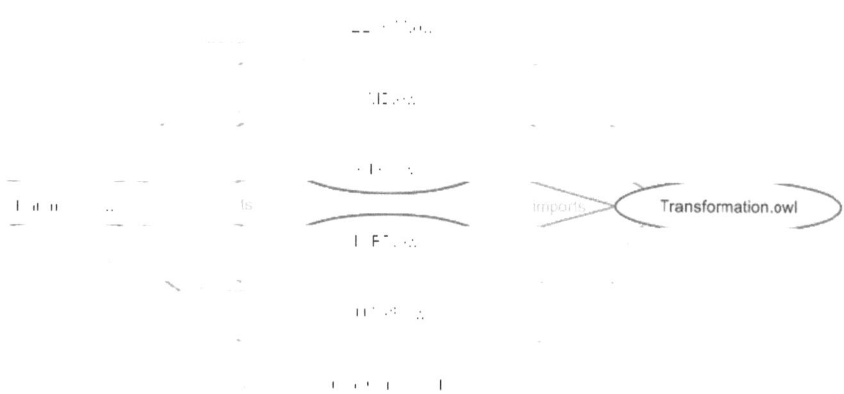

Fig. 3. Ontology relationships [1]

The explicit parts of ontology (Figure 3) are processed by reasoners. The inferred ontology contains relations between particular classes of taxonomies.

5 Conclusions

The HABITATS project built an environment that enables to share and combine data from various sources. The project results were validated through the HABITATS Reference Laboratory and pilot applications. On the basis of different pilots, HABITATS defined and tested harmonisation rules for spatial environmental data and designed the concept of Reference Laboratory as a tool for testing the interoperability and supporting unification of outputs cross different pilots.

References

1. Čepický, J., et al.: D3.3.2 Data models and transformation processes (2012)
2. Charvát, K., et al.: D4.2.2 HABITATS INSPIRE Networking Architecture (2011), http://www.slideshare.net/kcharvat/habitats-d422-final
3. Charvát, K., et al.: D4.3.2 HABITATS networking services and service toolkit (2013)
4. Čerba, O.: Ontology Of Cartographic Terminology. In: International Conference on Cartography and GIS (2012)

Dynamic Data Driven Ensemble for Wildfire Behaviour Assessment: A Case Study

Margherita Di Leo[1], Daniele de Rigo[1,2], Dario Rodriguez-Aseretto[1], Claudio Bosco[3], Thomas Petroliagkis[1], Andrea Camia[1], and Jesús San-Miguel-Ayanz[1]

[1] European Commission, Joint Research Centre, Institute for Environment and Sustainability, Via E. Fermi 2749, I-21027 Ispra (VA), Italy
{margherita.di-leo,daniele.de-rigo,dario.rodriguez, jesus.san-miguel}@jrc.ec.europa.eu

[2] Politecnico di Milano, Dipartimento di Elettronica e Informazione, Via Ponzio 34/5, I-20133 Milano, Italy

[3] Loughborough University, Department of Civil and Building Engineering Loughborough LE11 3TU, United Kingdom
C.Bosco@lboro.ac.uk

Abstract. Wildfire information has long been collected in Europe, with particular focus on forest fires. The European Forest Fire Information System (EFFIS) of the European Commission complements and harmonises the information collected by member countries and covers the forest fire management cycle. This latter ranges from forest fire preparedness to post-fire impact analysis. However, predicting and simulating fire event dynamics requires the integrated modelling of several sources of uncertainty. Here we present a case study of a novel conceptualization based on a Semantic Array Programming (SemAP) application of the Dynamic Data Driven Application Systems (DDDAS) concept. The case study is based on a new architecture for adaptive and robust modelling of wildfire behaviour. It focuses on the module for simulating wildfire dynamics under fire control scenarios. Rapid assessment of the involved impact due to carbon emission and potential soil erosion is also shown. Uncertainty is assessed by ensembling an array of simulations which consider the uncertainty in meteorology, fuel, software modules. The event under investigation is a major wildfire occurred in 2012, widely reported as one of the worst in the Valencia region, Spain. The inherent data, modelling and software uncertainty are discussed and preliminary results of the robust data-driven ensemble application are presented. The case study suitably illustrates a typical modelling context in many European areas – for which timely collecting accurate local information on vegetation, fuel, humidity, wind fields is not feasible – where robust and flexible approaches may prove as a viable modelling strategy.

Keywords: Wildfire behaviour, Forest fires, Integrated natural resources modelling and management, Semantic Array Programming, DDDAS.

1 Introduction

Among natural hazards, wildfires cause huge environmental, economic, social damages. In the Mediterranean area, a changing climate toward dryer and hotter

J. Hřebíček et al. (Eds.): ISESS 2013, IFIP AICT 413, pp. 11–22, 2013.

seasons seriously increases the fire risk. Wildfire information has been collected in Europe for a long time, with particular focus on forest fires. The European Forest Fire Information System (EFFIS) of the European Commission complements and harmonises the information collected by member countries and covers the forest fire management cycle [1]. Analysis and prediction of wildfire propagation is growingly supported by computational modelling. However, accurate wildfire behaviour modelling may require a large number of input data and parameters. The impossibility of reliably measuring all of them along with the numerical modelling complexity, simplifications and uncertainty (data and software uncertainty [2, 3]) are critical aspects to address. The transdisciplinary nature of the problem is the key driver for its intrinsic complexity.

The spatial extent of most wildfires involves local scales. However, the aim for the described fire information system to support the variety of conditions at continental scale leads the modelling architecture [4] to cover several transdisciplinary aspects [2] aside the direct life and health risks [5]. Fire-induced geomorphic and ecological transformations [6–8] may cause severe carbon emissions and alter biodiversity [9, 10] even due to habitat fragmentation [11–13]. They may also affect landslide susceptibility [14] and soil erosion [15, 16] with potential off-site impacts [17, 18] on water resources [19] (e.g. on floods [20, 21] and water quality [22]). The involved complexity and uncertainty suggested the architecture to be set within the context of integrated natural resources modelling and management (INRMM) [23].

Here we present a case study of a novel conceptualization based on a Semantic Array Programming (SemAP) [24, 25] application of the Dynamic Data Driven Application Systems (DDDAS) [26, 27] concept. The case study is based on a new architecture for adaptive and robust modelling of wildfire behaviour, proposed in [4]. We focus on the architecture modules for simulating wildfire dynamics under fire control scenarios. Rapid impact assessment of carbon emission and potential soil erosion is also shown. An array of simulations is ensembled considering the uncertainty in meteorology, fuel and software modules.

1.1 The European Forest Fire Information System (EFFIS)

EFFIS is a comprehensive environmental information system covering the full cycle of forest fire management from forest fire prevention and preparedness to post-fire damage analysis. The system provides information to over 30 countries in the European and Mediterranean regions, to the aim of harmonising the prevention and fire fighting [1].

Exploratory research is currently on-going to integrate fire behaviour models, with emphasis on extreme conditions, physical modelling, smoke dispersion and spotting related to forest fire occurrences. Another aspect relates the use of high resolution meteorological data from ENCWF [28] and Meteo France [29].

EFFIS exploits geospatial and computational modelling tools within a modelling paradigm aiming at robustness and semantic transparency [2, 3].

Forest fire occurrences are collected both on the basis of hot spots provided by NASA and of fire news inventoried by the operator using the dedicated web application implemented within the EFFIS portal, a tool to easily detect and geoparse the fire news [30]. The satellite images used to detect burnt areas are the daily Moderate-Resolution Imaging Spectroradiometer (MODIS) by NASA's Terra and Aqua satellites [31], of 250m resolution.

The procedure consists in a semi-automated process. Hot spots and FireNews points are displayed on a desktop GIS and drive the operator's attention on possible anomalies distinguishable on MODIS images. Where a possible anomaly is identified, the operator compares the image with MODIS images relative to previous days, in order to reveal the burnt area, visible as a garnet coloured spot. Once a fire is detected, it is digitised as a polygon and updated with all the relative information, e.g. extension, starting date, etc. If the fire is still active the day after, the polygon is updated with the new perimeter. Fire perimeters are stored in the EFFIS database. A fire's record stores burnt perimeters day by day, so as to enable the use for future needs and research purposes.

2 Materials and Methods

The case study applies the modelling architecture proposed in [4]. This architecture is designed to exploit information updates whenever available so as to timely adapt wildfire behaviour modelling to changing conditions (e.g. meteorology).

The Case Study. The analysed event refers to a major wildfire (also served as a benchmark in [32]) occurred near Valencia, the third largest city of Spain. The fire started on June 28 2012 in Dos Aguas and was under control on July 4, burning an area of 32424 ha. Our models' input is the 2nd July recorded burnt perimeter. The spread occurring in 24 hours is simulated.

Applying the Modelling Architecture. The modelling architecture of [4] relies on an adaptive control strategy [33] based on stochastic dynamic programming (SDP), which is widely applied in natural resources management [34–36]. Wildfire behaviour dynamics is spatially distributed. For example, the case study is characterised by highly variable local patterns of slope, aspect, fuel type, wind, humidity and rain. Unfortunately, SDP computational costs are exponential with the size of the vector of states $\mathbf{x}$ representing the environmental system. Since the SDP approach easily becomes intractable, approximated algoritms [37–39] are useful. [4,32] proposed one of those approximate techniques – Partial Open Loop Feedback Control (POLFC) [39] – for integrated wildfire behaviour modelling.

The dynamics of fire events may be represented as a data transformation model (D-TM) accounting for the evolution of the state $\mathbf{x}_t = \{x_{t,c_1}, x_{t,c_2}, \cdots\}$ which describes at time t the fire likelihood and severity for each spatial cell c. The evolution is modelled in the discrete time interval Δt with respect to the anthropogenic control $\mathbf{u}_t$, the disturbances $\xi_{t+\Delta t}$ (wind, humidity, rainfall), and the system characteristics (slope, aspect, fuel distribution, parameterised as θ_t):

$$\mathbf{x}_{t+\Delta t} = \;::\!\left| f(\theta_t, \mathbf{x}_t, \mathbf{u}_t, \xi^{\tau}_{t+\Delta t}) \right|^{sem}_{::} \Leftrightarrow \begin{cases} \mathbf{x}_{t+\Delta t} = f\left(\theta_t, \mathbf{x}_t, \mathbf{u}_t, \xi^{\tau}_{t+\Delta t}\right) \\ \Box\, sem(\mathbf{x}_{t+\Delta t}, f, \theta_t, \mathbf{x}_t, \mathbf{u}_t, \xi^{\tau}_{t+\Delta t}) \end{cases} \tag{1}$$

$$\text{where:} \quad \begin{array}{llll} t & \in \mathcal{U}^t & \mathbf{x}_t = \mathcal{U}^{\mathbf{x}}_t & \mathbf{u}_t \in \mathcal{U}^{\mathbf{u}}_t(\mathbf{x}_t) \\ \xi^{\tau}_{t+\Delta t(t,\mathbf{x}_t)} & \sim \phi(\cdot \,|\, I_\tau) & \in \mathcal{U}^{\xi}_t(\mathbf{x}_t, \mathbf{u}_t, I_\tau),\ \tau \in \mathcal{U}^t & \theta_t = \theta(\mathbf{x}_t, \mathbf{u}_t) \end{array}$$

Semantic checks *sem* are imposed as pre-, post-conditions and invariants on inputs, outputs and the D-TM module $f(\cdot)$ itself [3]. $\Box$ *sem* is a modal/deontic logic expression meaning: "it ought to be that *sem*". *sem* is a set of valid array-based semantic constraints following the *semantic array programming* paradigm [24,25] . Straightforward examples of semantic constraints are exemplified in the following as active links **::*constraint*::**. The disturbance vector $\xi^{\tau}_{t+\Delta t}$ represents the uncertain set of forecasted disturbances $\phi(\cdot \mid I_{\tau})$ and depends on $\mathbf{u}_t$, $\mathbf{x}_t$ and the available data-driven information I_{τ}. A POLFC control problem is proposed at each time t to mitigate (sub-optimal minimisation) the global fire costs:

$$u^t(\cdot) = \underset{\mathbf{u} \in \mathcal{U}^{\mathbf{u}}_{t,\,t_{\text{end}}}}{\arg\min} \left[\mathcal{C}^{\,\text{carbon emission},\,t} \quad \mathcal{C}^{\,\text{soil erosion},\,t} \quad \cdots \quad \mathcal{C}^{\,n,\,t} \right] \qquad (2)$$

$\mathbf{u}_{\cdot}$ are planned in the set of possible actions $\mathcal{U}^{\mathbf{u}}_{t,\,t_{\text{end}}}$ along the whole event lifespan $\mathcal{U}^t$. Since the controls are spatially distributed in wide raster grids, heuristic control options are explored and the uncertainty associated with their predicted effects is assessed with a robust ensemble strategy. Fig. 2 illustrates a simplistic example where the carbon emission and soil erosion costs are highlighted.

Design Diversity: The Meteorological Data. Three wind-forecast scenarios (fig. 1) were utilised as input for the model simulations, to explore the uncertainty associated with the meteorological models. Wind-forecast scenarios were based on a set of operational Numerical Weather Prediction (NWP) models with heterogeneous spatial and temporal resolution:

- The ECMWF (European Centre for Medium-Range Weather Forecasts) operational high-resolution single global deterministic model (ec_{16}), with a horizontal resolution of about 16 km [40–42]. The model is initiated on both the 00 and 12 UTC analysis fields reaching to an 11-hour forecast horizon with an archived time-step of 1 hour;
- The ECMWF EPS (Ensemble Prediction System) global deterministic Control component, with a horizontal resolution of about 32 km (ec_{32}). This single-model deterministic platform is initiated from a truncated analysis of ec_{16} while no errors are introduced to this initial field [43, 44];
- The DWD/Deutscher Wetterdienst (German Meteorological Service) operational single deterministic global icosahedral-hexagonal grid point model with a horizontal resolution of about 20 km [45, 46]. DWD model values have been retrieved at a regular 25x25 km grid (dw_{25}).

The forecasts have been spatially upsampled so as to disaggregate the average values of the coarse meteo-grids to the denser simulation grid.

The interpolation driven uncertainty (IDU) [47] has been explored by comparing an upsampling which preserves the local average invariance (LAI) [24] with an usual non-LAI upsampling (i.e. typically confusing the average value of each coarses-grid cell with the value of the associated centroid).

Design Diversity: The State Transition Function. An ensemble approach is proposed for mitigating the software uncertainty [2,3] associated with the state transition function (eq. 1). Although two particular free software simulators [32] are described, the proposed method is general and applicable to other simulators:

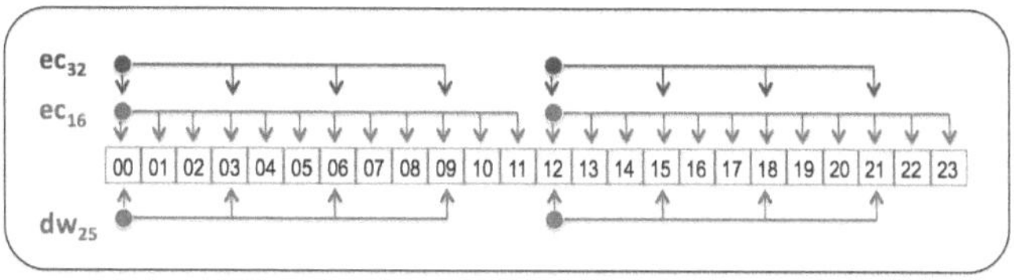

Fig. 1. Graphical representation of model analysis updates during a typical 24-hour interval. Number of dots corresponds to the number of updates, while arrows correspond to various time-steps utilized by model platforms.

- FireSim implements the library FireLib [48] and the Rothermel model [49] by means of a cellular-automata simulator. FireSim is a deterministic parametric and discrete event type simulator structured as a sequential set of D-TMs with a pipeline structure of four stages.
- The second simulator is implemented in GRASS GIS [50]. The spread function is based on the least cost path algorithm and simulates the elliptically anisotropic spread. It is implemented in the function *r.spread* [51], which takes as input the rate of spread (ROS) generated by the function *r.ros* [52]. This latter computes the ROS following the Rothermel model [49] and is based on the Fortran code by [53]. The direction of the maximum ROS is the vector sum of the forward ROS in wind direction and that in upslope direction. The obtained raster map layers serve as inputs for *r.spread.*

Both simulators take in input the standard fuel models defined by [54]. Fuel information was derived from the fuel type map of Europe developed by JRC [55]. The classification scheme adopted for the fuel map encompasses 42 fuel types representing the variety of fuel complexes found in the European landscapes. A cross-walk to the original set of 13 fire behaviour fuel models tabulated by [54] was also developed in order to obtain standard fuel parameters as input to the Rothermels fire spread model [56]. Because of the uncertainties in fuel model attributions and the non-univocal correspondence between fuel types and fuel models, multiple assignments were allowed. The expected proportions of fuel models extent within each fuel type were assigned based on expert knowledge. Equiprobable fuel maps have been generated for each run of the ensemble.

The two simulators are deterministic instances of eq. 1. For each spatial cell c and time t, the state $x_{t,\,c}$ simulated by an instance's run is converted to be a **::binary::**[1] flag to register whether c is burnt. Uncertainty analysis is implemented by ensembling multiple instance runs for different scenarios. Instances expect slope and wind speed input as **::nonnegative::**[2] layers while aspect and wind direction as **::angle::**[3] values. The weighted-quantile analysis of the ensemble

[1] `http://mastrave.org/doc/mtv_m/check_is#SAP_binary`
[2] `http://mastrave.org/doc/mtv_m/check_is#SAP_nonnegative`
[3] `http://mastrave.org/doc/mtv_m/check_is#SAP_angle`

requires all output **::matrix::**[4] layers to have the **::same_size::**[5] for generating a **::sortable::**[6] 3-dimensional array **::3-array::**[7].

3 Rapid Assessment of Wildfire Impacts

As stated in [4] wildfires are associated with high carbon emission and alteration of ecosystems and soil properties. Hydrology, land cover and soil characteristics are the factors that mainly influence the fire - geomorphic processes connection.

Wildfires can increase soil erosion rate by orders of magnitude. Wildfires can totally remove the protective surface layers, as the litter, and increase the hydrophobic condition at or below the ground surface [57–60].

For the rapid assessment of carbon emissions, soil erosion and other post-fire impacts as landslide susceptibility, a robust computational modelling is required [4, 14]. Within the present study a quick assessment of post-fire carbon emissions and soil erosion was carried out. The e-RUSLE model [15, 16] was applied in Dos Aguas for estimating post-fire soil erosion changes.

Combustion Carbon Emissions. Combustion in wildfires and associated carbon emissions vary depending on the local fire intensity and the characteristics of fuel (fuel moisture, share of dead and live fuel, share of woody and herbaceous fuel, ...) [61]. Specific burning efficiency and emission factors can be associated to classes of fuel and typology of fire (flaming or smoldering) so as for fuel models to be associated with static factors. However, the actual available fuel load is an essential information for a realistic carbon emission assessment. At European scale, a harmonised map of living forest carbon stock has recently been improved the standard methodology for IPCC Tier 1 level [62]. This map served as input for a preliminary, exploratory spatialisation of the carbon emission assessment.

Soil Erosion. The e-RUSLE approach is based on the joint use of a low data requirement model and an innovative technique for model inputs estimation.

The model's architecture is based both on semantic array programming [24, 25] and computational reproducibility [15], easing the integration of natural resources models. Furthermore the family of models to which e-RUSLE belong provide long term erosion averages and have been applied in many different climatic and environmental conditions all over the world [63–65].

Public available electronic datasets – European Soil Geographical Database (ESGDB) [66], Harmonized World Soil Database (HWSD) [67], Shuttel Radar Topography Mission (SRTM) [68], CORINE Land Cover (CLC) [69] and ENSEMBLES Observations gridded dataset (E-OBS) [70] – has been used for calculating the pre and post-fire soil erosion maps in the study site. Electronic archives are one of the main data source for research community, allowing accessibility to large volumes of data and having the capacity to preserve historical data [71]. The e-RUSLE factors have been calculated following the approach presented in [16, 72].

[4] `http://mastrave.org/doc/mtv_m/check_is#SAP_matrix`
[5] `http://mastrave.org/doc/mtv_m/check_is#SAP_same_size`
[6] `http://mastrave.org/doc/mtv_m/check_is#SAP_sortable`
[7] `http://mastrave.org/doc/mtv_m/check_is#SAP_3-array`

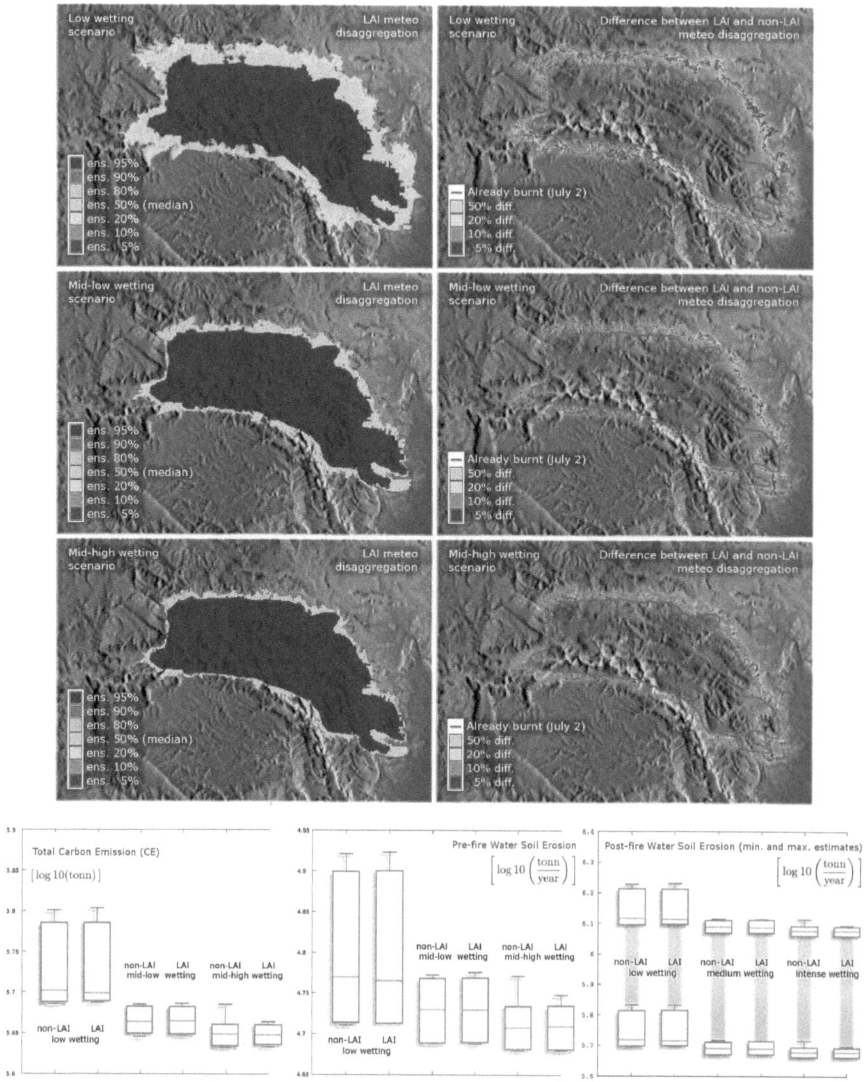

Fig. 2. Dos Aguas, Valencia – Spain. Wildfire behaviour simulations with ensemble prediction (July 3 2012). To illustrate, simplistic fire control scenarios are shown: low wetting (*top*); mid-low wetting (*middle*) and mid-high wetting (*bottom*). Interpolation driven uncertainty (IDU) is also assessed. Meteorological forecasts have been spatially disaggregated by either imposing local average invariance (LAI) [24] (*left*) or not (usual non-LAI disaggregation). The difference is shown at *right*. Below: rapid assessment of wildfire impacts. Total carbon emission and soil erosion by water (pre- and post-fire) have been estimated for each ensemble simulation of each scenario. Box-plots: box with quartiles 25%, 50% (red line), 75%; wiskers with 5% and 95% quantiles. Although the assessment considers the whole burnt area and not only the one which would have been burnt in the simulated day under the different scenarios, the non-LAI induced component of IDU is nonlinear and even its one-day effects may occasionally be non-negligible for the impact assessment of the whole burnt area.

Within the model, the Cover Management factor (C factor) and the soil erodibility factor (K factor) are strongly influenced by high-severity burns. In severe, slow-moving fires, the combustion of vegetative materials creates a gas that penetrating the soil profile causes the soil to repel water. This post-fire hydrophobic condition, reducing the soil permeability, is considered as one of the main causes of the increasing in runoff after a fire [17,58].

A rapid assessment of post-fire soil erosion is provided applying the DerBos nondimensionalised equation [4]. It describes a RUSLE based lower- and upper-bound for the expected soil erosion.

As stated in Larsen and McDonald [59], the algorithm for calculating K values are not consistent with our current understanding of erosion processes. High severity burns can increase sediment yields of 2-3 orders of magnitude [17] but considering the changes in soil permeability and organic matter content the maximum increase in K factor is around 100%. Despite the evident underestimation, we incresed the K factor passing from K_c^{pre} to K_c^{burnt} [4] of 100%. Because of the lack of information regarding post-fire C factor values and the link with fire intensity, we considered only the case of high severity burns. For C_c^{α}, C_c^{β} and C_c^{γ} we applied equispaced values in the range from 0.5 to 0.2 [59,73,74], representing the maximum and minimum degradation of the cover factor due to the fire.

4 Conclusions

A real case study (Valencia region, Spain) is presented applying the modelling architecture proposed in [4] as a new conceptualization for wildfire prediction [75]. The methodology constitutes an exploratory investigation for possibly expanding the ability of the European Forest Fire Information System (EFFIS) to support wildfire behaviour analysis and management. Transdisciplinary modelling integration – required by the EFFIS continental scale – has been exemplified by means of the rapid assessment of pre- and post-fire potential soil erosion along with carbon emissions. First results are presented on a robust ensemble for mitigating the deep uncertainty [4] associated with static information (fuel mapping), dynamic data forecast (provided by external meteorological models) and software uncertainty [2,3] (internal modules for simulating the wildfire behaviour). Design diversity has been applied to meteo forecast models (exploiting three different models for wind-forecast scenarios) and fire propagation models (ensembling the predictions of two independent implementations of the Rothermel model). The case study illustrates the preliminary feasibility of real time and robust rapid assessment of wildfire impacts under varying management options.

References

1. San-Miguel-Ayanz, J., Schulte, E., Schmuck, G., Camia, A., Strobl, P., Liberta, G., et al.: Comprehensive Monitoring of Wildfires in Europe: The European Forest Fire Information System (EFFIS). In: Tiefenbacher, J. (ed.) Approaches to Managing Disaster - Assessing Hazards, Emergencies and Disaster Impacts, ch. 5. InTech (2012)
2. de Rigo, D.: Behind the horizon of reproducible integrated environmental modelling at European scale: Ethics and practice of scientific knowledge freedom. F1000 Research (submitted, 2013)

3. de Rigo, D.: Software Uncertainty in Integrated Environmental Modelling: The role of Semantics and Open Science. Geophys. Res. Abstr. 15, 13292+ (2013)
4. de Rigo, D., Rodriguez-Aseretto, D., Bosco, C., Di Leo, M., San-Miguel-Ayanz, J.: An Architecture for Adaptive Robust Modelling of Wildfire Behaviour under Deep Uncertainty. In: Hřebíček, J., Schimak, G., Kubásek, M., Rizzoli, A. (eds.) ISESS 2013. IFIP AICT, vol. 413, pp. 367–380. Springer, Heidelberg (2013)
5. Richardson, L.A., Champ, P.A., Loomis, J.B.: The hidden cost of wildfires: Economic valuation of health effects of wildfire smoke exposure in southern California. J. For. Econ. 18(1), 14–35 (2012)
6. Herrando, S., Brotons, L., Llacuna, S.: Does fire increase the spatial heterogeneity of bird communities in Mediterranean landscapes? Ibis 145(2), 307–317 (2003)
7. Torre, I., Daz, M.: Small mammal abundance in Mediterranean post-fire habitats: A role for predators? Acta Oecologica 25(3), 137–142 (2004)
8. Di Piazza, G.V., Di Stefano, C., Ferro, V.: Modelling the effects of a bushfire on erosion in a mediterranean basin. Hydrol. Sci. J. 52(6), 1253–1270 (2007)
9. Moreira, F., Russo, D.: Modelling the impact of agricultural abandonment and wildfires on vertebrate diversity in Mediterranean Europe. Landsc. Ecol. 22(10), 1461–1476 (2007)
10. Acácio, V., Holmgren, M., Rego, F., Moreira, F., Mohren, G.M.J.: Are drought and wildfires turning Mediterranean cork oak forests into persistent shrublands? Agrofor. Syst. 76(2), 389–400 (2009)
11. Estreguil, C., Caudullo, G., de Rigo, D., Whitmore, C., San-Miguel-Ayanz, J.: Reporting on European forest fragmentation: Standardized indices and web map services. IEEE Earthzine 5(2), 384031+ (2012); 2nd quarter theme: Forest Resource Information
12. Estreguil, C., Caudullo, G., de Rigo, D., San-Miguel-Ayanz, J.: Forest landscape in Europe: Pattern, fragmentation and connectivity. EUR Scientific and Technical Research 25717(JRC 77295) (2013)
13. Estreguil, C., de Rigo, D., Caudullo, G.: Towards an integrated and reproducible characterisation of habitat pattern. Environmental Model. & Software (submitted, 2013)
14. Bosco, C., de Rigo, D., Dijkstra, T., Sander, G., Wasowski, J.: Multi-Scale Robust Modelling of Landslide Susceptibility: Regional Rapid Assessment and Catchment Robust Fuzzy Ensemble. In: Hřebíček, J., Schimak, G., Kubásek, M., Rizzoli, A. (eds.) ISESS 2013. IFIP AICT, vol. 413, pp. 321–335. Springer, Heidelberg (2013)
15. de Rigo, D., Bosco, C.: Architecture of a Pan-European Framework for Integrated Soil Water Erosion Assessment. In: Hřebíček, J., Schimak, G., Denzer, R. (eds.) Environmental Software Systems. IFIP AICT, vol. 359, pp. 310–318. Springer, Heidelberg (2011)
16. Bosco, C., de Rigo, D., Dewitte, O., Poesen, J., Panagos, P.: Modelling Soil Erosion at European Scale. Towards Harmonization and Reproducibility (in prep.)
17. Shakesby, R., Doerr, S.: Wildfire as a hydrological and geomorphological agent. Earth Sci. Rev. 74(3-4), 269–307 (2006)
18. Ice, G.G., Neary, D.G., Adams, P.W.: Effects of wildfire on soils and watershed processes. J. Forestry 102(5), 16–20 (2004)
19. Moody, J.A., Martin, D.A., Haire, S.L., Kinner, D.A.: Linking runoff response to burn severity after a wildfire. Hydrol. Proces. 22(13), 2063–2074 (2008)
20. Candela, A., Aronica, G., Santoro, M.: Effects of forest fires on flood frequency curves in a Mediterranean catchment. J. Hydrol. Sci. 50(2), 193–206 (2005)
21. Rulli, M.C., Rosso, R.: Hydrologic response of upland catchments to wildfires. Adv. Water Resour. 30(10), 2072–2086 (2007)
22. Smith, H.G., Sheridan, G.J., Lane, P.N.J., Nyman, P., Haydon, S.: Wildfire effects on water quality in forest catchments: a review with implications for water supply. J. Hydrol. 396(1-2), 170–192 (2011)

23. de Rigo, D.: Integrated Natural Resources Modelling and Management: minimal redefinition of a known challenge for environmental modelling. Excerpt from the Call for a shared research agenda toward scientific knowledge freedom, Maieutike Research Initiative
24. de Rigo, D.: Semantic Array Programming for Environmental Modelling: Application of the Mastrave Library. In: Int. Congress on Environmental Modelling and Software. Managing Resources of a Limited Plant, Pathways and Visions under Uncertainty, Sixth Biennial Meeting, pp. 1167–1176 (2012)
25. de Rigo, D.: Semantic array programming with Mastrave - introduction to semantic computational modelling (2012)
26. Darema, F.: Dynamic Data Driven Applications Systems: A New Paradigm for Application Simulations and Measurements. In: Bubak, M., van Albada, G.D., Sloot, P.M.A., Dongarra, J. (eds.) ICCS 2004. LNCS, vol. 3038, pp. 662–669. Springer, Heidelberg (2004)
27. Denham, M., Cortés, A., Margalef, T., Luque, E.: Applying a Dynamic Data Driven Genetic Algorithm to Improve Forest Fire Spread Prediction. In: Bubak, M., van Albada, G.D., Dongarra, J., Sloot, P.M.A. (eds.) ICCS 2008, Part III. LNCS, vol. 5103, pp. 36–45. Springer, Heidelberg (2008)
28. Hollingsworth, A., Engelen, R., Textor, C., Benedetti, A., Boucher, O., Chevallier, F., Simmons, A.J.: Toward a monitoring and forecasting system for atmospheric composition. Bull. Amer. Meteor. Soc. 89(8), 1147–1164 (2008)
29. Bouttier, F.: The Météo-France NWP system: Description, recent changes and plans. CNRM (2010)
30. Corti, P., San-Miguel-Ayanz, J., Camia, A., McInerney, D., Boca, R., Di Leo, M.: Fire news management in the context of the European Forest Fire Information System (EFFIS). In: Proceedings of Quinta Conferenza Italiana Sul Software Geografico e Sui Dati Geografici Liberi, GFOSS DAY 2012 (2012)
31. Barnes, W.L., Xiong, X., Salomonson, V.V.: Status of terra MODIS and aqua MODIS. Advances in Space Research 32(11), 2099–2106 (2003)
32. Rodriguez-Aseretto, D., de Rigo, D., Di Leo, M., Cortes, A., San-Miguel-Ayanz, J.: A data-driven model for large wildfire behaviour prediction in Europe. Procedia Computer Science 18, 1861–1870 (2013)
33. Lara, M., Doyen, L.: Sustainable Management of Natural Resources: Mathematical Models and Methods. Springer (2008)
34. Rani, D., Moreira, M.: Simulation Optimization Modeling: A Survey and Potential Application in Reservoir Systems Operation. Water Resour. Manag. 24(6), 1107–1138 (2010)
35. Ferreira, L., Constantino, M., Borges, J.G.: A stochastic approach to optimize Maritime pine (Pinus pinaster Ait.) stand management scheduling under fire risk. An application in Portugal. Ann. Oper. Res, 1–19 (2011)
36. Minas, J.P., Hearne, J.W., Handmer, J.W.: A review of operations research methods applicable to wildfire management. Int. J. Wildland Fire 21(3), 189+ (2012)
37. de Rigo, D., Rizzoli, A.E., Soncini-Sessa, R., Weber, E., Zenesi, P.: Neuro-dynamic programming for the efficient management of reservoir networks. In: Proceedings of MODSIM 2001, Int. Congress on Modelling and Simulation held in Canberra, Australia, vol. 4, pp. 1949–1954 (2001) ISBN: 0-867405252
38. de Rigo, D., Castelletti, A., Rizzoli, A.E., Soncini-Sessa, R., Weber, E.: A selective improvement technique for fastening neuro-dynamic programming in water resources network management. IFAC-PapersOnLine 16, 7–12 (2005)
39. Castelletti, A., de Rigo, D., Tepsich, L., Soncini-Sessa, R., Weber, E.: On-Line design of water reservoir policies based on inflow prediction. IFAC-PapersOnLine 17, 14540–14545 (2008)
40. Andersson, E., Thpaut, J.-N.: ECMWFs 4D-Var data assimilation system the genesis and ten years in operations. ECMWF Newsletter 115, 8–12 (2008)

41. Isaksen, L., Haseler, J., Buizza, R., Leutbecher, M.: The new Ensemble of Data Assimilations. ECMWF Newsletter 123, 15–21 (2010)
42. Miller, M., Buizza, R., Haseler, J., Hortal, M., Janessen, P., Untch, A.: Increased resolution in the ECMWF deterministic and ensemble prediction systems, ECMWF Newsletter No. 124 pp.10-16. ECMWF, UK (2010)
43. Molteni, F., Buizza, R., Palmer, T.N., Petroliagis, T.I.: The ECMWF Ensemble Prediction System: Methodology and validation. Q. J. R. Meteorol. Soc. 122, 73–119 (1996)
44. Buizza, R., Bidlot, J.-R., Wedi, N., Fuentes, M., Hamrud, M., Holt, G., Vitart, F.: The new ECMWF VAREPS. Q. J. Roy. Meteorol. Soc. 133, 681–695 (2007)
45. Majewski, D.: The new global icosahedral hexagonal grid point model GME of the Deutscher Wetterdienst. In: Proc. ECMWF Seminar on Recent Developments in Numerical Methods for Atmospheric Modelling, pp. 173–201. ECMWF, Reading (1998)
46. Majewski, D., Liermann, D., Prohl, P., Ritter, B., Buchhold, M., Hanisch, T., Paul, G., Wergen, W., Baumgardner, J.: The operational global icosahedral-hexagonal grid point model GME: Description and high resolution tests. Mon. Wea. Rev. 130, 319–338 (2002)
47. de Rigo, D., Guariso, G.: Rewarding Open Science: A Collaborative Review System for Semantically-Enhanced Free Software and Environmental Data Modelling (in prep., 2013)
48. Bevins, C.D.: FireLib user manual and technical reference. Systems for Environmental Management (1996)
49. Rothermel, R.: How to predict the spread and intensity of forest and range fires. Ogden, UT, USA: US Department of Agriculture, Forest Service, Intermountain Forest and Range Experiment Station (1983)
50. Neteler, M., Bowman, M.H., Landa, M., Metz, M.: GRASS GIS: A multi-purpose open source GIS. Environmental Modelling & Software 31, 124–130 (2012)
51. Xu, J., Lathrop, R.G.: GRASS GIS manual: r.spread. In: GRASS GIS 6.4.3svn Reference Manual (2006)
52. Xu, J.: GRASS GIS manual: r.ros. In: GRASS GIS 6.4.3svn Reference Manual (2009)
53. Andrews, P., LLoyd, P.: Fire modeling and information system technology. Int. J. Wildland Fire 10(4), 343–352 (2001)
54. Albini, F.A.: Estimating wildfire behavior and effects. Gen. Tech. Rep. INT-30. Ogden, Utah: Department of Agriculture, Forest Service, Intermountain Forest and Range Experiment Station (1976)
55. Camia, A., et al.: Development of the fuel type map of Europe. EUR Scientific and Technical Research (in prep.)
56. Rothermel, R.C.: A mathematical model for predicting fire spread in wildland fuels. Res. Pap. INT-115. Ogden, UT: U.S. Department of Agriculture, Forest Service, Intermountain Forest and Range Experiment Station (1972)
57. Certini, G.: Effects of fire on properties of forest soils: A review. Oecologia 143, 1–10 (2005)
58. DeBano, L.F.: The role of fire and soil heating on water repellency in wildland environments: A review. J. Hydrol. 231-232, 195–206 (2000)
59. Larsen, I.J., MacDonald, L.H.: Predicting post-fire sediment yields at the hillslope scale: testing RUSLE and disturbed WEPP. Water Resour. Res. 43(11), W11412 (2007)
60. Scott, D.F., Van Wyk, D.B.: The effects of wildfire on soil wettability and hydrological behavior of an afforested catchment. J. Hydrol. 121, 256–293 (1990)
61. Barbosa, P., Camia, A., Kucera, J., Libert, G., Palumbo, I., San-Miguel-Ayanz, J., Schmuck, G.: Assessment of Forest Fire Impacts and Emissions in the European Union Based on the European Forest Fire Information System. Developments in Environmental Science 8, 197–208 (2008)

62. de Rigo, D., Barredo, J.I., Busetto, L., Caudullo, G., San-Miguel-Ayanz, J.: Continental-Scale Living Forest Biomass and Carbon Stock: a Robust Fuzzy Ensemble of IPCC Tier 1 Maps for Europe. In: Hřebíček, J., Schimak, G., Kubásek, M., Rizzoli, A. (eds.) ISESS 2013. IFIP AICT, vol. 413, pp. 271–284. Springer, Heidelberg (2013)
63. Bosco, C., Rusco, E., Montanarella, L., Panagos, P.: Soil erosion in the Alpine area: risk assessment and climate change. Studi Trent. Sci. Nat. 85, 117–123 (2009)
64. Angima, S.D., Stott, D.E., ONeill, M.K., Ong, C.K., Weesies, G.A.: Soil erosion prediction using RUSLE for central Kenyan highland conditions. Agriculture Ecosystems and Environment 97, 295–308 (2003)
65. Kinnell, P.I.: Event soil loss, runoff and the Universal Soil Loss Equation family of models: A review. Journal of Hydrology 385, 384–397 (2010)
66. Heineke, H.J., Eckelmann, W., Thomasson, A.J., Jones, R.J.A., Montanarella, L., Buckley, B.: Land Information Systems: Developments for planning the sustainable use of land resources. Office for Official Publications of the European Communities. EUR 17729 EN (1998)
67. FAO, IIASA, ISRIC, ISSCAS, JRC: World Soil Database (version 1.1). FAO, Rome, Italy and IIASA, Laxenburg, Austria (2009)
68. Farr, T.G., Rosen, P.A., Caro, E., Crippen, R., Duren, R.: The Shuttle Radar Topography Mission. Reviews of Geophysics 45, RG2004 (2007)
69. European Environment Agency: Corine Land Cover 1990, 2000 and 2006 raster data version 15 (August 2011)
70. Haylock, M.R., Hofstra, N., Klein Tank, A.M.G., Klok, E.J., Jones, P.D., New, M.: A European daily high-resolution gridded dataset of surface temperature and precipitation. J. Geophys. Res. (Atmospheres) 113, D20119 (2008)
71. Panagos, P., Jones, A., Bosco, C., Senthil Kumar, P.S.: European digital archive on soil maps (EuDASM): Preserving important soil data for public free access. Int. J. of Digital Earth 4(5) (2011)
72. Bosco, C., Rusco, E., Montanarella, L., Oliveri, S.: Soil erosion risk assessment in the alpine area according to the IPCC scenarios. In: Threats to Soil Quality in Europe. EUR 23438 EN, pp. 47–58 (2008)
73. Cebecauer, T., Hofierka, J.: The consequences of land-cover changes on soil erosion distribution in Slovakia. Geomorphology 98, 187–198 (2008)
74. Terranova, O., Antronico, L., Coscarelli, R., Iaquinta, P.: Soil erosion risk scenarios in the Mediterranean environment using RUSLE and GIS: An application model for Calabria (southern Italy). Geomorphology 112, 228–245 (2009)
75. Rodriguez-Aseretto, R., Cortés, A., Margalef, T., Luque, E.: An Adaptive System for Forest Fire Behavior Prediction. In: 11th IEEE Int. Conf. on Comp. Sci. Eng., CSE 2008, pp. 275–282. IEEE (2008)

Automated Semantic Validation of Crowdsourced Local Information – The Case of the Web Application "Climate Twins"

Alexander Kaufmann[1], Jan Peters-Anders[1], Sinan Yurtsever[2], and Luca Petronzio[3]

[1] AIT Austrian Institute of Technology GmbH, Vienna, Austria
{alexander.kaufmann,jan.peters-anders}@ait.ac.at
[2] ATOS, Madrid, Spain
sinan.yurtsever@atosresearch.eu
[3] Telespazio
luca.petronzio@telespazio.com

Abstract. Climate Twins is a freely accessible Web application which provides easily comprehensible information on the projected climate developments in Europe at a regional level. Climate Twins is based on the idea that an impression of the future climate of a specific region can be received by indicating regions which have a similar present climate. Within the 7th EU FP project "TaToo" ontology-based functionalities have been added to the basic Climate Twins Web application in order to allow users to enrich the original information base - restricted to only two climate parameters ('temperature' and 'precipitation') - by themselves by adding any further geo-located information which they consider relevant in the context of a specific local climate (e.g. information on local vegetation and fauna). This kind of crowdsourcing information raises the problem of safeguarding the quality of the added information. In this paper the authors show how the domain ontology of Climate Twins can be used to semantically validate the coherence of new entries in order to prevent incorrect links.

Keywords: semantics, information enrichment, climate change.

1 Introduction

Climate change is often discussed on a rather abstract level. Raising awareness for what climate change actually means to people is important to bring the discussion from an academic level to the level of political action to induce necessary changes in the way we produce the goods and the way we behave as consumers. In order to visualize climate change on the regional level, the AIT Foresight & Policy Development Department has developed a Web application called "Climate Twins" which makes the projected changes of the climate manifest on a high spatial resolution by showing regions with a comparable climate today or in the recent past [5]. The similarity of climate conditions, however, is based on temperature and precipitation only. The picture of the climate in a region is much richer. Therefore it has been decided by

J. Hřebíček et al. (Eds.): ISESS 2013, IFIP AICT 413, pp. 23–30, 2013.

the Climate Twins publisher to to involve the user community of the Climate Twins Web application to enrich the regional climate information in a self-organized way. This required additional functionalities which have been developed in the 7th EU FP-project "TaToo" for which Climate Twins was one of three validation scenarios. An important issue of community-based information enrichment is the problem of safe-guarding quality when users can freely upload any information they want. In this paper we describe how we use semantics – the Climate Twins domain ontology – for the automated check of user entries to block incorrect links.

2 Climate Twins – A Web Tool for Visualizing Climate Change

Climate Twins is a regional climate simulation application, developed by the AIT Foresight & Policy Development Department, that is freely accessible on the Web [1]. The objective of Climate Twins is to offer a simple to use tool to people to get a vivid impression of the future development of the climate conditions in certain areas of Europe they are interested in. 'Simple to use' means that people do not have to be climate research experts for being able to get information from Climate Twins.

The basic idea of Climate Twins is to show people what the climate in their respective regions will look like in the future by showing other regions which have similar climate conditions today. For example, a person living in Austria's capital Vienna wants to know how the local climate will be two decades from now. The Climate Twins application returns those areas in Europe that have today similar climate conditions that are projected to be prevalent in the Vienna area in 20 years. The comparison of climate conditions in Climate Twins is based on two parameters: 'temperature' and 'precipitation'.

Climate Twins uses the temperature and precipitation data from the regional climate projection model COSMO-CLM (COnsortium for Small-scale MOdeling – Climate Local Model) at a 18 x 18 km resolution [2]. Projected data are available until 2100. Historical data are available back until 1961. Finding comparable areas is done by comparing the projected data of the climate parameters 'temperature' and 'precipitation' in the future with the respective present data. Those areas where future and present climate parameters are similar are then called "climate twin regions" [8].

For a particular Climate Twins query the user locates the grid cell he/she is interested in – the so-called "point of interest" (POI) on the map "Selection" in the Climate Twins Web application. Then he/she selects the future time period for the POI and the reference time period for all other grid cells (available time periods are the decades between 1961 and 2100). Further choices available to the user concern the climate parameters (temperature or precipitation or both), the similarity measures (proportional similarity or Hellinger coefficient), which part of the year to be compared (the whole year or one of the four seasons), the weighting of the climate parameters (if both are selected) and the similarity measure thresholds (the closer to 1, the less grid cells will be found with a similar climate). The results are then displayed in the "Results" map. There the user can easily see which places in Europe show similar climate conditions today as they are projected to be in the future at the POI [9].

3 Involving Users to Enrich the Information of Climate Twins

The project TaToo has provided several Web services that could be user to improve the functionality of Climate Twins with regard to information enrichment by involving its users. For this purpose three TaToo public services have been integrated: The tagging service enables users to add additional information on a particular Climate Twin region for the benefit of other users. The discovery service enables the search of additional information on a particular Climate Twin region. The evaluation service gives the users a possibility to evaluate the community-added resources. The evaluation functionality helps the users to distinguish valuable from weak resources added by other users. Originally, this has been the only way of pointing out good as well as weak resources and relied completely on self-regulation within the user-community.

If a user wants to upload the link to a certain resource he/she considers to contain useful climate-related information to be shared with other users the Add Info-tab within the Climate Twins interface has to be used (see Fig. 1). There the user has to select a particular location that fits best to the spatial relevance of the additional information. The user does this by clicking on an adequate Climate Twins raster cell. For this cell he/she gets a list of GeoNames [3] located in this cell from which a particular GeoNames location has to be selected. Then he/she has to fill in the link to the resource (i.e. the URL) and the title of the resource. He/she can also add a brief description and select an adequate topic from the Climate Twins domain ontology as a keyword of the resource. With pressing the submit-button the user uploads the URL of this additional resource and adds the corresponding meta-information to the TaToo knowledge base. It is from then on retrievable by other users of Climate Twins.

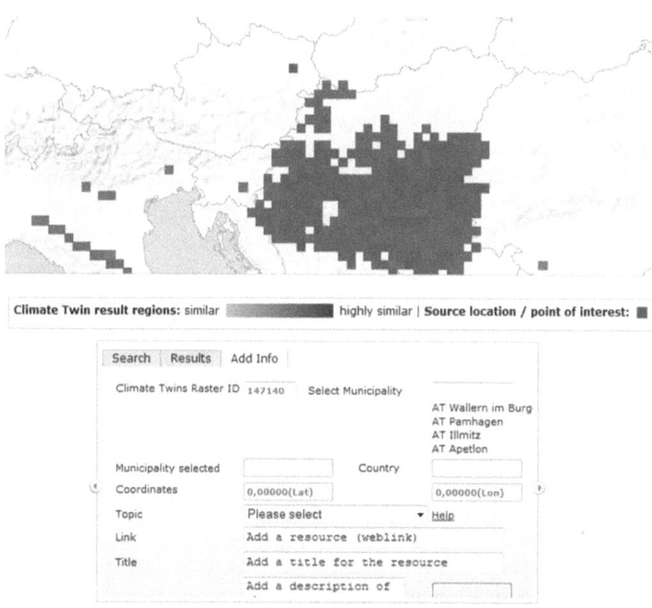

Fig. 1. Screenshot of the Climate Twins interface: part of the Results map and the Add Info tab

Fig. 1 shows the Add Info tab which the user has to fill to upload a new information to the knowledge base. The Add Info tab is right below the Results map showing the similar regions for a certain Climate Twins query. In this example the user has selected a Climate Twins raster cell in Burgenland at the eastern border of Austria. In this cell four GeoNames are present (Wallern, Pamhagen, Illmitz, Apetlon) from which the user has to select one.

Since there was no a priori-restriction of eligible resources in the original version of the Web application there was no guarantee that an added link points to a valuable information. There was no check whether an information is climate-related in any meaningful way or whether it has been geo-located correctly or whether a link has been correctly entered. The only process of quality assurance foreseen was user evaluation, i.e. self-regulation by the user-community itself.

Evaluating information can be done by either ranking its usefulness on a scale ranging from one to five stars or writing a comment or using both options. Other users see the average rating in the list of retrieved results (having used the TaToo-discovery function of Climate Twins). Furthermore if a user opens a link from the list of results in the dedicated viewer he/she finds more detailed evaluation information in the upper part of the viewer: the thread of comments and the individual ratings of the users before. It is also the place where the user enters his/her evaluation.

Of course, by relying on this kind of evaluation, it is impossible to block incorrect information. Other users can only rank it low and write critical comments, but the information remains in the knowledge base and cannot be removed.

4 Safeguarding Quality of User Entries by Means of Semantics

After having integrated the three TaToo services the new Climate Twins application with its added functionalities has been evaluated by test users from several research fields with experience in environmental issues (natural sciences, economics, geography, spatial planning, nature protection and engineering). Many comments expressed serious doubts whether self-regulation by rating and commenting could ever be sufficient to maintain an adequate quality level of the resources added by the user community. As a consequence, the idea was born to use the domain ontology for automated quality checks. Before the Climate Twins domain ontology was only used as a means to refine searches by ontology topics, so this opportunity to make more use of semantics seemed very attractive to the provider of Climate Twins.

The Climate Twins domain ontology has been constructed from two ontological resources: SWEET and reegle. The "Semantic Web for Earth and Environmental Terminology" (SWEET) is a large ontology developed by the Jet Propulsion Laboratory [4]. It covers almost any field of environmental science. So far, however, it lacks sophisticated relations between the concepts. "Reegle – the Clean Energy Info Portal" is an information portal for clean energy information [7]. It consists of country energy profiles including energy statistics, a clean energy search tool, a directory of relevant actors and stakeholders, a reegle blog and an extensive glossary of energy-related terms and concepts. Both sources could not be used directly as the Climate Twins domain ontology. SWEET is too broad, reegle is too detailed for the purpose of Climate Twins. As a consequence, a new much more focused preliminary domain ontology has been derived from these two sources (using only parts of the namespaces of the two ontologies), aligned to the TaToo bridge ontology and applied in the background of the Climate Twins application. The bridge ontology includes also GeoNames representing geographical space in the TaToo ontology framework [6].

The Climate Twins domain ontology is still preliminary. It is not yet covering all important concepts related to climate change, nor does it include GeoNames for all of Europe. Still it is sufficiently comprehensive to demonstrate the semantics-based automated quality check of user entries. In the following we describe the two cases where we have so far applied this method: geographically misplaced resources and resources where title and content description are inconsistent.

4.1 Misplacing a Resource

In this case a user uploads a link to a reasonable resource but geo-locates it incorrectly. Due to the fact that Climate Twins requires the selection of a particular GeoName, errors should occur less often than in the case of Web applications where resources can be located freely in geographical space (e.g. Google Earth). Nevertheless, it can still happen. A frequent source of mistakes are distinct places which share the same or a very similar name. Misplacing a resource in Climate Twins means that the user has linked it to a wrong GeoName. The logic of the automated check procedure in this case consists in the comparison of the GeoName selected for geo-locating the resource and any GeoName appearing in the title of the resource. If both are not identical, there needs to be an error and the entry is blocked.

In the following flowchart (see Fig. 2) this procedure is demonstrated with the example of the small village 'Andau' mixed up with the similarly sounding but distinct village 'Antau'. Both names are not only very similar, they are even located in the same Austrian province of Burgenland which makes such a mistake not unlikely.

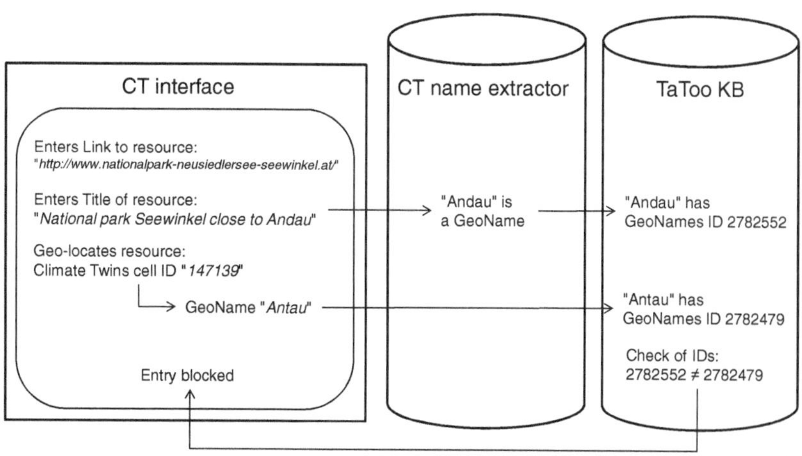

Fig. 2. Flowchart of the checking process for correct geo-location

In this example the user adds information to the national park "Seewinkel" in north-eastern Burgenland, close to Lake Neusiedl. The URL of this web resource is correct but not relevant here, because it does not contain a GeoName. He also adds a title. This entry matters, because there is a GeoName in it. Within the Climate Twins application there is a procedure to extract names from any text the user enters into the Add Info tab. One of these text elements in the title is the GeoName 'Andau'. Then the user makes a mistake while choosing the corresponding Climate Twins raster cell. He ends up in a cell containing the GeoName 'Antau' and selects it for geo-locating the information. Both GeoNames are sent to TaToo where the GeoNames IDs of both places are compared. Since they are not the same, the entry is blocked. Andau and Antau are distinct GeoNames, both cannot be true for the same information, so the entry must contain an error.

4.2 Selecting a Topic that is Incorrectly Related

In this case there occurs an inconsistency between two terms. Technically speaking the user chooses two disjoint ontology concepts for the same entry. This means there are two ontology individuals describing the same resource but are not accordingly related. Of course, for being able to run such a check procedure the underlying ontology has to contain all the required relations between the concepts as object properties.

The following example demonstrates this type of check: A Climate Twins user wants to add an information on animals typical of a peculiar Austrian region, the steppe region at its border to Hungary. What is wrong is that she links it to a wrong climate zone, i.e. a climate where a steppe cannot exist.

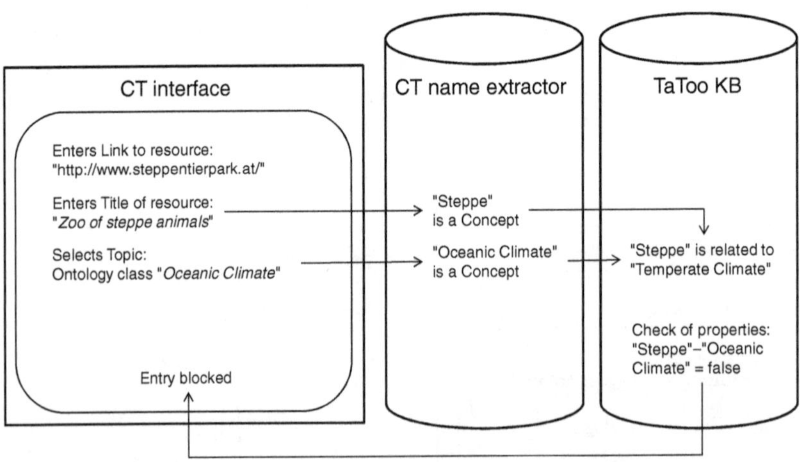

Fig. 3. Flowchart of the checking process for consistent terms

The user enters the URL of the respective zoo and the title. In both entries the word 'steppe' appears. In the title it is a distinct word, so it can easily be extracted and sent to TaToo as an ontology concept. Furthermore, the user selects one of the ontology individuals as a topic, i.e. a keyword indicating the content of the added information.

Here she chooses 'oceanic climate'. TaToo checks the object properties. In the ontology 'steppe' cannot be related to 'oceanic climate'. They are disjoint classes. This is due to the fact that a steppe can only exist in temperate and continental climates. With typing 'steppe' in the title and selecting 'oceanic climate' as topic the user has wrongly established a de facto-relation between the ecosystem 'steppe' and an incompatible climate zone. Since this relation cannot be true, the entry is blocked.

5 Conclusions

Semantics have the potential to provide a good means to check automatically the quality of freely uploaded information by users of the Climate Twins application. The Climate Twins publisher has no chance to do this personally, so an automatic procedure was essential. As the two examples described above show, the approach works basically. It is a first step, however, and the number of possible checks has to be extended a lot and some technical issues have to be tackled in order to be of real practical value in safeguarding the quality of user entries. In the following we list some major issues:

The name extraction procedure used in the Climate Twins application is still very simple. In the present version it cannot handle phrases. Regarding the first example of misplacing a resource this means that a lot of place names cannot be extracted, because they consist of more than one word. For example, there are two towns in Austria with the name 'Waidhofen', one with the extension 'an der Thaya', the other with 'an der Ybbs'. The name of the river is crucial for identifying the place unambiguously. The present name extractor cannot distinguish such places and hand over the correct GeoName to the TaToo system. A more powerful name extraction procedure is therefore needed.

The domain ontology must get more comprehensive. This is not only about adding concepts, which, of course, is important for its own sake. It is more important, however, to increase the internal complexity of the ontology. Formulating new specific relations between ontology concepts will allow for more types of checks. In the second example above, we only dealt with the relation between climate and vegetation. No other relations between ontology classes have been defined so far. Of course, this is an almost open-ended task with a lot of tricky issues to be tackled regarding the definition of relevant and correct properties which correctly link the concepts under all circumstances. This is more difficult than it seems at first glance, because what is obviously a correct relation in one context can be wrong in another.

Climate Twins is still indicated as beta-version. The main reason is a still weak usability. Much of it concerns deficiencies regarding its user interface: More intuitive titles and labels are needed, a more detailed map to make it easier for Climate Twins users to orient themselves. Furthermore the two maps now should be collapsed into one map for both selection of the POI and viewing the results. But usability concerns also the new automated checks. User-friendliness in this context means that a user who has made one of the errors for which an automatic check exists receives meaningful feedback. For now a wrong entry is simply blocked. In the future it is planned that there are more meaningful error messages which contain the reason why an entry has been blocked. In the first case of a misplaced resource this could be something like: "Your information is probably misplaced. Please, check the location again." Overall, usability has to be improved a lot, before we can expect that self-organized information enrichment by Climate Twins users will take off.

Acknowledgements. The research leading to these results has received funding from the European Community's Seventh Framework Programme (FP7/2007-2013) under grant agreement nr. 247893.

References

1. Climate Twins Viewer: AIT Climate Twins Viewer v3.0, http://foresight.ait.ac.at/projects/tatoo/
2. COSMO-CLM: Consortium for Small-scale Modeling – Climate Local Model, http://www.cosmo-model.org/
3. GeoNames: GeoNames geographical database, http://www.geonames.org
4. Jet Propulsion Laboratory, California Institute of Technology: SWEET - Semantic Web for Earth and Environmental Terminology, http://sweet.jpl.nasa.gov
5. Loibl, W., Peters-Anders, J., Züger, J.: Climate Twins – a Tool to Explore Future Climate Impacts by Assessing Real World Conditions: Exploration Principles, Underlying Data, Similarity Conditions and Uncertainty Ranges. Geophysical Research Abstracts 12, EGU2010-12149 (2010)
6. Pariente, T., Fuentes, J.M., Sanguino, M.A., Yurtsever, S., Avellino, G., Rizzoli, A.E., Nešić, S.: A Model for Semantic Annotation of Environmental Resources: The TaToo Semantic Framework. In: Hřebíček, J., Schimak, G., Denzer, R. (eds.) ISESS 2011. IFIP AICT, vol. 359, pp. 419–427. Springer, Heidelberg (2011)
7. Renewable Energy & Energy Efficiency Partnership (REEEP): reegle - Clean Energy Info Portal, http://www.reegle.info
8. Ungar, J.: A Comparative Analysis of Region Pairs Matching Current and Future Climate Conditions. Diploma thesis, University of Vienna, Department of Geography and Regional Research (2010)
9. Ungar, J., Peters-Anders, J., Loibl, W.: Climate Twins – An Attempt to Quantify Climatological Similarities. In: Hřebíček, J., Schimak, G., Denzer, R. (eds.) ISESS 2011. IFIP AICT, vol. 359, pp. 428–436. Springer, Heidelberg (2011)

Information System Efficiency as an Attribute in Environmental Information Systems

Miloš Koch and Zuzana Chvátalová

Brno University of Technology, Faculty of Business and Management,
Kolejni 4, 612 00 Brno, Czech Republic
{koch,chvatalova}@fbm.vutbr.cz

Abstract. In terms of contemporary civilization development one of the key aspects of environmental protection is the effort to avoid wasting resources. In the case of environmental information systems there is a necessity to strive for such systems, which deliver the highest benefits with the lowest possible consumption of resources (material, energy, as well as human, etc.). Therefore, every information system must be regarded comprehensively, i.e. as a set of mutually interactive subsystems (components, parts). The experiences confirm that the information system, parts of which are not mutually balanced, leads to the lower effectiveness and efficiency of the system as a whole in contrast with a system which may be managed in such a way that all its parts are maintained at a comparable level. Thus, the effectiveness of environmental/information systems must be harmonized as a whole. This paper presents "the HOS method" developed in the Faculty of Business and Management of the Brno University of Technology. This method enables the primary assessment of the balance of organizations' information systems. The HOS analysis is based on a considering of the basic subsystems: Hardware, Orgware and Software (hence the title HOS), but also the following subsystems: Peopleware, Dataware, Customers, Suppliers and Information System Management. The understandable and transparent HOS diagrams for assessing the balance of an information system are presented in a short case study where authors submit also some selected statistics about their current survey results obtained by means of empirical measurement of approximately four hundred chosen organizations from the Czech Republic and Slovak republic. The research is realized using the Zefis portal.

Keywords: Balance, Efficiency and Effectiveness, Environment, Environmental Information System, Evaluation, HOS Analysis and HOS Diagram, Information and Communication Technologies, Information system, Subsystems.

1 Introduction

Information system/s (IS) and *information and communication technologies* (ICT) are now an indispensable part of the everyday life of people on Earth and environmental information systems are their important subset [5], [6]. Since the early beginnings across centuries up until the substantially accelerated present, the philosophy of calculations, "counting mechanisms", functionality and even the production itself of

J. Hřebíček et al. (Eds.): ISESS 2013, IFIP AICT 413, pp. 31–43, 2013.

mechanical counting aids, are formed and limited by the needs and possibilities of the real world, of concrete problem-solving (at present, problems of a multiple character), and above all, connected with technical progress. The implementation of IS and ICT in its whole development thus unmistakably corresponds with the challenges posed by material, space, time, cost and energy connected with their environmental aspects, and furthermore, it has been challenged by the users' literacy, the organization and direction, the data accessibility, and by ensuring the safety, and many other factors. With the passing of time, it is necessary to understand this field of human activities as a complex system, i.e., a set of mutually interconnected elements, which reacts by certain behavior to the inputs (i.e., transforms the inputs into outputs) and serves a certain goal. At the same time, there are objective requirements that can be attached to them (e.g. from the viewpoint of the sustainability of development, more in [13]) by the external world.

Assessment of IS has been done from various viewpoints, on different levels of difficulty. It is frequently determined by conditions and according to the purpose which the assessment is supposed to serve. Many authors deal with this [4], [7], [11], [12]. Let us name the following authors of interesting works or projects: R. Weber [14] (the evaluation of theories within the information systems,) R. Gomez and S. Pather [3] (the experiences of ICT evaluation – not only the measurable tangible and quantifiable benefits of ICT but also the intangible benefits of ICT are more important from a developmental perspective). Among Czech experts dealing with IS evaluation let us mention professor J. Voříšek [15].

At the Faculty of Business and Management of the Brno University of Technology, the HOS method is being developed. This method reflects that it is necessary to understand whole IS as a complex of means and activities. This means that the IS is a developed system which is formed by subsystems. The HOS method assesses the level of balance of these subsystems, i.e., the balance of the whole IS (balanced systems are more effective than unbalanced). In dealing with this issue more widely, the concepts of system, its efficiency and effectiveness play a basic part.

At first, the HOS analysis was based on considering three basic subsystems of the IS: *Hardware, Orgware* and *Software* (hence the title HOS), and later, these areas were assigned as well: *Peopleware, Dataware, Customers, Suppliers, Management IS* [8]. The HOS diagrams for assessing the efficiency of a every information system are usefully comprehensible. The scale on the HOS diagram (indicates the level of efficiency of the IS) shows a balance of the IS by using the evaluation of its subsystems, identifying the level of the whole IS and the recommended level for the given IS.

2 Aspects of Information System Efficiency Evaluation

The basic problem in the HOS method is to determine *the subsystems*, i.e., which parts of the IS are appropriate for being considered, and how to evaluate their level. Based on a long-term validation of the importance of individual subsystems we can finally select eight basic subsystems (parts, components, areas, elements) of the IS. We consider the eight subsystems presented in Table 1 to be the components of the IS. We evaluate the balance which supports its efficiency.

Table 1. The subsystems in the HOS method (Source: Authors´ elaboration)

The monitored subsystems of the information system using the HOS method			
Abbreviation	**Level**	**Subsystem**	**Description of the evaluation of the respective subsystem**
HW	L_1	*Hardware*	Level of the technical equipment (hardware products) of the given organization;
SW	L_2	*Software*	Monitoring of features, complexity of use, operation and controls of the program equipment (software products) of the given organization;
OW	L_3	*Orgware*	Rules for the operation of information systems, the recommended operating procedures, security rules;
PW	L_4	*Peopleware*	Level of computer literacy of users and primarily their duties and responsibility with respect to a given information system;
DW	L_5	*Dataware*	Availability, management and security data sets; their usefulness in the organization processes;
CU	L_6	*Customers*	The term "customer" can be seen as a real customer, such as a user of electronic commerce (e-commerce), citizen looking for information or as any organization employee that needs an information system and its outputs to work;
SU	L_7	*Suppliers*	The term "supplier" is meant in the sense of someone who ensures the operation of the given information system; if it is a system whose operation and support are provided by another organization, the concept of supplier is understood in the usual sense; if the organization provides the operations or support of information systems directly, then the term "contractor" means just these workers;
MA	L_8	*Management IS*	The quality of information systems management in relation to information strategy, the consistency of application of the rules and the perception of the end-user information system;

3 The HOS Method

To use the HOS method we can focus on information systems (e.g., environmental information systems).

Let us consider $n \in N$ (N is a set of natural numbers) and the information system (IS) is composed of individual subsystems S_i, i=1,…,n (we can also call them as components, parts, elements, areas, etc.).

The fundamental concept of the HOS method is based on:

- the evaluation of individual subsystem levels L_i, i=1,…,n of the information system IS, where the level L_i is for the subsystem S_i, i=1,…,n and
- the identification of the subsystem/s with "the worst" level/s. This/these subsystem/s has/ve a negative impact on (i.e., reduce) the whole IS level L_w.

Let us imagine that the IS consists of its subsystems and their interrelationships, and it has a defined way of its behavior. If the level of its individual subsystems is different (we understand "level" in terms of the degree to which a considered subsystem conforms to our needs). The question what the level of the IS will be as a whole arises from this debate. The HOS method is primarily designed to find weaknesses in the IS, and therefore it is necessary to evaluate the level of the weakest link. This means we assume that the IS is as good as its weakest subsystem, i.e. see (1):

$$L_w = \min\{L_1,.., L_m\}, \tag{1}$$

where L_w is the level of the whole IS,
L_i and L_j are the levels of subsystems S_i and S_j,
$1 \leqq m \leqq n$: $\forall i, j \in \{1,.., m\}, i \neq j : L_i \neq L_j$.

The aim of the HOS method is assessing the level of the key subsystems of the IS and to determine whether the levels of these subsystems are mutually comparable (i.e., they have the same or a similar level). Then we can say that it is a so-called *a balanced information system*, otherwise we say it is *an unbalanced information system*, resp. *a heavy unbalanced information system.* Unbalance of the IS usually leads to the inefficiency of the whole IS, because the system inefficiencies increase, for example, the total cost. Thus, it is obvious that inefficient IS as a whole lead to difficulty in the optimal use of resources, and therefore they burden nature with its excessive utilization and pollution (manufacturing, energy, emissions, waste, biodiversity, etc.).

Let us visualize the situation as an example of the two different compared information systems IS_1, IS_2 with four subsystems S_1, S_2, S_3, S_4 (see Fig. 1).

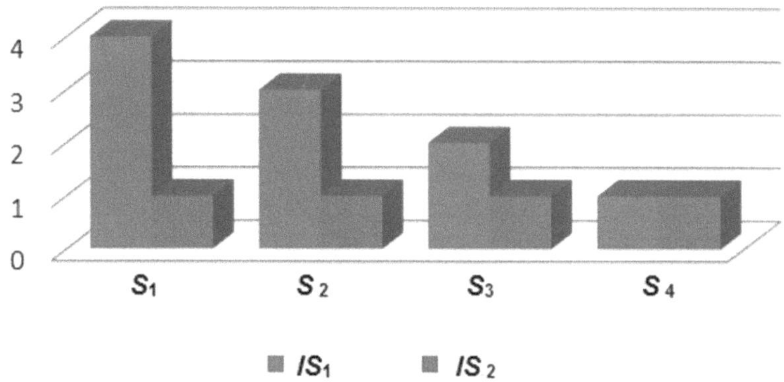

Fig. 1. Subsystem levels of compared IS_1 and IS_2 (Source: Authors´ elaboration in Microsoft Excel according to [10])

Whole levels of both information systems IS_1, IS_2 are equivalent, i.e., we can write (2):

$$L_{w_1}(IS_1) \cong L_{w_2}(IS_2)\cdot \qquad (2)$$

The histogram in Fig. 1 presents this situation: we apply (1) and we can say that (2) holds. In terms of the usefulness of information systems IS_1 and IS_2 in the organization we have to say: they are comparable, because the same degree of their whole levels is given as a level of the weakest subsystems of each of them (in both cases, it is the level of the subsystem S_4). However, from the histogram is possible to deduce that the total costs of the information system IS_1 will be substantially higher than the information system IS_2, i.e., we can write (3):

$$TC_{w_1}(IS_1) > TC_{w_2}(IS_2)\,, \qquad (3)$$

where $TC_{w_1}(IS_1)$, $TC_{w_2}(IS_2)$ are the total costs of the information systems IS_1, IS_2.

This source of inefficiency in terms of sustainable development definitely cannot be supported. Studies indicate that the improper implementation of IS and ICT leads to the considerable environmental pollution as a result of direct or indirect emissions (mainly CO_2, SF_6, etc.), direct and indirect energy consumption, material consumption, water consumption, increase of the amount of (hazardous) waste, etc. It is also necessary to include investment and non-investment costs to protect the environment (recycling, optimization of organization and management, compliance with environmental laws), etc. We believe that the current sustainable development and the respect for the environment are a significant asset and an essential fact to the performance and the success of each organization.

3.1 The Monitored Subsystems in the HOS Method

After long-term analyses we selected eight basic subsystems of the IS, see Table 1.

3.2 The Subsystem Level

The first step in evaluating the HOS method is an assessment of the level of individual subsystems, briefly *the subsystem level.* Each subsystem is rated on a four-point scale as 1 – *bad*, 2 – *rather bad*, 3 – *rather good*, 4 – *good.*

We even tried a more precise evaluation using the whole range of real interval <1;4>. But it turns out that, due to the explanatory character of this method and the way of self-evaluation by the user, a more accurate evaluation is unnecessary[1].

The assessment of each subsystem can be made by qualified assessment experts (but it excludes the use of methods for the initial assessment by the organization itself), or using control questions for each subsystem by using a questionnaire. The HOS method uses the control issues, ten questions for each subsystem. Their formulation is based on checking the most common weaknesses. For the purpose of the questions we used the expert opinion [1]. Currently, this set of questions is being modified. This modification is based on the feedback evaluation and other concrete experience with surveys in organizations.

The subsystem levels of the IS are plotted as the boundary of the light gray irregular octagon in the sample chart in Fig. 2, Fig. 3 and Fig. 4 (more precisely, the intersection of the octagon boundary and all eight axes).

3.3 The Whole Level

The whole level of the IS is determined by its weakest link, see (1). The whole level of the IS is plotted as the boundary of the gray regular octagon in the sample chart in Fig. 2, Fig. 3 and Fig. 4 (more precisely, the intersection of the octagon boundary and all eight axes).

Let L_w be the level of the whole information system *IS* and L_i, $i = 1,..,n$ the levels of the subsystems S_i of the information system *IS*.

We consider the information system *IS a balanced information system* provided that all its eight subsystems are at the same level or, at most, three of its subsystem levels are different from the others by no more than one evaluation point[2], see (4). In this case of we can assume that it is the IS with the optimal efficiency ratio: benefits to costs.

$$[\sum_{i=1}^{n}(L_i - L_w) \leq 3] \wedge [\forall i \in \{1,..,n\} : L_i - L_w \leq 1] \tag{4}$$

An unbalanced information system is such IS that does not satisfy the first part of the condition (4), but the each subsystem level (evaluation done on each axis) does not exceed a difference of one evaluation point than has of the whole IS level 's, see (5).

$$[\sum_{i=1}^{n}(L_i - L_w) > 3] \wedge [\forall i \in \{1,..,n\} : L_i - L_w \leq 1] \tag{5}$$

---

1 It is possible to access in this way: If someone chooses to assess this using the real number in the range <1;4>, the required argument can be rounded down to an integer. We can use the Microsoft Excel command: *INT* (*number*).

2 *Note*: In (4), (5), (6) for $\forall i \in \{1,..,n\} : \min(L_i - L_w) = 0$.

A heavy unbalanced information system is such IS that does not satisfy these conditions (3) and (4) because in the IS the subsystem exists, one which has a level (evaluation on axis) exceeding a difference of one evaluation point compared with what the whole level is (note, that it does not depend on the value of the expression $\sum_{i=1}^{n}(L_i - L_w)$), see (6).

$$\exists i \in \{1,..,n\} : L_i - L_w > 1 \tag{6}$$

3.4 The Recommended Level

The recommended level of IS is based on the importance of the IS which an organization attaches to it. If the IS is indispensable for the organization's activities, then the recommended level is 4, i.e., *good.* For IS, without, which the activity of the organization is possible, albeit with great difficulties, the recommended level is 3, i.e., *rather good.* If an organization can exist without the analyzed IS and this state of affairs brings to the organization only very few or no problems at all, the recommended level is 2, i.e., *rather bad.* In this case, the question arises whether the IS has any sense for the organization at all, and whether the costs incurred are proportional to the benefit. The recommended status must be understood as the minimum required level. The recommended level of the IS is plotted as the bold black curve (the regular octagon) in the sample chart in Fig. 2, Fig. 3 and Fig. 4 (more precisely, the intersection of the octagon boundary and all eight axes).

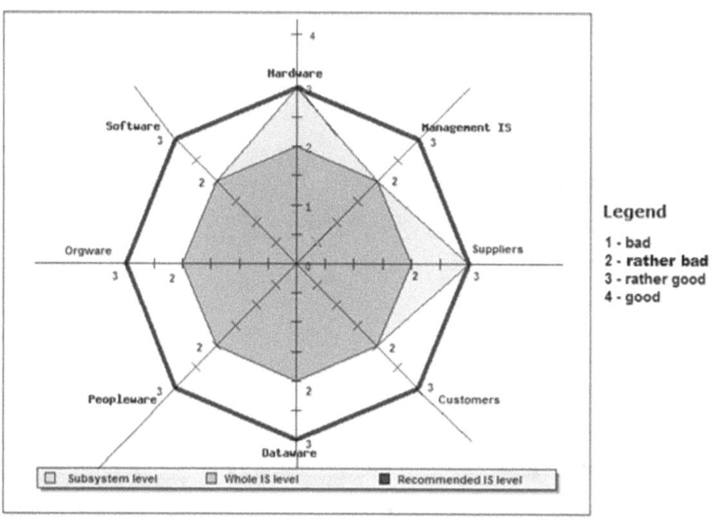

Fig. 2. The HOS diagram for the balanced IS (Source: Authors´ elaboration [10])

4 The HOS Diagram – Case Study

In the following examples will present the HOS analysis of the IS balance of four organizations from our research using the Zefis portal [10]. We have selected organizations that submit environmental reporting. The results are calculated in the Maple system. Many useful instructions we gain in [2], [8], [9].

4.1 The Balanced Information System

In Fig. 2 is presented the sample of the balanced IS's evaluation of one selected organization from the Zefis portal [10]. The HOS analysis of the evaluation of this IS is shown in detail in Tab. 2 and Tab. 3.

Table 2. Subsystem levels of the IS using the HOS method (Source: Authors´ elaboration [10])

	SUBSYSTEMS							
	HW	SW	OW	PW	DW	CU	SU	MA
HOS Diagram Fig. 2	*Light grey irregular octagon boundary*							
Level	L_1	L_2	L_3	L_4	L_5	L_6	L_7	L_8
	3	2	2	2	2	2	3	2

Table 3. The evaluation of the IS using the HOS method (Source: Authors´ elaboration [10])

	HOS Diagram Fig. 2	According to Condition/s	IS
Whole Level	*Grey regular octagon boundary*	(1) $L_w = \min\{3,2\} = 2$	2
Currrent Status		(4) $\sum_{i=1}^{8}(L_i - L_w) = 2 \le 3$ $\forall i \in \{1,..,8\} : L_i - 2 \le 1$	*Balanced*
Recommended Level	*Bold black regular octagon*	--	3 *Rather good*

Recapitulation: The monitored IS's whole level is equal to 2. Only two subsystems (HW and SU) are at a level which differs from the whole IS level (namely concerning the admissible 1 point evaluation), thus (4) is fulfilled and therefore the IS is *balanced.* Its recommended level is equal to 3 (rather good).

4.2 The Unbalanced Information System

In Fig. 3 we present the sample of the unbalanced IS's evaluation of one organization from the Zefis portal [10]. The HOS analysis of the evaluation of the IS is shown in detail in Tab. 4 and Tab. 5.

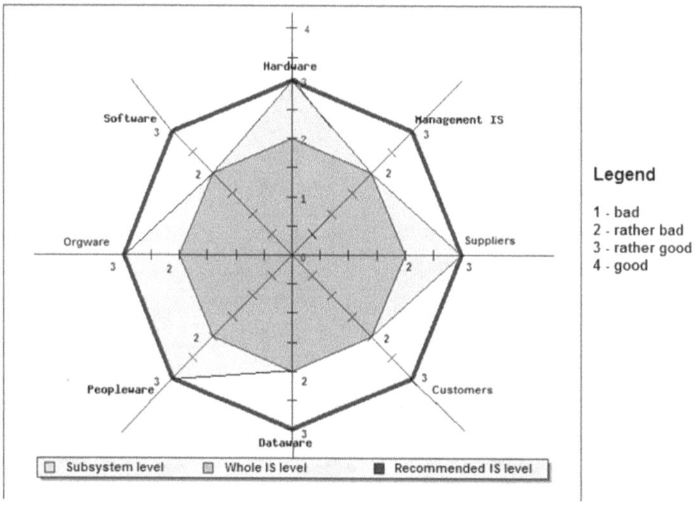

Fig. 3. The HOS diagram for the unbalanced IS (Source: Authors´ elaboration [10])

Table 4. Subsystem levels of the IS using the HOS method (Source: Authors´ elaboration [10])

	SUBSYSTEMS							
	HW	SW	OW	PW	DW	CU	SU	MA
HOS Diagram Fig. 3	*Light grey irregular octagon boundary*							
Level	L_1	L_2	L_3	L_4	L_5	L_6	L_7	L_8
	3	2	3	3	2	2	3	2

Table 5. The evaluation of the IS using the HOS method (Source: Authors´ elaboration [10])

	INFORMATION SYSTEM		
	HOS Diagram Fig. 3	**According to Condition/s**	**IS**
Whole Level	*Grey regular octagon boundary*	(1) $L_w = \min\{3,2\} = 2$	2
Currrent Status		(5) $\sum_{i=1}^{8}(L_i - L_w) = 4 > 3$ $\forall i \in \{1,..,8\} : L_i - 2 \leq 1$	*Unbalanced*
Recommended Level	*Bold black regular octagon*	--	3 *Rather good*

Recapitulation: The monitored IS's whole level is equal to 2. Half of the subsystems (HW, PW, OW and SU) are at a level which differs from the whole IS level, (namely concerning the admissible 1 point evaluation), i.e. (4) is not partly fulfilled, thus (4) is not valid, (5) is true and therefore the IS is *unbalanced*. Its recommended level is equal to 3 (rather good).

4.3 The Heavy Unbalanced Information System

In Fig. 4 we present the sample of the heavy unbalanced IS's evaluation of one organization from the Zefis portal [8]. The HOS analysis of the evaluation of the IS is shown in detail in Tab. 6 and Tab. 7.

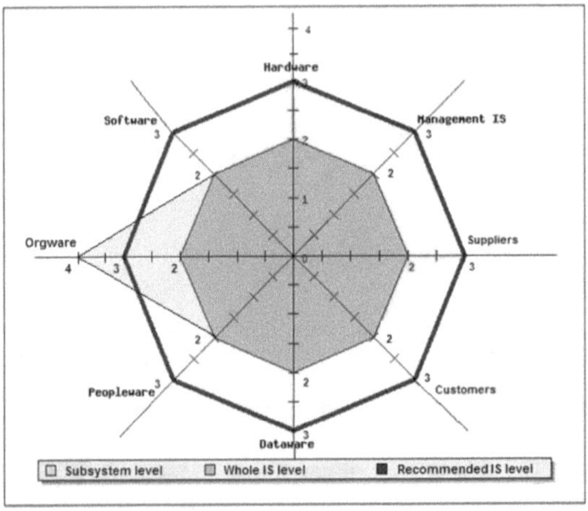

Fig. 4. The HOS diagram for the heavy unbalanced IS (Source: Authors´ elaboration [10])

Table 6. Subsystem levels of the IS using the HOS method (Source: Authors´ elaboration [10])

	SUBSYSTEMS							
	HW	**SW**	**OW**	**PW**	**DW**	**CU**	**SU**	**MA**
HOS Diagram Fig. 4	*Light grey irregular octagon boundary*							
Level	L_1	L_2	L_3	L_4	L_5	L_6	L_7	L_8
	2	2	4	2	2	2	2	2

Table 7. The evaluation of the IS using the HOS method (Source: Authors´ elaboration [10])

	INFORMATION SYSTEM		
	HOS diagram Fig. 4	**According to condition/s**	**IS**
Whole Level	*Grey regular octagon boundary*	(1) $L_w = \min\{4,2\} = 2$	2
Currrent Status		(6) $\sum_{i=1}^{8}(L_i - 2_w) = 2 \leq 3$ $\exists i = 3 : L_3 - 2 = 2 > 1$	*Heavy unbalanced*
Recommended Level	*Bold black regular octagon*	--	3 *Rather good*

Recapitulation: The monitored IS's whole level is equal to 2. It is sufficient that the level of one subsystem (OW) differs by more than one evaluation point from the value of the whole IS level; i.e., (4) and (5) are not valid, thus (6) is true and therefore the IS is, heavy unbalanced. Its recommended level is equal to 3 (rather good).

5 The Outcomes and Results

The Zefis portal [10] offers organizations (free of charge, anonymously) to evaluate the efficiency and balance of their IS using the HOS method (by means of questionnaires).

The HOS method has been tested on a sample of 425 organizations from the Czech Republic and Slovak Republic in the period from 2010 until now (the state of affairs as of March 4, 2013). Fig. 5 presents the aggregated average results of the HOS method. We apply the evaluation criteria (1), (4), (5), (6) then the Fig. 6 represents the sample of monitored organizations with environmental information systems. The Fig. 6 presents the situation in terms of how given organizations understand the importance of their IS, and especially, whether the organizations would function even without the IS.

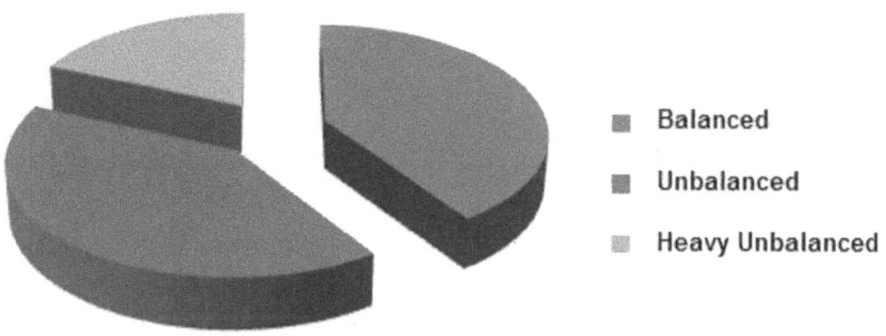

Fig. 5. The proportions of organizations by criteria: a/an balanced / unbalanced / heavy unbalanced state of the monitored IS (Source: Authors´ elaboration in Microsoft Excel according to [10])

Let us summarize the results. This output shows that, of the relevant sample of organizations: approximately its two fifths (40.9 %) have a balanced IS. Approximately the same portion of the sample (40.5 %) has an unbalanced IS. The rest (18.6 %), i.e., less than approximately one fifth of the sample, has a heavy unbalanced IS. The first two partial results are quite optimistic, because also the part of organizations that has an unbalanced IS has the ability to constructive way to remedy the situation. Organizations with a heavy unbalanced IS probably need to have the costs in addition for the more complicated deciding whether A total renovation of their information systems will be appropriate for them or not.

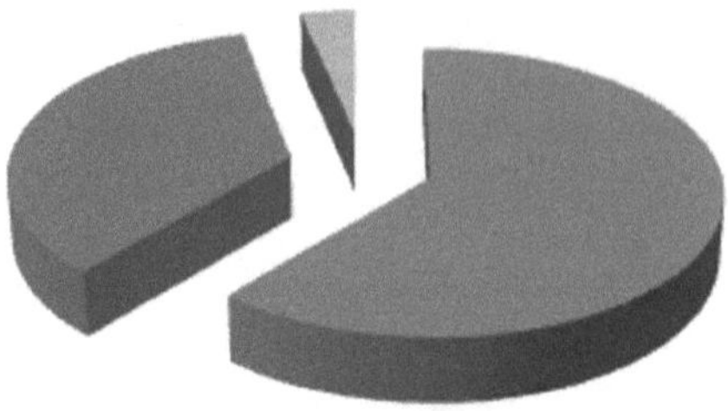

Fig. 6. The proportion of organizations by criteria: the importance of the IS for the organization (Source: Authors´ elaboration in Microsoft Excel according to [10])

In our sample survey of meonitored organizations responded, approximately 61.9 % of the organizations could not function in any case whatsoever without the IS, approximately 34.8 % of them could function only partially, with difficulties, and only approximately 3.3 % of the organizations believe that without this IS they could work with little or no difficulty.

6 Conclusion

The HOS method is one of methods implemented at the Zefis portal [10]. This portal is designated mainly for organizations from the Czech and Slovak Republics, for the assessment of their IS including environmental information systems (EIS). (The portal is partly accessible in English as well).

Experience and response organizations using the HOS method and the Zefis portal have brought positive feedback. The base method and guidance of Zefis portal provide an effective tool to perform very rapid initial assessment of the status of their IS including EIS. In turn, leads to a reduction of costs and not only for the organization itself, but also from the perspective of environmental protection. Another important – though hidden – attribute is the fact that the HOS method is forcing the organization to the reflection and self-reflection, how to improve a particular subsystems of their IS. This leads to an increase in its overall usefulness and effectiveness for these organizations, as well as for the environment. The HOS method on the Zefis portal shall in addition, when viewing the results of the user suggestions to improve their IS in various subsystems. This can contribute to more effective use of information management systems, which ultimately leads to reduce wasted resources and unnecessary burden for the environment.

Limiting the resource wasting and an unnecessary burdening of the environment is the only possibility of survival of not only of organizations, but also of civilization as such.

Acknowledgments. This paper was supported by grant FP-S-13-2148 'The Application of ICT and Mathematical Methods in Business Management' of the Internal Grant Agency at Brno University of Technology.

References

1. Dovrtěl, J.: Selected Aspects of the Effectiveness of Information Systems (Dissertation in Czech). BUT FBM, Brno (2004)
2. Gander, W., Hřebíček, J.: Solving Problems in Scientific Computing Using Maple and Matlab, 4th edn. Springer, Berlin (2004)
3. Gomez, R., Pather, S.: ICT Evaluation: Are We Asking the Right Questions? The Electronic Journal of Information Systems in Developing Countries 50, 1–14 (2012), http://www.ejisdc.org/ojs2/index.php/ejisdc/article/view/879
4. Grover, V., Jeong, S.R., Segars, A.H.: Information Systems Effectiveness: The Construct Space and Patters of Application. Information and Management 31, 177–191 (1996)
5. Hilty, L., Page, B., Hřebíček, J.: Environmental Informatics. Environmental Modelling & Software 21, 1517–1518 (2006)
6. Hřebíček, J., Pillmann, W.: eEnvironment: Reality and Challenges for eEnvironment Implementation in Europe. In: Hřebíček, J., Schimak, G., Denzer, R. (eds.) ISESS 2011. IFIP AICT, vol. 359, pp. 1–14. Springer, Heidelberg (2011)
7. Cha-Jan Chang, J., King, W.: Measuring the Performance of Information Systems: A Functional Scorecard. Journal of Management Information Systems 22, 85–115 (2005)
8. Chvátalová, Z., Hřebíček, J.: Modeling and Simulation Utility Functions with Maple. Mendel Journal Series 1, 552–557 (2012)
9. Chvátalová, Z., Hřebíček, J.: Computational Finance and Finance Economics with Maple. International Journal of Mathematical Models and Methods in Applied Sciences 7, 541–550 (2013)
10. Koch, M.: ZEFIS: Online Assessment for your Information System. BUT FBM, Brno (2013), http://www.zefis.cz/
11. Miller, J., Doyle, B.A.: Measuring the Effectiveness of Computer-Based Information Systems in the Financial Services Sector. MIS Quarterly 11, 107–126 (1987)
12. Molnár, Z.: Effectiveness of Information Systems, 1st edn. Grada Publishing, Praha (2000) (in Czech)
13. Soukopová, J., Hřebíček, J., Štencl, M., Trenz, O.: Integration of Economic, Environmental, Social and Corporate Governance Performance and Reporting in Enterprises. Acta Universitatis Agriculturae et Silviculturae Mendelianae Brunensis 59, 157–166 (2011)
14. Weber, R.: Evaluating and Developing Theories in the Information Systems Discipline. Journal of the Association for Information Systems13 (2012), http://aisel.aisnet.org/jais/vol13/iss1/2
15. Voříšek, J.: Strategic Management of Information System and System Integration. Management Press, Praha (2002) (in Czech)

PROMETHEE-GAIA Method as a Support of the Decision-Making Process in Evaluating Technical Facilities

Alena Kocmanová, Marie Dočekalová, and Jiří Luňáček

Brno University of Technology, Faculty of Business and Management, Kolejni 4, 612 00 Brno, Czech Republic
{kocmanova,docekalova,lunacek}@fbm.vutbr.cz

Abstract. This paper describes the application of PROMETHEE-GAIA methodology in a multiple criteria analysis to rank potential environmental investments in mineral-processing companies. The intent of the paper is to identify best technical facilities on the basis of preferential relations between a set of variants. The method of Total Cost Analysis (TCA) was chosen to define the criteria. The economic and environmental costs, as well as the benefits of these technical facilities, were determined by means of this method. PROMETHEE is one of the methods in the Multi Criteria Analysis (MCA) category. The MCA, as the name implies, deals with the evaluation of a number of variants by several criteria. The technical facility was selected by a comparative analysis involving five influential parameters (Investment Costs, Annual Operating Costs, Operating Income, Administrative Costs and Disposal Fees, Economic and Environmental Benefits). As expected, the analysis resulted in a preferential ranking of these technical facilities.

Keywords: total cost analysis, multicriteria analysis, PROMETHEE, environmental technology, technical facilities.

1 Introduction

Decision-making is one of the most important activities in managing a private company, although decisions must likewise be made in public administration and in self-governing entities. The ability to make a quick and correct decision is the key to success in any field on any level, be it in business circles, in a national arena, or on an international scale. Decisions are made about financial strategies, investment projects, public government contracts, and a multitude of other endeavors. Public tenders, even though governed by law, would have much fewer problems in defending the outcome if they employed a multiple criteria approach. The impact of these decisions on the entire society should be thoroughly scrutinized. Rather than relying on purely economic considerations, or a subjective judgment, it is necessary to incorporate some additional criteria. This is particularly true when awarding public contracts, for example when trying to choose the right environmental investment, ecological innovation

J. Hřebíček et al. (Eds.): ISESS 2013, IFIP AICT 413, pp. 44–53, 2013.

or technology, it is necessary to consider also the environmental and the social aspects, along with the economic criteria.

A multi criteria variant analysis represents a valuable tool in reaching rational decisions about environmental investments in the light of multiple criteria and conditions of risks and uncertainty. In a variant evaluation by multiple criteria, it is recommended to use more than one method to verify the sensitivity of the preferential order of variants depending on the method used. If the task is to select an optimal environmental investment, it is necessary to take into account all costs associated with the investment inclusive of the economic, environmental, and social benefits. A good method for this kind of evaluation of environmental investments is Total Cost Analysis (TCA), which selects the optimal environmental investment by means of multiple criteria analysis.

The intent of this paper is to select the optimal environmental investment (technical facility) for the production of cement clinkers, glass, basalt, and ceramic products in companies processing minerals, using the methods of Total Cost Analysis (TCA) and Preference Ranking Organization for Enrichment Evaluation (PROMETHEE).

2 Theoretical Approach

The 1990s witnessed an emerging trend in the Czech Republic of investing heavily in environmental protection. Per Czech Statistical Office, CZK 83.3 billion was spent on in 2011 to protect the environment, or about 2.2% of the GDP. The environmental investments jumped 10% from the previous year, to CZK 24.8 billion. When environmental investments are pondered, methodological nuances may be crucial. These investments may be inspired by administrative actions, voluntary standards (eco-labeling), stricter product requirements or increasing environmental costs. They are usually accompanied by lack of information about cost-effective solutions. To evaluate an investment, we should know, first of all, if the project is even realistic, and, if so, which of the proposed alternatives is the best. Environmental investments must take into account the non-financial aspects as well, such as the impact on the quality of air, climate, water (surface and underground), soil, etc. The evaluation of an investment (technical equipment) in pursuit of environmental objectives lends itself to multiple criteria-based decision making methods, because these projects tend to be complex and multifaceted [13]. The proposed methods guarantee the inclusion of all relevant data [14]. The goal of such evaluation is done primarily to identify the best option and to rank all options from best to worst. The best variant is usually a compromise [18].

For the purposes of making decisions with multiple criteria, the tasks can be itemized in various ways according to different authors. A breakdown is possible according to the task-solving goal and the information associated with it [2]. Authors Fiala, Jablonský and Maňas [8] describe a division based on the manner of defining the set of acceptable variants. Fotr, Dědina and Hrůzová [9] introduce other types of itemization: by the structure of the problem to be resolved, by the information about the states of the world and the consequences of variants, by the nature of the decision subject viewed as a time factor or the level of control. Finally, Triantaphyllou [18] emphasizes the division mainly by the type of input data.

3 Methodology

Arriving to decisions about environmental investments through multiple criteria analysis involves several steps, from formulating the possible versions of a decision to actually adopting it. The objective of the multi criteria variant evaluation is, in most cases, a selection of the optimal (compromise) variant and creating a platform for complex decision-making processes typical for environmental investments. The multi criteria analysis is essentially a mathematical model, the application of which should lead to a selection of the optimal variant from a group of variants feasible in a given situation. Optimal variants are expressed by means of a set of criteria that serve to select the optimal variant. Choosing the right criteria is an important step towards an objective assessment of all variants, and so is the determination how much weight should be given to each criterion in accordance with its importance.

A decision to proceed with a certain environmental investment should consider the requirements of Best Available Techniques (BAT). At the EU level, BAT for selected industrial and agricultural activities are defined and quantified in Best Available Techniques Reference Documents (BREF).

From the financial perspective, environmental investments may be evaluated by identifying all direct and indirect costs, the economic and environmental benefits, and applying a multiple criteria analyzing method like TCA or PROMETHEE. The latter is applicable to a variety of decision problems in diverse fields [1]. PROMETHEE belongs (together with ELECTRE) to a group of outranking methods [17] and it is based on pairwise comparisons [10]. This method was presented for the first time by Jean-Pierre Brans in 1982, see e.g. [4-6]. Together with its descriptive complement, it is known as the PROMETHEE & GAIA methods.

3.1 Total Cost Assessment (TCA) Method

The TCA method includes all relevant costs and benefits related to an environmental investment with the exception of social costs, or externalities [12].This method was developed in an organization named Tellus Institute in Boston. It has been described in foreign literature and modified in various ways [7], [11-12]. The costs are classified as direct, indirect, conditional, and difficult-to-quantify. The most important part of the cost analysis is the necessity to determine all inputs and outputs, to identify all relevant costs and benefits resulting from the environmental investment expressed in monetary units or described qualitatively. The difficulty of itemizing the different types of costs, an essential input in TCA, depends on the probability of their occurrence in the future. The final decision needs to consider all potential factors that may affect the overall effectiveness of a given environmental investments and define their mutual relations. It is therefore imperative to assess both the quantitative and the qualitative criteria.

3.2 PROMETHEE Method I, II

The PROMETHEE-GAIA method is a multicriteria decision method developed by J. P. Brans and B. Mareschal [4-5], [15]. The PROMETHEE-type methods are based on a paired comparison of variants, progressively in terms of all the criteria. The result of this comparison is an expression of preferential intensity between the pairs of variants evaluated with regard to all criteria.

$$P(a_i, a_j) = P\lfloor f(a_i) - f(a_j) \rfloor = P(d), P(d) \in (0,1) \tag{1}$$

A higher *P(d)* value indicates a higher degree of preference of the variant ai over the variant aj. A double-sided preference is described by function $H(d) \in$ <-1,1>, defined as:

$$H(d) = \begin{bmatrix} P(a_i, a_j, d \geq 0 \\ P(a_i, a_j, d \leq 0 \end{bmatrix} \tag{2}$$

PROMETHEE recognizes six types of simple functions to express the preferential strength, the so-called generalized criteria, where the parameter q always signifies the breadth of the preferential area and the significance of the parameter p varies with the type of criteria. The calculation consists in selecting an appropriate generalized criterion a for each criterion, and calculating the values of preferential functions for each criterion a of all pairs of variants:

$$P(a_i, a_j), h = 1,2,...k, i, j = 1,2,...p \tag{3}$$

Also calculated is a multi-objective preferential index π $(a_i, a_j) \in \langle 0, 1 \rangle$, which measures the preferential strength of variant a_i over variant a_j for all criteria:

$$\pi(a_i, a_j) = \sum_{h=1}^{k} v_h P_h(a_i, a_j) \tag{4}$$

The outflow of $F'(a_i)$ and inflow $F'(a_j)$ for all variants is determined as:

$$F'(a_i) = \sum_{a_j \in A} \pi(a_i, a_j) \tag{5}$$

$$F'(a_i) = \sum_{a_j \in A} \pi(a_j, a_i) \tag{6}$$

This is followed by an evaluation by PROMETHEE I, and possibly by PROMETHEE II, which works with net flows.

$$F(a_i) = F^+(a_i) - F^-(a_i) \tag{7}$$

3.3 GAIA Method

The method called Geometrical Analysis for Interactive Aid (GAIA) can be described as a graphic resource for analyzing MCA tasks in continuation of the PROMETHEE category of methods. GAIA method is based on the assumption that each variant aris not characterized by criteria values, but by the so called vector of mono-criteria flows $S_i\,(a_r),\ i=1,2,...,k$, which are defined as follows:

$$S_i(a_r) = \frac{1}{p-1}\sum_{s=1}^{p}\left[p_i(a_r,a_s) - p_i(a_s,a_r)\right] \tag{8}$$

Each variant can therefore be represented in a k-dimensional vector space by an R^k vector

$$q_r = [\ (S_1(a_r),\ S_2(a_r),\ \dots\dots\dots\dots,\ S_k(a_r)] \tag{9}$$

If the number of criteria is k = 2, which is not a typical situation, then all variants can be displayed in one plane. Otherwise, the representation must be in a space of a higher dimension. The model for selecting optimal environmental investments for mineral processing companies was processed by Visual PROMETHEE Academic Edition 2013 software (http://www.promethee-gaia.net/index.html).

4 Results and Discussion

The multiple criteria analysis serves to select the optimal environmental investment for companies processing minerals, taking into account the economic and environmental parameters of the equipment. The resultant ranking provides the rationale for a decision whether the environmental investment should be implemented using public funds. In this paper, the most effective drier was chosen among the eight available and pointed with a_i on the basis of six criteria pointed with f_j. The multi criteria decision problem will be formulated as follows: $max\ \{(f_1(a),\ \dots f_k\,(a))\ \vert\ a \in A\},\ A=\{a_1,\ \dots,\ a_n\}$ is a finite set of variants, $\{f_1,\dots,f_k\}$ is a set of evaluation criteria. A methodical solution in selecting the optimal environmental investment is based on the TCA method which involved six established criteria:

- f_1 - Investment expenditure, CZK
- f_2 - Annual operating costs, CZK
- f_3 - Operating income, CZK
- f_4 - Administrative cost and disposal fees, CZK
- f_5 - Economic benefits, CZK
- f_6 - Environmental benefits, CZK

The TCA evaluation method encompasses all relevant costs and benefits from the company's viewpoint, but it does not include the social costs, i.e. the externalities (Kennedy, 1998). The next step was the application of the PROMETHEE multi criteria decision method. The model for the evaluation of future environmental investments contains an expert estimate of the set criteria, expressed in the Czech currency (CZK), see Tables 1-4 below.

Table 1. Environmental investment rated by given criteria – costs and income

Environmental investment	*Investment Cost*		*Operational cost*	*Operational income*
	Life expectancy	Equipment cost		
Unit	Years	CZK	CZK/year	CZK/year
	Max	Min	Min	Max
Instalation of wet gas cleaner	20	5,000,000	100,000	0
Replacement of burner	10	1,000,000	20,000	100,000
Replacement of dry electrostatic filters	20	10,000,000	50,000	15,000
Replacement of hose filters	20	40,000,000	10,0000	0
Replacement of exhaust filters	20	250,000,000	30,000	0
Replacement of burner	10	35,000,000	50,000	100,000
Electrostatic separator	15	6, 000, 000	50,000	0
Replacement of adsorbers (cascade&hose filters, electrostaticseparators)	15	10,000,000	40,000	0

Table 2. Environmental investment rated by given criteria – administrative costs and fees

	One-time administrative costs and fees		
Environmental investment	Permit	Documentation	Reporting duty
Unit	CZK	CZK	CZK
		Min	
Installation of wet gas cleaner	30,000	20,000	30,000
Replacement of burner	30,000	10,000	30,000
Replacement of dry electrostatic filters	30,000	15,000	30,000
Replacement of hose filters	30,000	50,000	30,000
Replacement of exhaust filters	30,000	10,000	30,000
Replacement of burner	30,000	15,000	30,000
Electrostatic separator	30,000	15,000	30,000
Replacement of absorbers (cascade&hose filters, electrostaticseparators)	30,000	15,000	30,000

Table 3. Environmental investment rated by given criteria – economic benefits

	Economic benefit	
Environmental investment	Better image	Productivity improvement
Unit	CZK/year	CZK/year
	Max	
Installation of wet gas cleaner	50,000	0
Replacement of burner	50,000	30,000
Replacement of dry electrostatic filters	50,000	0
Replacement of hose filters	300,000	0
Replacement of exhaust filters	150,000	0
Replacement of burner	150,000	30,000
Electrostatic separator	150,000	0
Replacement of adsorbers (cascade&hose filters, electrostaticseparators)	150,000	0

Table 4. Environmental investment rated by given criteria – environmental benefits

	Environmental benefits		
Environmental investment	Impact on air and climate	Impact on soil	Impact on ecological stability
Unit	CZK/year	CZK/year	CZK/year
		Max	
Instalation of wet gas cleaner	0	10,000	50,000
Replacement of burner	10,000	0	0
Replacement of dry electrostatic filters	10,000	10,000	50,000
Replacement of hose filters	20,000	0	5,000,000
Replacement of exhaust filters	20,000	0	5,000,000
Replacement of burner	60,000	0	500,000
Electrostatic separator	10,000	0	500,000
Replacement of adsorbers (cascade&hose filters, electrostaticseparators)	15,000	0	500,000

The same weight (0.1667) is assigned for each criterion. The PROMETHEE II provides complete ranking, see Fig. 1. It is based on a balance of two preferential flow. The upper half of the scale (green) corresponds to a positive Phi score and the bottom half (in red) to negative score. Investments 2 (Replacement of burner) is above all investments while investments 1 (Instalation of wet gas cleaner) and 4 (Replacement of hose filters) are the worst.

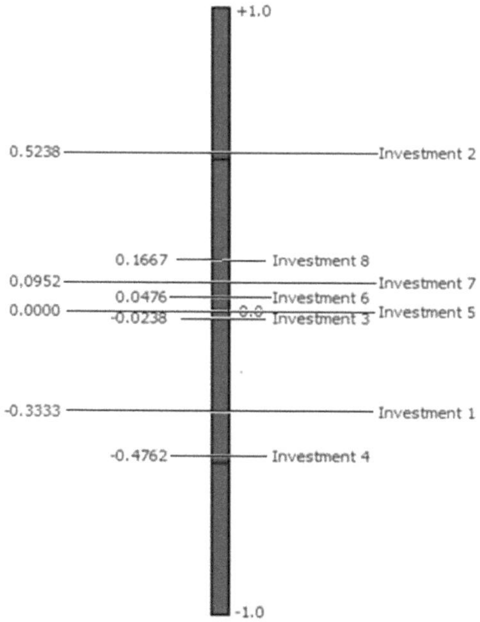

Fig. 1. The PROMETHEE II complete ranking

In the Table 5 PROMETHEE Flow Table we can see Phi, Phi+, Phi- score. Investments are ordered by PROMETHEE II complete ranking. The positive flow expresses how much an alternative is dominating the other ones, and the negative flow how much it is dominated by the other ones.

Table 5. PROMETHEE Flow Table

Environmental investments	*Phi*	*Phi+*	*Phi-*
Investment 2	0.5238	0.7381	0.2143
Investment 8	0.1667	0.4762	0.3095
Investment 7	0.0952	0.4286	0.3333
Investment 6	0.0476	0.4524	0.4048
Investment 5	0.0000	0.4286	0.4286
Investment 3	-0.0238	0.4048	0.4286
Investment 1	-0.3333	0.2619	0.5952
Investment 4	-0.4762	0.1905	0.6667

It is clear, that PROMETHEE II is influenced by weights assigned to criteria. Walking Weight allows to adjust the weights and to observe the resulting changes. If we modify weights of Annual operating costs at 21% and Administrative cost and disposal fees at 21%, than other weights will change to 14%. Investment 2 still dominates the others and investment 5 (Replacement of exhaust filters) has positive Phi score. Fig. 2 shows the PROMETHEE Network which presents PROMETHEE I partial ranking. Investments are nodes and preferences are indicated by arrows.

Investment 2 is significantly preferred to other investments. Investments 6 (Replacement of burner), 5 (Replacement of exhaust filters) and 3 (Replacement of dry electrostatic filters) are incomparable and very close to each other.

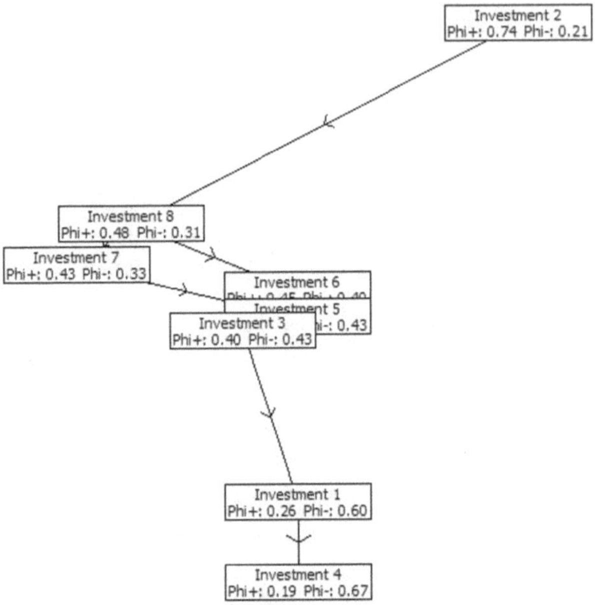

Fig. 2. PROMETHEE Network

5 Conclusions

When evaluating environmental investments (e.g. technical facilities) it is very important to identify and consider all types of costs, potential cost savings and other benefits associated with the investment, because this is the only way to properly evaluate the effectiveness of investments. The method of Total Cost Analysis was chosen to define the criteria – investments costs, operational costs, operational income, administrative costs and fees, economic benefits (better image and productivity improvement), environmental benefits (impact on air and climate, impact on soil and impact on ecological stability). After collecting and analyzing data follows an evaluation of investment efficiency. To evaluate 8 technical facilities by 6 criteria, the PROMETHEE as one of the methods in the multi criteria analysis was employed. The result of the method proposed in this paper is the final ranking of investments on the basis of which a qualified decision can be made.

Acknowledgements. This paper is supported by The Czech Science Foundation. Name of the Project: Construction of Methods for Multifactor Assessment of Company Complex Performance in Selected Sectors. Reg. Nr. P403/11/2085.

References

1. Behzadian, M., Kazemzadeh, R.B., Albadvi, A., Aghdasi, M.: PROMETHEE: A comprehensive literature review on methodologies and applications. European Journal of Operational Research 200, 198–215 (2010)
2. Brožová, H., Houška, M., Šubrt, T.: Modely pro vícekriteriální rozhodování. CREDIT, Praha (2003)
3. Brans, J.P.: L'ingénierie de la décision: élaboration d'instruments d'aide à la décision. La méthode PROMETHEE. Presses de l'Université Laval (1982)
4. Brans, J.P., Mareschal, B.: Promethee V: MCDM problems with additional segmentation constraints. INFOR 30, 85–96 (1992)
5. Brans, J.P., Mareschal, B.: PROMETHEE methods. In: Multiple Criteria Decision Analysis: State of the Art Surveys, pp. 163–186. Springer, New York (2005)
6. Brans, J.P., Vincke, P.: A preference ranking organisation method: The PROMETHEE method for MCDM. Management Science 31, 647–656 (1985)
7. Cokins, G.: Aktivity-Based Cost Management. An Executive's Guide. John Wiley & Sons. Inc., New York (2001)
8. Fiala, P., Jablonský, J., Maňas, M.: Vícekriteriální rozhodování. VŠE, Praha (1994)
9. Fotr, J., Dědina, J., Hrůzová, H.: Manažerské rozhodování. EKOPRESS, Praha (2003)
10. Greco, S. (ed.): Multiple criteria decision analysis: state of the art surveys, vol. 78. Springer (2004)
11. Kennedy, M.L.: The Cost of Changing: Total Cost Assessment of Solvent Alternatives. The Massachusetts Toxic Use Reduction Institute. University of Massachusetts Lowel (1994)
12. Kennedy, M.L.: Total Cost Assessment for Environmental Engineers and Managers. John Wiley & Sons. Inc., New York (1998)
13. Kiker, G.A., Bridges, T.S., Varghese, A., Seager, T.P., Linkov, I.: Application of multicriteria decision analysis in environmental decision making. Integrated Environmental Assessment and Management 2, 95–108 (2005)
14. Lahdelma, R., Salminen, P., Hokkanen, J.: Using multicriteria methods in environmental planning and management. Environmental Management 26, 595–605 (2000)
15. Macharis, C., Springael, J., De Brucher, K., Verbeke, A.: Promethee and AHP: the design of operational synergies in multicriteria analysis. Strengthening PROMETHEE with ideas of AHP. European Journal of Operational Research 153, 307–317 (2004)
16. Mareschal, B., Brans, J.P.: Geometrical representations for MCDA. The GAIA module. European Journal of Operational Research 34, 69–77 (1988)
17. Munier, N.: A Strategy for Using Multicriteria Analysis in Decision-making: A Guide for Simple and Complex Environmental Projects. Springer, Dordrecht (2011)
18. Triantaphyllou, E.: Multi-criteria decision making methods, pp. 5–21. Springer US (2000)

Incorporate Sensor Data and Dynamic Modeling for Real Time Control of Sewer Systems

Erik de Rooij and Klaas-Jan van Heeringen

Deltares, P.O. Box 177, 2600 MH Delft, The Netherlands
{erik.derooij,klaasjan.vanheeringen}@deltares.nl

Abstract. We discuss how Delft-FEWS is applied to incorporate real-time sensor data and dynamic modeling to allow Real Time Control (RTC) of three sewer systems in the area of the Hoeksche Waard, the Netherlands. The purpose of this project was to reduce sewerage spilling to the surface water by optimizing the available storage in the sewer systems. In this paper we will present how Delft-FEWS makes use of Simple Object Access Protocol (SOAP) webservices to exchange real-time data with multiple sewer systems and their different Supervisory control and data acquisition (SCADA) systems. We will also explain how we combine this with dynamic models, allowing for central automatic control of multiple sewer systems.

Keywords: Delft-FEWS, Real Time Control, Webservices, SOAP, SCADA systems, dynamic modeling, Sewer Systems, Sensor Web.

1 Introduction

Over the last decade Delft-FEWS has been applied in a wide range of operational water management and forecasting systems. In this case study Delft-FEWS is used for the first time to integrate dynamic models with automatic system control. This introduces new challenges concerning; real-time availability of sensor data, fail safe mechanisms in case of network interruptions and fast response times to allow for automatic control. In this paper we discuss how we tackled these challenges, experimenting with different mechanisms for exchanging sensor data between the SCADA systems that are operated at the municipalities and the Waste Water Treatment Plants (WWTP) and Delft-FEWS.

1.1 Description Delft-FEWS

Delft-FEWS is a proven, real time software infrastructure for operational water management and forecasting. The system is a sophisticated collection of modules designed for building an operational water management system customized to the specific requirements of an organization. The philosophy of the system is to provide an open shell system for managing the operational management process [1-2]. This shell incorporates a comprehensive set of general data handling utilities, while provid-

J. Hřebíček et al. (Eds.): ISESS 2013, IFIP AICT 413, pp. 54–61, 2013.

ing an open interface to a wide range of forecasting models. The modular and configurable nature of the system allows it to be used both in rudimentary systems and in highly complex systems. Delft-FEWS can be deployed as stand-alone environment, as a fully automated client-server application, or as a web-service component within another forecasting system.

1.2 Case Study Hoeksche Waard

For this case study the regional water authority Hollandse Delta and five municipalities (Binnenmaas, Cromstrijen, Korendijk, Oud-Beijerland and Strijen) (see Fig. 1) are exploring ways to increase the efficiency of the waste water treatment process and to improve water quality [4]. Together with Deltares and Delft University of Technology they started three pilot projects aimed to centralize automatic control of the pumps in the sewer systems of the municipalities and the pumps that transfer sewage water to WWTP.

Each of the municipalities and the WWTP operate their own SCADA system. These systems collect the data from the sensors within the sewer system and display this to the operators. From these displays the operators can monitor and control their part of the system. The SCADA systems of the different municipalities and WWTPs are not connected. By introducing Delft-FEWS as a central control component we hope to connect the SCADA systems and achieve the following objectives.

The main objective of this project is to reduce combined sewerage overflow (CSO) volumes from the sewer systems to the surface water. This objective is achieved by optimizing the remaining available storage within the sewer system during storm events. Another objective is the reduction of pumping costs and waste water treatment costs [3].

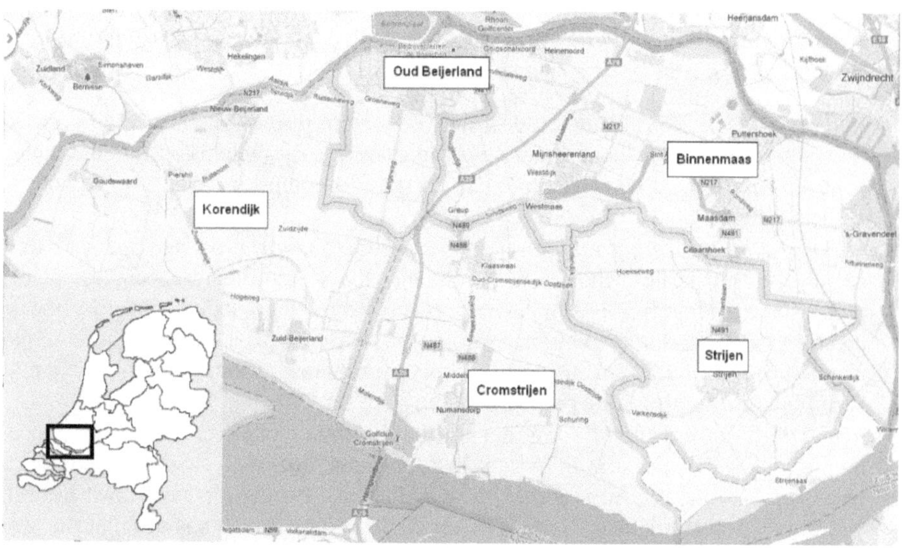

Fig. 1. Hoeksche Waard area with the five municipalities

2 Centralizing Control by Creating a Sensor Web

The Netherlands have an extensive sewer network that collects waste water from households, industries and urban overland flow. To transport the waste water through the system towards the WWTPs an extensive network of sewer pumps is maintained. The capacity of the sewer system is limited and during storm events the system can fill up, resulting in harmful sewage spills to surface waters.

Each municipality is responsible for maintaining and operating the pumps within its boundaries, while the regional water authority is responsible for operating the WWTPs. In general the pumps work independently and do not take into account the levels within the rest of the system. Once the sewage exceeds the local start level the pump will commence operation and will continue until the level drops below the local stop level. Each sewer pump is linked to a SCADA system from which it can be monitored and controlled. The SCADA systems of the municipalities and WWTPs are not linked together.

By linking the SCADA systems of the municipalities and WWTPs together and by introducing a dynamic modeling component it becomes possible to optimize the sewage flow throughout the whole system and at the same time allows for an automated centralized control of the system.

To accomplish this webservices are used to link the SCADA systems together and Delft-FEWS is introduced as dynamic modeling component to be the central information system. Data is collected from each SCADA system and is fed to Delft-FEWS, where a RTC plugin calculates the optimal pump settings. These settings calculated from Delft-FEWS are sent back to the SCADA systems in order to control the sewer pumps.

2.1 Automatic Control Requirements

The introduction of the 'automatic control' aspect results in extra requirements to ensure the safety of the sewer system and pumps:

- *Fail over mechanism*: In the case of network interruptions, or on receiving incorrect pump settings the local system must react to this by switching from automated control back to local control to prevent damage to pumps or cause overflowing to surface waters.
- *Secured against external interference*: The system must be protected against external malevolent parties who try to 'hack' into the system and take control of the pump stations. Firstly this should be prevented by using adequate security measures such as: user authentication, limit access to known IP-addresses and communication over HTTPS. Secondly the fail over mechanism should kick in as soon as harmful control settings are passed to a pumping station.
- *Real-Time data availability*: In order for the dynamic models to produce any meaningful control output it is important that the input data is as close to real-time as possible. Also the return period of the calculated control parameters must be well within the system response time. If either of these requirements is not met the system will fail over. Experience shows that a response time equal to or less than 30 seconds is required.

3 Creating a Sensor Web with Webservices

As described earlier in this paper, the SCADA systems of the municipalities and WWTPs reside on individual networks not connected in any way. Each municipality is responsible for maintaining and operating its own SCADA system. None of the municipalities are very eager to open up their system to outsiders and relinquish control. In order to create a Sensor Web from these individual and closed systems it is necessary to minimize the required changes and risks to the local systems. By using webservices the impact to the local systems is kept to a minimum:

- Open a communication port over which all HTTP or HTTPS traffic is sent,
- Install a software component locally on each individual SCADA system that enables the SCADA system to interact with the Sensor Web.

During the course of the project two mechanisms are implemented for transferring data between the SCADA systems and the central controller Delft-FEWS (see Fig. 2):

1. SCADA data is sent to a webservices hosted by Delft-FEWS (FewsPiService).
2. Delft-FEWS interacts with a webservice hosted on the side of the SCADA system.

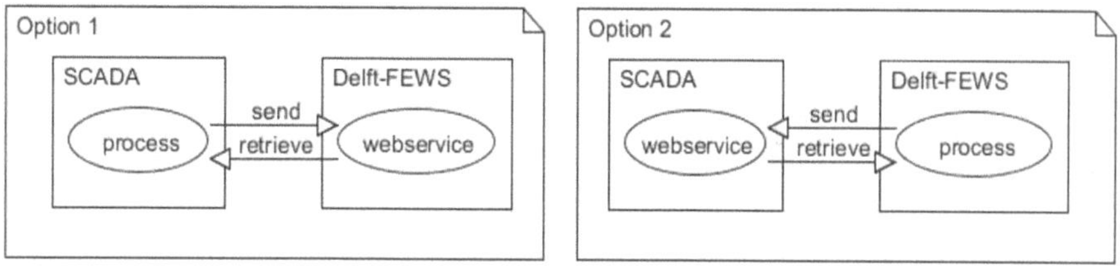

Fig. 2. Overview transfer mechanisms

Both of these mechanisms have to be able to interact with the following two SCADA system types:

- Control Maestro: Produced by Elusions SAS (www.elusions.com) and uses OLE for Process Control (OPC) technology.
- Flygt (RTiem): Produced by Xylem Nederland (www.xylemwatersolutions.com).

3.1 Webservice Hosted by Delft-FEWS

At the start of the project a decision is made to make use of the webservice provided by Delft-FEWS, the FewsPiService[1], which exchanges data using the Published Interface (PI) XML format[2]. On the SCADA system side this requires the installation of a JAVA process that can exchange data between the SCADA system and the Delft-

[1] http://publicwiki.deltares.nl/display/FEWSDOC/Fews+PI+service
[2] http://publicwiki.deltares.nl/display/FEWSDOC/The+Delft-Fews+Published+interface+%28PI%29

FEWS webservice. This process is called the Data Transfer Service (DTS) and requires a local configuration file containing information about the data to transfer and at what frequency the data should be transferred. For example one type of data can be scheduled every second while the other is scheduled once a day, at a certain time. Besides the sending of SCADA data, the DTS also sends log messages produced during data retrieval. Furthermore during network interruptions the DTS stores the data locally in order to re-send it once the network connection is re-established. A schematic representation of this setup can be seen in Fig. 3.

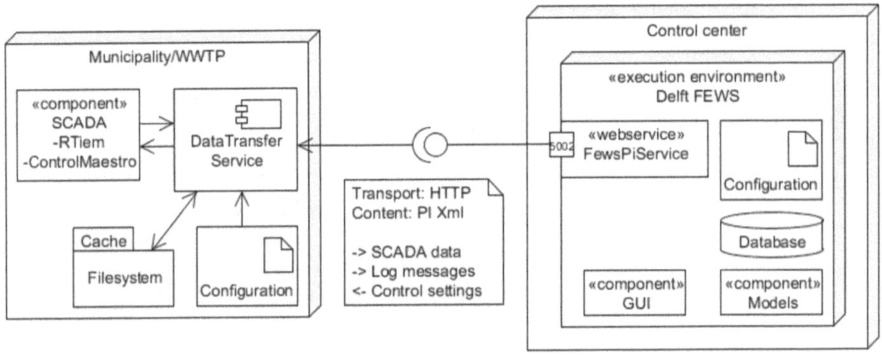

Fig. 3. Schematic representation of webservice hosted by Delft-FEWS

Advantages of this approach:

- At the time of implementation, this approach had the advantage of smaller response times, due to the Delft-FEWS system not being able to cope well with model runtimes less than one minute.
- Because data is sent at regular intervals it is possible to build a continuous historical archive of the SCADA data. Even during network interruptions the data is cached and is sent later on. This is convenient hence SCADA systems do not always store their historical data.
- By being able to tune the transfer frequencies it is possible to reduce network load and the amount of queries on the SCADA system.
- It is easy to manage and send log messages as each process has a dedicated destination.

Disadvantages of this approach:

- Because the DTS process does not run synchronously with the running of the dynamic models, data often arrives slightly too late for the model run or is retrieved slightly before completion of the model run. Resulting in the output control settings lagging in time.
- Because the configuration of the DTS is located on the local SCADA systems, updates always require the assistance of the system operator of the municipalities. This always results in delays.
- It is not possible to transfer data using HTTPS, as the Delft-FEWS webservice does not support this.

3.2 Webservice Hosted on SCADA System

Later on in the project a decision is made to transfer the control of data exchange to Delft-FEWS. This became possible due to improvements in performance of Delft-FEWS allowing it to run the models as frequent as two to three times a minute. The main reasons for wanting to transfer control are:

- Synchronizing data exchange with model runs.
- Transferring control of configuration to Delft-FEWS

For this change a PI webservice is created that is hosted in a Tomcat web container installed on the SCADA system. Now each time Delft-FEWS performs a model run data is first retrieved from the webservice on the SCADA system, then the dynamic model is run and finally the results are sent back to the SCADA system. For each SCADA system an implementation of the new PI webservice is required. Delft-FEWS does not have any dependencies with the SCADA systems as it communicates through the interface hosted by the PI webservice. Now all configurations concerning data transfer and scheduling is controlled by Delft-FEWS. A schematic representation of this setup can be seen in Fig. 4.

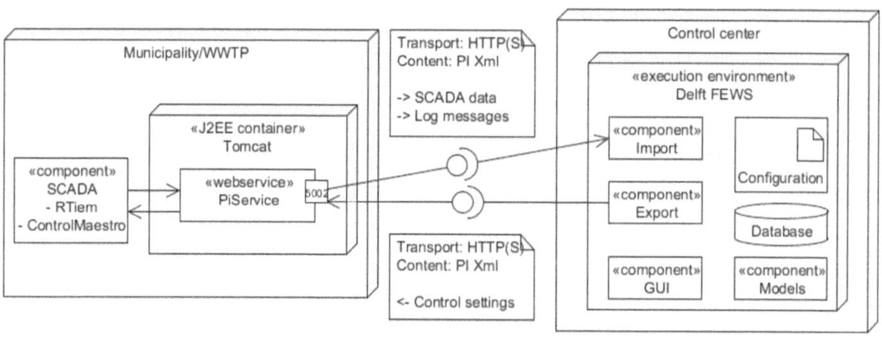

Fig. 4. Schematic representation of webservice hosted on SCADA system

Advantages of this approach:

- Model run always uses the latest data and returns data as soon as it becomes available.
- Configuration resides within Delft-FEWS. It is even possible to update the webservice component through the manager console of Tomcat.
- It is possible to communicate over HTTPS using the security configuration available within Tomcat.
- It is still possible to tune the transfer frequencies by scheduling the import process at different rates.

Disadvantages of this approach:

- There is no longer a caching mechanism. So whenever the network connection is interrupted a gap will appear in the historical archive. If the SCADA system allows historical queries of the data the gap can be filled later during a manual import run.
- Because it is more flexible to query the SCADA system for historical data (if supported by SCADA), it introduces the risk of OutOfMemory exceptions when querying a too large period. At present no mechanism has been built into the system that splits the results into smaller segments. The only security against this is to report the requested data volume is too large and to return only a partial result set.
- It is harder to manage log messages because logged information has to be stored until client retrieves this explicitly in a separate call. If not managed this could lead to memory issues.

4 Project Results

Prior to this project, the Hoeksche Waard sewer system consisted of numerous individual sensor networks. By linking these networks a greater insight was gained in the actual working of the network as a whole. This revealed that the designed specifications of the sewer system did not correspond to the actual situation.

The improved monitoring resulted in a better quantification of the CSO. According to the improved model results there was 48% more CSO than estimated. However because of these new findings it was also possible to implement some simple and cheap improvements.

The goal to reduce pumping costs was not achieved. As a result of the main goal, to reduce emissions, it was not possible to operate the sewer pumps in a more energy efficient way. The only possible solution for reducing pumping costs would be to reduce the inflow of rainwater into the sewer system.

Furthermore flow measurements showed that about two times more water than estimated is entering the sewer system and is being processed by the WWTP's. This leaves a lot of room for improvements and cost reductions.

5 Conclusions and Future Challenges

This project demonstrates how relatively small adjustments to existing SCADA systems allows them to be converted into centrally controlled operating systems. Currently this project is running as an operational system and plans exist to expand the functionality to other municipalities.

During this project we found that when using a Sensor Web there are inherently going to be network issues or partial system failures. It is therefore important to have a robust backup mechanism in place to assure the integrity of the vital system components.

When transferring SCADA data, package size is usually not a concern, as the amount of data transferred is small. But when it is possible to query larger amounts of data, memory management will become a concern. For the future it would be good to investigate mechanisms that support the transfer of large volumes of data, either by segmenting the results or by streaming them.

Acknowledgments. This project was supported by grants from Agentschap NL, The Netherlands. This agency provided a subsidy from the "Innovatieprogramma Kaderrichtlijn Water" of the Dutch Ministry of Transport, Public Works and Water Management. This work is part of the project "Integrale Sturing Afvalwaterketen en watersysteem Hoeksche Waard".

References

1. Werner, M.G.F., Dijk, M., van Schellekens, J.: Delft-FEWS: An open shell flood forecasting system. In: 6th International Conference on Hydroinformatics, pp. 1205–1212. World Scientific Publishing Company, Singapore (2004)
2. Gijsbers, P.J.A., Werner, M.G.F., Schellekens, J.: Delft-FEWS: A proven infrastructure to bring data, sensors and models together 650. In: Proceedings of the iEMSs Fourth Biennial Meeting: International Congress on Environmental Modelling and Software, Barcelona, Catalonia, vol. 1, pp. 28–36 (2008)
3. van Loenen, A., van Heeringen, K.-J., van Nooyen, R., van Velzen, E.: Leveraging existing infrastructure for central automatic control of multiple sewer systems. In: 10th International Conference on Hydroinformatics, Hamburg, Germany (2012)
4. van Nooijen, R.R.P., Kolechkina, A.G., van Heeringen, K.-J., van Velzen, E.: Implementation of central control for multiple sewer systems. In: 12th International Conference on Urban Drainage, Porto Alegre, Brazil, pp. 10–15 (2011)

Data Fusion in the Environmental Domain

Hylke van der Schaaf[1], Mike Kobernus[2], Markus Falgenhauer[3], Jasmin Pielorz[4], and Kym Watson[1]

[1] Fraunhofer IOSB, Fraunhoferstr. 1, 76131 Karlsruhe, Germany
hylke.vanderschaaf@iosb.fraunhofer.de,
kym.watson@iosb.frauhofer.de
[2] NILU - Norsk institutt for luftforskning, Instituttveien 18, Kjeller, Norway
mike.kobernus@nilu.no
[3] Austrian Institute of Technology, Donau-City-Str. 1, 1220 Vienna, Austria
markus.falgenhauer@ait.ac.at
[4] Ubimet GmbH, Dresdner Str. 82, 1200 Vienna, Austria
jpielorz@ubimet.com

Abstract. Air quality and air pollution have a very large impact on human health. The sensitivity to different pollutants varies per person, therefore it is important that citizens can get personalised air quality information. The Personal Environmental Information System (PEIS) aims at delivering just that. The PEIS takes sensor data from several data providers and employs a service-oriented architecture to deliver these observations to the user through a smartphone application. The PEIS also uses scientific models to fuse the sensor data and create new, derived observations. To make scientific models that fuse sensor data fit better in a service-oriented architecture, a software framework called Fusion4Decision was developed. This framework is based on Open Geospatial Consortium standards and allows scientific models written in languages like MATLAB or R to be available as a web service.

Keywords: air quality, open geospatial standards, environmental modelling, fusion.

1 Introduction

From the moment of our birth until our very last breath, we spend our lives immersed in a mixture of gases and aerosols known as the air or atmosphere. Unsurprisingly, our well-being is closely connected to the characteristics of the air: some of the gases and chemicals suspended therein are toxic, others may cause allergic reactions, and our bodies are also sensitive to changes in for instance humidity or oxygen levels, air temperature and pressure. Moreover, sensitivity to these substances and changes is very different from one person to another. Despite this, the information on the state of atmosphere available today is entirely generic, and by no means personalized. We intend to address this lack of individualized information through the development of a Personal Environmental Information System (PEIS).

J. Hřebíček et al. (Eds.): ISESS 2013, IFIP AICT 413, pp. 62–70, 2013.

The PEIS aims at enhancing the human to environment interaction by providing personalized air quality, pollen and meteorological data for chosen locations and personalized alerts and notifications about changes in atmospheric conditions [15,14]. The PEIS also enables users to keep a personal health diary and submit their own environmental observations to the system. Pilot projects currently run in Oslo (Norway) and Vienna (Austria).

The PEIS is a distributed system with a mobile application interface. The data used by the PEIS is supplied by several data providers. Sensor data is available for air quality, meteorology and pollen. These data are presented to the user as they are and they are fed into models. To allow the PEIS to use custom models written in different languages for data fusion, the Fusion4Decision (F4D) system was developed. This allows domain experts to write models in their favourite language and then make those models available in a service oriented architecture, using the OGC SPS standard interface.

This paper provides an example on how heterogeneous data sources and models in the environmental domain can be combined in one language-independent service that agrees with OGC standards [7]. In section 2, we first give an overview over the PEIS mobile application and explain how it relates to the EU project ENVIROFI [2], which is followed by a detailed description of its architecture in section 3. After discussing the different data sources that have to be taken into account in the scope of environmental applications, section 4 lays out how the data fusion service works in case of the PEIS. Finally, we conclude with a discussion of the resulting service and provide a future outlook.

2 Personal Environmental Information System

The PEIS is one of three end user applications to come out of ENVIROFI, a co-funded research project within the Future Internet Public Private Partnership (FI-PPP) programme of the EU's Seventh Framework Programme (FP7). The project was dedicated to the environmental usage area of the Future Internet [10]. It explored environmental enablers (applications for collecting and processing environmental data) and provided environmental sector requirements to FIWARE, the FI-PPP core platform project [12,16]. These were deployed in collaboration with large stakeholder communities in three scenarios, specifically in the terrestrial, atmospheric and marine environments, and laid the foundation for an environmental observation web. In this way, ENVIROFI helps Europe tackle the grand societal challenges of climate change, environmental degradation, and sustainable growth.

The PEIS is an innovative application for smartphones which accesses data and services from multiple providers and combines them into a single system with an intuitive and easy to use interface. It gives users more control over their environment as to what information they track (e.g., particulate matter, ozone, humidity, etc.), how it is presented (GIS overlays as in Fig. 1, tabular data, etc.) and how they are alerted to changes in the environment. The PEIS system provides users with a fully customisable, interactive application that enables the user to track air borne pollutants, pollen, and meteorology in an entirely personalized manner [15,14]. Through the creation of profiles, the user can specify just what air quality, pollen and meteorological data should be tracked. Threshold levels can be set for any of these tracked components and notifications triggered when threshold levels are surpassed, alerting the user to the change.

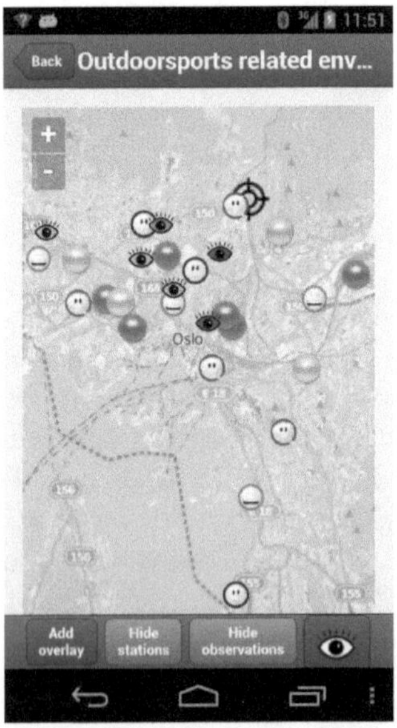

Fig. 1. PEIS map-view, showing stations and user observations

Fig. 2. Health observations and calender view in the PEIS

For most people, this type of system may be more than is required for general, day to day interest in air quality. However, for some, this type of configurable tracking and notification mechanism is key to their health, since sensitivity to airborne pollutants is very different from one person to another and in many cases is highly dependent on the individuals patterns of behaviour or their health. For those who are more at risk the ability to monitor current air quality in a very localized environment is extremely useful. In addition, with air quality forecasts (which use a sophisticated modelling system predicting trends in air quality in the very near future), at risk users can make plans for two or even three days in advance (the system provides up to three days forecasts, with daily updates). This is a far cry from typical air quality data provided to most users which is usually only the current air quality index for the nearest city, often based on the highest relative concentration value measured for any component in the city [5,13]. In contrast, the PEIS provides air quality data, specific to the users' requirements, in the area where he is currently located, in near real time. Since the smartphone supplies the user's GPS coordinates, it is possible to not only provide current data for that exact location, but to also store the user's location and create an exposure value for the user every 24 hours. In addition, users can report their own observations, including reports on their current health situation. These are stored in the system, and any additional sensor data available from the phone (such as air pressure, for example) can be stored at the same time. This essentially turns the user into a sensor node in the network. Once the user has been providing their own observations for a while, it is possible to view these in a calendar (Fig. 2), which can aid in spotting potential patterns in the user's health. For example, headaches occurring at times of changing barometric pressure.

3 Architecture

The system is implemented as a service-oriented architecture based on open standards. Central to the architecture is the Mobile Data Acquisition Framework (MDAF [11]), created by the Austrian Institute of Technology (AIT), which manages observations and keeps track of observations relevant to the specific user's needs (Fig. 3).

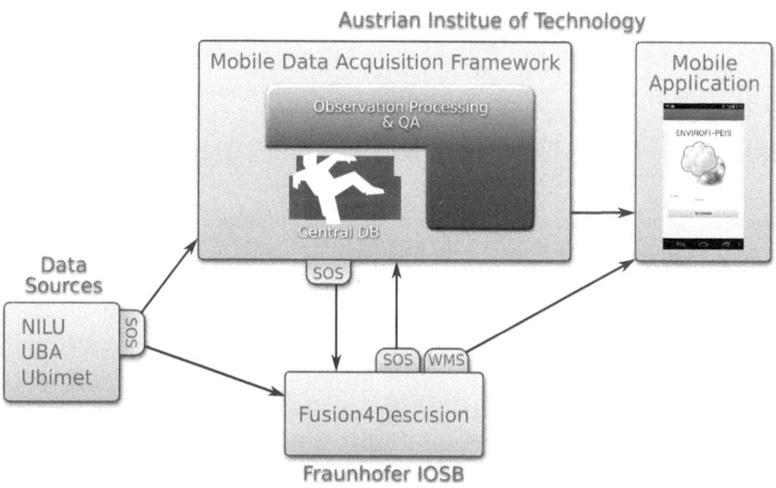

Fig. 3. PEIS Architecture overview

Air quality data, both current and predictive, are provided by the Norwegian Institute for Air Research (NILU) and The Federal Environment Agency (UBA), while meteorological data is provided by Ubimet. Each data provider has its own web services for accessing its data. This data is accessed directly by MDAF, but is also fed into models to create GIS map overlays. These overlays are in turn accessed by the PEIS using a Web Map Service (WMS). The execution of models is described in more detail in the next section.

Internally MDAF uses GeoCouchDB. CouchDB [1,6] is a document-based, open source, NoSQL database that combines advantages of an easy to use document-orientated database with the scalability of a relational database. It uses JavaScript as query language and MapReduce with HTTP REST as API. GeoCouchDB is a special development of CouchDB for location aware applications. As documents within the GeoCouchDB we use GeoJSON strings which are JSON strings with additional geographic information.

To enable cross platform compatibility the phone application is developed using PhoneGap [3] (Fig. 4). Phonegap is a free and open source framework that allows the creation of mobile apps using HTML, CSS and JavaScript and it provides a link to device-specific APIs. The supported platforms are: iPhone, Android, Windows Phone, BlackBerry and many more.

Fig. 4. The PEIS application software stack

On top of Phonegap, we use Sencha Touch [4] which is a flexible HTML5 mobile application framework to enable a quality user experience. It uses the model-view-controller architecture pattern to structure the application.

4 Standardised On-Demand Model Execution

Part of the core functionality of the PEIS is the processing of data from environmental sensors. These can be both, quantitative as well as qualitative sources, and can be broken down into three basic groups: (i) standardized, gauged sensors, e.g. from professional-grade weather stations measuring meteorological conditions or advanced measurement stations for pollutant concentrations, (ii) non-standardized sensors, e.g. supplied by home-grown weather stations, as well as their mobile phones which can be fitted with basic sensor systems (measuring air pressure for example) and finally, (iii) subjective observations from the end users themselves in the form of their personal perceptions on either the environment (weather, pollution, pollen, etc.) or how this affects them directly, such as having a headache, being dizzy, sneezing, etc. Using all of the data generated by this heterogeneous mix of sensors is not straightforward. What exactly was measured? Who measured it? How reliable is the data and how can it be accessed? The required sensor data is often only available through proprietary, non-standard interfaces, complicating the use of data from different data providers in the same model.

To make use of these heterogeneous data specialized models are required. These models are created by domain experts [17]. Since domain experts are not programmers, and since mathematical modelling is best done with specialized tools, the models needed for processing the available data are often created in packages like MATLAB or R. As a result the models are not easily deployable as a service on the internet and thus not available to (client) applications.

The Open Geospatial Consortium (OGC) is a non-profit, international voluntary consensus standards organization that is leading the development of standards for geospatial and location-based services [7]. The standards are applied to a wide range of domains. The primary objective of the standards is to support information discovery, access, processing, fusion and decision-making. The relevant OGC standards with respect to sensors and sensor data are the OGC Sensor Web Enablement (SWE) standards, which allow sensor data to be accessed in a uniform way over the Internet with web services and XML encodings.

The SWE services of relevance to this paper are: (a) Sensor Observation Service (SOS) to access sensor observations with queries filtering on the phenomenon (property) and the spatial and temporal domains of the observations, (b) Sensor Planning Service (SPS) to parametrise, schedule and execute assets such as sensors. The main operations of the SPS are DescribeTasking (to get the tasking parameters), GetFeasibility (to ascertain if the asset can be tasked with the given parameters) and Submit (to actually execute the task). During the execution the operations GetStatus and Cancel are available.

The SWE services are accompanied by two fundamental information models:

1. The Sensor Model Language (SensorML) which describes the static metadata of a sensor such as make, model, operator and measurement procedure. It can also describe the parameters that a sensor accepts if a sensor can be tasked using a SPS request.
2. The Observation & Measurement Model (O&M Model) which describes observations and the relation between an observation, the associated feature of interest (what was observed) and the associated sensor (how it was observed).

To allow the publication of models as a service, with a well-defined interface, Fraunhofer IOSB has designed the Fusion4Decision architecture. In this architecture, different types of models, called Fusors, implemented in languages like R, MATLAB, Python or C, can be started and tracked using a SPS and the results of the model execution can be retrieved using a SOS. In other words, the model is treated as a sensor and the model results are considered to be new observations. When an application needs a certain model to run on a specific set of data, the application can task the F4D SPS server to execute the model. During the execution of the model the application can track the progress of the execution. Once the model has finished, the application can access the resulting observation through the SOS and use it for further processing, or display it to the user. The requirement for the Fusors is that they are implemented in a way so they can be started by a Java application, that they accept their parameters in a single SensorML document and that they return a single observation document. The returned document can link to other generated files for increased efficiency.

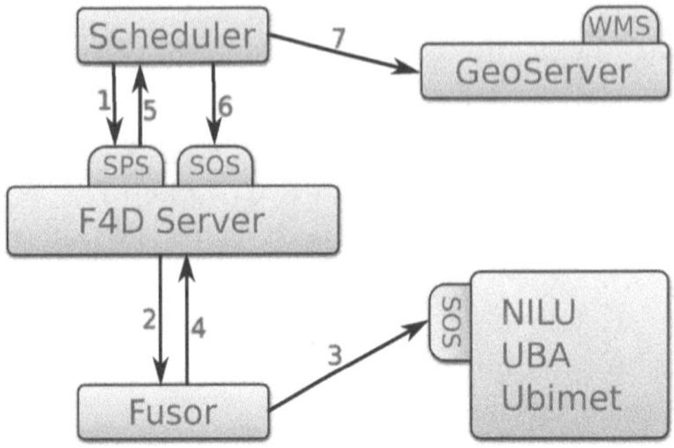

Fig. 5. Use of the F4D architecture in ENVIROFI

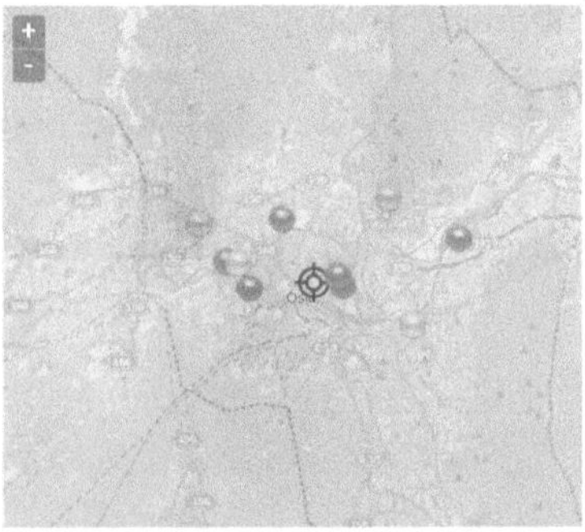

Fig. 6. A map of Oslo with an overlay showing the concentration, interpolated from measurements at the 10 stations shown

In the current prototype of the PEIS the Fusion4Decision system was successfully used to turn point measurements of air pollutants, gathered at fixed measuring stations, into overlay maps of pollutant concentrations covering the cities of Oslo and Vienna. Figure 5 describes this process in detail. When the client that wants to schedule a task (the Scheduler) determines a new model execution is needed, it requests a new task by calling the GetFeasibility and Submit operations of the SPS interface of the F4D server (1). The F4D Server starts the requested Fusor with the supplied parameters (2). The Fusor requests data from various SOS servers (3), and calculates the requested results. The Fusor returns the results to the F4D server (4) which makes them available in its SOS interface. The F4D server then sends a notification to the client that scheduled the task (5) or the client that regularly polls the F4D server to monitor the progress of the execution of the task using the GetStatus operation. The Scheduler then fetches the results from the SOS service of the F4D server (6) and pushes the results into the GeoServer (7), such that they become available for the PEIS to use.

The currently used underlying models use Bayesian maximum entropy calculations to interpolate sensor data from the different measurement stations (Fig. 6). The Bayesian Maximum Entropy method is able to consider soft sensor data (e.g. the sensor value lies in an interval) and additional phenomenological knowledge in the form of models. The results are statistics quantifying the uncertainty of the spatiotemporal interpolation given the uncertainty of the available information.

The overall BME fusion method is structured in three stages [8,9]:

1. prior stage: Consideration of general physical and scientific knowledge $\mathcal{G}$ about the spatio-temporal properties of the phenomenon of interest. It is what is known before experience with the specific situation is applied. The prior probability distribution of the so-called random field of the phenomenon is determined using the maximum entropy (ME) principle, i.e. it is the most uninformative (unbiased) probability distribution given only $\mathcal{G}$.
2. meta-prior stage: Consideration of case-specific hard and soft data of the phenomenon of interest. This information is denoted by $\mathcal{S}$ (for specific knowledge) and is based on observations and measurements. Hard data refers to values believed to be accurate. Soft data is accompanied by uncertainty information such as a probability distribution for the value range.
3. posterior stage: Processing (fusion) of the available knowledge $\mathcal{G}$ and $\mathcal{S}$ of the prior and meta-prior stages respectively to make a probabilistic map of the phenomenon for a given set of spatiotemporal points (typically a grid). The map is a statement of the general knowledge $\mathcal{G}$ relative to the case-specific knowledge $\mathcal{S}$ and is derived using Bayesian conditional probabilities.

The knowledge $\mathcal{S}$ is represented as an observation collection described with the O&M model and including uncertainty information in UncertML. The map resulting from the posterior stage is represented as a coverage with associated uncertainty information. The BME models are implemented in MATLAB and compiled as a Java program for integration into the F4D architecture. Models written in other languages, such as R, can be integrated as well.

5 Conclusions and Recommendations

Future work will include combining these personal health reports with atmospheric condition measurements to create personalized health profiles. This will help to develop a closer relationship between PEIS users and their environment, from both, an atmospheric observation and health observation perspective. Raising public consciousness of their environment and its potential impact upon them is a key goal of the project as this could lead to improved decision making on the side of the users, which will lead to improved health and better knowledge in turn. These objectives are just the tip of the iceberg, however, since for systems like the PEIS that can include sensors at the user level, the potentiality is huge. Crowd sourcing, humans as sensors and even bio-monitoring is still in its infancy and the future will see more and more micro sensors deployed and contributing to our understanding of both the environment and human health impact. The PEIS and the Fusion4Decision system that supports it will play a role in this future as a platform for data collection, processing and dissemination.

Acknowledgments. The research leading to these results has received funding from the European Community's Seventh Framework Programme (FP7/2007-2013) under grant agreement n° 284898 (ENVIROFI).

References

1. Apache couch DB (April 2013), `http://couchdb.apache.org/`
2. ENVIROFI: The environmental observation web and its service applications within the future internet (June 2013), `http://www.envirofi.eu/`
3. PhoneGap (April 2013), `http://phonegap.com/`
4. Sencha touch: Build mobile web apps with html5 (April 2013), `http://www.sencha.com/products/touch`
5. AIRNow: Air quality index (aqi) - a guide to air quality and your health (April 2013), `http://www.airnow.gov/index.cfm?action=aqibasics.index`
6. Anderson, C., Lehnardt, J., Slater, N.: CouchDB: The Definitive Guide: The Definitive Guide. O'Reilly Media (2010)
7. Botts, M., Percivall, G., Reed, C., Davidson, J.: OGC® sensor web enablement: Overview and high level architecture. In: Nittel, S., Labrinidis, A., Stefanidis, A. (eds.) GSN 2006. LNCS, vol. 4540, pp. 175–190. Springer, Heidelberg (2008)
8. Christakos, G.: Modern spatiotemporal geostatistics, 2nd edn. Oxford University Press, New York (2001)
9. Christakos, G., Bogeart, P., Serre, M.L.: Temporal GIS: advanced functions for field-based applications. Springer (2002) ISBN 3540414762, 9783540414766
10. FI-PPP: The future internet public-private partnership programme (April 2013), `http://www.fi-ppp.eu/`
11. Havlik, D., Kutschera, P., Geyer, J., Egly, M.: Geospatial service interface and encoding for mobile applications. In: Arndt, H.-K., Knetsch, G., Pillmann, W. (eds.) Man-Environment-Bauhaus: Light up the Ideas of Environmental Informatics, Proceedings of 26th International Conference on Informatics for Environmental Protection, Dessau, Germany, pp. 413–421 (2012) ISBN: 978-3-8440-1248-4
12. Havlik, D., Schade, S., Sabeur, Z.A., Mazzetti, P., Watson, K., Berre, A.J., Lorenzo Mon, J.: From sensor to observation web with environmental enablers in the future internet. Sensors 11(4), 3874–3907 (2011)
13. Klopfer, M. (ed.): EO2HEAVEN, Mitigating Health Risks. The EO2HEAVEN Consortium (2013), `http://eo2heaven.org`
14. Kobernus, M., Havlik, D., Van der Schaaf, H., Pielorz, J., Falgenhauer, M.: A quest for affordable personalized atmospheric exposure estimates. In: Proceedings of the 26th International Conference on Informatics for Environmental Protection (2012)
15. Kobernus, M., Pielorz, J.: ENVIROFI D2.3.2: Functional and Organisational Specification for PIS Pilot II (May 2012), `http://envirofi.eu/Downloads/PublicDeliverables/tabid/4983/Default.aspx`
16. Schade, S., Fogarty, B., Kobernus, M., Schleidt, K., Gaughan, P., Mazzetti, P., Berre, A.-J.: Environmental information systems on the internet: A need for change. In: Hřebíček, J., Schimak, G., Denzer, R. (eds.) ISESS 2011. IFIP AICT, vol. 359, pp. 144–153. Springer, Heidelberg (2011)
17. Watson, V., Watson, K.: Design of a software framework based on geospatial standards to facilitate environmental modelling workflows. In: Seppelt, R., Voinov, A.A., Lange, S., Bankamp, D. (eds.) 6th International Congress on Environmental Modelling and Software (iEMSs). iEMSs (July 2012), `http://www.iemss.org/society/index.php/iemss-2012-proceedings` ISBN: 978-88-9035-742-8

Approaching Cross-Domain Search in Environmental Applications – Towards Linked Data

Gerald Schimak[1], Luca Petronzio[2], and Tomas Pariente Lobo[3]

[1] AIT Austrian Institute of Technology GmbH, Vienna, Austria
gerald.schimak@ait.ac.at
[2] Telespazio Spa, Rome, Italy
luca.petronzio@telespazio.com
[3] Atos Spain, Madrid
tomas.parientelobo@atosresearch.eu

Abstract. TaToo – Tagging tool based on a semantic discovery framework, a project funded by the European Commission provides a web-based solution for easy and accurate discovery as well as tagging of environmental resources. The novelty relies on a semantic framework integrating different domain ontologies in a multi-domain and multilingual context. The underlying ontology framework, comprises besides the different domain ontologies (e.g. related to climate change, agro-environmental and anthropogenic impact domains) also concepts and methods to establish a mapping between the domain ontologies and so-called minimal environmental resource model (MERM). Together this forms a suitable and usable bridge ontology allowing a cross-domain discovery by using aligned ontologies concepts from different domains. The clear advantage for the end-user is that he is now able to find relevant information stemming from other domains, (like from impact of pollutant, climate change or temperature on human health) that he would not have found before but would be even more important to him that the ones only from his domain of expertise. The TaToo semantic framework extends cross-domain search evolving towards the Linked Data initiative by providing a linking functionality. The cross-domain search can be extended including in the search results also linked resources.

Keywords: semantics, information enrichment, cross-domain, search, discovery, annotation.

1 Introduction

Today's Holy Grail is the development of cross boundaries (cross-domain, cross-organisations) usage of information, especially including search & discovery, interpretation, access, tagging, reliability, trustworthiness plus all the necessary steps needed of information enrichment for a broader usage of information objects.

The appeal of cross-domain searching is obvious. A simple example: Why, after all, should the text of, say, Shakespeare's Hamlet only be retrieved alongside other texts – an artificial constraint? Why not alongside related resources from other domains such as sound recordings of performances of the play or digital images of the costumes?

J. Hřebíček et al. (Eds.): ISESS 2013, IFIP AICT 413, pp. 71–84, 2013.

Why not allowing someone else, e.g. a third party, to give his impression on an information object or to link it with some other information items of importance to him or to his/her community. This would really pave the way to a broader and better understanding of information often given by highly specified experts and improve its usage.

The problem so far can be described as a tragedy – as discussed in Havlik's "Tragedy of meta-information" [1] and also identified by ORCHESTRA (http://www.eu-orchestra.org/), an FP6-Integrated Project. It is that the "appropriate" meta-information depends on the intended use, which in turn can never be known in advance. For example, the set of meta-information required by a scientist interested in climate change may be very different from the meta-information required for automatic report generation in the context of "Clean Air For Europe" (CAFE) directive. Thus hundreds or thousands of meta-data structures have been defined and put in place.

Macolm Polfreman's article [2] brings it to the point: "The problem is that the archive, library, gallery, and museum communities have developed frequently incompatible metadata structures of their own". But it does not make sense just to blame on them as nearly everyone on the globe did it that way (and more or less it was discussed as good practice to provide meta-data).

Some promising approaches like the Dublin Core are widely used but have some weakness in loosing potentially information by their tight structuring of meta-information [3] and its involved mapping process.

So we had the idea to overcome all this tragedy of meta-data definition, acquisition and burden of collecting and entering them, by developing a semantic based system that allows to identify e.g. similarity by using ontologies going hand in hand with producing a way of cross-domain processing (searching, discovering, tagging) of information pieces (e.g. understood as resources in our naming conventions, which are data, services, models, etc.) and let doing the work of information enrichment (more or less the former entering of meta-information) peoples, experts, or communities interested in a specific topic touching various domains (e.g. like climate change or air pollution or its impact on human health).

In the TaToo project (Tagging Tools based on a Semantic Discovery Framework, Contract Nr. 247893; www.tatoo-fp7.eu) we stressed this issue specifically, and developed a semantic framework that allows the alignment of ontologies of different domains (like human health, agro-environmental, climate change) up to a bridge ontology (covering the necessary and minimal concepts suitable for the environmental domain). Following this approach we have been able to facilitate cross-domain search on the one hand and on the other to allow integration of additional domains respectively their domain ontology. This last point is still a piece of hard work, but at the end we can provide an added value by extending the cross domain search to the user, as a key service, that can enrich users experience significantly.

2 The Semantic Challenge

As introduced before TaToo's goal is semantic tagging and searching of environmental resources. TaToo aims to capitalise on the principles of the Semantic Web using ontologies as the underlying model for tagging and searching resources. The main objectives of the ontologies within the TaToo Framework are the following [4]:

- Allow formal tagging and searching of environmental resources;
- Allow contextual cross-domain tagging and searching of environmental resources;
- Foster multilingualism issues;
- Implement an extensible and more accurate searching mechanism.

To achieve this objectives one of the major challenges was the development of these ontologies. TaToo decided to base the development applying existing standards (W3C standards) and (re-)use proven methodologies like the NeOn methodology [5] and tools such as the NeOn Toolkit [6] as much as possible. But also to avoid the use of complex semantic for tagging and discovery.

2.1 The TaToo Ontology Framework

One can easily imagine a key challenge when applying semantics is the conceptualization and formalization of ontologies. As mentioned above the NeOn methodology for engineering ontologies was adopted [7]. This methodology provides guidance for all aspects of the ontology development, focusing on collaborative aspects on designing ontologies, the reuse of existing resources (ontological and non-ontological), and the further maintenance and evolution of the ontologies. In order to find out the common parts (terms, expressions, vocabulary) across the different domains (climate change, health, agro-environmental, etc.) covered by TaToo in its Validation Scenarios [8-10] several workshops, for introducing and collecting competency questions, were held.

Thereof, the main conclusions drove towards the definition and implementation of a common minimal model to describe environmental resources and annotations, and a semantic infrastructure where different domain ontologies shall be relatively easy plugged in; furthermore three main issues (stemming from an extensive requirements analysis [11]) impacted on the shape of the TaToo Ontology Framework, namely: i) take into account multilingual aspects; ii) allow formal tagging and searching environmental resources; and iii) allow cross-domain search.

With respect to the multilingual approach of the ontologies The final decision made in TaToo was to use the RDFSchema metadata rdfs:label and to add multilingual metadata storage in the TaToo Ontologies (including domain ontologies) [12].

Formal tagging and searching of heterogeneous and disperse environmental resources implies to define an ontology resource model. For practical reasons formal tagging in opposition to informal tagging should use a common and well-defined minimal model. For this purpose the Minimum Environmental Resource Model (MERM) was introduced. MERM is the cornerstone of the TaToo Ontologies, meaning that all resources annotated within TaToo contain tags according to the MERM model.

This approach implies that common elements of the domain ontologies have been abstracted and mapped to a common and really simple ontology in order to allow semantic interoperability. Within TaToo this common ontology is the so-called Bridge Ontology.

A high level description of the TaToo Ontology Framework can be seen in Fig. 1 below:

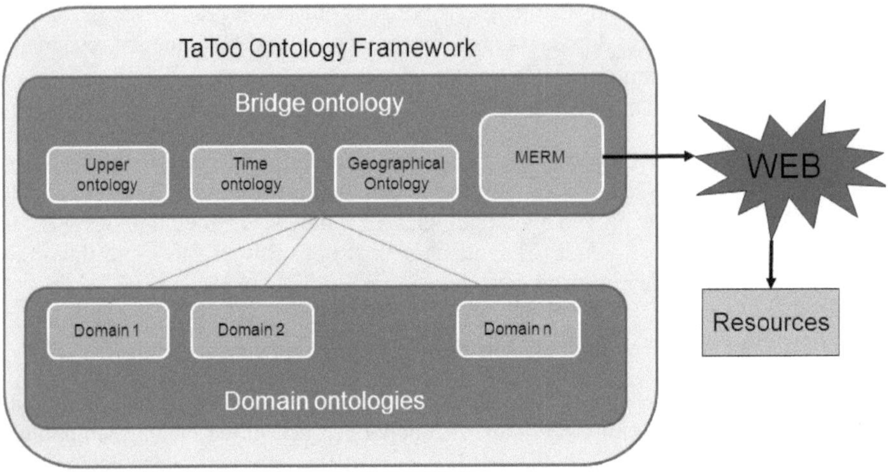

Fig. 1. High-level overview of the TaToo Ontology Framework

The ontologies defined in the Fig. 1 are as following:

- Minimal Environmental Resource Model (MERM): MERM is the ontology that gives structure to the annotations of a given resource, hence an effort to identify a minimal model that, without limiting the expression of specific domains, acts as a reference for past and future applications in the domain.
- MERM specifies concepts and properties that allow the description of what is important to say about a resource common to the environmental domain. MERM is extensible in order to allow specialization on different sub-domains, if needed.
- TaToo Bridge Ontology (BRIDGE Ontology): The BRIDGE Ontology main objective is to map elements coming from different domains fostering easier cross-domain integration. It acts as a bridge or hub where elements of different domains are mapped together. The BRIDGE Ontology also provides a unified way to annotate entities common to most of the sub-domains of the environmental domain. In this sense, the BRIDGE Ontology contains MERM plus a set of common domain elements, such as time and geographical ontologies. Domain ontologies can map to these common elements in order to achieve cross-domain semantic interoperability.
- Domain ontologies: A set of ontologies from the different sub-domains that are mapped to the common elements of the BRIDGE Ontology.

For the conceptualization of MERM three main classes are defined: Resource, AccesInfo, and Annotation.

- Resource is the main class for MERM. The Resource class includes annotation management information (author, date of creation, etc.). A resource presents a set of annotations describing what the resource is about and annotations used to discover the resource;
- ResourceAccessInfo is about accessing the resource. We assume that a resource has a single AccessInfo. The content of the access information depends on the type of the resource. Although access information for a web document would be different from access information for a web service, in TaToo Ontology we assume a single way to access to any resource; through its URI;
- Annotation describes what a resource is about according to the people tagging the resource. It is therefore an opinion about the resource, which might or not be true. The content of an annotation depends on the type of the resource (an annotation for a web document would be different from an annotation for a web service). Saying that "some resource is related to some domain concepts" would be the most basic type of annotation in TaToo. Resources can be tagged and discovered taking annotations as basis;
- Evaluations are also part of the MERM model, allowing users to evaluate both resources and resource annotations according to various evaluation criteria;
- Linking part of the MERM model enables TaToo resources to be linked by two types of links. The first type refers to the so-called typed links defined by a selected set of well-established properties such as rdfs:seeAlso, rdfs:isDefinedBy, dcterms:hasPart, dcterms:isPartOf, etc. The second type refers to similarity links that link two similar TaToo resources. The MERM model introduces its own property (merm:similarTo) that describes the similarity links.

As an implementation convention, within TaToo all domain elements subject of annotation must be subclasses of the bridge:Topic class.

The BRIDGE Ontology imports MERM, plus the environmental BRIDGE elements (GeoNames [14], and NUTS ontologies W3C Time [15]).

More details on the relation between MERM annotations and domain elements is managed using some BRIDGE Ontology elements can be found in [4].

2.2 Procedure to Add New Ontologies to the TaToo Framework

There is no single standard ontology or shared vocabulary that encompasses the entire environmental domain, so the TaToo Ontology Framework faced the challenge of being capable of dealing with annotations from different domain ontologies while enabling a certain degree of cross-domain discovery capabilities, and at the same time offering a unified, extensible and relatively simple ontology framework.

The main purpose of the TaToo Domain Ontologies is that of offering new domain elements (individuals) as topics (individuals of subclasses of bridge:Topic) to allow semantic annotation and search of resources following the MERM model. This implies that the focus of TaToo is not only in interoperability issues, but also of allowing a simple an effective way of adding new domain elements for the annotation and discovery, taking into account cross-domain considerations.

Cross-domain mappings are the most complex issue when adding a new ontology. Allowing direct mappings between different domain ontologies could potentially end-up in a really unmanageable network of interconnected ontologies. To avoid this issue, we propose to abstract the mappings in a way that all classes or individuals subject of mapping will be available at the bridge ontology. The requisites for introducing a new domain ontology into TaToo are the following:

- The new domain ontology must be an OWL2 RL ontology [13];
- The classes of individuals that can potentially be used for annotation (topics) must be made explicitly subclasses of the bridge:Topic class of the bridge ontology;
- The domain ontologies must provide individuals of the topics (subclasses of bridge:Topic) that will be used for annotation. TaToo does not allow annotation of resources using classes in order ease the inference process and get an optimal performance.

3 Information Enrichment Implementation

The TaToo Framework enables experts as well as arbitrary users to share trusted and reliable information and to allow easy discovery and semantically enhanced tagging [16] of existing environmental resources.

The TaToo Framework has gone through several short-end software lifecycles to progressively design and implement the core business logic, the extension of functionalities and finally the consolidation of the framework together with the enrichment of available tools and integration of Linked Data initiative. The framework demonstrates also the use of the ontology framework (as already discussed above) engineered during the TaToo project as a parallel implementation activity and how its integration within the implemented software framework enables cross-domain search and tagging, Linked Data extension, similarity between resources and multilingual aspects. Having in mind also to design a flexible and robust system the TaToo Framework architecture is composed of four different tiers:

- Presentation tier: provides tools to the end users to access the TaToo Knowledge Base;
- Public Service tier: provides Web Services interfaces to directly access the framework functionalities;
- Business tier: implements the core business logic of the framework, in particular semantic reasoning and harvesting;
- Cross tier: includes security and administration services.

3.1 Presentation Tier

The Presentation tier contains all the components the user can take advantage of to access the functionality provided by the TaToo Framework, in particular the tagging, discovery, evaluation and geo-referentiation functionalities.

The implementation of a portal [17] as a first entry point and showcase to access the TaToo Framework has been motivated by the TaToo vision of providing easy access to users, thus only requiring a browser and an Internet connection wherever the users are located, with no requirement to the user to install other software or plug-ins.

The reference implementation of the TaToo Portal is based on Liferay Community Edition technology (http://www.liferay.com/).

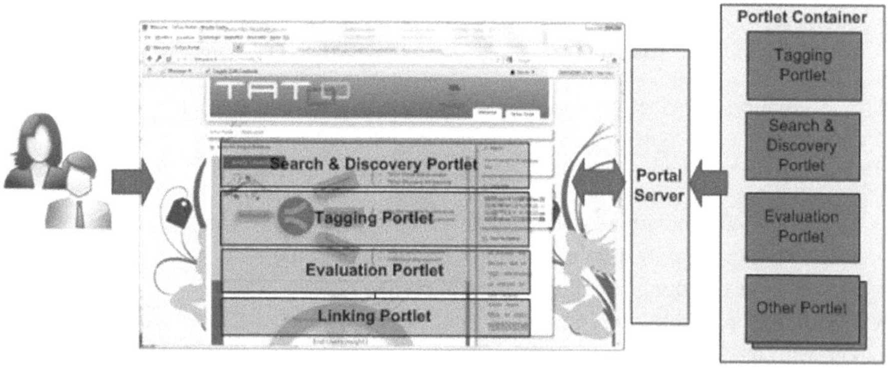

Fig. 2. TaToo Web Portal

The implementation of the final version of the TaToo Portal have been influenced by several live and online evaluation and validation surveys performed by end users that have been conducted during Scientific Conferences or by invitation sent by the TaToo Project Consortium.

3.2 Public Service Tier

The Presentation tier provides the Web Services interfaces (SOAP) in order to perform semantic Tagging, Discovery, Evaluation and Linking operations on TaToo resources.

The Tagging Services expose the tagging functionality of the TaToo System through its public interfaces. The Tagging Services allow to:

- Register new resources;
- Associate annotations with resources;
- Retrieve annotations associated with resources;
- Delete annotations.

The Discovery Services expose the public discovery functionality to User Components. They support semantic search and discovery of annotated resources. The discovery process is query driven, allowing the user to selected certain terms from an ontology that are then used for the semantic search.

In particular the following search operations are available:

- search: This is the main discovery operation, allowing the client to filter by topic, by annotation types, and by several resource and annotation metadata according to the TaToo Ontology Framework. The client may also decide whether the search will also retrieve similar resources and/or decide if they would like to perform cross-domain search;
- rectangleSearch: Main operation for performing GeoNames location-based queries within a rectangle delimited by two GeoNames coordinates;
- nutsSearch: As in the previous case, this operation is intended to perform location-based queries, but based on NUTS regions (from the NUTS ontology).

The Evaluation Services provide necessary functionalities for the evaluation of the TaToo-managed resources and the resource annotations. The functionalities of the Evaluation Services could be grouped into the following four categories:

- Creating new evaluations of the resources and the resource annotations;
- Retrieving existing resource and annotation evaluations;
- Removing existing resource and annotation evaluations, and;
- Generating and retrieving average evaluation scores of the resources and resource annotations.

The TaToo linked data functionalities are implemented through the Linking Services. In particular, they provide the following list of functionalities:

- Adding a predefined set of typed links between TaToo resources;
- Adding similarity links between similar TaToo resources;
- Retrieving a list of resources linked to a given resources;
- Retrieving a list of similar resources of a given resource;
- Removing the typed links and similarity links set between TaToo resources.

The Linking chain allows to "link" resources by a predefined set of properties available from the TaToo Ontology framework.

3.3 Business Tier

The Business tier is composed of the following set of components [18]:

- The Semantic Processor;
- The Harvester and Connectors, and;
- The Tagging, Discovery, Evaluation and Linking processors (where the business logic of operations resides).

The Semantic Processor is a Building Block made of a set of Core Components. It offers the infrastructure to access to the TaToo Semantic Repository and ontologies allowing semantically enhanced tagging, discovery, evaluation and linking functionalities.

The Semantic Processor provides functions to:

- Retrieve ontologies on the basis of user-defined properties like domain or context;
- Store and retrieve semantic annotations for resources (RDF-Triples);
- Search for semantic annotations;
- Semantic reasoning based on the existing ontologies.

The semantic functionality is realised by the following components composing the Semantic Processor internal architecture:

- The Ontology Manager to manage access to the TaToo Ontologies;
- The Semantic Framework to manage the interaction with the Semantic Repository and the Reasoner;
- Reasoner to infer new knowledge from the available RDF-Triples & ontologies and to check for incongruent (or inconsistent) information while managing ontologies;
- Triplestore (Semantic Repository) which is a data base for the storage and retrieval of RDF-Metadata.

The Resource Harvester is the Core Component responsible for harvesting meta-information from available resources that could be data (catalogues), web services, web pages, etc. A more and more large meta-information set is essential to improve the process of searching for resources returning valuable results to the requesting user.

As result of the final implementation phase the following Harvester connectors have been implemented and are currently harvesting resources via the:

- GENESI-DEC Connector - harvests OGC meta-information from GENESI-DEC catalogues, only those were pertaining to TaToo domains are then stored in the repository;
- JRC Connector – harvests meta-information from JRC Validation Scenario catalogues, generated in the context of the TaToo project;
- Masaryk University Connector – harvests meta-information from Masaryk University Validation Scenario, generated from the legacy RDF;
- Linked Data Connector – harvests resources pertaining to the TaToo domains that are part of Linked Data;
- Web Site Connector – harvests meta-information stored as RDFa in any Web Site available. As a proof of concept the TaToo Project Web Site has been harvested and stored as a resource.

The Linked Data Connector provides harvesting capabilities to the TaToo Framework to crawl and harvest Linked Data resources. The connector is able to integrate TaToo resources annotated in the Knowledge Base repository with linked resources from the Semantic Web. The connector presents then two main advantages, largely increases the knowledge base stored in the TaToo repository, and plugs the TaToo resources in the Linked Open Data cloud.

The Linked Data Connector is based on an already existing web crawler adapted to traverse and harvest content from the Linked Data web. The first action of the Linked Data connector harvesting workflow is then to crawl resources from the Linked Web. The implemented connector filters the result triples based on the link types supported in TaToo. Once the Linked Data connector receives the filtered set of triple statements, it directly stores the triples set using the TaToo Linking library.

The final step of the Linked Data workflow is to perform a second round parsing of the harvested LOD (Linked Open Data) cloud datasets to check if there are meaningful annotations in the context of TaToo. In this case a specific mapping is performed and resources are also semantically annotated with concepts from TaToo Domain Ontologies.

3.4 Cross-Domain, Similarity and Multilinguality Samples

The aim of this section is to provide a few views on how using the TaToo Portal as an entry point and how it is possible to obtain results from cross-domain searches, or extend the semantic search results by using Linked Data and similarity, and finally how to demonstrate how multlinquality got implemented.

The Cross-domain search feature is implemented in TaToo as optional, the framework leaves to the user the choice whether to perform a search only in its own domain or extend it to other domains.

The User can access the TaToo Portal and operate a search through the Search tools depicted in Fig. 3.

There the user has used a concept from the Regional Climate Change domain selected in the Ontology Tree above in the portal and then in the search results are displayed, properly ranked, resources from the Regional Climate Change domain together with resources from the Earth Science (OGC – Open Geospatial Consortium) domain, annotated with a different concept from the GENESI-DEC Domain Ontology that is aligned with the concept used to search from the Regional Climate Change Domain Ontology.

In the same search it was also selected the "Extension" option that triggers the usage of Linked Data. The search results include then together with the resources found thanks to their semantic annotation the linked resources, that means bound together by using one of the provided set of links, or properties, within TaToo.

Fig. 4 shows one of the linked resources found in the previous search.

Fig. 3. Cross-domain search results

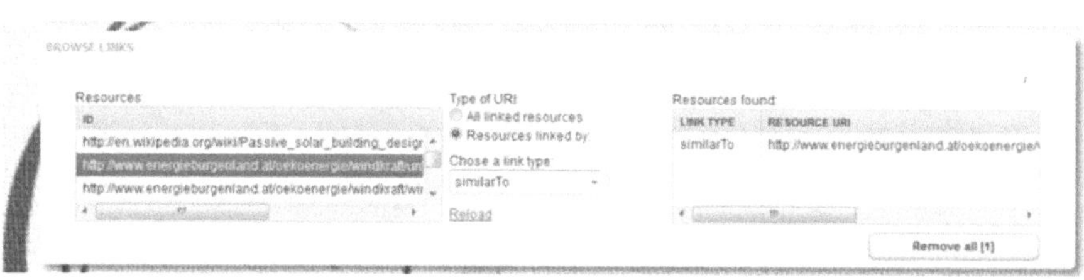

Fig. 4. Linked Resources search results

Finally, thanks to the Ontology engineering combined with the Software implementation best practices the multilinguality has been integrated within the framework at different levels:

- User Interface available in different languages;
- Ontology terms available in different languages;
- Tagging and Discovery using different languages.

Fig. 5 shows the Tagging Portlet translated in Czech language. Both User Interface elements and Anthropogenic Impact Domain Ontology are available in the selected language and can be used to search and annotate.

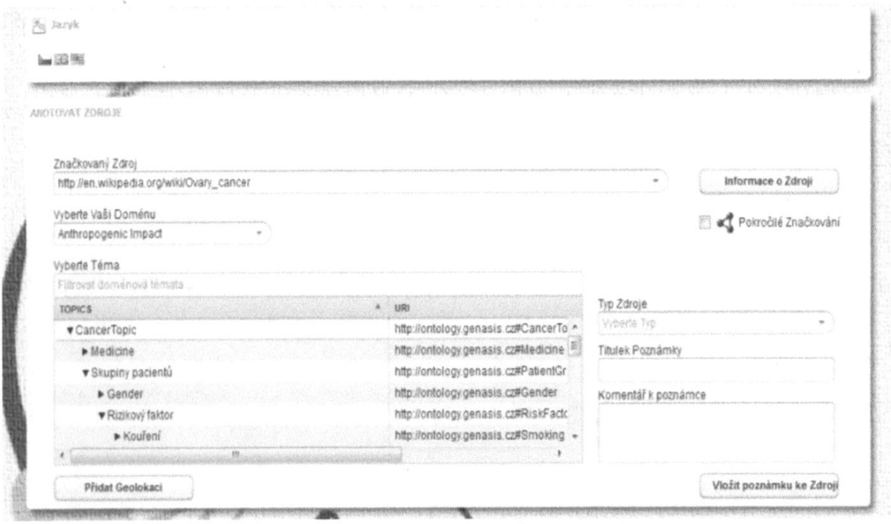

Fig. 5. Multilinguality

4 Conclusion

In this article we stressed the importance as well as the challenges of a semantic framework to reach to a satisfying extent cross-domain search and discovery. Still there remains the burden of doing alignments of ontologies of the different domains to the common part of all of them which is the bridge ontology. A task that is not easily done in couple of minutes.

But, what is gained by the TaToo approach is that we can extend the system by allowing to include new domains respectably domain ontologies and we do not loose information, in the contrary we enrich information to an object by additional tagging using a well-defined environment. We can link and reference to other (similar) resources. All this facilitates interoperability and understanding of cross domain issues.

Note: The majority of the software (e.g. public services) developed by the project are open source and can be requested by the TaToo Consortium Partners. The project documentation (deliverables) is available on the TaToo WebPage (www.tatoo-fp7.eu).

Acknowledgements. The research leading to these results has received funding from the European Community's Seventh Framework Programme (FP7/2007-2013) under Grant Agreement Number 247893. The authors would like to thank all the Partners of the TaToo Consortium for their valuable inputs and successful execution of the project.

References

1. Havlik, D., Schimak, G.: The Tragedy of Meta Information. In: Geophysical Research Abstracts. EGU General Assembly 12 (2010)
2. Polfreman M.: Cross-domain searching of library, archive and museum metadata – some considerations. Museums and Images JISC-FAIR Cluster Group - Cross Domain Searching Issues Paper; Document Date: July 1 (2004)
3. Woodley, M.: Crosswalks: the path to universal access. In: Introduction to Metadata: Pathways to Digital Information, Getty Information Institute, Los Angeles (2000), http://www.getty.edu/research/institute/standards/intrometadata/2_articles/woodley/
4. Pariente Lobo, T., Fuentes, J.M., Sanguino, M.A., Yurtsever, S., Dihe, P., Nesic, S., Avellino, G., Petronzio, L.: Semantic Framework Implementation Prototype, Deliverable 4.1.3 of TaToo project, Public Document (2012), http://www.tatoo-fp7.eu
5. Gómez-Pérez, A., Motta, E., Suárez-Figueroa, M.C.: NeOn Methodology in a Nutshell. NeOn-Book (2010), http://www.neonproject.org/nw/NeOn_Book
6. NeOn-Toolkit: The NeOn Toolkit, http://neon-toolkit.org/
7. Suárez-Figueroa, M.C., et al.: NeOn D5.4.1. NeOn Methodology for Building Contextualized Ontology Networks (February 2008)
8. Kaufmann, A., Peters-Anders, J.: Refined Scenario Definition – Case 1- V3, Deliverable 5.1.7 of TaToo project, Public Document (2012), http://www.tatoo-fp7.eu
9. Donatelli, M., Fanchini, D.: Refined Scenario Definition – Case 2- V3, Deliverable 5.1.8 of TaToo project, Public Document (2012), http://www.tatoo-fp7.eu
10. Kubásek, M., Hřebíček, J.: Refined Scenario Definition – Case 3- V3, Deliverable 5.1.9 of TaToo project, Public Document (2012), http://www.tatoo-fp7.eu
11. Schimak, G., Božić, B., Kaufmann, A., Peters-Anders, J., Dihé, P., Pariente Lobo, T.: The TaToo Semantic Case – Requirements, impacts and applications. In: Pillmann, W., Schade, S., Smits, P. (eds.) Proceedings of the 25th EnviroInfo 2011: Innovations in Sharing Environmental Observation and Information, vol. 2, pp. 832–847. SHAKER Verlag (2011)
12. Dihe, P., Avellino, G., Petroncio, L., Pariente, T., Fuentes, J.M., Yurtsever, S., Rizzoli, A., Nesic, S., Božić, B., Schimak, G.: Semantic Service Environment and Framework Architecture V3, Deliverable 3.1.3 of TaToo project, Public Document (2012), http://www.tatoo-fp7.eu
13. OWL 2 Web Ontology Language, http://www.w3.org/TR/owl2-profiles/#OWL_2_RL
14. GEONAMES: GeoNames ontology, http://www.GeoNames.org/ontology
15. OWL-Time, OWL time ontology, http://www.w3.org/TR/owl-time/

16. Avellino, G., Lobo, T.P., Schimak, G., Rizzoli, A.E., Dihé, P.: Applying semantics in the environmental domain: The TaToo project approach. In: Pillmann, W., Schade, S., Smits, P. (eds.) Proceedings of the 25th EnviroInfo 2011: Innovations in Sharing Environmental Observation and Information, vol. 2, pp. 848–855. SHAKER Verlag (2011)
17. Avellino, G., Petronzio, L., Lobo, T.P., Lopez, J.M.F., Yurtsever, S.: The TaToo Portal – Tagging and Discovery within TaToo. In: Pillmann, W., Schade, S., Smits, P. (eds.) Proceedings of the 25th EnviroInfo 2011: Innovations in Sharing Environmental Observation and Information. SHAKER Verlag (2011)
18. Dihé, P., et al.: An Architecture for the Semantic Enhancement of Environmental Resources. In: Hřebíček, J., Schimak, G., Denzer, R. (eds.) ISESS 2011. IFIP AICT, vol. 359, pp. 372–384. Springer, Heidelberg (2011)

ENVIROFI - Bringing Biodiversity to the Future Internet

Katharina Schleidt[1], Nina Laurenne[2], Andrea Giacomelli[3], and Denis Havlik[4]

[1] Umweltbundesamt Austria GmbH
katharina.schleidt@umweltbundesamt.at
[2] Semantic Computing Research Group (SeCo), Aalto University
firstname.lastname@aalto.fi
[3] Consiglio Nazionale delle Ricerche
a.giacomelli@iia.cnr.it
[4] Austrian Institute of Technology
Denis.Havlik@ait.ac.at

Abstract. In order to meet the requirements for observational biodiversity data, new sources of data must be enabled; for this purpose, new tools will be required. In order to effectively implement such tools, standardized building blocks such as the enablers defined by the FI will be very valuable. Thus, it is of utmost urgency that the requirements posed by the biodiversity sector are clearly structured and made available for implementation of environmental enablers within the FI. The ENVIROFI project has gone to great lengths so provide these requirements for various environmental domains; of specific relevant to this paper being the biodiversity domain. We hope to successfully contribute to the future of e-Environment in this manner.

1 Introduction

In the last decade, the pervasiveness of the Internet has become such that a world without online access to available resources has become unthinkable. At present, the internet consists of loosely linked documents and web applications. The Future Internet (FI) [1], an initiative of Organisation for Economic Co-operation and Development (OECD), seeks to expand this to the internet of People, Things and Services.

Information and communication technologies (ICT) for the environment generate huge amounts of environmental observations. At the same time, FI technologies and mobile applications empower amateur naturalists to provide as well as consume data. These data must be harmonized, exchanged between various stakeholders and processed - often in real time.

The ENVIROFI project, funded by the European Commission under the Seventh Framework Program (FP7) Future Internet Public Private Partnership (FI-PPP) Programme, addressed the needs and use cases of the environmental usage area within the FI-PPP context.

The project envisions a European Observation Web for the environment enabled by the FI. With this in mind, it explored the next generation of an ICT infrastructure which will enable global, decentralized, interactive, and dynamic environmental applications; integrating the data, services, and models from governments, the private sector, and other observation data.

J. Hřebíček et al. (Eds.): ISESS 2013, IFIP AICT 413, pp. 85–97, 2013.

Within the ENVIROFI project, Work Package 1 "Bringing Biodiversity into the Future Internet" focused on the use of Future Internet technology for survey, analysis, quality assurance, persistence and dissemination of biodiversity data in the following areas:

- Requirements for mobile data entry devices for recording biodiversity occurrence data.
- Requirements for context aware quality assurance of reported data, automated identification support and intelligent crowdsourcing of biodiversity occurrence data.
- Standards based data models for observational data as well as a semantic backbone for biodiversity data.

This paper analyses the requirements to the Future Internet of a mobile biodiversity occurrence information system, and proposes solutions required for the creation of such a system.

2 Requirements on Mobile Devices

2.1 Non-functional Requirements

As an initial step, an analysis of the requirements for mobile data entry devices for recording biodiversity occurrence data was performed. Traditionally, biodiversity data stems from organized surveys performed by trained professionals.

In many cases the results of the data survey are initially collected in paper logbooks and then later transferred to electronic media due to constraints of using mobile devices in the field. One of the reasons for the reluctant uptake of mobile devices was the prohibitive cost of professional devices such as the handheld devices created by companies such as Trimble[1]. Through the advent of powerful and inexpensive mobile devices, it is now possible to create affordable solutions for mobile biodiversity data entry.

Due to the increased financial pressures on the environmental sector, it is often not possible to obtain the necessary amount of observational data by relying only on professional survey data; thus, this data must be complemented by data provided by experienced amateur naturalists. While these potential data providers often possess a great deal of knowledge over certain areas of biodiversity, they are less likely to invest large amounts of time and energy in repetitive activities such as performing fledged professional surveys. Consequently, these users must be offered a possibility to report only part of the required observational data and encouraged to improve the overall survey quality at their own pace. Ideally, the same application should be presented in various levels of complexity for different user groups, using simpler terms and requiring fewer types of observations for less experiences users.

Data provision should be bidirectional, with the user able to interactively obtain information about habitats and relevant biodiversity in a specific area. In addition to this

1 http://www.trimble.com/

being a strong support for professional users in their work, this opens the scope of the mobile applications for a wider community of amateurs, allowing them to gain valuable information on the area of interest.

As many relevant areas of interest for the biodiversity sector are remote areas, one cannot assume that the mobile application will have constant connectivity. The application must remain fully functional during the duration of a field trip and opportunistically exchange information with the central database in the areas with good network coverage. This means that:

1. The user must be able to select a specific area in which they will be working or trekking in advance while they still have full network connectivity and all relevant data for this area must be cached for later use when in the field.
2. All observational data provided by the user in the field must be stored locally and uploaded once network connectivity is available.
3. The feedback on observations entered by user (e.g. assessing of the plausibility by other users or background services) can only be provided asynchronously.

A further difficulty often encountered during protracted use of mobile devices is the battery duration. Using current technology only about 4 hours of continuous use are possible, severely constraining the scope of use for mobile applications. While the mobile application can be designed to keep energy requirements to a minimum, the techniques used often interfere with the usability of the system; thus, we are hoping that new technological developments will provide a new generation of devices with the required battery duration as required for professional field work.

2.2 Main Usage Patterns

The biodiversity survey applications must support following types of use

- Output:
  (a) View locations of existing species and habitat occurrence records;
  (b) View detailed information on observations on existing species and habitat occurrence records;
  (c) View further information on species or habitats (nice-to-have).
- Input
  (a) Provide a new species or habitat occurrence record with location;
  (b) Provide new observations on existing species and habitat occurrence records;
  (c) Support functionality for species and habitat identification.
- Administrative
  (a) Define and cache areas of interest for later offline use in the field;
  (b) Download species and habitat occurrence data;
  (c) Provide feedback on existing and new species or habitat occurrence records and observations;
  (d) Standardized data models for access to species or habitat occurrence records and observations;
  (e) Standardization of species and habitat nomenclature.

These usage patterns are found in all participatory sensing applications, but some adaptation will be required for an optimal user experience on mobile devices.

3 Quality Assurance of Biodiversity Observations

Quality assurance of biodiversity data has always been a challenge. In contrast to other areas of environmental data provision, where the focus is on physical and chemical measurements, biodiversity data has always depended on observational data provided by humans, with all subjective errors entailed. While in some cases these observations are complemented by physical specimens that can be archived for future reference, there is a multitude of reasons why this is not often possible, including:

- endangered species should not be further endangered by removal of specimens, be they entire individuals or only parts (i.e. leaves or branches);
- certain species are dangerous, and thus should only be collected by professionals;
- some species can only be observed at a distance.

In addition, storage of specimens is costly. Many institutions currently tasked with the storage and conservation of scientific specimens are struggling to maintain their current collections; some have been closed, with other institutions struggling to find storage opportunities for the existing material. Thus, adding vast amounts of additional specimens as would result from wider data provision to these already overburdened institutions is not an option. New mechanisms for assuring the correctness of field observations must be found.

To date, biodiversity occurrence data is provided with one unique species identification. Upon later inspection of the data, this identification is often shown to be wrong; a new identification is provided replacing the old one. However, there is no guarantee that the new identification is actually correct; by removing the old identification originally provided, valuable information can get lost. Thus, we propose allowing for the provision of multiple identifications for one species occurrence record, whereby each species identification is assigned a level of plausibility. While the species identification with the highest plausibility will be displayed to the user as the default identification, it should be possible to view the other identifications provided if desired.

3.1 Context Aware Quality Assurance

One approach for assuring the quality of biodiversity observation data is a context aware approach. Through the mobile devices used for the provision of biodiversity observations, the system can report the exact location of the user at the time of reporting the observation, as well as the time at which the observation was first reported to the system.

By comparison of the species reported by the user with existing background information such as known species distributions a plausibility of the reported species existing in the area reported can be calculated. However, species distribution information is only available for certain species, is not always completely reliable, and is prone to change due to additional factors such as human pressures and climate change. Thus, additional spatial mechanisms must be used for the calculation of the plausibility of a biodiversity occurrence record.

Various additional mechanisms identified for extending the possible scope for the calculation of species occurrence plausibility include:

- comparing the biogeographical region the species is known to occur in with the biogeographical region the species observation was reported from;
- comparing the habitat the species is known to occur in with the habitat the species observation was reported from;
- using the distribution of a species known to prefer a similar habitat to determine the plausibility that the reported species could occur in the region reported;
- using Food-Webs together with the distributions of other species known to be linked to the reported species via a food-chain.

In addition to the spatial mechanisms listed above, temporal information can also be valuable in determining the plausibility of a species occurring at a specific place and time. Based on phenological information pertaining to the species reported such as migratory patterns, the date of emergence of leaves and flowers or the date of leaf coloring and fall in deciduous trees, the system can determine the plausibility that the reported species could actually be sighted at the place and time reported.

A further mechanism for determining the plausibility of a specific species occurrence record utilizes knowledge on common mistakes made in species identification. While one would assume that these types of mistakes only pertain to amateur naturalists, systematic misidentifications are also common amongst scientists and often propagated from one generation of scientists to the next. Using local knowledge on common misidentifications, the system can show the user reporting a commonly misidentified species images of the other more plausible species; this can help the user decide on the final identification of the individual organism sighted, as well as providing further plausibility information for the biodiversity occurrence record.

3.2 Automated Identification Support

Various types of automated support in the identification of species have been explored both within ENVIROFI as well as in other related projects. Organizations such as NatureGate [2] in Finland provide well organized identifications keys utilizing the power of IT-based systems to allow a user to provide those species characteristics. Institutions such as the Laboratoire Informatique & Systématique [3] provide generic identification key software [4] that can be configured by providing SDD [5] encoded species characteristics. As this area has been well explored in other projects, it was not directly considered within ENVIROFI; however, such functionality would be essential for a complete production system.

The two areas of automated identification support explored within the ENVIROFI project pertain to the automatic recognition of species based on images or sounds recorded. For the identification of images, as the provision of a species identification based on an entire individual is not currently technically possible, a fairly sound identification can be provided based on the image of an individual leaf of a tree.

Of these two approaches, image recognition was prototyped pertaining to leaves of trees. Identification services have been implemented that analyze the shape and coloration of a leaf image provided and return a ranked list of possible species identifications. This list of possible identifications can then be provided to the user together with further images representative for this species, allowing the user to decide which of the possible species identifications returned best corresponds to the individual tree they are trying to identify.

While it is possible to identify some species using only leaf images, for the correct identification of many species, further information will be required. In some cases this complementary information could be provided through additional images of other parts of the tree such as flowers or fruits. In other cases more complex morphological information will be required; for this approach it would be advisable to integrate the leaf recognition service with an identification key for the provision of such information.

Regardless of the approach chosen for automated identification support, which can provide valuable support in the identification process, it is imperative that the final decision on the identification of the individual sighted is left to the human user.

3.3 Crowdsourcing of Biodiversity Occurrence Data

Traditionally, biodiversity occurrence data stems from organized surveys performed by trained professionals. Due to the increased financial pressures on the environmental sector, the necessary resources are often not available to perform the survey activities required for the comprehensive assessment of the state of biodiversity.

At the same time, with the advent of various networked platforms and social media, grass-roots activities triggered by non-professionals with an interest in a specific area are becoming a widespread alternative to existing organized governmental and scientific activities. This can provide a valuable complement to existing data and information available from conventional sources. At the same time such activities can serve to sensitize the general population to various issues pertaining to biodiversity, both by allowing them to actively engage in the survey process as well as by providing the users with new information directly pertaining to their current interest through context sensitive information provision.

However, when dealing with crowd-sourced data, it is difficult to ascertain the reliability of the data provided. While these potential data providers often possess a great deal of knowledge over certain areas of biodiversity, it is difficult to determine the reliability of the data they provide without some sort of certification of their individual track-record; thus, the approach of storing the user's reliability rating has been selected. In an initial phase, only one reliability rating will be stored together with the user data; however, in production use the reliability rating must be defined at a far more fine-grained level, providing information on the user's level of proficiency pertaining to different areas of biodiversity. Proficiency categories can be defined based on the structure of the taxonomic hierarchy, using the taxonomic ranks Order, Family or Genus for the definition of a proficiency rating. A user having a high proficiency rating for certain plant families may have a very low proficiency rating pertaining to animals. Depending on the species identified, the relevant user rating should be taken into account in determining the plausibility of a species identification.

An additional factor in determining the plausibility of a species identification can be provided by utilizing the user's own trust in their identification skill. While in many cases the user will be sure of the correct identification they are providing, in other cases the user will be aware of the fact that they are not confident of the identification being provided. By allowing the user to add a subjective confidence rating to their identification, further information on the plausibility of this identification can be obtained.

Various factors can be utilized to determine a user's reliability rating. An initial value can be set depending on the context through which the user has registered with the system. A user registering via an accredited organization or providing proof of higher education in a specific relevant area is provided a higher reliability rating than an anonymous user. However, the user's reliability rating can be subsequently modified depending on the accuracy of data provided by this user; if the identifications provided with observation records by this user are often corrected by subsequent users, the user's reliability rating will be reduced. Conversely, if the identifications provided with observation records by this user are regularly confirmed by more experienced users, the user's reliability rating will be raised.

In order to perform these adjustments on a user's reliability rating, methods of providing subsequent feedback to existing biodiversity occurrence records must be provided. The following options have been proposed:

- **Re-identification of existing biodiversity occurrence records in the field:** if a user comes across an individual in the field that has an incorrect identification, the user may provide a different identification for this individual.
- **Provision of test images at the end of the identification process:** once a user has completed the identification process for an individual, the user is shown six existing images, whereby these images fall into three categories as follows:
  - images with a high plausibility rating for the species the user just identified;
  - images which are definitely not the species the user identified (but may be commonly misidentified as this species);
  - images which have a low plausibility rating for the species the user just identified.

  If the user correctly selects the images that have a high plausibility rating for the species identified and rejects the images that are definitely not this species, the plausibility that the new observation provided is correct can be raised. At the same time, if the user has shown proficiency in the identification of this species based on the known material, their choices pertaining to the occurrence records with a low plausibility rating can be used to either raise or lower the plausibility for these occurrence records as well.

- **Feedback via a web portal:** while mobile devices are very useful for user interaction in the field, their use is limited due to the small display size as well as difficulties in data entry. Thus, a web portal allowing the user to browse through existing biodiversity occurrence records and provide feedback on the occurrence data provided is a valuable complement to the mobile applications that are the focus of this work.

While there will never be full certainty in the identification of biodiversity occurrence records stemming from observations, especially when dealing with data of diverse provenance, through the use of feedback mechanisms as described it should be possible to provide data of a sufficient quality for scientific and administrative use. Due to the plausibility information provided for each species identification as well as the reliability ratings attached to each data provided, the end user of the data can filter the available data to assure the necessary level of data quality is provided.

4 Data and Information Models

4.1 Standards Based Data Models

There are several international standards for the structuring, storage and sharing of biodiversity data available; most of these stem from TDWG Biodiversity Information Systems [6], a standards organization specialized in the provision of standards pertaining to biodiversity. The two TDWG standards pertaining to biodiversity occurrence records are:

- **Darwin Core (DwC):** The Darwin Core is body of standards. The Darwin Core is primarily based on taxa, their occurrence in nature as documented by observations, specimens, and samples, and related information.
- **Access to Biological Collections Data (ABCD):** The ABCD Schema is a comprehensive standard for the access to and exchange of data about specimens and observations (a.k.a. primary biodiversity data). The ABCD Schema attempts to be comprehensive and highly structured, supporting data from a wide variety of databases.

While these standards have served as an inspiration for the work in ENVIROFI, they have several drawbacks. The following deficits guided us in our decision not to utilize these existing standards:

- **ISO Compatibility:** The TDWG standards are not compatible with Geography Markup Language (GML) [7] and the wider family of ISO spatial standards mostly defined by the Open Geospatial Consortium (OGC) [8]. As these standards form the backbone of most modern environmental data standards, specifically those defined in the implementation of the INSPIRE Directive (2007/2/EC) [9] in Europe, ISO compliance was deemed essential.
- **Lack of Flexibility:** while the ISO standards for observational data are very generically defined to guarantee a high level of flexibility, TDWG standards are usually custom tailored to a specific field. The following areas required higher flexibility than the TDWG standards:
  - **Provision of Multiple Identifications:** for the various quality assurance mechanisms described above, it is necessary to be able to provide multiple identifications with integrated plausibility levels for individual biodiversity occurrence records:

– **Dynamic Definition of Properties:** due to the differing user groups to be served by the ENVIROFI Biodiversity Application, different observational parameters describing different properties observed on an individual must be supported.

Based on these considerations, the observational data model defined within EN ISO 19156 Observations & Measurements (O&M) was selected as the basis for this work. This data model foresees dynamic parameterization of the properties provided, while the TDWG standards would require definition of specialized schemata for this purpose. In addition, as the identification of an individual is seen as an observation on this individual, multiple identifications for one individual can be provided as differing observations on this individual.

The core of the O&M Schema is the Observation class, which provides a result (association Range) of a specified property (association Phenomenon) determined through the application of a specified procedure (association ProcessUsed) on an explicit object (association Domain) together with relevant metadata such as observation and result times or responsible parties.

The object the properties are being ascertained on, the target of the Domain association has the role featureOfInterest (FoI) for the Observation. Special care must be taken in modeling the FoI, as often properties that should be the result of an observation on the FoI are modeled as explicit attributes of the FoI. An example is the species identification, which is often considered to be an integral part of the object being observed; however, as the identification initially assigned to this individual is often incorrect, it must be seen as an observation on the individual. We have chosen to refer to the FoI defining the individual the observations are being made on as a Species Point (SP), and relegating all further information to the status of observations on this SP.

The following properties have been identified as relevant to tree survey:

Table 1. Tree Survey Properties

Inventory	**Classifications**
Tree number	Social Position (Kraft)
Identification	Status
Species Identification	Tree Status Forest Inventory
Location	Tree Status Site Inventory
Relative Position	Barkbeetle Infestation
Geographic Coordinates	Wind Damage
Length Properties	Light Changes
Breast Height Diameter (BHD)	Tree damage (1-n)
Tree Height	**Comments**
Crowning Height	Comment
Deadwood Height	
Crown Diameter	

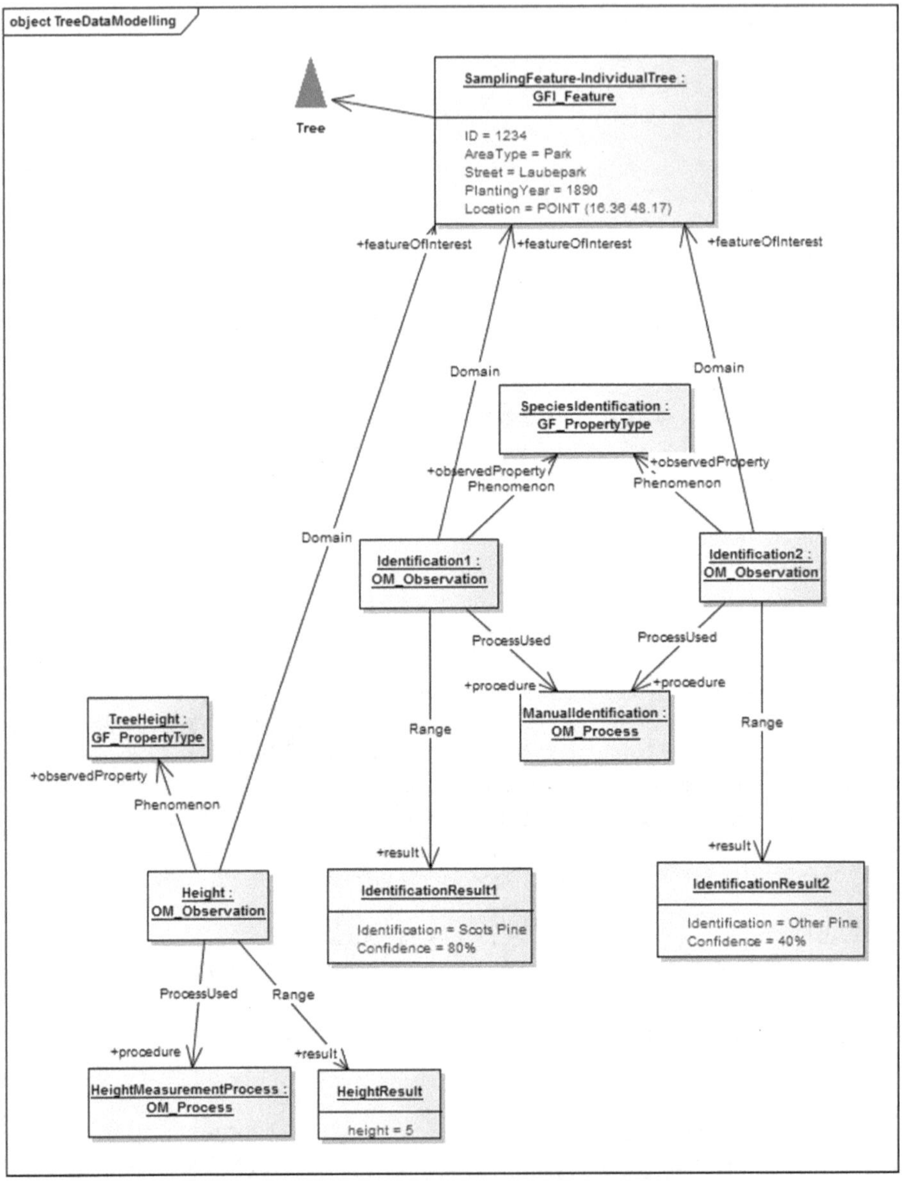

Fig. 1. Individual Species Point with Multiple Identifications

As these properties are dynamically assigned, adding new properties to the system is only a matter of configuration; new parameters can be easily extended. The same is true for the procedure, which is the target of the observation association ProcessUsed; various procedures can be dynamically defined and referred to by observations.

In the example below we provide a simplified sketch of the ENVIROFI Biodiversity data model, based on the O&M schema. In this example we show how an individual, in this case an individual tree in a forest, can have two conflicting identifications with different levels of plausibility as expressed by the confidence attribute in the result.

4.2 Semantic Backbone

A recurring problem with the analysis of biodiversity data stems from inconsistencies in the nomenclature used for the identification of species and habitats. Historically, different names have been established for the same concept, the same name been reused for different concepts, or the primary concepts used to describe organisms modified over time. In order to overcome this problem, various projects such as ALTER-Net [10] or SEEK [11] and SONet [12] have proposed the use of ontologies for structuring environmental data.

Within ENVIROFI, it was proposed to utilize the power the TaxMeOn Species Ontology [13] developed of ontologies for the following purposes:

- Structuring and storage of taxonomic information. This should include both scientific and common names
- Storing additional species information as required for context aware quality assurance (i.e. biogeographical regions of occurrence, types often wrongly identified)
- Semantic mapping between species lists (both common and scientific species names from national lists)

The Taxonomic Meta-Ontology (TaxMeOn) is for managing the scientific and vernacular names of organisms. The model consists of three parts that serve different purposes and can be used independently, too. Our focus is here in species lists which forms vary from a simple name list to more information containing list for a professional use.

TaxMeOn defines classes and properties that can be used for building name ontologies for creating semantic metadata for describing observational data. The model is based on RDF(S) and some features of the OWL, and it contains 42 classes (excl. subclasses of the taxonomic ranks) and 36 properties. The key classes for managing biological names are the following:

1. A **scientific name** which is binominal and consists of a genus name and a species name. In the Linnaean name system taxa are organized hierarchically (a species, genus, family etc.). The class TaxonInChecklist expresses a scientific name and the hierarchy is expressed using the relation isPartOfHigherTaxon.
2. An **author name** of a scientific name is a reference to the original publication where the original description of a taxon was first published. The scientific name authorship is expressed using the relation hasScientificNameAuthorship. The author names are often abbreviated, for example Linnaeus may be referred to L. or Linn. The model supports abbreviations. The usage of a reference to an original publication is limited to biology and regulated by the nomenclatural codes, and it differs from the ordinary way of referring to a scientific publication.
3. A **reference** consists of three subclasses: a checklist, a scientific publication and other source. For instance, the latter can be used if an unofficial note needs to be

referred to. The class Reference allows associating scientific publication to a name and specifying the checklist that is used.
4. The class **Nomenclatural code** specifies the set of rules that are applied to a scientific name. The nomenclature of plants and animals differ from each other.
5. According to the nomenclatures a scientific name can have different **statuses**, for example the status of a scientific name can be rejected.
6. **Taxonomic status** indicates the current state of a name, for example a name can be a valid name or a synonym of a valid name (the subclass of TaxonomicStatus).
7. The scientific names are easily misspelled and incorrect spellings have often spread in literature. The **SpellingStatus** is used for indicating the orthographic status of a name, e.g. misspelling.
8. The **VernacularName** is a class for the common names of organisms in different languages. It is supported to specify a name as an accepted name or an alternative name.
9. **Taxonomic rank** expresses the hierarchal level of a taxon (a species, a genus, a family etc.; subclasses of the TaxonomicRank).

By providing species identifications through reference to the ontology instead of textual entries, clarity can be achieved in the explicit meaning of the identification. The additional information available to the individual species can be utilizes for context aware quality assurance, provision of additional information on specific species as well as support in merging species occurrence data stemming from other sources.

5 Conclusions

In order to meet the requirements for observational biodiversity data, new sources of data must be enabled. For this purpose, new tools will be required; in order to effectively implement such tools, standardized building blocks such as the enablers defined by the FI would be very valuable. Thus, it is of utmost urgency that the requirements posed by the biodiversity sector are clearly structured and made available for implementation of environmental enablers within the FI.

The ENVIROFI project has gone to great lengths so provide these requirements for various environmental domains; of specific relevant to this paper being the biodiversity domain. We hope to successfully contribute to the future of e-Environment in this manner.

The ideas discussed in this paper were also partially developed and tested in the field. More technical information on the overall application architecture and on the specific enablers facilitating the development of FI-enabled VGI applications can be found in [14] and [15] respectively.

Acknowledgements. We thank the ENVIROFI (FP7 – 284898) project consortium for the lively discussions we had as well as the EC for providing the necessary funding for this work. This paper is based on our common findings within the ENVIROFI Project.

References

1. European Future Internet Portal, http://www.future-internet.eu/
2. Official Homepage of NatureGate, http://www.luontoportti.com/suomi/en/tekijat/
3. Official Homepage of Laboratoire Informatique & Systématique, http://www.infosyslab.fr/
4. Official Homepage of Laboratoire Informatique & Systématique Identification Key Software, http://www.identificationkey.fr/
5. Official Homepage of the Structured Descriptive Data (SDD) TDWG Standard, http://www.tdwg.org/standards/116/
6. Official Homepage of TDWG Biodiversity Information Standards, http://www.tdwg.org/
7. Official Homepage for GML, http://www.opengeospatial.org/standards/gml/
8. Official Homepage for OGC, http://www.opengeospatial.org/
9. INSPIRE Directive Legal Text, http://eur-lex.europa.eu/LexUriServ/LexUriServ.do?uri=OJ:L:2007:108:0001:0014:EN:PDF
10. Official Homepage for the ALTER-Net Project, http://www.alter-net.info/
11. Official Homepage for the SEEK Project, http://seek.ecoinformatics.org/
12. Official Homepage for the SONet Project, https://sonet.ecoinformatics.org/
13. Tuominen, J., Laurenne, N., Hyvönen, E.: Biological Names and Taxonomies on the Semantic Web – Managing the Change in Scientific Conception. In: Antoniou, G., Grobelnik, M., Simperl, E., Parsia, B., Plexousakis, D., De Leenheer, P., Pan, J. (eds.) ESWC 2011, Part II. LNCS, vol. 6644, pp. 255–269. Springer, Heidelberg (2011)
14. Usländer, T., Berre, A., Granell, C., Havlik, D., Lorenzo, J., Sabeur, Z., Modafferi, S.: The Future Internet Enablement of the Environment Information Space. In: Hřebíček, J., et al. (eds.) ISESS 2013. IFIP AICT, vol. 413, Springer, Heidelberg (2013)
15. Havlik, D., Egly, M., Huber, H., Kutschera, P., Falgenhauer, M., Cizek, M.: Robust and trusted crowd-sourcing and crowd-tasking in the Future Internet. In: Hřebíček, J., et al. (eds.) ISESS 2013. IFIP AICT, vol. 413, Springer, Heidelberg (2013)

Acquisition and Representation of Knowledge for Atmospheric New Particle Formation

Markus Stocker[1], Elham Baranizadeh[2], Amar Hamed[2], Mauno Rönkkö[1], Annele Virtanen[2], Ari Laaksonen[2,3], Harri Portin[4], Mika Komppula[4], and Mikko Kolehmainen[1]

[1] Environmental Informatics Group, Department of Environmental Science, University of Eastern Finland, P.O. Box 1627, 70211 Kuopio, Finland
{markus.stocker,mauno.ronkko,mikko.kolehmainen}@uef.fi
[2] Aerosol Physics Group, Department of Applied Physics, University of Eastern Finland, P.O. Box 1627, 70211 Kuopio, Finland
{elham.baranizadeh,amar.hamed,annele.virtanen}@uef.fi
[3] Finnish Meteorological Institute, P.O. Box 503, 00101 Helsinki, Finland
ari.laaksonen@fmi.fi
[4] Finnish Meteorological Institute, P.O. Box 1627, 70211 Kuopio, Finland
{harri.portin,mika.komppula}@fmi.fi

Abstract. Sensors are used in environmental science to monitor an increasingly large multitude of properties of real world phenomena. An important scientific aim of such monitoring is more accurate and more complete understanding of phenomena, with respect to, e.g., their formation, development, or interactions. Properties and phenomena may be, for instance, mass or concentration and particulate matter or eutrophication, respectively. Typically, measurement data must undergo considerable processing in order to become useful to a scientific aim. We outline the architecture and implementation of an ontology-based environmental software system for the automated representation of knowledge for real world situations acquired from measurement data. We evaluate and discuss the system for the automated representation of knowledge for situations of atmospheric new particle formation. Such knowledge is acquired from measurement data for the particle size distribution of a polydisperse aerosol, as measured by a differential mobility particle sizer.

Keywords: Knowledge representation, new particle formation, ontology, situation theory, machine learning.

1 Introduction

Atmospheric New Particle Formation (NPF) and the growth of newly formed particles have been well documented in a wide variety of environments all over the world [1]. Aerosol particles are known to influence quality of life, for instance by affecting human health [1]. Newly formed nano-sized particles can grow, through condensation and coagulation processes, and directly effect on Earth's radiation balance by scattering sunlight. Indirectly, their potential to

J. Hřebíček et al. (Eds.): ISESS 2013, IFIP AICT 413, pp. 98–108, 2013.

grow large enough to act as Cloud Condensation Nuclei (CCN) and their possible activation to cloud droplets results in more scattering of radiation. It is known that the scattering of radiation has a cooling effect on the climate [2]. However, the magnitude of indirect effects remains the single largest uncertainty in current estimates of anthropogenic radiative forcing [2], leading to large uncertainties in the calculations of future climate change.

The study of NPF relies on methods for the identification and characterization of these atmospheric events. At the base of such methods is the measurement of particle size distribution for polydisperse aerosols. Of specific interest are particles with diameter size ranging 10×10^{-9} to 10×10^{-6} m. The resulting data, as measured over the course of a day at a specific location, can be visualized to assess the presence of NPF [3]. Quantitative methods have been developed to extract, from measurement data, basic NPF characteristics, such as particle growth and formation rates [3].

Different classifications have been proposed in order to characterize NPF [3–5]. The criterion established by Hamed *et al.* [4] is based on event clarity. According to the criterion, nucleation event classes 1, 2 and 3 indicate strong, intermediate, and weak nucleation event (E) days, respectively. Days during which no particle formation is observed are classified as non-event (NE). Days that are neither E or NE are called undefined class, or class 0. The tasks of NPF detection and characterization are typically carried out visually by experts [4].

We outline the architecture and implementation of an ontology-based environmental software system aimed at the automated representation of knowledge for real world situations acquired from measurement data. In translating measurements to observations, the system performs a semantic enrichment of (heterogeneous) measurement data. Observations are consistent with the terminology defined by the Semantic Sensor Network (SSN) ontology [6–8]. Given observations, the system acquires and represents knowledge for real world situations. Situations are consistent with the terminology defined by the Situation Theory Ontology (STO) [9]. The STO has been used to represent airport security situations [10], in methods for situation identification [11], and for intelligent sensor network information fusion [12]. We evaluate and discuss the presented system for situations of NPF acquired from data for the particle size distribution of a polydisperse aerosol, as measured by a differential mobility particle sizer.

The presented system can support domain experts in the detection and characterization of NPF. Moreover, it has the potential for full process automation. More generally speaking, by committing to the terminology and semantics explicitly formalized in ontologies, the system provides a unified framework for the representation of sensor observations and real world situations acquired from observations. The system can be adapted to specific domains by extension, specifically by extending ontologies with domain knowledge and software components with domain specific application logic [13, 14].

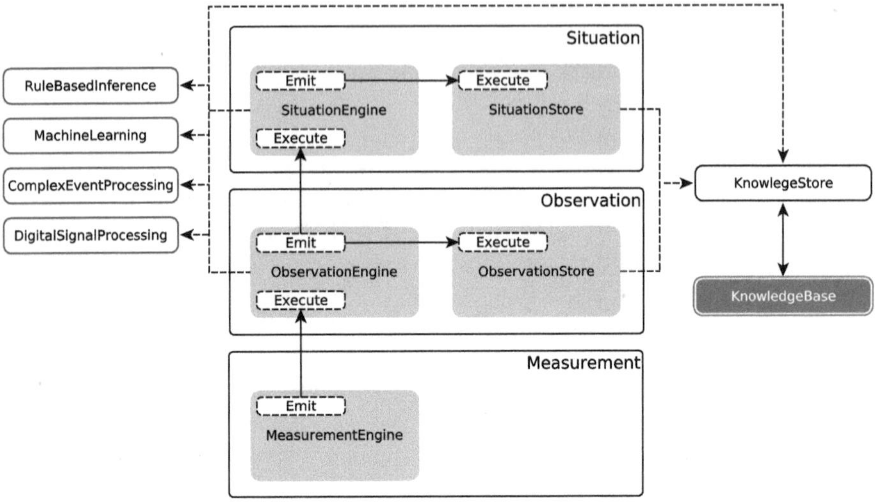

Fig. 1. System architecture for the representation of situational knowledge acquired from sensor measurement data showing the three layers of measurement, observation, and situation as well as the main components and modules, and their interactions

2 Materials and Methods

Materials: We used expert-labeled sensor measurement data for the particle size distribution of a polydisperse aerosol, in the 7 to 800 nm diameter size range as measured between January 2008 and December 2011 by a Differential Mobility Particle Sizer (DMPS) located at the Puijo semi-urban measurement station in Kuopio, Finland [15]. A DMPS consists of a Differential Mobility Analyzer (DMA) and a Condensation Particle Counter (CPC); the particles of a polydisperse aerosol (source) are first classified according to diameter size by the DMA and then counted by the CPC [16]. The DMPS measures the particle number concentration [cm^{-3}] for 40 discrete diameter sizes, on average 5 samples per hour. Daily measurement data results in a data matrix $m \times n$, where m is the number of samples (typically approx. 120) and $n = 40$ is the number of discrete diameter sizes. Daily measurement data is persisted in a text file.

We developed a software architecture for a system aimed at automated, near real-time, distributed, and continuous acquisition of sensor measurements; translation of measurements into semantically enriched observations; representation and persistence of observations; automated acquisition of situations from observations; representation and persistence of situations. Figure 1 provides an overview of the architecture. The core of the architecture consists of three layers that build on each other: the measurement, observation, and situation layers. Layers consist of components, including the measurement, observation, and situation engines as well as the observation and situation stores. Components communicate over (TCP/IP) streams. Modules implement computational services, such as for machine learning or complex event processing, needed by components. The knowledge base is a third-party system.

The architecture is implemented on top of Storm, a distributed real-time computation system for the processing of streams of data.[1] A Storm topology consists of nodes and streams. Nodes may be of two types: (1) a source of streams or (2) a consumer of input streams as well as data processor and, possibly, source of new streams. Components and streams map to Storm nodes and Storm streams, respectively.

At the measurement layer, a measurement engine implements the software logic necessary to acquire data from a sensor, and process it into measurements. In its simplest form, a measurement is a tuple consisting of a time stamp and a measurement value. Measurements are forwarded to Storm streams. At the observation layer, an observation engine subscribes to streams of measurements and processes measurements to observations, consistent with the SSN ontology. This step amounts to a semantic enrichment of measurements. Observations are forwarded to streams. Still at the observation layer, an observation store may subscribe to streams of observations and require the knowledge store to persist observations. At the situation layer, a situation engine subscribes to streams of observations. A situation engine implements (one or more) knowledge acquisition tasks and makes use of computational services provided by modules. Knowledge for situations acquired from observations is represented consistent with the STO and persisted by the situation store.

The STO captures key aspects of the situation theory developed by Barwise and Perry [17] and extended by Devlin [18]. The theory formalizes the semantics of situations by means of the expression $s \models \sigma$ (read "s supports σ") meaning that the infon σ is "made factual" by the situation s. According to the definition by Devlin, the object $\ll R, a_i, \ldots, a_m, i \gg$ is a well-defined infon if R is an n-place relation and $a_1, \ldots, a_m$ ($m \leq n$) are objects appropriate for the argument places $i_1, \ldots, i_m$ of R, and if the filling of argument places $i_1, \ldots, i_m$ is sufficient to satisfy the minimality conditions for R, and $i = 0, 1$ is the polarity. Minimality conditions "determine which particular groups of argument roles need to be filled in order to produce an infon" [18]. The polarity is the 'truth value' of the infon. If $i = 1$ then the objects $a_1, \ldots, a_m$ stand in the relation R; else the objects do not stand in the relation R. Parameters, denoted as $\dot{a}$, make reference to *arbitrary* objects of a given type. For instance, $\dot{l}$ and $\dot{t}$ typically denote parameters for arbitrary objects of type spatial location and temporal location, respectively. Anchors are a mechanism to assign values to parameters. Hence, the parameter $\dot{t}$ may anchor the value for the current time.

We used the WURVOC Ontology of units of Measure (OM) [19] to model quantities and dimensions, in particular at the observation layer. We used the OWL 2 Web Ontology Language (OWL 2) [20], the Resource Description Framework (RDF) [21] and Protégé[2] to manage the ontologies. The knowledge base is

[1] `http://storm-project.net/`

[2] `http://protege.stanford.edu`

implemented by the Stardog RDF database.[3] We used WEKA [22] for machine learning. Software was implemented in Java.

Table 1. The key characteristics of training datasets for NPF detection and NPF characterization. The table provides an overview of the mapping from (expert) label (Hamed *et al.* [4]) to (training) class. It includes the number (#) of samples. Note that, in order to construct approximately balanced training classes, we limited the number of samples per training class to 160 and 50 in NPF detection and NPF characterization, respectively.

		Learning tasks			
		NPF detection		NPF characterization	
Label	#	Class	#	Class	#
0	113				
4	531	NE	160		
NE	200				
1	14			C1	14
2	47	E	159	C2	47
3	98			C3	50

Methods: The expert-labelled sensor measurement data was used to train and evaluate the performance of Multi-Layer Perceptron artificial neural network (MLP) classifiers for the two tasks of NPF detection and NPF characterization. NPF detection is a 2-class classification of days into event days (E) and non-event days (NE). NPF characterization is a 3-class classification of event days (E) into events of type class 1 (C1), class 2 (C2), or class 3 (C3). The multivariate daily DMPS measurement data was transformed to a vector by means of Singular Value Decomposition (SVD). Together with the label, such vectors formed a labelled dataset which we used to automatically generate training datasets. Table 1 provides an overview of the key characteristics of training datasets. MLP classification performance was evaluated using 10-fold cross validation, meaning that the training dataset was partitioned into 10 disjoint and equal sized folds, and for each fold a classifier was trained using the other 9 folds and then tested on the fold. Intermediate results were averaged. For both NPF detection and NPF characterization the MLP networks consisted of 40 input neurons and one hidden layer with 21 hidden neurons. The number of output neurons was 2 and 3 for NPF detection and NPF characterization, respectively. The learning rate and momentum were set to 0.3 and 0.2, respectively.

To accommodate domain knowledge we extended from both the SSN and STO upper ontologies. Listing 1.1 provides an overview. Specifically, we extended the SSN ontology to accommodate our domain specific sensing device, namely an individual instance of the `DifferentialMobilityParticleSizer` class, which was modeled as a subclass of `ssn:SensingDevice`. The DMPS implements a process of `ssn:Sensing` which is characterized by the unit of measure $[\text{cm}^{-3}]$. Moreover,

[3] `http://stardog.com/`

Listing 1.1. Subset of axioms used to accommodate domain knowledge. The set of terminological axioms is followed by assertional axioms. We show the modelling of the concentration of particles with diameter 7.0 nm and omit the similar modelling of the concentration of particles for the remaining 39 diameter sizes. Situations of NPF support infos with npfe-relation and event class as relevant individual. Individual names such as f16 are abbreviated random UUIDs.

```
DifferentialMobilityParticleSizer ⊑ ssn:SensingDevice
PolydisperseAerosol ⊑ ssn:FeatureOfInterest
ParticleConcentration ⊑ ssn:Property
NewParticleFormation ⊑ sto:Situation
EventClass ⊑ sto:RelevantIndividual
rdfs:domain(forParticleDiameter , ParticleConcentration)
rdfs:range(forParticleDiameter , om:diameter)

# The DMPS located at Puijo
DifferentialMobilityParticleSizer(f16)
# implements a process of sensing
ssn:Sensing(b5b)
ssn:implements(f16 , b5b)
# characterized by the unit of measure [cm^-3]
DUL:isCharacterizedBy(b5b, om:reciprocal_cubic_centimetre)
# The concentration of particles
ParticleConcentration(0f1)
# for a particle diameter
om:diameter(a2b)
forParticleDiameter(0f1 , a2b)
# of length 7.0 nm
om:Measure(2b2)
om:value(a2b, 2b2)
om:numerical_value(2b2, 7.0)
om:unit_of_measure_or_measurement_scale(2b2, om:nanometre)
# is observed by the DMPS
ssn:observes(f16 , 0f1)
# and is a property of the polydisperse aerosol at Puijo
PolydisperseAerosol(01d)
ssn:hasProperty(01d, 0f1)

# Situations of NPF support infons with npfe-relation
sto:Relation(npfe)
# and a relevant individual for the NPF event class
EventClass(c1)
EventClass(c2)
EventClass(c3)
```

we extended `ssn:Property` with the subclass `ParticleConcentration` to model particle number concentration for discrete diameter sizes. Finally, we extended `ssn:FeatureOfInterest` with the subclass `PolydisperseAerosol` to model the polydisperse aerosol at Puijo. Thus, the DMPS observes the particle number concentration for 40 discrete diameter sizes ranging 7-800 nm, which are properties of the polydisperse aerosol at Puijo. Finally, we extended `sto:Situation` with the subclass `NewParticleFormation` to represent knowledge about situations of interest to our domain. Such situations support npfe-relation infons with temporal and spatial locations as well as the related NPF event class.

3 Results and Discussion

Correctly classified instances, as computed by WEKA, resulted to be approximately 73% and 54% for NPF detection and NPF characterization, respectively. Precision (and recall) figures are 0.730 (0.744), 0.737 (0.723), 0.444 (0.286), 0.524 (0.468), and 0.567 (0.680) for the classes NE, E, C1, C2, and C3, respectively. Precision, and in particular recall, are notably low for C1. The class is misclassified as either C2 or C3 approximately three times out of four. This performance can be explained by the very low number of training samples for the class C1 (14). Furthermore, recall of events in NPF detection is an issue because at 72.3% the system will not retrieve many events of NPF. Overall, the performance of NPF detection is satisfactory while the performance of NPF characterization is mediocre.

The results suggest that the current implementation is not mature enough for full automation in a real-time context. However, the system may support domain experts in the detection and characterization of NPF and simplify the manual task to a machine assisted visual confirmation step, e.g. via a web application. Still, it is important to improve classification performance, an aim in future work. We have performed some preliminary experiments which resulted in approximately 80% correctly classified instances for NPF detection. The performance may be improved by including other attributes in classification, such as total daily particle concentration, average daily solar flux, or average daily SO_2 concentration. Both solar flux and SO_2 are known to correlate with events of NPF [1, 4]. Moreover, instead of using SVD on daily DMPS measurement data and use the resulting vector in classification, we may compute certain futures from daily DMPS measurement data and use those in classification. Such features include particle growth rate and formation rate [3]. This approach may also be interesting because it could extend the (prevalently data driven modelling) architecture shown in Figure 1 with modules for mechanistic modelling.

In the presented system, knowledge provided by different system parts is managed by a knowledge base. Both domain experts and software commit to use a common knowledge representation language as well as shared upper and domain ontologies. Specifically, knowledge provided by domain experts (Listing 1.1) is formalized and extends upper ontologies, at the observation layer for information about the measurement infrastructure and the observed real world phenomena and at the situation layer for information about real world situations that are of interest to the domain. Such information is used by the components of the architecture in order to "learn" about the domain. On the other hand, system components also commit to use the (same) knowledge representation language and extended terminologies in the (automated) representation of knowledge acquired at different layers.

At the observation layer, measurements are semantically enriched and represented as observations [14]. Given an average of approximately 4800 measurements per day and a time interval of 4 years, the observation engine generates around 80 million RDF statements, which are persisted in the knowledge base. SPARQL [23] can be used to query observations according to, e.g., sensing devices, features, properties, time. Results can be visualized as tables, (multivariate) time series plots, heat maps, or in other form. Summary statistics can be easily computed for data generated by a heterogeneous sensor network that measures a range of properties of real world features. We have demonstrated the flexibility of SPARQL to domain experts at the University of Eastern Finland and Finnish Meteorological Institute. First impressions underscore how a knowledge base that manages measurement data for sensing devices deployed in the field would be an invaluable repository and a considerable improvement compared to current measurement data management. In addition to sensor measurement, experts have also highlighted the possibility of integrating measurement data acquired (manually) in a laboratory.

At the situation layer, situations are acquired from observations. Obviously, the number of RDF statements generated by the situation engine is far lower than the number of statements generated by the observation engine. We argue that this is generally the case. SPARQL can be used for flexible querying of situational knowledge acquired from one (or more) heterogeneous sensor network(s). Situations can be augmented or refined with new or more accurate knowledge, either manually or automatically. For instance, in the presented use case, situations support the infon $\ll npfe, \dot{c}, \dot{l}, \dot{t}, 1 \gg$, where $\dot{c}$, $\dot{l}$, and $\dot{t}$ are parameters for the event class, spatial location, and temporal location, respectively. For a specific situation, acquired by NPF detection, the value to be anchored to the parameter $\dot{c}$ for the event class is automatically assessed, by NPF characterization. However, because this assessment is not sufficiently reliable, a domain expert should review, and may correct, the automatically represented information. This amounts to manual refinement of automatically represented situations with more accurate information. Finally, of interest to situational knowledge are intuitive visualization techniques. Real world situations are typically located along spatial and temporal dimensions. Aside the visualization of situations on an interactive map, we may also visualize situations on an interactive timeline.

We selected the ontology approach, rather than classical relational databases, and build our system on semantic web technologies for several reasons. Domain ontologies developed in Protégé can be tested for satisfiability and consistency. We can, thus, check whether an ontology is meaningful [24]. Knowledge bases offer powerful reasoning services. Specifically, Stardog supports several reasoning levels, including RDFS, QL, RL, EL, and DL.[4] With the ontology approach we can focus on the modelling of domain knowledge and semantics, and leave the data modelling to the knowledge base. We build our system on several, readily available, ontologies, including the SSN ontology, the STO, and the OM ontology. In future work we will also include the SWEET ontology,[5] in particular the concepts `chem:Substance` and `aero:Aerosol` (among others) and GeoNames[6] for (qualitative) spatial modelling (e.g. sensor location). Such terminologies can support the modelling of domain knowledge, since they provide an organization of relevant generic concepts and relations. Moreover, they can guide the design and implementation of software systems. In fact, we aligned our system implementation to relevant concepts and relations of both the SSN ontology (e.g. `ssn:SensingDevice` and `ssn:observedBy`) and STO (e.g. `sto:Situation` and `sto:supportedInfon`). Hence, the system is generic and its implementation can be reused across domains. Concrete applications [13, 14] can be implemented by extending specific system components and modules, as long as domain knowledge is aligned to the discussed upper ontologies. In the ontology approach, the integration and extension of terminologies designed by third parties is relatively straightforward. Finally, we also aim at showing that scalable solutions for the discussed problem can be designed and implemented using technologies other than classical relational databases.

4 Conclusions

For the domain of aerosol science and the study of New Particle Formation (NPF), we have presented a software system that supports the automated processing of measurement data into observations, consistent with the Semantic Sensor Network ontology, as well as the automated representation of situational knowledge, consistent with the Situation Theory Ontology, for events of NPF, acquired from observations by means of machine learning. Results show that, currently, machine classification performance is insufficient to allow for reliable automated NPF characterization. However, the system can guide domain experts with an estimate, and the automatically represented knowledge can be improved by expert refinement. The knowledge base integrating observations made by a heterogeneous sensor network and situational knowledge for real world phenomena is promising for its potential value to domain experts. In future work we aim at improving classification performance for both NPF detection and NPF characterization and further integration of measurement data, in particular for sensors other than a differential mobility particle sizer. Moreover, we plan to add an intermediate layer, named *derivation*, between the observation and situation layer of the presented architecture. This layer will support the translation of SSN observations to observations of *datasets* as well as the application of algorithmic transformations to datasets. The result of such transformations are datasets.

[4] `http://www.w3.org/TR/owl2-profiles/`

[5] `http://sweet.jpl.nasa.gov/ontology/`

[6] `http://www.geonames.org/`

Datasets can be persisted in the knowledge base. Example transformations include gap filling, outlier detection/removal, spatio-temporal interpolation, or digital signal processing. We are currently evaluating the use of the RDF Data Cube Vocabulary [25] at the derivation layer.

Acknowledgements. The Finnish Measurement, Monitoring and Environmental Efficiency Assessment research program for supporting the publication. Clark & Parsia LLC for a Stardog developer license.

References

1. Kulmala, M., Vehkamäki, H., Petäjä, T., Dal Maso, M., Lauri, A., Kerminen, V., Birmili, W., McMurry, P.: Formation and growth rates of ultrafine atmospheric particles: a review of observations. Journal of Aerosol Science 35(2), 143–176 (2004)
2. Solomon, S., Qin, D., Manning, M., Chen, Z., Marquis, M., Averyt, K., Tignor, M., H.L., M.: Climate Change 2007: The Physical Science Basis. Contribution of Working Group I to the Fourth Assessment Report of the Intergovernmental Panel on Climate Change. Cambridge University Press, Cambridge (2007)
3. Dal Maso, M., Kulmala, M., Riipinen, I., Wagner, R., Hussein, T., Aalto, P., Lehtinen, K.: Formation and growth of fresh atmospheric aerosols: eight years of aerosol size distribution data from SMEAR II, Hyytiala, Finland. Boreal Environment Research 10(5), 323–336 (2005)
4. Hamed, A., Joutsensaari, J., Mikkonen, S., Sogacheva, L., Dal Maso, M., Kulmala, M., Cavalli, F., Fuzzi, S., Facchini, M., Decesari, S., et al.: Nucleation and growth of new particles in Po Valley, Italy. Atmospheric Chemistry and Physics 7(2), 355–376 (2007)
5. Vana, M., Ehn, M., Petäjä, T., Vuollekoski, H., Aalto, P., de Leeuw, G., Ceburnis, D., O'Dowd, C.D., Kulmala, M.: Characteristic features of air ions at Mace Head on the west coast of Ireland. Atmospheric Research 90(2), 278–286 (2008)
6. Compton, M.: What Now and Where Next for the W3C Semantic Sensor Networks Incubator Group Sensor Ontology. In: Proceedings of the 4th International Workshop on Semantic Sensor Networks (SSN11). CEUR-WS, pp. 1–8 (2011)
7. Janowicz, K., Compton, M.: The stimulus-sensor-observation ontology design pattern and its integration into the semantic sensor network ontology. In: The 3rd International Workshop on Semantic Sensor Networks, pp. 7–11 (2010)
8. W3C Semantic Sensor Network Incubator Group: Semantic Sensor Network Ontology. Technical report, W3C (2009)
9. Kokar, M.M., Matheus, C.J., Baclawski, K.: Ontology-based situation awareness. Inf. Fusion 10(1), 83–98 (2009)
10. Fenza, G., Furno, D., Loia, V., Veniero, M.: Agent-based Cognitive approach to Airport Security Situation Awareness. In: Proceedings of the, International Conference on Complex, Intelligent and Software Intensive Systems, CISIS 2010, pp. 1057–1062. IEEE Computer Society (2010)
11. De Maio, C., Fenza, G., Furno, D., Loia, V.: Swarm-based semantic fuzzy reasoning for situation awareness computing. In: 2012 IEEE International Conference on Fuzzy Systems (FUZZ-IEEE), pp. 1–7 (June 2012)

12. Doulaverakis, C., Konstantinou, N., Knape, T., Kompatsiaris, I., Soldatos, J.: An Approach to Intelligent Information Fusion in Sensor Saturated Urban Environments. In: 2011 European Intelligence and Security Informatics Conference (EISIC), pp. 108–115 (September 2011)
13. Stocker, M., Rönkkö, M., Kolehmainen, M.: Making Sense of Sensor Data Using Ontology: A Discussion for Residential Building Monitoring. In: Iliadis, L., Maglogiannis, I., Papadopoulos, H., Karatzas, K., Sioutas, S. (eds.) Artificial Intelligence Applications and Innovations, Part II. IFIP AICT, vol. 382, pp. 341–350. Springer, Boston (2012)
14. Stocker, M., Rönkkö, M., Kolehmainen, M.: Making Sense of Sensor Data Using Ontology: A Discussion for Road Vehicle Classification. In: Seppelt, R., Voinov, A., Lange, S., Bankamp, D. (eds.) 2012 International Congress on Environmental Modelling and Software: Managing Resources of a Limited Planet: Pathways and Visions under Uncertainty, Sixth Biennial Meeting, Leipzig, Germany, pp. 2387–2394. International Environmental Modelling and Software Society (iEMSs) (July 2012)
15. Leskinen, A., Portin, H., Komppula, M., Miettinen, P., Arola, A., Lihavainen, H., Hatakka, J., Laaksonen, A., Lehtinen, K.: Overview of the research activities and results at Puijo semi-urban measurement station. Boreal Env. Res. 14, 576–590 (2009)
16. Kulkarni, P., Baron, P.A., Willeke, K.: Aerosol Measurement: Principles, Techniques, and Applications. Wiley (2011)
17. Barwise, J., Perry, J.: Situations and attitudes. Bradford books. MIT Press (1983)
18. Devlin, K.: Logic and information. Cambridge University Press (1995)
19. Rijgersberg, H., Wigham, M., Top, J.: How semantics can improve engineering processes: A case of units of measure and quantities. Advanced Engineering Informatics 25(2), 276–287 (2011)
20. W3C OWL Working Group: OWL 2 Web Ontology Language. W3C Recommendation, W3C (December 2012)
21. Manola, F., Miller, E.: RDF Primer. Technical Report W3C Recommendation, W3C (2004)
22. Hall, M., Frank, E., Holmes, G., Pfahringer, B., Reutemann, P., Witten, I.H.: The WEKA Data Mining Software: An Update. SIGKDD Explorations 11 (2009)
23. Prud'hommeaux, E., Seaborne, A.: SPARQL Query Language for RDF. Technical Report W3C Recommendation, W3C (2008)
24. Baader, F., Calvanese, D., McGuinness, D., Nardi, D., Patel-Schneider, P.: The Description Logic Handbook: Theory, Implementation and Applications, 2nd edn. Cambridge University Press (2007)
25. Cyganiak, R., Reynolds, D.: The RDF Data Cube Vocabulary. W3C Working Draft, W3C (March 2013)

The Future Internet Enablement of the Environment Information Space

Thomas Usländer[1], Arne J. Berre[2], Carlos Granell[3], Denis Havlik[4], José Lorenzo[5], Zoheir Sabeur[6], and Stefano Modafferi[6]

[1] Fraunhofer IOSB, Germany
thomas.uslaender@iosb.fraunhofer.de
[2] SINTEF, Norway
arne.j.berre@sintef.no
[3] European Commission, Joint Research Centre, Italy
carlos.granell@jrc.ec.europa.eu
[4] Austrian Institute of Technology (AIT)
Denis.Havlik@ait.ac.at
[5] AtoS, Spain
jose.lorenzo@atosresearch.eu
[6] University of Southampton IT Innovation Centre, Faculty of Physical Sciences and Engineering, United Kingdom
{zas,sm}@it-innovation.soton.ac.uk

Abstract. This paper motivates the enablement of the Future Internet to become a highly functional service platform supporting the design and the operation of software applications in the Environmental Information Space. It reports on the experience made by the European research project ENVIROFI as one of the usage area projects within the Future Internet Public-Private Partnership programme. It describes the software components (environmental and specific enablers) which are required to connect with the domain-independent capabilities (generic enablers) of the Future Internet core platform for geospatially and environmentally-driven applications.

Keywords: Future Internet, ENVIROFI, generic enablement, specific enablement, environment information space, Open Geospatial Consortium, OGC.

1 Motivation

Starting from proprietary software frameworks and the societal need for interoperable environmental services, the object- and service-orientated architecture design approaches of the past have paved the way to open, standardized and distributed design of the Environmental Information Space (EIS) [1]. The emerging Future Internet (FI) platform aims to establish an even more generic set of software building blocks which simplify the building of cross-domain applications.

With the mainstream trends towards virtually centralized computing resources (cloud computing) combined with the predominant role and extension of the Internet

J. Hřebíček et al. (Eds.): ISESS 2013, IFIP AICT 413, pp. 109–120, 2013.

technology towards an Internet of Things (IoT), Content, Services and People, EIS stakeholders are facing a disruptive technology shift which will affect the established technologies as well as the business models of EIS. The environmental data generated by smart low-cost sensors, also known as "Things" in the IoT context, is getting increasingly ubiquitous instead of being offered as distinguished "products" by service providers in the environmental domain sectors. Smart processing capabilities, e.g. the provision of aggregated environmental data as thematic layers in Geographic Information Systems (GIS), can be used on demand. The intention to combine huge amounts of environmental sensor-based observations with unstructured or semi-structured information of Web documents (e.g. environmental reports, images, videos) as well as with the human observations posted in social media transforms the design requirements of future EIS and the analysis of the existing sources of environmental information into "Big Data" management requirements.

In order to profit from these development, EIS stakeholders (e.g. designers, implementers) have to face the challenging task of integrating the legacy and future EIS applications into the emerging Future Internet framework. They need to understand the current and upcoming user requirements [2], often expressed by thematic experts of a particular domain, and to map these requirements onto highly generic FI software building blocks. In order to assure their re-use and sustainable implementations in context of the FI Enablement, such building blocks shall be specified and developed in an agile way, feature simple and well-defined services interfaces, and follow existing standards whenever possible. This sets the requirements for building the new generation of EIS to higher grounds once again while agile development methodologies and new domain-independent FI software building blocks have to be embraced.

Past experiences with the EIS design have stressed the fact that the majority of environmental information objects have geospatial references. This is reflected in the European regulations on information sharing and spatial data infrastructures such as INSPIRE, as well as in the widespread use of standards promoted by the Open Geospatial Consortium (OGC) within the environmental usage area. Hence, the building blocks of such an open and FI-enabled EIS have to support the geospatial processing as well as the existing open standards for geospatial data and service interfaces.

The rest of the paper is organized as follows. Section 2 of this paper introduces the FI-PPP programme and its environmental usage area project ENVIROFI. Three environmental pilot applications pursued by ENVIROFI are introduced in Section 3, followed by the presentation of the ENVIROFI Conceptual Architecture and the discussion of the generic and specific enablers in sections 4 and 5 respectively. The encountered technical challenges and the lessons learnt during the project are discussed in section 6. Finally, section 7 summarizes the vision of the future development of an FI-enabled EIS and environmentally enabled FI.

2 The Future Internet Public-Private Partnership

The European Commission (EC) launched the Future Internet Public-Private Partnership programme (FI-PPP) with the idea of enabling a broad range of Internet applications, including those of the environmental knowledge domain. The FI-PPP programme shall advance a shared vision for common pan-European technology platforms and their implementation, as well as the integration and harmonization of the relevant policy, legal, and regulatory frameworks. As set in the Digital Agenda for Europe[1], these frameworks are prerequisites for realizing a common European digital market economy with inclusive access to knowledge and awareness to societal issues. Phase 1 of the FI-PPP started a set of interrelated projects, addressing different use cases in vertical domain sectors. They cooperated with the ICT Core Platform which is developed in the FI-WARE project (see http://www.fi-ware.eu/).

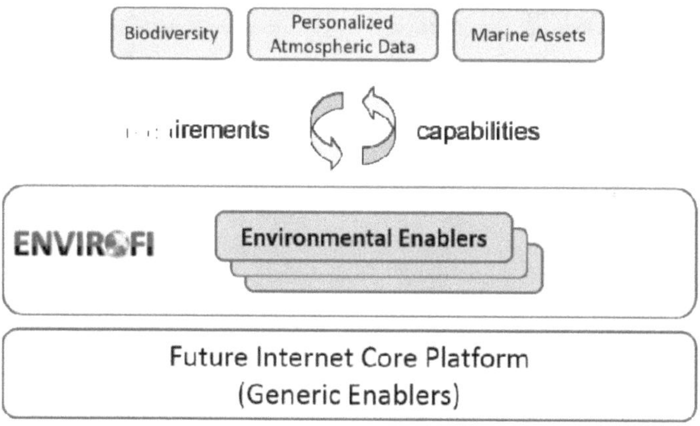

Fig. 1. ENVIROFI Approach to Design an Environmental Observation Web

As illustrated in Fig. 1 the FI-WARE platform is an ICT platform based on generic functional building blocks: the so called Generic Enablers (GE). These are used by the usage area projects in order to validate and demonstrate the new generation FI enabled environmental applications. Furthermore, the project provides development benchmarks for innovation schemes and guiding new market sector Enterprise buy-in FI technologies. ENVIROFI represents the environmental usage area within phase 1 of FI-PPP (see http://www.envirofi.eu/). It explores the so-called environmental enablers, i.e. reusable building blocks for collecting and processing environmental data, and provides environmental sector requirements to FI-WARE. Thus, ENVIROFI lays the foundation for an FI-enabled Environmental Observation Web, which will help Europe to tackle the grand challenges of climate change, socio-economic pressures on the environment and sustainable development.

ENVIROFI's vision is to establish an Environmental Observation Web in which all environmental data from sensors, citizens and models become available through the Internet in a standardized and usable format. ENVIROFI specifically works on three environmental application areas: biodiversity, personalized atmospheric data and marine assets.

[1] http://ec.europa.eu/digital-agenda/

3 Pilot Applications

The identification of the specific building blocks and enablers for the environment has been guided by an agile service-oriented analysis and design methodology [1, 3] of the EIS requirements inherited from existing European initiatives such as Copernicus[2] and INSPIRE[3], and the more specific requirements of the following pilot applications.

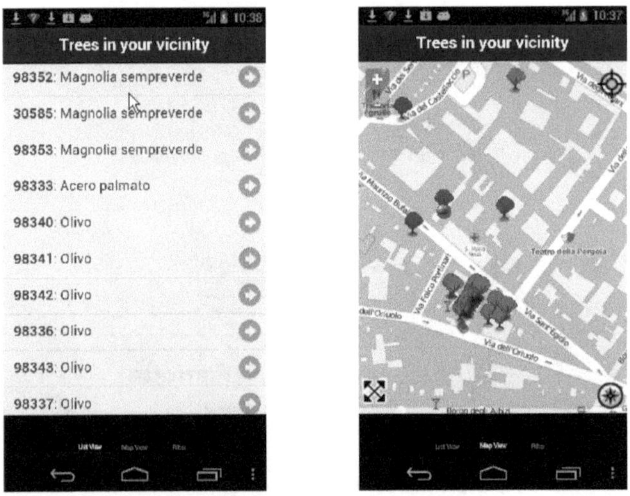

Fig. 2. Screenshots of the BIO app: The left hand side screenshot shows a list of trees around a given user's location. The right hand side screenshot shows such information on a map of the City of Florence (Italy).

1. **Bringing Biodiversity into the Future Internet**: This pilot focuses on the usage of FI technology for the survey, analysis, quality assurance, persistence and dissemination of biodiversity data. The Biodiversity Survey Application (BIO App) aims to support the monitoring of terrestrial biodiversity. This enables users from a wide variety of backgrounds to: (i) Access data on interesting biodiversity occurrences which are recorded at their geographic location; (ii) Provide data on biodiversity occurrences which they observe; and (iii) Provide feedback on the correctness of existing and new recorded biodiversity occurrence in the system. In order to allow data provisioning by users, they may either add information to existing entries on individual plants or animals, or provide information on specimen which are not contained within the system. Although the BIO App operates in any geographical region, sample data sets in the cities of Vienna (Austria) and Florence (Italy) (see Fig. 2) allow users to test with large data sets, based on open data tree layers which are made available.

---

[2] Previously known as GMES (Global Monitoring for Environment and Security), see `http://copernicus.eu/`

[3] Infrastructure for Spatial Information in the European Community, see `http://inspire.jrc.ec.europa.eu/`

2. **Personal Information System for Air Pollutants, Allergens and Meteorological conditions**: This pilot focuses on the interaction with the environment by creating a series of integrated tools that comprise the backbone of the Personal Information System for Air Pollutants, Allergens and Meteorological conditions which potentially affect human health and well-being. The Personal Environmental Information System (PEIS) is a fully customizable, interactive mobile application enabling users to monitor information on weather conditions and atmospheric exposures and related hazards from their own perspective. PEIS was tested for the cities of Oslo (Norway) and Vienna (Austria) (see Fig. 3).

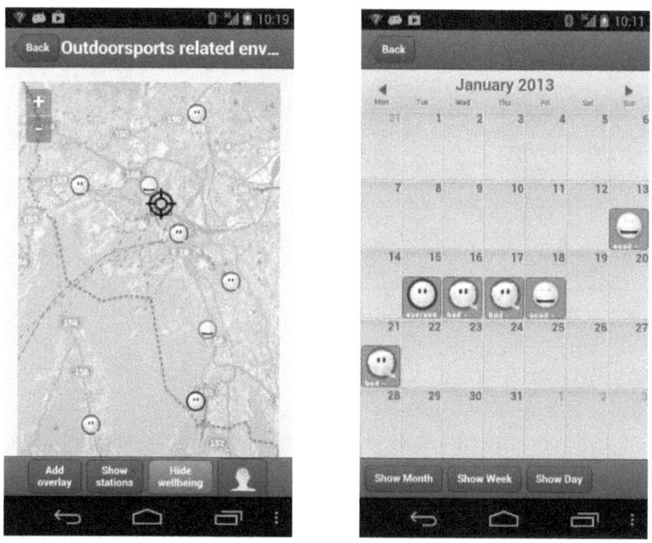

Fig. 3. Screenshots of the PEIS app: The left hand side screenshot shows a map of health and environmental conditions at outdoor sport facilities in the city of Oslo (Norway). The right hand side screenshot shows a personalized health calendar.

3. **Future Internet Collaborative Usage of Marine Environmental Assets**: This pilot focuses on the collaborative use of FI technology for Marine Environmental Assets. The Marine Asset decision Support Tool (MAST) is a Web portal that mashes up a number of freely available information and data feeds to monitor operational weather conditions in the Galway Bay, Ireland, promoting situational awareness among offshore marine operators. Through the intelligent fusion of real-time observation data with model predictions, the portal can be used to identify monitor sea-state conditions, water quality and the encountered risks for the marine industry to operate at sea. These include the aquaculture, fishing and leisure industries (see Fig. 4).

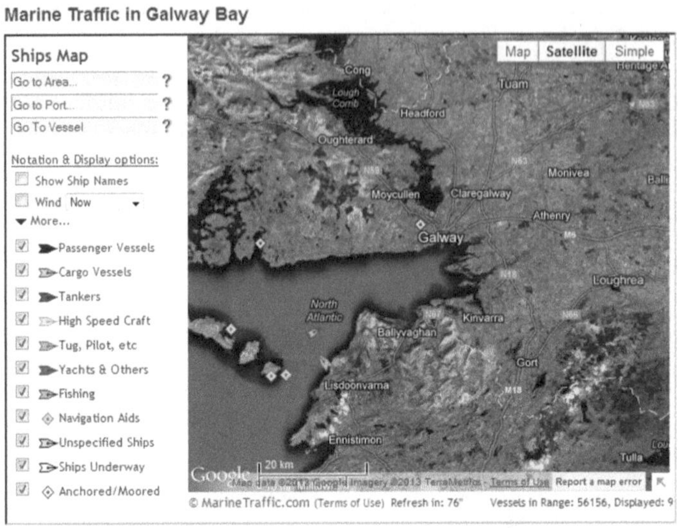

Fig. 4. Screenshots of the MAST portal: it shows real time conditions collected from various observation platforms around Galway Bay

4 Conceptual Architecture

The FI-WARE technical chapters are: Cloud Hosting, Data/Context Management, Internet of Things (IoT) Services Enablement, Applications/Services Ecosystem and Delivery Framework, Security and Interface to Networks and Devices (I2ND). They are highly independent from each other, and may heavily differ in terms of the adopted standards, technologies, data models, data representations and interaction patterns. They are application and use case agnostic to the level where none of the semantics that is implicitly available in the Use Case projects has been a-priori known to the FI-WARE GEs. See the list of GEs in the FI-WARE catalogue in [5]. In addition, the FI-WARE architecture supports the co-existence of various Web service technologies, as well as the resource-oriented, request/reply and event driven architectural styles.

The ENVIROFI conceptual architecture was built by profiling the various GEs according to the needs of the environmental domain, and also by identifying and specifying the functions specific to EIS applications. Furthermore, as the environmental domain heavily relies on generic geospatial standards, it was essential in the ENVIROFI conceptual architecture design (see Fig. 5) to accommodate these standards on top of the GEs, or better, integrate them among the GEs cluster.

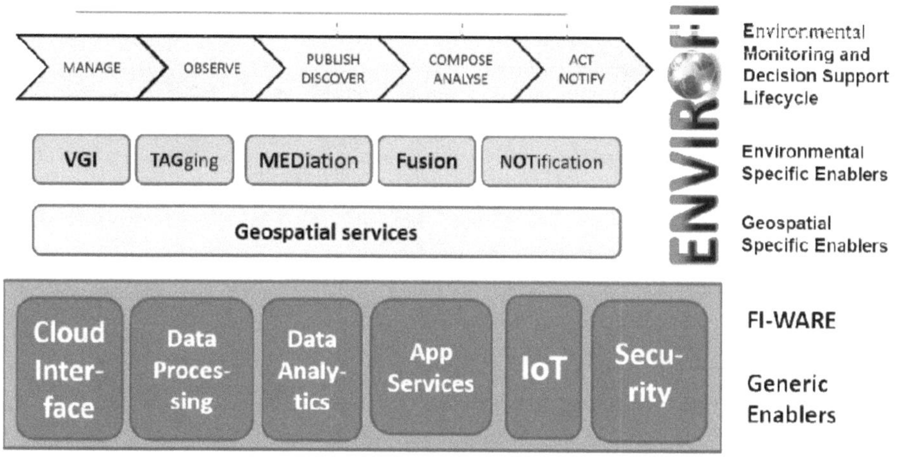

Fig. 5. ENVIROFI Conceptual Architecture

The specification of the ENVIROFI Conceptual Architecture [4] results from the work and experience of European research projects such as ORCHESTRA[4], SANY[5], EO2HEAVEN[6], SWING[7], ENVISION[8], EuroGEOSS[9] and TRIDEC[10], and relies upon the following pillars:

- Reference models originally specified by standardization organizations such as ISO, OGC, OMG, OASIS and The Open Group, and tailored to the environmental domain [1].
- Profiles of geospatial services and information models specified by ISO and OGC.
- Specifications and implementations of architectural styles, design patterns and (partial) solutions.

A generic Environmental Monitoring and Decision Lifecycle based upon the INSPIRE lifecycle perspective served as a framework for the generalization of use case functionality from the various environmental pilot applications. ENVIROFI contributed with extensions to standardization under CEN /TC289 TR 15449 on Spatial Data Infrastructure, Reference model [6] and Service Centric view [7], and to ISO/TC211 19119 Geospatial Services [8] and further input to the related OGC Topic 12 on Service Architecture. ENVIROFI has also contributed to examples of the use of an RM-ODP modeling approach by use case-based specification templates.

---

4 Open Architecture and Spatial Data Infrastructure for Risk Management.
5 Sensors Anywhere.
6 Earth Observation and Environmental Modelling for the Mitigation of Health Risks.
7 Semantic Web Services Interoperability for Geospatial decision making.
8 ENVIronmental Services Infrastructure with ONtologies.
9 European approach to GEOSS (Global Earth Observation System of Systems).
10 Collaborative, Complex and Critical Decision-Support in Evolving Crisis.

5 Geospatial and Environmental Enablers

In order to cope with the domain-independent geospatial aspects, ENVIROFI explicitly identified a set of specific enablers (SE) that address geospatial data provisioning and storage. These geospatial enablers encompass known geospatial information models (e.g. the OGC Observation and Measurements Model) and services (e.g. the OGC Sensor Observation Service). Hence, they may also be reused in other usage areas that deal with geo-referenced data processing, e.g. risk and crisis management [12].

The environmental SEs have been identified under five major categories which are derived from very distinct enablement functionalities[11]:

1. **Harvesters, Connectors and Mediators**: These constitute a collection of brokers, connectors and mediator services which support protocols and data models. The models are specialized in the environmental domain. This thematic class of specific enabler facilitates easier interoperability between specific enabler based services, using agile and flexible service implementations.
2. **Volunteered Geographic Information**: This specializes in geo-referenced observation and data sampling the environmental usage area. The services in this thematic class provide ways to record and archive geo-tagged measurements for later use by other classed of specific enablers. The enablers in this class are designed to support crowdsourcing and crowdtasking of environmental measurements, performed by a large-scale community of volunteering environmental observers.
3. **Semantic Tagging Tools**: These tools and services provide support for semantic enrichment of environmental data streams and sources. This thematic class includes environmental domain ontology support, harvester services and linked data services, which allow uncertainty annotation of existing measurement resources.
4. **Fusion Tools for Heterogeneous Data Sources**: These are composed by environmental data fusion services, which may operate at different semantic levels. This thematic class includes pre-processing, feature extraction, situation assessment and forecasting services, with those for preparing and aggregating environmental data into formats which are suitable for use by end users and FI downstream services.
5. **Event detection and notification services**: These include services, which provide a variety of notification mechanisms, which are compatible with the environmental geospatial standards and protocols.

Fig. 6 below illustrates an example of a KML file which is produced as output using a chain of environmental enablers, which led to the fusion of environmental observations (in this example sea surface temperature) using three semantically conflicting and asynchronous heterogeneous data sources such as in situ, space-borne and model data [10].

---

[11] See the list of SEs in the ENVIROFI catalogue [9].

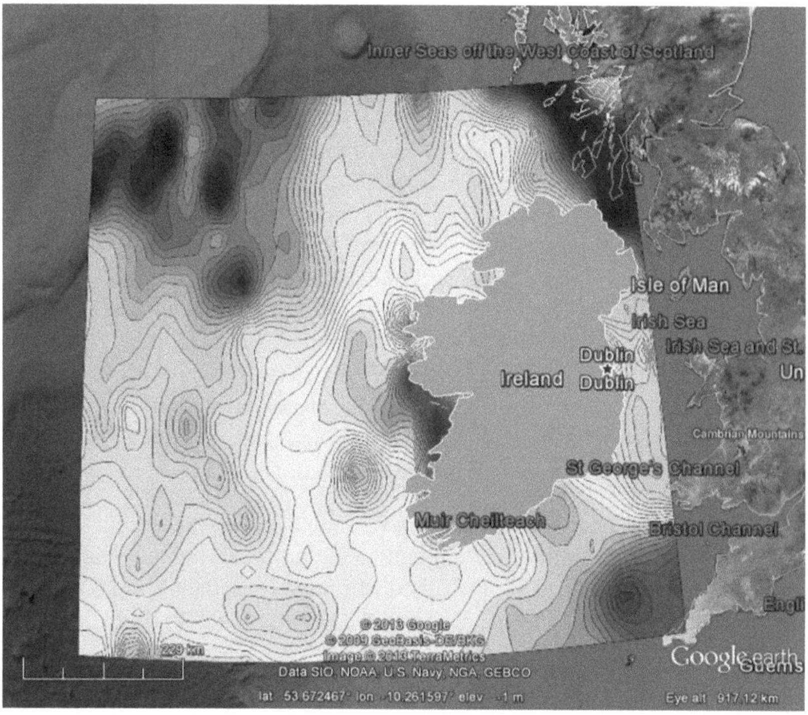

Fig. 6. Future Internet enablement of water quality monitoring in the marine environment, using generic data fusion from multiple observation sources

The above mentioned categories of specific enablers for the environmental domain are also designed as application agnostic as much as possible. They offer the possibility to extend their usage as specific enablers for the environmental domain in its broadest sense and beyond.

6 Generic Enablers and Cross-Domain Geospatial Environments

Many of the existing and upcoming GEs from the FI-WARE Cloud Hosting, Data/Context management, and Security chapters are suitable for "filling the (standardization) gaps" in the environmental usage area. Others, most notably the GEs of the IoT chapter to some extent overlap with some types of OGC services.

The main challenge in adopting the GEs for real-world environmental domain applications is the harmonization and integration across the GE chapters. A potential customer needs one contact institution and one technically harmonized platform to take sustainable design decisions. Unfortunately, the initial release of the FI-WARE platform featured GEs at very different levels of maturity, and as a whole lacked harmonization and integration across the GE chapters. For example, the GEs did not support the authentication and authorization functionality of the identity management GE. This is expected to change with the second release of the software and the introduction the FI-WARE Open Innovation Laboratory (OIL) later in the year 2013.

Another important challenge of the FI-WARE platform is to address the weak support for geospatial data and processing. This is most notable at the level of event processing, likely to slow down the uptake of the FI-WARE event processing framework in the environmental usage area, and cannot be fully solved without extending the GE functionality. The underlying OMA-NGSI[12] data model is generic enough to allow the embedding of geospatial information, but the FI-WARE architecture does not define the way to do this. Likewise, the architecture does not prevent geospatial processing, but no standardized geospatial processing functions are defined for existing FI-WARE GEs from the Data/Context management Chapter.

It is also important to understand that FI-WARE neither defines a standardized way for encapsulation of the processing services, nor a consistent set of standard data retrieval services. To a certain extent, this deficiency is compensated by the Publish/Subscribe Context broker GE and the Mediator GE [5]:

- The Mediator GE is defined as a middleware application responsible for providing interoperability among different communication protocols and among different data models. However, the ability to mediate between standardized OGC services used in the environmental usage area and the services used in other domains were not tested in ENVIROFI.
- The Pub/Sub Context Broker GE allows "Content providers" to register and publish information in form of the OMA-NGSI compliant events ("Context Entities"). "Context consumers" can subscribe to registered data sources and consume this information. The Pub/Sub context broker GE is essentially an event broker, but the Context Entities remain cached on the server until pre-defined "expiration time" and can be accessed over RESTful service interface. We have successfully used this property of Pub/Sub context broker to exchange data between the Environmental Georeferenced Observation service SE [9] and the Wirecloud Application Mashup GE in an extension of the biodiversity pilot.

In order to cope with the domain-independent geospatial aspects, ENVIROFI explicitly identified a set of enablers that address geospatial data provisioning and storage. These geospatial enablers encompass the geospatial information models (e.g. the OGC Observation and Measurements Model) and services (e.g. the OGC Sensor Observation Service). Hence, they may also be reused in other usage areas that require geo-referenced data and data processing, e.g. security, transport and logistics, health, risk and crisis management and so forth. With the exception of the application-specific models, most of the ENVIROFI enablers provide application and even usage area neutral functionality which involves observing, publishing, accessing and analyzing geospatially and temporally referenced events and measurements. Fig. 7 below illustrates the potential usage of ENVIROFI SEs in various cross-domain usage areas

---

[12] Open Mobile Alliance Next Generation Services Interface, see `http://technical.openmobilealliance.org/Technical/release_program/ngsi_v1_0.aspx`

of the evolving Future Internet. Indeed, further use cases and scenarios may be described to benefit from the FI-WARE and ENVIROFI architecture, and clearly put in value the enablers and technologies being developed over the project [11].

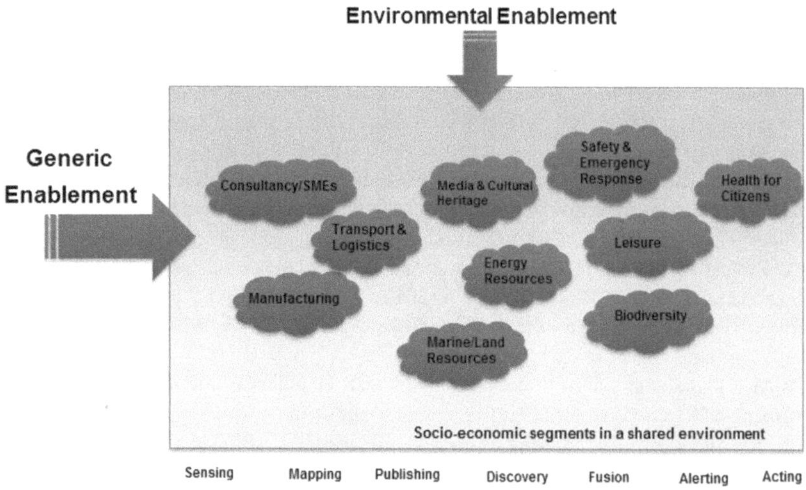

Fig. 7. Vision of the Environmental Enablement in Future Internet Cross-domain Areas

7 Conclusion

Despite the current weak support for geospatial data and processing, the ENVIROFI project team considers most of the FI-WARE GEs as generic enough and welcomes the extensions of currently available software building blocks. Many of the existing and upcoming GEs are suitable for "filling the (standardization) gaps" in the environmental usage area. The combination of GEs and environmental SEs could thus be used to build or support many cross-domain scenarios in the FI-PPP large-scale trials in phase 3 and beyond.

Together with SEs from other usage areas, the FI-WARE platform has the potential of becoming one of the important commodities and technology accelerators for European Small to Medium Enterprise organizations. In this sense it will be a major contributor in the growth and expansion of ICT-driven enterprises in the ever growing global digital market. However, the real value and success of FI-WARE generic enablement will greatly depend on its ability to deliver working, intuitive and easy-to-use enablers. These should not only connect to the specific enablers of the usage areas but also take on board the geospatial and intelligent data processing requirements in FI experimentations under the various phases of the FI-PPP and beyond. Only then, the creation of sustainable ICT driven ecosystems with robust business and value chains will be achieved.

References

1. Usländer, T.: Service-oriented Design of Environmental Information Systems. PhD thesis of the Karlsruhe Institute of Technology (KIT). KIT Scientific Publishing (2010), http://digbib.ubka.uni-karlsruhe.de/volltexte/1000016721
2. Schade, S., Fogarty, B., Kobernus, M., Schleidt, K., Gaughan, P., Mazzetti, P., Berre, A.: Environmental Information Systems on the Internet: A Need for Change. In: Hřebíček, J., Schimak, G., Denzer, R. (eds.) ISESS 2011. IFIP AICT, vol. 359, pp. 144–153. Springer, Heidelberg (2011)
3. Usländer, T., Batz, T.: How to Analyse User Requirements for Service-Oriented Environmental Information Systems. In: Hřebíček, J., Schimak, G., Denzer, R. (eds.) ISESS 2011. IFIP AICT, vol. 359, pp. 161–168. Springer, Heidelberg (2011)
4. ENVIROFI Consortium (Ed.). D4.2 Environmental Architecture. Deliverable D4.2 of the FP7 project ENVIROFI Work Package 4 (2012), http://www.envirofi.eu/Downloads/PublicDeliverables/tabid/4983/Default.aspx
5. FI-WARE Catalogue. All information, documentation and tools you need as a developer to start using a Generic Enabler (GE) Implementation, http://catalogue.fi-ware.eu/
6. CEN/TC289 – TR 15449 Geographic information – Spatial Data Infrastructures – part 1 Reference model (November 2011)
7. CEN/TC289 – TR 15449 Geographic information – Spatial Data Infrastructures – part 4, Service Centric view (May 2012)
8. ISO/TC211 – ISO 19119 PT/EC revision draft (May 2013)
9. ENVIROFI Catalogue. Central repository for implementations of Specific Enablers (SE) for the environmental Usage Area within the Future Internet Public Private Partnership programme (FI-PPP), http://catalogue.envirofi.eu/
10. Sabeur, Z., Chakravarthy, A., Bashevoy, M., Modafferi, S.: Structured Multi-level Data Fusion and Modelling of Heterogeneous Environmental Data for Future Internet Applications. Geophysical Research Abstracts, Vol. 15, EGU2013-6311-2. EGU General Assembly (2013)
11. ENVIROFI Consortium (Ed.). D6.2.3 Report on Community Consultation. Deliverable D6.2.3 of the FP7 project ENVIROFI Work Package 6 (2013), http://bscw.fi-ppp.eu/pub/bscw.cgi/d17436/ENVIROFI%20D6.2.3_Report_on_Community_Consultation_III.pdf
12. Berre, A.J., Usländer, T., Schade, S.: Identification and Specification of Generic and Specific Enablers of the Future Internet – Illustrated by the Geospatial and Environmental Domain. In: Abramowicz, W., Llorente, I.M., Surridge, M., Zisman, A., Vayssière, J. (eds.) ServiceWave 2011. LNCS, vol. 6994, pp. 278–289. Springer, Heidelberg (2011)

A Model Integration Framework for Assessing Integrated Landscape Management Strategies

Jared M. Abodeely[1], David J. Muth[1], Joshua B. Koch[1], and Kenneth M. Bryden[2]

[1] Idaho National Laboratory, Idaho Falls, ID USA
{Jared.Abodeely,David.Muth,Joshua.Koch}@inl.gov
[2] Iowa State University, Ames, IA USA
kmbryden@iastate.edu

Abstract. Nitrogen application is a standard practice for maximizing productivity of an agronomic system. The challenge is that many commercial scale agricultural systems are inefficient in utilizing the nitrogen that is applied. Therefore, understanding the impact of land management practices on nitrogen use inefficiencies within the agroecosystem is critical. This paper presents an integrated model that quantifies the impact of various land management practices on specific agroecosystem units. This integrated model is composed of the Wind Erosion Prediction System (WEPS), the Revised Universal Soil Loss Equation, Version 2 (RUSLE2), the Soil Condition Index (SCI), and the daily CENTURY model, DAYCENT. The integrated model was used to determine the impact of land management strategies on greenhouse gas emissions and nitrate leaching in a 60.5 ha field in Webster County, Iowa, USA. It was found that nitrogen use efficiency can vary significantly across a field and that integrated land management strategies can reduce overall nitrogen losses.

Keywords: integrated model, soil organic carbon, greenhouse gas emissions, nitrate leaching.

1 Introduction

In some agricultural systems a significant amount of nitrogen fertilizer is required for the land to reach its maximum productivity potential. However, these agricultural systems are often inefficient in their utilization of the nitrogen, and much of the nitrogen is lost through greenhouse gas emissions and nitrate leaching. The U.S. Environmental Protection Agency has determined that in 2011 agricultural systems accounted for 8% of the total greenhouse gas emissions in the United States and that 69% of the total N_2O emissions were due to agricultural practices [1]. This is significant because N_2O has a global warming potential 310 times greater than CO_2. Nitrate leaching is another concern as nitrogen leaving a field through runoff or tile drainage enters river systems and migrates into open water. This is evident in the northern Gulf of Mexico where hypoxia is occurring due to increased concentrations of nitrogen delivered from the Mississippi River system [2].

Nitrogen losses in agroecosystems can be attributed to several factors including land management practices, crop rotation, drainage, soil organic matter levels, and climate patterns [3]. This paper presents an integrated model composed of several

J. Hřebíček et al. (Eds.): ISESS 2013, IFIP AICT 413, pp. 121–128, 2013.

agronomic models for simulating environmental processes that quantify the impact of various land management practices on specific agroecosystem units. Using the integrated model the potential of integrated landscape management strategies to improve nitrogen use efficiency and mitigate nitrogen losses from the agroecosystem are examined. Two land management scenarios are assessed: 1) a traditional Midwest row cropping rotation (corn-soybean) and 2) an integrated landscape management with a corn-soybean rotation with a rye clover cover crop and switchgrass in low production areas. The study is performed using high-fidelity soil, landscape, and grain yield data for a 60.5 ha field in Webster County, Iowa, USA.

2 Integrated Model

Agronomic models provide the ability to assess a broad range of soil, climates, and land management practices. However, these models are often disparate in nature and focused on a specific environmental process. Muth and Bryden [4] addressed the challenges of integrating agronomic models through development of a model integration framework. Initial research efforts for the integrated model were focused on sustainable residue removal with models used by the US Department of Agriculture – Natural Resources Conservation Service for developing land management plans. These models include the Wind Erosion Prediction System (WEPS) [5], Revised Universal Soil Loss Equation, Version 2 (RUSLE2) [6], and the Soil Conditioning Index (SCI) [7]. The integrated model predicted sustainable residue removal rates based on wind- and water-induced soil erosion and qualitative soil organic carbon constraints. The integrated model was used to investigate sustainable residue removal at the national scale [8] and at the subfield scale [9-10].

In this paper, this integrated model has been extended to include the biogeochemical model DAYCENT [11]. The DAYCENT model quantifies exchanges of carbon and nitrogen between the soil, vegetation, and atmosphere, enabling assessment of the impacts of land management decisions on soil organic carbon, greenhouse gas emissions, and nitrate leaching.

The data flow in the integrated model is shown in Fig. 1. The model uses high-fidelity datasets including light detection and ranging (LiDAR) for slope and elevation ($0.5m^2$), grain yield data taken from a yield monitor of a combine during grain harvest ($5m^2$), and the Soil Survey Geographic (SSURGO) Database (10m – 100m). The integrated model is initialized with the selection of an agroecosystem unit and land management practice. Soil and climate data are retrieved based on the selected location. The soil, climate, and land management data are then assembled into native file formats for each of the integrated models. The WEPS model is initialized and simulated to produce wind-induced soil erosion. The RUSLE2 model is then initialized and simulated to determine the water-induced soil erosion. The soil erosion values are then passed to the SCI model and considered in determining the qualitative soil organic carbon value. Also considered in the SCI calculation are organic matter and field operations. The system is then determined to either be sustainable or unsustainable using total soil erosion and soil organic carbon criteria. The DAYCENT model is then initialized and simulated to predict and quantify the long-term impacts on soil organic carbon, greenhouse gas emissions, and nitrate leaching.

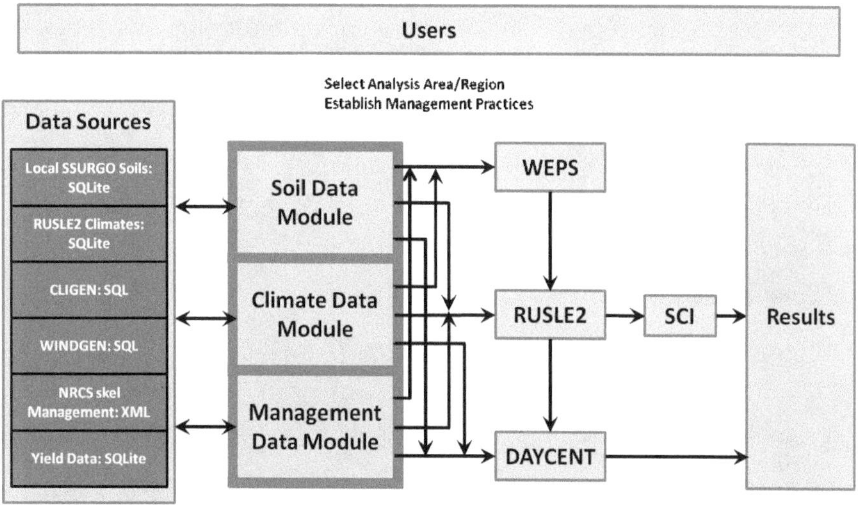

Fig. 1. Model integration framework used to assess environmental processes

One of the challenges in the integrated model is using consistent yield data across all the models since WEPS and RUSLE2 use yield as an input, and DAYCENT predicts yield based on soil, climate, and land management practices. To account for this difference, a yield calibration module is utilized with DAYCENT to enable yield consistency across the integrated model. While this enables a more accurate assessment of the land management decisions, it increases the simulation time due to the parameter adjustment methodology employed by the yield calibration module and because the process occurs for each year of the simulation. The WEPS, RUSLE2, and SCI simulations are less than a minute while the DAYCENT simulation/calibration process is several minutes.

3 Methodology

3.1 Field Discretization

The differences in computational runtime between the various models impacts the fidelity the agroecosystem unit can be analyzed. The relatively quick runtime of the WEPS, RUSLE2, and SCI models enables them to be simulated at every yield data point. However, simulation of the DAYCENT model with yield calibration at every yield data point is computationally expensive. To address this challenge, a discretization methodology was developed that captures the variability of the soil, slope, and yield at a scale that is appropriate for understanding the environmental processes while enabling utilization of the DAYCENT model. A 10 m × 10 m grid was determined to be the optimum resolution for (1) representing existing agricultural equipment used in the field and (2) aggregating the data to a level that is representative of the field.

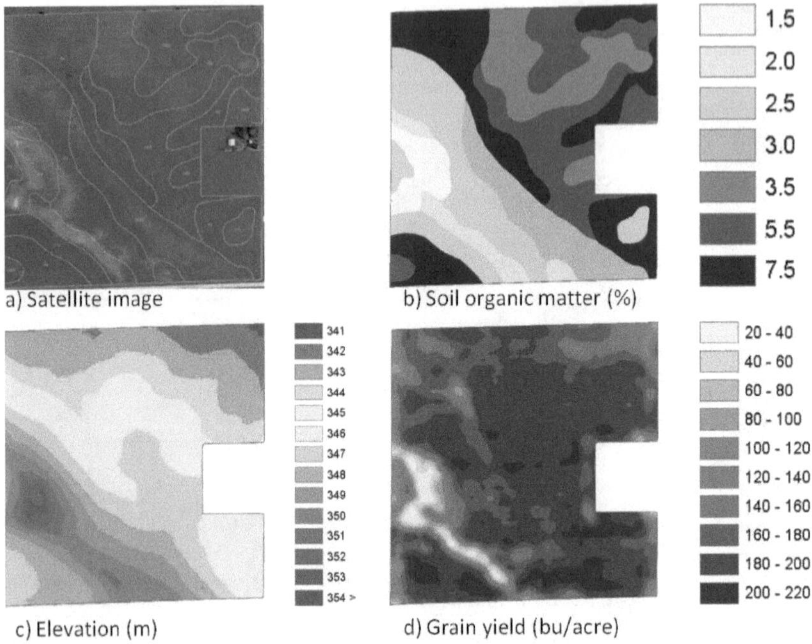

Fig. 2. Soil and grain yield characteristic for a field in Webster County, IA, USA. a) SSURGO soil map, b) soil organic matter (%), c) elevation (m), and d) grain yield (bu/acre)

3.2 Case Study

The integrated model was applied to a 60.5 ha field in Webster County, Iowa, USA. Fig. 2 shows the high-fidelity data utilized to assess the field. Fig. 2a shows a satellite image of the field with the SSURGO soil map overlaid. The field is composed of seven soil types. The variation in soil type is noticeable from the satellite image. Fig. 2b shows the soil organic matter levels of the various soil types found in the field. Soil organic matter retains and recycles nutrients, improves soil structure, enhances water exchange characteristics, and sustains microbial life in the soil [12] and pro vides an indicator of productivity [13]. Fig. 2c shows the field elevation using LiDAR data. Slope values required by the integrated model are determined using this elevation data. Fig. 2d shows the 2010 corn grain yield distribution using yield data gathered during harvest from the grain combine. More than 24,000 yield data points were taken during the 2010 grain harvest. Corn grain yield varies across the field from 1.25 Mg/ha (20 bushel/acre) on the high elevation areas to almost 15 Mg/ha (220 bushel/acre) in the lower elevation areas. The field was discretized into 6048 grid cells, as shown in Fig. 3. Grain yield and slope data points were averaged within each grid cell to generate a representative value. In addition, the dominant soil type was determined based on area coverage within the grid cell. These values were used within the integrated model by DAYCENT.

The integrated model was used to compare the impacts of two land management scenarios—a traditional Midwest row crop rotation and an integrated landscape management strategy on nitrogen use efficiency in the agroecosystem. The traditional row crop rotation is a corn-soybean rotation assuming 2010 corn grain yield and 1.48 Mg/ha soybean yield. A fertilizer application of 0.235 Mg_N/ha was applied during the spring in the corn year. A reduced tillage regime and no residue harvest are also assumed. For the integrated landscape management scenario, sections of the field with low productivity are taken out of row crop production and replaced with switchgrass. Switchgrass has been identified as a potential biomass resource for biofuel production and can potentially improve carbon sequestration, nutrient recovery from runoff, and soil remediation [14-15]. Approximately 35% of the field is converted from traditional row crops to switchgrass. The areas selected are based on the productivity of the land and the ability to manage the land with existing equipment without inconveniencing the farmer. The switchgrass stand is assumed to have a two-year establishment period followed by productive years where 80% of the biomass is harvested in the fall. Each year a fertilizer application of 0.101 Mg_N/ha is applied. The remaining area of the field still in row crop production assumes the same fertilizer application and reduced tillage with the addition of a rye clover cover crop. In addition, residue was removed from the row crop area at the highest sustainable rate. Each scenario is simulated for 10 years to compare the impacts of the two land management scenarios on nitrogen use efficiency and losses.

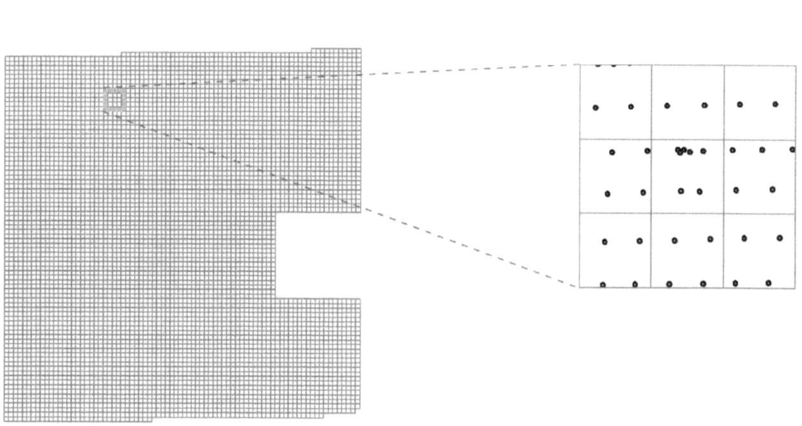

Fig. 3. Discretized field with an expanded view of a 30m x 30m section of the field with grain yield data points

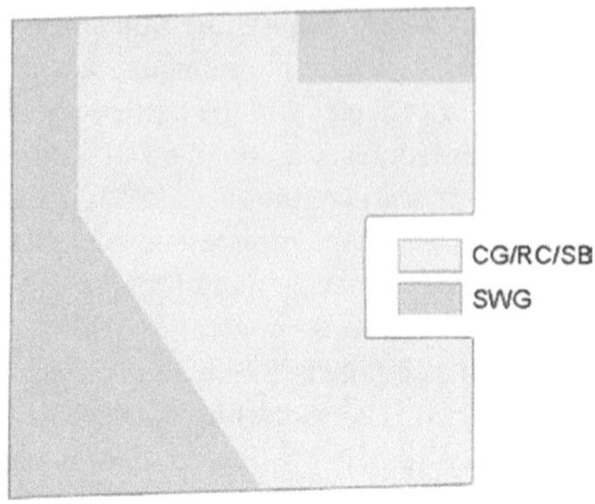

Fig. 4. Field representation of integrated landscape management

4 Results

Fig. 5 shows the losses that occur in each of the land management strategies. Fig. 5a,b show the agroecosystem unit losses due to N_2O gas fluxes. It can be seen that when calculating the N_2O flux that the integrated model is sensitive to soil texture. Soils with higher organic matter tend to have higher N_2O flux. Although crop rotation selection can reduce N_2O emissions, the general trend of higher organic matter leading to high N_2O emissions is consistent. The integrated land management strategy reduces the greenhouse gas emissions relative to the traditional row crop system from the field from 1.46 Mg_N to 0.87 Mg_N over 10 years. Fig. 5c,d show the nitrogen losses due to

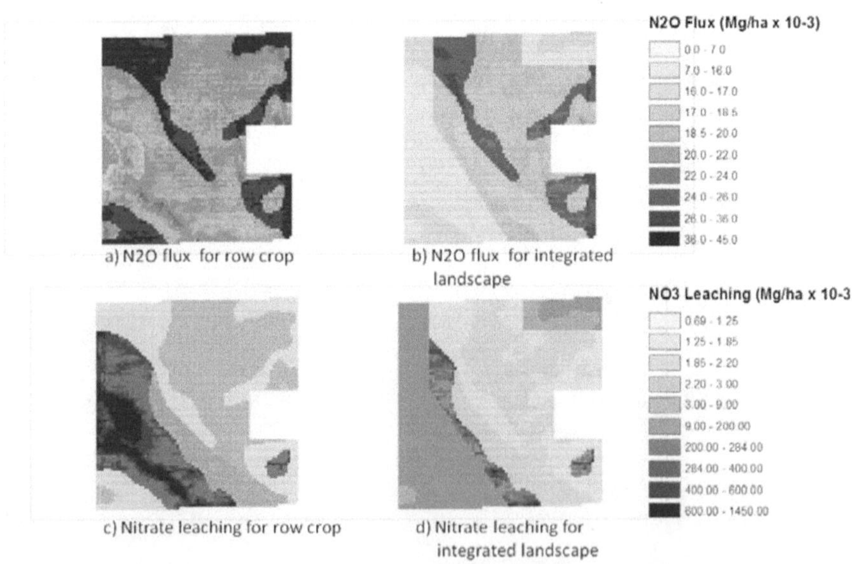

Fig. 5. Nitrogen losses for the row crop and integrated landscape managements

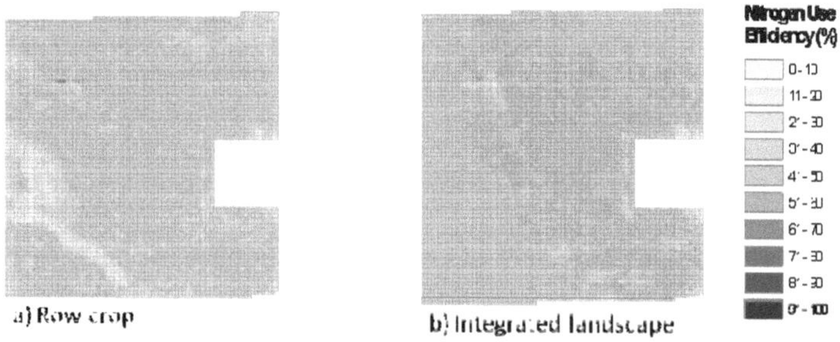

Fig. 6. Nitrogen use efficiency for the row crop and integrated landscape managements

nitrate leaching. Implementation of switchgrass reduces total nitrate leaching relative to the traditional row crop management by 4.83 Mg_N over 10 years, a 63% decrease.

Fig. 6a and 6b show the nitrogen use efficiency in the agroecosystem unit for both land management scenarios. As shown the integrated landscape significantly increases the efficiency of nitrogen use in the low production areas. Overall, the integrated land management strategy increases nitrogen use efficiency by 6% relative to the traditional row crop scenario, reducing nitrogen losses relative to the traditional row crop scenario by 6.48 Mg_N over a ten-year period.

5 Conclusions

The integrated model was used to investigate a traditional row crop rotation and an integrated landscape management scenario in a field with diverse soil and topographical characteristics. The integrated landscape management replaced low production areas in a traditional row crop rotation with switchgrass and added a rye cover crop where the traditional row crop was maintained. The integrated landscape improved nutrient use efficiency resulting in reduced nitrogen loss from the agroecosystem unit. The integrated land management strategy reduced N_2O emissions and nitrate leaching by 40% and 63%, respectively, resulting in a net nitrogen loss reduction of 0.11 Mg_N/ha. In addition, these nitrogen losses have an economic value associated with them and as markets for energy feedstocks begin to grow, the integrated model can enable decisions towards determining land management strategies that improve the economics, environment, and total biomass production.

Acknowledgements. The authors gratefully acknowledge the funding support from DOE's Office of Biomass Programs, as well as significant support from all partners in the DOE Biomass Regional Feedstock Partnership Program.

References

1. U.S. Environmental Protection Agency: Inventory of U.S. Greenhouse Gas Emissions and Sinks: 1990-2011, `http://www.epa.gov/climatechange/Downloads/ghgemissions/US-GHG-Inventory-2013-Main-Text.pdf`

2. Rabalais, N.N., Turner, R.E.: Hypoxia in the northern Gulf of Mexico: Description, causes and change. In: Coastal Hypoxia: Consequences for Living Resources and Ecosystems, vol. 58, pp. 1–36. AGU, Washington, DC (2001)
3. Dinnes, D.L., Karlen, D.L., Jaynes, D.B., Kaspar, T.C., Hatfield, J.L., Colvin, T.S., Cambardella, C.A.: Nitrogen Management Strategies to Reduce Nitrate Leaching in Tile-Drained Midwestern Soils. Agron. J. 94, 153–171 (2002)
4. Muth, D.J., Bryden, K.M.: An integrated model for assessment of sustainable agricultural residue removal limits for bioenergy systems. Environ. Modell. Softw. 39, 50–69 (2013)
5. Official NRCS-WEPS Site. Wind Erosion Prediction System, http://www.weru.ksu.edu/nrcs/wepsnrcs.html
6. Revised Universal Soil Loss Equation, Version 2 (RUSLE2), http://fargo.nserl.purdue.edu/rusle2_dataweb/RUSLE2_Index.htm
7. Soil Conditioning Index (SCI), http://soils.usda.gov/sqi/concepts/soil_organic_matter/som_sci.html
8. Muth, D.J., Bryden, K.M., Nelson, R.: Sustainable agricultural residue removal for bioenergy: a spatially comprehensive US national assessment. Appl. Energy 102, 403–417 (2013)
9. Muth, D.J., McCorkle, D.S., Koch, J.B., Bryden, K.M.: Modeling Sustainable Agricultural Residue Removal at the Subfield Scale. Agron. J. 104, 970–981 (2012)
10. Muth, D.J., Bryden, K.M.: A Conceptual Evaluation of Sustainable Variable-Rate Agricultural Residue Removal. J. Environ. Qual. 41, 1774–1786 (2012)
11. Abodeely, J.M., Muth, D.J., Bryden, K.M.: Integration of the DAYCENT Biogeochemical Model within a Multi-Model Framework. In: 6th International Congress on Environmental Modelling and Software, pp. 1287–1294. Leipzig, Germany (2012)
12. Wilhelm, W.W., Johnson, J.M.-F., Karlen, D.L., Lightle, D.T.: USDA-ARS: Corn stover to sustain soil organic carbon further constrains biomass supply. Agron. J. 99, 1665–1667 (2007)
13. Bauer, A., Black, A.L.: Quantification of the Effect of Soil Organic Matter Content on Soil Productivity. Soil Sci. Soc. Am. J. 58, 185–193 (1994)
14. Bransby, D.I., McLaughlin, S.B., Parrish, D.J.: A review of carbon and nitrogen balances in switchgrass grown for energy. Biomass Bioenergy 14, 379–384 (1998)
15. Keshwani, D.R., Cheng, J.J.: Switchgrass for bioethanol and other value-added applications: A review. Bioresour. Technol. 100, 1515–1523 (2009)

E-SMART: Environmental Sensing for Monitoring and Advising in Real-Time

Matthew Aitkenhead, David Donnelly, Malcolm Coull, and Helaina Black

The James Hutton Institute, Craigiebuckler, Aberdeen, AB15 8QH, UK
{matt.aitkenhead,david.donnelly,malcolm.coull,
helaina.black}@hutton.ac.uk

Abstract. Smart monitoring, using real-time environmental sensing with links to server-side data processing/modeling, allows progression from data acquisition to useful information generation. The use of modern technology such as mobile phones to provide imagery and other types of data along with GPS-derived coordinates enables researchers and stakeholders to integrate ground-based observations with existing datasets. We have developed an infrastructure linking mobile communications, server-side processing and storage of data and imagery, and field-based access to existing spatial datasets. This infrastructure has been used for the development of a number of mobile phone apps (applications) and web-based applications, and has proved useful for stakeholders in agriculture, science and policy. In addition to giving information on the capacity development, we demonstrate useful applications relating to the upload, interpretation and integration of data (e.g. automated interpretation of soil profile imagery, carbon content estimation from soil colour) while focusing on the technical aspects of the underpinning system.

Keywords: Environment, mobile phones, apps, monitoring, soil, models.

1 Introduction

1.1 What Is Needed in the Modern World

There are many ways in which information about the environment can assist land managers and policymakers in making decisions about appropriate management actions. Much of the current research in the environmental sciences is focussed not only on what processes and relationships there are within the natural world, but also on how we can improve our ability to observe the environment. Several factors are important in terms of how useful information is in this area, including:

- Accuracy – observations and measurements must be accurate, in order to be of use. In order to achieve this accuracy, a certain level of expertise is required in either making the observations or in interpreting them.
- Relevance – there are many different characteristics of the environment that a manager or policymaker might be interested in, and observations that do not match

J. Hřebíček et al. (Eds.): ISESS 2013, IFIP AICT 413, pp. 129–142, 2013.

the requirements are of little use. Making sure that the user's specific requirements are identified and satisfied is therefore important in terms of utility and cost-effectiveness.

- Spatial distribution – sometimes, a single observation is all that is required. Often however, multiple observations are needed to gain an insight into the spatial distribution of the parameter of interest. The pattern of observations is very much dependant upon what the user is looking at.
- Timeliness – some characteristics of the environment such as topography do not change very much over time, while others such as weather can change very rapidly. More recent observations can be much more useful to the user than older ones, although repeating the observations implies increased cost and effort.

Modern technology has much to offer in the field of environmental monitoring, and mobile phone apps (applications) in particular have the potential to allow land managers access to existing datasets and to make their own observations rapidly and cheaply. The number of different sensors embedded into a modern mobile phone, such as camera, microphone, accelerometer and GPS device, enables it to act as a platform for multiple different types of observations [1]. Another benefit of mobile phone apps in environmental monitoring is that they allow the user to make the observations where they are most relevant, rather than being forced to rely on maps and datasets that have been acquired for different purposes.

Here we demonstrate a framework for the development of mobile phone-based environmental monitoring apps, and highlight three examples of the way in which this framework has been used to produce apps for land managers in Scotland.

1.2 The State of the Art

A number of different mobile phone apps are available for environmental monitoring. These fall into two different categories, for observation recording and data access. The manner in which each app operates is different, as is the data that is either recorded or accessed by the user. Examples of 'observation' apps include species identification [2-4] and water quality monitoring [5]. This type of app requires a certain level of knowledge on the part of the user, in order for them to produce reliable observation data. It also often requires monitoring equipment to be used, for example pH meters, which can provide measurements that the mobile phone is not equipped to take. Another type of app that allows users to record their own information is the 'mapping' app, which can enable land managers, for example, to record the positions at which samples have been taken for later analysis [6, 7]. This type of app is useful for recording and retrieving the spatial distribution of observation points in relation to fields or other landscape features. However, it still requires the user to get additional (often costly and time-consuming) analysis carried out. There are also apps that process user observations and provide decision support for agricultural and environmental management, for example in crop pest monitoring [8], crop yield prediction [9, 10] and farm management [11, 12].

Examples of apps that allow the user to access existing data include interfaces for soil maps (e.g. [13-15]) and more general environmental map information (e.g. [16]). These apps use the GPS function of the phone to pass the user's location to a server where the map is stored. This information is then passed back to the user. The value of this kind of app is that it can supply the user with existing information, usually for free, that is from a reliable source. However, the observations that have been recorded do not always match those that are required by the user, and even if they do then they may be several years old. For some applications this is not an issue, but as mentioned above, some soil and landscape characteristics, for example land cover or soil moisture content, can change rapidly.

1.3 Data Requirements

The data requirements for monitoring are highly parameter-dependent. Some measurements, such as the elevation of a particular location, can be acquired through direct detection of elevation using a mobile phone's GPS functions. More complex examples, such as presence of specific plant species, would required either sophisticated image analysis from photographs of the environment or direct data input from an expert user. There are also many different ways in which each monitoring challenge can be tackled, along a spectrum of reliance upon user expertise. For true environmental monitoring that is consistent, reliable and can be carried out by land managers who are not likely to be experts in the relevant scientific fields, there needs to be a high degree of automation built in. Each case will have specific data requirements, and will also require ways in which this data is processed.

1.4 Model Requirements

Leading on from the data requirements of automated environmental monitoring is the need to interpret the information acquired through mobile phone sensors and turn it into something relevant for the user. Again, this interpretation will be highly application-specific, and so if an organisation is looking to develop more than one app, or even a suite of apps that can be used for environmental monitoring, then it is more cost-effective to develop a generalisable data interpretation approach that can be used in a range of different ways. This implies the use of modern statistical or artificial intelligence-based approaches that can be applied to a wide range of data types and input/output combinations.

1.5 Personnel Requirements

The development of an app for environmental monitoring is a highly multidisciplinary task, requiring expertise in several different scientific and technical areas. These include the development of the app itself, which requires specific programming skills, and the development and management of datasets and models that will interpret sensor data and produce information for the user. App development also implies a focussed project mentality enabling personnel to work closely together.

2 Methods

The work described here is not a standard scientific experiment, but rather the development and implementation of a framework for linking mobile phone technology with models for environmental monitoring. As such, the methods give a description of how this was achieved. We have divided the methodology into three main areas. Firstly, the organisation structure of the framework is presented. This includes the staffing and skills required, and the information and workflow arrangements set up to ensure success. Secondly, we cover the data framework used to integrate the various components of a successful system. Three examples are given of different types of monitoring system that we have implemented successfully. Thirdly, we cover the challenges that were overcome during the course of the ESMART project [17], and how they were tackled at the James Hutton Institute [18].

2.1 Organisational Structure

In order to successfully develop apps for environmental monitoring, a number of key staff roles are required. These do not all have to be different members of staff, and it may actually be beneficial if at least one of these roles (the project coordinator) is taken by a member of staff who is also fulfilling one of the other roles. This ensures that two of the main challenges mentioned later, namely keeping project expectations realistic and realising the potential of the project team, are successfully handled. The key roles are as follows:

- App developer – while all of the roles identified for the development of environmental monitoring apps are important, this one is perhaps the most obvious for successful app development. The developer is the person who designs and implements the 'front end' of the app, and who ensures that basic information input/output between the app and any server-side processing is handled. The development of apps for different platforms requires skills in different development environments, and it is possible that more than one developer will be required. The app developer also needs to have a certain level of understanding about the environmental characteristic or characteristics that are being monitored or measured. Without this understanding, it is difficult to produce a user interface and background processing chain that is relevant for the requirements of the project. For many of the apps mentioned in the Introduction, the majority of the relevant data to be passed to the user is actually held on servers, and an understanding of the extraction and handling of this data is also important.
- Data manager – as mentioned above, environmental monitoring that is capable of adding value greater than simply data entry or access in the field will require the handling and manipulation of existing environmental data. Unless the user simply wants to record observations that are linked to geographical location, there will be some element of integrating new information with existing datasets, usually with a spatial nature (i.e. digital maps). Therefore, at least one member of the development team needs to be familiar with the nature and handling of digital maps and

databases. In addition to digital maps, other forms of data may be required. Tabular databases of records are one example that was also used throughout the ESMART project. The database manager for an app development project therefore needs to be able to be able to set up and handle database schemas that can hold new observations generated either by the user or by models that are generating information automatically.

- Modeller – for nearly all of the apps mentioned in the Introduction, there is an emphasis either on data recording for future use, or access to existing datasets, with the main utility for the user in both cases that the app allows these activities to be done in the field. However, while entering or accessing data from the location of interest is no doubt more useful than via a PC from another location, there are two limitations from this situation that are important to consider here: (1) that any information entered by the user does not immediately benefit them, and (2) that any data accessed from existing datasets is dependent for its relevance upon how old it is. What we are adding through the inclusion of a modelling aspect is the ability for the user to accomplish two things at once: add new data to an existing framework and get some new and relevant information back out that is derived from their inputs, while minimising the amount of effort and required expertise on the part of the user. The role of the modeller is therefore to develop software models that can provide the user with new and useful information using a combination of their own input and existing data. A priority here is to simplify the requirements placed on the user and make the computers do the work.
- Project coordinator – the development of mobile phone apps that can provide 'added value' in response to user inputs requires a high degree of integration between the visible user interface, the running of customised models (either through server-side processing or within the mobile phone itself), and the mechanisms involved in passing data back and forth between interface, models and any secondary processing/preparation of data. Even in a situation where all of the individual components can be implemented, a coordinating overview of the app framework is required.

2.2 Data Framework

The data infrastructure of the ESMART project has a conceptually similar appearance to that of the organisational structure. The intention of the project is to provide a framework within which mobile phone apps for environmental monitoring may be developed. Each app has broadly similar data requirements, namely (A) access to spatial datasets, (B) models that take inputs from either the user or from these spatial datasets (or both), and (C) software that links the user interface, the database servers and the models. In order to satisfy these requirements, we have set up project-specific servers and made sure that Institute technical staff members are familiar with, and capable of handling, our server requirements. We have also developed a portfolio of spatial datasets of Scottish environmental characteristics, organised appropriately and split into subsets that allow more rapid access, which can be used to provide information for models. These datasets include soil, topographic, land cover, climatic and geological maps, each at a spatial scale of 100m and using a common coordinate system. For each app that is developed, a schema is generated that contains a copy of these datasets adjusted as required for the specific purposes of the app.

In addition to the servers and the spatial datasets stored on them, we have developed the capacity to produce rapid and flexible environmental models for predicting the outputs of interest. We have found that a neural network modelling architecture is sufficiently flexible, generalizable and rapid to satisfy our requirements in this area. Each neural network model is trained as appropriate, using a common training methodology and architecture. Additional processing of information, for example image analysis of photographs provided by the user, is carried out using custom subroutines designed as required. For communication between the various components of an app (user interface, spatial datasets, server-based models), we have found that the programming language PHP is suitable. PHP can be used to pass information from the user interface to the servers, and also to encode the neural network models and other processing subroutines. Figure 1 shows a schematic of the data infrastructure as it relates to the operation of each app, with the boxes representing specific components of the app and the lines representing the flow of operation between individual components.

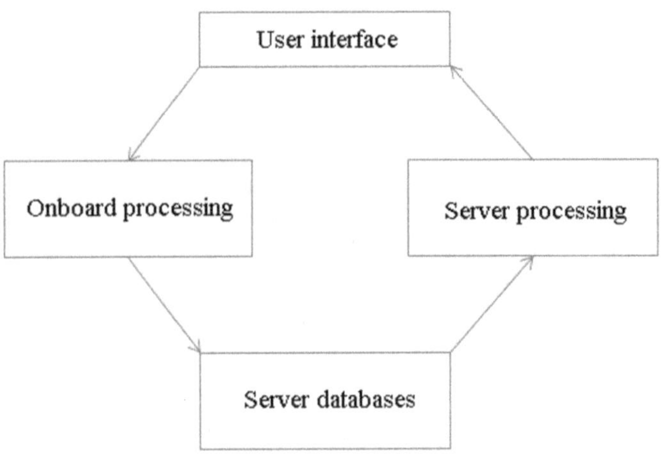

Fig. 1. Generalised data infrastructure of an app developed within the ESMART project

The model represented in Figure 1 contains four components, but these components overlap to a certain extent, and the activities taking place within each box vary between apps. In order to give a clearer depiction of the data flow and processing within any one app, we have provided in Figure 2 a conceptual diagram of the SOCiT app shown in Section 3.3. This diagram contains more detail than Figure 1, but would not serve as a replacement, as it contains several data flow and processing components specific to the app. Where appropriate, we have labeled each component of the SOCiT app in relation to the four main components shown in Figure 1.

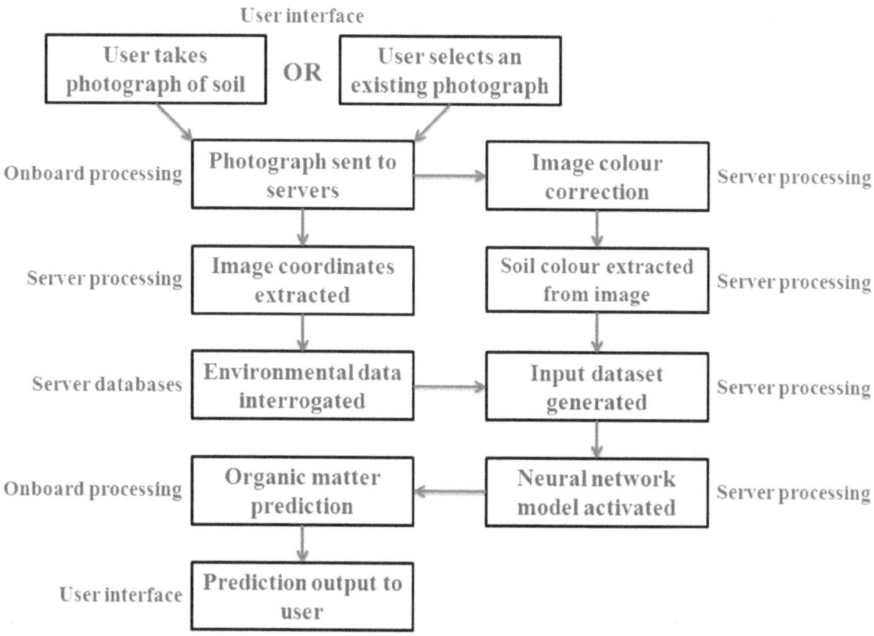

Fig. 2. Data infrastructure of the SOCiT app developed within the ESMART project

2.3 Challenges Encountered

2.3.1 Keeping Expectations Realistic

There is a certain air of urgency about the adoption of mobile phone technology by environmental agencies and research organisations in the UK. This is very definitely a fast-paced field, and there is a feeling that everyone is trying to 'get ahead of the competition'. To a certain extent this is appropriate – the first group that manages to produce useful, cost-effective apps for land managers and environmental scientists is likely to benefit more than those following. However, there are risks involved in simply trying to be first, rather than best. A shoddily-produced or unreliable app is a highly visible target for criticism, and does nothing to improve the reputation of those producing it. Any app that will be used by non-expert stakeholders in the field must be both attractive and simple to use but must also produce something new and useful – farmers and other stakeholder groups often spread information about new developments, both good and bad, by word of mouth.

An understanding of what is realistically possible is also necessary. Many apps that can be downloaded for free have been produced at great expense by expert production teams who intend to recover their costs through advertising revenues. Project funding in the environmental sciences is of an entirely different nature, and it is important for both developers and funders to appreciate this.

2.3.2 Understanding the Potential

Running counter to the challenge of keeping expectations realistic is the need to ensure that the full potential of the technology is utilised. There are three main characteristics of modern mobile phone technology that provide avenues for developers to exploit:

- Geolocation – most modern mobile phones have in-built GPS technology almost equal to GPS devices. The inclusion of the user's location as information allows an app potential access to available spatial datasets. However, there is also scope for returning mapped data to the user and to models that can use this to develop new information about the location of interest.
- Processing power – it is several years since the 1GHz barrier was passed in terms of mobile phone processor speed, and it is standard for memory capacity to exceed 1GB. This enables the device to handle sophisticated and computationally intensive processes. There is therefore the potential to generate new information through modelling or image/data analysis.
- Database access – environmental agencies and research establishments either hold, or have access to, spatial datasets and databases in digital form. The practicalities of accessing this information from a mobile phone are no more complex than accessing them from a website.
- Sensor technology - an important aspect of most, if not all modern mobile phones is their function as cameras, which could potentially be augmented by the inclusion of additional 'snap-on' sensor devices.

2.3.3 Institutional Silos

While the compartmentalising of research areas has not been as much of an issue within the James Hutton Institute as it could have been, there are potential hurdles that must be overcome for successful development of environmental monitoring apps. A wide number of different areas of expertise are required in the development of any app that does more than simply record information or regurgitate it upon demand. In addition, the development of an app is not the only prerequisite to success; also important are the publicising and dissemination of the app, the provision of long-term support for users and the handling of any legal, administrative and reporting requirements. Successful implementation therefore requires good communication and as staff with the required skills are likely to be working from within a number of different research, technical support and administrative 'branches' of an organisation, there are challenges in bringing them together.

2.3.4 IP Issues

When an organisation develops an app for mobile phones, the intellectual property of the app is normally fairly clear-cut – the organisation simply holds the intellectual property rights to the app and allows users to access it under licence. However, this situation is complicated when the app makes use of information from other sources, such as digital maps or remote sensing imagery, which has been acquired by the organisation under licence. If the organisation is constrained to the use of this data in a particular way, for example the licence was provided for purely research purposes, this will likely prevent the organisation from selling the app for profit. And if the project was funded by an external funding body, then the situation becomes even more complicated – the intellectual ownership of the app may be split between the funders and the organisation that produced it, or entirely with one of the two. And what of the outputs from such an app? If a farmer enters information about one of their fields, and this information is processed and produces an output, who owns the results? Is the organisation that funded the app development allowed access to this new information, or the organisation that produced the app?

It is advisable for agreement to be sought on these issues prior to the development phase. Separate agreement needs to be reached on the intellectual ownership and usage rights of (A) underlying datasets, (B) the app itself, (C) any information supplied by the user to the app, and (D) outputs generated and passed back to the user.

2.3.5 Project Administration

As pointed out above, the success of an app depend on more than how well the software is designed. The frameworks within which Apple and Android app developers operate are significantly different from one another, but there are commonalities between the two. One example of these is in relation to contractual obligations and intellectual property, as mentioned above. For an organisation to be able to publish apps, there must be some kind of legal contract between the organisation and those hosting and supplying the app to end-users. This contractual link is more obvious with Apple developers than for Android, but exists for both. The signing of contracts and other legal agreements is almost by definitions a responsibility of senior staff, who need to have the authority to agree that the organisation will be bound by certain legal requirements. However, the role of senior staff is unlikely to also include day-to-day management of an app development project, which needs to be carried out by a tightly focussed group with high levels of internal communication. The development and production of a mobile phone app is generally a focussed, relatively short-term project. In this sense, it is different from a lot of environmental research and can be funded through small, directed project budgets. However, there are two considerations for the long-term, that of user support and software licensing. The intention will undoubtedly be for the app to be used over the long-term, and the development team should expect to have to provide user support and further application development past the timeline of project funding. It is impossible, particularly in the current financial climate, to guarantee even a small funding source for this kind of work in perpetuity, and so the organisation responsible for app development needs to ensure that there is some kind of allocation made from underpinning capacity funding. Additionally, further future costs will be incurred due to annual license renewal for software and app developer accounts.

3 Results

Here we present and discuss three of the apps that have been produced by the James Hutton Institute, either directly under the ESMART project or with its support. Each app has certain common features, but is distinct in the manner of its use and information flow, and the stakeholder groups at which it is targeted.

3.1 SIFSS

SIFSS (Soil Indicators for Scottish Soils) is an app that allows the user to find out what soil type is in their area, to explore the characteristics of around 600 different Scottish soils, to discover the differences in soil characteristics between cultivated and uncultivated soils and to examine a range of key indicators of soil quality. It is the only app that gives access to the full dataset held from the Soil Survey of Scotland.

SIFSS allows the user to use their device's GPS or to use a place name or postcode, or just browse using the image map on their device to choose a location in Scotland to investigate information about soil. This information includes pH, soil carbon, N, P, K etc. directly from the James Hutton Institute soils database (see Figure 3).

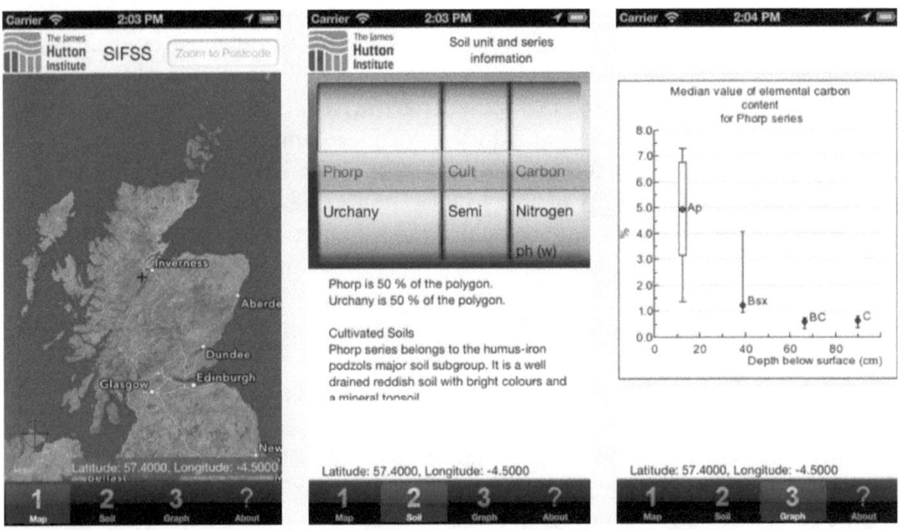

Fig. 3. Screen shots from the SIFSS iPhone/iPad app

3.2 iDee

The iDee project aims to directly involve people and communities in the River Dee catchment in North-East Scotland in environmental recording. Using the iDee app and website, the user can submit their own records of river conditions in the Dee catchment in the form of a photograph of the river together with simple assessments of water clarity, flow speed and algae cover. An interactive map allows the user to choose and view records submitted by other people to see how conditions in the catchment change over time.

The iDee app was created for the Dee Catchment Partnership by the James Hutton Institute, and while it is currently only available for Apple devices (Figure 4), the idee.hutton.ac.uk web site can be used on any device.

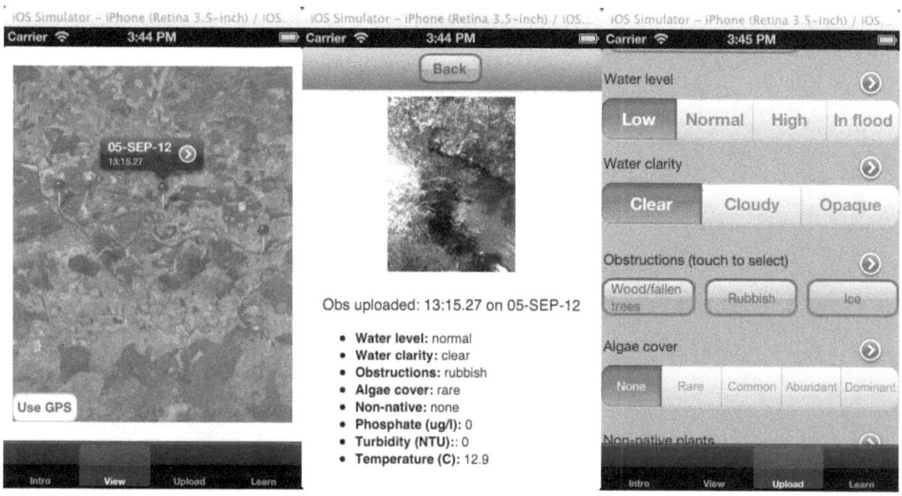

Fig. 4. Screen shots from the iDee iPhone/iPad app

3.3 SOCiT

SOCiT is a free app aimed at farmers, land managers and other land users who want to know how much carbon is in the soil. It is available for download on iTunes. The app uses information about the user's position to access existing digital maps of environmental characteristics, such as elevation, climate and geology. Combining this information with data extracted automatically from a photograph of the soil of interest, it uses a mathematical model to predict topsoil organic matter and carbon content (Figure 2). The app has been designed to work in any rural agricultural area of Scotland. The user also requires an image colour correction card that is supplied directly from the James Hutton Institute for free upon request. Figure 5 provides a series of screen shots from this app.

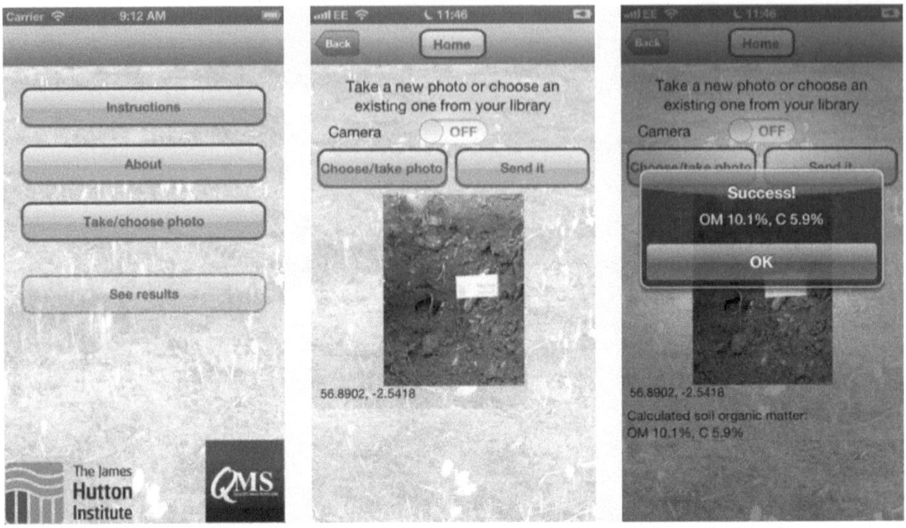

Fig. 5. Screen shots from the SOCiT iPhone/iPad app

3.4 Ongoing Projects

Each of the above apps has been successfully developed for iPhone and iPad. Current market share of apps is shifted towards Android, and there is an intention to develop each of the apps mentioned here, and any future apps, for this wider set of users. The additional skills required are available within the James Hutton Institute, and development of these apps is in hand.

Additional apps are also being proposed, to make use of existing models and data resources developed and held at the James Hutton Institute. Now that we have successfully implemented a framework through which increasingly useful mobile phone apps can be developed for land managers and environmental scientists, the only real obstacle is finding and allocating the resources to allow us to do so.

4 Discussion

Specific infrastructure and skill requirements exist for the successful implementation of a mobile phone app that can be used for environmental monitoring. Putting this infrastructure in place and ensuring that members of staff have the necessary skills represents a significant effort, both economically and in terms of project management. However, once an organisation has all of the key components in place, it is in a position to repeat initial successes in this area. In the last year, the James Hutton Institute has put in place the technical, data-management and personnel resources to allow it to produce apps that do more than allow the standard map referencing or data entry from the field that is the industry standard. We have demonstrated this potential with three apps for iPhone/iPad, the most sophisticated of which (SOCiT) gives the user an estimate of soil organic matter content based on information about the location and soil imagery. The intention is to develop more apps, for both Apple and Android platforms that are capable of producing new information about the user's environment and that facilitate environmental monitoring.

5 Conclusions

We have demonstrated a successful framework for the development of useful mobile phone apps for environmental monitoring. This framework depends for its success on the identification of key roles and the creation of a focussed, results-oriented group in relation to the implementation of each app. It also relies on an organisational infrastructure that is sufficiently flexible to accommodate the technical requirements of specific apps, and that contains an appropriate mix of different types of staff expertise.

Acknowledgements. The authors would like to thank Mr Andrew Thorburn, Mr Donald Barrie and Professor Colin Campbell from the James Hutton Institute for their assistance with the work carried out during the ESMART project. We would also like to acknowledge the funding support provided by QMS (Quality Meat Scotland) for the development of the SOCiT app.

References

1. Lane, N.D., Miluzzo, E., Lu, H., Peebles, D., Choudhury, T., Campbell, A.T.: A survey of mobile phone sensing. IEEE Communications Magazine 48(9), 140–150 (2010)
2. Graham, E.: What's invasive (2013), http://whatsinvasive.com/
3. Terrestrial Ecosystem Research Network (TERN) Transects for Environmental Monitoring and Decision Making (TREND) app (2013), http://www.trendsa.org.au/mobile_app
4. European Environment Agency (EEA) NatureWatch invasive species app (2013), http://www.eea.europa.eu/mobile
5. Information for Community Oriented Municipal Services (iComms) Water Quality Reporter (2013), http://www.icomms.org/
6. Farmlogic Soil Test Pro app (2013), http://www.farmlogic.com/tools/mobile_apps/mobile_soil_sampling.aspx
7. PrecisionEarth: PrecisionEarth soil sampling & recording app. (2012), http://precisionearth.com/
8. University of Nebraska-Lincoln Aphid Speed Scout (2012), http://real.unl.edu/iOS/AphidSpeedScout/
9. Optimizer, Optimizer corn yield app (2013), http://www.optmzr.co/
10. Precision Planting: Yieldcheck app (2013), http://precisionplanting.com/
11. CropTrak iCropTrak farm management app (2012), http://www.icroptrak.com/
12. Stringybark software Farm Manager app (2012), http://stringybarksoftware.com/

13. United States Department of Agriculture Natural Resource Conservation Service (USDA NRCS) Soil Web Survey (2013), http://casoilresource.lawr.ucdavis.edu/soilweb/
14. Commonwealth Scientific and Industrial Research Organisation (CSIRO) SoilMapp for iPad: soil information at your fingertips (2013), http://www.csiro.au/en/Organisation-Structure/Flagships/Sustainable-Agriculture-Flagship/SoilMapp-for-iPad.aspx
15. British Geological Survey (BGS) mySoil App (2013), http://www.bgs.ac.uk/mysoil/
16. National Geographic Information System (NGIS) Natural Environment Map in Taiwan iOS app (2011), http://www.arcgis.com/home/item.html?id=9143a461e1da461f91ce7dac677e9ae8
17. Aitkenhead, M.: The E-SMART project (2013), http://www.hutton.ac.uk/research/groups/information-and-computational-sciences/esmart
18. The James Hutton Institute (2013) Institute website, http://www.hutton.ac.uk/

Identifying Smart Solutions for Fighting Illegal Logging and Timber Trade

Ioannis N. Athanasiadis[1,2], Despina Anastasiadou[2], Kostas Koulinas[2], and Fotis Kiourtsis[3]

[1] Democritus University of Thrace, Xanthi, Greece
ioannis@athanasiadis.info
[2] Information Technologies Institute, Centre for Research and Technology Hellas, Thessaloniki, Greece
[3] General Directorate of Forests & Rural Affairs, Decentralized Administration of Macedonia and Thrace, Thessaloniki, Greece

Abstract. We investigate how smart Information and Communication Technologies (ICT) solutions can be used for combating illegal logging and timber trade. We put together techniques from agile requirements engineering to propose a methodology for identifying user stories and associated risks and priorities and via a collaborative, participatory, single day workshop, named *inception workshop*. We present our findings from the first application of the method, with the active involvement of the relevant stakeholders, i.e technical and domain experts, which concluded in seven user stories.

Keywords: Agile methods, requirements engineering, forestry, illegal logging, illegal timber trade, user stories, participatory methods, inception workshop, smart solutions, RFID, low cost sensors.

1 Introduction

Illegal logging and timber trade is currently a major concern among all countries world wide [1], while it is acknowledged as a complex problem that involves not only environmental dimension, but also economical and societal ones. While until very recently illegal logging was not an important issue in Greece [2], nowadays it has been intensified due to the economic crisis [3, 4]. In this paper, we investigate how smart Information and Communication Technologies (ICT) solutions can be used for combating illegal logging and timber trade. Following an analysis of the whole logging and timber trade lifecycle, performed with participatory methods and agile techniques, we have identified the weak links, which could potentially be improved by technical solutions.

With the active involvement of the relevant stakeholders, we identified seven user stories for strengthening modus operandi, each one of which introduces one or more ICT technologies. These include electronic documents; barcode and RFID (Radio-frequency identification) tagging; low-cost, low-power sensor networks for motion and volume detection; remote sensing and satellite imaging;

J. Hřebíček et al. (Eds.): ISESS 2013, IFIP AICT 413, pp. 143–153, 2013.

integrated service-oriented systems. We also identified associated risks and priorities for the implementation pilot of these user stories.

The rest of this paper is organized as follows: Section 2 presents methods and tools, i.e. the problem at hand related to illegal logging and timber trade (par. 2.1), agile methods for requirements engineering (par. 2.2), and our methodology for identifying and prioritizing solutions (par. 2.3). Section 3 demonstrates the methodology application and our key findings, while section 4 presents with a discussion of our experiences with the process.

2 Methods and Tools

2.1 The Problem at Hand: AITOLOS Project

The AITOLOS project aims to foster cross-border collaboration and coordinate the activities of Greece and the former Yugoslav Republic of Macedonia against illegal logging and timber trade. Specifically, AITOLOS contributions are towards the following directions:

(a) Assess current legal framework, and define specific recommendations for enhancing legal and regulatory framework in order to fully address legal aspects of illegal logging/trade.
(b) Develop policy guidelines for addressing illegal logging, with a special focus on cross-border timber trade.
(c) Develop an operational plan across participating border regions for implementing policy/guidelines, including the technical specifications, stakeholders roles, monitoring mechanism, timetable and budget.
(d) Engage in communication and dissemination activities at regional,national and European level, presenting developed policy guidelines and regional operational plan.
(e) Two implementation pilots, one per country, in order to assess project results in real world situations. Pilots will run in forest offices of Goumenissa and Bitola.

The project duration is 15 months (2012-2013), and its consortium comprise of the Decentralized Administration of Macedonia & Thrace, Public Enterprise '*Macedonian Forests*', Centre for Research and Technology Hellas (CERTH), Technological Education Institute of Serres (TEIS), and National Association of Private Forest Owners.

One of the major challenges of AITOLOS project was to identify smart ICT solutions to be incorporated into the everyday work of national forest protection services, to facilitate and support traditional processes of forest management, and to protect forest ecosystems against illegal logging and trade; proving that modern technologies can be successfully incorporated into everyday forestry service duties. Identification, prioritization, and risk estimation for building a potential solutions portofolio was a major challenge for the AITOLOS project. Also, selecting which of the potential solutions can be demonstrated as the project pilot was further constrained by our limited amount of resources. To tackle these challenges, a bidirectional knowledge shift between forest and computer scientists was required. To develop a common language and achieve consensus among technical and domain scientist, we adopted agile methods originating in software engineering, and put them in action in a single-day, intensive, *'inception' workshop*.

2.2 Agile Methods for Identifying Key Requirements

Capturing user requirements consist a moving target in projects that are ill-defined, or fast-changing, thus the past years agile approaches as Extreme Programming or Scrum have been proposed for capturing requirements informally, i.e. as users stories.

Agile methods for requirements engineering are based on four principles [5]: *a*) put customers in the center of attention, i.e. involve actively stakeholders through out the process; *b*) employ early verification and validation techniques, i.e by developing acceptance tests for user stories, or use cases; *c*) identify early non-functional requirements and analyze them with customers; *d*) manage change and incorporate it in the process of requirements engineering.

Cao and Ramesh, in [6], presented an empirical study that evaluates agile requirements practices in 16 organizations, and reveals that intensive communication between the developers and customers as the most important one.

Inspired by agile requirements practices, we developed a methodology for engaging end-users early in the process, and applied it in AITOLOS project technical solution identification.

2.3 The Inception Workshop Approach

An *inception workshop* aims to help project participants (i.e. both technology and domain experts) to identify key requirements of the project technical solution. This can be achieved through eliciting initial user stories, which will be subsequently developed through iterations. Domain and technology experts interact in a single day intensive workshop, in order to come to a common understanding as to the scope of the project technical solution.

Participant roles participants are domain and technology experts, and a facilitator. Domain experts participate in order to present their application domain and present problems in their everyday work practice. Technology experts come to the workshop for understanding better the problem domain and specify the technical solution The facilitator is responsible for running the workshop according to schedule and keep discussions focused.

The *inception workshop* is organized in three sessions. Figure 2.3 presents a typical agenda of an inception workshop.

The first session focuses on *domain understanding*, in our case forestry and illegal logging. The goal of this session is to detail the current processes followed, and increase the understanding of the technical experts. Domain experts share their knowledge in a **story-telling** fashion: They communicate their knowledge

30'	Welcome - The workshop process	Facilitator
90'	'*Storytelling*' (Domain understanding session)	Domain experts
30'	Short Break	
90'	'*Technology roadshow*' (Explore technology opportunities)	Technical experts
60'	Long Break	
60'	'*Synthesis I*' (Define first user stories)	All
30'	Short Break	
60'	'*Synthesis II*' (Set priorities and identify risks)	All

Fig. 1. Typical agenda of an inception workshop (activities with durations)

by sharing experiences and anecdotes. In a collaborative team environment, domain experts stand up to tell the story of their every day work, in a procedural manner, while technology experts intervene with clarifying questions. The facilitator needs to keep the storyline progressing, and prevent the domain experts from diving into details.

The second session focus is on understanding technology opportunities and identifying technology solutions. Through a brainstorming process, technical experts '*put guns on the table*', and exemplify how technical solutions might be used for fighting illegal logging. Existing or new solutions are presented informally by technical experts, in order to increase domain experts awareness of the capabilities of ICT systems, and their application in the domain. In a **technology roadshow**, technical experts present solutions in the form of examples, not only related to the domain at hand, but from other domains as well. Domain experts are free to ask questions, so that they understand what technology tools may offer, and the facilitator is responsible for moderating jargon language and reminding technical experts to use simple examples from known applications to the domain experts.

The final, third, session aim is to produce a **synthesis** of the two views, the one of the domain experts with that of technology experts. In a collaborative team environment, domain and technology experts identify together *initial user stories*. Agile requirements methods consider user stories as very high-level definition of requirements, containing just enough information so that the developers can produce a reasonable estimate of the effort to implement it [7]. In our case, initial user stories aim to produce very high-level requirements of technical solutions for the project at hand. While identifying those, we also set priorities in terms of effectiveness, risk and cost.

3 Demonstration: The AITOLOS Inception Workshop

For identifying the potential technical solutions of the AITOLOS project (see par. 2.1), we organized the first AITOLOS inception workshop, in Thessaloniki, Greece, on January 25, 2013, with the participation of all Greek project partners. Approximately 12 domain experts from two forest services (Kilkis and Goumenissa), and 6 technology experts from CERTH and TEIS participated with the main goal of answering the question: 'How smart ICT solutions may be applied in forest service practice in order to combat illegal logging and timber trade?'. We had a very vibrant and interesting discussions in the form of an 'inception workshop'. In the following paragraphs, we present our findings from the workshop. *Storytelling* and *Technology roadshow* sessions are presented with annotations from bibliography, to strengthen the opinions that have been revealed during the workshop. *Synthesis* session presents our results in terms of the technical solutions identified, presented here as user stories, annotated with our estimates in terms of risks and costs.

3.1 Storytelling

During the storytelling session, forest scientists presented the whole lifecycle of wood, from the forest to the timber market. Illegal logging activities include the harvest, transportation, purchase or sale of timber in violation of national laws. The harvesting procedure itself may be illegal, including using corrupt means to gain access to forests; extraction without permission or from a protected area; the cutting of protected species; or the extraction of timber in excess of agreed limits [8].

We went through five steps in the timber logging and trade value chain: *a*) Timber pre-marking, *b*) Felling and pilling, *c*) Delivery and transport, *d*) Storage (and trade) *e*) Cross-border timber trade. We highlighted differences in the process between fuel wood and technical wood. Our findings are summarized in Figure 2.

Issues related to illegal logging were presented, and verified that 'two types of illegal logging can be clearly distinguished in South-East Europe countries: (a) poverty driven; and (b) commercial. These types of illegal logging generally take place in different forest areas, are carried out by different offenders, and have different impacts. For example, poverty-driven illegal logging is typically carried out in remote locations, out of sight of the state forest management and law enforcement bodies' [9]. Domain experts provided with evidence from their areas and detailed incidents of illegal logging and timber trade of the two types. We concluded that in Greece, poverty-driven illegal logging typically involves fuel-wood, while commercial illegal logging is related to both fuel-wood and technical wood.

At the same time, we realized that Greece is not an exception to the UN report observation that in South-East Europe many countries have weak forest information systems [10]. Most forest services in Greece are largely doing their business with 'pen and paper', and digitization of processes and documents are

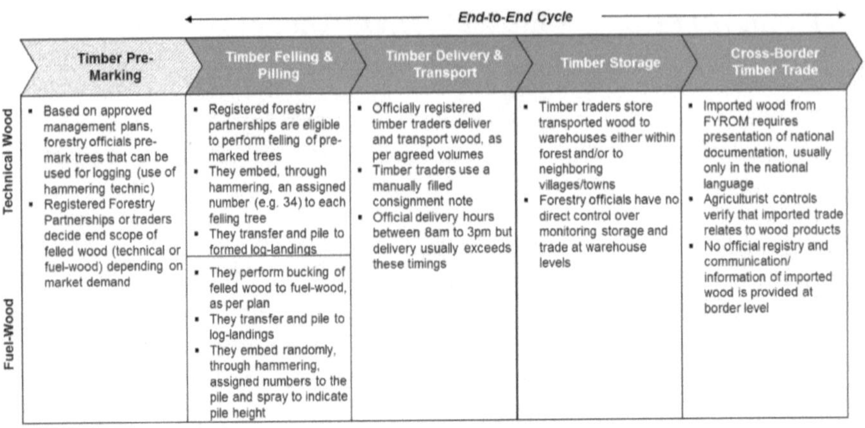

Fig. 2. Workshop notes on timber logging and trade value chain

still in its infancy. Wood value chain in Greece remains undeveloped and several obstacles remain related to official forest maps, forest data recording and cadastral, adaptation of EU specifications, and producing official forest maps, forest data recording, rapid adaptation to EU specifications [11].

3.2 Technology Roadshow

During the *technology roadshow* session, technical scientists with different backgrounds presented relevant technologies that could be potentially used in improving the situation with illegal logging and timber trade. Among those are the followings:

Central systems for issuing permits: A recent Interpol/UNEP report [1] recommends to centralize nationally the issuing of permits for land clearance whether for logging, plantations or ranching, and permits for road transport of timber with strong anti-counterfeit measures.

DNA fingerprinting: The use of ambient sonic spray ionization mass spectrometry as phytochemical markers [12], and DNA fingerprinting [13–17] has been reported for identifying timber origin, and verifying timber supply chains integrity, especially for high-valued timber. Such forestry forensics tools can be proved effective as a means against illegal logging, however the effects of population substructure and inbreeding [18] need to be taken under account.

Market control mechanisms include the expansion of market inspection competencies to ensure the complete control of documentation accompanying shipments of wood, including statements of origin and logging permits [9].

Forest certification: Chain of custody control measures ensures that illegally sourced wood cannot be processed in plants, as some proof of origin certification is required. Forest governance and overall forest preservation can be improved through the introduction of certificates for sustainable forest management and chain of custody certificates [9].

GIS systems and databases: Field computers, GPS receivers and GIS databases can ensure the constant monitoring of wood flow and so prevent illegal trade [9].

Marking logs with RFID tags: Pulping compatible, passive UHF transponder made of a wood composite material was developed for marking of logs together with robust RFID readers for the harvester and saw mills [19].

Satellite imagery at multiple temporal and spatial resolutions can effectively increase transparency in the forestry sector by revealing the rate and extent of deforestation on an annual basis and identifying potential areas of illegal logging [20].

Machine vision: Wood identification using anatomical characteristics can be automated through machine vision, as reported in [21], that could be used for fraud detection.

Technology experts presented the above-mentioned technologies, giving examples from forestry applications abroad and from scientific literature, along with examples from other domains with which our domain experts were familiar with. These included market retail, taxation and fiscal systems, electronic tickets, parking stations, and fire protection.

3.3 Synthesis: User Stories, Risks, Priorities

The previous two sessions increased confidence of our technology experts with illegal logging and timber trade issues, and our domain scientists with modern ICT technologies, and their potential in forest protection. Following a longer break were informal discussions continued, we reconvened in collaborative session where we started shaping solution ideas into the following user stories. The order of presentation is not significant.

S1: Electronic documents with security characteristics. All documents involved in timber transportation (from the log landing areas to the market) becomes digital, to avoid fraudulent documents. This involves both logging permits and timber transportation notes (which accompany trucks from log landing areas to timber warehouses). New document templates incorporate security features in the form of barcode, QR code, or e-signatures, which are instantly verifiable. Using low cost smart devices, security features can be verified online, via accessing web addresses over the web.
Risk assessment and priority: This solution is easy to implement and resembles an electronic ticketing mechanism with several control points. Legal enforcement and information security infrastructure are the main challenges of this solution.

S2: Smart tags for timber marking. Currently, timber marking is performed with hammers that mark wood with forest office, or forester code. This is a weak point that can be significantly improved by introducing unique tags

for all wood marked by hammer. This can be achieved by using plastic barcode or RFID tags, or RFID nails from pulping compatible material. This solution is ideal both for proving technical wood log origin, and for timber certification.
Risk assessment and priority: RFID technology has been extensively applied in market applications, however there is little experience with timber marking. In this respect, reading capacity of equipment needs to be verified in the environmental conditions of log landing areas. At the same time, there are operational challenges related to who applied the tags, who bears the cost of the application and how to check tags when logs are loaded on trucks. Last but not least, information security and cryptographic characteristics of the information encoded in the tags needs to be studied in depth.

S3: Cross-border uniform, multilingual reports. Forms attached to timber imported (or exported) are standardized, and all texts appear in both local languages, and English. This solution removes language barriers at inspection points and could be achieved by developing a common standardized template to be shared between the neighboring countries.
Risk assessment and priority: While seems simple, due to the different policies and regulations of the two countries, achieving consensus might be a problem. Enforcement and orchestrating all involved authorities in both countries could be troublesome too. Also, when exported in processed form, timber becomes a marketable product. This implies that European common market rules apply, and forest regulations do not apply any longer, rather is a matter of custom inspection and international trade.

S4: Truck load detection. Manual estimation of timber volume loaded on trucks is another weak point of the value chain. Using an one-stop gate, equipped with low cost, low power sensors to calculate automatically wood volume loaded on a truck. This could be done for example by two depth cameras getting two viewpoints of a truck: top and rear, and detecting the contour of the volume loaded.
Risk assessment and priority: While today there are several cheap sensors and this solution seems easy to deploy, our estimation was that it is gonna be hard to calibrate with accuracy, and truck type knowledge might be needed. To the best of our knowledge there are no similar applications, at least reported in literature, so a prototype is needed to evaluate the proof of concept.

S5: Truck motion detection. While effective monitoring of timber transport and access control to forest areas have been identified as a key to combat illegal logging, this is a very difficult task, as forests are not closed and a dense network of dirt roads and paths exist. However, as a forest service we would like to see implemented a typical fleet management scenario, that uses GPS/GSM network to detect truck routes, and report them online. Similar approaches might use RFID tags or automatic vehicle identification (AVI) at forest gates to patrol entrance/exit from certain checkpoints.
Risk assessment and priority: While this solution could be extremely effective, transporters consent would be required and privacy issues may arise.

Also implementation in forest conditions are not well studied, and poor signal coverage may affect efficiency.

S6: Satellite imaging. This story considers the application of image processing algorithms on satellite images to detect areas of intervention and protection. Such tools have been applied extensively in forest fire monitoring. In our case, we identified two scenarios. The operational scenario, where we use images on demand, in order to identify illegal logging at the time it happens, and the strategic scenario, where the same techniques may apply on low-cost, historical images, to estimate forest degradation due to illegal logging and prioritize future intervention actions.
Risk assessment and priority: We need to keep a balance between image cost and resolution needed. A high-resolution, on demand image cost is extremely expensive, thus the operational scenario seem not viable, at least for fuel wood related illegal activities.

S7: Systems integration. In the current situation, information doesn't flow across the borders, neither between the various national services involved in the inspection of timber trade value chain (i.e customs, taxation, police and forest authorities). As a forest service we need an online system that links customs with forest services, and can notify forest services on vehicle movements across borders, and report on volume and type of wood, origin and destination, etc.
Risk assessment and priority: While this seems from a technical point of view a relatively straightforward solution, we identify risks related to data content, system interoperability and standardization, and the implementation of an appropriate information security infrastructure.

4 Discussion

The *inception workshop* helped a lot our project members to develop a common language and understand better each other expertise. It was also proven helpful in building team spirit due to the participatory sessions, where interactions brought our team closer.

We also managed to identify early in the project, what is more important to our own users, rather than considering generic literature cases or an average hypothetical user.

Additionally, before and after the workshop we conducted a survey to evaluate participant perceptions. Currently we process the survey results, and in general we identified that there was a knowledge shift in both domain and technical scientists. Also, participant feedback on the process was very positive.

We identified that a main issue during our AITOLOS inception workshop was the large number of participants. This made discussions last longer than expected, but on the other hand, it was an important factor in producing seven user stories in a single day. Future plans include organizing a second inception workshop with rest of the project partners, and we intend to repeat the same process with half of the participants, reducing the total number of participants to ten (i.e 3-4 technical experts, 5-6 domain experts, 1 facilitator).

Last, but not least, the user stories we identified with the inception workshop are *epic*, in the sense that they can be broken down to smaller, more targeted stories. However, this was rather expected as the goal of the inception workshop was not to identify certain features to be implemented, rather it was to identify systems to be subsequently specified in detail.

Acknowledgments. Authors would like to thank the personnel of the Forest Service Offices of Goumenissa and Kilkis, and the technical experts from Centre for Research and Technology Hellas, and Technological Education Institute of Serres for their participation to the inception workshop.

This work have been partially supported by the IPA cross-border programme 'Greece-The former Yugoslav Republic of Macedonia 2007-2013' under grant AITOLOS, which is co-funded by the European Union and by national funds of the participating countries.

More about the project at `http://www.aitolos.eu`.

References

1. Nellemann, C.: INTERPOL Environmental Crime Programme. In: Carbon, G. (ed.) Black Trade: Illegal Logging, Tax Fraud and Laundering in the Worlds Tropical Forests. A Rapid Response Assessment. United Nations Environment Programme, GRID-Arendal (2012), `www.grida.no`
2. Ministry of Rural Development and Food: National report of Greece concerning the illegal logging and the trade of the illegally produced forest products. Illegal logging and trade of illegally-derived wood products in the UNECE region: Causes and extent, UNECE & FAO (2004), `http://www.unece.org/fileadmin/DAM/timber/docs/sem/2004-1/full_reports/Greece.pdf`
3. Papachristou, H.: Greek austerity fuels illegal logging, ministry says. News article, Reuters (2012), `http://www.reuters.com/article/2012/09/18/greece-logging-idUSL5E8KIKAY20120918` (Published online on) (September 18, 2012)
4. Lowen, M.: Greece crisis: Poor turn to illegal logging to survive. News Article, BBC (2013), `http://www.bbc.co.uk/news/world-europe-21227572` (Published online on January 28, 2013)
5. Eberlein, A., Leite, J.: Agile requirements definition: A view from requirements engineering. In: Proceedings of the International Workshop on Time-Constrained Requirements Engineering (TCRE 2002), pp. 4–8 (2002)
6. Cao, L., Ramesh, B.: Agile requirements engineering practices: An empirical study. Software, IEEE 25, 60–67 (2008)
7. Cobb, C.G.: Making Sense of Agile Project Management. John Wiley and Sons (2011)
8. Marijnissen, C., Ozinga, S., Richards, B., Risso, S.: Facing Reality: How to halt the import of illegal timber in the EU. Technical report, FERN - Greenpeace European Unit - WWF European Policy Office (2004)
9. Markus-Johansson, M., Mesquita, B., Nemeth, A., Dimovski, M., Monnier, C., Kiss-Parciu, P.: Illegal Logging in South Eastern Europe. REC Working Paper, Regional Environmental Center for Central and Eastern Europe (2010)
10. UN: The European Forest Sector Outlook Study II 2010-2030. Technical report, UNECE, FAO (2011)

11. Koulelis, P.P.: Greek timber industries and wood product markets over the last century: Development constraints and future directions. Annals of Forest Research 54, 229–240 (2011)
12. Cabral, E., Simas, R., Santos, V., Queiroga, C., Da Cunha, V., De Sa, G., Daroda, R., Eberlin, M.: Wood typification by Venturi easy ambient sonic spray ionization mass spectrometry: The case of the endangered Mahogany trees. Journal of Mass Spectrometry 47, 1–6 (2012)
13. Lowe, A., Wong, K.N., Tiong, Y.S., Iyerh, S., Chew, F.T.: A DNA method to verify the integrity of timber supply chains; Confirming the legal sourcing of merbau timber from logging concession to sawmill. Silvae Genetica 59, 263–268 (2010)
14. Nowakowska, J.: Application of DNA markers against illegal logging as a new tool for the Forest Guard Service. Folia Forestalia Polonica, Series A 53, 142–149 (2011)
15. Jolivet, C., Degen, B.: Use of DNA fingerprints to control the origin of sapelli timber (Entandrophragma cylindricum) at the forest concession level in Cameroon. Forensic Science International: Genetics 6, 487–493 (2012)
16. Degen, B., Ward, S., Lemes, M., Navarro, C., Cavers, S., Sebbenn, A.: Verifying the geographic origin of mahogany (Swietenia macrophylla King) with DNA-fingerprints. Forensic Science International: Genetics 7, 55–62 (2013)
17. Ganopoulos, I., Aravanopoulos, F., Madesis, P., Pasentsis, K., Bosmali, I., Ouzounis, C., Tsaftaris, A.: Taxonomic Identification of Mediterranean Pines and Their Hybrids Based on the High Resolution Melting (HRM) and trnL Approaches: From Cytoplasmic Inheritance to Timber Tracing. PLoS ONE 8 e60945 (2013)
18. Tnah, L., Lee, S., Ng, K., Faridah, Q.Z., Faridah-Hanum, I.: Forensic DNA profiling of tropical timber species in Peninsular Malaysia. Forest Ecology and Management 259, 1436–1446 (2010)
19. Häkli, J., Jaakkola, K., Pursula, P., Huusko, M., Nummila, K.: UHF RFID based tracking of logs in the forest industry. In: RFID 2010: International IEEE Conference on RFID, pp. 245–251 (2010)
20. Fuller, D.: Tropical forest monitoring and remote sensing: A new era of transparency in forest governance? Singapore Journal of Tropical Geography 27, 15–29 (2006)
21. Hermanson, J., Wiedenhoeft, A.: A brief review of machine vision in the context of automated wood identification systems. IAWA Journal 32, 233–250 (2011)

Using Smartphones to Profile Mobility Patterns in a Living Lab for the Transition to E-mobility

Francesca Cellina[1], Anna Förster[1], Davide Rivola[1], Luca Pampuri[1], Roman Rudel[1], and Andrea Emilio Rizzoli[2]

[1] SUPSI - University of Applied Sciences of Southern Switzerland
{francesca.cellina,anna.foerster,davide.rivola,luca.pampuri, roman.rudel}@supsi.ch

[2] IDSIA, USI-SUPSI, Switzerland
andrea@idsia.ch

Abstract. The diffusion of electric vehicles is currently the most promising opportunity to reduce the dependency on fossil fuel in the mobility sector. However, the adoption of the electric vehicles seems to be still hindered by psychological and behavioural barriers. Thus, in order to understand and to foster the transition towards more sustainable mobility styles, it becomes essential to adopt an inter-disciplinary approach. In this framework, the e-mobiliTI project was launched in late 2012 in Southern Switzerland. It aims at understanding the potential for transition in the mobility system at the local level, with a special focus on electric mobility. The project builds upon a small living lab made up of around twenty families, who will be monitored in all their trips through smart mobile devices, in order to get quantitative data, and through focus groups, in order to get qualitative data and perceptions.

Here we discuss the major challenge in the initial stage of the e-mobiliTI project, that is the gathering of reliable and high-quality data on users' behaviour. We describe the automatic tracking system, and the data processing and the qualitative assessment approach and comment on the overall performances of the living lab experiment.

Keywords: personal mobility tracking, smartphones, mobility living lab, mobility transition.

1 Introduction and Motivation

The diffusion of electric vehicles is currently the most promising opportunity to reduce the dependency on fossil fuels in the mobility sector and to pave the way for the transition to a more sustainable society.

In the last few years technology made remarkable progresses. In Switzerland, nowadays more than 80% of the demand of personal mobility activities and patterns could be covered by electric vehicles: the 2010 Swiss Transport and mobility census in fact indicates that 84% of the trips are shorter than 10 kilometers and that 73% of them are shorter than 5 kilometres [1]. Currently there are no monetary incentives at the national level for electric vehicles, since the Swiss

J. Hřebíček et al. (Eds.): ISESS 2013, IFIP AICT 413, pp. 154–163, 2013.

government believes their introduction should only be driven by the market [2]. However, in Canton Ticino, located in Southern Switzerland, the government is on the verge to approve an incentive scheme for electric and hybrid vehicles, allowing a reduction of the vehicle sale price up to 10'000 CHF.In addition, the network for public charging stations for electric vehicles is already densely distributed all over the urbanised areas. The charging infrastructure will be furthermore supported by a private initiative, named "EVite", which is planning to install up to 250 public fast-charge stations in Switzerland. Finally, the cost of electric vehicles has worldwide dramatically decreased, up to the point that in a few years it is expected to become equivalent to conventional fossil fuelled vehicles (see for example [3]).

Therefore, all the conditions seem to favour a transition to electric mobility, but this is going to be a large scale and long-term process. Transitions of this type are determined by a wide set of factors, acting together. In the case of electric vehicles, their adoption seems to be still hindered by psychological and behavioural barriers (e.g. issues such as the breadth of the autonomy range, the availability and diffusion of recharge points, the performance, see [4], [5], [6], [7]). Thus, in order to understand and, at a later stage, to foster the transition towards more sustainable mobility styles, it becomes essential to adopt an inter-disciplinary approach, able to integrate psychology, sociology, cultural and spatial planning aspects, alongside the technology and engineering aspects.

Among various factors, consumer behaviour is one of the most influential ones, at least as important as technology ([8]). Adopting a socio-technical perspective ([9], [10]) and an actor-based approach allows a deeper and more effective understanding of this fundamental phenomenon. The living lab research approach ([11]) offers promising perspectives, since it allows real life users to explore and experience new technologies in complex, real-world settings.

In the remainder of this paper, first we present the e-mobiliTI project context (Section 2), then we describe the smart-phone *app* we developed to track user behaviour and collect user feedback (Section 3). We then quickly describe the approach for profiling user behaviour and understanding the drivers for change (Sections 4), and finally we critically discuss the major challenges we will face during the project (Section 5).

2 The e-mobiliTI Living Lab Experiment

The e-mobiliTI project is funded and coordinated by the University of Applied Sciences of Southern Switzerland (SUPSI) and it is supported by regional authorities, with the final aim of elaborating policy guidelines, incentives, procedures and strategies for a transition in the mobility sector at the urban level.

2.1 Background and Approach

The e-mobiliTI project adopts an inter-disciplinary, bottom-up approach, aimed at exploring the potential effects and benefits of innovative mobility technologies and options through experimentation and direct interaction with the end-users.

Similar socio-technical experiments were held in the past, in the framework of the Strategic Niche Management (SNM) approach, which was developed in the late Nineties at the European level and was successfully applied in the transportation sector ([12], [13]). More recently, similar approaches started being indicated as *Living lab* processes, intended as inter-disciplinary and user-centric methodologies, that, involving real users in actual living environments, allow to *develop, deploy and test new technologies and strategies capable of responding to the complex and evolving needs of the present complex society* [1].

The SNM and Living lab methodologies are especially effective in systems offering environmental and social benefits which are undervalued on the market [13], in particular by single individuals, as in the transportation sector. Successful experiences in this sector were for example gathered in Denmark since 2009, with the TryAnEV/Insero E-mobility project [2]. For less recent experiences, one can for example refer to [14].

2.2 Research Objectives and Expected Outcome

Project activities are being developed in two main phases (see Figure 1): in Spring 2013 the families involved in the living lab are monitored in order to identify their *present mobility patterns and styles*, while in Spring 2014 they are monitored while testing and experiencing *new mobility options*, namely electric bicycles and cars as well as car-sharing (*Mobility*), bike-sharing (*Velopass*) and season tickets for public means of transportation (*Arcobaleno*).

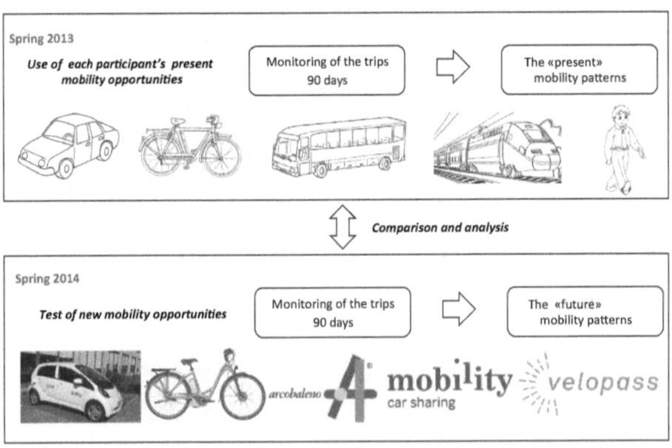

Fig. 1. The two phases of the e-mobiliTI experiment

Providing the participants with the new mobility options, e-mobiliTI removes the present main barrier to the diffusion of electric cars and bicycles, that is their

[1] MIT living lab website `http://livinglabs.mit.edu/`, retrieved on April 24th, 2013.
[2] Nordic Sustainable Logistics Network website `http://www.noslone.com/News/Details.aspx?id=70`, retrieved on May 13rd, 2013.

price, and creates a niche in which we can analyse the participants behaviour with respect to the other barriers, such as: range, recharge and performance for the electric cars; physical effort and safety for e-bikes; status-symbol and rigidity for vehicle sharing; rigidity and punctuality for public transport.

Comparing the mobility patterns and styles observed during the two monitoring phases allows us to test the following hypotheses:

1. barriers affecting the use of electric vehicles are mainly psychological and can be removed through experience - or, in other words: using an electric vehicle or bicycle does not require major sacrifices or reductions in the present single individuals well-being;
2. having an electric vehicle or bicycle at one's disposal acts as a leverage for a general transformation of the individual mobility patterns (slow mobility, public transport and use of shared mobility services), instead of merely acting as a pure substitute of conventional vehicles and bicycles.

In order to identify the mobility patterns and styles of the participants in the living lab, during the two phases we rely on both quantitative automatic data-gathering techniques and on group discussions and single interviews for the analysis and review of qualitative data. Automatic data-gathering is performed by means of a specifically developed smart-phone application *(app)* that relies on GPS tracks and on Bluetooth sensors placed on vehicles and bikes to identify transport modes [15]. The *app* posts GPS and Bluetooth data to a centrally hosted database, where data are post-processed in order to make them available to further processing by using data classification techniques aiming at the identification of mobility patterns and styles. Such users profiling also considers information gathered through qualitative discussion groups, mainly focus-groups: the latter suggests hypotheses on styles and patterns to be verified by means of the data mining classification. From the computer science point of view, thus, the e-mobiliTI project has the further objective of effectively integrating the quantitative information produced by the smart mobile devices to the qualitative information produced by the focus groups.

2.3 Selection of the Participants

For the setting of the experiment, we selected the Lugano region in the Southern Switzerland Ticino Canton, i.e. an urban conurbation of around 135'000 inhabitants. It is particularly interesting because, even though it is often congested and characterised by frequent traffic jams, it already offers alternatives to traditional private owned fossil fuelled vehicles, such as an efficient public transportation network and car-sharing and bike-sharing services.

Selecting the participating families is a key and delicate activity. We performed it by means of a public open call, promoted via the local mass-media (newspapers, television, radio) and via flyers distributed all over the city: therefore, participants come from a process of self-nomination of the interested people.

The public call was open to generic *families*, i.e. either *single individuals* or *couples* or *couples with (older) children*. More than 350 candidates answered the call: a much higher number than we initially expected, reflecting a very high interest also in the common people for new mobility technologies and options. The selection process aimed at identifying an homogeneous group of subjects based on the place of living, on the demographic features of the family (single, couple, couple with (older) children) and on the present use of mobility options. This process allowed us to identify a set of *sixteen families*, for a total number of *thirty persons* actively involved.

3 Smart Mobile Devices

In order to capture as precisely as possible and as many as possible trips from our users, we turned our attention to smartphone based gathering. This approach turned out to be a bless and a curse at the same time: a bless, because we are able to gather really a vast amount of data from different users for long periods of time; a curse, because the usage of the smartphone based application proved to be quite complex for many users. In this section, we detail the general architecture of our solution and present the main functionalities of the application.

3.1 General Architecture

The general architecture of our system, which we also call the *e-mobiliTI app*, is presented in Figure 2. It consists of the smartphone based application, which communicates to the central server to send the gathered data and to receive the daily reports. The application gathers GPS and Bluetooth data every 1-3 minutes, depending on how fast the user is moving (1 min when moving fast). If the smartphone is not moving at all, the GPS and Bluetooth sampling is interrupted to save the battery. This data is sent regularly to the central server, which stores the data. Once a day, at a user-defined time, the server evaluates all available data, automatically identifies moving segments of the user together with the used vehicle and sends these segments in the form of a daily report back to the smartphone. There, the user is notified and can review all segments and correct them, where necessary.

Additionally to the smartphone (currently running Android OS), each user's car is equipped with a Bluetooth tag. Any Bluetooth device can be used, but the system requires that the tag is always on. A good option is the stick'n'find Bluetooth tag[3] or the Blue Solar car kit from MrHandsFree[4], which are always on and are very energy efficient.

3.2 Automatic Recognition of Trip Segments

One of the most important functionalities of the system is its ability to recognise the movements of the users automatically and to break them down into segments. According to the commonly used terminology, we define:

[3] http://www.sticknfind.com

[4] http://www.mrhandsfree.com

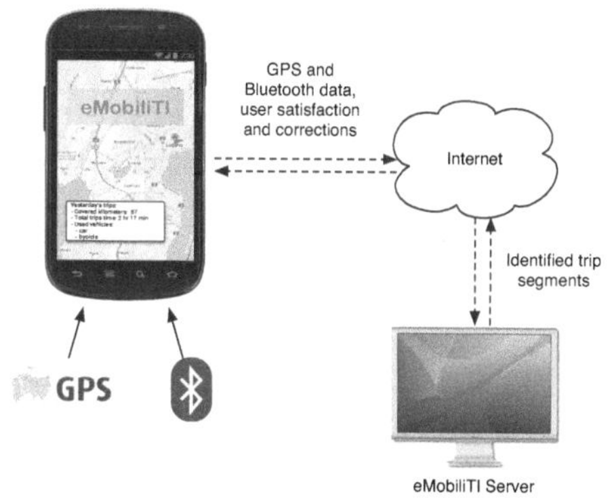

Fig. 2. The general architecture of our smartphone based solution. The e-mobiliTI app gathers GPS and Bluetooth data and sends them to the central server. The server analyses the data daily and sends to the user all identified trip segments. The user can thus make corrections to the properties of the trip and give her overall satisfaction level, which are again sent to the server for backup.

- **Trip:** a movement of the user starting at her home and ending also there. There are also other possible definitions of a trip, however this one is the most well defined and non-ambiguous one.
- **Segment:** part of a trip, which can be associated with a single vehicle used (a bus, a car, on foot, etc.), separated by stop points. A trip can consist of a single segment, but usually it consists of more than one segment.
- **Stop:** it is a humanly defined stop point on a trip, corresponding to some short or long interrupt to the trip. For example, stopping at the gas station, to chat with a friend, at the bakery etc. However, it also includes long stops, such as for whole day work in the office.

Thus, a trip is a sequence of alternating stops and segments, where both the start and the end places are at home. Our system is able to recognise all there levels of the trip: the stops, the segments and the trips themselves. The algorithm is based on the following steps:

- Recognize stop points by identifying GPS points very close to each other (at most 100 meters apart) for at least 2 minutes.
- Segment the data into stops (as identified above) and segments.
- Validate the segments with some minimum requirements for distance and velocity. Delete all failed segments by merging them together.
- Validate the stops between the segments again by deleting them first and analysing the newly formed segments. If the segment is a valid one (see above), the stop remains deleted. If the segment does not fulfil the requirements any more, put back the stop.

The vehicle is recognised by thresholds on the mean and maximum speeds in a segment. The Bluetooth data is used to recognise the own cars of the users and to differentiate them from public transport or other cars.

The developed algorithm is quite flexible and is able to identify short stops (such as changing the bus at a bus stop), while at the same removing artefacts such as stopping at the traffic light. Of course, not all artefacts are removed and there remain a certain number of segments, which are shorter than expected.

3.3 Graphical User Interface and Functionalities

The *app* graphical user interface for the review of the trips by the users is presented in Figure 3. The left-most picture (a) shows the daily survey with the general daily satisfaction level and note and all identified segments for that day. (b) presents the details of one segment, with start and end times, total distance, a map with start and end places and the recognised vehicle. Also, the start and end places can be associated with labels (e.g. home, office, school, bus stop, etc.) and the user can also enter her motivation for the trip, the number of people she is travelling with and the satisfaction for the segment only. Once the user completes the report, the updated data is sent back to the server for final storage. Of course, the user can change also old surveys and segments or complete the survey later than the predefined time.

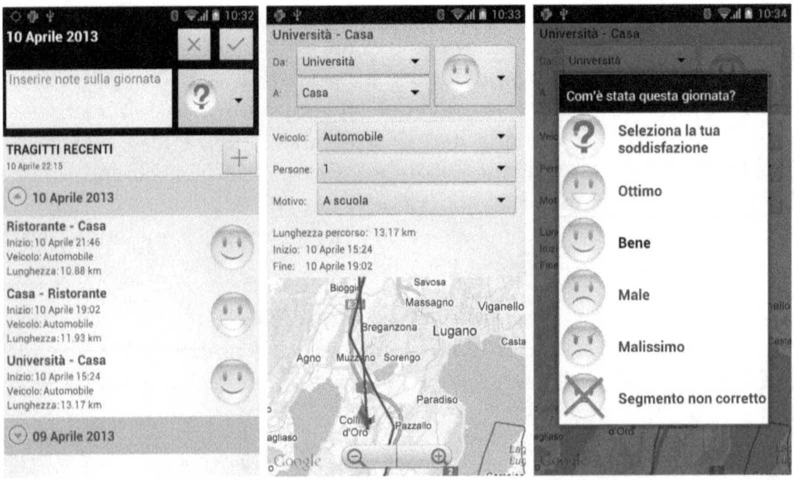

Fig. 3. The graphical user interface to review the movements of the day and to insert details and satisfaction level

3.4 Practical Notes

The implementation on the smartphone is straightforward, but requires a lot of additional functionalities to make it user-friendly at one side and robust on the other side. The main challenge comes from the very different technological skills of the users and their attitude towards the experiment.

In the implementation we also considered the following requirements: when powering on the phone the app had to be automatically started; GPS and Bluetooth had to be also turned on at start-up and kept active; the notification of the alarm signal for the daily report had to remain loud even when the smartphone was silenced; failed data transmission to the servers because of lack of signal had to be queued up for later re-send; all entered data needed to be saved without additional user confirmation; the phone settings were corrected not to allow data roaming, but to always allow 3G data connection (for cost efficiency), and not to use any energy saving options (as they are faulty on many Android phones).

4 User Profiling and Evaluation

The collected data on the various mobility choices of all participants in the experiment will then be processed in order to discover and understand how they make their daily personal mobility choices. Given the limited size of the sample, this part of our study is aimed at understanding how a specific person chooses a specific mobility option as a function of various possible influencing factors, such as distance, tightness of schedule, reason for the trip, number of intermediate stops, the weather conditions and so on. For each user, her daily trips will be analysed, looking for patterns, such as the choice of a given mean of transport, or a combination of means, to a specific purpose. We will therefore build a mobility profile for each user, and we will then also look for similarities among all users, on the basis of their typology.

During the second phase of the project, the same users will be monitored, but this time their travel options will be wider, as they will have the choice of opting for public transport, shared mobility and electric mobility options. The data collected during this part of the experiment will also be used to create the users' profiles. The new profiles will be then compared with the old ones, i.e. those obtained before the introduction of new mobility options, and the objective of the comparison will be the identification of the most relevant factors influencing a transport choice change, provided a given degree of profile similarity among the users.

5 Discussion and Conclusions

The *e-mobiliTI* project relies on the strong involvement of the users, the key components of the living lab, in order to evaluate how a transition to sustainable mobility styles could happen. Such a project is therefore fraught with problems which are common to this type of projects, namely: users' motivation to actively participate in the project can dwindle; automated data collection covers only quantitative data, while qualitative data collection presents some challenges; finally, the size of the sample is small, and thus the general validity of the output of the project, the policy guidelines, can be criticised.

The users's attitude during the project is probably the most important challenge, since it cannot be tackled from a purely technical point of view. Literature on the living lab approach in fact implicitly assumes that the users are cheap or unpaid contributors, motivated by the hope that their participation will solve their problems or lead to better policy-making [16]. Will our living labs achieve and sustain the necessary level of user mobilisation and cooperation in the absence of stronger incentives [17]? Will we be able enough in motivating the participants as active co-creators of the transition towards a new mobility system? This challenge is particularly relevant during the first e-mobiliTI monitoring phase, when they are asked to use their own mobility options, thus being little interested in getting reports of their trips. We expect that probably, when in the second monitoring phase they will be allowed to try new mobility options, their curiosity respect to the daily survey will increase.

Provided that users keep an high involvement in the project and keep the e-mobiliTI app active at all times and regularly collect data, certain types of information, such as the reason for the trip, the number of people travelling together and the overall satisfaction for the trip, can only come from personal information enrichments, voluntarily entered by the participants to the e-mobiliTI living lab. For this reason, the *app* provides the users with a daily report of their segments, which, besides offering them the possibility to correct erroneous trips, allows them to insert the cited qualitative information. However, this is not enough to grasp the full reasons for their behaviour. That is why we believe that physical, personal contact during focus groups and single interviews is essential to fully understand their mobility patterns and styles.

Finally, different users may face different "real-world" situations, or even interpret them in different ways, according to their personal perception and value systems [18]. Thus, in order to get enough different perspectives, it is important to involve a large enough number of users, which, in some cases, might be burdensome and costly. In our experiment, due to limits in the available budget, the number of participating families is very small: therefore, no general representativeness can be attributed to the mobility patterns and lifestyle observed within the e-mobiliTI living lab. The project in fact cannot provide results capable of representing the whole Lugano region inhabitants behaviour and perceptions. While we acknowledge this limitation, on the other hand the information gathered during the project lifetime can provide a starting point for the niche characterisation and suggest the initial actions to be taken to favour the transition towards different mobility lifestyles.

References

1. Office fédéral de la statistique (OFS), Office fédéral du développement territorial (ARE): La mobilité Suisse - Rèsultats du Microrecensement Mobilité et transports 2010. Technical report, Neuchàtel et Berne (2012)
2. IA-HEV: Hybrid and electric vehicles annual report. Technical report, IEA (2012)
3. World Bank (PRTM Management Consultants): The China New Energy Vehicles Program: challenges and opportunities. Technical report, Bejing (2011)
4. Everett, A., Burgess, M., Harris, M., Mansbridge, S., Lewis, E., Walsh, C., Carrol, S.: Initial Findings from the Ultra Low Carbon Vehicle Demonstrator Programme: How quickly did users adapt? - A review of preliminary usage and perception data from the first three months of vehicle trials. Technical report (2011)

5. Turrentine, T., Garas, A., Lentz, J., Woodjack, J.: The UC Davis MINI E Consumer Study. Technical Report Research Report UCD-ITS-RR-11-05, 2011, Institute of Transportation Studies, University of California, Davis (2011)
6. Deloitte: Gaining traction - A customer view of electric vehicle mass adoption in the U.S. automotive market. Technical report (2010)
7. Cocron, P., Bühler, F., Neumann, I., Franke, T., Krems, J., Schwalm, M., Keinath, A.: Methods of evaluating electric vehicles from a user's perspective - the MINI E field trial in berlin. IET Intelligent Transport Systems 5(2), 127–133 (2011)
8. Geels, F., Kemp, R., Dudley, G., Lyons, G.: Automobility in transition? A socio-technical analysis of sustainable transport. Routledge (2011)
9. Schot, J.: Towards new forms of participatory technology development. Technology Analysis & Strategic Management 13(1), 39–52 (2001)
10. Geels, F., Schot, J.: Typology of sociotechnical transition pathways. Research Policy 36(3), 399–417 (2007)
11. Higgins, A., Klein, S.: Introduction to the living lab approach. In: Tan, Y.H., et al. (eds.) Accelerating Global Supply Chains with IT-Innovation, Springer, Heidelberg (2011)
12. Kemp, R., Schot, J., Hoogma, R.: Regime shifts through processes of niche formation. the approach of strategic niche management. Technology Analysis & Strategic Management 10(2), 175–195 (1998)
13. Kemp, R., Truffer, B., Harms, S.: Strategic niche management for sustainable mobility. In: Rennings, K., Hohmeyer, O., Ottinger, R. (eds.) Social costs and sustainable mobility: strategies and experiences in Europe and the United States, pp. 167–188. Physica-Verlag, Heidelberg (2000)
14. Hoogma, R., Kemp, R., Schot, J., Truffer, B.: Experimenting for sustainable transport: the approach of strategic niche management. Spon Press, London (2002)
15. Förster, A., Cabrini, M., Garg, K., Giordano, S.: Understanding and optimizing human mobility with smart phones. In: Poster at the International Conference on ICT for Sustainability, Zurich, Switzerland (2013)
16. Mulder, I., Stappers, P.J.: Co-creating in practice: results and challenges. In: Collaborative Innovation: Emerging Technologies, Environments and Communities (Proceedings of the 15th International Conference on Concurrent Enterprising, ICE 2009, Leiden, The Netherlands, June 22–24, Centre for Concurrent Enterprise, Nottingham (2009)
17. Dutilleul, B., Birrer, F.A., Mensink, W.: Unpacking european living labs: analysing innovation's social dimensions. Central European Journal of Public Policy 4(1), 60–85 (2010)
18. Bergvall-Kåreborn, B., Ihlström Eriksson, C., Ståhlbröst, A., Svensson, J.: A milieu for innovation–defining living labs. In: 2nd ISPIM Innovation Symposium, New York (2009)

Robust and Trusted Crowd-Sourcing and Crowd-Tasking in the Future Internet

Denis Havlik, Maria Egly, Hermann Huber, Peter Kutschera, Markus Falgenhauer, and Markus Cizek

AIT Austrian Institute of Technology GmbH, Austria
{Denis.Havlik,Maria.Egly,Hermann.Huber,Peter.Kutschera, Markus.Falgenhauer,Markus.Cizek}@ait.ac.at

Abstract. A great majority of the EU citizens already owns a cellular phone. An increasing part of these phones are smartphones with a broadband internet connection. This growing network of smart internet enabled devices could act as a dense sensing network, as well as a tool for individual informing and tasking of mobile citizens and volunteers.

In order to fully harvest this new resource, we need to understand its rules and develop adequate tools. This paper lists some of the peculiarities of ad-hoc volunteer networks supported by smartphones. It furthermore compares the capabilities and limitations of these networks with other means of observation gathering and activity coordination. This includes: (1) a reflection on the motivation for users participation; (2) human and technical limitations of smartphone-enabled volunteer networks; (3) legal and ethical challenges; (4) reliability and usability issues; as well as (5) issues related to trust and quality of information.

The second part of the paper presents our experiences with design and prototypic development of the tools supporting volunteer efforts in the field of environmental monitoring, e-health and crisis management. This development is presented in a wider scope of the "Future Internet Public Private Partnership" research programme.

Finally, the paper summarizes our findings and recommendations for further developments.

Keywords: crowdsourcing, crowdtasking, human sensors, local situation awareness, volunteered geographic information, future internet, mobile applications.

1 Motivation

According to a recent comScore whitepaper [1], the smartphone penetration has surpassed the 50% mark in US, Canada and the EU5 (UK, France, Germany, Spain and Italy) during 2012. Moreover, the mobile media already accounts for 37% of the total time spent online and for a majority of usage in categories well suited for use on smartphones and tablet PCs: maps, weather, music and social networking. These devices combine (almost) "always on" broadband connectivity, ad-hoc local

J. Hřebíček et al. (Eds.): ISESS 2013, IFIP AICT 413, pp. 164–176, 2013.

connectivity, significant storage capacity and computing power with intuitive user interfaces and a number of built-in as well as add-on sensors for various purposes.

Numerous researchers have been testing the usability of smartphones as platforms for mobile sensing, in the recent past [2-3]. From the point of view of user interaction, this resulted in two distinct application classes: "opportunistic sensing" [4] applications, where the user merely allows an application to report (interpreted) readings of selected sensors, and the participatory sensing applications [5], where users are an integral part of the sensing and quality assurance loop ("human sensors", "crowd-sourced quality assurance"), and actively provide information on their environment.

In this paper, we shall consider crowdsourcing of observations, data interpretation, dissemination and (crowd-) tasking as integral part of the mobile Volunteered Geographic Information (VGI) applications and present a prototype of the Mobile Data Acquisition Framework (MDAF) which simplifies the task of developing such applications. The word "observation" is used in-line with the definition of the Open Geospatial Consortium, to indicate geospatially and temporary referenced data reported by hardware sensors, human sensors or automated processes (nowcasts, forecasts, virtual sensors, and indicators).

The paper starts with a discussion of the challenges and opportunities of mobile VGI applications (section 2). Section 3 presents the MDAF framework and section 4 discusses MDAF's ability to answer these challenges. Finally, section 5 "conclusions" presents the main lessons learned in MDAF development.

2 Challenges and Opportunities of Mobile VGI Applications

This section discusses some "peculiarities" of VGI applications as compared to a classical monitoring system where all sensors and networking equipment are maintained and calibrated by a central authority.

For the start, let us consider some challenges directly resulting from the technology used in VGI applications: inaccurate sensors, unreliable networks, sub-optimal positioning of the smartphones for taking measurements and the need to recharge the battery on (at least) a daily basis. On the whole, smartphones aren't optimal sensing systems, and in reality there are only three reasons for (mis)using them in this way: (1) the phones are already out there and we don't have to pay for them; (2) they are at the appropriate location to provide information on phone users and their surroundings; and (3) the phone users can provide us with information we could not easily get from hardware sensors.

In opportunistic sensing application, the users merely need to be sufficiently motivated to install an application and let it use the smartphone resources. On the other hand, the participative applications rely on users as active providers of observations or "human sensors"[1]. In some areas, human sensing surpasses the capability of hardware sensors and automated algorithms for sensor data interpretation: recognising and interpreting combinations of visual, audible, and to some extent also olfactory pat-

---

[1] In the simplest case, users may only be asked to use their own judgement to decide when to start recording and assure the optimal positioning of the smartphone.

terns and anomalies is easy for humans and very difficult for hardware sensing systems. On the other hand, humans are notoriously bad at some tasks which can be easily performed by hardware sensing systems:

- Humans get easily tired and bored, and are notoriously bad at performing repetitive activities.
- Reported data will often depend on the individual sensitivity (e.g. some people are sensitive to weather changes or pollen, others aren't), cultural background (is a playground full of children noisy or music to ones ears?) and even on the current mood of the reporter.
- Finally, the humans are not good at quantification of observations (how loud is "very loud" in decibel? Which temperature corresponds to "very cold"?)

Fully fledged analysis of the participants' motives and the ways to enhance the motivation, as presented in e.g. [6], is beyond the scope of this paper. Nevertheless, it is important to keep in mind that the *users' motivation* is crucial for success of all VGI applications. Some key concepts influencing the users' motivation for participation in voluntary activities are summarized below:

- Participants can have very different reasons for being both constructive (providing observations, helping with quality assurance) and destructive (providing misinformation, misusing the system for spamming or for the personal gain).
- Direct or indirect personal gain, including both "monetary gain" and the possibility to improve own knowledge (intellectual gain) are very strong motivations –for constructive and for the destructive contributions.
- Another very strong motivation is the "fun factor" that originates from using the application.
- Knowledge that own contributions will be acknowledged, appreciated, seen by the peers and used for the well-defined and "good" purpose also greatly enhances the user motivation.
- The ease of use of the system and active solicitation of the information are crucial for attracting the participants with relatively low motivation levels.
- Concerns about misuse of users' observations and privacy-related information may lower the willingness of the citizens to participate.

As mentioned above, the participants can be motivated for both constructive and destructive contributions. In addition, the "quality" of the human sensors and of the inexpensive sensors built-in smartphones is a-priory unknown, may vary over time, and in-lab calibration is often unpractical.

Consequently, the quality of the information in VGI applications is usually estimated based on the combination of statistic methods and human judgment. In many cases, the application also takes users "reputation" into account – a numeric value indicating how likely it is that a certain user will contribute high quality observations. Some methods for calculating the users' reputation are discussed in [7].

3 MDAF Framework

3.1 Vision and Functional Architecture

Mobile Data Acquisition Framework (MDAF) is a set of mobile and backend software components designed to simplify the task of implementing mobile VGI applications – independent on the contents and scope of these data. Our long-term goal is to develop an application framework which allows loosely organized groups to conduct measurement campaigns and to analyse and discuss the results online. For the reasons explained in section 2, the framework must also: (1) support automated and manual tasking of the volunteers; (2) provide mechanisms for interpreting the meaning and quality of the observations; and (3) assure that the privacy and security of the users and the data transferred and stored in the framework is handled carefully according to legal constraints.

The use of MDAF from the mobile user perspective is illustrated on Fig. 1 below.

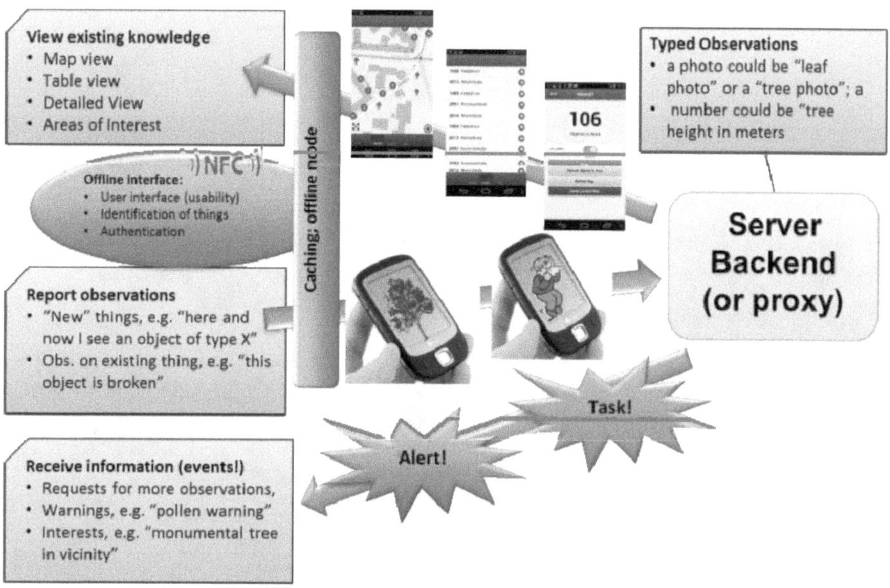

Fig. 1. MDAF mobile use cases

Users can: (1) view existing knowledge; (2) report new objects of interest[2] (OOI) and new observations on existing OOIs; and (3) receive alerts and event notifications on changes and requests for performing certain tasks pertinent to their areas on interest (AOI)[3]. The "offline mode" and "offline interface" are necessary to improve the overall reliability and usability of the system:

2 OOI is a specialization of "thing" in internet of things and "feature" in OGC context.

3 AOI is, as a word suggests, an area that the user is interested in.

- "Offline mode" is a consequence of the requirement that the application remains fully functional in slow and unreliable networks. As a welcome side-effect, the pre-loading and caching of the information pertinent to user-defined AOIs lowers the requirements on network bandwidth and latency during field trials.
- "Offline interface" refers to the requirement that users can provide input by means other than the smartphone GUI interface, e.g. using some of the built-in sensors to either directly provide the observation, or to record the user's input.

From the perspective of a service using the data collected and stored within the framework, MDAF must provide a way to: (1) combine the observations received from mobile users with data from external sources; (2) quality assure and interpret the resulting data heap; and (3) assure that the results can be queried and accessed by any service that needs them. For the sake of efficiency, the framework must both generate events, which can be used to trigger external processing services for each incoming observation and provide a service interface for accessing the observation ("pull") at a later time. Main MDAF use cases from the backend perspective are illustrated in the Fig. 2 below.

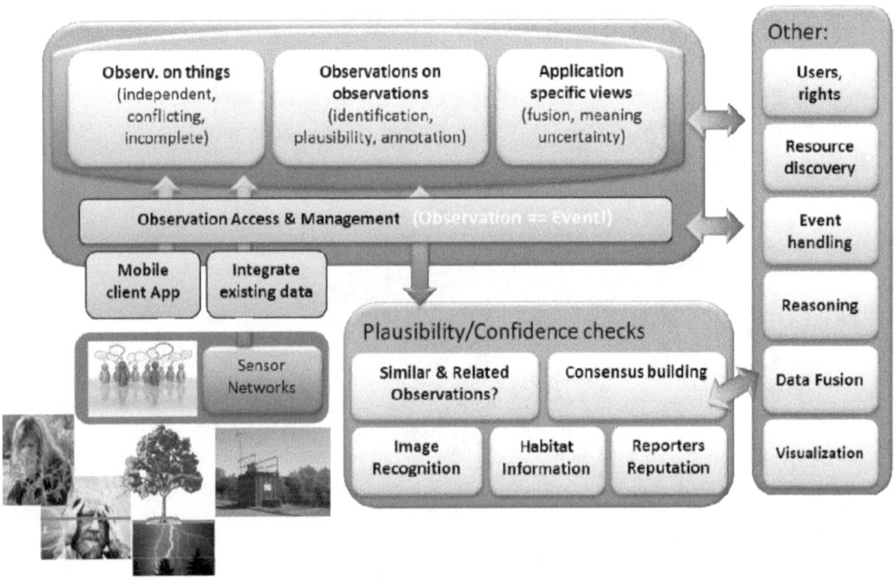

Fig. 2. MDAF backend use cases

This figure also illustrates the way MDAF organizes information and interacts with other services:

- Most of the data is considered "observations" – geospatially and temporary annotated information on some real-world or virtual "objects of interest". The observations can originate from our own application, from third party observation networks or from some processing service.

- "Objects of interest" themselves can be either simple data objects representing the real world objects[4] or other observations. The observations which refer to existing observations are called "observations on observations". They provide additional information on quality and meaning of the initial observations. This information can be provided by users as well as by automated processes.
- The framework does not prevent the users and automated processes from publishing incomplete or conflicting observations and a-priory does not make any judgments on the quality of this information. The task of interpreting the resulting observation heap and producing a useful view on this data is considered application specific, delegated to external processes and the results are stored in "application specific views".

The importance of this data model for VGI applications can be easily illustrated bythe example of "consensus building" (Fig. 3). If several users are asked to report observations on a single OOI, the most likely result is a non-consistent heap of observations. The fact that all these observations can peacefully co-exist within MDAF allows us to decide which of the observations are most plausible at application level, possibly taking into account some additional information, such as users' reputations, observations reported by automated processes, statistical analysis of the nearby reports or the probability of reported observations in a local habitat.

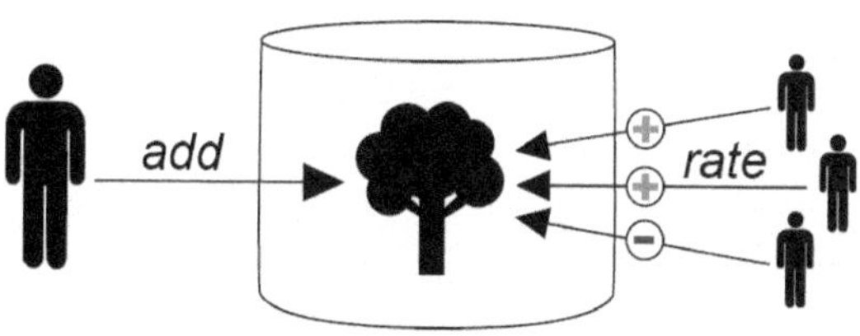

Fig. 3. Consensus Building

3.2 Implementation Architecture

Our initial idea was to follow the ideas of the SANY SensorSA architecture and reuse the existing OGC service interfaces, most notably the Sensor Observation Service (SOS) for serving of the observations [8, 9]. However, the initial tests with smartphones indicated that the combination of (Geo)JSON and RESTful service interfaces is more appropriate for mobile applications than the XML/OGC-style web service combination used by SOS. In addition, we needed a service that would run both on a backend and on mobile phone and synchronize the data to the users' device as needed.

CouchDB [10] offered exactly the type of functionality we needed: (1) a geospatially-enabled database which can be deployed both on a smartphone and on a backend; (2) RESTful http interface and a native support for (Geo)JSON; (3) Robust multi-master "eventually consistent" synchronization protocol allowing two-way partial synchronization of the data that fulfils the query criteria; and (4) Built-in event queue which can be used to trigger the sending of appropriate messages to users and other services.

---

[4] In simplest case, the OOI will contain a unique ID and a reference to object type.

MDAF implementation architecture is illustrated in Fig. 4.

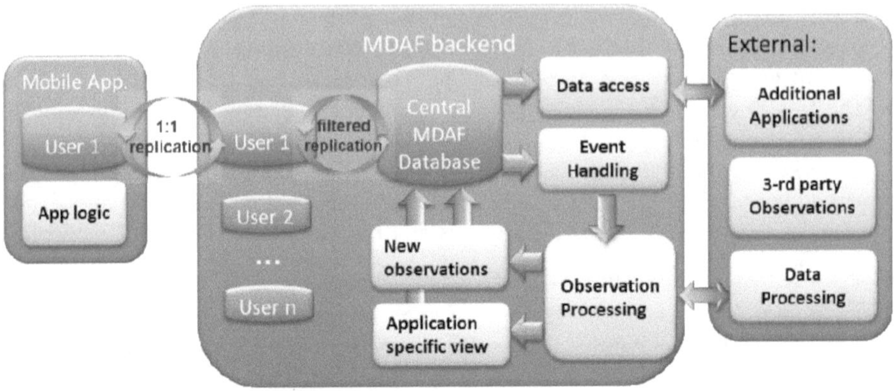

Fig. 4. MDAF implementation architecture

- The MDAF backend maintains a 1:1 replica of the user database on each mobile device. Every database is connected to the central MDAF database via filtered replication. This architecture assures the eventual consistency between all databases.
- CouchDB generates internal events for each database change. MDAF event-hander monitors these events and triggers appropriate observation processing modules.
- MDAF observation processing modules are self-contained pieces of software which implement most of the MDAF server-side application logic. They usually receive new observations from the event handler, generate additional observations based on processing results and store them on the central database from where they can be distributed to mobile users. In many cases the modules act as connectors and the actual processing is delegated to external services.

Our main concern on the mobile application side was to assure portability. In order to achieve this goal, we opted for a hybrid development model where most of the application logic is realized using standardized platform independent technology. Platform-specific libraries are only used to access the underlying hardware.

As illustrated in Fig. 5, the hardware access is handled by PhoneGap/Apache Cordova. The application functionality build upon Cordova is offered by the Sencha Touch mobile application framework. An evaluation of the usability of the Cordova/Sencha Touch combination for cross-platform mobile application development can be found e.g. in [11].

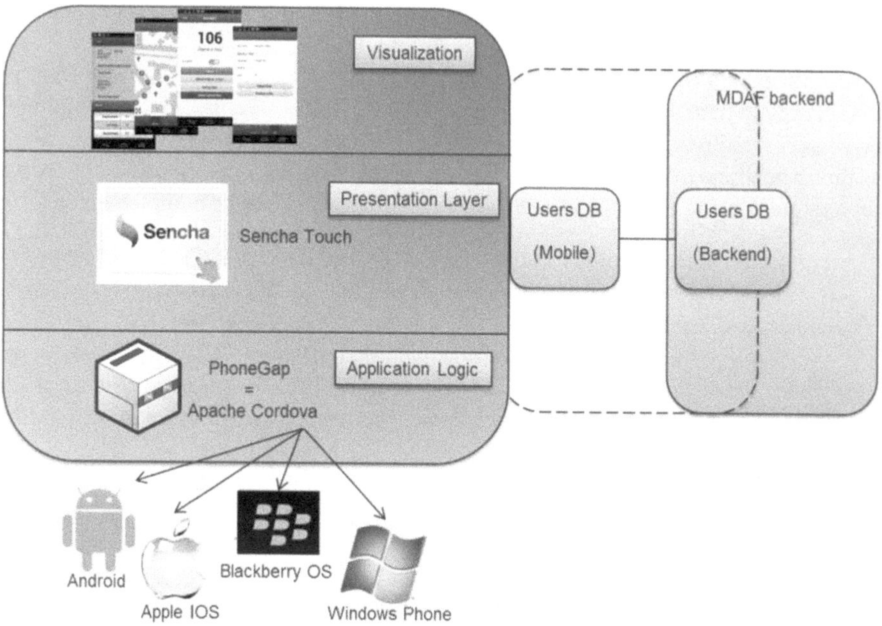

Fig. 5. Technologies used in MDAF mobile app development

3.3 Integration in ENVIROFI and FI-Ware Architecture

The power of the MDAF framework originates from its ability to reuse the data and functionality offered by the third party data access and processing services. Within the scope of ENVIROFI, our primary goal was to assure that the MDAF framework can be used in combination with Generic Enablers (GEs) developed within FI-Ware project, as well as in combination with the environmental Specific Enablers (SEs) developed within ENVIROFI [12]. In fact, the MDAF framework is an integral part of the ENVIROFI architecture and implements three SEs:

- MDAF backend service is the reference implementation of the Environmental Georeferenced Observation Collection Service.
- The caching database deployed on user devices is the reference implementation of the Environmental Georeferenced Observation Proxy Service.
- Two mobile applications (see section 4.1) are considered implementations of the Environmental-Georeferenced Observation App.

The interaction of other enablers and other third party applications and services is achieved through implementation of the appropriate data access interfaces and observation processing/connector modules (see Fig. 4). The simplest type of modules acts as clients to third party data access services and can be used to import the data from external sources into MDAF. So far, the data can be imported from: OGC Web

Feature Service [13], OGC SOS [14], Shapefiles and proprietary RESTful interface of the NILU air quality service.

More complex modules provide read-only or read-write access to MDAF observations. Primary data access to MDAF observation is the RESTful CouchDB interface. This interface allows retrieval of observations corresponding to specific geo-locations (within Bounding Box), observation IDs, OOI IDs, filtering on various object properties and publication of new observations. In addition, we have also implemented:

- A connector to TS-Toolbox and SUDPLAN's implementation of the Sensor Observation Service [15]. This allows us to provide access to MDAF observations over standardized SOS interface.
- A connector to ENVIROFI Environmental Image Sample Classification Service SE [16]. This allows us to automatically submit all observations of the "leaf image" type to this SE as well as to integrate the results (list of species with similar leafs) as additional observations.
- A connector to Google Cloud Messaging [17] provides an alternative channel (in addition to CouchDB replication mechanism) for pushing alerts and task requests to MDAF users.
- Finally, a two-way connector for exchanging events (observations) between MDAF and the Publish/Subscribe Context Broker GE [18] facilitates integration of MDAF in the Future Internet applications.

4 Development Status and Outlook

The development of MDAF started in Q4 2012, with the intention to support the development of the "conceptual prototypes" of the future internet enabled environmental applications. In the course of the project, the complexity and maturity of the MDAF framework grew way beyond a simple proof of concept. Nevertheless, further effort is required to reach the maturity level which would allow us to use this framework in real world applications.

4.1 ENVIROFI Pilots

Within ENVIROFI, MDAF has been used in two sample applications concentrating on participatory sensing and on the human/environment interaction:

- The Personal Environmental Information System (PEIS) application allows users to track information on airborne pollutants, pollen as well as meteorology and correlates these with personal well-being information. Additionally, it enables the user to share current local environmental conditions as described in [19, 20]. The main user interface elements and use cases of the PEIS application are illustrated in Fig. 6.

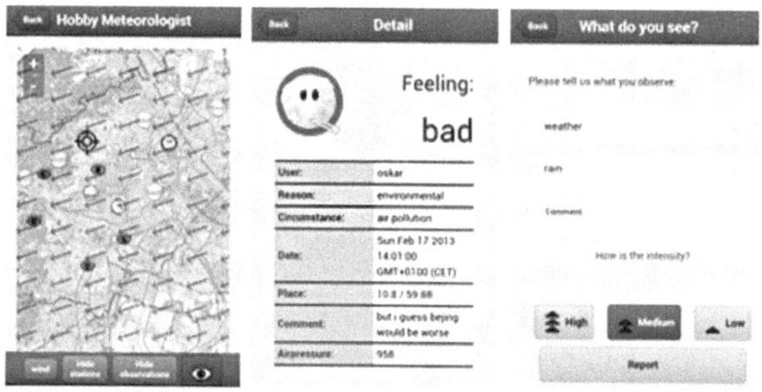

Fig. 6. PEIS enables users to relate personal well-being to environmental observations

- The "biodiversity" application allows users to receive and publish information on biodiversity in area(s) they are interested in [21]. The current prototype has been designed and developed in cooperation with the Long Term Ecological Research (LTER) Austria [22] and specializes in the objects of the type "tree".

Both applications can be downloaded from ENVIROFI catalogue and tested on android smartphones. PEIS app is developed only to a proof of concept level, whereas the biodiversity prototype is almost completed in the terms of features.

Comparison of the MDAF-related features of the pilots is shown in the Table 1.

Table 1. MDAF-supported functionality in ENVIROFI pilots

	Biodiversity pilot	**PEIS pilot**
Observations from external sources	Tree databases of Vienna, Tuscany and LTER Eisenwurzen (WFS, shapefiles)	Air quality in Vienna, Stockholm and Karlsruhe (SOS, NILU RESTful)
Observations from mobile users	Various observations on trees, in-line with requirements of the LTER-Austria.	Barometric sensor reading and own reactions on environmental conditions
Observations generated by SEs	Leaf identification	Air quality coverage
Service interfaces	CouchDB RESTful	CouchDB, OGC SOS
Offline GUI	Assign NFC tag to OOI; identify OOI by NFC tag	-
Events	CouchDB, FI-Ware, Google cloud messaging	CouchDB
Filtering	By area, numeric parameters, text	Threshold
Key functionality implemented by FI-Ware GEs	Authentication; image store; event handling; web GUI interface	-

	Biodiversity pilot	PEIS pilot
Areas of Interest	User-defined AOIs	predefined regions: Vienna, Stockholm, Karlsruhe
Alerting	New observations in users AOIs. Can be sorted by time and by distance from the user.	Threshold for various air quality parameters
Tasking	Request additional observations (partially implemented)	-
Crowdsourced quality assurance	"consensus" (partially implemented)	-
Automated quality assurance	Leaf identification; common name/Latin name; probability within habitat (partially implemented)	-
Offline mode	Background maps, tree types and status within user-defined areas of interest	-

4.2 Beyond ENVIROFI

So far, the MDAF framework has been developed and tested primarily in the context of environmental observations and the Future Internet Public-Private Partnership initiative. Due to combination of time constraints, resource constrains and low maturity of the FI-Ware platform, we were able to integrate and test only a small subset of the FI-Ware GEs. In spite of the weak support for geospatial data and processing, which has been identified as main shortcoming of FI-Ware by the ENVIROFI team [23], our analysis indicated several GEs with "high potential" for future VGI applications.

Likewise, the ENVIROFI experience has thought us that the use of external data fusion and processing services can greatly improve the overall quality and usability of VGI applications. However, most of the algorithms used by these services are highly application specific and therefore not likely to ever become part of the core MDAF offering.

With ENVIROFI project approaching its finalization, our mid-term goal is to use MDAF to facilitate active citizen participation in applications combining environmental awareness with topics such as human health/well-being, traffic, energy or crisis management.

Our development priorities are therefore to: (1) develop robust and generic reputation, crowdtasking and consensus building modules; (2) stimulate the volunteer's motivation by various incentives (see [6]); (3) improve the overall quality, performance and scalability of the framework; (4) consolidate and standardize the MDAF data model and service interfaces; (5) improve the overall security of the application and assure the delicate information (e.g. information related to users health or sightings of rare species) is adequately protected per design; (6) further develop the concept of "offline use" in case of network failure towards fully autonomous pear to pear exchange of information over ad-hoc networks; (7) simplify the task of developing new application to the level where this can be done by users with no software development experience; and (8) test the promise of OS-independent development by actually rolling out the application on non-Android devices.

5 Conclusions

The MDAF framework demonstrates the feasibility of developing robust and trusted crowd-sourcing and crowd-tasking applications in the context of Future Internet. Some of the concepts and technologies tested in MDAF and validated in the ENVIROFI applications are novel and have the potential for greatly enhancing the functionality offered by future mobile VGI applications. For example:

- The use of document-oriented CouchDB database with native http RESTful interface allows greater flexibility at the level of data models.
- The hybrid development model where most of the application is written in HTML5/Javascript lowers the cost of cross-platform development.
- The CouchDB replication mechanism and the possibility to run the same database on server backend and on the mobile device greatly improve usability of the application on slow and unreliable networks, offsets workload from the central server and minimizes the requirements on bandwidth and latency during field work.
- User-defined Areas of Interest completely eliminate the need for disclosing the current users' position to the central server, while still assuring the messages relevant for the user (e.g. alerts pertinent to his or her AOIs).
- The data model explained in section 3.1 allows MDAF-based applications to provide a consistent and easy to understand views on observations, even if the underlying dataset contains incomplete and conflicting observations.

Based on the experiences with development of the biodiversity and PEIS applications, we are confident that MDAF provides excellent answers to most of the challenges mentioned in section 2 of this paper and will develop into a fully-fledged framework for VGI applications. However, it is important to understand that the MDAF development is not yet finalized and that some important features (e.g. fully-fledged trust and reliability management, incentives for participating volunteers and secured data access) will have to be developed in the follow-up projects.

Acknowledgments. The research leading to these results has received funding from the European Community's Seventh Framework Programme (FP7/2007-2013) under grant agreement n◦ 284898 (ENVIROFI).

References

1. Lipsman, A.C.A.: Mobile Future in Focus (2013), `http://www.comscore.com/Insights/Presentations_and_Whitepapers/2013/2013_Mobile_Future_in_Focus`
2. Lane, N.D., Miluzzo, E., Lu, H., Peebles, D., Choudhury, T., Campbell, A.T.: A Survey of Mobile Phone Sensing. In: IEEE Communications Magazine, vol. 09, pp. 140–150. IEEE Computer Society Press (2010)
3. Lane, N.D.: Community-Aware Smartphone Sensing Systems. In: IEEE Internet Computing, vol. 16(3), pp. 60–64. IEEE Computer Society (2012)
4. Campbell, A.T., Eisenman, S.B., Lane, N.D., Miluzzo, E., Peterson, R.A.: People-centric urban sensing. In: Proceedings of the 2nd Annual International Workshop on Wireless Internet (WINCON). ACM (2006)

5. Burke, J., Estrin, D., Hansen, M., Ramanathan, N., Reddy, S., Srivastava, M.B.: Participatory Sensing. In: Workshop on World-Sensor-Web (WSW 2006): Mobile Device Centric Sensor Networks, ApplicationsWoodrow. ACM (2006)
6. Coleman, D.J., Georgiadou, Y., Labonte, J.: Volunteered Geographic Information: The Nature and Motivation of Produsers. International Journal of Spatial Data Infrastructures Research 4, 332–358 (2009), Joint Research Centre of the European Commission
7. Bishr, M.: Trust & Reputation Models for Human Sensor, Ph. D. Thesis, University of Münster, Germany (2011)
8. Havlik, D., Schade, S., Sabeur, Z.A., Mazzetti, P., Watson, K., Berre, A.J., Mon, J.L.: From Sensor to Observation Web with Environmental Enablers in the Future Internet. Sensors 11, 3874–3907 (2011)
9. Klopfer, M., Simonis, I. (eds.): SANY - An Open Service Architecture for Sensor Networks. SANY Consortium (2009)
10. Anderson, J.C., Lehnardt, J., Slater, N.: CouchDB: The Definitive Guide. O'Reilly Media, Inc. (2010)
11. Heitkötter, H., Hanschke, S., Majchrzak, T.A.: Evaluating Cross-Platform Development Approaches for Mobile Applications. In: Cordeiro, J., Krempels, K.-H. (eds.) WEBIST 2012. LNBIP, vol. 140, pp. 120–138. Springer, Heidelberg (2013)
12. ENVIROFI Consortium: Catalogue of Specific environmental enablers and prototype applications (2013), `http://catalogue.envirofi.eu/enablers`
13. OGC Consortium: Web Feature Service (2010), `http://www.opengeospatial.org/standards/wfs`
14. OGC Consortium, Sensor Observation Service (2007), `http://www.opengeospatial.org/standards/sos`
15. Kutschera, P., Bartha, M., Havlik, D.: SUDPLAN's Experiences with the OGC-Based Model Web Services for the Climate Change Usage Area. In: Hřebíček, J., Schimak, G., Denzer, R. (eds.) ISESS 2011. IFIP AICT, vol. 359, pp. 589–604. Springer, Heidelberg (2011)
16. Middleton, S.E.: Environmental Image Sample Classification Service (2013), `http://catalogue.envirofi.eu/enablers/environmental-image-sample-classification-service`
17. Google: Google Cloud Messaging (2013), `http://developer.android.com/google/gcm/index.html`
18. Márquez, F.G.: Publish/Subscribe Context Broker - SAMSON Broker (2013), `http://catalogue.fi-ware.eu/enablers/publishsubscribe-context-broker-samson-broker`
19. Kobernus, M., Pielorz, J.: D2.3.2 Functional and Organisational Specification for PEIS Pilot II, Public deliverable of the ENVIROFI project. ENVIROFI Consortium (2012)
20. Van der Schaaf, H., Kobernus, M., Falgenhauer, M., Pielorz, J., Watson, K.: Data fusion in the environmental domain. In: Hřebíček, J., et al. (eds.) ISESS 2013. IFIP AICT, vol. 413, Springer, Heidelberg (2013)
21. Schleidt, K., Egly, M., Havlik, D., Schade, S., Giacomelli, A., Mazzetti, P., Laurenne, N., Tuominen, J., Middleton, S.E., Arbab-Zavar, B.: D1.3.2 Functional and Organisational Specification for Biodiversity Pilots II, Public deliverable of ENVIROFI project. ENVIROFI Consortium (2012)
22. Umweltbundesamt Austria: Austrian Long Term Research Network (2013), `http://www.umweltbundesamt.at/umweltsituation/lter_allgemein`
23. Usländer, T., Berre, A., Canut, C.G., Havlik, D., Lorenzo, J., Sabeur, Z., Modefferi, S.: The Future Internet Enablement of the Environment Informaiton Space. In: Hřebíček, J., et al. (eds.) ISESS 2013. IFIP AICT, vol. 413, Springer, Heidelberg (2013)

Mapping of Illegal Dumps in the Czech Republic – Using a Crowd-Sourcing Approach

Miroslav Kubásek

EnviWeb s.r.o., Březová 352/6, 637 00 Brno, Czech Republic
miroslav.kubasek@enviweb.cz

Abstract. The paper describes an environmental project ZmapujTo (www.ZmapujTo.cz) which is intended to fight against illegal waste dumps in the Czech Republic using a modern, effective and widespread platform. As a supporting tool a smartphone/tablet reporting application ZmapujTo has been developed. This freely available mobile application enables users to report illegal dumps and also overloaded of municipal waste containers. The project ZmapujTo is intended for all people who want to live in a clean environment in their cities, villages or countryside and additionally want to raise awareness of unacceptable environmental behaviour of people.

Users of this GIS based mobile application can report illegal dumps quickly and easily by using their smartphones/tablets. Thus, everybody can use this interactive public web form to report illegal dumps in the Czech Republic. The objective of this project is to contribute to reduce environmental pollution caused by illegal dumps in the Czech Republic by making available a modern, effective and widespread platform for reporting and monitoring of illegal dumps.

Keywords: illegal dumps, mapping, reporting, GPS, mobile application, crowd-sourcing.

1 Introduction

Illegal dump sites have always been a trouble endangering not only local environment but also human health. It is estimated that currently 100 million tons of illegally waste is dumped world-wide [1].Therefore, people should not be indifferent to that and report to competent and responsible authorities if they find an illegal dump site in order to take actions. The illegal trafficking and dumping of waste heavly affects our environmental, economy and human well being. The costs for cleaning-up illegal dump sites are extremely high, especially if this involves multiple countries. For example, approximately 134,000 tons of waste was illegally dumped in a large gravel pit in North East Europe and the extraction and transport of this waste costs 160 Euros per ton which added up to over 21 million Euros. It is even more costly when the waste is considered to be hazardous. The cost rises to about 300 Euros or more per ton [2].

J. Hřebíček et al. (Eds.): ISESS 2013, IFIP AICT 413, pp. 177–187, 2013.

For example, the situation in United Kingdom is monitored by councils across England to reveal their illegal dumping statistics, to compare relative performance, and to identify opportunities for improvement. In 2012, 732,052 illegal dumps on public land have been reported, 9.1% less compared to previous year (805,320) according to the AnyJunk Fly Tipping Report 2012 [3]. £33m have been spent by the councils on illegal dumping clearance; but only 0.6% of illegal dumping incidents were successfully prosecuted.

During the last decade, in the Czech Republic, several thousand cases of illegal dumping have been reported. For example, the average amount of accumulated waste in the years 2007 to 2009, put to illegal dumps in the forests was about 8 thousand m^3. The average annual area coverd by illegal dumps for example in the forests is approximately 3,4 hectares and 250 tons. The average annual disposal costs for waste from illegal dumps in the forest is about 40 thousand Euros. In 2010, it 830 illegal dumps have been registered in the forests of the Czech Republic, their total area reaches almost 5,4 hectares with 280 tons of waste. Disposal costs exceeded 40 thousand Euros [4].

An even worse situation is the illegal dumping of waste in municipallity areas in the Czech Republic. In 20011 130 thousand tons of waste have been disposed as reported by the the Waste Management Information System of the Czech Republic (2013) [5] operated by the Czech Environmental Agency (CENIA).

The cost of the disposal of illegal dumps exceeded 20 million Euros. In addition to that, it it can be assumed that many other cases of domestic illegal dumping have not been reported because of insufficient inspection. Local municipal governments confronted with such illegal dumping issues have been working hard to take countermeasures.

An important approach for environmental protection suitable for all regions is to promote the enforcement of municipal policies assuming the residents' improved awareness and deliberate participation. To understand visible changes to the environment such as illegal dumping, in particular, it is urgently needed to create a mechanism and/or tools for a close collaboration between the residents and administration in the Czech Republic.

The progress in information and communication technologies (ICT) in recent years has extended to the environment surrounding geographic information systems (GISs). These GISs are integrated into professional spatial analysis tools conventionally used by research institutes and technical consultants up to familiar information utilization tools using general PCs in USA [1], Japan[6-8] and Australia. The Regional Illegal Dumping Database in Illegal Dumping project is being managed by Hunter and Central Coast region in New South Wales [9] to compare and analyse data collected by councils, from dumping sites. All incidents are entered into a database, which is linked to a GIS that is able to produce locations' maps for each council area.

The "e-japan" Strategy formulated by the Japanese Government in 2001 advocates the introduction of ICT to the administration, aiming at the establishment of an infrastructure for the world's most advanced communications environment by 2005. "GIS Action Program 2002-2005" based on the e-japan Strategy includes the plan for effectively using GISs in various governmental fields to streamline administration and to provide higher-quality administrative services [7].

"Let's do it 2008" project was a campaign that began in Estonia that was led by one of the inventors of Skype - Ahti Heinla and the founder of Microlink and Delfi- Rainer Nolvak. They put cutting-edge ICT and 40,000 volunteers to work Saturday to clean-up the tiny Baltic Sea state of Estonia. Heinla and Nolvak used special software based on Google Earth, positioning software for mobile phones and mobile phones with GPS to map and take images of illegal garbage dumps across the country. The innovative software also brings the massive garbage collection campaign virtually into living rooms where Estonians can follow its real-time progress, organisers said. The project aims to recycle 80 percent of the estimated 10,000 tonnes of garbage in illegal dumps[1].

The European Commission has supported the EuropeAid Project "Waste Governance – European neighbourhood and partnership instrument (ENPI) East" (Azerbaijan, Armenia, Belarus, Georgia, Moldova, Russia, and Ukraine) with "Completed Pilot Region Inventories of Existing Illegal/Non-Compliant Disposal Sites"[2]. One key sub-result whose achievement will ultimately lead to the achievement of the main result is the development of an existing register of illegal waste disposal sites through, firstly, a desk-based review of the existing inventory and then it's updating through a series of field visits and stakeholder consultation throughout the pilot region.

1.1 Utilization of ICT in Dealing with Environmental Issues

A Ddescription of ICT related projects in the area of environmental respectively waste problems in the Visegrad countries are published in the report "Visegrad Countries: Environmental Problems and Policies" [10]. Authors are describing good and bad practices of ICT technology utilization. On the base of existing social and environment-related systems operating in the Visegrad countries region the report concluded, that several crucial common problems exist. Such as:

Poor visibility - the most common and very serious disadvantage. Reporting systems are not advertised properly, thus people do not know that they can use them.

Poor or even lack of internationalization - reporting systems are usually provided in one local language either translated partially or incorrectly. Lack of translation into a commonly recognized international language (e.g., English) makes it difficult or even impossible for foreigners to find such a system and/or use it.

Limited scope - many systems have been built to serve only one particular purpose or to help in solving problems of problems only of small regions. Some of them are discontinued after having solved the problem.

Lack of integration with other systems - systems shows almost no traces of integration with other systems, especially with systems funded by the government.

---

1 Map of the garbage dumps can be seen at: `http://www.teeme2008.ee/?op=body=55ar/mas`

2 See web site `http://www.wastegovernance.org/index_eng.html`

These general problems pointed out in this analyse should be taken into account when preparing or extending social system dealing with reporting of environmental issues.

2 Project ZmapujTo

The project ZmapujTo[3] dealing with the mapping illegal dump sites was started in the middle of 2011, the web site (http://www.ZmapujTo.cz) have been put in operation during summer 2012. The initial aim of the project was to motivate citizens to start creating a map of illegal dump sites in their surroundings

At the end of 2012 we put the administrative part of the system for municipalities into operation and we tested its functionality in different pilot territories. The administration function has been fully operational in May 2013 and we are now trying to involve as many municipalities, town districts and other organizations responsible for disposal of illegal dump sites as possible.

The main aim of this project is the possibility to inform the public on the issue and economic drawbacks of illegal dump sites. The project also envisages a more efficient use of municipality finance resources used for the disposal of dump sites; furthermore it will also contribute to catch the illegal waste dumping people or companies. ers.; and finally, it will contribute to their prevention.

An active involvement the administration of municipalities in the project has many advantages. The municipality makes the citizens aware that it takes the issue of illegal dump sites serious and it also saves its expenses of dedecting new dump sites in its territory. Moreover, a timely dedection of a new illegal dump site reduces the expenses of its disposal. The GIS application allows for an easy observation of the sites even in difficult terrains, including the plot number and the owner. An appropriate municipal authority can thus gain maximum information about the illegal dump site without seeing the site personally.

Timely disposed illegal dump sites can also prevent environmental impact and damage to a larger extent (soil and water contamination, hazard of fire etc.), it improves the safety of human health, provides prevention of possible injuries (children, animals etc,), and prevents further uncontrollable extension of illegal dump sites (see Figure 1).

To enable a detailed viewing of reported illegal dumps and their surrounding areas we have included into an interactive map several advanced GIS technologies. The first one is the Google Street View, which enable users to get a very detail overview of the situation around the illegal dump. Google Street View is a feature of Google Maps that provides 360° panoramic street-level views and allows users to view parts of selected cities and their surrounding metropolitan areas at ground level. Google Street View displays photos that have been previously taken by a camera mounted on a car, and can be navigated using either the arrow keys on the keyboard or by using the mouse to click on arrows displayed on the screen. Using these devices, the photos can be viewed in different sizes, from any direction, and from a variety of angles[4].

[3] English equivalent of Czech words "Zmapuj To" is "Map It".

[4] Source: `http://en.wikipedia.org/wiki/Google_Street_View`

Panoramio is the second feature, which enables geolocation-oriented photo sharing. The goal is to allow users to learn more about a given area by viewing the photos that other users have taken at that place[5].

Next feature enables the integration of a cadastral map of the Czech Republic[6]. This feature can simply identify an owner of a specific plot so that the municipal officer can directly contact the owner.

This interactive map is also enhancement by a set of layers that enable users to adjust the map according to their needs. A user can select from an ortofoto layer either displaying satellite images, hybrid layers or displaying a mixture of normal and satellite views and terrain layers or displaying a physical map based on terrain information. Furthermore the map can also integrate three-part map layers e.g. Bing maps, Yahoo maps or Open Street maps. The interactive map also integrates a heat map (graphical representation of data where the individual values contained in a matrix are represented as colours). This heat map enables visualizing the frequency of illegal dumping.

Fig. 1. ZmapujTo approach

5 Panoramio is reachable under web address `http://www.panoramio.com`

6 See `http://nahlizenidokn.cuzk.cz/`

Thanks to the linkage of the project ZmapujTo to local governments of the Czech Republic as well as to forest owners, PLA administrators, etc., the person who owns the specific site can quickly react and take action.

The ZmapujTo project also aims to inform the public about the dangers related to illegal dump sites, motivate the public to report dump sites, but also offers municipalities and other organizations a tool for administration of all these reports.

2.1 Reporting Tools

The primary way of sending reports about illegal dump sites is by using the mobile phone application ZmapujTo. Currently, this application is available Android and iPhone operating systems and everybody can download it from the project web site or via Google Play[7] and Apple iTunes[8].

The user can take a photo of an illegal dump site with his smart phone and marks its exact location in a map. The GPS in the smart phone determines the position approximately and the user can then change it so that it fits the reality. Then the user selects the size of an illegal dump site, specifies what type of waste is there and sends the report (Figure 2).

Fig. 2. Screenshot of the mobile device application

[7] `https://play.google.com/`

[8] `https://itunes.apple.com`

All reports are anonymous, but the user can also insert optionally a comment or his contact data. After checking if the report is genuine (elimination of spam or experimental/erroneous reports), all reports are displayed in the interactive map at the web site and the related municipalities can use this information to take actions.

Alternatively, the user can use the web interactive form to report an illegal dump site. The interactive form can be opened in a computer with a common web browser, or in a mobile device (smartphone, tablet). The address for the web browser is http://new.zmapujto.cz.

2.2 Interactive Map

After the mentioned verification of reports, all reported illegal dump sites appear in an interactive map, available at web address http://skladky.zmapujto.cz (Figure 3.). This map allows filterning for new reports, reports about to be removed or the already removed dump sites. At the bottom part of the interactive map a list of the latest reported illegal dump sites is displayed. The map has also a function that can show a list of the dump sites closest to a selected location, including information about the distance from/to the site.

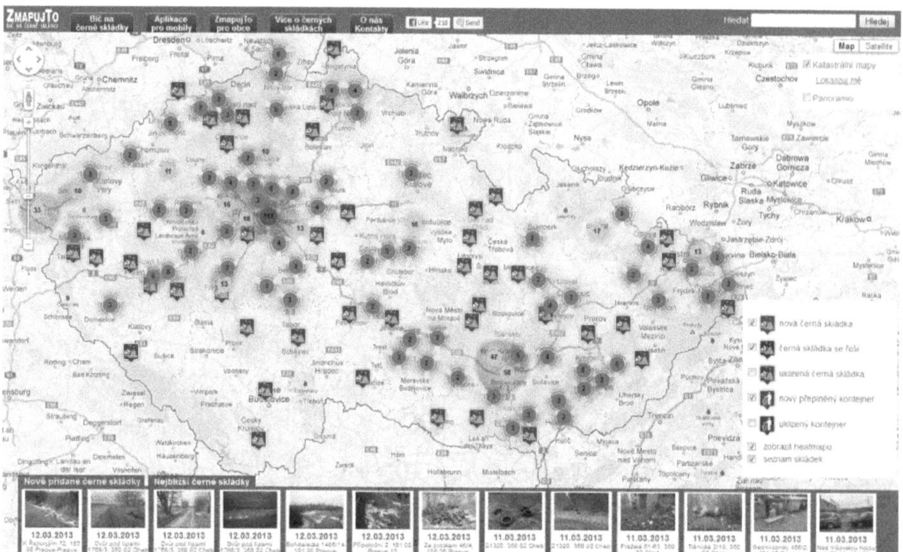

Fig. 3. Print screen of the interactive map with illegal dump sites

The interactive map has several layers (including satellite images), as well as the map layer of the cadastre of real estates, which allows for an easy identification of the owner of related property. It is forssen that in the future, more layers can be integrated (e.g. points of waste collection, legal dump sites, contaminated places, etc.) from the Czech National Geoportal[9] using INSPIRE standards. This map is also equipped with

[9] `http://geoportal.gov.cz/web/`

the Google Street View technology which allows viewing the place using panoramic photos taken in the street. Thanks to this, a user can gain a very detailed view where the illegal dump site is without even going there.

2.3 Illegal Dumps Management

An important part of the project is the administration and management of illegal dump sites, which allows municipalities or other interested organizations to get involved in the project. The aim of this part of application is to provide a simple tool providing information to municipalities about illegal dump sites reported in its district (Figure 4.). The administrator of the municipality allotted to a specific dump site can change its status – e.g. when they are preparing to dispose it, it has been removed, comments can be added, or reports concerning the expenses of its disposal. If a new illegal dump site appears within the district, the administration system of application sends an email to the administrator of the municipality. The administrator can also use the statistics of the district and a widget, which can be placed on the municipality web site and inform thus the visitors of the municipality web site that the municipality is involved in the project, how many dump sites have been reported, removed or what expenses have been related to its disposal.

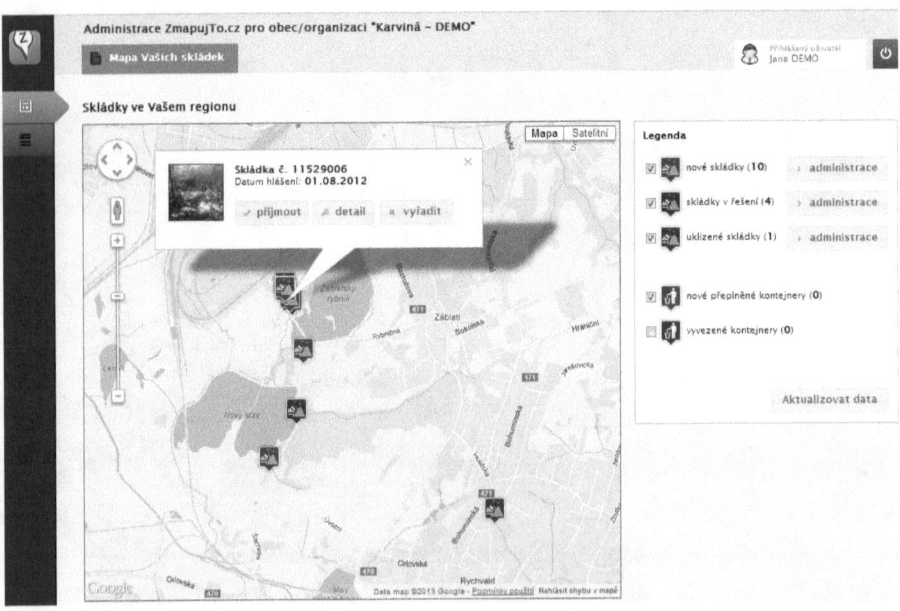

Fig. 4. Screenshot of the administration application for illegal dump sites

Currently, there are 29 municipalities and institutions from the entire Czech Republic involved in ZmapujTo project. We are collecting their experience using the information system, their comments concerning improvements and we are expanding the service based on these feedbacks. There is also a DEMO version of the administration application available so that a potentially interested person can try this system.

One of the most active system users are the officials of the city district Praha 10, whose are using this system intensively. They have even been able to involve city policemen, whose now report illegal dump sites and overfilled garbage bins as a part of their daily guarding walks. Within two months, there have been 371 illegal dump sites reported and 294 of them have been disposed.

2.4 Illegal Dumps Analyse

Since July 2012 1780 illegal dumps have been reported and more then 400 dumps have been cleaned (State to 15. July 2013). The Figure 5 shows the aquivalent area representation of the reported dumps. The greatest part of dumps has a medium size (44%). Small sized dumps are illegal dumps which can be cleaned and transported by hand, the medium sized dumps are dumps up to $5m^3$ and above this volume it is recognized as a big dump. The small amount of overloaded containers is due to the fact that this type of reporting has stared much later (since January 2013).

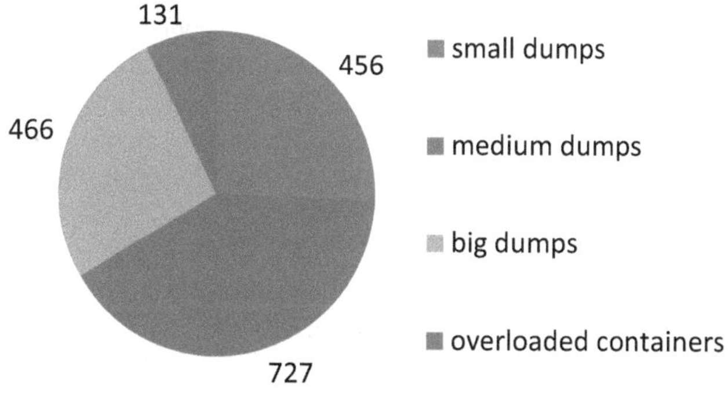

Fig. 5. Representation of illegal dumps

During the reporting process the user has to specify in more detail the waste type. The user can select from 12 waste categories which cover the whole range of possible wastes. He can assign one or more categories in one report. As can be seen in Figure 6, the most common type of an illegal dumped waste is household waste and plastics. More than ten percent of these dumps still contains construction and demolition waste but also automotive waste (tires, batteries etc.).

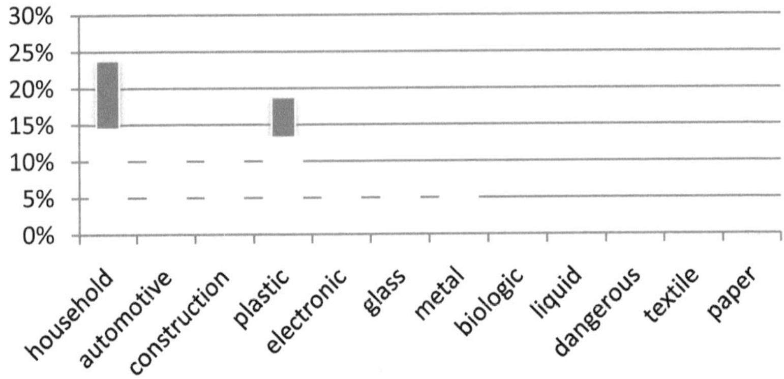

Fig. 6. Distribution of waste per type

3 Conclusion

The ZmapujTo project has been launched a year ago. During its existence over 1700 illegal dump sites in the Czech Republic have been reported and mapped in the system. The crowd-sourcing approach and the developed mobile application for reporting and mapping facilitated the dedection of illegal dumps. Furthermore the public awareness was heavily increased. 29 pilot municipalities of the Czech Republic are already involved in this project and over 400 illegal dump sites have been disposed since the last year. [10]

We believe that the ZmapujTo project is vital and has a great potential in the future also taking into account its possible usage in neighbouring States of the Czech Republic, taking advantage of the INSPIRE GEOPORTAL[10]. We are looking forward to others users interested in getting involved either local governments or organizations responsible for elimination of illegal dump sites.

References

1. Hanfman, E.: A Comprehensive Assessment of Illegal Waste Dumping. Water Health Educator (2009), http://www.waterhealtheducator.com/upload/Illegal%20Waste%20Dumping%20Article.pdf
2. Europol. Europol warns of increase in illegal waste dumping (2011), https://www.europol.europa.eu/content/simplenews/europol-warns-increase-illegal-waste-dumping-1057
3. AnyJunk Fly Tipping Report 2012 (2012), http://www.anyjunk.co.uk/fly-tipping-report-2012

[10] http://inspire-geoportal.ec.europa.eu/

4. Fechtnerová, A.: The issue of illegal dumping on land designated for forestry. Lesnická práce - časopis pro lesnickou vědu a praxi 3 (2011), http://www.silvarium.cz/lesnicka-prace-c-3-11/problematika-cernych-skladek-na-pozemcich-urcenych-k-plneni-funkci-lesa
5. Waste Management Information System of the Czech Republic (2013), http://www1.cenia.cz/www/odpady/isoh
6. Morita, A., Takagishi, S.: GIS-based Support Systems for the Reduction of Illegal Dumping. In: Proceedings 22nd Annual Esri International User Conference–Exploring the New World of GIS!, San Diego, California, U.S.A, July 8-12 (2002), http://proceedings.esri.com/library/userconf/proc02/pap1172/p1172.htm
7. Tasaki, T., Kawahata, T., Osako, M., Matsui, Y., Takagishi, S., Morita, A., Akishima, S.: A GIS-based zoning of illegal dumping potential for efficient surveillance. Waste Management 27, 256–267 (2007)
8. Ichinose, D., Yamamot, M.: On the relationship between the provision of waste management service and illegal dumping. Resource and Energy Economics 33, 79–93 (2011)
9. Illegal Dumping project (2013), http://www.hccrems.com.au/Programs/Environmental-Compliance/Sub-projects/Illegal-Dumping.aspx
Šauer, P., et al.: Visegrad Countries: Environmental Problems and Policies. Published by CENIA, Prague (2013)
10. Kubásek, M., Hřebíček, J.: Crowdsource Approach for Mapping of Illegal Dumps in the Czech Republic. International Journal of Spatial Data Infrastructures Research (to be pubished 2013)

A Mobile Application for Reporting Disease Incidents

Hylke van der Schaaf[1], Ente Rood[2], and Kym Watson[1]

[1] Fraunhofer IOSB, Frauhoferstraße 1, 76131 Karlsruhe, Germany
{hylke.vanderschaaf,kym.watson}@iosb.frauhofer.de
[2] Koninklijk Instituut voor de Tropen, Amsterdam, The Netherlands
E.Rood@kit.nl

Abstract. The current procedure for the reporting of cholera cases in Uganda contains many manual steps across several levels of the health infrastructure. Because of this there is a large chance of errors in the information flow, possibly delaying the signalling of an outbreak. The lack of accurate and complete data also hinders research into the spread of cholera. To improve the cholera reporting an application called Dira (Disease Incidence Reporting Application) has been developed for mobile devices, that allows the field registration of patients to be done quickly, easily and accurately. By entering the data directly on an electronic device there is no longer the need for separate digitization steps. By transferring case data directly from the hand-held device to a central server at the Ugandan Ministry of Health, the reliability of the data can be increased and the time to launch a response to an outbreak can be decreased.

Keywords: cholera,health,reporting,mobile application.

1 Introduction

EO2HEAVEN (Earth Observation and ENVironmental modelling for the mitigation of HEAlth risks, EU FP7 244100, [1]) has developed methodologies, correlation models, spatial data services (using OGC standards) and applications supporting the main activities in environmental health. EO2HEAVEN work is organized in three case studies: a) environmental effects on allergies and cardiovascular diseases in Saxony, Germany, b) environmental challenges to health caused by air pollution from petrochemical plants in Durban, South Africa, and c) investigating the impact of climate variables on cholera outbreaks in Uganda.

The research in the cholera case study requires accurate records of cholera cases. Currently, cases of cholera are recorded on paper at the cholera treatment centres. These forms are then digitized and sent, once a week, by email and hard-copy, to the District Health Office. The District Health Office gathers the reports of the district, digitizes them if needed, and sends the aggregated data to the Ministry of Health. This whole process takes up to two weeks, hindering a timely response to a cholera outbreak. The process also involves many manual

J. Hřebíček et al. (Eds.): ISESS 2013, IFIP AICT 413, pp. 188–195, 2013.

copy actions, making it very error-prone and making the resulting cholera case data unreliable. Since a patient record, once sent, can not be updated the cholera case data available at the Ugandan Ministry of Health is often incomplete or erroneous.

To improve the cholera reporting in Uganda, an application called Dira (Disease Incidence Reporting Application) has been developed for mobile devices, that allows the field registration of patients to be done quickly, easily and accurately. By entering the data directly on an electronic device there is no longer the need for separate digitization steps, while by transferring case data directly from the hand-held device to a central server at the Ministry of Health, the reliability of the data can be greatly increased and the time to launch a response to an outbreak can be greatly decreased.

2 Health Surveillance and Health Information Management

Public health surveillance aims at the collection, analysis, interpretation, use and dissemination of data for guiding decision making to prevent and control the occurrence of disease and injury [6]. The fundamental principle of such a system is to provide valid, accurate, unbiased and timely data to decision makers to plan interventions. As the process of collecting detailed health information can become very cost inefficient when conducted on a small temporal and spatial scale and will compromise the quality of the data collected, most surveillance systems rely on aggregated data collated at district levels and reported monthly to national authorities [5]. The utility of surveillance data, collected on various time intervals, can serve various objectives for the assessment of and the response to the occurrence of disease. The immediate detection of emerging outbreaks, for example, will rely on a prompt and timely reporting of newly detected cases to monitor the progression of an epidemic and to monitor the effectiveness of instantaneous interventions. At the same time, health system monitoring and resource allocation in National health policies will require quality data available over more extended time periods.

3 Cholera Surveillance and Reporting in Uganda

In 2000 the Ugandan ministry of Health adapted the Integrated Disease Surveillance and Response (IDSR) strategy with the intention to achieve an integrated and intervention oriented surveillance system as outlined in the International Health Regulations [8]. This framework uses various surveillance indicators (core and process) to monitor trends in the occurrence of communicable diseases and to evaluate the effectiveness of program interventions [4]. The IDSR integrates the systemic collection and dissemination of a number of diseases, previously monitored by different vertical disease programs across all layers of the health care system, including cholera. The core functions of this system are to 1) detect cases in a timely fashion, 2) record cases according to quality standards, 3) to report cases transparently using an adequate communication network and 4) to analyze the available data for response planning and epidemic mitigation [5].

An evaluation of the IDSR system functions conducted in 2008, showed a considerable improvement of surveillance system functions over the period 2000-2007 [4]. Both the completeness of reports as well as the timeliness of monthly reporting increased substantially (respective increase over period 2001-2007 of report completeness: 49% – 85%; and timeliness: 50% – 77%). In contrast to the increase in monthly reports sent to the National surveillance department, the consistency of weekly reports submitted showed a short upsurge from 49% in 2001, to 96% in 2004, but diminished again to 53% by the year 2007. The authors conclude that this reduction can be attributed to reduced financial investments by the National government leading to reporting gaps.

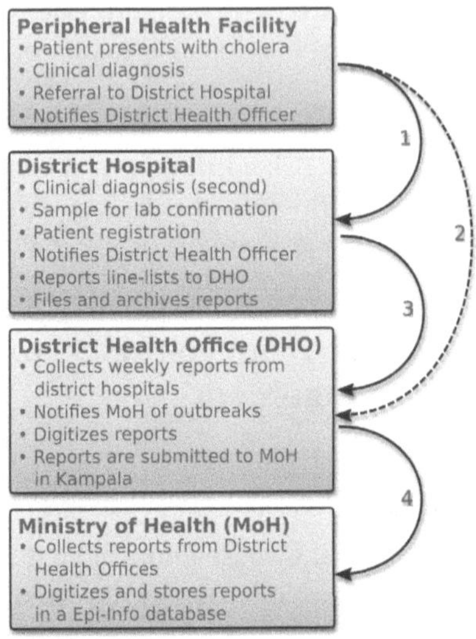

Fig. 1. Current cholera reporting procedure: 1.New cholera cases are directly reported from the peripheral health centers to the district health office. 2.Weekly cholera reports are sent to the DHO to trace epidemic outbreaks. 3.New Cholera cases are directly reported from district health office to the MoH. 4.Monthly reports are compiled at the DHO and sent to the MoH to monitor cholera prevalence. 5.MoH reports to the WHO and other international initiatives (not shown).

A descriptive assessment of the cholera surveillance system in Uganda, conducted in the Kasese district in April 2012 as part of the EO2HEAVEN project (unpublished data) showed several quality measures in place including the confirmation of the completeness and timeliness of reports. However, at this point, limited effort had been invested to perform regular data quality checks or further data analysis within the district. Data validation of reports is a core activity by the Ministry of Health though this is not consistently done due to inadequate resources. Post outbreak evaluation is standard practice that the Ministry of Health recommends for all affected districts but this too is subject to availability of resources to conduct the evaluation [3].

The current data flow and digitization process has not been standardized across different levels of the health infrastructure. Capacity to digitize information and to perform basic data analysis or visualization tasks is limited. Likewise the current staff lacks basic knowledge of data management and analysis. Even though weekly reports and time series graphs are produced from the data, the lack of data quality checks and critical evaluation of the data limits the efforts to interpret the results. As a result, data managers have little to no sense of ownership and perceived value of their work for others, resulting in a limited accountability. The stratification of the health information system over multiple administrative levels confines the prompt dissemination of information. This also introduces errors in the information flow, potentially delaying the signalling of an outbreak when it occurs. Even though alternative procedures exist to flag initial cases for follow up investigations, the protocols are fuzzy and the system is prone to human error. Moreover the lack of resources limits the efforts for efficient active surveillance. It appears that the flow of information from the peripheral health facilities to the ministry of health is hampered by lack of tools to collect, store and disseminate this information. The reporting system relies on various consecutive steps as visualized in Fig. 1.

Based on the requirements identified, EO2HEAVEN has developed a standardized data collection protocol and associated data collection tool to facilitate data collection and provide access to health information across all levels of the health infrastructure. The tool will benefit both health practitioners working in the field as well as health authorities working at the national ministry of health as it 1) Reduces the number of human errors induced by the current reporting system 2) Reduces the amount of resources spent on data management and dissemination 3) Improves response time, 4) Increases the effectiveness of surveillance by facilitating the exchange and interpretation of information at a sub-district level and 5) Facilitates communication across all levels of the health system.

4 Dira: A Mobile Application for Incident Reporting

Dira, Disease Incidence Reporting Application is an application for tablet PCs and smartphones running the Android platform. It is developed to improve the accuracy and speed of recording of cholera cases in third world countries. The main screen (see Fig. 2) of the application is the listing of the entered cases. From here cases can be directly viewed and searched. All other functionality is directly available from the menu.

4.1 Data Entry

The application has an easy-to-use interface for entering patient data. It allows for the collection of primary patient data such as name and home town, symptoms and epidemiological information, such as the patients primary water source, and which villages he or she has visited in the week before falling ill.

To increase the ease of entering village names, to reduce variations in spelling of village names, and to insure that the correct village is entered when multiple villages with the same name exist in different counties, the application has a database of village names, sorted by sub-county, county and district. When entering a village name, the user is prompted for the district, county, sub-county and village name. If the user is certain of the spelling of the name of the village, he can directly type in the village name. The system will give as-you-type

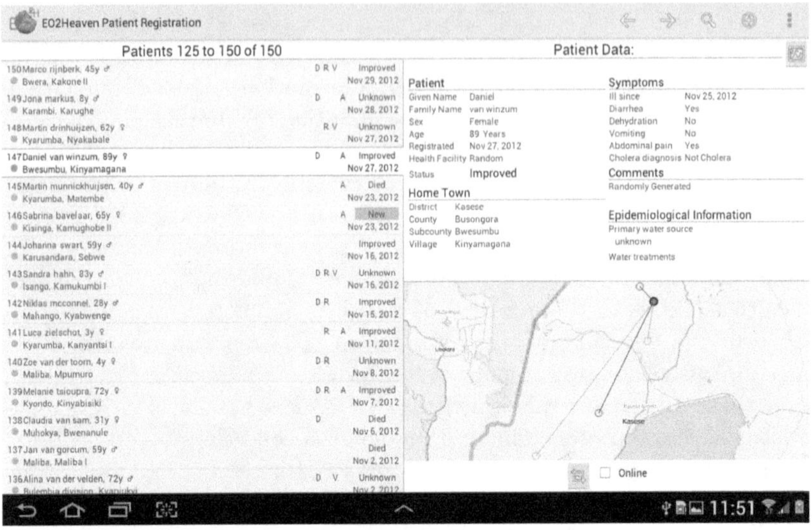

Fig. 2. Dira main screen

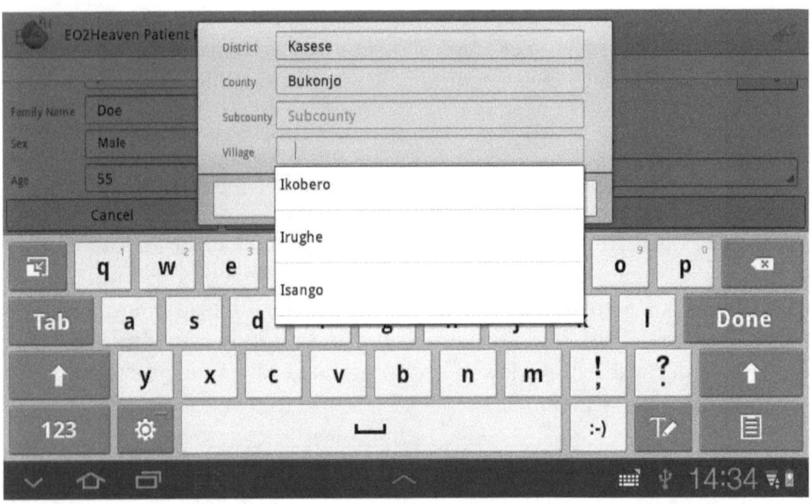

Fig. 3. Entering a village name

suggestions based on all known villages. When the user selects one of these suggestions, the district, county and sub-county names will be automatically filled in. If the user is not certain of the village name, he can drill-down by starting with the district, county or sub-county. Once one of these fields is filled in, the suggestions for the other fields will be limited to entries matching the already completed fields (see Fig. 3). The user can always override the suggestions by typing in the full village name.

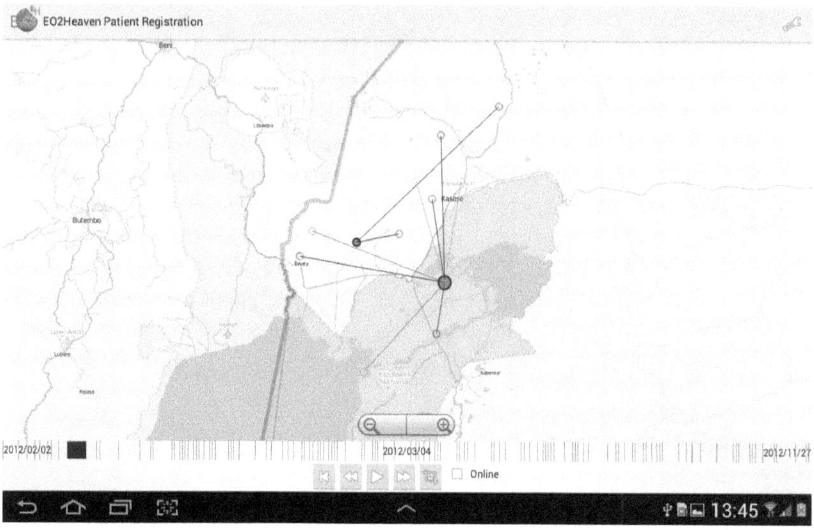

Fig. 4. Display of cases on a map, animated in time

When editing a case, before the changes are stored, the user is presented with an overview page highlighting all changed fields in the patient record. This in-between page allows the user to confirm that he did not make any accidental changes while working with the input form.

4.2 Data Visualisation

The application has several ways to visualize the data. It can show graphs of cases aggregated per week, for several time periods (running or previous month, last 6 or 12 months, all). It can show individual cases on a map, together with the villages that the patient has visited in the week before he became ill. It can also show all, or a subset of cases on a map, animated in time, giving a visual overview of the progress of an outbreak (see Fig. 4).

4.3 Data Transmission

To gather the data in a central place the application can send the patient data to a central server over any available internet connection, e.g. Wi-Fi or GSM. The application has been made robust for field work with intermittent internet connectivity. If the data transfer is interrupted, the application will automatically resend the data from the point of failure. The application will also automatically resend patient records that have been updated, meaning patient status can be updated even after the patient record has been sent to the central server, leading to a more complete data set.

4.4 Data Output

It is important for the uptake of the application that its use can seamlessly integrate into the existing workflows at the local level, with respect to data collection and archiving. To facilitate this, the application can export the data records available on the device in various file formats, like CSV for importing into existing patient record systems, and PDF for easy printing.

Since internet connectivity is not always available in Uganda, these exports will still allow the old case-reporting methods to be used, by sending printouts from the local treatment centres through the District Health Offices, to the Ministry of Health. This will give the users the confidence that even if there is no internet connectivity, their data-collecting efforts are not in vain.

4.5 Implementation Details

The platform used to implement Dira is Android. The choice for Android was made because of the large diversity of devices available with this platform, from low-cost phones to high-performance tablets. Some ebook-readers with an e-ink display also run Android and these e-readers could be very suitable for data collection in the field, as their screens do not become hard to read in very bright sunlight, and they use very little power. We have not yet been able to test any e-readers.

For field work it seems that 7 inch tablets are optimal. Devices of this size are easy to hold in one hand, which is important when used while standing or walking, yet they are still large enough to give a good overview when using forms or maps.

The application is implemented as a native Android application, making it very resource efficient. This also allows it to fully utilize the internationalization functionality of the platform and gives it the ability to easily specify different layouts for different device sizes and portrait/landscape orientations. For storing patient data on the device, the OS-provided SQLite database is used. This database also stores village data that is used to give autocomplete / as-you-type suggestions when entering a patients home town or visited locations. For graphing functions are implemented with the AchartEngine library [7]. For the map display functions, the Mapsforge library [2] is used, which can work with off-line vector maps. This allows the map functionality to work fully without internet connectivity.

5 Conclusions and Recommendations

The Dira application was received very well by local health workers when they tried it out at two EO2HEAVEN workshops in Uganda. Users especially appreciated the ease of use of the application. A key feature is its ability to work completely off-line with no internet access. Feedback from these workshops will be used to further improve the application. It is important that the application is smoothly integrated in the current organisational workflows. The vision is to incorporate this application into an early warning system for different diseases in developing countries.

Acknowledgments. The research leading to these results has received funding from the European Community's Seventh Framework Programme (FP7/2007-2013) under grant agreement n° 244100 (EO2HEAVEN)

References

1. EO2HEAVEN: Earth observation and environmental modelling for the mitigation of health risks (june 2013), http://eo2heaven.org/
2. mapsforge: free mapping and navigation tools (june 2013), http://code.google.com/p/mapsforge/
3. Lukwago, L.: personal communication (February 2012)
4. Lukwago, L., Nanyunja, M., Ndayimirije, N., Wamala, J., Malimbo, M., Mbabazi, W., Gasasira, A., Nabukenya, I.N., Musenero, M., Alemu, W., Perry, H., Nsubuga, P., Talisuna, A.: The implementation of integrated disease surveillance and response in uganda: a review of progress and challenges between 2001 and 2007. Health Policy and Planning 28, 30–40 (2013)
5. Nsubuga, P., White, M.E., Thacker, S.B., Anderson, M.A., Blount, S.B., Broome, C.V., Chiller, T.M., Espitia, V., Imtiaz, R., Sosin, D., Stroup, D.F., Tauxe, R.V., Vijayaraghavan, M., Trostle, M.: Disease Control Priorities in Developing Countries chap. Public Health Surveillance: A Tool for Targeting and Monitoring Interventions, pp. 997–1015. World Bank, Washington (2006)
6. Thacker, S., Berkelman, R.: Public health surveillance in the united states. Epidemiol Rev. 37, 1–18 (1988)
7. The 4ViewSoft Company: AChartEngine (june 2013), http://www.achartengine.org
8. WHO: International Health Regulations (2005). WHO (2008)

"My Environment" – A Dashboard for Environmental Information on Mobile Devices

Thorsten Schlachter[1], Clemens Düpmeier[1], Rainer Weidemann[1],
Wolfgang Schillinger[2], and Nina Bayer[2]

[1] Karlsruhe Institute of Technology (KIT)
Institute for Applied Computer Science
Hermann-von-Helmholtz-Platz 1, 76344 Eggenstein-Leopoldshafen, Germany
{thorsten.schlachter,clemens.duepmeier,rainer.weidemann}@kit.edu
[2] Baden-Wuerttemberg State Institute for Environment, Measurements and Nature
Griesbachstraße 1, 76185 Karlsruhe, Germany
{wolfgang.schillinger,nina.bayer}@lubw.bwl.de

Abstract. This paper describes a universal approach to the development of a cross-platform and multi-functional environmental mobile application. According to the Pareto principle [1] only common use cases are implemented and can be described in a lightweight description format instead of being explicitly programmed. These use cases include information about the environment ("my environment"), reporting of environmental data ("crowd sourcing"), and environmental experience ("electronic nature guide").

1 Introduction

Every month users of mobile devices are facing many new applications. Those can be used to access information of their interest in almost any place and situation using context information such as current position and time. People use apps for localized weather forecasts, travel information (including delays), news, stock quotes, etc., and maybe even provide information by themselves, implicitly by allowing apps to send location and speed (traffic flow) information back to service providers or explicitly by reporting their own observations via a crowdsourcing application.

At least in Europe authorities are committed to actively provide environmental information to the public [2]. For that reason many environmental information systems (EIS) do implement web frontends, web services, or even service oriented architectures and it is relatively easy to use these for the implementation of mobile frontends, e.g. mobile apps.

The range of environmental information provided varies from unstructured data, like text documents and media files, up to structured and semi-structured information, like tabular data, spatial data, metadata, time series, charts, and also includes specialized information for specific applications. Nevertheless, although data is very diverse some recurring usage patterns can be observed.

In [3] and [4] a descriptive approach of a universal mobile application accessing EIS based on extended OpenSearch descriptions, outlined in section 2, was presented.

J. Hřebíček et al. (Eds.): ISESS 2013, IFIP AICT 413, pp. 196–203, 2013.

In [5] and [6] we added our first lessons learned in the field of cross platform development for mobile devices, summarized in section 3.

Within the rest of this paper we would like to add our ideas of an integrated display of environmental information for our app. We additionally present an extension of our use cases, now including environmental reporting and environmental experience for both, the public and professionals.

2 No Programming Required!

In order to avoid programming for new fields of application the most important concept for this environmental app is a description format, called "target system description" based on the OpenSearch description format. The OpenSearch description format allows for the description of URL patterns for dynamic access to web systems. Originally the OpenSearch descriptions format was designed for access to web interfaces of search engines and mainly supplies a syntactic description for such queries. The semantics of its parameters is rigid. However, OpenSearch descriptions can be extended. For the purpose of use for a mobile environmental application, the following features have been added:

- Freely definable URL parameters.
- Semantic classification of these parameters, i.e. a rudimentary description of their semantics, for instance, the geographical position of a place in a certain format.
- A description of the sources of these parameters, e.g. automatically determined from a GPS component or by user selection from a given list of values.
- Structural presentation of the results, e.g. mapping of attributes.
- (HTML-)Templates for the display of certain result types.
- Content-dependent grouping information.

Such extended OpenSearch descriptions supply the necessary information for the registration of an information source in an application, for the generation of queries, and for the visualization of received data.

The template mechanism can also be applied to HTTP POST requests, which enables the app to write data to various destinations, e.g. REST services. With added HTTP basic authentication as well as OAuth (2.0) authorization flexible services for the storing of reported data can be described.

A "target system catalog" format, which extends the RSS 2.0 specification, is used for the description of multiple target systems, allowing the distinction of target systems, e.g. for different customers (Federal States) using the same app, and also the grouping of related target systems, e.g. different types of nature reserves or conservation areas.

3 HTML5-Based Cross-Platform Approach

Instead of developing independent native apps for each particular target system (possibly with the use or consideration of the appropriate specification), cross platform development aims to deploy one application without (or at least without major) changes to multiple target platforms.

This is often done by combining an unchanged common core of the application (top box in Fig. 1) with platform-dependent parts provided by cross development frameworks.

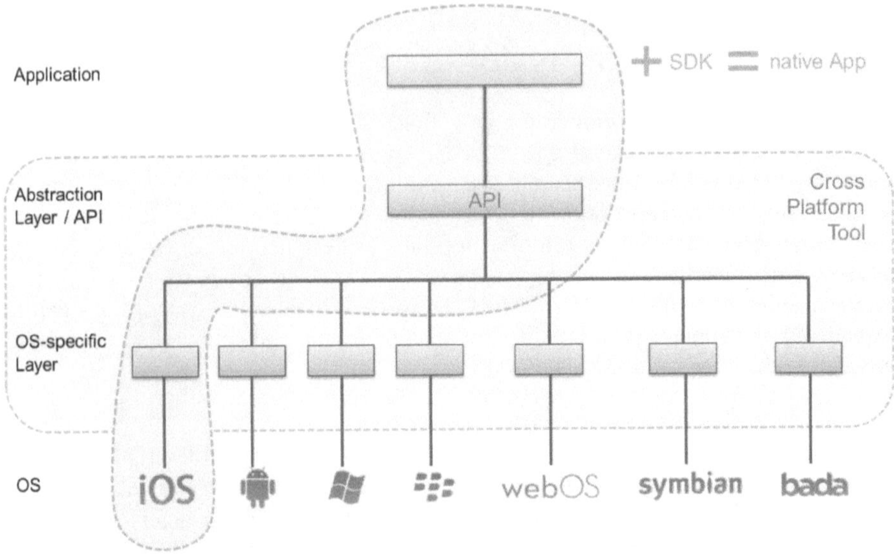

Fig. 1. Deploying an iOS-application using a cross-platform development tool

Such a combination then can be delivered for a particular platform. In Fig. 1 the OS specific "package" of an application, platform-independent and platform-dependent parts, is shown by the darker highlighted area.

The cross-platform development tool (brighter highlighted area) provides an application programming interface (API) against which the application can be programmed. On its backend it provides a specific implementation for each target platform (OS) that translates abstract commands into native functionalities of the system. According to the respective programming interface manufacturers of cross-platform development tools have to implement the OS-specific layer for each supported operating system once. This is especially easy if well-established features, commonly available for all or at least many target systems, have to be implemented. More specific features may be emulated by the interface, in the worst case, they are not supported.

Some manufacturers of cross platform development tools fall back on pre-existing technologies. One possible technology of note is Java, which is available for all popular mobile platforms.

A similar approach relying on widely available technologies is based on Web technology. The central idea is to run the actual application within a Web browser component which is already available on all platforms.

With the help of a cross platform Web development tool (and possibly using a platform-specific software development kit (SDK)) the generated native app contains all of the necessary resources for the application, in particular at least one HTML page, any associated graphics and CSS files as well as one or more JavaScript libraries that contain the abstract interface and the OS-specific implementations of the API. The display of the graphical user interface creates the browser in the form of HTML components, whose appearance can be controlled using Cascading Style Sheets (CSS).

The application logic is implemented with the help of the libraries in the form of JavaScript functions that may also provide functionality that is not standard in many Web browsers such as Multi-touch controls.

After a first prototype for the Android platform was fully implemented in Java, a second prototype based on Web technologies (HTML5) was developed to compare native and cross platform approaches. The second prototype was able to proof that a cross platform approach for the app is more feasible and so the first productive version of the app is based on Javascript-based Web technologies and a cross platform toolkit providing native functionalities through a common Javascript API.

The used PhoneGap/Cordova framework for cross development supports the particular use of components of different operating systems (including iOS, Android, Windows Mobile and BlackBerry) in a portable way. I.e. PhoneGap is able to access the camera, contacts, the file system, accelerometers, and GPS.

JQuery Mobile (JQM) provides a fully touch-optimized and asynchronous interaction with mobile browser-based applications. JQM runs on all commonly used browsers (like Firefox, Chrome, Opera and Safari), but in detail the display may show browser- and platform-dependent differences.

The Google Maps API is used for the display of maps and other data which include geospatial representations.

4 An Environmental Dashboard

The first prototypes of our mobile application behaved like a collection of single mini applications for querying EIS. Data provided by one data source has been displayed after a user request and has not been put in context of information provided by other data sources.

Our actual approach combines data and displays from several data sources within a kind of dashboard, dynamically changing when context changes or data sources provide new data (Fig. 2). The user may register for certain types of information and/or certain data sources. The app dynamically queries these using context information like sensor data (GPS, radio cell) or default settings provided by the user.

For example, spatial data from multiple data sources are merged into a single map view with different layers for each data source and the user has the possibility to individually switch single layers or groups of layers on or off, and thus display only the relevant data important to him.

Other areas of the dashboard can show other types of information, e.g. a media carousel can combine previews of all media objects collected from several data sources, and there still may be areas for other specialized mini apps, e.g. display of water levels or air pollution and the respective forecasts.

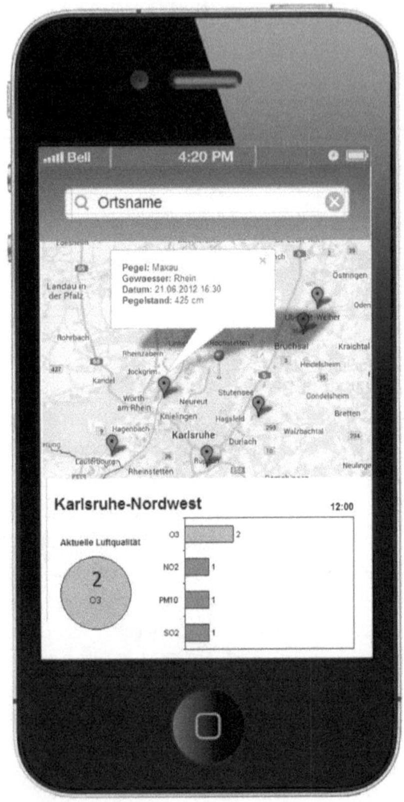

Fig. 2. Environmental dashboard with preview map and air quality data

Starting at the dashboard the user has additional possibilities to navigate to a more sophisticated query form, a more detailed display of a single item, or even to the original information system. We believe that in most cases a quick overview with a reduced complexity at the dashboard will meet the user's needs. I.e., a user frequently may be interested in the actual ozone levels before going jogging. With sports in mind he rarely is interested in querying a time series of ozone data at his location, even if he could do so.

Nevertheless the user is provided with functions to formulate complex queries or to filter larger hit lists. The navigation philosophy "dashboard/map – search/filtering – detail view – alarm/navigation/reporting" is applied within all use cases of the app. Key views and click paths are shown in Fig. 3.

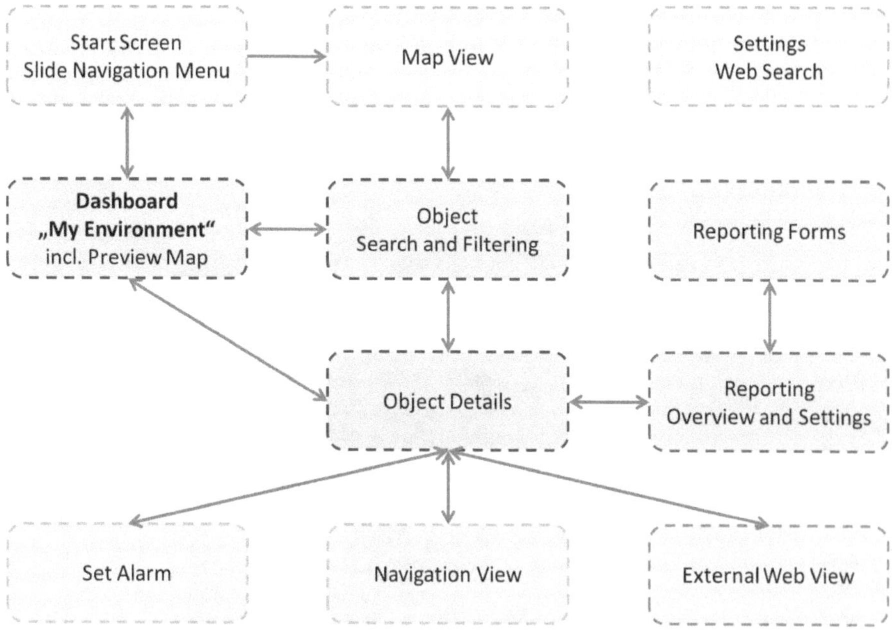

Fig. 3. Key views and click paths for the use cases "information" and "reporting". (Views for the use case "environmental experience" are not included.).

5 Beyond "Pure" Environmental Information

In addition to the environmental information dashboard, the app implements two other use cases, "reporting of environmental information" as well as "environmental experience" (electronic nature guide).

In the majority of cases reporting of environmental information follows a pattern of two. "Ad hoc reporting" aims to capture spontaneous observations, i.e. to take some photos or a video of a certain species, tag them with some additional information (e.g. locality, timestamp and some user input), and upload them to a server. In contrast, "systematic reporting" mostly does not start from scratch but aims to verify or complement existing information. This means a user has to be enabled to query (search and filter) existing objects, to view object details, and to supplement these details. For example, an employee moves to multiple measuring points and checks their condition. So he may be provided with additional routing information, and, of course, with the existing data for each station. In both cases there may be need for caching collected data offline on the device and for a delayed upload (to an EIS or to the cloud).

Many functions of the app are implemented using cloud services. For example, routing information is provided by the Google Directions API [7], and place names are determined using the Google Geocoding API [8].

From the perspective of EIS cloud services can help to close the gap between available and published data by providing infrastructure, storage, services, or even whole applications. For example spatial data easily can be (re-)published using tools like Google Maps for Business [9] when original systems do not provide standardized interfaces or are not powerful enough to cope with a large number of simultaneous requests. Structured or binary data may be stored in the cloud using services like Google Fusion Tables or Google Drive and the respective APIs [10, 11]. The discussion of the use of cloud services will be subject of another publication.

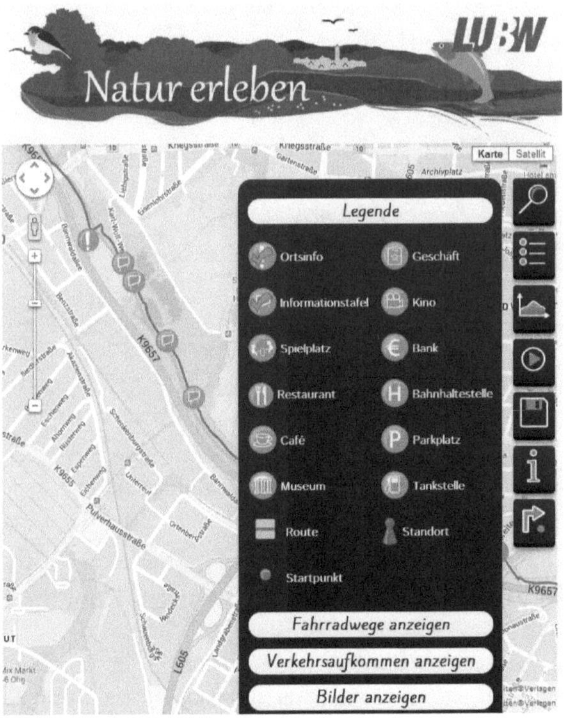

Fig. 4. Display of environmental and touristic information along a given track creating an environmental experience for the user. (Prototype by N. Bayer).

"Environmental experience" is our third use case, which has been prototypically implemented as a web application (Fig. 4) [12]. It combines aspects of environmental information and environmental reporting with the navigation along a given track such as a nature trail, as well as "learning by playing" elements in the manner of geocaching [13].

The application enables the user to navigate along the nature trails and provides him additional information (text, media, technical information). Besides fixed predetermined routes, there are also geocaching routes, where the user only knows the starting point of the route and is provided with basic additional information/criteria like the length of the course. The route itself is described by geocaching waypoints which arise from the repeated answering of questions.

6 Conclusion and Outlook

With three applications within a single app, description instead of programming, and cross platform capabilities our project "Meine Umwelt" ("My Environment") has the capability to cover many standard requirements for environmental mobile applications.

However it has room for improvements in the future. At this time information is actively pulled from EIS. Push services should be enabled in order to avoid needless time and energy consuming requests.

The use case "environmental experience" is still in an early stage and needs to be improved. And many more EIS have to be connected to the app so that the user can be informed comprehensively on all his relevant environmental issues.

References

1. http://en.wikipedia.org/wiki/Pareto_principle
2. Directive 2003/4/EC of the European Parliament and of the Council of 28, on public access to environmental information and repealing Council Directive 90/313/EEC (January 2003), http://eur-lex.europa.eu/LexUriServ/LexUriServ.do?uri=CELEX:32003L0004:EN:NOT
3. Schlachter, T., Düpmeier, C., Geiger, W., Weidemann, R., Ebel, R., Tauber, M., Zetzmann, K.: Concept of a universal mobile application accessing environmental information systems. In: EnviroInfo 2011: 25th Internat. Conf. on Environmental Informatics, Ispra, IT, October 5-7 (2011)
4. Schlachter, T., Düpmeier, C., Geiger, W., Weidemann, R., Ebel, R., Tauber, M., Zetzmann, K.: LUPO mobil. Ein Schichtenmodell zur Auswahl und Nutzung von Umweltdiensten auf mobilen Endgeräten."; In: Mayer-Föll, R. [Hrsg.] UIS Baden-Württemberg. F+E Vorhaben KEWA Phase VI 2010/11; KIT Scientific Reports KIT-SR 7586; pp. 33-42 (2011)
5. Schlachter, T., Weidemann, R., Ebel, R., Schillinger, W., Zetzmann, K.: Building Mobile Environmental Apps Using Web and Cloud Technologies. In: EnviroInfo 2012: 26th Internat. Conf. on Environmental Informatics, Dessau, DE, August 29-31 (2012)
6. Schlachter, T., Düpmeier, C., Weidemann, R., et al.: LUPO mobil - Nutzung von Webtechnologie zur Entwicklung plattformübergreifend einsetzbarer, mobiler Umwelt-Anwendungen"; In: Weissenbach, K. [Hrsg.] UIS Baden-Württemberg. F+E Vorhaben MAF-UIS Phase I 2011/12; KIT Scientific Reports KIT-SR 7616; pp. 59-70 (2012)
7. https://developers.google.com/maps/documentation/directions/
8. https://developers.google.com/maps/documentation/geocoding/
9. http://www.google.com/enterprise/mapsearth/products/mapsengine.html
10. https://developers.google.com/fusiontables/
11. https://developers.google.com/drive/
12. Bayer, N.: Konzept und prototypische Realisierung eines mobilen Umweltführers Bachelor thesis, Hochschule Karlsruhe – Technik und Wirtschaft (2013)
13. http://www.geocaching.com

Estimating Impacts of Environmental Interventions in Monitoring Programs Requires Conceptual Data Models and Robust Statistical Processing
(Position Paper)

Ladislav Dušek[1,2], Jana Klánová[2], Jiří Jarkovský[1], Jakub Gregor[1,2], Richard Hůlek[2], Ivan Holoubek[2], and Jiří Hřebíček[1]

[1] Institute of Biostatistics and Analyses, Masaryk University, Kamenice 126/3, 625 00 Brno, Czech Republic
{dusek,jarkovsky,gregor,hulek,hrebicek}@iba.muni.cz
[2] Research Centre for Toxic Compounds in the Environment, Masaryk University, Kamenice 753/5 (A29), 625 00 Brno, Czech Republic
{klanova,holoubek}@recetox.muni.cz

Abstract. This paper discusses main problems associated with evaluation of performance and impact of long-term environmental programs. Lack of data standards, incompleteness of archived datasets and insufficient statistical power were identified as main limits in functionality of monitoring networks. To avoid these failures, environmental programs should be designed with inception to incorporate data management as their integral part. Especially in global programs, local and regional data managers should invest significant proportion of their effort to handle documentation in terms of standardized coding, data formats, metadata coding and consistency of records over time. Up-to-date trends in building knowledge-based infrastructures are illustrated using example of monitoring of atmospheric pollution by persistent organic pollutants (POPs). Conceptual model usable to facilitate the integration and analysis of data on POPs concentrations is introduced with its multilayer hierarchy of entities (POPs as nomenclature classes, couples "observation – measurement" as content classes). Robust set of statistical methods for processing of time series of concentration data is discussed from the viewpoint of practical implementation within running monitoring programs. It consists of the following components: baseline pollution estimates, uncertainty analyses, spatial extrapolations, effect size estimates, time trend identification and quantification. Development of tools supporting standardized environmental data management is rapidly expanding field of science which results in the following challenges for applied informatics and statistics: log-term sustainability of information systems, data-related metadata coding and archiving, tools for automated integration and reporting of data.

Keywords: environmental monitoring, persistent organic pollutants, data model, data standards, statistics.

J. Hřebíček et al. (Eds.): ISESS 2013, IFIP AICT 413, pp. 204–221, 2013.

1 Introduction – Assessing Impact of Environmental Programs

For the past several decades, there has been considerable effort to establish representative environmental monitoring networks as a response to increasing pollution by hazardous compounds released from various anthropogenic sources. The effort has resulted in more or less specialized monitoring programs, designed to cover various environmental matrices. Recently adopted international environmental policies make it imperative for most of the developed countries to upgrade their national monitoring programs, to keep them sustainable and to share exported data in continental and global level. However, the evaluation of various policies launching environmental interventions still remains a rather tentative domain, which is often reasoned as a consequence of the complexity of the problem, multiplicity of stakeholders to be addressed or by prevalence of multilevel uncertainties [1-2].

Nevertheless, increasing spatial and temporal scales of environmental programs strongly stimulate needs to compare their performance and to evaluate their reachable cost/benefit ratio. Special emphasis is placed on standards related to the quality of source data which form the real base of any functional monitoring [3-4]. Due to recently sharply growing availability of data from networks covering the whole continents or entire globe, the task of analyses performed at larger scales has become feasible, among others due to affordable computational costs and capacity of information systems [5-6]. Many complex methods, models and scenarios have been employed in the assessment of outcomes of environmental programs [7-11]; however, the process of evaluation still basically depends on applied stochastic procedures. From this viewpoint, we can distinguish two principal approaches in the evaluation of environmental programs, regardless of their content and focus, which corresponds to the common classification of statistical techniques:

1. Descriptive evaluation, typically aimed to extract unbiased estimates of proper summary statistics, mostly generated from varying or incomplete database records.
2. Inferential evaluation which is concerned with testing hypotheses and with making predictions or other inferences about examined environment. In this case, the main topic is generalization of outcomes from monitoring programs to a larger scale.

Although both approaches play important role, only inferential methodology can contribute to the impact assessment of environmental interventions:

1. Spatial comparison and/or comparability of different monitoring programs or data sources. Information on inter-regional or inter-continental differences in pollution is helpful in identification of contamination sources.
2. Assessment of time-related trends in pollution which can be attributed to a specific source (in case of increasing trend) or to a particular effective prevention or intervention (in case of decreasing trend). Statistically significant decreasing trend in pollution can be accepted as proof of impact of the examined program.

Impact assessment is becoming obligatory component of current environmental programs, which, however, must be intrinsically incorporated in the study plan from their very beginning. Otherwise we must solve many problems like selection bias, insufficient comparability of sites or impact of historical events which coincided with the evaluated program and thereby provided an alternative explanation for the observed trends. To avoid these problems we need properly designed monitoring plans which can transform the collation of data into robust information systems enabling standardized comparative analyses of sites and regions. The main goals of this paper are to outline statistical methods useful in such analyses, to discuss robust approaches which can overcome weaknesses associated with historical environmental data and to document rules to be adopted in prospectively planned monitoring studies. Main recommendations will be illustrated using example of monitoring of air pollution by persistent organic pollutants (POPs).

2 Current Limits in the Evaluation of Environmental Programs

One of the main causes of difficulties in the evaluation of environmental programs is inadequately designed databases which then come into a conflict with desired endpoints. These contradictions form an essential basis of many problems that cannot be solved retrospectively once the data are obtained in inadequate form. To avoid it, proper standards including statistical power planning should be implemented in obligatory guidelines of any "evidence-based" monitoring program. This results in challenges for current development in this field.

Most problems limiting information value of the data arise from the heterogeneity of input data ranging from laboratory biotests to multilevel epidemiologic observations. There is no "universal" truth in the outcomes of various environmental programs. Progress increasingly requires standardized access to multi-disciplinary information resources, including chemical, geological, meteorological, epidemiologic and demographic data. Each broadly ranged ecological study must adopt both following scenarios [12-13]:

1. retrospective exploitation of data sources and their description in discovery process;
2. prospective arrangement enabling effective electronic data capture in future.

Another source of problem is proper selection of territories where the monitoring is conducted. Especially for long-term programs it might be difficult to keep the network sustainable in sites with stable environmental conditions. In this sense, the information sustainability of the programs means that they are able to gather data in continuously comparable sources (sites), matched in a paired (repeated measures) design. The network cannot randomly assign data obtained from individual sites in past to recent outcomes from different sites or to some randomly selected control (background) sites. Growing number of studies is not accompanied by adequate progress in information background and prospectively optimized plans [14-15]. It inevitably results in publishing of non-consistent outcomes with ad hoc data management support. To discover such broadly heterogeneous data we need consensus on data and metadata standardization, but this alone is not enough. We need sufficiently comprehensive conceptual models, advancing development of formal ontologies over environmental data capture systems.

Although there are some usable standardizing concepts already published (Ecological Metadata Language) [16-17], they are not extensively used in practice. Environmental data collection is still subject to research in the informatics field. Especially in global monitoring networks, we need standardized documentation of data in the concept of metadata. Metadata may contain handling instructions for data, such as details of its format, information about its quality, and information required by the user to determine its accessibility for different modes of use.

To conclude, widespread support for ontology-based approaches is not implemented in the field of ecological risk assessment [18]. However, interest in developing ontologies has been growing, because new synthetic environmental analyses increasingly rely on access to a broad range of cross-disciplinary data sources and monitoring studies. The effective system should encompass not only structure and content of such data repositories, but also hierarchical architecture and mutual relationships among components [19–21].

3 Persistent Organic Pollutants (POPs) as Model

To illustrate the role of standardized data management in environmental studies, we took persistent organic pollutants (POPs) as a proper model namely due to the following reasons:

- POPs have become recently intensively studied due to their properties which represent remarkable risk for ecosystems and human population (persistence, bioaccumulation, carcinogenicity, genotoxicity).
- Clear nomenclature of POPs serves as suitable classification system which facilitates development of formal concepts based on well defined entities.
- POPs serve as a proper model for global monitoring studies because these pollutants are in focused attention of several widespread monitoring programs which mostly deal with air sampling as driven by the Global Monitoring Plan (GMP) of the Stockholm Convention (SC; negotiated under the auspices of the United Nations Environmental Program, UNEP) [22]. Any contribution to the standardization of these monitoring activities can be extensively utilized by the running active programs which have been recently overviewed by Klanova and Harner [23]. Moreover, more than twelve years after the adoption of the Stockholm Convention, the questions on its effectiveness prioritize development of functional IT infrastructures with capability to address both recent and historical time trends in POPs fate in the environment.
- The global framework established by the Stockholm Convention is vital and targets specific groups of POPs [24]. The decisions by the Conferences of the Parties to the Basel, Rotterdam and Stockholm conventions on enhancing cooperation and coordination among the three conventions invite Parties to coordinate their efforts when implementing the three conventions to ensure close cooperation among relevant ministries and programs at the national levels. The Conference of the Parties initiated „the establishment of arrangements to provide itself with comparable monitoring data on the presence of the [POP] chemicals as well as their regional and global environmental transport" [25].

4 Conceptual Data Model for Environmental Monitoring of POPs

Here we proposed simplified conceptual model that should broaden our capability for understanding the validity, content and relevance of the data coming from environmental monitoring (Table 1). The model is based on hierarchically layered architecture providing different levels of classifiers or properties of homologous entities as well as scoring of data origin. The model works with three principle layers:

1. **Entities** defined on the basis of internationally standardized nomenclature systems. The level is linked with classifiers, i.e. given properties extending the nomenclature and filtering homologous groups of entities (in this example: POPs).
2. **Observation – measurement** level and its descriptors, focused namely on time & space coordinates, methodical attributes, measured endpoints, reference benchmarking of their value and validity scoring.
3. **Content** identification describing employed measures, units and precision estimates.

Important principles applied in the model construction are further summarized as follows:

1. Reduction of the number of object properties. When necessary, the set of attributes can be expanded before specifically designed data discovery. Too many object properties cannot be utilized efficiently in retrospective exploitation of the resources.
2. Any relevant data discovery must reflect heterogeneity of experimental and methodical approaches. That is why type of the study or data resource is obligatory attribute among the observation descriptors.
3. A measured value cannot be interpreted without reference to a defined/known measurement standards or reference benchmarks. Both internal reference norms (e.g. self-benchmarking of time series data) and external benchmarks (e.g. background concentration levels or limits, hygienic norms, detection limit of applied method) can be applied.
4. Descriptors of measures must fulfill obligatory measurement standard, i.e., the units, scales and lists of attributes defining origin of the measures (e.g., examined matrix, sampling methods, etc.)

The model template enables the IT tools to integrate various data resources. The relevance of the integration process relies on the ability to determine if two values (studies) are compatible, not only in time and space coordinates. Description of model levels in Table 1 implies multivariate interactions of classifiers extending the nomenclature. The most important added value of the model is the capability to determine whether two data sets can be either fully or partially merged or mutually related (correlated) once they are discovered. During the decision process the system undertakes important steps in each level of its architecture:

1. The system must control relevance and compatibility at the taxonomic level (nomenclature) and in space & time localization of data resources
2. Identified data resources must be assessed whether and at what semantic resolution the data are compatible (level of classifiers and/or extending descriptors like type of the study, etc.)

Table 1. Conceptual model proposed for environmental monitoring focused on persistent organic pollutants (POPs)*

CONCEPTUAL MODEL - LEVELS	DEFINITION & COMMENT
RESOURCE IDENTIFICATORS	
	Obligatory descriptors identifying institution (project) which guarantees the data (mostly also as owner of the resource content). In already closed resources, the identification is supplied with overall time/ space description.
OBJECTS – KEY ENTITIES	
NOMENCLATURE	Internationally used nomenclature of POP compounds [23-24]. It allows selection of individual compounds and their groups.
OBJECT CLASSIFIERS	Categorized classifiers derived from external (encyclopedic) sources of information. Classifying criterion is linked to the individual compounds and/or to their groups. Classifiers typically define groups but can be used also for scoring of individual records.
Attributes (physical properties, carcinogenicity, etc.)	Attributes extending recognition of nomenclature classes, coded as no/yes/suspected. Code is directly interlinked with individual compounds. The information can be extracted from accessible international databases.
Reference concentration values	Internal (time series analysis, background values) and/or external reference benchmarks. The classifier is coupled with given entity (compound), typically with direct link to matrix sampled and method used.
OBSERVATION – MEASUREMENT (OM)	
TIME & SPACE COORDINATES	Obligatory attributes, also proposed as inescapable items of any data standard.
STUDY TYPE (design)	Study type (list): Long-term environmental monitoring / Short-term environmental monitoring / Case studies / Screening.
PROBLEM STUDIED (exposure)	Problem studied (list): Accident, short-term exposure / Long-term exposure / Random inspection (survey of some area) / Examination of background (reference) site
METHODICAL ATTRIBUTES Measured entities Matrix Experimental unit Sampling methods Analytical methods	Obligatory identification of observation – measurement, necessary for interpretation of measured values. Measured entities select nomenclature items which are examined in given environmental matrix (soil, sediment, water, air, biota). Experimental unit identifies context of measured values (micro-samples within site, site – single sample, site – mixed sample, sample mixed across sites). Sampling and analytical methods fulfill minimized list of items which follows standardized norms and guidelines.
CONTENT	
Measures Units Precision measures	Content of the resource, in case of POPs mostly concentration levels in internationally standardized unit scales. Precision measures include sample variability (in concentration units) or detection limits of performed analytical methods.

* Example based on the solution adopted in the Global Monitoring Plan of the Stockholm convention [22], [24], [31]

Finally, the measurement standards for the mutually related environmental endpoints have to be controlled for compatibility (units, scales, reference benchmarking, methodical origin).

5 Projection of Conceptual Model to Data Standards

The quality of conceptual model determines its utility for assisting in data discovery and information searching. However, the applicability strongly depends on quality of description and content of processed data sets. That is why we should insist on minimized obligatory database components and their descriptors. Minimized data model as standard can be used both retrospectively (scoring of validity of discovered resources) and prospectively (when designing new data capture system). Proposed conceptual model intrinsically encompasses these obligatory items: institution (origin of data), time & space coordinates, type of resource, examined entities (compounds), measures and methodical descriptors (values and units, matrices, methods).

The system allows for any type of reasonable extension; additional properties may be added on demand. However, minimized data standard ensures accessibility of key information namely in the measurement level of the model; i.e. when and where measurements were recorded, who recorded each measurement, the methodology of measurements, study design and aims. In this way the model can improve data visibility for search engines and enables greater levels of automation of common data transformation and integration.

Proper conceptual model also contributes to widely recommended discovery of data according to the concepts they really represent [26-27]. Ontology should represent the knowledge in a domain of interest, defined via the terminology (concepts, nomenclature) used within the domain and the properties and relationships among domain objects [16-17]. This concept is fully implemented in the model proposed here; the nomenclature baseline is extended by selected descriptors with defined dependencies. It is a formal framework for observational studies where we adopted structured approach recognizing key entities (nomenclature classes) in the 1st level and their characteristics (classifiers). Second level consists of measurements and their characteristics, i.e. validity criteria, origin of data, etc. Third level covers content identification, namely values and units, scales.

6 Impact of Conceptual Model on Standardized Data Management

Implementation of conceptual data models increases the usage value of monitoring programs and facilitates their sustainability. To keep the information system for long-term monitoring activities sustainable, it is necessary to develop some strategy for handling the incorporation of consecutively gathered data into the archive. That is why the description of data, methods and sites in standardized metadata are vitally important – it is the only way how to track the changes in sites, regional networks, etc. [28]. In any evidence-based environmental monitoring, the data structure must be well defined, but flexible enough to reflect a wide range of possible hypotheses. Of course, such system cannot be constructed retrospectively, on demand of running analyses. Baseline standardization through conceptual data model positively impacts upon analytical procedures, namely in the following four fields:

- **Integration of disparate terminology** which facilitates standardized management of metadata, digital libraries and specific tools working with data.
- **Hierarchical structure advances the data analysis.** The conceptual model intrinsically distinguishes hierarchy of levels and descriptors which facilitates implementation of tools focused on data analysis and knowledge mining. The position of nomenclature entities and measurements can be used to denote a wide range of entity characteristics (nominal or ordinal measures). Using the hierarchy of descriptors we can easily decide whether the data are useful for a particular analysis.
- **Conceptual model supports robust comparison of values.** Regarding data analysis, very important attribute is incorporation of measurement level and its characteristics. Validity criteria reflect some precision measures as well as reference values or protocol standards. A measured value cannot be interpreted and analyzed without reference to a defined measurement standard.
- **Stratified analyses and integration of different data sources.** Hierarchical relationships among nomenclature classes and descriptors also potentiate development of automated SW tools for comparison of values using different strata. For example we can summarize concentrations of some pollutant according to site locations because the sites and their load provide context for observation of contamination. The sites and matrices thus provide a context for measurement of POPs exposure. Different summaries can be then interlinked using various time frames. The concept facilitates data integration, reasoned by compatibility of interlinked values.

7 Environmental Programs Require Robust Statistical Processing

From the viewpoint of statistics, environmental monitoring can be characterized as processing of heterogeneous data leading to probabilistic estimation of some uncertain (prospective approach) or on the other hand relatively certain (retrospective approach) risk event. Environmental studies are complex and require processing of highly diverse data. Even if we can recently get standardized, structured data fully accessible for analyses, it is often difficult to select the best approach how to

- extract information from data collected in past, mostly prior to standardization;
- realize later analytic steps which assume some specific input or data aggregation from the preceding measurements;
- merge different data sets collected in different times ;
- mutually relate various measured factors.

To illustrate the role of statistical methods typically employed in analyses of long-term monitoring studies we use already mentioned Global Monitoring Plan (GMP) of the Stockholm Convention (SC) which is focused on specific groups of POPs [22], [24-25]. The GMP gathers relevant global data on POPs concentrations in collection campaigns consecutive in time. The volume and diversity of the data generated by the global campaigns necessitated a centralized approach to quality assurance, data management and robust statistical processing. The principal methodical goals are to:

- summarize the GMP data into a comprehensive report, database and on-line reporting tool,
- propose methodology for the effectiveness evaluation of the existing data to be used as baseline estimates against future concentration data which can be measured,
- recommend a strategy for further environmental monitoring efforts to assess temporal and spatial trends in POPs concentrations.

Following six methodical steps were recommended for the statistical processing of the GMP data with respect to objectives of the GMP [24-25] (Table 2):

A. Exploratory Statistics

1. Data and metadata pre-processing.
2. Summary statistics of the baseline concentrations of POPs.
3. Analysis of uncertainty.

B. Inferential Statistics

4. Power analysis.
5. Stochastic identification of the time trends.
6. Quantification of time trends.

7.1 Data and Metadata Pre-processing

Correct definition of data is an unavoidable prerequisite of all subsequent statistical analyses in any monitoring. Only reliably reported values can be accepted for any spatial or temporal comparisons. The prior evaluation procedures guarantee comparability of different samples, especially from the viewpoint of site type, matrix, sampling method, time span and sampling frequency. Heterogeneity in these factors might dramatically increase the uncertainty in the final outcomes. The pre-processing procedures also limit the impact of uncontrolled covariates and thus reduce the risk of a false detection of a trend or the risk of neglecting truly significant changes.

Metadata as "data describing data" represent extremely important standardizing component, mainly in long-term programs where historical datasets have been planned to be merged with the recent ones. Each data set should be supplied with the following minimum metadata profile:

- site identification (coordinates, name, ID, responsible data manager)
- site typology (background vs. impacted, type of anthropogenic influence, etc)
- information value of the program (type: long-term, occasional, ad hoc study, etc.)
- information value of data (primary data vs. aggregated values, type of aggregation)
- sampling strategy and frequency, applied design
- quality control (scoring of usability of records)

Table 2. Components of robust set of statistical methods recommended for long-term monitoring studies*

COMPONENT	ASSUMPTIONS – OUTCOMES
I. Data and metadata pre-processing	
Assumptions	Recorded primary data in the database (not only e.g. annually aggregated values); accessibility of variability measures for any type of central summary statistics
Outcomes	Initial data filtering according to the objective entities such as site-matrix type and analyzed compounds. The filters also check/verify completeness of the primary database records (sample distribution profile, reported sampling frequency, no. of detected LOQs, LOQs handling rules). Identification and exclusion of proved extreme or unreliable values (e.g. checking their quantile position in the sample distribution function, probabilistic reconstruction of sample distribution, etc.).
II. Summary statistics and baseline estimates	
Assumptions	Correct coding of primary data (direct sampling records) or aggregated values with proper variability measures
Outcomes	Baseline estimates of quantitative values (here concentration of POPs fro give time period. Two approaches are principally recommended for an assessment of the concentration summary statistics: a) Median estimate reported together with a 5th-95th percentile range b) Geometric mean estimate based on the log-transformed annual averages with a corresponding 95% confidence interval
III. Uncertainty analysis	
Assumptions	Correctly spatially references measured values, accessible variability measures for any kind of summarized statistics
Outcomes	Information on intra-regional and inter-regional homogeneity Spatial and temporal benchmarking of the sites (or other defined territories) Check of merging of data from different sites, time periods, etc. Sensitivity analysis tracking changes in information outcome when the database is changed (updated, split, etc.)
IV. Power analysis - effect size estimates	
Assumptions	Evidence – based design of the program: i.e. known sampling strategy, type of collected data, primary end-point and estimate of expected ranges in measured values. Possibility to test the power calculation in a pilot study performed on available primary data sets.
Outcomes	Effect size estimate: quantification of the minimum detectable difference as a base for relevant estimates spatial differences, changes over time and, if possible, the time trends. Defined credibility of a program for future use of its data.
V. Time trend identification	
Assumptions	At least 4-5 consecutive time points (typically years) reported for the examined time series. Possibility to filter out confounding or masking components like seasonality.
Outcomes	Statistically proved time trend and its direction.

Table 2. (*continued*)

VI. Time trend quantification	
Assumptions	Consecutive time points with known variability behind used values (e.g. annually aggregated averages). For relevant estimates, at least 4-5 consecutive time points are preferred rather than two point differencing.
Outcomes	Time-related difference in the measured value, expressed with the following attributes: - difference as an absolute value expressed in the concentration units - relative (%) change expressed as an index of the value detected in the initial time point (year) - 95% confidence interval of the time-related difference - p value of the trend test - corresponding minimum detectable annual difference

* Example customized for time series of POPs concentrations in the Global Monitoring Plan of the Stockholm convention [22, 24, 31]

7.2 Summary Statistics of the Baseline Concentrations of POPs

Annually aggregated POPs concentrations calculated as arithmetic means of the primary values were recommended for quantitative analyses. The quantitative estimates of concentration values in a specific historical period are called baseline estimates. Variability of baseline values can be quantified on a local, regional or global level, merging appropriate data sets. Pooling of underlying data (both primary and aggregated), however, has to be supported by an uncertainty analysis (see Chapter 7.3). Non-parametric tests like Mann-Kendall U test or Kruskal-Wallis test are recommended for the inter-regional comparison of the POPs concentrations. Parametric tests like ANOVA models or analysis of covariance can be applied only after effective normalizing transformation of the concentration estimates.

7.3 Analysis of Uncertainty

The decision makers working with the outcomes of POPs monitoring meet widely varying types of information, each type being associated with its own uncertainties. Local occasional measurements reveal different distributional profiles than long-term, regularly sampled time series. Although the POPs concentrations are mostly collected as a part of some large monitoring network, they must be correctly spatially referenced, since they apply to a specific territory (site). The uncertainty measures therefore mostly reflect spatial and time-related variability [29].

As data reported to the global monitoring networks are typically generated by a variety of programs from various background sites, they have to be inspected for an intra-regional and inter-regional homogeneity in the annually averaged values. Graphically, the regional variability can be reported as their intra-regional 5th-95th percentile range. Sample distribution functions of the regional samples can be then compared and tested by proper robust method (Kolmogorov-Smirnov test, Kruskal-Wallis test). The same applies for geometric means of the averaged concentrations and their 95% confidence intervals.

Another test of relevance is also an examination whether the information outcome has changed if the dataset was improved or degraded. Multimodal distributional patterns, which often indicate the influence of some "effect modifying factor", can be proved and distinguished from unimodal sample distribution [30]. The homogeneity should be also assessed in the time trend analyses (i.e. presence and the same direction of the trend changes and the annual difference). A year-to-year difference can be compared among the time-series based on the individual sites. Such variability can be expressed as standardized year-to-year difference or as coefficient of variation (expressed in %). Application of the time-related regression models and their residuals is possible as well. In accessible time series, homogeneity (or non-homogeneity) in a year-to-year variance indicate representativeness and stability of the identified time trends. The time series reported from various sites can be merged for more powerful trend change analysis only if their homogeneity was proved.

7.4 Power Analysis

A power analysis is an obligatory step defining magnitude of changes reliably detectable by the statistical methods. Power analysis minimizes risk of the wrong conclusion, misinterpretation or confusing generalization of the observed values. Although its value is widely known, it seldom receives attention in the environmental literature. However, it is essentially vital for establishing program credibility, for supporting effectiveness analyses, and for creating a permanent record of its data for future use.

The relevance of data should be scored mainly with respect to a required decision-making process. In case of global POPs monitoring it is evidently the power of collected data to distinguish substantially increased or decreased POPs levels in the examined matrices and territories in time. Therefore, any time trend analysis must be accompanied with power analysis and the identified trends must be always reported together with corresponding minimum detectable difference. The power analysis estimates a minimum difference between two annually aggregated concentration values detectable by paired t-test on log-transformed data ($\alpha = 0.05$ and $\beta = 0.20$). Appropriate non-parametric alternative like Wilcoxon-rank-sum test or Mann-Kendall test can be used as well, especially when the analysis is based on primary, rather than normalized concentration data.

7.5 Stochastic Identification of the Time Trends

The time trends are identified via qualitative test of statistical significance of the time-related changes observed in the consecutive measurements. At least five consecutive annually aggregated concentration values are required when assessing time trends using one of the following robust techniques:

- The Daniel's test as an application of the Spearman's rank correlation coefficient between the concentration values and corresponding time ranks.
- The Mann-Kendal test as a non-parametric test for detection of a trend in the time series based on binary coding of the changes in measurements consecutive in time.

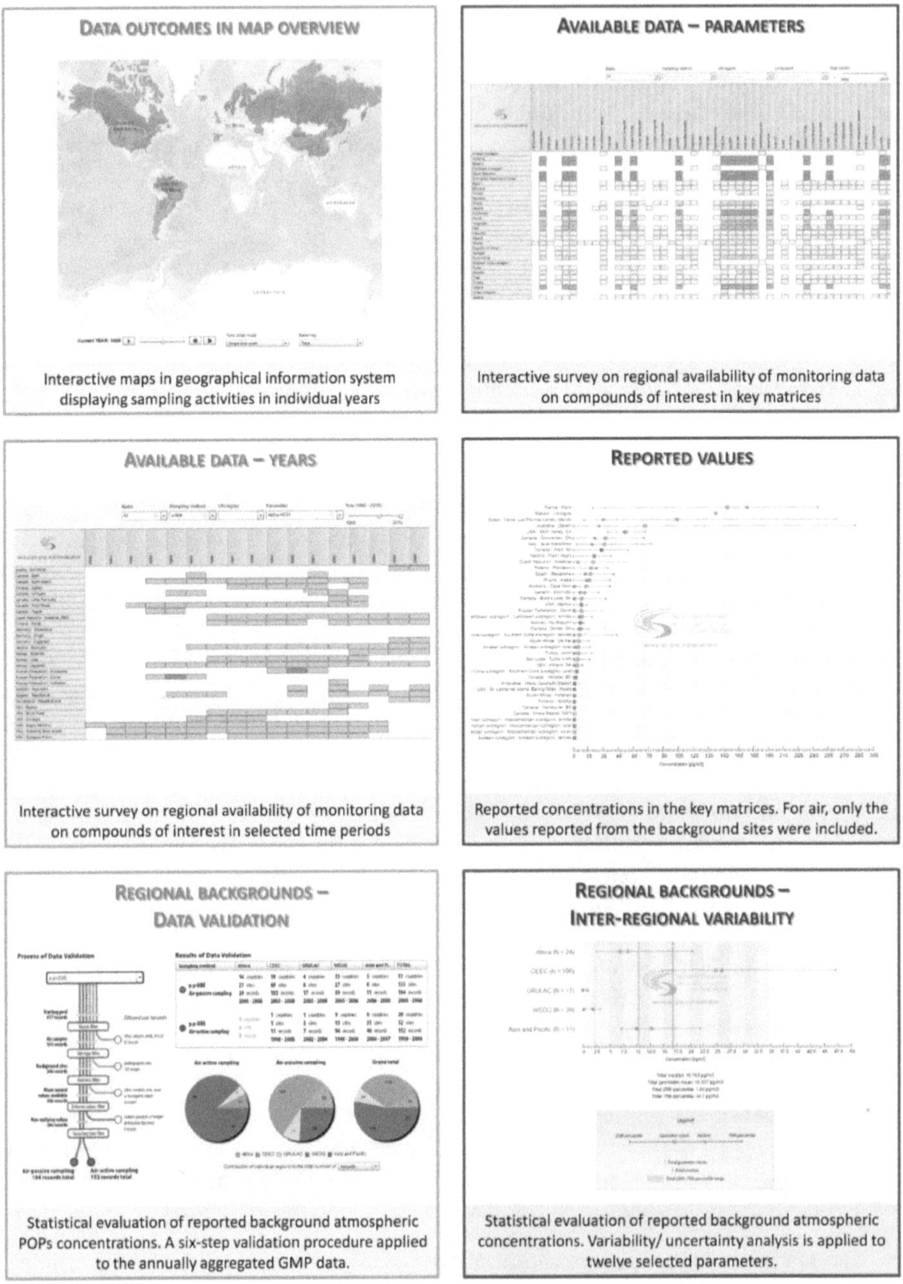

Fig. 1. Global Monitoring Plan of the Stockholm Convention on Persistent Organic Pollutants: implemented reporting and on-line retrieval of data. Available from: www.pops-gmp.org [31].

Direction of the time trend (concentration values increasing or decreasing in time) has to be recorded whenever it is confirmed as statistically significant. In addition, any concentration change over time should be reported in the same way, although there is no exact statistical significance behind. Both statistically significant and non-significant time changes over time must be correctly quantified in the reports and marked with p value generated by appropriate tests (see Chapter 7.6).

7.6 Quantification of Time Trends

Quantification of a time trend should be performed whenever the proper statistical tests confirm significant and consistent time-related differences in the POPs concentrations. Quantified trend means a difference $\Delta = y1 - y2$, where y1 respectively y2 correspond to the annually aggregated concentration values recorded in two different consecutive years. We should be aware of the fact that the recent trend estimates based on just two points („start–end") differencing can be biased by random errors or fluctuations behind such occasionally measured values. Recent trends should be also related to historical changes occurred in the past because historical trends determine observable changes in more recent periods.

8 Example of Standards Implemented in Global Monitoring of Atmospheric Pollution by POPs: Data Repository with Automated Data Integration and Reporting

The illustrated conceptual model (Table 1) as well as robust statistical processing and reporting of POPs concentration data (Table 2) have already been implemented in a functional data repository and SW toolkit forming IT infrastructure of the Global Monitoring Plan of the Stockholm Convention as defined in Chapter 3 of this paper [22], [24-25]. The methodology, data outcomes and user methodical guide are published in the portal www.pops-gmp.org [31].

The information system distinguishes object entities (nomenclature items) and enables users to stratify accessible measures (content of resources) across a set of classifiers and methodically important attributes. Maintenance of data in an online system is stable and associated with allocated resources for maintaining global connectivity, timely migration to new hardware and inevitable content updates.

The system serves as an example of IT background of global infrastructure for global monitoring and data management. The created data repository is capable of active accumulation of datasets from various sources and ensures their long-term accessibility. Adoption of standards documenting content of the databases, both at the level of data and metadata, enhances usability of the data even for most demanding endpoints, such as long-term trend assessment.

In the centralized data repository, the metadata description is inextricably linked to the datasets, archived and managed in conjunction with the data. The coding of site and data description finally leads to a knowledge-based system which involves gathering and encoding of various types of missing values, values lower than limits of quantification, thesauri, classification rules. Consequently, it allows for implementation of some codes driving automated pre-processing and reporting of imported data which acts upon this coded additional information.

Finally, on-line working interactive communication and reporting level allows users to keep control over the data and to use them, e.g., for regional purposes. The system is equipped with flexible interactive on-line browsing and reporting instruments which are customized for the primary datasets in the data warehouse. Development of such interactive retrieval of data required high degree of familiarity with the data structure and its interpretation; the usage of reports, however, is intuitive and does not limit users, even those without any experience in the field. The outcomes of the reports are documented in Figure 1.

9 Conclusions and Future Challenges

In this study, we summarized main problems associated with the assessment of performance and impact of long-term environmental programs. Lack of data standards, incompleteness of archived datasets and insufficient statistical power were identified as main limits of functionality of monitoring networks. It is evident that we cannot carry out adequate assessment of the large scale systems with techniques that have been successful in smaller systems with limited heterogeneity [32].

Progress in environmental monitoring, which has been currently advanced by molecular and genomic techniques, need to be supported by development of novel type of IT infrastructures [33-34]. These tools should reflect more universal standards of environmental metadata that go beyond the scope of a particular program or case study [35]. Another inevitable priority is the sustainability of the long-term monitoring programs which does not mean only archiving of gathered datasets. The challenging dimension of sustainability, which is not yet adequately reflected in the literature, includes storing and merging of different datasets in a way maintaining their usage value and information potential [36]. Completed data should remain available and usable after the end of projects. The projects must ensure that the data are captured in a structured way, with necessary metadata description [37]. Nowadays, such a goal is best achieved by utilizing an electronic database coupled with appropriate procedures and data flow facilitated by active participation of data managers and scientists [38].

However, there are still many barriers remaining, especially in global monitoring networks. Data heterogeneity, geographic dispersion of sampling places, funding limitations, incompleteness in time series and other factors often lead to localized, "file-based" solutions [39]. Building centralized data-based archives capable of ensuring retrospective collection of data from these partial repositories represents another priority of current strategies for environmental monitoring. Emphasized attention to the structured collection and correct coding of metadata as well as the primary data must become a key component of either functional approach, not only in eco-informatics [40].

Acknowledgements. This work has been supported by the project TB010MZP058 "Development of the system for spatial evaluation of the environmental contamination".

References

1. Mickwitz, P.: A framework for evaluating environmental policy instruments: Context and key concepts. Evaluation 9, 415–436 (2003)
2. Crabbé, A., Leroy, P.: The handbook of environmental policy evaluation. Earthscan, London (2008)
3. Goodchild, M.F.: Communicating the results of accuracy assessment: metadata, digital libraries, and assessing fitness for use. In: Mowrer, H.T., Congalton, R.G. (eds.) Quantifying Spatial Uncertainty in Natural Resources: Theory and Applications for GIS and Remote Sensing, pp. 3–15. CRC Press (2000)
4. Rönnbäck, B.I., Nordberg, M.L., Olsson, A., Östman, A.: Evaluation of environmental monitoring strategies. Ambio 32, 495–501 (2003)
5. Gray, J., Liu, D., Nieto-Santisteban, M., Szalay, A., DeWitt, D., Heber, G.: Scientific Data Management in the Coming Decade. CTWatch Quarterly 1, 17–26 (2005)
6. Brunt, J.W., McCartney, P., Baker, K., Stafford, S.G.: The future of eco-informatics in long-term ecological research. The International Institute of Informatics and Systemics IIIS 7. In: Proceedings of the 6th World Multi-Conference on Systematics, Cybernetics and Informatics, Orlando, FL, USA, July 14–18, pp. 367–372 (2002)
7. Dardari, D., Conti, A., Buratti, C., Verdone, R.: Mathematical evaluation of environmental monitoring estimation error through energy-efficient wireless sensor networks. IEEE. Trans. Mob. Comput. 6, 790–802 (2007)
8. Svenfelt, A., Engström, R., Hojer, M.: Use of explorative scenarios in environmental policy-making-Evaluation of policy instruments for management of land, water and the built environment. Futures 42, 1166–1175 (2010)
9. Mermet, L., Bille, R., Leroy, M.: Concern-Focused Evaluation for Ambiguous and Conflicting Policies: An Approach From the Environmental Field. Am. J. Eval. 31, 180–198 (2010)
10. EPA (Environmental Protection Agency): EPA requirements for quality assurance project plans. EPA QA/R-5, United States Environmental Protection Agency, Washington, DC (2001)
11. EPA (Environmental Protection Agency): STORET, EPA's largest computerized environmental data system (2008), `http://www.epa.gov/storet/`
12. Michener, W.K., Brunt, J.W.: Ecological Data: Design, Management and Processing. Blackwell Science, Oxford (2000)
13. Jones, M.B., Schildhauer, M., Reichman, O.J., Bowers, S.: The New Bioinformatics: integrating ecological data from the gene to the biosphere. Ann. Rev. Ecol. Evol. Syst. 37, 519–544 (2006)
14. Elmagarmid, A., Rusinkiewicz, M., Sheth, A.: Management of Heterogeneous and Autonomous Database Systems, vol. 4. Morgan Kaufmann, San Francisco (1999)
15. Grossman, D.A., Frieder, O.: Information Retrieval: Algorithms and Heuristics. Springer, Heidelberg (2004)
16. Darwin Core: Darwin Core Schema (version 1.3), a draft standard of the Taxonomic Database Working Group (TDWG), `http://wiki.tdwg.org/DarwinCore`
17. DCMI: DCMI Metadata Terms, `http://www.dublincore.org/documents/dcmi-terms`
18. Huang, P.S., Shih, L.H.: Effective environmental management through environmental knowledge management. Int. J. Environ. Sci. Tech. 6, 35–50 (2009)
19. Madin, J., Bowers, S., Schildhauer, M., Krivov, S., Pennington, D., Villa, F.: An ontology for describing and synthesizing ecological observation data. Int. J. Ecol. Informatics 2, 279–296 (2007)

20. Williams, R.J., Martinez, N.D., Golbeck, J.: Ontologies for ecoinformatics. J. Web Semant. 4, 237–242 (2006)
21. Hale, S.S., Hollister, J.W.: Beyond data management: how ecoinformatics can benefit environmental monitoring programs. Environ. Monit. Assess. 150, 227–235 (2009)
22. Stockholm Convention on Persistent Organic Pollutants (POPs): Interim Secretariat for the Stockholm Convention, United Nations Environmental Programme (UNEP) Chemicals: Geneva, Switzerland (October 2001), `http://chm.pops.int`
23. Klanova, J., Harner, T.: The challenge of producing reliable results under highly variable conditions and the role of passive air samplers in the Global Monitoring Plan. Trends Anal. Chem. 46, 139–149 (2013)
24. Secretariat of the Stockholm Convention: UNEP Report of the First Expert Meeting to update the Guidance on the Global Monitoring Plan for Persistent Organic Pollutants (2010), `http://chm.pops.int/Programmes/GlobalMonitoringPlan/Meetings/GMP1stExpertMeeting2010/tabid/760/ctl/Download/mid/3261/language/en-US/Default.aspx`
25. United Nations Environment Programme: Guidance on the Global Monitoring Plan for Persistent Organic Pollutants. Secretariat of the Stockholm Convention on Persistent Organic Pollutants, Geneva (2007)
26. Berkley, C., Jones, M.B., Bojilova, J., Higgins, D.: Metacat: a schema-independent XML database system. In: Proc. of the 13th Intl. Conf. on Scientific and Statistical Database Management. IEEE Computer Society Press (2001)
27. Borgida, A.: Description logics in data management. IEEE Trans. Knowl. Data Eng. 7, 671–682 (1995)
28. Cook, R.B., Olson, R.J., Kanciruk, P., Hook, L.A.: Best practices for preparing ecological data sets to share and archive. Bulleting of the Ecological Society of America 82, 138–141 (2001)
29. Cornélis, B., Brunet, S.: A policy-maker point of view on uncertainties in spatial decisions. In: Shi, W., Fisher, P.F., Goodchild, M.F. (eds.) Spatial Data Quality, pp. 168–185. Taylor & Francis (2002)
30. Hartigan, J.A., Hartigan, P.M.: The Dip Test of Unimodality. Ann of Stat. 13, 70–84 (1985)
31. Hůlek, R., Jarkovský, J., Borůvková, J., Kalina, J., Gregor, J., Šebková, K., Schwarz, D., Klánová, J., Dušek, L.: Global Monitoring Plan of the Stockholm Convention on Persistent Organic Pollutants: visualization and on-line analysis of data from the monitoring reports. Masaryk University (2013), `http://www.pops-gmp.org/visualization`
32. Christensen, S.W., Brandt, C.C., McCracken, M.K.: Importance of data management in a long-term biological monitoring program. Environ. Manage. 47, 1112–1124 (2011)
33. Emerson, K.J., Merz, C.R., Catchen, J.M., Hohenlohe, P.A., Cresko, W.A., Bradshaw, W.E., Holzapfel, C.M.: Resolving postglacial phylogeography using high-throughput sequencing. Proc. Natl. Acad. Sci. USA 107, 16196–16200 (2010)
34. Sork, V.L., Waits, L.: Introduction: contributions of landscape genetics approaches, insights, and future potential. Mol. Ecol. 19, 3489–3495 (2010)
35. Metzger, K.J., Klaper, R., Thomas, M.A.: Implications of informatics approaches in ecological research. Ecological Informatics 6, 4–12 (2011)
36. Ozmen-Ertekin, D., Ozbay, K.: Dynamic data maintenance for quality data, quality research. Int. J. Inform. Manage. 32, 282–293 (2012)
37. Michener, W.K.: Meta-information concepts for ecological data management. Ecol. Inform. 1, 3–7 (2006)

38. Horsburgh, J.S., Tarboton, D.G., Piasecki, M., Maidment, D.R., Zaslavsky, I., Valentine, D., Whitenack, T.: An integrated system for publishing environmental observations data. Environ. Modell. Softw. 24, 879–888 (2009)
39. Bertzky, M., Stoll-Kleemann, S.: Multi-level discrepancies with sharing data on protected areas: What we have and what we need for the global village. J. Environ. Manage. 90, 8–24 (2009)
40. Michener, W.K., Jones, M.B.: Ecoinformatics: supporting ecology as a data-intensive science Trends Ecol. Evol. 27, 85–93 (2012)

Tools for Collection, Analysis and Visualization of Data from the Stockholm Convention Global Monitoring Plan on Persistent Organic Pollutants

Jakub Gregor[1,2], Richard Hůlek[1,2], Jiří Jarkovský[1], Jana Borůvková[2], Jiří Kalina[1,2], Kateřina Šebková[2], Daniel Schwarz[1], Jana Klánová[2], and Ladislav Dušek[1]

[1] Institute of Biostatistics and Analyses, Masaryk University,
Kamenice 126/3, 625 00 Brno, Czech Republic
{gregor,hulek,jarkovsky,kalina,schwarz,dusek}@iba.muni.cz
[2] Research Centre for Toxic Compounds in the Environment, Masaryk University,
Kamenice 753/5 (A29), 625 00 Brno, Czech Republic
{boruvkova,sebkova,klanova}@recetox.muni.cz

Abstract. The Global Monitoring Plan for persistent organic pollutants is an important component of the effectiveness evaluation of the Stockholm Convention and its main objective is assessment of long-term changes in POPs concentrations in core matrices – ambient air and human tissues (milk, blood). This paper summarizes results of activities of the Research Centre for Toxic Compounds in the Environment and the Institute of Biostatistics and Analyses, Masaryk University, Czech Republic, which have been performed on the basis of the mandate given by Global Coordination Group for GMP and Secretariat of the Stockholm Convention: content analysis of the GMP monitoring reports published in 2009, on-line visualization tool for browsing and analyzing collected data from the monitoring reports, and proposal of a design of future data collection campaigns.

Keywords: Stockholm Convention, Global Monitoring Plan, POPs, database, data collection, visualization.

1 Background

Multilateral environmental agreements on chemicals management have brought requests for developing instruments that are able to collect or evaluate data and establish and/or optimize risk management related to increasing levels of chemicals in the environment. Such instruments can be employed on both national and international level.

Development and adjustment of systems for the collection, analysis, and visualization of environmental data must cope with relatively high heterogeneity of collected data, e.g. in terms of data sources (institutions, projects, purposes), different matrices (ambient air, water, soil, sediments, human tissues), or chemical parameters (isomers, degradation products). It is therefore important to strictly define data structure and code lists to ensure reliability of all collected and analysed data and provide detailed

J. Hřebíček et al. (Eds.): ISESS 2013, IFIP AICT 413, pp. 222–229, 2013.

guidance and support to all users participating in the data collection process. Furthermore, sufficiently complex conceptual models, advancing development of formal ontologies for environmental and epidemiological data acquisition systems are needed [1,2].

The Global Monitoring Plan (GMP) for persistent organic pollutants (POPs) is an important component of the effectiveness evaluation of the Stockholm Convention (SC) and its main objective is the assessment of long-term changes in POPs concentrations in core matrices – ambient air and human tissues (milk, blood). A methodological guidance document specifying the methods to be used and compounds to be monitored was published in 2007[3]. The first global data collection occurred in 2008 and the results were published in five regional monitoring reports in 2009.

Based on the tasks identified by the Global Coordination Group for GMP and the Secretariat of the Stockholm Convention, the Research Centre for Toxic Compounds in the Environment (RECETOX) and the Institute of Biostatistics and Analyses (IBA MU), Masaryk University, Czech Republic, performed a content analysis of the GMP reports[4], prepared electronic tools for storage and on-line visualization of data published therein[5], and proposed a comprehensive IT solution for future data collection campaigns.

2 Content Analysis of GMP Monitoring Reports

The first step in the analysis of the regional GMP reports was to determine the relation of all reported parameters to the Stockholm Convention. The classification strictly followed scope of the Stockholm Convention and the GMP Guidance document and their amendments in time. Considerations were also given to the reported data content and relevance of existing records for further data collection (Fig. 1).

1. 12 initial POPs included in the Stockholm Convention in 2001, their congeners, isomers and degradation products specified in the GMP Guidance document (2007) – 58 parameters;
2. Additional 10 POPs listed in the Stockholm Convention in 2009 and 2011 and specified in the updated GMP Guidance document (version 2009) – 7 parameters;
3. All other compounds, their sums and toxic equivalents related to the Stockholm Convention but not specified in any of the GMP Guidance documents – 84 parameters;
4. Compounds found in the GMP reports but not related to the Stockholm Convention (i.e. PAHs) – 22 parameters.

The analysis revealed several serious challenges related to data standardization in the GMP report. The reports suffered from the lack of standardized taxonomy for POPs, their isomers, transformation products and summations. Heterogeneity of the data was further enhanced by reporting various toxic equivalents (TEQ) (based on WHO TEF values from various years) rather than concentrations of the individual polychlorinated dibenzo-p-dioxins (PCDDs), Polychlorinated dibenzofurans (PCDFs) and polychlorinated biphenyls (PCBs). Unclear identification of units, time and spatial scales of the reported concentrations as well as insufficient specification of aggregated data belonged to other frequently identified drawbacks. The review, however, resulted in conclusion that the available GMP data can be used for baseline statistical processing. Data can be analysed as annually aggregated time series or at least as relevant point estimates, prepared for comparison with the next data collection campaign.

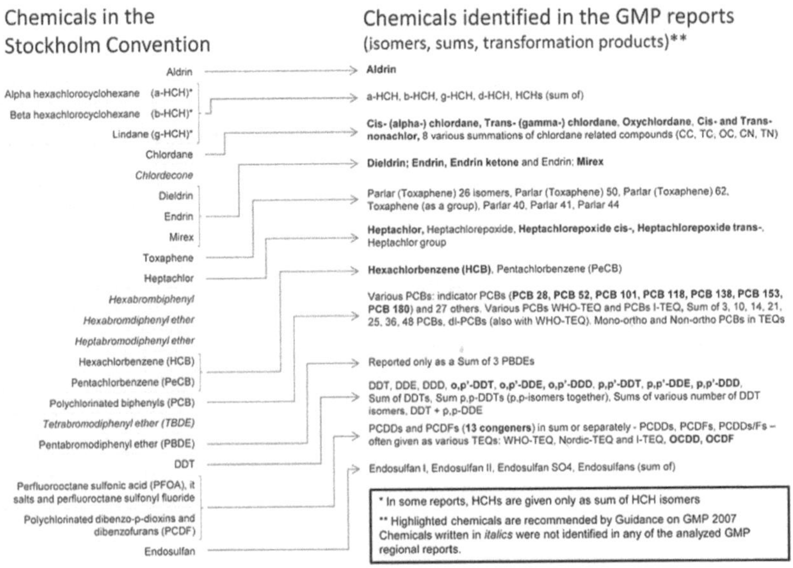

Fig. 1. Overview of chemicals identified in GMP reports

3 GMP Database

Data from the GMP regional reports were transformed into a form that was suitable for both database development and analytical processing. Data had to be extracted from textual form, tables, citations and charts into a common standardized data structure with shared property lists (matrices, chemicals, units…) in several consecutive steps:

1. **Identification of Collected Data.** First of all, an overall inspection of all regional GMP reports was performed. The main goal was to identify reported matrices, measured chemicals (compounds), used values and aggregation characteristics as well as sampling frequency and time ranges.
2. **Datasets Volume Identification.** Determination of volume (range and size) of datasets was equally important, because an appropriate storage technology had to be chosen. A relational database engine was used as the main storage solution in order to enable fast sorting, aggregation and selection.
3. **Design of a Common Data Structure.** For the purposes of the data extraction from particular regional reports, a common data structure was used allowing for maintenances of all collected values – both directly measured and aggregated.
4. **Extraction of Data from Reports into a Common Data Structure.** During the process of data extraction from regional reports, a number of problems had to be

solved. All missing information had to be looked up in the supporting text of the reports or was completed based on the context knowledge. Most of the problems were caused by the non-existent standardized form for data collection.

5. **Data Validation.** The database entries were double-checked after the digitalization. Manual validation was applied particularly to problems with text labels and variables, which could not be automatically solved by analytic approaches. Second step of validation included analytical tools to identify missing values, extreme values, and duplicities.
6. **Data Pre-processing.** Some records in GMP reports contained one value for multiple years, or countries (sub-regions). Such values were individually assigned to all covered countries and/or years.
7. **Data Aggregation.** For the purposes of analytical outcomes from regional reports, annual aggregation of values was used. Aggregations were calculated and prepared using specialized statistic software (SPSS). Import of re-calculated data into the database was done by automated procedures to ensure data consistency and integrity. Different aggregation approaches were used for:

- Primary and aggregated data obtained from the GMP reports
- Different matrices (air and human tissues)

8. **Data Import.** MS Excel sheets were used as data transfer medium and imported into MSSQL Server Database by means of standard embedded functions and features. Microsoft tools and data files were used in all steps to ensure maximal compatibility during the whole process of data transfer and processing.

Aggregated values originating in GMP reports are stored in a relational database consisting of 10 basic entities: UN Region; Country; Site; Matrix; Compound; Parameter; Record; Aggregation; Percentile; Unit.

4 On-Line Data Visualization Tool

In addition to the content analysis of GMP monitoring reports and transferring the published data into the database, an on-line data visualization tool was developed to allow easy searching, browsing, and analysing the GMP data in their annually aggregated form (www.pops-gmp.org). The browser provides easily accessible information on the performance of monitoring programmes in individual countries and/or regions sorted by matrices, time, and compounds. The database underlying the visualization tool was supplemented with additional data from other public sources, in particular from large environmental monitoring programmes, as some conclusions of the regional monitoring reports refer to them (EMEP[1], AMAP[2], IADN[3], MONET[4]). The visualization comprises of three descriptive and three analytical tools:

---

[1] European Monitoring and Evaluation Programme.

[2] Arctic Monitoring and Assessment Programme.

[3] Integrated Atmospheric Deposition Network.

[4] Passive Ambient Air Monitoring Network.

1. **Monitoring Overview.** An interactive map displaying monitoring activities in the individual years.
2. **Available Data – Parameters.** Regional availability of monitoring data on compounds of interest in key matrices.
3. **Available Data – Years.** Regional availability of monitoring data in selected time periods.
4. **Reported Values.** Reported concentrations in the key matrices. For air, only the concentration values reported from the background (urban, suburban, rural, remote, mountain and polar) sites were included.
5. **Regional Backgrounds – Data Validation.** A 6-step validation procedure of data from ambient air monitoring.
6. **Regional Backgrounds – Inter-regional Variability.** Statistical evaluation of reported background atmospheric concentrations.

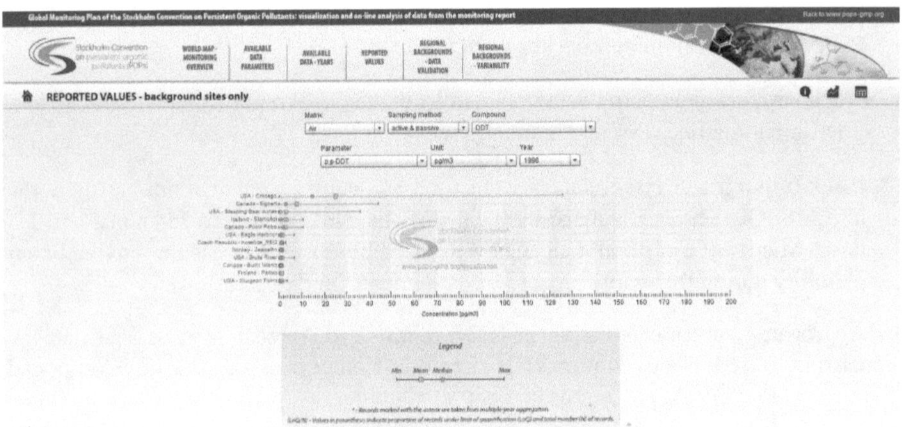

Fig. 2. On-line data visualization of GMP data – tool 4: reported values. Available at www.pops-gmp.org.

The visualization tools were developed using Adobe Flash and ArcGIS technologies. The user-defined outputs are available for download as a graphic file (PNG format); the underlying data from tools 4–6 are also available in text format which is easy to import to MS Excel.

The website has been designed particularly for purposes of the Stockholm Convention effectiveness evaluation by UNEP, as well as by environmental professionals and lay public.

5 Design of Upcoming Data Collection Campaigns

Considering the drawbacks identified in the 2008 GMP reports, an updated design of data collection and processing for future data collection campaigns was proposed by RECETOX and IBA MU and approved by experts from the GMP Global Coordination Group and Regional Organization Groups. The proposed design includes, among others, an electronic data capture system, a defined parametric data structure for both primary and aggregated data, and predefined code lists for most of the items to ensure maximum standardization and uniformity of reported data to allow reliable analysis and reporting.

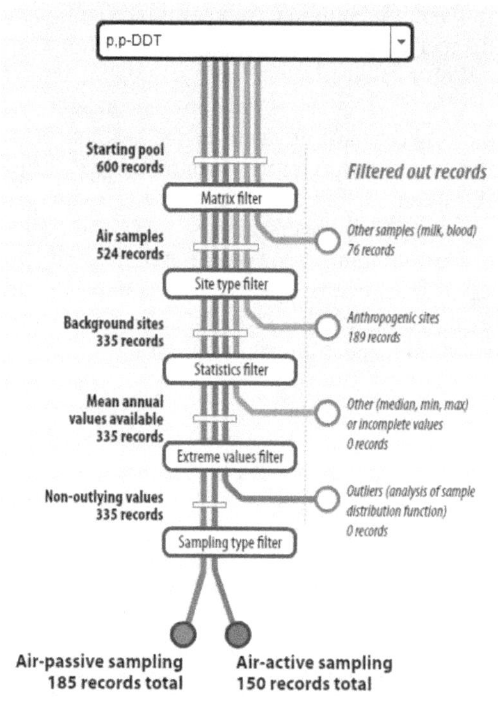

Fig. 3. On-line data visualization of GMP data – tool 5: a 6-step validation procedure of data from ambient air monitoring. Available at www.pops-gmp.org.

Structure of the data capture system and the code lists are adjusted to data from monitored environmental matrices (ambient air, human tissues) and to both primary and aggregated values. Options to check and correct the GMP1 data, should need be, are also included.

There are two principal ways for data recording into the database, each with specific data fields:

1. Data sets of primary POPs concentration data in a given site and at given time points; these records can be also uploaded automatically through standardized import MS Excel sheets.
2. Data sets of aggregated data on mean (median) annual POPs concentrations with proper variability measures (minimum, maximum, percentiles, standard deviation).

The GMP data warehouse has been built on an on-line web-based information system with a central data repository to collect relevant data (TrialDB system). TrialDB system was developed in cooperation with the Yale University [6-8].

Complex generic environment for on-line data management is recommended for the development of on-line databases with multiple touch-points collecting primary data. Multi-tier architecture (client - application server - database server) is the most commonly used approach. A common web browser (i.e. Internet Explorer) is employed as a client in that architecture. Internet connection is then necessary for all participating users. Web browser is available as default equipment in all personal computers; thus there is no need to install any specialized software on the users' workstations. Users can access all functions of the system (data manipulation, entry, editing, viewing, reporting, analyses, etc.) through their web browser.

Communication between the client and the server is always realized via secured (encrypted) https protocol (128-bit encryption is used), and data security is guaranteed in a standardized manner: system administrator assigns access login and password to individual users. Access level can be specified separately for each account. The legal safeguarding is defined in contracts between the users and the provider.

The main advantages of the centralized on-line solution are:

- Regular monitoring by trained staff;
- Security and availability for sharing among clients (when permitted);
- No need of upgrades, because changes in the web application are made centrally;
- Monitoring the progress of the project and addressing potential problems as soon as possible.

The proposed system is equipped with a number of useful functionalities for data reporting, handling and evaluation. In addition, a track changes tool displays recent changes in a form or group of forms, a data validation tool provides a list of incomplete records and missing items and some other useful tools are available. A standardized workflow for data reporting from involved monitoring programmes and subsequent approval by Regional Organization Groups has been set up to allow effective completion of regional monitoring reports in late 2014.

6 Conclusions

The GMP data available in monitoring reports from 2008 can be used for baseline statistical processing. It is, however, necessary to design a standardized and parametric data collection system to allow reliable assessment of time trends in concentrations of POPs listed in the Stockholm Convention. As recommended by the Global Coordination Group and updated Guidance on GMP [9], the data should be obtained mainly from large environmental monitoring programmes and surveys (EMEP, AMAP, GAPS, IADN, MONET, WHO human milk surveys).

Acknowledgements. This work has been supported by the project TB010MZP058 "Development of the system for spatial evaluation of the environmental contamination".

References

1. Dušek, L., Hřebíček, J., Kubásek, M., Jarkovský, J., Kalina, J., Baroš, R., Bednářová, Z., Klánová, J., Holoubek, I.: Conceptual Model Enhancing Accessibility of Data from Cancer–Related Environmental Risk Assessment Studies. In: Hřebíček, J., Schimak, G., Denzer, R. (eds.) ISESS 2011. IFIP AICT, vol. 359, pp. 461–479. Springer, Heidelberg (2011)
2. Jarkovský, J., Dušek, L., Janoušová, E.: Is On-Line Data Analysis Safety? Pitfalls Steaming from Automated Processing of Heterogeneous Environmental Data and Possible Solutions. In: Hřebíček, J., Schimak, G., Denzer, R. (eds.) ISESS 2011. IFIP AICT, vol. 359, pp. 486–490. Springer, Heidelberg (2011)
3. United Nations Environment Programme: Guidance on the Global Monitoring Plan for Persistent Organic Pollutants. Secretariat of the Stockholm Convention on Persistent Organic Pollutants, Geneva (2007)
4. Klánová, J., Dušek, L., Borůvková, J., Hůlek, R., Šebková, K., Gregor, J., Jarkovský, J., Kohút, L., Hřebíček, J., Holoubek, I.: The initial analysis of the Global Monitoring Plan (GMP) reports and a detailed proposal to develop an interactive on-line data storage, handling, and presentation module for the GMP in the framework of the GENASIS database and risk assessment tool. Masaryk University, Brno (2012)
5. Hůlek, R., Jarkovský, J., Borůvková, J., Kalina, J., Gregor, J., Šebková, K., Schwarz, D., Klánová, J., Dušek, L.: Global Monitoring Plan of the Stockholm Convention on Persistent Organic Pollutants: visualization and on-line analysis of data from the monitoring reports. Masaryk University (2013), `http://www.pops-gmp.org/visualization`
6. Nadkarni, P.M., Brandt, C., Frawley, S., Sayward, F.G., Einbinder, R., Zelterman, D., Schacter, L., Miller, P.L.: Managing attribute–value clinical trials data using the ACT/DB client-server database system. J. Am. Med. Inform. Assoc. 5, 139–151 (1998)
7. Nadkarni, P.M., Brandt, C.M., Marenco, L.: WebEAV: automatic metadata-driven generation of web interfaces to entity-attribute-value databases. J. Am. Med. Inform. Assoc. 7, 343–356 (2000)
8. Nadkarni, P.M., Marenco, L.: Easing the transition between attribute-value databases and conventional databases for scientific data. In: Proc. AMIA Symp., pp. 483–487 (2001)
9. United Nations Environment Programme: Guidance on the Global Monitoring Plan for Persistent Organic Pollutants. Secretariat of the Stockholm Convention on Persistent Organic Pollutants, Geneva (2013)

GENASIS System Architecture
On the Way from Environmental Data Repository towards a Research Infrastructure

Richard Hůlek[1,2], Jiří Jarkovský[1], Miroslav Kubásek[1], Jakub Gregor[1,2], Jiří Hřebíček[1], Ladislav Dušek[1], Jana Klánová[2], Kateřina Šebková[2], Jana Borůvková[2], and Ivan Holoubek[2]

[1] Institute of Biostatistics and Analyses, Masaryk University, Kamenice 126/3, 625 00 Brno, Czech Republic
{hulek,jarkovsky,dusek,kubasek,gregor,hrebicek}@iba.muni.cz
[2] Research Centre for Toxic Compounds in the Environment, Masaryk University, Kamenice 753/5 (A29), 625 00 Brno, Czech Republic
{klanova,sebkovaboruvkova,holoubek}@recetox.muni.cz

Abstract. GENASIS (Global ENvironmental ASsessment and Information System) is the web environmental information system which is an environmental data repository that provides comprehensive information on chemical contamination of the environment by Persistent Organic Pollutants (POPs). GENASIS combines data from long-term environmental monitoring programmes operated by RECETOX with validated data from partner institutions, and provides data archives, data management services and analytical processing of data. In the past few years the GENASIS system has undergone rapid development and grew up from the data repository into a scientific data infrastructure.

Keywords: GENASIS, information system, web portal, system architecture, environmental monitoring, POPs, Stockholm Convention.

1 Introduction

The environment is significantly impacted by human activities. For example the occurrence of many dangerous diseases such as respiration problems or various types of cancer is directly affected by environmental conditions, particularly by artificial contamination of the environment by toxic chemical compounds. Tracking current state and evaluating long-term trends of chemical contamination is subject of interest of many monitoring programs over the globe [1-2].

Information and communication technology (ICT) tools for collection and storage of environmental data, ensuring their quality and security as well as tools suitable for statistical processing and evaluating of data are needed, preferably within an environmental information system. Such an information system (IS) plays an important part in decision making processes, effectiveness evaluation, research activities as well as in education.

J. Hřebíček et al. (Eds.): ISESS 2013, IFIP AICT 413, pp. 230–239, 2013.

GENASIS (Global ENvironmetal ASessment and Evaluation System) provides data repository capacities optimized for monitoring of important elements of environment (ambient air, surface water, soil, sediments or even human tissues), statistical processing tools directly connected to data repository, reporting tools and an information and publication platform in form of web portal. All its ICT tools, features and services are wisely integrated together to create its own "ecosystem" [1-2].

The GENASIS system and its software components are subject of this paper which describes the overall system architecture, individual component's design and features and the way they are interconnected and used.

The GENASIS system is developed by two cooperating institutions: The Research Centre for Toxic Compounds in the Environment (RECETOX) and the Institute of Biostatistics and Analyses (IBA), Masaryk University, Czech Republic.

Since the first time the idea of the GENASIS system occurred, it has been intended to serve not only for internal purposes of RECETOX but also for a broader group of users from the Stockholm Convention parties [3] and a wider range of activities. This vision impacted all decisions about the system architecture, its components and the way the system is being developed.

2 GENASIS System Architecture

The GENASIS system architecture is inspired by several design patterns and technology standards: modular design, serviced oriented architecture (SOA) [4] and a data warehouse principle [5].

Modules of GENASIS are designed to gather similar sort of system functionality into one system component with a defined interface, so that the inner implementation would be hidden inside of the module and could be updated or changed by different module implementation which supports the same required interface.

Services in terms of SOA are independent system components which provide some kind of functions with additional value (services) to other components. Services must have well defined interfaces and usually communicate through standardized communication protocols and data formats.

The data warehouse is a specific form of relational database which is optimized for complex queries upon large amounts of data integrated together from various sources.

The GENASIS system architecture (Fig. 1) consists of:

- **Data repository** is the core of the whole system and is composed of a database of primary data and a data warehouse.
- **Controllers and engines** are server side software components which ensure data quality and security and other features. A *data warehouse engine* transforms for example all updated data when any changes are performed on primary data. Design of internal controllers and engines was implemented to ensure strict separation of concerns.
- **Services** allow controlled and secured access for particular users or other system components to stored data (and data services); spatially oriented services are used for spatial analysis, computations and for map publishing (spatial services); services focused on data management provide features for importing data and to setup security constrains (data management services); features of advanced statistical computations are provided by computational services.

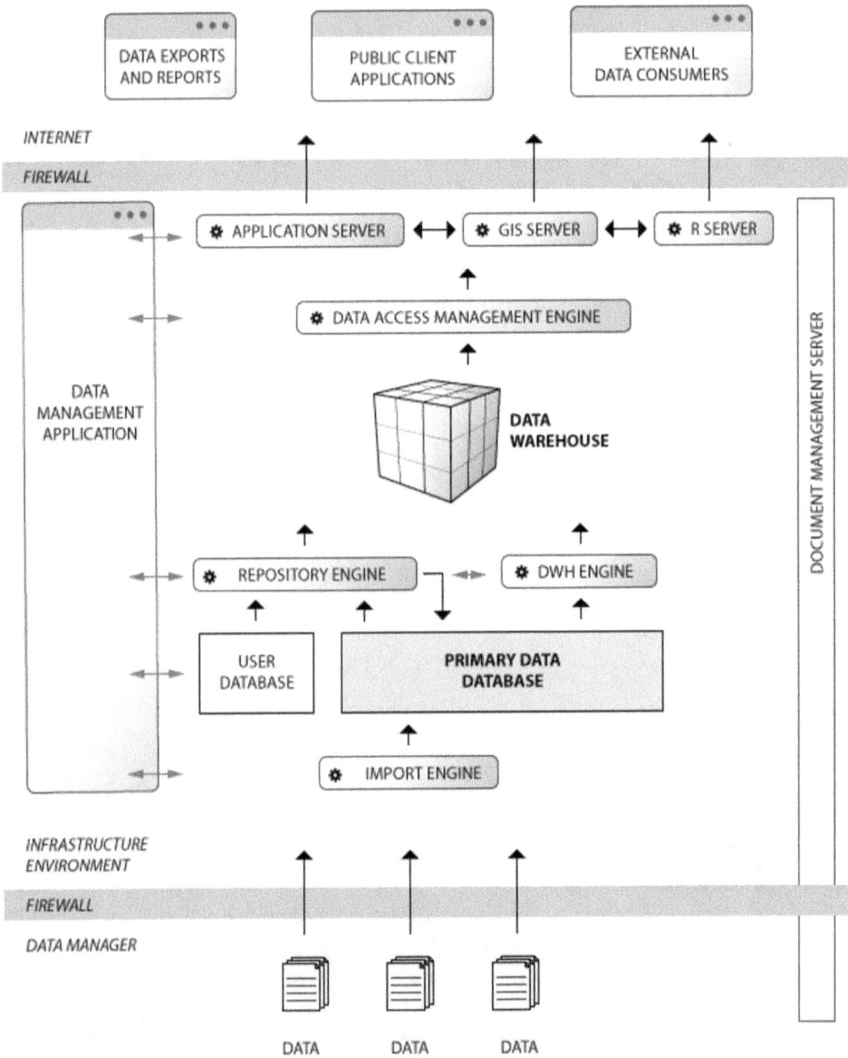

Fig. 1. GENASIS system architecture (Source: Authors)

- **Client applications** offer visualization of data stored in the data repository and enable users exploring the content of the database, perform basic evaluation analysis and export their outputs for further use.
- **Admin Panel** is used by data managers for data import, configuration of security constrains and export of raw - data into data sheets.

- **Supporting tools** such as document management system are used to maintain project documentation, workflows and to archive data packages provided directly by individual data providers.

Individual components of the GENASIS infrastructure are described in detail in the following subchapters.

2.1 Data Providers and Data Validation

Data validation is an important process in gathering data from various environmental monitoring networks from different data providers. Before importing to the database the data must be validated, its nomenclature and code lists harmonized as well as transformed into a data format which is supported by the import engine [2], [6], [8].

All data from data providers are saved in a protected archive before any modifications are made.

Individual steps of data preparation are performed manually by specialized staff.

2.2 Import Engine

The import engine is the system component which ensures proper import of data to the primary database. It is controlled via an admin panel application through which data sheets in a standardized format are passed in. Imported sheets are parsed to elementary records and further processed. The main responsibilities of the import engine are:

- Data transfer from data manager's PC to the main server infrastructure;
- Format validation of data from the input;
- Error detection and proper notification handling;
- Storing data into the database.

The import engine is capable of handling tens of thousands records in a row; thus, the data managers can import data very efficiently. Data are imported into the database in batches so that they could be processed as groups. The overall process of the data import is divided into several stages:

- Import of samples;
- Samples imported, waiting for confirmation;
- Samples confirmed by guarantee;
- Import of values;
- Values imported, waiting for confirmation;
- Values confirmed by guarantee;
- Import batch cancelled;
- Import batch confirmed – data are published and cannot be modified any more.

Only data from confirmed import batches are used in the primary database and other components of the GENASIS system.

Import engine is implemented in a way that all operations with data are transactional – it means that a principle "all or nothing" is applied and any single error leads to cancellation of the whole dataset import. This guarantees integrity of all imported data.

2.3 Primary Data Database (Repository)

The primary database – called repository as well – is the core of the system. It is designed to work as a storage capacity of heterogeneous data from environmental monitoring networks. Data structures allow storing data from ambient air, soil, water, plants, sediments and other types of environmental matrices from both long-term and short-term monitoring programs of the Stockholm convention [3].

Data are organized into a logical hierarchy: Institution – Project (Monitoring network) – Site – Matrix – Sample – Parameter – Value – Unit. Beside this core structure there are data structures to provide more information about classification, categorization and detailed description of data. Samples are described according to the matrix, sampling method, analysis method and other factors. Sampling sites are located by spatial coordinates, described in text by the expert who performed measurements and classified by a set of parameters such as pollution source descriptors, geography, pedology or hydrology categorization. Measured values are stored with proper information on limit of quantification (LOQ) and the way it was determined.

Uniform storage data structures and standardized code lists ensures standardization of data formats and their content so that comparability of individual records would be ensured.

2.4 User Database

The user database serves as a helper database of the system. It contains information about user accounts and access rights. While nearly all operations in the system are registered, thanks to the user database it is possible to track authors and editors of records and their changes. The user database is important in connection with data access management module where it provides a flexible access control list (see detailed description in chapter 2.8).

2.5 Repository Engine

The repository engine expands functionality of the primary database. It is controlled by an admin panel application and integrates procedures over the primary database such as quality assurance heuristics and transformation procedures.

Quality assurance heuristics are single purpose server side computer programs which attempt to identify potential quality leaks which cannot be discovered during the phase of data import (such as duplicate records, missing records in sequences etc.).

Transformation procedures convert values stored in the database into transformed parameters. A good example of such transformation is converting concentration of a particular dioxin compound to its toxic equivalency (TEQ) or calculating sum of indicator PCBs (polychlorinated biphenyls). Transformations are performed directly upon the database so that the data managers would not be required to calculate additional parameters manually and import them separately.

Repository engine scripts are designed to run in batches and their execution is scheduled by a central update time table or is controlled by the admin panel application.

2.6 Data Warehouse

The data warehouse database contains transformed and pre-processed data prepared to be delivered to the end users of the system by any of the top level system applications or services (chapters 2.9, 2.10, 2.11). It integrates data from various monitoring programs into one common data structure optimized for reporting and data analysis and it represents a platform for decision-support data.

Analytical codes are added to elementary records to enable a broader range of selections and filters. Complex calculations and aggregations are calculated in advance in order to speed up results delivery.

The concept of the data warehouse enables integration of external data sources regardless its physical data structure.

Thanks to separation of the data warehouse database from the rest of the repository, it is easy to perform changes in the primary database without affecting the data warehouse database.

2.7 Data Warehouse Engine

The data warehouse engine connects the primary data database and the data warehouse. It is responsible for its creation and performs updates of the data warehouse based on changes in the primary database. Update procedures are scheduled by a central scheduling mechanism or are executed by the admin panel application.

Since the primary database and the data warehouse have different data structures, update procedures of the data warehouse engine implement all necessary transformations, calculations and aggregations.

Procedures implemented in the data warehouse engine are designed to be executed in batch mode because these tasks usually take a longer time to be executed.

2.8 Data Access Management Engine

Since the GENASIS system and its infrastructure has been used for various purposes and it contains data from more than 20 large monitoring programs as well as internal research experiments to date, there are various requirements to control access to individual record sets. Not all data are suitable for public use.

A custom data access management was implemented in the GENASIS system to meet specific needs of data protection. It is controlled by the admin panel application and it creates an access control list (ACL) for each user based on rules which arise from logical data structure, access scopes (private data / protected data / public data), data ownership and combinations of these factors.

The access control lists are partially pre-processed in advance (when some changes in repository occur) to speed up results delivery when large amounts of data are queried.

2.9 Application Server and Data Services

The application server provides an environment for various applications which are built on top of the GENASIS infrastructure. Applications typically access the data warehouse database or rarely primary database and perform further data transformations and calculations and pass results to the presentation tier of particular application's user interface. The application server provides pre-configured connectivity to other components of the system (such as GIS server or R server) as well as the appropriate computation resources.

2.10 GIS Server and Spatial Services

A GIS server extends functions of the GENASIS system by adding support for spatial data visualizations, map distribution and performing spatial analysis in real-time. The GIS server also allows linking additional information to stored data based on spatial references.

2.11 R Server and Computation Services

The R (R Core Team) server platform [6] was integrated into the GENASIS infrastructure to make use of its computations capabilities. Algorithms already implemented in free software programming language R are used instead of custom-created implementations of non-trivial statistical tests and models.

The integration of the R server improves the level of delivered services and speeds up the development of new analytical tools of the GENASIS system.

The R server consumes data from the data warehouse as well as from the primary data database and delivers results to various components of the GENASIS system. Analytical applications make use of the R server computation services for example to calculate time series statistical tests; the repository engine uses the R server component for parameter transformations.

2.12 Publicly Available Applications

Once various data from different sources are validated and imported into GENASIS repositories it is desired to analyse them in terms of existing patterns and trends. An important additional value of the GENASIS system consists in analytical and interpretation tools which are directly connected to the data repository and perform statistical computations in real time.

The GENASIS on-line data browser [7] offers visualization of data from both national and international POPs monitoring projects of the Stockholm convention. Among the most important projects are Košetice EMEP station, MONET network and others. At present (June 2013), the database contains POPs data from 20 projects, 4 continents, 58 countries, 295 compounds and over 606,000 records collected during more than 20 years (first records date 1988) of continuous POPs monitoring. Monitored compounds cover most of these listed under the Stockholm Convention on Persistent Organic Pollutants [3].

In contrast to real-time general purpose web based statistical and interpretation tools, there are case studies [8] which deal with a specific topic or a selected part of the data repository. Similarly to the data browser they allow for online analysis of selected data, which are processed and presented in a structure and level adjusted to the particular purpose and study and are supplemented with further information on a given topic, methodology, and interpretation of obtained results.

2.13 External Data Consumers – Interoperability

Thanks to service-oriented and modular architecture the GENASIS system is capable of sharing data with external data consumers at various levels. The system can provide direct access to primary or aggregated data from the whole repository or parts of it. Besides data alone, results of analyses such as trend tests or maps of spatial distribution of contamination by particular compounds can be also provided.

2.14 Data Exports

Data exports in standardized format are an important function of the GENASIS system. Data exports are used in research, scientific publications and for both national and international reporting activities of Stockholm Convention CEE Regional POPs Centre [9].

2.15 Admin Panel

The admin panel is a standalone application connected to the GENASIS server infrastructure and is used by data managers to manage data in the GENASIS database and surrounding system components.

Through the admin panel data managers are able to:

- Import data into the database;
- Execute update procedures in the data warehouse engine;
- Execute transformations and quality assurance heuristics in the repository engine;
- Manage access to data and maintain user accounts;
- View and export data.

2.16 Document Management Server

The document management server works beside the GENASIS system itself but it still can be considered as a part of the whole infrastructure. The document management system is not (yet) connected with any other GENASIS system component programmatically but plays an important role in the overall process.

In the document management server there is stored the whole project documentation, content of code lists, all metadata and description of used standards as well as configuration properties needed for proper system set up. There is also an data archive containing the data as provided by the data providers. The document management server has the advantage against regular file system folders of flexible access control management and version tracking of all documents.

2.17 Firewalls and Security Components

The GENASIS system infrastructure is protected across several levels. GENASIS infrastructure is operated in Masaryk University Brno, Czech Republic [9]. The university's subnet is monitored and there are two levels of firewalls in front of each server. All activities and services are provided in accordance with ISO certifications EN ISO 9001:2009, ISO/IEC 20000-1:2006, and ISO/IEC 27001:2006.

3 Conclusion

The GENASIS system is used not only for internal purposes of RECETOX activities; it represents a research infrastructure available for use by other institutions, hosting data from different monitoring programs and providing its features and services.

Core services of the GENASIS infrastructure were used during development of tools for collection, analysis and visualization of data from the Stockholm Convention Global Monitoring Plan on Persistent Organic Pollutants [10]. Infrastructure of the GENASIS system is also used for the purpose of national inventories on persistent organic pollutants in the Czech Republic to meet requirements given by the Stockholm Convention on Persistent Organic Pollutants [3].

Acknowledgements. This work was supported by the TACR project no. TB010MZP058 "Development of the system for spatial evaluation of the environmental contamination".

References

1. Holoubek, I., Dušek, L., Klánová, J., Kubásek, M., Jarkovský, J., Baroš, R., Komprdová, K., Bednářová, Z., Hůlek, R., Hřebíček, J.: GENASIS Information System: A Global Environmental Assessment of Persistent Organic Pollutants. In: Hřebíček, J., Schimak, G., Denzer, R. (eds.) ISESS 2011. IFIP Advances in Information and Communication Technology, vol. 359, pp. 480–485. Springer, Heidelberg (2011)
2. Hůlek, R., Kubásek, M., Gregor, J.: Data Management for Environmental Monitoring. In: 26th International Conference on Informatics for Environmental Protection, EnviroInfo 2012, Dessau, Germany, vol. 2012, pp. 631–637 (2012)
3. Stockholm Convention on Persistent Organic Pollutants (POPs): Interim Secretariat for the Stockholm Convention, United Nations Environmental Programme (UNEP) Chemicals: Geneva, Switzerland, http://chm.pops.int
4. Arsanjani, A., Zhang, L.J., Ellis, M., Allam, A., Channabasavaiah, K.: A service-oriented reference architecture. IT Professional 9(3), 10–17 (2007)
5. Chaudhuri, S., Dayal, U.: An overview of data warehousing and OLAP technology. SIGMOD Record (ACM Special Interest Group on Management of Data) 26(1), 65–74 (1997)
6. R Core Team: R: A language and environment for statistical computing, http://www.R-project.org/
7. Jarkovský, J., Dušek, L., Kubásek, M., Kohút, L., Klánová, J., Hůlek, R., Gregor, J., Šebková, K., Borůvková, J., Hřebíček, J., Holoubek, I.: On-line data browser for environmental monitoring and associated information systems. Masaryk University (2011), http://www.genasis.cz/

8. Komprdová, K., Komprda, J., Sáňka, M., Hájek, O., Hůlek, R., Jarkovský, J.: Spatial modeling of pesticides concentrations and pools. Masaryk University (2011), http://www.genasis.cz/case-studies/pops-spatial-modeling/
9. Stockholm Convention CEE Regional POPs Centre, http://recetox.muni.cz/index-en.php?pg=regional-pops-center
10. Masaryk University Brno, http://www.muni.cz
11. Hůlek, R., Jarkovský, J., Borůvková, J., Kalina, J., Gregor, J., Šebková, K., Schwarz, D., Klánová, J., Dušek, L.: Global Monitoring Plan of the Stockholm Convention on Persistent Organic Pollutants: visualization and on-line analysis of data from the monitoring reports. Masaryk University (2013), http://www.pops-gmp.org/visualization/

Integration of R Statistical Environment into ICT Infrastructure of GMP and GENASIS

Richard Hůlek, Jiří Kalina, Ladislav Dušek, and Jiří Jarkovský

Research Centre for Toxic Compounds in the Environment
Kamenice 735/5, 625 00 Brno, Czech Republic,
Institute of Biostatistics and Analyses,
Kamenice 126/3, 625 00 Brno, Czech Republic
{hulek,dusek,jarkovsky}@iba.muni.cz, kalina@mail.muni.cz
http://recetox.muni.cz,
http://www.iba.muni.cz

Abstract. The Global Monitoring Plan (GMP) was established as a tool for providing a worldwide overview of Stockholm Convention (SC) compliance by monitoring and evaluation of SC 22 persistent organic pollutants (POPs) concentration levels and their trends. In order to evaluate a dataset on POPs concentrations from the initial GMP campaign, it was essential to use advanced statistical methods which are not incorporated in commonly used database languages. Instead of a complete realization of these methods in the main programming language, in which the application is developed, this language was used only as an interface to the server version of the powerful statistical software R. The involvement of the R language into the environmental pollution data assessment infrastructure of the Global Environmental Assessment Information System (GENASIS) adopted for the GMP data makes easier to avoid disambiguities in data analysis and brings a powerful tool for advanced statistical analysis and visualization of GMP and GENASIS data.

Keywords: R, statistical computing, web application, system architecture, Stockholm Convention, POPs, GMP, GENASIS.

1 Background

The Stockholm Convention (SC) was signed by 128 parties in 2001 in order to protect both the environment and human health from persistent organic pollutants (POPs). It has been in force since 2004, changing an approach to POPs all around the world by the elimination or a significant restriction of their manufacturing and application. Most of the production and use of listed compounds was abandoned, but a huge amount of POPs, although still slowly decreasing in general, is still present in the environment. Thus a detailed and reliable analysis of the still developing situation is needed for the evaluation of SC effectiveness.

An arrangement of the Global Monitoring Plan (GMP) was established as a tool for providing a worldwide overview of the Stockholm Convention (SC) compliance by monitoring and evaluation of SC 22 persistent organic pollutants (POPs) concentration levels and their trends.

J. Hřebíček et al. (Eds.): ISESS 2013, IFIP AICT 413, pp. 240–252, 2013.

The results of GMP reporting are assessed and visualised via an international part of the Global Environmental Assessment Information System (GENASIS) which encompasses a wider set of projects on environmental contamination by chemicals, namely POPs. Run in the cooperation between the Research Centre for Toxic Compounds in the Environment (RECETOX) and Institute of Biostatistics and Analyses (IBA) at Masaryk University, the GENASIS system combines expertise, validated data from partner institutions, inputs from regular environmental monitoring programs and provides data repository, analytical tools and data management crowned by online web visualization platform [1].

A prime related data mining and a further treatment on the dataset of POPs concentrations from the GMP and other projects needs to employ several advanced statistical methods, which are not incorporated in any of the commonly used database languages neither are standard distributions of main languages. Thus, several different possibilities come under consideration on how to treat the task of a sufficiently robust and effective development and running of statistical computations.

There are several options to use instant solutions on database level (i.e. Oracle DBMS_STAT_FUNC package [11], Oracle Data Mining (ODM) extension [12], PostgreSQL MADlib library [13] and a few others) which enable several data mining functionalities as native SQL functions.

Another possibility is to involve some of the wide range of main programming language statistical libraries such as PHP Statistics extension [14] or Commons Math for Java [15], but usually they are less efficient than SQL language extensions.

However, both of these solutions have certain shortcomings given by either a limited scope (in particular, a small number of incorporated statistical techniques and options of their customization) or a high price. Nevertheless, even the most expensive ready-made packages are principally limited by the set of functions that were included by their creators and the development of a new function is thus impossible or very difficult.

Similar imperfections occur when trying to create one's own main programming language statistical library. The development, optimization and validation of algorithms and their perpetual modifications and improvements represent a rather complicated way to achieve satisfactory results of environmental data mining, especially in the changing environment of different programming languages and different requirements of infrastructure projects.

Finally, all of the above-mentioned solutions are part of an area that is easily accessible for experts from the IT field but generally unintelligible for (bio)statisticians, who are primarily involved in environmental data mining and evaluation.

These conclusions led us to a decision to prefer the use of a main programming language (PHP) only as an interface to the server version of a powerful statistical software R, especially in the scope of further statistical analyses carried out at the Institute of Biostatistics and Analyses in the framework of the GENASIS project. The involvement of the R language into the environmental pollution data assessment infrastructure makes the development considerably more efficient not only by engaging a wider group of experts into the process of data analysis and makes available a powerful tool, but also facilitates to avoid ambiguities in data, allowing an advanced visualization and analysis of environmental data.

1.1 R

R is the name of a computational environment based on the S language developed in 1970's in Bell Laboratories. It comprises an extensive pool of statistical techniques, which can be easily extended via a wide range of packages available online through the Comprehensive R Archive Network (CRAN). Due to the fact, that R is available as a Free Software under the terms of the Free Software Foundation's GNU General Public Licence in source code form [2], it achieved enormous popularity in recent years and the number of its users is estimated to exceed 2,000,000 worldwide.

R compiles and runs on a wide variety of UNIX platforms, Windows, MacOS and related systems. The environment itself is built in the R language, wherefore any extensions of existing and development of new functions can be conducted easily recurrently from basic elements of the language up to advanced statistical techniques.

Another advantage of R lies not only in the fact, that it is widely used in the statistical community, but due to its open-source character, it allows to share an enormous amount of experience, methods and ideas suitable for statistical computations. In the field of environmental sciences, a lot of different projects were solved using the R language and numerous methods were published in a form of R packages [15].

On the other hand, the R environment has one substantial disadvantage: from the very beginning, it was constructed as a single-user system intended primarily for a desktop use; therefore its use for solving multi-user tasks is rather complicated and has certain limitations, as well as a partial incompatibility of R environment data structures and formats used during data transmission between the clients and server.

Although the use of R as a tool providing advanced statistical techniques solves all problems listed above and allows the use of an almost unlimited set of algorithms and functions for calculations, server implementation remains quite challenging with regrds to its desktop character. Therefore, in terms of efficiency, R should be implemented into computational infrastructure in its widest meaning in the frame of projects solved at IBA. Thus, its use should not be confined to the processing of data from GMP, but it should be used in all applications under the information system GENASIS in the widest possible context.

2 Integration of R into ICT Infrastructure

The integration of R into web applications is not a trivial task. A rather complex structure of the R language suggests for such form of integration that will be both most usable across applications and flexible enough to deal with various requirements on computing tasks. It should also meet many other criteria that are primarily implied by the general characteristics of web applications and their development process.

2.1 Requirements

Web applications operate on a client – server communication architecture. All the application logic is located at the server. Individual users send requests to a central server through their web browsers (clients). Requests are being interpreted and evaluated. The output is processed and sent back to the client computer in the form of response (so called the request – response communication). It is obvious that the R software integration needs to be at the server side due to the fact that today's trend is to create a so called thin client that does not perform any computational operations and displays only server outputs. On the other hand, the thick clients contain a substantial portion of application logic. This kind of architecture (communication model client-server, where client means thin client) supports a wide range of client devices from conventional desktop computers or laptops to recently highly popular mobile phones and tablets where R support would not be possible.

Another important characteristic, which determines a certain set of requirements for the integration method, is the fact that web applications are typically accessed simultaneously by multiple users. Multi-user environment is the main difference from the other applications, where R is run locally on a desktop computer or laptop. Due to the existence of multiple users, the server integration of R has to support concurrent processing in multiple processes so that required system response time would be met. Concurrent processes have to be isolated among each other and not to share for example program variables or other resources.

Using R on local computers program scripts, helper libraries and R packages are written in a way where the output of particular computing functions prints result to a standard or error output (print on a screen), saves a file to a local hard drive or returns a value with pointer to the system memory. Such a different approach to returning result values is not fully applicable in the server environment because, e.g., sharing data among servers on hard drives is not always possible.

In general the particular form of R software integration into the server environment should support maximum flexibility and effective use in different web applications. For that reason it is efficient to integrate it as a component in commonly used ICT infrastructure and a set of helper communication code libraries rather than building web applications with one-off R support solutions.

2.2 Methods of Integration

R can be effectively embedded with other programming languages and environments. Implementations of interfaces to R exist for most of the main programming languages such as Java [3], C++ or Python [4]. The interface for communication with C and Fortran is already implemented in R. A module for Apache2 web server RApache (mod_r) [5] is the important technology for the development of web applications, which enables the R code to be run directly from the web server environment.

There are various approaches to R software integration into web applications. They differ mainly in the general concept.

First, an extension for R support in a main programming language, in which the application is developed, can be used. The main programming language is used to create the application logic, access to data sources, data transformation and the definition of application design. The interface then allows for calling the computation and statistical functions directly from the given programming language. Such solution is simple and relatively easy to manage. There are, however, limited possibilities of scaling and moreover, when further web applications are developed using different programming languages it is necessary to implement the whole R support again.

The integration of the R support on a more general level is much more flexible, so that the access to computation functions would not be limited by using one or another programming language or platform – i.e. the choice of integration in a form of web services. Web service in general play an important role in building service-oriented systems and infrastructures. The communication between individual system components runs by means of the universal and widely supported communication protocol (e.g., Hypertext Transfer Protocol – HTTP), which is also used for sending requests between the components for processing individual tasks. A concrete internal design (implementation) of the components remains fully hidden behind a strictly defined programme interface (API), through which the components communicate to each other. Making the computation services of R language accessible in a form of web services, which can be called via web interface by a system of unique addresses (URL), will enable a flexible development of web applications, which might use computation possibilities easily and independently on the used programming language, platform or form of internal implementation. The separation of computation and application server also facilitates system scaling, security, or, e.g., caching. This is because the computation backend is integrated within the infrastructure at one place only and system resources can be assigned effectively, so that the sufficient system response would be ensured and the resources would be effectively used.

During the integration of R to the GENASIS system infrastructure we preferred the inclusion of a computation server as a separated component communication on a SOA principle [6].

2.3 Computation Backend Architecture

Computation backend server is accessible through a central proxy which enables load balancing and – if implemented – caching functionality as well. Thanks to using the load balancing proxy it is easy to scale the computation backend just by adding additional resources (servers).

Proxy, cache and computation servers are situated in a protected infrastructure environment behind a firewall which protects them from threats from the Internet. Services of computation backend are accessible via client applications or via direct calls of particular R functions and scripts mapped to URL addresses.

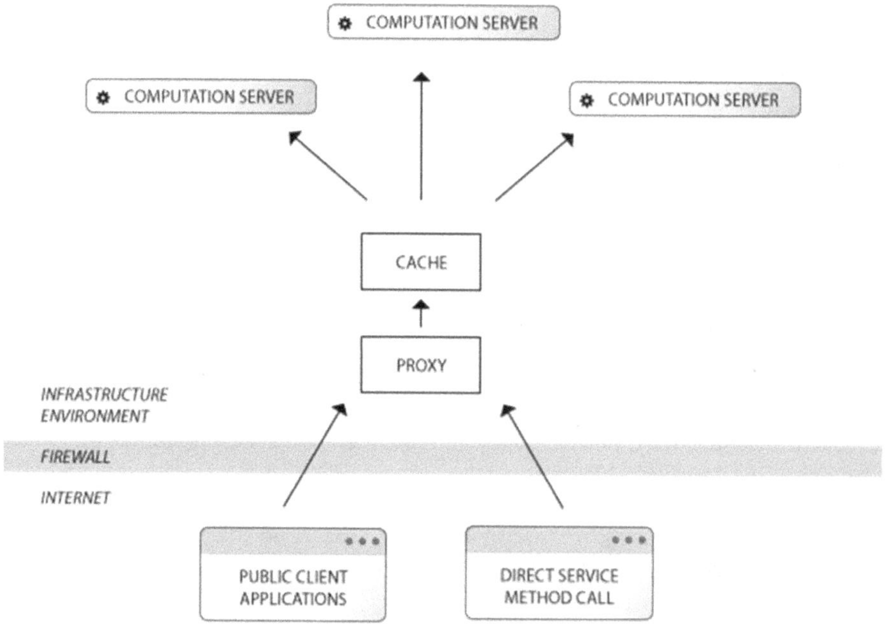

Fig. 1. Computation backend architecture

2.4 Technologies Used

Due to a very quick progress in recent years there are technologies available which enables integration of R into server environment.

RApache is a module extension of popular Apache2 web server enables executing R scripts on the server. It supports multiuser access by running isolated processes for the each request. Since Apache2 is HTTP server, RApache communicates with its clients over HTTP as well. RApache works very well together with R package Brew [7] which enables mixing R and HTML code together in one document. Such approach speeds up the development rapidly.

RApache and Brew are sufficient technologies to create web applications directly just by writing R and HTML code. But for more complicated applications it is more reliable to build web application in the main language such as PHP or Java and to use R functions published via RApache as remote method calls.

Following this approach, one can very easily realize there are many repeating steps: mapping functions to particular URL addresses, reading parameters from request, transforming results into the requested format before sending a response back to the client and many more.

To implement computation backend a much smarter solution can be used: OpenCPU [8]. It is built on the top of the RApache. It offers REST (Representational State Transfer) API with an automated mapping of URL addresses onto available R functions and custom scripts. OpenCPU transforms the output into various formats such as the plain text, JSON or as a downloadable picture.

Client applications build on the top of the GENASIS ICT infrastructure are usually developed in PHP, HTML and JavaScript or Apache Flex framework.

2.5 Details of Client-Server Communication

The communication between web applications and the computational server is realized via the HTTP protocol. Request parameters are passed into R scripts using GET and POST methods. To create a request on the client side of the application, which desires to call R-based computation backend, it is necessary to create HTTP request programmatically, which means namely: targeting the request to a proper URL address; serializing all data structures (simple variables, data arrays or complex objects as well as binary files) into a transferable format and making them ready to be sent as parameters.

When a request is sent to computation, the backend application must wait until the result is calculated and sent back in the server response. Results must be unserialized, data structures must be converted into a proper format of the target language and passed into program variables.

In order to resolve all of the above mentioned tasks, a custom code library which makes it easier to communicate PHP web applications and R-based computational backends has been developed. The library helps to convert all data and data structures into JSON format, resolves minor (but important) differences of implementation of JSON in PHP and R and helps to create proper HTTP request using the curl library available in PHP language. All binary data are encoded by base64 [9] algorithm in order to be easily transferrable over the text-based HTTP protocol. Since the JSON implementation on the R side (JSONIO [10]) is a significant consumer of system resources, the helper library in addition offers an optimized way to transfer large amounts of data to the computation backend.

3 Results

3.1 GENASIS ICT Infrastructure

The R-based computation server is an important part of the GENASIS system architecture. It provides R capabilities in the form of web services accessible via web interface to any application developed in this infrastructure.

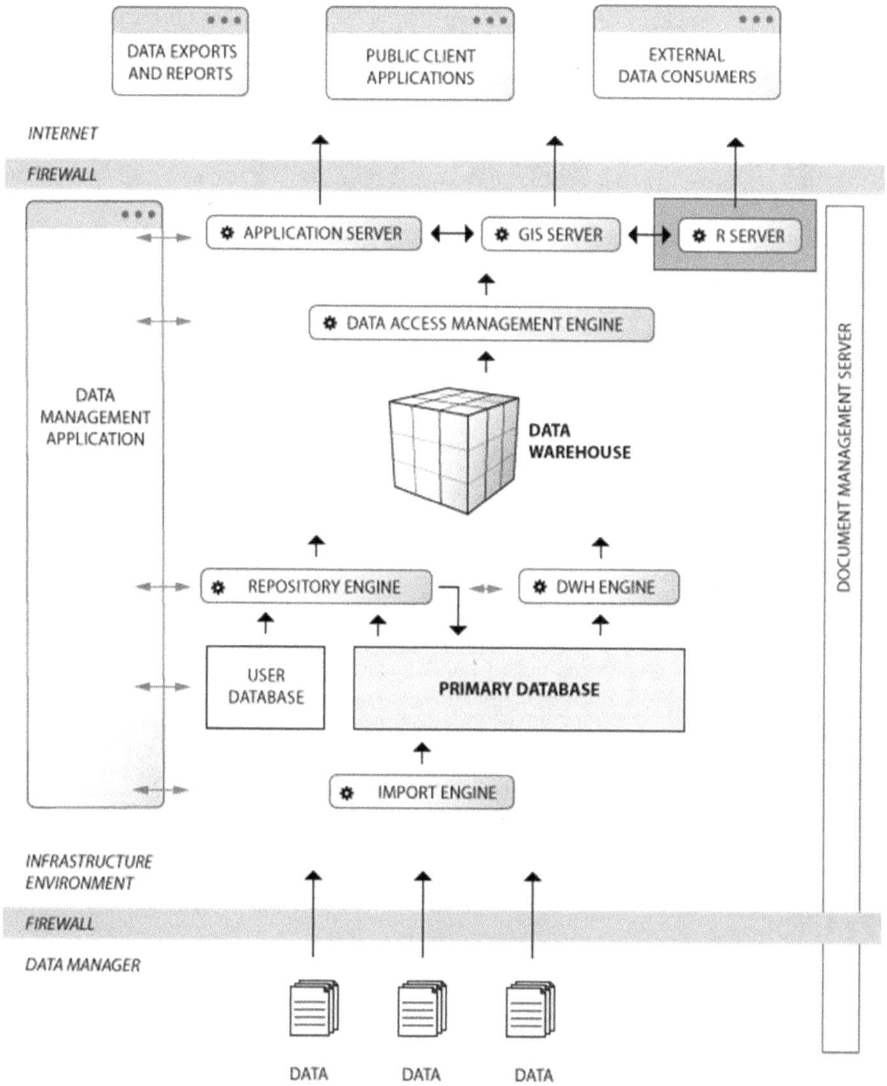

Fig. 2. GENASIS ICT infrastructure

3.2 Using of R-Based Computation Services

Environmental issues, including the tasks of air pollution, are in the centre of R users attention from the very first stages of its development. Over time, numerous packages containing functions relevant for the use in the evaluation of environmental data were published. GENASIS system procedures and, in particular, the evaluation of data from the GMP is largely implemented using functions included in the packages supplied with the standard R distribution (base, compiler, datasets, graphics, grDevices, grid, methods, parallel, splines, stats, stats4, tcltk), so descriptive statistics, correlation and linear trends etc. do not require additional functions. Nevertheless, the situation in time series analysis and outlier exclusion is more complicated and requires the use of some extending packages.

Several R packages relevant to GMP and GENASIS statistical tasks are listed in Table 1.

Table 1. R packages relevant to GMP and GENASIS computations

Name	Content	Year published
forecast	Methods and tools for displaying and analysing univariate time series forecasts including exponential smoothing via state space models and automatic ARIMA modelling.	2009
Kendall	A modest package for Kendall rank correlation and Mann-Kendall trend test computations, both used in GMP trend analysis.	2005
dyn	Another package for time series regression. The regression functions to be used with time series including specifications that may contain lags, diffs and missing values.	2005
openair	Tools to analyse, interpret and understand air pollution data. Data are typically hourly time series and both monitoring data and dispersion model output can be analysed. Many functions can also be applied to other data, including meteorological and traffic data.	2010
tseries	Package originally created for time series analysis and computational finance, but useful also in environmental applications.	1999
mvtnorm	Package allowing to compute multivariate normal and t probabilities, quantiles, random deviates and densities used in linear trend confidence intervals.	2000
car	Package accompanying a book of J. Fox and S. Weisberg, An R Companion to Applied Regression contains i.a. the Cooks distance used in outlier exclusion.	2001

For the purpose of a reliable estimation of the mean overall POPs concentrations and related statistical measures, it was necessary to apply a six-step validation procedure on the annually aggregated database of GMP I records (which contains either mean, median, minimum, maximum or any combination of these statistics). This procedure was developed in order to maintain a high predictive value of the GMP records and to avoid any bias in estimates of the background concentration values. It consists of the elimination of useless data records (inappropriately aggregated data, incomplete values, outliers) and employs several R base functions (both parametric and non-parametric) such as min, max, median, quantiles, geometric and arithmetic mean, standard deviation and several data transformations) as well as advanced methods of outlier exclusion and description as χ^2 testing or Cook's distance involved in R package *car*.

Further steps in GMP and GENASIS data analysis could be divided into 5 sections employing differnet kinds of R functions:

Baseline concentration – a set of descriptive statistics used to an elementary description of all the data available for each parameter (compound or set of compounds). The qualitative data include the number of entries and the time span of their recording (measurement), quantitative non-parametric statistics embraces the median and the pair of percentiles (5th and 95th). In terms of parametric description, the geometric mean seems to be the best description of the central tendency, which corresponds approximately to a lognormal distribution of the concentration of pollutants in the air. The confidence interval of the geometric mean was also computed, using log-normal distribution of values and concept of geometric standard deviation. As additional statistics, the arithmetic mean and standard deviation were also determined.

Detectable difference was expressed by two different concepts. The first alternative was the minimal annual concentration decline (increase) statistically significantly different from 0 based on standard t-test procedure on mean annual differences on all reasonable sites (i.e. on sites, where at least two concentration values in different times were available, the mean difference was computed as the total difference between initial and final value divided by the length of the time series. Such a value was repeated according to this length.).

The second alternative of detectable difference considers the minimal statistically significant slope of concentration line using a computation of a confidence interval for linear regression (or exponential regression in the case of logarithmic transformation). This computations requires the use of multivariate normal distribution parameters included in the package *mvtnorm*, quantile functions of normal and log-normal distributions and linear regression modelling tool, which are really complicated to implement properly without R.

Identification of time trend – several different methods are being used for the identification of statistically significant time trend, from which non-parametric Mann-Kendall test was chosen for GMP due to its robustness and low demands on the length of time series and its data distribution.

The test compares the relative magnitudes of concentration rather than the data values themselves for each site individually. One benefit of this test is that the data do not need to conform to any particular distribution. Moreover, data reported as not-detected can be included by assigning them a common value which is smaller than the smallest measured value in the data set. The procedure assumes that only one data value exists per time period, which is ensured by annual aggregations on sites. The final statistics of Mann-Kendall test S is normalized and compared with the critical value of normal distribution.

There is also the possibility of employing other methods of trend identification such as the non-parametric Spearman's rank correlation trend test, parametric Pearson's correlation trend test and again the linear and exponential models. Moreover, short-term predictions of time series are in progress, employing several advanced time series modelling methods (exponential smoothing and ARIMA models). For this section, not only the *Kendall* package, but also the time series modelling packages were used as well as correlation functions included in the base set of R functions.

Trend quantification – in the case of trend quantification, two different approaches are implemented, using parametric and non-parametric statistical techniques. The least square method based on linear and exponential regression models is employed for the parametric estimation of mean annual decrease/increase of POPs concentrations in the first case, whereas the non-parametric approach uses the Theil-Sen estimator. This robust linear regression method provides the value of trend slope as a median value of slopes of all pairs of time points inside the time series. The computed slope value corresponds well to the Mann-Kendall trend test, since the Kendall correlation coefficient between the measurements and modeled values converges to zero. The intercept of the Theil-Sen model is computed likewise as a median of differences between (without any intercept) modeled and measured values.

After a cancellation of *mblm* package from the R CRAN repository, the best option for employing the Theil-Sen estimator is to use the more complex and embracing package *openair*.

Visualization techniques – the current visualization techniques within the GENASIS and GMP web platforms comprise interactive histograms of concentrations, time series plots, trends visualisations including confidence intervals, detectable alternative dependence plots (power analysis), outliers exclusion plots, autocorrelation plots and several others. With the presence of plot drawing libraries as graphics or several more specific such as lattice, there are many possibilities to draw the whole spectrum of these plots within the R environment. Nevertheless, data transfer of the results in the form of jpeg/png files and restricted possibilities of plots graphic design resulted in the variant, that R is used only for precomputation of the plots, while the plots themselves are generated inside the main programming language by an appropriate graphic library.

4 Discussion

Although the implementation of the server version R environment is far from being a trivial task, after its completion it became the essential part of the environmental pollution data infrastructure, which provides a practically unlimited space for statistical computations and visualization of results.

There are several effects induced by the possibility to run statistical computations inside the web applications in the R environment:

- The R environment offers a huge pool of ready-made packages for almost all applications in the field of environmental statistics.
- The open source character and recurrent form of the R environment allows an easy creation of new functions and packages for sharing experiences both inside and outside the institution.
- The separation of computing and application layers of infrastructure allows a larger number of experts to participate in the development of statistical algorithms, without having to completely control the main programming language environment - especially (bio)statisticians without a deeper in-sight into IT technologies.
- Using the proven functions and procedures makes it easier to avoid ambiguities in the data analysis and saves a substantial part of the development capacity used for validation and correction processes.
- In the case of large volumes of data processing, the stability of the system is at a significantly higher level due to the use of optimized algorithms in the R environment.

Undoubtedly the high increase of development efficiency, worth taking the difficulties associated with the implementation of R into the data infrastructure.

Acknowledgment. This work was supported by the TACR project No. TB010-MZP058 "Development of the system for spatial evaluation of the environmental contamination".

References

1. GENASIS environmental data repository, `http://www.genasis.cz/index-en.php`
2. R Core Team. R: A language and environment for statistical computing, `http://www.r-project.org/`
3. rJava: Low-level R to Java interface. R package version 0.9-4, `http://cran.r-project.org/web/packages/rJava/index.html`
4. rpy, a robust Python interface to the R Programming Language `http://rpy.sourceforge.net/rpy.html`
5. Horner J.: Embedding R within the Apache Web Server: Whats the Use? `http://biostat.mc.vanderbilt.edu/wiki/pub/Main/RApacheProject/paper.pdf`
6. Arsanjani, A., Zhang, L.J., Ellis, M., Allam, A., Channabasavaiah, K.: A service-oriented reference architecture. IT Professional 9(3), 10–17 (2007)

7. Framework for Report Generation. R package version 1.0-6, http://cran.r-project.org/package=brew
8. OpenCPU, https://public.opencpu.org/pages/
9. Tools for base64 encoding. R package version 0.1-1, http://cran.r-project.org/package=base64enc
10. Serialize R objects to JSON, JavaScript Object Notation. R package version 1.0-3, http://cran.r-project.org/package=RJSONIO
11. Oracle Database PL/SQL Packages and Types Reference 102 DBMS_STAT_FUNCS, http://docs.oracle.com/cd/B19306_01/appdev.102/b14258/d_stat_f.htm
12. Oracle Data Mining, http://www.oracle.com/technetwork/database/options/advanced-analytics/odm/index.html
13. MADlib, http://madlib.net
14. PHP Statistics, http://www.php.net/manual/en/intro.stats.php
15. CRAN Task View: Analysis of Ecological and Environmental Data, http://cran.r-project.org/web/views/Environmetrics.html

GENASIS: On-Line Data Browser for Environmental Monitoring and Associated Information Systems

Jiří Jarkovský[1], Ladislav Dušek[1], Miroslav Kubásek[1], Richard Hůlek[1,2], Jakub Gregor[1,2], Jiří Hřebíček[1], Jana Klánová[2], Kateřina Šebková[2], Jana Borůvková[2], and Ivan Holoubek[2]

[1] Institute of Biostatistics and Analyses, Masaryk University, Kamenice 126/3, 625 00 Brno, Czech Republic
{jarkovsky,dusek,kubasek,hulek,gregor,hrebicek}@iba.muni.cz
[2] Research Centre for Toxic Compounds in the Environment, Masaryk University, Kamenice 753/5 (A29), 625 00 Brno, Czech Republic
{klanova,sebkova,boruvkova,holoubek}@recetox.muni.cz

Abstract. The current environmental research is producing large quantities of data on all environmental matrices and number of chemical and biological endpoints. However, there is a paradox in this situation that we suffer from lack of representative data, since their production has not been accordingly supported by development of systems that make them accessible for analysis. The introduced GENASIS data browser aims in its current version on the problem of persistent organic pollutants and the analysis, visualization and interpretation of data from their monitoring networks. The development of the GENASIS involves problems of the so called information pyramid: monitoring (obtaining environmental data), data validation (including uncertainty quantification), data analysis, and methods of aggregation of data from different sources, model generalization and interpretation of results, and their presentation. Within the frame of the GENASIS data browser, we introduce a user-friendly system for the visualization and analysis of contamination of all environmental compartments by persistent organic pollutants and evaluation of actual POPs contamination, its long-term trends and seasonal fluctuations. Project outcomes are useful as information source both for lay public and experts, as well as for the process of the Stockholm Convention implementation.

Keywords: Stockholm Convention, POPs, database, data analysis, online visualization.

1 Background

The current environmental research is producing large quantities of data on all environmental matrices and number of chemical and biological endpoints. Production of a high amount of data has been currently typical for all fields of human activities and concerns also environmental research and monitoring. Thousands of projects have

J. Hřebíček et al. (Eds.): ISESS 2013, IFIP AICT 413, pp. 253–261, 2013.

been in progress on the national and international level, which generate data that must be administered, processed, analyzed, and interpreted. However, there is a paradox in this situation that we suffer from lack of representative data, since their production has not been accordingly supported by development of systems that make them accessible for analysis [1-2]; we currently face the "Data rich but information poor" problem when based on these quantities of data only small part of information is utilized for interpretation or decision support.

The introduced GENASIS data browser [3] aims in its current version on the problem of persistent organic pollutants and the analysis, visualization and interpretation of data from their monitoring networks. The project exists within the frame of the Stockholm Convention [4] tools and its objectives are availability of a user-friendly system for the visualization and analysis of contamination of all environmental compartments by persistent organic pollutants and evaluation of actual Persistent Organic Pollutants (POPs) contamination, its long-term trends and seasonal fluctuations.

The GENASIS project utilizes data from national and international monitoring networks to obtain as-complete-as-possible set of information and a representative picture of environmental contamination by POPs [5]. Project outcomes should be useful as an information source both for lay public and experts, as well as for the process of the Stockholm Convention implementation. The system has been developed while taking into account requirements of the international projects aimed on collection and sharing of environmental information in Europe, such as the Shared Environmental Information System (SEIS) jointly with the INSPIRE Directive, the Single Information Space in Europe for the Environment (SISE) and the Global Monitoring for Environment and Security (GMES) initiative [6-7].

2 Persistent Organic Pollutants and the Stockholm Convention

Persistent organic pollutants (POPs) remain in the center of scientific attention due to their slow rates of degradation, their toxicity, and potential for both long-range transport and bioaccumulation in living organisms. This group of compounds covers large number of various chemicals from industrial products like PCBs through pesticides like HCH and DDT till by-products of combustion or chemical processes like PAHs. Due to high importance of POPs the Stockholm Convention on Persistent Organic Pollutants (POPs) entered into force on the May 17, 2004, and has currently 179 signatory parties [4]. The main objective of the Stockholm Convention (SC) is to protect human health and the environment from persistent organic pollutants by reducing or eliminating their releases into the environment. Parties to the Stockholm Convention are required to develop National Implementation Plans to demonstrate their implementation of the SC obligations including arrangements necessary for effectiveness evaluation of adopted measures. The data on POPs monitoring and their utilization in decision support is thus a high priority task in the process of the SC implementation.

3 Data Sources for GENASIS System

The GENASIS data browser utilizes the available POPs data from the monitoring networks to obtain as-complete-as-possible set of information and a representative picture of environmental contamination by POPs. The system is developed with taking into mind requirements of the international projects aimed on collection and sharing of environmental information in Europe like Single Information Space in Europe for the Environment (SISE) and the Shared Environmental Information System (SEIS) jointly with the INSPIRE Directive and the Global Monitoring for Environment and Security (GMES) initiative [7]. The development of the GENASIS involves problems of the so called information pyramid: *monitoring* (obtaining environmental data), *data validation* (including uncertainty quantification), *data analysis*, and *methods of aggregation of data* from different sources, *model generalization* and *interpretation of results*, and *their presentation.*

The GENASIS browser [5] offers visualization of data from both national and international monitoring projects. Among the most important are Košetice EMEP station, MONET network and others. The database contains data from 5 projects, 4 continents, 38 countries, 98 compounds and over 500,000 records collected during more than 20 years of continuous monitoring. Range of monitored compounds covers most of those listed under the Stockholm Convention.

4 System Description

GENASIS (Global ENvironmetal ASessment and Evaluation System) provides data repository capacities optimized for monitoring of important elements of environment (ambient air, surface water, soil, sediments or even human tissues), statistical processing tools directly connected to data repository, reporting tools and information and publication platform in form of web portal. All tools, features and services are wisely integrated together to create its own "ecosystem".

In the core of system is implemented robust data storage capacity. Data storage must be able to handle big amounts of heterogeneous data and provide fast results for requests of analytical tools and application services which provide data to other system components. Data storage is provided by database of primary data. Compute-intensive data transformations and calculations are preprocessed and stored in data warehouse to provide results as fast as possible. This solution meets general speed requirements and is characterized by good scalability.

On the top of the robust data storage infrastructure is built a set of services with Service Oriented Architecture (SOA) design pattern in mind. Individual services are focused on providing specialized functionality and provide its own interfaces so it is possible to combine and interconnect them in applications which are built over this system layer. These architectural design patterns were extended by integration of additional technologies: ArcGIS Server provides extension for spatial operations and map publishing, server side R is used for advanced statistical computations.

All of these core features are further used in interconnected applications of GENASIS system. These applications provide analytical tools directly connected to data repository and perform statistical computations in real time (Fig. 1). All GENASIS applications are publicly available so anyone can access and evaluate data and use the outputs for his needs. In addition to general data browser, which is described in this paper, subsets of data are used also for specialized case studies [8].

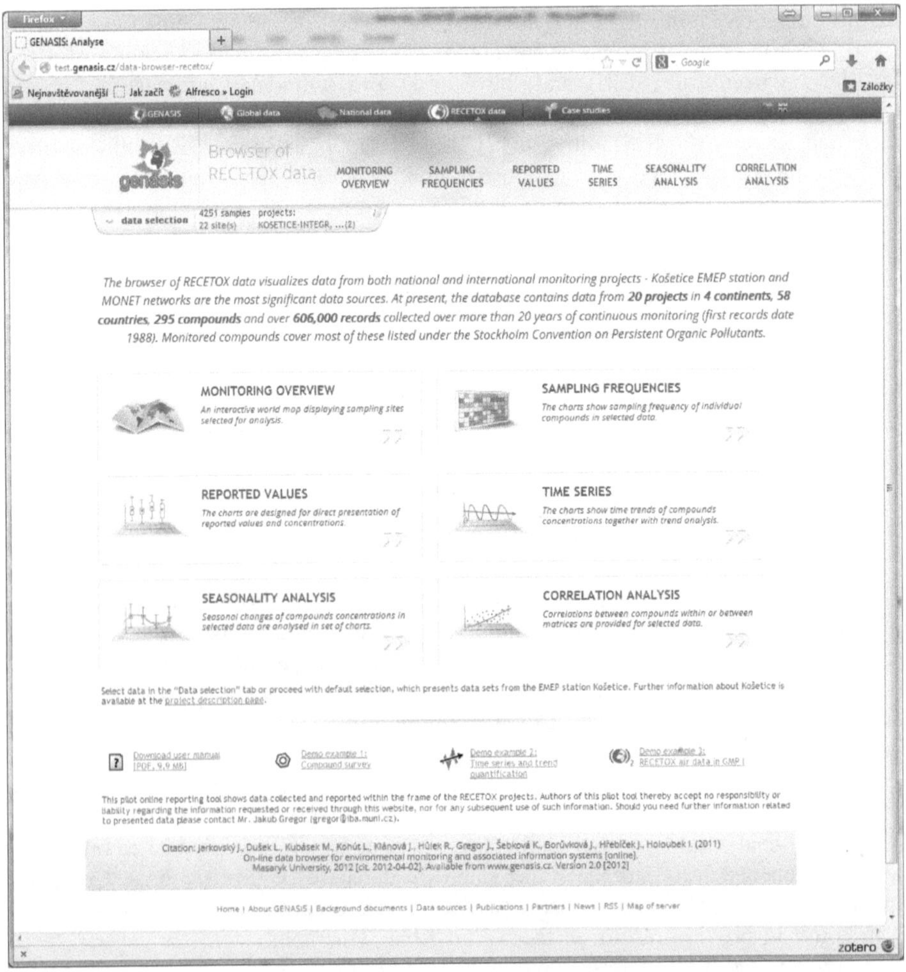

Fig. 1. Entry page of the GENASIS system (Source: www.genasis.cz)

5 Data Analysis Tools

The primary step of the analysis is the selection of the data using the basic set of characteristics (several predefined selections are also provided): Matrix; Type of monitoring; Project/monitoring network; Time range; Region; Parameter/compound; Individual sites (Fig. 2). The selected data can be analyzed in the following ways covering all standard requirements for the environmental research in this field.

Fig. 2. Selection module of the GENASIS system

Monitoring overview is an interactive world map displaying sampling sites selected for analysis (Fig. 3).

Fig. 3. Overview of the selected sites

Sampling frequencies are charts showing sampling frequency of individual compounds in selected data; the analytical module allows description of single compounds, comparison of compound on several sites or comparison of sampling frequency of multiple compounds (Fig. 4).

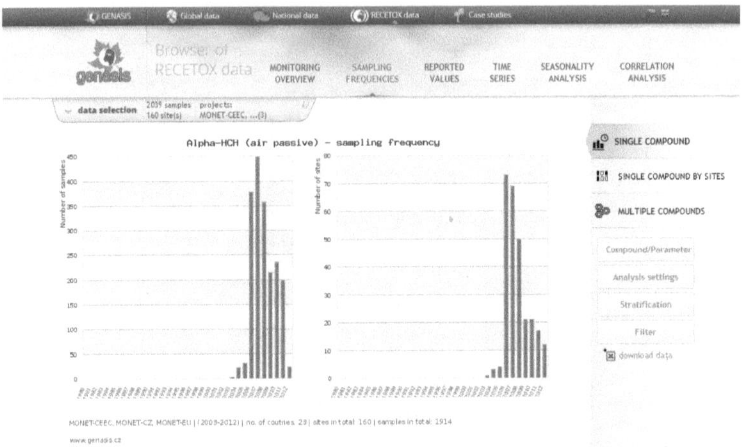

Fig. 4. Overview of sampling frequencies

Reported values charts are designed for direct presentation of reported values and concentrations; several types of charts (histograms, box and whisker plots and charts for individual sites) are provided (Fig. 5).

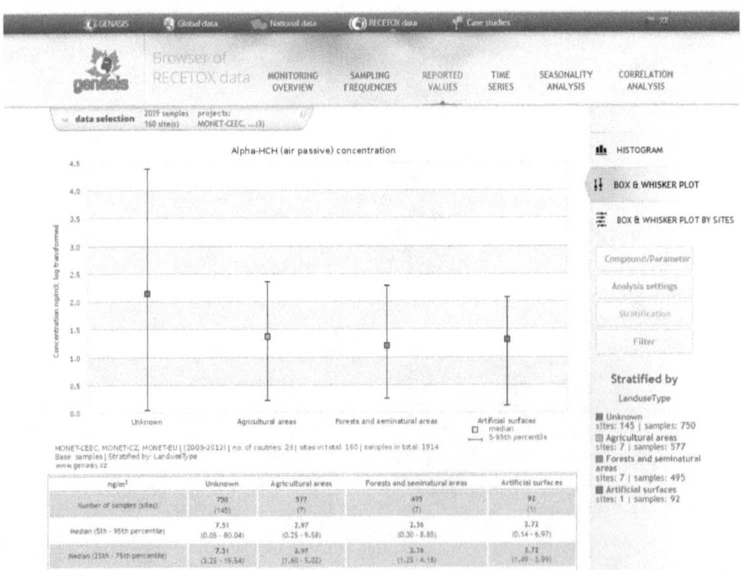

Fig. 5. Visualization of measured concentrations

Time series analysis shows time trends of compounds concentrations together with trend analysis (Fig. 6). The visualization of time series is supported by smoothing and time series data aggregation; trend analysis uses both non parametric (Daniels test, Mann Kendall test) and parametric approach (trend estimate using linear regression model with least square method). Seasonality analysis presents seasonal changes of compounds concentrations in selected data; several levels of time aggregation are possible.

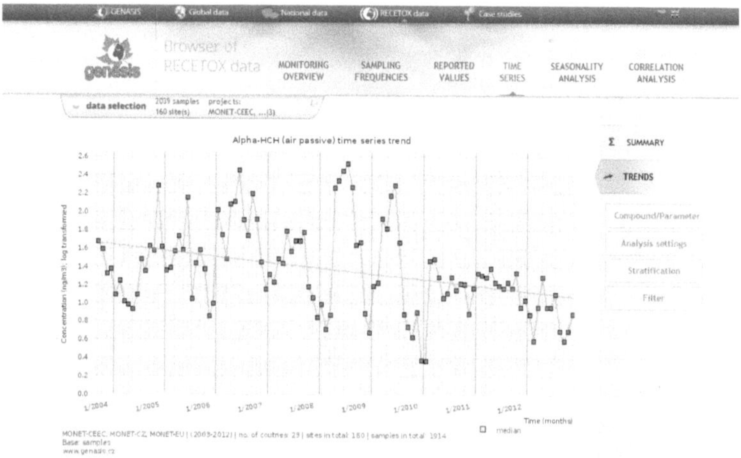

Fig. 6. Time series and trend analysis

Correlation analysis analyses correlations between compounds within or between matrices for selected data; both parametric and nonparametric correlation coefficients are available (Fig. 7).

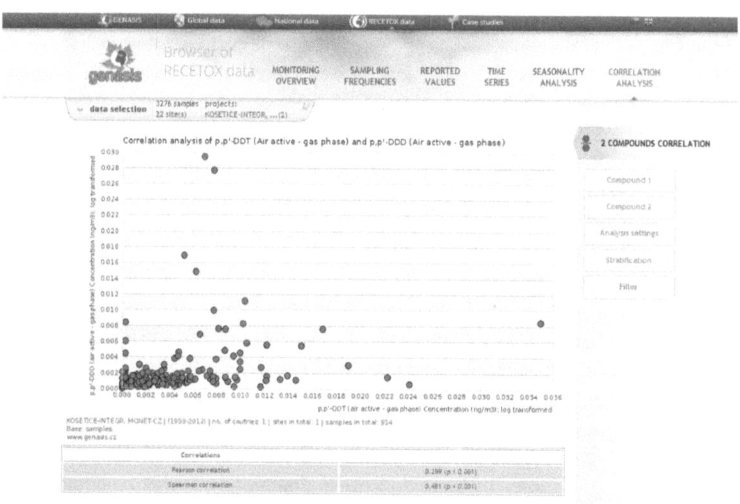

Fig. 7. Correlation analysis

All the analyses share also additional abilities allowing focused analyses:

- Filtering and splitting of selected data according to set of sampling sites characteristics; due to dynamic structure of the database these sampling sites characteristics can be easily enriched by newly available data.
- Several types of data transformation are available allowing user to work with various statistical distribution of data.
- Both parametric and nonparametric descriptive statistics and trend tests are provided.
- Export of the data for charts is available.

6 Conclusion

Within the frame of the GENASIS data browser [9], we introduce a user-friendly system for the visualization and analysis of contamination of all environmental compartments by persistent organic pollutants and evaluation of actual POPs contamination, its long-term trends and seasonal fluctuations. Project outcomes are useful as information source both for lay public and experts, as well as for the process of the Stockholm Convention implementation.

Acknowledgements. This work has been supported by the project TB010MZP058 "Development of the system for spatial evaluation of the environmental contamination".

References

1. Luis, G.B.: Principles for evaluating health risks in children associated with exposure to chemicals. Environmental health criteria 237, WHO Library Cataloguing-in-Publication Data. World Health Organization 2006 (2006)
2. JRC ISPRA: 2nd edition of the Technical Guidance Document (TGD) on Risk Assessment of Chemical Substances following European Regulations and Directives (2007)
3. Holoubek, I., Dušek, L., Klánová, J., Kubásek, M., Jarkovský, J., Baroš, R., Komprdová, K., Bednářová, Z., Hůlek, R., Hřebíček, J.: GENASIS Information System: A Global Environmental Assessment of Persistent Organic Pollutants. In: Hřebíček, J., Schimak, G., Denzer, R. (eds.) ISESS 2011. IFIP AICT, vol. 359, pp. 480–485. Springer, Heidelberg (2011)
4. Stockholm Convention on Persistent Organic Pollutants (POPs): Interim Secretariat for the Stockholm Convention, United Nations Environmental Programme (UNEP) Chemicals: Geneva, Switzerland, `http://chm.pops.int`
5. Kubásek, M., Hřebíček, J., Kalina, J., Dušek, L., Urbánek, J., Holoubek, I.: Global Environmental Assessment Requires Global Functional Searching Engines: Robust Application of TaToo Tools. In: Hřebíček, J., Schimak, G., Denzer, R. (eds.) ISESS 2011. IFIP AICT, vol. 359, pp. 507–518. Springer, Heidelberg (2011)
6. Hřebíček, J., Pillmann, W.: Shared Environmental Information System and Single Information Space in Europe for the Environment: Antipodes or Associates. In: Proceedings of the European conference of the Czech Presidency of the Council of the EU TOWARDS eENVIRONMENT. Opportunities of SEIS and SISE: Integrating Environmental Knowledge in Europe, pp. 447–458. Masaryk University, Brno (2009)

7. Hřebíček, J., Pillmann, W.: eEnvironment: Reality and Challenges for eEnvironment Implementation in Europe. In: Hřebíček, J., Schimak, G., Denzer, R. (eds.) ISESS 2011. IFIP AICT, vol. 359, pp. 1–14. Springer, Heidelberg (2011)
8. Komprdová, K., Komprda, J., Sáňka, M., Hájek, O.: Spatial Soil Modeling of Organochlorine Pesticides, Their Pools and Volatilization Fluxes. In: Hřebíček, J., Schimak, G., Denzer, R. (eds.) ISESS 2011. IFIP AICT, vol. 359, pp. 500–506. Springer, Heidelberg (2011)
9. Kubásek, M., Hřebíček, J., Kalina, J., Dušek, L., Urbánek, J., Holoubek, I.: Global Environmental Assessment Requires Global Functional Searching Engines: Robust Application of TaToo Tools. In: Hřebíček, J., Schimak, G., Denzer, R. (eds.) ISESS 2011. IFIP AICT, vol. 359, pp. 507–518. Springer, Heidelberg (2011)

Development of a Catchment Management Tool to Assess Environmental Risk from Nutrient Loadings Using Open Source GIS

Ian Packham, Eva Mockler, and Michael Bruen

UCD Dooge Centre for Water Resource Research, School of Civil, Structural and Environmental Engineering, University College Dublin, Belfield, Dublin 4, Ireland
{i.packham,eva.mockler,michael.bruen}@ucd.ie

Abstract. A Catchment Management Tool (CMT) is being developed for the Irish Environmental Protection Agency (EPA) that will allow River Basin Managers to evaluate the environmental risk from organic and inorganic nutrient loadings due to various land uses and human activities. The CMT was built using Open Source GIS (Geographical Information System) software, to facilitate development and widen the research user-base of the tool. The user interface of the CMT is designed to be flexible to allow local knowledge to be included in the system as well as the possibility to try what-if scenarios in relation to environmental assessment. Contaminant loadings are calculated from publicly available data and hydrogeologically susceptible areas (HSAs) are formed by combining soil and geological GIS layers. Further combinations of Loadings and HSAs allow Critical Source Areas to be delineated that identify areas contributing significant amounts of a given contaminant to any selected water body.

Keywords: Catchment Management, Critical Source Areas, Open Source GIS Software, Contaminant Loadings, Hydrogeology, Susceptibility.

1 Introduction

In response to the European Union's Water Framework Directive (WFD) requirements that River Basin Districts meet good quality status by 2015, the Irish Environmental Protection Agency (EPA) has commissioned a Catchment Management Tool (CMT) to assist in planning its activities. The CMT is one of the outputs of the Pathways Project, a multidisciplinary collaborative project that aims to improve understanding of hydrological and hydrogeological pathways of water-borne nutrient transport. The knowledge gained from this and other recent projects will inform the CMT and in particular the evaluation of environmental risk due to nutrient loadings.

This paper begins by describing the development of the Catchment Management Tool using Open Source Geographical Information System (GIS) software. Open Source software was chosen during the design and development stage of the CMT for fast, accessible ongoing software development and to widen the potential research user-base of the tool. Then the paper describes the data sources and methods used to

J. Hřebíček et al. (Eds.): ISESS 2013, IFIP AICT 413, pp. 262–270, 2013.

compute the nutrient loadings for a chosen catchment and thus how the various information sources are combined to produce Critical Source Areas of environmental risk. The paper concludes with potential further applications of the CMT, such as varying loadings due to land-use change and other what-if scenarios. The flexible nature of the CMT is designed to help river basin managers select appropriate measures to achieve good WFD status.

1.1 Critical Source Areas

Critical Source Areas are defined as those areas that contribute a disproportionally high pollutant loading to the receiving river, and occur where high loadings coincide with hydrogeologically susceptibility areas (HSAs). HSAs are those areas that are sensitive to a given contaminant (such as the nutrients nitrogen or phosphorous) in relation to a given receptor (such as groundwater or stream/lake). The coincidence of a sensitive area with the likelihood of high loading of that contaminant indicates a possible environmental risk. When defined in this way, critical source areas can be delineated using GIS layers and appropriate factors.

2 Software Design

2.1 Free and Open Source Software for Geospatial

Open Source software often allows for a flexible development environment because the software in question is often designed to be easy to set up and interact with other software [1]; in contrast to proprietary software that is often designed as a stand-alone system. The Open Source Geospatial Foundation (OSGeo [2]) is the umbrella organization that supports collaborative projects and organizes the annual international Free and Open Source Software for Geospatial (FOSS4G) conference. As well as the projects overseen by OSGeo, there are a large number of other GIS projects listed at [3] and [4] and reviewed, for example, in [5].

The GIS desktop viewer chosen for the Pathways CMT was Quantum GIS (QGIS) [6] which can view multiple sources of GIS data, such as ESRI shapefiles, Raster files and databases. The principal data source is the PostgreSQL [7] database which can be spatially enabled using PostGIS [8]. The advantage of using such a database is that the available data can be processed and stored for viewing, in this case via the QGIS Python plugin architecture. Working copies of data can be saved as individual projects and updated to simulate separately different scenarios. Other Open Source software combinations are available, but the QGIS / PostGIS pairing was seen as the most flexible option for the Pathways project.

2.2 Three Levels of Interaction

To allow for different levels of prior knowledge and GIS experience for the users, a range of levels of interaction is provided in the user interface, see Fig. 1.

Level 1: Source Layers. At the first level of interaction, the user can choose the basic source data to view as GIS-layers for the catchment of interest, such as soil properties, geology and land-use as well as potential point sources of contaminants. This data is pre-processed so the user can easily view data of interest with minimal interaction; summary reports and statistics are readily available.

Level 2: Static Analysis Layers. At the next level of interaction, users can choose a specific catchment and view in detail the loadings and static critical source areas formed by combining the Level 1 layers. Initially the data is created automatically from default data and parameters, but the user can choose to update this data, if they have more local knowledge or experience of the actual processes involved.

Level 3: Dynamic Analysis Layers. The final level of interaction requires the user to input hydrological data specific to the catchment. The more complicated lateral flows are modeled using a rainfall-runoff model, producing dynamic critical source areas.

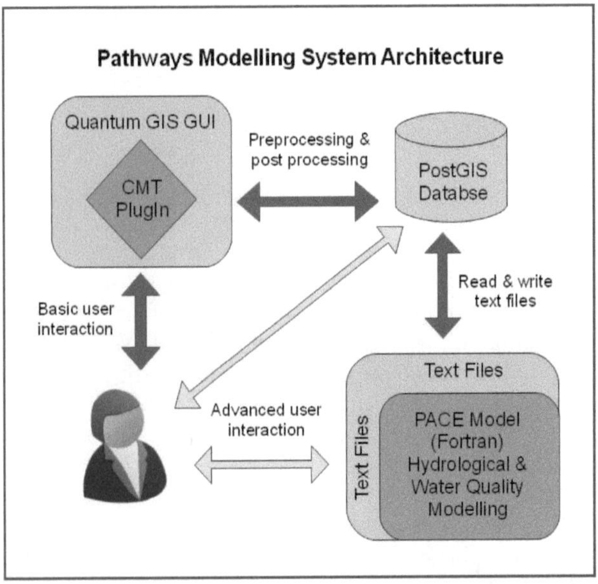

Fig. 1. CMT user interaction diagram

2.3 Flexible Software Design

The software was designed as a loosely coupled system so that individual modules can be developed independently and replaced if necessary in the future. The QGIS software is used to view the GIS data; Fig. 2 shows an example, on the left hand side is the layer legend. A plugin written in the Python language lets the user choose which data to load from the PostGIS database (right hand side of Fig. 2). The Python code uses the QGIS application programming interface (API) to define different types of data needed for the CMT plugin and loads this via the database API into QGIS. Summary data and statistics of the relevant layers are easily computed in the plugin. The data has been preprocessed to speed up access for Level 1 Source Layers but also to allow combinations of data sources to be intersected "on the fly" (created dynamically) for Level 2 Static Analysis Layers. New data tables are created in the database using SQL (structured query language) and PostGIS spatial functions.

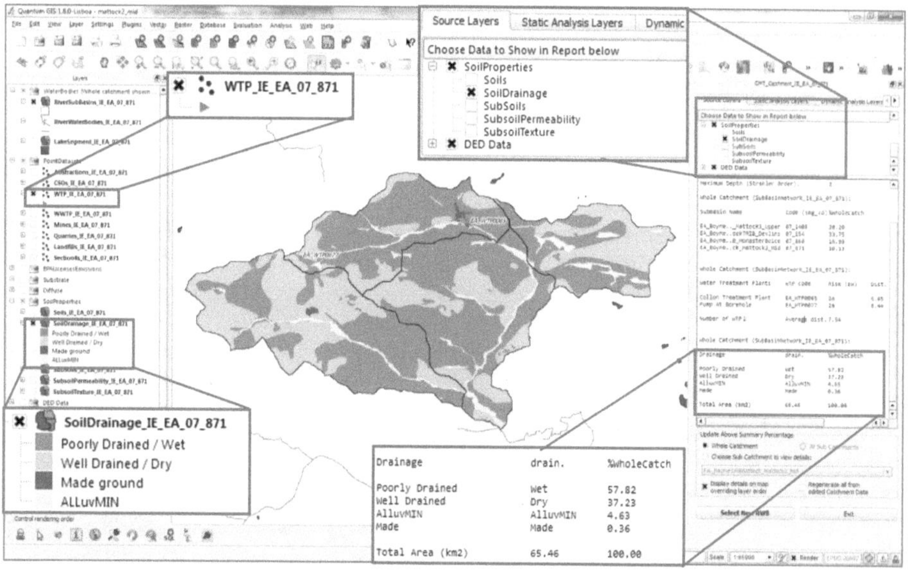

Fig. 2. Screenshot of the CMT – note usual GIS tools are provided by QGIS, the plugin to the right allows the user to choose data layers and view summary statistics of those layers below

Some tables in Level 2 are copies of Level 1 source layers and can be edited by the user. Functionality to change the underlying data is available in QGIS and many other desktop GIS systems. The user might modify the data because, for example, they might know that the in-built default loading values are not appropriate and should be changed or want to simulate changes in, for instance, land-use. After updating the data source, all the other related data tables can be updated from one button click and the outcome of those changes will be immediately seen.

For Level 3, the Pathways Computational Engine (PACE) has been written as a standalone FORTRAN package [9]. Communication with the CMT and database is undertaken only in the form of reading and writing text files. This loose coupling means that the Computational Engine can be developed in parallel with the GIS component and can be amended without affecting the rest of the CMT.

3 Evaluating Environmental Risk

3.1 Source Layers

Geological and soil mapping information were supplied by the EPA and GSI (Geographical Survey of Ireland) in digital form. Derivatives of the EPA / Teagasc (Irish Agriculture and Food Development Authority) soil maps [10] include soil drainage and subsoil permeability. Other relevant maps include aquifer type, groundwater vulnerability [11], Corine land-cover, slope and lakes, a number of these were combined to make the national recharge map [12]. Point sources such as landfills, water treatment plants and registered IPPC (integrated pollution prevention and control) facilities are also included as well as possible hazards such as abstractions and quarries. These maps are all available to view in Level 1 of the CMT (Fig. 2).

3.2 Calculating Nutrient Loadings

In the CMT, contaminant loadings to land surfaces are calculated from publicly available data from the Irish Central Statistics Office (CSO). Population figures and the number of houses and type of sewage system that serves each house [13] are available. Land use and the number of type of animals in an Electoral District is provided by the Census of Agriculture [14] which will soon be updated from 2000 to 2010 data. This data is then used to calculate nutrient (Nitrogen and Phosphorous) loading to different hydrological pathways using the Groundwater Task Team (GTT) [15] methodology, modified for Irish settings by the EPA.

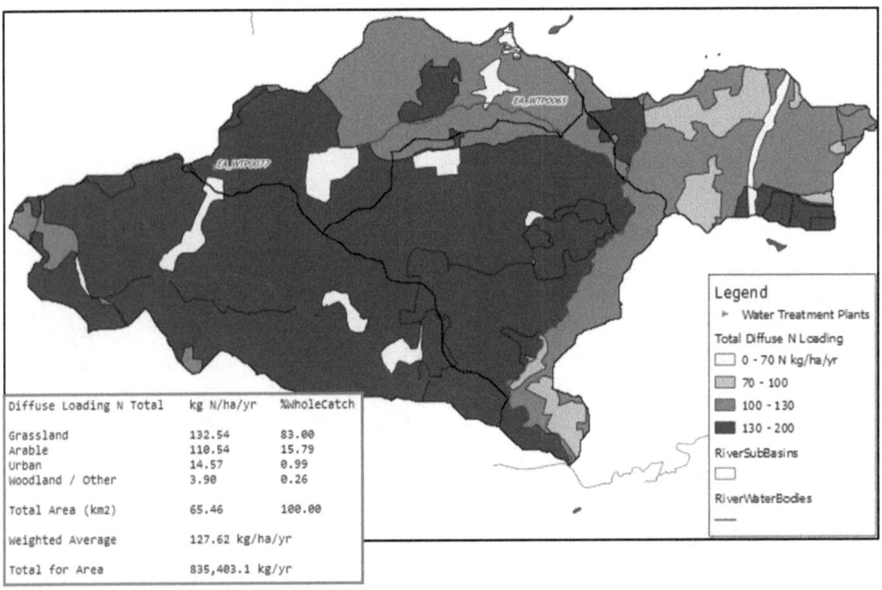

Fig. 3. Example of total input Nitrate loading to the Mattock_mid catchment, on the border of Co Louth and Co Meath in Ireland; inset shows averages broken down by land-use

The hydrological pathways are modeled at river basin sub-catchment level (average area 13 km^2), so intersection of these with the Irish Electoral District (average area 20 km^2) gives a coarse variation in loadings across the chosen catchment. An example output of the loadings calculated for nitrate by the CMT is shown in Fig. 3, including a statistical summary.

The GTT methodology also models, in an empirical way, the processes of the nutrient going through the soil and to the particular receptor that is required to model attenuation processes. For example leaching in Grassland Pastures is calculated using NCycle_IRL [16]. Leaching of Nitrogen is related to soil drainage or texture, so the combination of the loadings map (Fig. 3) and the soil drainage map (Fig. 2) are used to create a loading map of N from soil drainage (not shown).

3.3 Creating Hydrogeologically Susceptible Areas (HSAs)

Hydrogeologically susceptible areas are those areas that are sensitive to a given nutrient reaching a given water body receptor. Creating these layers requires identifying the soil or geological properties that are relevant through literature review, expert opinion or empirical analysis. For example, for nitrogen to groundwater, the combination of subsoil permeability and depth to bedrock is used to determine ranges of susceptibility. The EPA has published a risk-based methodology for single house treatment system that includes a number of susceptibility factors [17]; they also consider denitrification in bedrock for the Groundwater pathway. These factors are being reviewed and modified for use in the CMT.

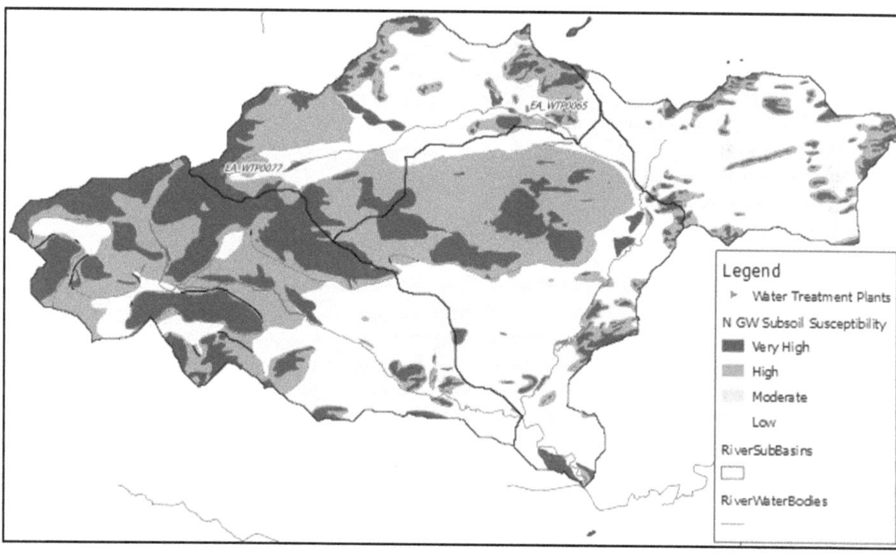

Fig. 4. Hydrogeologically Susceptible Areas (HSAs) of Nitrogen to the Groundwater receptor for the Mattock_mid catchment

Fig. 4 shows an example of the HSAs derived for the Mattock catchment, in the North East of Ireland, for Nitrogen to groundwater. Very High and High susceptibility relate to areas of extreme vulnerability (subsoil less than 3m subsoil thick) or highly permeable subsoils. These categories can be given quantitative risk values which in effect are the attenuation rates of nitrogen to groundwater.

Susceptibility maps for phosphorous to groundwater can be created using the same procedure. Susceptibility to surface water receptors is more problematic as no temporal variation is considered when using the GIS layers. Multiple receptors might also exist in the catchment (in terms of lakes and streams), so only the loading, mobilization and delivery to the local surface water receptor is being considered. The computational engine in Level 3 of the CMT has the capabilities to model the lateral flow and temporal dynamics required for generic surface water receptors.

3.4 Critical Source Areas (CSAs)

By combining the Loadings information with the Hydrogeological Susceptibility information it is possible to create Critical Source Area (CSA) maps for each contaminant. When intersecting the maps, the loadings rates from soil drainage can be multiplied by the attenuation rates or other factors derived from the susceptibility maps. The amount of loading that reaches groundwater also needs to take into account as well as the proportion of effective rainfall involved.

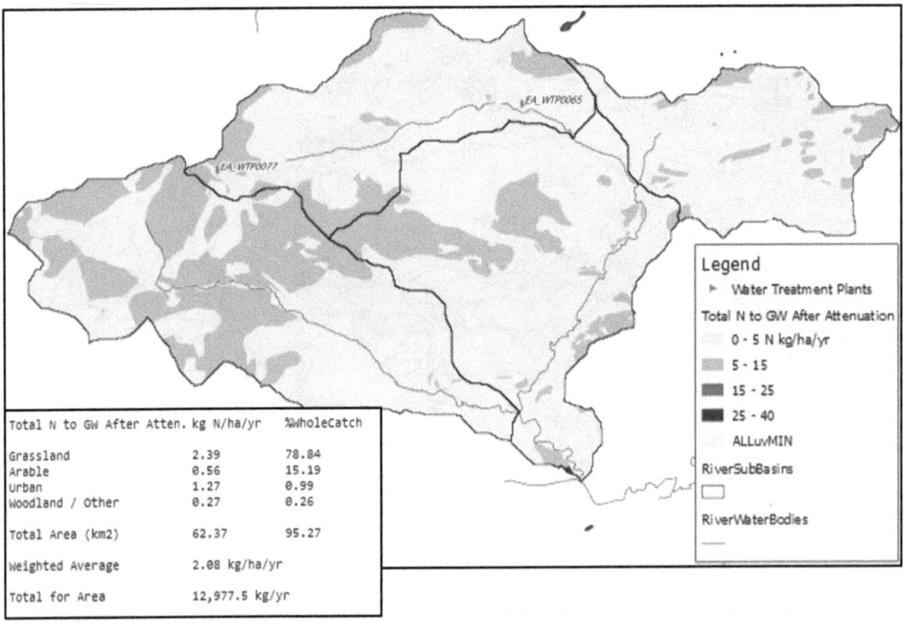

Fig. 5. Critical Source Areas for Nitrogen to Groundwater receptor from diffuse sources; this is final loading after attenuation for Mattock_mid catchment; inset shows averages

The national recharge map [12] is available within the CMT which includes effective rainfall and recharge coefficient. The latter coefficient determines the amount of effective rainfall that will reach the groundwater depending on the hydrogeological category of the area. Most of the Mattock catchment shown in the figures is underlain by a "non-productive" aquifer, that is, it does not have the storage capacity or transmission capability to supply more than a few houses. This is simulated in the Recharge Map by a "cap" that limits the amount of water that the aquifer will accept.

The recharge cap has an additional effect on the model output as it is assumed that excess rainfall is effectively rejected by the aquifer. In the CMT, a modified recharge coefficient is used that takes into account the recharge cap for non-productive aquifers. Fig. 5 shows the Critical Source Areas to groundwater after combining loadings with soil drainage, the attenuation factors and modified recharge coefficients.

4 Further Applications and What-If Scenarios

The CMT is designed to be user friendly by pre-processing as much of the available data as possible so an initial overview can be evaluated by the user or River Basin District Manager very quickly. If the Manager is aware, from local knowledge, of inaccuracies in the underlying loadings data or subsequent output, then changes to the input data and parameters are allowed because copies of the original data can be saved under a different project name. This flexibility also allows for what-if scenarios to be generated as the user can change the loadings data to simulate an increase in, for instance, inorganic fertilizer or leakage at a waste water treatment facility, for example. Also a change in land-use could be simulated by allowing changes to the copy of the Corine land-cover layer. So a change in arable land to urban could be simulated or vice versa. It might also be possible to simulate the effects of climate change by adjusting meteorological inputs or changing the properties of the soil or geology layers and analyze the outcome in the Critical Source Areas.

5 Conclusions

A Catchment Management Tool (CMT) is under development for the Irish EPA. The user interface of the CMT is designed to be flexible to allow local knowledge to be included in the system as well as the possibility to try what-if scenarios in relation to environmental assessment. Contaminant loadings are calculated from publicly available data and hydrogeologically susceptible areas (HSAs) are formed by combining soil and geological GIS layers. Further combination of Loadings and HSAs allow Critical Source Areas to be delineated that identify high risk areas from a given contaminant to a receptor of interest.

This paper has described the software architecture of the CMT and the data sources required for the loadings and geological susceptibility layers. Examples of critical source area maps produced from the CMT were shown. Changes in those maps due to user interaction can be used to simulate the changes in environmental risk due to possible future scenarios. Initial feedback suggests the tool will be useful for the EPA and River Basin District Managers; further user testing and validation of the loading and attenuation rates is required before final release.

Acknowledgments. The Pathways Project is an EPA STRIVE funded project (2007-W-CD-1-S1), the collaborative partners are Queens University Belfast, University College Dublin and Trinity College Dublin. The authors would also like to thank the project Steering Committee for invaluable discussion and advice.

References

1. Sanz-Salinas, J.-G., Montesinos-Lajara, M.: Current Panorama of the FOSS4G Ecosystem. Novatica X(2), 43–51 (2009)
2. Open Source Geospatial Foundation, `http://www.osgeo.org`
3. Open Source GIS, `http://www.opensourcegis.org`
4. Free and Open Source GIS, `http://www.freegis.org`
5. Steiniger, S., Bocher, E.: An Overview on Current Free and Open Source Desktop GIS Developments. Int. J. of Geographical Information Science 23(10), 1345–1370 (2008)
6. Quantum GIS, `http://www.qgis.org`
7. PostgreSQL, `http://www.postgresql.org`
8. PostGIS, `http://www.postgis.org`
9. Mockler, E., Bruen, M.: A variable-structure catchment model as the engine for a water quality decision support tool facilitating scientific debate and collaboration. In: Seppelt, R., Voinov, A.A., Lange, S., Bankamp, D. (eds.) International Congress on Environmental Modelling and Software - iEMSS 2012 (2012)
10. Fealy, R.M., Green, S., Loftus, M., Meehan, R., Radford, T., Cronin, C., Bulfin, M.: Teagasc EPA Soil and Subsoils Mapping Project-Final Report. Volume I, Teagasc, Dublin (2009)
11. DELG/EPA/GSI: Groundwater Protection Schemes. Department of the Environment and Local Government, Environmental Protection Agency and Geological Survey of Ireland (1999)
12. Hunter Williams, N., Misstear, B., Daly, D., Johnston, P., Lee, M., Cooney, P., Hickey, C.: A National Groundwater Recharge Map for Ireland. National Hydrology Conference (2011)
13. CSO, Small Area Population Statistics (SAPS) (2006), `http://census.cso.ie/census/ReportFolders/ReportFolders.aspx`
14. CSO, Census of Agriculture 2000 Detailed Results: (2000), `http://www.cso.ie/px/pxcoa2000/database/censusofagriculture2000/censusofagriculture2000.asp`
15. Groundwater Task Team: Cumulative Nitrogen and Phosphorus Loadings to Groundwater, produced by ENTEC on behalf of the Scottish Environment Protection Agency, the Environment Agency (England & Wales), the Environmental Protection Agency and the Northern Ireland Environment Agency (2010)
16. del Prado, A., Scholefield, D., Brown, L.: NCYCLE_IRL: A model to predict N fluxes in Irish grasslands. Environmental Protection Agency, Ireland (2005)
17. EPA: A risk-based methodology to assist in the regulation of Domestic Waste Water Treatment Systems. Environmental Protection Agency, Ireland (2013)

Continental-Scale Living Forest Biomass and Carbon Stock: A Robust Fuzzy Ensemble of IPCC Tier 1 Maps for Europe

Daniele de Rigo[1,2], José I. Barredo[1], Lorenzo Busetto[1], Giovanni Caudullo[1], and Jesús San-Miguel-Ayanz[1]

[1] European Commission, Joint Research Centre, Institute for Environment and Sustainability, via E. Fermi 2749, I-21027 Ispra (VA), Italy
[2] Politecnico di Milano, Dipartimento di Elettronica e Informazione, via Ponzio 34/5, I-20133 Milano, Italy

Abstract. Forest ecosystems play a key role in the global carbon cycle. Spatially explicit data and assessments of forest biomass and carbon are therefore crucial for designing and implementing effective sustainable forest management options and forest related policies. In this contribution, we present European-wide maps of forest biomass and carbon stock spatially disaggregated at 1km x 1km. The maps originated from a spatialisation improvement of the IPCC methodology for estimating the forest biomass at IPCC Tier 1 level (IPCC-T1). Using a categorical map of ecological zones within the mapping technique may originate boundary effects between the ecological zones. This may induce undue artifacts in the outcomes, as evident in previously published maps generated with the IPCC-T1 methodology. Here we present a novel method for IPCC-T1 biomass mapping which mitigates these artifacts. We propose the use of a fuzzy similarity map of the FAO ecological zones computed by estimating the relative distance similarity (RDS) of each grid-cells climate and geography with respect to the FAO ecological zones. A robust ensemble approach was used to merge an array of simple models with spatially distributed fuzzy set-membership. This allowed the boundary artifacts to be reduced, while mitigating the impact of model semantic extrapolation. The chain of semantically enhanced data-transformations is described following the semantic array programming paradigm. Preliminary results obtained from the application of this novel approach are presented along with a discussion of its impact on the derived maps.

Keywords: Ecological Zones, Living Forest Biomass, Living Forest Carbon Stock, IPCC Tier 1, Relative Distance Similarity, Robust Fuzzy Ensemble, Semantic Array Programming.

1 Introduction

Forest ecosystems play a key role in the global carbon cycle and are considered large and persistent carbon sinks [1]. Spatially explicit data and assessments

J. Hřebíček et al. (Eds.): ISESS 2013, IFIP AICT 413, pp. 271–284, 2013.

of forest biomass and carbon are therefore of great importance for the design and implementation of effective sustainable forest management options and forest related policies. In complex, wide scale problems, the effectiveness of the science-policy interface might depend on the availability, understandability and multiplicity of estimations, even for correctly assessing the extent and sources of critical uncertainties [2–7]. In this context, European-wide maps of forest biomass and carbon stock spatially disaggregated at $1\,km \times 1\,km$ (LAEA grid) were recently developed [8], exploiting the IPCC Tier 1 level methodology described in [9]. This is a cost-effective approach for biomass and carbon mapping over large geographical regions where gaps in accurate biomass information exist. It is considered appropriate for continental or global studies [10–12]. According to this methodology [8], above ground living biomass B_c^{ag} for a given cell c of a generic regular grid may be estimated by aggregating the biomass contributions by forest type `typ` [13] – coniferous (con) and broadleaved (bro) – and management type `man` – semi-natural forest (sn) and plantations (pl):

$$B_c^{ag} = \sum_{\substack{\mathtt{man}\in\{sn,pl\}\\ \mathtt{typ}\in\{con,bro\}}} B_c^{ag,\mathtt{man},\mathtt{typ}} \tag{1}$$

The contributions $B_c^{ag,\mathtt{man},\mathtt{typ}}$ are determined on the basis of two factors. First, the fraction $fc_c^{\mathtt{man},\mathtt{typ}}$ of the cell's area covered by each combination {`man`,`typ`}. Second, the average above-ground biomass $b_c^{ag,\mathtt{man}}$ associated to semi-natural forest and plantations in the ecological zone (e.g. boreal, temperate, subtropical; mountain vs. steppe etc.; from hereafter, EZ) to which c belongs (see tab. 1):

$$B_c^{ag,\mathtt{man},\mathtt{typ}} = b_c^{ag,\mathtt{man}} \cdot fc_c^{\mathtt{man},\mathtt{typ}} \tag{2}$$

Below-ground biomass B_c^{bg} and carbon stock (above- C_c^{ag} and below-ground C_c^{bg}) can be successively estimated as:

$$B_c^{bg} = \sum_{\substack{\mathtt{man}\in\{sn,pl\}\\ \mathtt{typ}\in\{con,bro\}}} B_c^{ag,\mathtt{man},\mathtt{typ}} \cdot \lambda_c^{\frac{bg}{ag}} = B_c^{ag} \cdot \lambda_c^{\frac{bg}{ag}} \tag{3a}$$

$$C_c^{ag} = \sum_{\substack{\mathtt{man}\in\{sn,pl\}\\ \mathtt{typ}\in\{con,bro\}}} B_c^{ag,\mathtt{man},\mathtt{typ}} \cdot \lambda_c^{\frac{C}{B},\mathtt{typ}} \tag{3b}$$

$$C_c^{bg} = \sum_{\substack{\mathtt{man}\in\{sn,pl\}\\ \mathtt{typ}\in\{con,bro\}}} B_c^{ag,\mathtt{man},\mathtt{typ}} \cdot \lambda_c^{\frac{C}{B},\mathtt{typ}} \cdot \lambda_c^{\frac{bg}{ag}} = C_c^{ag} \cdot \lambda_c^{\frac{bg}{ag}} \tag{3c}$$

where $\lambda_c^{\frac{bg}{ag}}$ is the ratio between the below-ground and the above-ground living forest biomass and $\lambda_c^{\frac{C}{B},\mathtt{typ}}$ is the ratio between tonns of living forest carbon and tonns of biomass (carbon fraction). Approximately, $\lambda_c^{\frac{C}{B},\mathtt{typ}}$ describes both above-ground and below-ground conditions, depending on the forest type. Both factors vary with the EZ to which c belongs (tab. 1).

For their pan-European IPCC-T1 application, [8] used as inputs the CORINE land cover map of 2006 [14] to determine the fraction of surface occupied by coniferous and broadleaved forests within $1\,km \times 1\,km$ cells of a pan-European grid. Each cell was associated with an EZ, following the map of Global Ecological Zones for the Global Forest Reporting [15,16]. Biomass and carbon stock for each cell were finally computed using eqs. 1–3. This method relies on the simplistic assumption that within a given EZ the coefficients of tab. 1 are constant.

Table 1. Above-ground forest biomass, ratio below- vs. above-ground biomass and carbon fraction of forest biomass in European semi-natural forests or plantations. Values from [17] (from [9] where marked with *); d.u. denotes dimensionless units.

Ecological zone ez	$b^{ag,sn,ez}$ $\left[\frac{tonn}{ha}\right]$	$b^{ag,pl,ez}$ $\left[\frac{tonn}{ha}\right]$	$\lambda^{\frac{bg}{ag},ez}$ d.u.	$\lambda^{\frac{C}{B},con,ez}$ d.u.	$\lambda^{\frac{C}{B},bro,ez}$ d.u.
Subtropical humid forest	180	180	0.24	0.47	0.47
Subtropical dry forest	130	90	0.28	0.47	0.47
Subtropical steppe	130*	con : 20 bro : 80	0.28*	0.47	0.47
Subtropical mountain system	135	95 (40-150)	0.27*	0.47	0.47
Temperate oceanic forest	120	200	0.29	0.51	0.48
Temperate continental forest	120	200	0.29	0.51	0.48
Temperate steppe	120*	n/a	0.29	0.51	0.48
Temperate mountain system	130	200	0.29	0.51	0.48
Boreal coniferous forest	50	40	0.39	0.51	0.48
Boreal tundra woodland	50*	25	0.39	0.51	0.48
Boreal mountain system	45	40	0.39	0.51	0.48
Polar	50*	25*	0.39*	0.47*	0.47*

The variability of the estimated quantities within each EZ is therefore only driven by the varying percentage of forest area within the cells, and evident artefacts in the output maps are present on the ecozones' boundaries (see also fig. 3).

In order for these effects to be mitigated, we propose an innovative fuzzy similarity map of EZs as an alternative input dataset so as to associate the different areas of Europe with more consistent average biomass amounts and carbon fraction coefficients. This fuzzy set map was computed by estimating the relative distance similarity (RDS) [18] of each grid-cells climatic and geographic conditions from the distribution of values typically observed within the FAO EZs. An unsupervised robust ensemble approach was used to merge an array of simple models with spatially distributed fuzzy set-membership. This allowed the boundary artifacts to be reduced, while mitigating the impact of model semantic extrapolation. The chain of semantically enhanced data-transformation models (D-TM) is described following the *semantic array programming* paradigm [19,20].

The theoretical basis and the preliminary results obtained from the application of this novel approach are presented along with a discussion of its impact on the derived maps.

2 Materials and Methods

The methodology to determine the fuzzy-set memberships of each $1\,km \times 1\,km$ spatial cell (on a LAEA regular grid, INSPIRE [21,22] compliant) to each EZ is briefly explained in the next subsection. It is based on a concise array-based mathematical notation for a series of D-TMs, implemented as array programming [23] modules in the GNU Octave/MATLAB programming language [24]. The adopted paradigm of semantic array programming [19,20] is supported by the Mastrave library and is defined by a modularisation of autonomous tasks with an intense effort toward their most concise generalization and reusability. Each module is semantically constrained with array-oriented compact invariants, pre- and post-conditions. Some straightforward constraints are exemplified in the following with the notation **::*constraint*::** and an associated active link.

From Crisp Ecological Zones to a Fuzzy Similarity Map. FAO produced an expert-based EZ map considering not only bioclimatic variables, but also potential and national vegetation maps [16]. The FAO classification does not match exactly with the Holdridge one, based exclusively on bioclimatic data [25], and in general is not meant to be fully reproduced as a computational model.

FAO EZs embed a broad semantic understanding of climatic specificities such as the mass elevation effect (MEE), for which altitude influences vegetation less "on larger mountains than on smaller ones" [16]. Our method exploits the FAO crisp map as input for training a fuzzy EZ classification with a set of N^C bioclimatic covariates (tab. 2). Besides temperature and precipitation derived factors, geographic ones are also used – elevation range and solar irradiation – so as to better discriminate e.g. polar, alpine and boreal EZs. Evapotranspiration is not used [30]. MEE is considered with neighbourhood analysis (spatial average moving windows) in two covariates. For each ecozone ez, the fuzzy set-membership map $\mu_{\cdot}^{ez}$ is derived in three steps with intermediate D-TMs (eqs. 4a-c).

The spatial extent $\mathcal{U}^S$ is partitioned in N^{EZ} crisp EZ sets $\mathcal{S}^{ez}$, $ez \in \mathcal{U}^{EZ} \equiv [1, N^{EZ}]$. However, an unclassified spatial buffer of 2 km along the boundary between different EZs accounts for the rasterisation error and leads to the set of classified grid cells $\tilde{\mathcal{U}}^S = \bigcup_{ez} \mathcal{S}^{ez} \subseteq \mathcal{U}^S$. The RDS analysis [18] (Mastrave library) associates to each pair composed by a set $\mathcal{S}^{ez}$ and a spatial cell c a similarity score RDS_c^{ez}. Given a training set $\mathring{\mathcal{U}}^S \subseteq \tilde{\mathcal{U}}^S$ of classified cells, RDS_c^{ez} expresses the **::possibility::**[1] $\in [0,1]$ for c to belong to ez based on the similarity of each i-th **::nonnegative::**[2] bioclimatic covariate C_c^i from the covariate's values in $\mathcal{S}^{ez} \cap \mathring{\mathcal{U}}^S$, aggregated $\forall i \in [1, N^C]$ with the RDS operator Ω:

$$\mathit{RDS}_c^{ez} = \max_{\nu \in \mathcal{S}^{ez}} \underset{i \in [1,N^C]}{\Omega} \frac{\min(C_c^i, C_\nu^i)}{\max(C_c^i, C_\nu^i)} \tag{4a}$$

The fuzzy set-membership μ_c^{ez} of c to $\mathcal{S}^{ez}$ is estimated by means of a robust ordinal algorithm. The algorithm has been designed so as for μ_c^{ez} to be

[1] `http://mastrave.org/doc/mtv_m/check_is#SAP_possibility`
[2] `http://mastrave.org/doc/mtv_m/check_is#SAP_nonnegative`

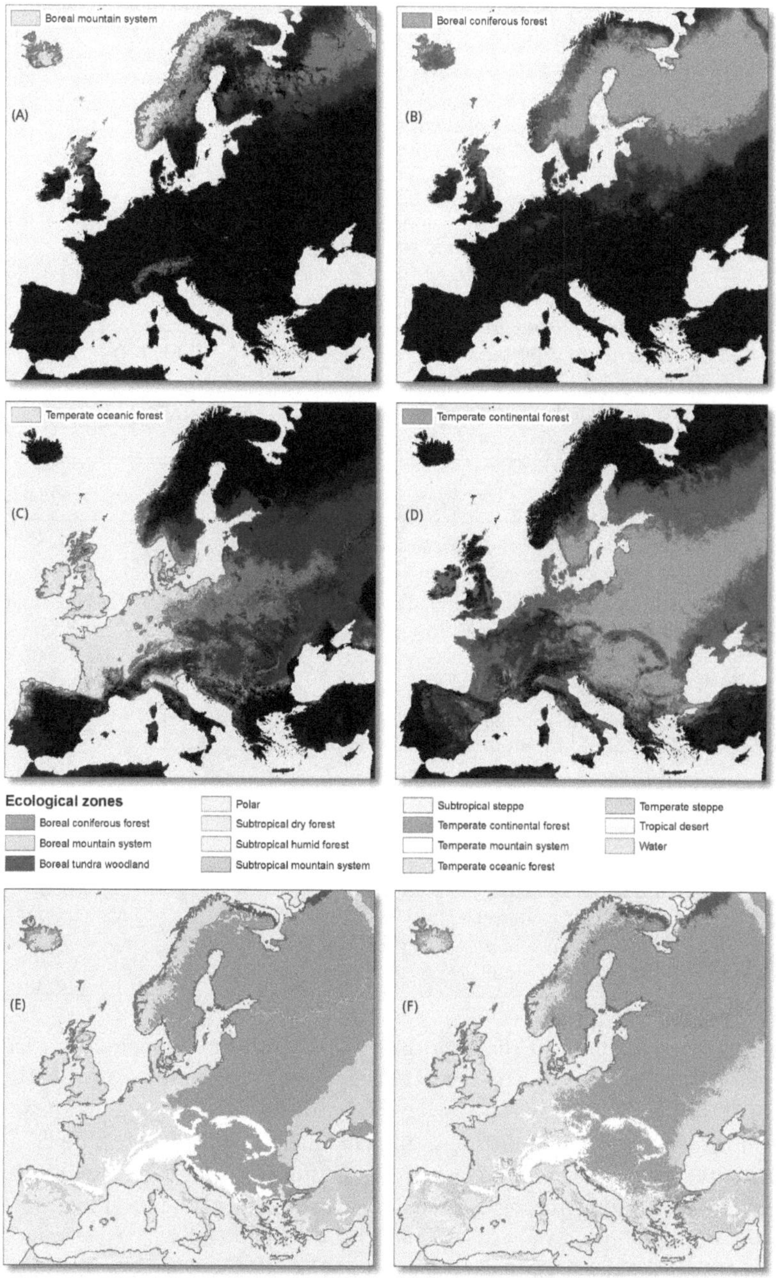

Fig. 1. Preliminary maps of fuzzy set-membership (A-D) of four ecological zones EZ, and comparison between the original FAO map (E) and a defuzzified map obtained by selecting among the fuzzy EZ sets the dominant (the highest) value of set-membership (F), for each spatial cell. Red lines denote the crisp boundary of the original FAO EZs.

Table 2. Bioclimatic variables and derived covariates C_c^i used for computing the fuzzy set-membership μ_c^{ez} of each spatial grid cell c to each ecological zone ez. Quantities averaged with a spatial moving window of $3\,km{\times}3\,km$ are denoted with $^{3\times3}$. Temperature in degrees Celsius; precipitation in mm; solar irradiations in $W\,h\,m^{-2}$; elevation range in m. T_c: annual avg. temperature. T_c^m, $\lfloor T \rfloor_c^m$, $\lceil T \rceil_c^m$: avg., min. and max. temperature of the month m. P, P_c^m: annual total precipitation and that of the month m.

Variable code	Description of the bioclimatic variable	Reference equation
S_c^{light}	Total potential solar irr. of the 6 lighter months [26]	
S_c^{dark}	Total potential solar irr. of the 6 darker months [26]	
T_c^{cold}	Average temperature of the coldest month	$\min_m T_c^m$
T_c^{warm}	Average temperature of the warmest month	$\max_m T_c^m$
ΔT_c	Mean of monthly temperature range	$1/12 \sum_m \lceil T \rceil_c^m - \lfloor T \rfloor_c^m$
T_c^{tundra}	Nordenskiöld index [27]	$T_c^{\text{warm}} + 0.1 T_c^{\text{cold}} - 9$
P_c^{dry}	Total precipitation of the driest month	$\min_m P_c^m$
P_c^{wet}	Total precipitation of the wettest month	$\max_m P_c^m$

Covariate code	Description of the bioclimatic covariate $C_c^i, i \in [1, 12]$	Reference equation
1. $\lceil \Delta S \rceil_c^{3\times3}$	$3\,km \times 3\,km$ seasonal variation of pot. solar irr. (normalised by semestral minimum)	$\dfrac{S_c^{\text{light},3\times3} - S_c^{\text{dark},3\times3}}{S_c^{\text{dark},3\times3}}$
2. S_c	Annual potential solar irradiation	$S_c^{\text{light}} + S_c^{\text{dark}}$
3. eT_c^{tundra}	Nordenskiöld exponential index	$\exp(\,T_c^{\text{tundra}}\,)$
4. T_c^{+100}	Shifted average annual temperature	$T_c + 100\,°C$
5. $T_c^{\text{cold}+100}$	Shifted average temp. of the coldest month [16]	$T_c^{\text{cold}} + 100\,°C$
6. $\lceil \Delta P \rceil_c$	Seasonal variation of monthly precipitation (normalised by minimum: driest month)	$\dfrac{P_c^{\text{wet}} - P_c^{\text{dry}}}{P_c^{\text{dry}}}$
7. $\lfloor \Delta P \rfloor_c$	Seasonal variation of monthly precipitation (normalised by maximum: wettest month)	$\dfrac{P_c^{\text{wet}} - P_c^{\text{dry}}}{P_c^{\text{wet}}}$
8. $T^{100}lP_c$	Shifted temperature per rain magnitude order	$T_c^{+100} / \log_{10}(\,P_t\,)$
9. $ER_c^{3\times3}$	$3\,km \times 3\,km$ average elevation range	
10. $N_c^{m \geq 10\,°C}$	Number of months with $T_c \geq 10\,°C$ [16]	$\sum_m [T_c^m \geq 10]$
11. $N_c^{m\text{dry}}$	Number of dry months [16]	$\sum_m [P_c^m \leq 2T_c^m]$
12. $S\Delta T_c$	Potential irradiation by montly temp. range (simplified avg. monthly solar entransy flux [28, 29])	$S_c \cdot \Delta T_c$

invariant to any monotonically non-decreasing, rank-preserving transformation of the RDS scores. This means that the sole semantic requirement for the similarities RDS_c^{ez} is to be **::sortable::**[3]. The second D-TM implements this mathematical invariance by computing an intermediate quantity, η_c^{ez}. The EZs whose similarity is above the median are ranked and the ranking is scaled in $[0, 1]$ so that the maximum among the scaled values η_c^{ez} is 1. This mathematically describes the semantic concept of ensuring each spatial cell c to belong with certainty to at least one EZ $ez^{*,c}$ (i.e. c is in the kernel of $\eta_c^{ez^{*,c}}$). In case two or more EZs in c share the maximum RDS score, c would belong to the kernel of each of them. The median threshold also ensures c not to belong to the farthest EZs $ez_i^{\varnothing,c}$. For example, it would otherwise be surprising for a spatial cell in the

[3] `http://mastrave.org/doc/mtv_m/check_is#SAP_sortable`

subtropical dry forest to be also substantially characterised by the polar zone. The described array transformation $f^{\eta} : \mathit{RDS} \mapsto \eta$ is equivalent to the following:

$$\begin{bmatrix} \cdots \\ \eta_c^{ez} \\ \cdots \end{bmatrix} = f^{\eta}(\mathit{RDS}_c) = \begin{bmatrix} \cdots \\ \dfrac{\max\left(0, \sum\limits_{\eta \in \mathcal{U}^{EZ}, \eta \neq ez} [\mathit{RDS}_c^{ez} \geq \mathit{RDS}_c^{\eta}] - \left\lfloor \dfrac{N^{EZ}-1}{2} \right\rfloor\right)}{N^{EZ} - 1 - \left\lfloor \dfrac{N^{EZ}-1}{2} \right\rfloor} \\ \cdots \end{bmatrix} \quad ez \in [1, N^{EZ}] \tag{4b}$$

where the Iverson bracket notation is used [31]. Depending on the number of EZs, the relative importance of the best score $\eta_c^{ez^*}$ may be noticeably diluted by the remaining ones. For example, considering 12 EZs and the ranking r of the 6 best fitting ones (i.e. the above-median ones), $\eta_c^{ez^*}$ would only account for 29% of the total $(6/\sum_{r=1}^{6} r)$. The third D-TM stabilises this behaviour by imposing the weight of the dominant EZ to be 50% with a simple power transformation:

$$\mu_c^{ez} = \left(\eta_c^{ez}\right)^k \qquad \text{such that} \quad 2 \cdot \left(N^{EZ}\right)^k = \sum_{r=1}^{N^{EZ}} r^k \tag{4c}$$

Table 3. Comparison between the crisp FAO EZ $\mathcal{S}^{ez}$ and their proposed extended fuzzy version μ_c^{ez}. A subset $\mathring{\mathcal{U}}^S$ of the classified grid cells has been used, uniformly distributed in $\tilde{\mathcal{U}}^S$ with approximately $1/16^{th}$ of the available cells. The number of cells of $\mathring{\mathcal{U}}^S$ within (*true set*) and outside (*false set*) a given $\mathcal{S}^{ez}$ is listed along with the mean square error (for all cells; only *true set* cells; only *false set* cells) between μ_c^{ez} and $\mathcal{S}^{ez}$.

Ecological zone ez	Overall	Within the FAO EZ		Outside the FAO EZ	
	$\underset{c \in \mathring{\mathcal{U}}^S}{\text{mse}}$	$\sum\limits_{c \in \mathring{\mathcal{U}}^S} [c \in \mathcal{S}^{ez}]$	$\underset{c \in \mathring{\mathcal{U}}^S \cap \mathcal{S}^{ez}}{\text{mse}}$	$\sum\limits_{c \in \mathring{\mathcal{U}}^S} [c \notin \mathcal{S}^{ez}]$	$\underset{c \in \mathring{\mathcal{U}}^S \setminus \mathcal{S}^{ez}}{\text{mse}}$
Subtropical humid forest	0.0072	748	0.009	566558	0.0072
Subtropical dry forest	0.026	63836	0.063	503470	0.022
Subtropical steppe	0.031	3246	0.0027	564060	0.032
Subtropical mountain system	0.039	30308	0.037	536998	0.04
Temperate oceanic forest	0.068	79124	0.018	488182	0.076
Temperate continental forest	0.11	164777	0.021	402529	0.15
Temperate steppe	0.041	31513	0.0031	535793	0.043
Temperate mountain system	0.042	26715	0.054	540591	0.041
Boreal coniferous forest	0.055	134513	0.014	432793	0.068
Boreal tundra woodland	0.029	3515	0.0056	563791	0.029
Boreal mountain system	0.039	24809	0.029	542497	0.04
Polar	0.015	4202	0.0089	563104	0.015

Fuzzy Ecozones for Biomass and Carbon Stock Ensembles. The final D-TM of eq.4c ensures μ_c^{ez} to show a series of desirable properties for unsupervised fuzzy ensemble applications. First, the aforementioned semantic properties of η_c^{ez} also hold for μ_c^{ez}. Second, a weighted median of quantities of choice (e.g.

EZ biomasses or carbon stocks), done using as weights the fuzzy scores μ_c^{ez}, would always select the quantity corresponding to the dominant EZ in c. This provides a very intuitive and coherent way for estimating defuzzified (crisp) ensembles: selecting a very robust ensemble operator such as the weighted median automatically implies a crisp ensemble.

On the other hand, a weighted average would instead use all the above-median best fuzzy scores and would merge the corresponding associated quantities (but not the ones having worst scores, thus automatically behaving as a robust statistics). Classic intermediate robust statistics as the trimmed mean would show intermediate behaviour, as the intuition would expect. Furthermore, jack-knife resampling techniques might safely be applied by iteratively removing one fuzzy EZ per time. In the worst case, the removal would decrease the overall sum of remaining μ_c^{ez} to not less than 50%, thus ensuring a quite robust statistical resampling. Fig. 2 shows the application of weighted average ensembles to estimate forest living biomass and carbon stock maps.

3 Results and Discussion

Maps of fuzzy set memberships obtained for four EZs are shown as an example in fig. 1 (A-D). The similarity with a given EZ can be nonnegligible even very far from the original FAO spatial domain of that zone (the red boundary in the maps). The reconstructed crisp EZ map obtained by selecting the dominant fuzzy EZ agrees with the FAO map where sudden spatial climatic changes occur (e.g. mountains) but highlights areas with smoother transitions (fig. 1 E and F). Results obtained by using the fuzzy EZ maps as the basis for the computation of forest biomass and carbon stock are shown in fig. 2. In particular, fig. 2 (A,B) show the average amount of below-ground and above-ground biomass per unit surface ($b_c^{ag,sn}$) estimated across Europe for semi-natural forests (irrespective of the current $fc_c^{sn,\mathrm{typ}}$) by transforming the fuzzy EZs into the corresponding fuzzy biomass models. It must be stressed that, without the use of the fuzzy EZ maps, computing the two maps with IPCC-T1 methods would have led to a piecewise constant approximation, with localised sudden spatial changes along the EZ boundaries.

Fig. 2 (C,D) show instead the total living carbon stock of forests ($C_c^{ag} + C_c^{bg}$) estimated with eqs. 1–3 by taking into account the membership values for the different EZs. Due to the lack of gridded high-resolution information regarding the management type of forests in Europe, the maps represent two very different scenarios. The first scenario (fig. 2 C) assumes all forests to be semi-natural, while in the second one all forests are assumed as plantations(fig. 2 D). An example of the improvements introduced by the use of the fuzzy EZ maps for the computation of the IPCC-T1 biomass maps is finally shown in fig. 3.

The detailed analysis referring to the southern part ofNorway and Sweden highlights how the evident artefacts introduced in the above-ground forest living biomass by [8] and related to the crisp boundaries between EZs (see panel B) are substantially reduced in the improved map (see panel D). Although preliminary,

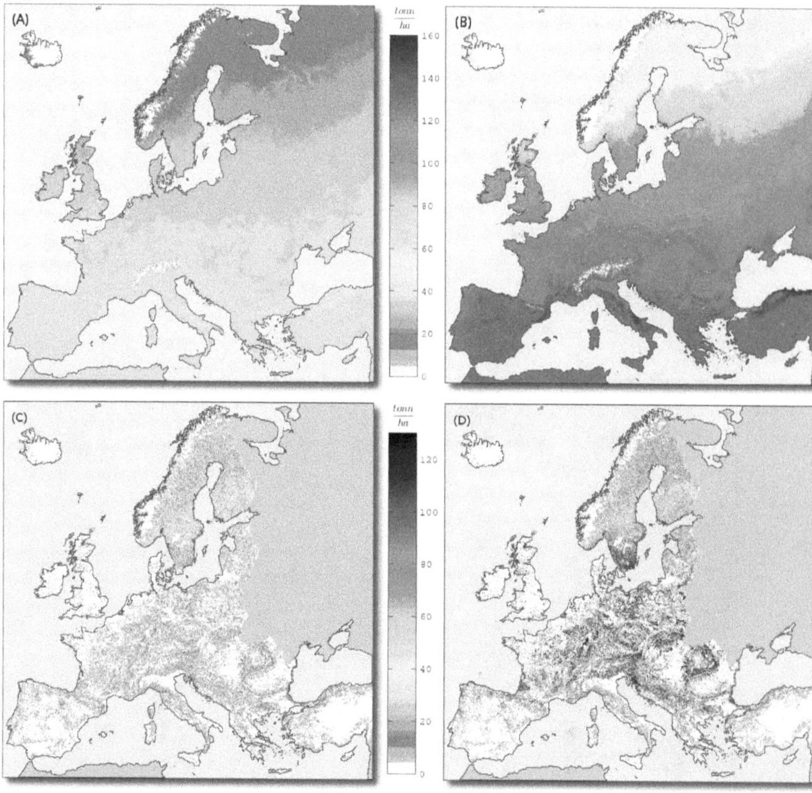

Fig. 2. Improved maps obtained exploiting the fuzzy EZ maps as an input for the IPCC TIER-1 mapping methodology. (A,B) Average amount of below ground ($b_c^{bg,\mathrm{sn}}$) and above-ground ($b_c^{ag,\mathrm{sn}}$) biomass per unit surface for semi-natural forests. (C,D) Total living carbon stock of forests ($C_c^{ag} + C_c^{bg}$) estimated considering all forests as semi-natural (C), and all forests as plantations (D).

these results suggest the proposed methodology for deriving fuzzy EZ maps to be beneficial in mitigating undue artifacts in IPCC-T1 large-scale biomass and carbon mapping. It must however be remembered that, although the quantities are disaggregated at $1\,km \times 1\,km$ spatial resolution in the resulting maps, the intrinsic limitations of the IPCC-T1 methodology (i.e., the use of constant coefficients to describe the typical characteristics of forests over huge spatial extents) suggest that local patterns should be considered as merely indicative.

Nevertheless, the method might prove helpful for interpolating spatial gaps in more detailed maps. The exemplified fuzzy ensemble modelling strategy – whose D-TM detailed computational description [32] is reproducible [33–35] and entirely based on free software [36] and data [37, 38] – lies within a family of semantically-enhanced ensemble techniques [39–42] which might also serve as reference for merging multiple partially independent estimations of biomass and carbon stock [43] and deriving new improved maps. Several factors, neglected in

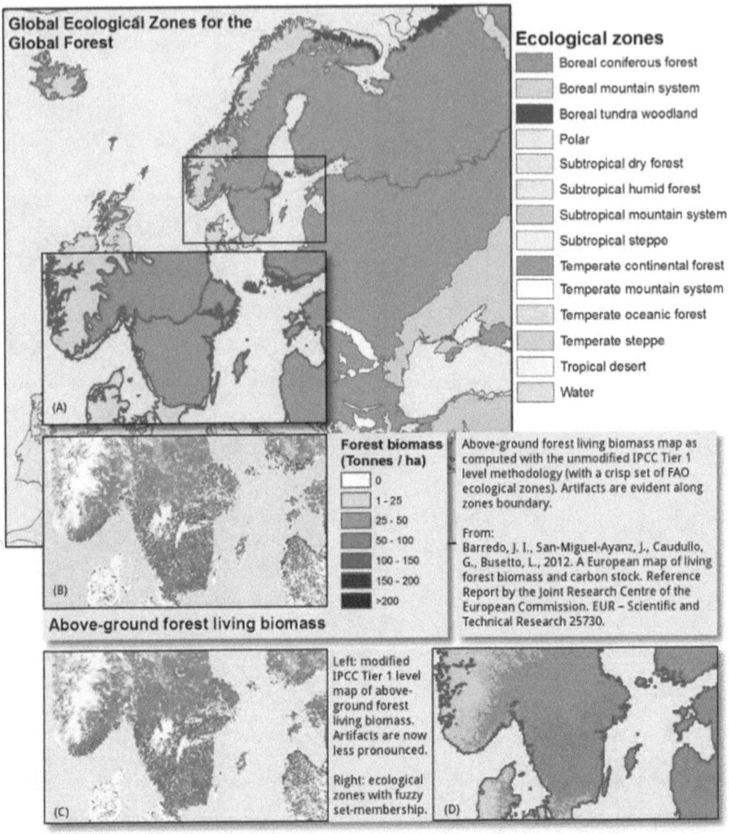

Fig. 3. Example of improvements introduced with the use of the fuzzy EZ maps for the IPCC-T1 modelling of above-ground forest living biomass in Europe. The area surrounding the southern part of Norway and Sweden is highlighted as it provides an example of artifacts (B) generated along the sharp boundary between FAO ecological zones as a result of the implementation of the unmodified IPCC-T1 [8]. The artefacts are substantially reduced in the improved map (C), thanks to the fuzzy approach used for the determination of the several coefficients used for biomass computation. (Note: The combined spatial fuzzy set-membership is here approximately rendered (D) by merging the colours associated to the different EZs in the legend of the FAO map).

a simplified IPCC-T1 approach, might be considered [44–46]. The disaggregation by forest type might be improved by introducing information on pan-European forest tree species distribution [47]. While fragmentation is correlated with carbon stock alteration [50, 51], its relationship with forest cover is complex [48, 49].The post-disturbance regeneration of vegetation might sometimes preserve the fractions $fc_c^{\mathtt{man,typ}}$, but it can alter biomass and carbon stock [50, 52]. In Europe, around 65000 wildfires per year burn approximately 500,000 hectares of wildland and forests [53]. The effect of fires and other disturbances also impacts on the canopy cover and thus on the non-living component of the carbon stock – here not modelled – by means of soil erosion [54–56]. Recent advances in European soil data [57], soil erosion [58, 59] and its relationship with wildfires might help to complement the analysis of changes in living and non-living forest carbon [41, 42].

4 Conclusions

A pan-European similarity analysis has been performed – at 1x1 km spatial grid resolution – by comparing each grid-cell climate and geography with respect to the FAO ecological zones. A rank-based mapping of the relative distance similarity (RDS) for each spatial cell c to the eco-zones EZ in the crisp classification system of FAO leaded to a robust fuzzy ecological classification. The resulting array of simple models offered a spatially distributed fuzzy set-membership of the ecological zones which extends the FAO system in the European continent by moving from crisp categorical classes to a semantic fuzzy composition of them.

The use of the derived novel fuzzy EZ maps as an input dataset for the computation of pan-European forest biomass and carbon maps using the IPCC Tier 1 methodology highlighted their usefulness for modelling applications relying on EZ maps as categorical input datasets. A noticeable characteristic of the modelling approach for deriving the fuzzy EZ maps is also that it entirely relies on geographic and climate information routinely estimated by global and regional climate models. This makes the method also suitable for modelling applications in which the spatial shift of ecological zones and of eco-zone dependant quantities need to be assessed under climate change scenarios.

References

1. Pan, Y., Birdsey, R.A., Fang, J., Houghton, R., Kauppi, P.E.: A Large and Persistent Carbon Sink in the Worlds Forests. Science 33(6045), 988–993 (2011)
2. Nabuurs, G.J., van Putten, B., Knippers, T.S., Mohren, G.M.J.: Comparison of uncertainties in carbon sequestration estimates for a tropical and a temperate forest. Forest Ecology and Management 256(3), 237–245 (2008)
3. Bonan, G.B.: Forests and climate change: Forcings, feedbacks, and the climate benefits of forests. Science 320(5882), 1444–1449 (2008)
4. Eurostat: Forestry in the EU and the world: a statistical portrait
5. Seebach, L., McCallum, I., Fritz, S., Kindermann, G., Leduc, S., et al.: Choice of forest map has implications for policy analysis: A case study on the EU biofuel target. Env. Sci. & Policy 22, 13–24 (2012)
6. de Rigo, D.: Behind the horizon of reproducible integrated environmental modelling at European scale: ethics and practice of scientific knowledge freedom. Submitted to F1000 Research (2013)
7. Food and Agriculture Organization of the United Nations: Global forest resources assessment 2010: Main report. (2010) ISBN: 9789251066546
8. Barredo, J.I., San-Miguel-Ayanz, J., Caudullo, G., Busetto, L.: A European map of living forest biomass and carbon stock. Reference Report by the Joint Research Centre of the European Commission. EUR – Sci. Tech. Res. 25730 (2012)

9. Ruesch, A., Gibbs, H. K.: New IPCC tier-1 global biomass carbon map for the year 2000. Carbon Dioxide Information Analysis Center, Oak Ridge National Lab (2008)
10. Koh, L.P., Ghazoul, J.: Spatially explicit scenario analysis for reconciling agricultural expansion, forest protection, and carbon conservation in indonesia. Proc. Natl. Acad. Sci. U.S.A. 107(24), 11140–11144 (2010)
11. West, P.C., Gibbs, H.K., Monfreda, C., Wagner, J., Barford, C.C., et al.: Trading carbon for food: Global comparison of carbon stocks vs. crop yields on agricultural land. Proc. Natl. Acad. Sci. U.S.A. 107(46), 19645–19648 (2010)
12. Nelson, E., Sander, H., Hawthorne, P., Conte, M., Ennaanay, D., et al.: Projecting global Land-Use change and its effect on ecosystem service provision and biodiversity with simple models. PLoS ONE 5(12), e14327+ (2010)
13. Kempeneers, P., Sedano, F., Seebach, L.M., Strobl, P., San-Miguel-Ayanz, J.: Data fusion of different spatial resolution remote sensing images applied to Forest-Type mapping. IEEE Trans. Geosci. Remote Sens. 49(12), 4977–4986 (2011)
14. EEA: CORINE land cover technical guide Addendum 2000. European Environment Agency, Technical report No 40, Copenhagen (2000)
15. Food and Agriculture Organization of the United Nations: Global Ecological Zoning for the Global Forest Resources Assessment 2000 - Final Report. Food and Agriculture Organization of the United Nations, Forestry Department, Rome, Italy (2001)
16. Food and Agriculture Organization of the United Nations: Global ecological Zones for FAO forest reporting: 2010 update. For. Resour. Assess. Work. Paper 179 (2012)
17. Aalde, H., Gonzalez, P., Gytarsky, M., Krug, T., Kurz, W. A., et al.: Forest Land. IPCC Guidelines for National Greenhouse Gas Inventories, Prepared by the National Greenhouse Gas Inventories Programme. The Intergovernmental Panel on Climate Change (IPCC), vol. 4, ch. 4, p. 83 (2006)
18. de Rigo, D.: Relative distance similarity as multivariate supervised or unsupervised ensemble interpolation (in prep. 2013)
19. de Rigo, D.: Semantic Array Programming for Environmental Modelling: Application of the Mastrave Library. In: Int. Congress on Environmental Modelling and Software. Managing Resources of a Limited Plant, Pathways and Visions under Uncertainty, Sixth Biennial Meeting, pp. 1167–1176 (2012)
20. de Rigo, D.: Semantic array programming with Mastrave - introduction to semantic computational modelling (2012)
21. European Parliament: Directive 2007/2/EC of the European Parliament and of the Council of 14 march 2007 establishing an infrastructure for spatial information in the european community (INSPIRE). Official J. Eur. Union 50(L 108), 1–14 (2007)
22. European Commission: Commission regulation (EC) no 1205/2008 of 3 december 2008 implementing directive 2007/2/EC of the European Parliament and of the Council as regards metadata. Official J. Eur. Union 51(L 326), 12–30 (2008)
23. Iverson, K.E.: Notation as a tool of thought. Commun. ACM 23(8), 444–465 (1980)
24. Eaton, J.W.: GNU Octave and reproducible research. J. Process Contr. 22(8), 1433–1438 (2012)
25. Holdridge, L. R.: Life zone ecology. Tropical Science Center, Costa Rica. 206p. (1967)
26. Hofierka, J., Suri, M., Huld, T.: GRASS GIS manual: r.sun. In: GRASS Development Team, 2013. GRASS GIS 6.4.3svn Ref. Manual. Open Source Geospat. Found. (2007)
27. Nordenskiold, O., Mecking, L.: The geography of the polar regions. Amer. Geogr. SOC. Spec. Publ. 8, 359 (1928)
28. Guo, Z.-Y., Zhu, H.-Y., Liang, X.-G.: Entransy A physical quantity describing heat transfer ability. Int. J. Heat Mass Transfer 50(13-14), 2545–2556 (2007)

29. Xu, M.: The thermodynamic basis of entransy and entransy dissipation. Energy 36(7), 4272–4277 (2011)
30. Milly, P.C.D., Dunne, K.A.: On the hydrologic adjustment of Climate-Model projections: The potential pitfall of potential evapotranspiration. Earth Interact 15(1), 1–14 (2011)
31. Knuth, D.E.: Two notes on notation. Amer. Math. Monthly 99(5), 403–422 (1992)
32. de Rigo, D., Barredo, J.I., Busetto, L., Caudullo, G., San Miguel-Ayanz, J.: Extending FAO Ecological Zones with Pan-European Fuzzy Geoclimatic Similarity. Environmental Data and Data Transformation Models Review (in prep., 2013)
33. Mesirov, J.P.: Accessible reproducible research. Science 327(5964), 415–416 (2010)
34. Nature: Devil in the details. Nature 470(7334), 305–306 (2011)
35. Kleiner, K.: Data on demand. Nature Climate Change 1(1), 10–12 (2011)
36. Stallman, R.M.: Viewpoint: Why "open source" misses the point of free software. Commun. ACM 52(6), 31–33 (2009)
37. Murray-Rust, P.: Open data in science. Serials Review 34(1), 52–64 (2008)
38. Stodden, V.: Trust your science? open your data and code. Amstat News 21–22 (2011)
39. de Rigo, D., Guariso, G.: Rewarding Open Science: A Collaborative Review System for Semantically-Enhanced Free Software and Environmental Data Modelling (in prep., 2013)
40. Bosco, C., de Rigo, D., Dijkstra, T., Sander, G., Wasowski, J.: Multi-Scale Robust Modelling of Landslide Susceptibility: Regional Rapid Assessment and Catchment Robust Fuzzy Ensemble. In: Hřebíček, J., Schimak, G., Kubásek, M., Rizzoli, A. (eds.) ISESS 2013. IFIP AICT, vol. 413, pp. 321–335. Springer, Heidelberg (2013)
41. de Rigo, D., Rodriguez-Aseretto, D., Bosco, C., Di Leo, M., San-Miguel-Ayanz, J.: An Architecture for Adaptive Robust Modelling of Wildfire Behaviour under Deep Uncertainty. In: Hřebíček, J., Schimak, G., Kubásek, M., Rizzoli, A. (eds.) ISESS 2013. IFIP AICT, vol. 413, pp. 367–380. Springer, Heidelberg (2013)
42. Di Leo, M., de Rigo, D., Rodriguez-Aseretto, D., Bosco, C., Petroliagkis, T., Camia, A., San-Miguel-Ayanz, J.: Dynamic Data Driven Ensemble for Wildfire Behaviour Assessment: A Case Study. In: Hřebíček, J., Schimak, G., Kubásek, M., Rizzoli, A. (eds.) ISESS 2013. IFIP AICT, vol. 413, pp. 11–22. Springer, Heidelberg (2013)
43. Mitchard, E.: A Comparison of Tropical Carbon Maps. Ecometrica, Natural Environment Research Council (NERC) (2013)
44. Ciais, P., Schelhaas, M.J., Zaehle, S., Piao, S.L., Cescatti, A., et al.: Carbon accumulation in european forests. Nature Geosci. 1(7), 425–429 (2008)
45. Mascarelli, A.: Earth's carbon sink downsized. Nature (2012)
46. Fiorese, G., Guariso, G.: Modeling the role of forests in a regional carbon mitigation plan. Renewable Energy 52, 175–182 (2013)
47. de Rigo, D., Caudullo, G. Amatulli, G., Strobl, P., San-Miguel-Ayanz, J.: Modelling tree species distribution in Europe with constrained spatial multi-frequency analysis (in prep.)
48. Estreguil, C., Caudullo, G., de Rigo, D., Whitmore, C., San-Miguel-Ayanz, J.: Reporting on European forest fragmentation: Standardized indices and web map services. IEEE Earthzine 5(2), 384031+ (2012); 2nd quarter theme: Forest Resource Information
49. Estreguil, C., Caudullo, G., de Rigo, D., San-Miguel-Ayanz, J.: Forest landscape in Europe: Pattern, fragmentation and connectivity. EUR – Sci. Tech. Res. (JRC 77295) (2013)
50. Malhi, Y., Baldocchi, D.D., Jarvis, P.G.: The carbon balance of tropical, temperate and boreal forests. Plant, Cell and Environment 22(6), 715–740 (1999)
51. Houghton, R.: Aboveground forest biomass and the global carbon balance. Global Change Biology 11(6), 945–958 (2005)
52. Myneni, R.B., Dong, J., Tucker, C.J., Kaufmann, R.K., Kauppi, P.E., et al.: A large carbon sink in the woody biomass of northern forests. Proc. Natl. Acad. Sci. U.S.A. 98(26), 14784–14789 (2001)

53. Rodriguez-Aseretto, D., de Rigo, D., Di Leo, M., Cortés, A., San-Miguel-Ayanz, J.: A data-driven model for large wildfire behaviour prediction in Europe. Procedia Computer Science 18, 1861–1870 (2013)
54. Lal, R.: Soil erosion and the global carbon budget. Env. Intl. 29(4), 437–450 (2003)
55. Lal, R.: Forest soils and carbon sequestration. For. Ecol. Manage. 220, 242–258 (2005)
56. Hurteau, M.D., Brooks, M.L.: Short- and long-term effects of fire on carbon in US dry temperate forest systems. BioScience 61(2), 139–146 (2011)
57. Panagos, P., Jones, A., Bosco, C., Senthil Kumar, P.S.: European digital archive on soil maps (EuDASM): Preserving important soil data for public free access. Int. J. of Digital Earth 4(5) (2011)
58. de Rigo, D., Bosco, C.: Architecture of a pan-european framework for integrated soil water erosion assessment. In: Hřebíček, J., Schimak, G., Denzer, R. (eds.) ISESS 2011. IFIP AICT, vol. 359, pp. 310–318. Springer, Heidelberg (2011)
59. Bosco, C., de Rigo, D., Dewitte, O., Poesen, J., Panagos, P.: Modelling Soil Erosion at European Scale. Towards Harmonization and Reproducibility (in prep.)

Geo-analysis of Landscape Level Degradation and Natural Risk Formation under Uncertainty

A Case Study of Selected Czech Urban Watercourses

Radka Báčová, Petr Kubíček, Jiří Jakubínský, Eva Svobodová, and Vladimír Herber

Department of Geography, Faculty of Science, Masaryk University, Brno, Czech Republic
rada.ba@mail.muni.cz, kubicek@geogr.muni.cz,
{jakubinsky,svobodova}@mail.muni.cz, herber@sci.muni.cz

Abstract. Rivers and streams in urban areas are losing their natural environmental qualities. A lack of river management and insufficient interest on the part of municipal authorities has become the norm, especially in the case of small watercourses. Here, an interdisciplinary approach to landscape-level degradation is applied to selected catchments in the Czech Republic (the streams known as the Lačnovský and the Leskava). This approach includes basic geographical approaches to the quantification of the relationship between anthropogenic pressure and ecological stability. The methods employed include hy-dromorphological field research, identification of changes in land use, and an inventory of anthropogenic landforms.

The existence of urban stream syndrome was established in model catchments. Parameters for these areas were compared with values for susceptibility to natural risk and preliminary results show a spatial pattern consistent with the areas at most risk also being prone to flooding. The observations were subjected to cartographic visualization and enriched with thematic and temporal uncertainty features, thus refining the results and highlighting any ambiguity.

Keywords: degradation of landscape, cartographic visualization, uncertainty, natural risk, river landscape, the Leskava stream, the Lačnovský stream.

1 Introduction

River landscapes have always supported human development, providing conditions and functions without which today's settlements could not exist. The floodplains along rivers, in particular, have been an ideal environment for life. Step by step, people come to inhabit this landscape and to exploit it. This progress intensified with population growth. New ways of life developed and became more sophisticated, than unnatural processes began to affect river landscapes. Primary ecological values started to decline. Anthropogenic pressure, directly or indirectly, currently influences approximately 30% of Earth's surface. Tockner [17] and others used the epithet "the double-edged face of floods" when describing the hazards and human benefits of flood plains. They had in mind the conflict between constantly-growing population density in floodplains and the increase in flood-related economic losses. Today, about

J. Hřebíček et al. (Eds.): ISESS 2013, IFIP AICT 413, pp. 285–293, 2013.

half of Europe's population and even 70% of the inhabitants of Japan live on flood-plains. This means that such areas are among the landscapes most altered by human activity and that natural flood plains are disappearing. The environmental problems that result from inappropriate land use changes have adverse effects on human society, at their most obvious in urbanized areas. Flood risk reduction is needed for them, but with existing scientific methods this is impossible for small watercourses.

In our study, a relationship between the landscape degradation of small urban wa-tercourses and local predisposition to the occurrence of natural risks (floods) is assumed. The main aim of the project is the identification and verification of the relationship between landscape degradation and the propensity to natural risk formation, by the use of geographical analysis and cartographical visualization [16], the identification of the most degraded areas and their comparison with values for natural risk susceptibility under uncertainty.

2 Methods and Data

The basic approaches to geographical research appropriate to an interdisciplinary approach were employed in the catchment areas selected, the streams known as the Leskava and the Lačnovský. These consisted of hydromorphological field-work and an inventory of man-made landforms with the purpose of evaluating the relationship between anthropogenic pressure and ecological stability. These were supplemented by a study of land use changes and identification of uncertainty in the geographical data.

The model catchments were selected because they were comparable in terms of prevailing conditions, as areas around which human activities have been intensive. These streams are typical representatives of the small urban watercourses of the Czech Republic. More than half of their catchment areas are modified by human activities, and they flow through city/suburban areas – the Leskava stream through Brno, Moravia's capital in the south of the province, and the Lačnovský stream through the city of Svitavy in the north.

2.1 Hydromorphological Field Survey

Fluvial geomorphology is one of the critical indicators of river landscape degradation. Figure 1 shows a diagram of the channel of the watercourse (cross-section) and morphometric characteristics measured for the purposes of analysis. The following pa-rameters were measured on the streams monitored: channel width (e.g. distance be-tween opposite banks [ChW], riverbed width [RW], countersink of the channel [C], slope of the banks (average of both sides) [S] and the channel material making up the shores and bottom of the watercourse [ChM]. All these parameters were measured with a laser distance meter. These characteristics were recorded at predefined profiles that best represented the morphological status of the channel in a given stream segment. Each segment had relatively homogeneous properties in terms of stream morphology.

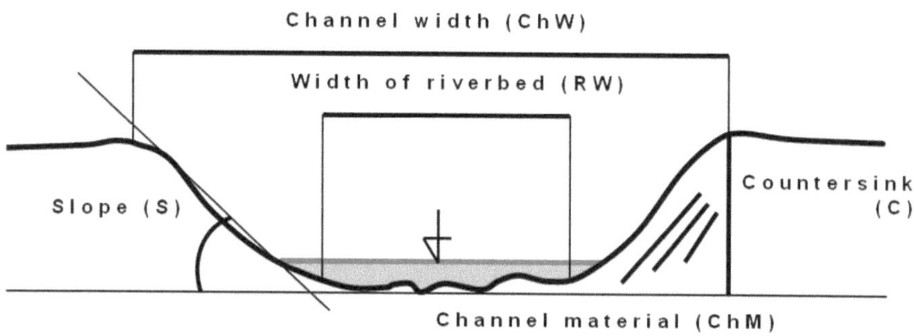

Fig. 1. Morphometric parameters measured in the channel

The acquired data was entered into the calculation of a pair of indicators that provide information on the status of channel morphometry and its patency: the channel morphology index (CMI) and the channel capacity coefficient (CCC) were further calculated. Both indicators may serve as sources of information about the potential for flooding, since they include a number of variables that may impact upon flooding processes. The aim of the CCC calculation is to identify the critical locations within the hydrographical network, where a flood can easily occur in the event of high discharge. Generally the higher the CCC, the lower the capacity of the channel and the greater the risk of flooding [7], [8].

The principle indicator calculation is represented by equations 1 and 2:

$$CCC = AW1 + BW2 + CW3 + DW4 + EW5 + FW6 \tag{1}$$

where the importance of variables and weights is: A … undersized bridge (W1 = 1.5); B … barrier within the channel (W2 = 1.2); C … reduced channel capacity (W3 = 1.4); D … narrow inundation area (W4 = 1.3); E … unsuitably situated buildings (W5 = 1.0); F … alternation of artificial and natural channel (W6 = 1.1).

$$CMI = (ChW - RW + C) * S/100 * ShM \tag{2}$$

Variables as Figure 1.

2.2 Inventory of Anthropogenic Landforms

The second method employed was to take an inventory of man-made landforms in model catchments. These were classified into nine categories on the basis of genesis [1]: mining landforms, industrial landforms, agricultural landforms, urban landforms, transport landforms, water landforms, funeral landforms, pilgrimage landforms, and recreational and sports landforms. We observed surface and subsurface man-made landforms [14], both of which may modify natural processes in the event of flooding.

The level of the natural process of flood modification was expressed as two weights, the first of which classified identified landforms into four categories, based on their effects on the process:

- Positive effect (0) – man-made landform built for flood protection. Its main function is retention of water in the landscape (e.g. water reservoir, pond, dam) and reduction of flood (e.g. levee, protective wall).
- Without effect - minimal negative effect (1) – behaviour of these landforms does not significantly affect the process of floods. However, they can change the hydrological regime and accelerate or decelerate runoff water from the landscape (e.g. mine, sand pit, wildlife crossing (ecoduct), agricultural terrace, sewage disposal plant, graveyard).
- Moderate negative effect (2) – landforms that can accelerate the flow of water in the watercourse (e.g. man-modified watercourse, canal). These may constitute a barrier to a flood or a source of sediments and debris that reduce the profile of the riverbed (e.g. mining heap, agricultural heap, industrial heap, bridge construction, waste dump). Land-forms that significantly reduce water infiltration and retention in the landscape are included here (e.g. industrial estate, urban planes, transport platforms, parking-sites, agricultural platforms).
- Very negative effect (3) – undersized bridge, weir, spillway and sewage network in the category of man-made landforms, strongly influencing the process of floods. These create barriers and radically change the flow of the flood.

These categories are valid for point (e.g. weir) and areal landforms (e.g. polder). However, the linear types are affected by their orientation with reference to the watercourse. They may fall into any of the categories; this principal is illustrated in Figure 2.

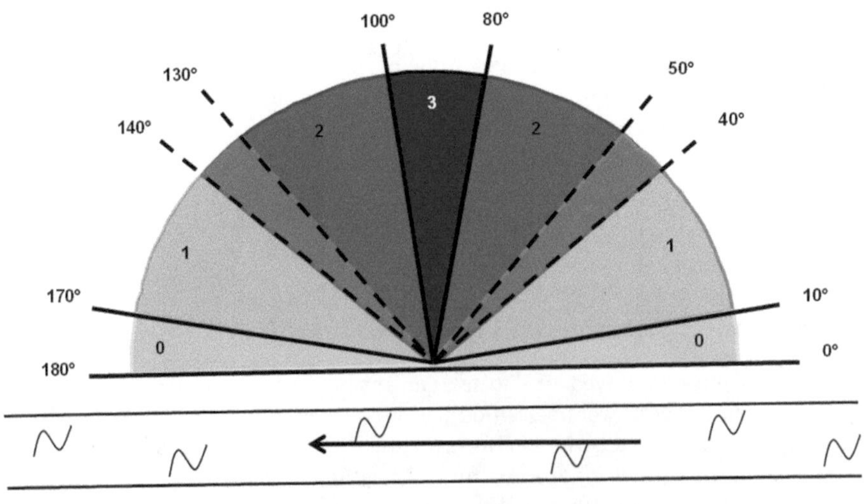

Fig. 2. Position of linear landform with reference to watercourse and effects on the process of flooding

The second of the weights specifies the location of anthropogenic form within the catchment area of the stream. We defined three zones – up to 5 metres from the watercourse (3), river landscape area beyond 5 metres from the watercourse (2)

remainder of the catchment beyond the river landscape (1). All these weights were identified on the basis of general experience in flood risk and catchment management, but they contain a certain degree of uncertainty, which is discussed below.

2.3 Identification of Uncertainty in Geodata

Uncertainty is generally considered as a quantitative or a qualitative variable characteristic of the spatial, typological or temporal value of a given reference attribute and may be identified at all stages of modelling reality. For this paper we have selected a definition arising out of a number of scientific studies [5], [12], [13], [15], [18] and [20], in which uncertainty is an indicator of systematic and random errors, together with accuracy, and vagueness (see Fig. 3.). However, the components of the entire concept have pure definitions that may be used with different meanings from author to author. Our purposes are best served by adoption of inherent and epistemic uncertainty-like components.

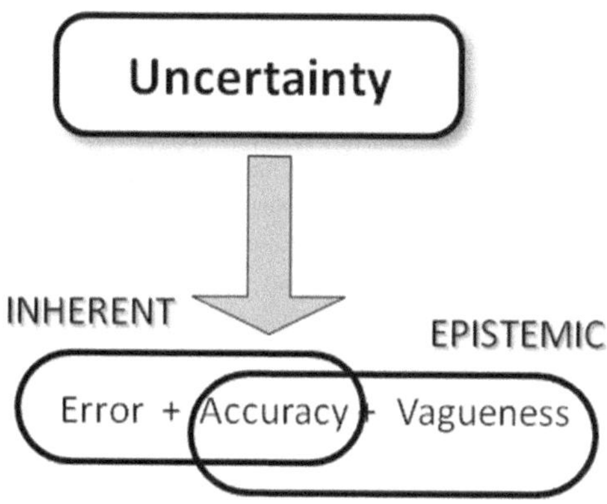

Fig. 3. Conceptualization of uncertainty in geographical data (Based on [12], [20])

This approach to data quality description involves significant spatial variability. A visualization method was therefore employed to identify the spatial pattern. A visualization of inherent uncertainty (measurable and precisely defined) appeared for the purposes of this study to be a suitable approach to supporting critical thinking and the identification of epistemic uncertainty, which is more subjective. Epistemic uncertainty depends largely on the individual user's knowledge and experience. Qualification was prioritized over quantification of uncertainty expressed with probability in this case study. Thus expert assessment of uncertainty was applied where quantification was not possible.

Most of the available visualization techniques focus strictly on a single factor, with respect to only a few selected parameters of uncertainty. For each parameter there are a number of recommendations as to the most appropriate method of communication of the phenomenon, but only a few of them are empirically based. Examples of these may be found in the literature, e.g. [2], [3], [9], [10], [11], [13]. Cartographic visualization of uncertainty is a challenge for cartographers [4], and in this case study the possibilities of multidimensional form have been explored. Two visualization methods were employed – comparatives and combined maps [12].

The system of uncertainty identification was discussed between geographers and cartographers. The objective was to evaluate thematic and temporal uncertainty together, with no accentuation of individual components. Thematic uncertainty was determined as the role of man-made landforms in the natural process of flood and corresponds with the three distance zones from the watercourse (above). The uncertainty grows with increasing distance because we are not able to make a determination of the impact of anthropogenic landforms in the watershed catchment. Thematic uncertainty was identified in the selected landforms in which experience was lacking of their impact in modelling floods. Temporal uncertainty was fixed in landforms where temporal change is obvious, e.g. the highest temporal uncertainty was reflected at a farm track, the lowest at a bridge construction.

3 Results and Discussion

The above analyses enable identification of the most degraded areas that may affect natural flood processes. Both hydromorphological indexes, CCC (channel coefficient capacity) and CMI (channel morphology index), define a level of natural state degra-dation conditioned by human activities in homogenous sections of the watercourse. The situation is at its worst in permanently occupied urban areas where people co-habit with streams. The anthropogenic landforms inventory supports these hypothe-ses.

Figure 4 clearly shows the river network and it easy to confirm that the most degraded area is river landscape, also the most vulnerable. The hexagonal net map is used as the main cartographic visualization method; it can take the ideal neighbourhood of every hexagon subarea into account. The area of these polygons is 1 ha, but changing this parameter may result in a different pattern. The efficiency of the techniques employed needs to be discussed. The rare visualization of single landforms might be more accurate, but its ability to be entered into future analyses is reduced.

The data quality is characterized by level of uncertainty. The reliable areas appear in the illustration in Fig. 5. Maps of comparatives (Fig. 5, upper picture) are created by two different maps, one showing the hexagonal net maps from the previous case and the other showing the bivariate map of temporal and thematic uncertainty by changing size and colours of symbols. The growing thematic uncertainty of the watershed basin with the most confident areas near the watercourse is clearly visible. Combined maps (Fig. 5, lower picture) show the same situation in one presentation. This is one typical example, where the most degraded hexagon (in the left part) in dark red is disputed, with a high level of temporal uncertainty. Its temporal change is highly variable and it is therefore not permanent enough to enter analysis. One future plan is to verify the benefit of qualitative and quantitative expressions of uncertainty. The authors remain, meanwhile, convinced that quantitative definition would be more confusing.

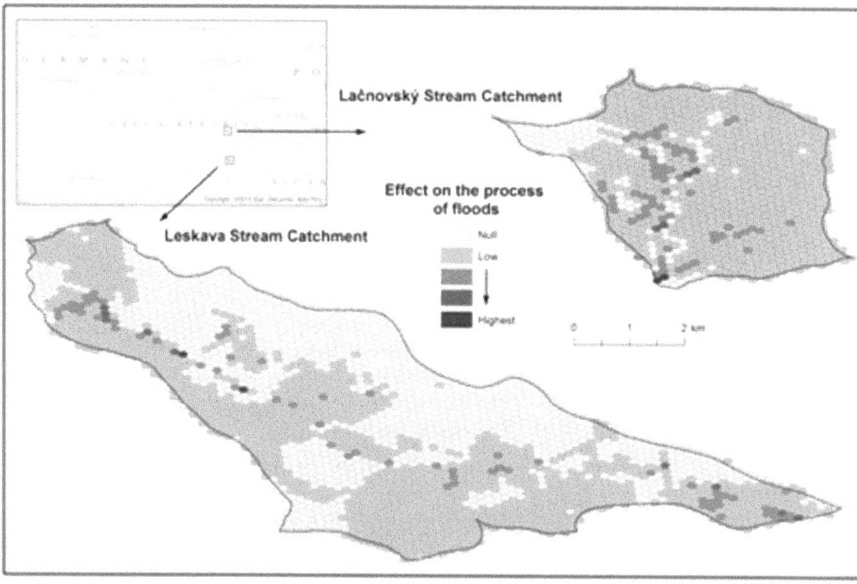

Fig. 4. The most degraded areas

These results confirmed the existence of urban stream syndrome, characterized by negative impact of running waters in urban areas. These are more prone to flood formation and increased risk of damage, since the incoming flood is of swift onset and the catchment is less able to cope with higher rainfall through infiltration. Other characteristics of urban stream syndrome are given in [6], [19].

Our preliminary results confirmed a relationship between the spatial pattern of flood risk and the pattern of degraded areas. We diagnosed that the most degraded territory corresponds with the potential zone of flooding (in the case of the Leskava stream), or with historically flooded areas (in the case of the Lačnovský stream). Therefore the contribution of this study may be summarised as a positive tool for defining flood risk areas in small watercourses, where the standard approach of the flood risk area definition based on risk matrix cannot be applied. The future development of this theme is an expansion of the model areas and possible verification of our results in other natural conditions in the Czech Republic.

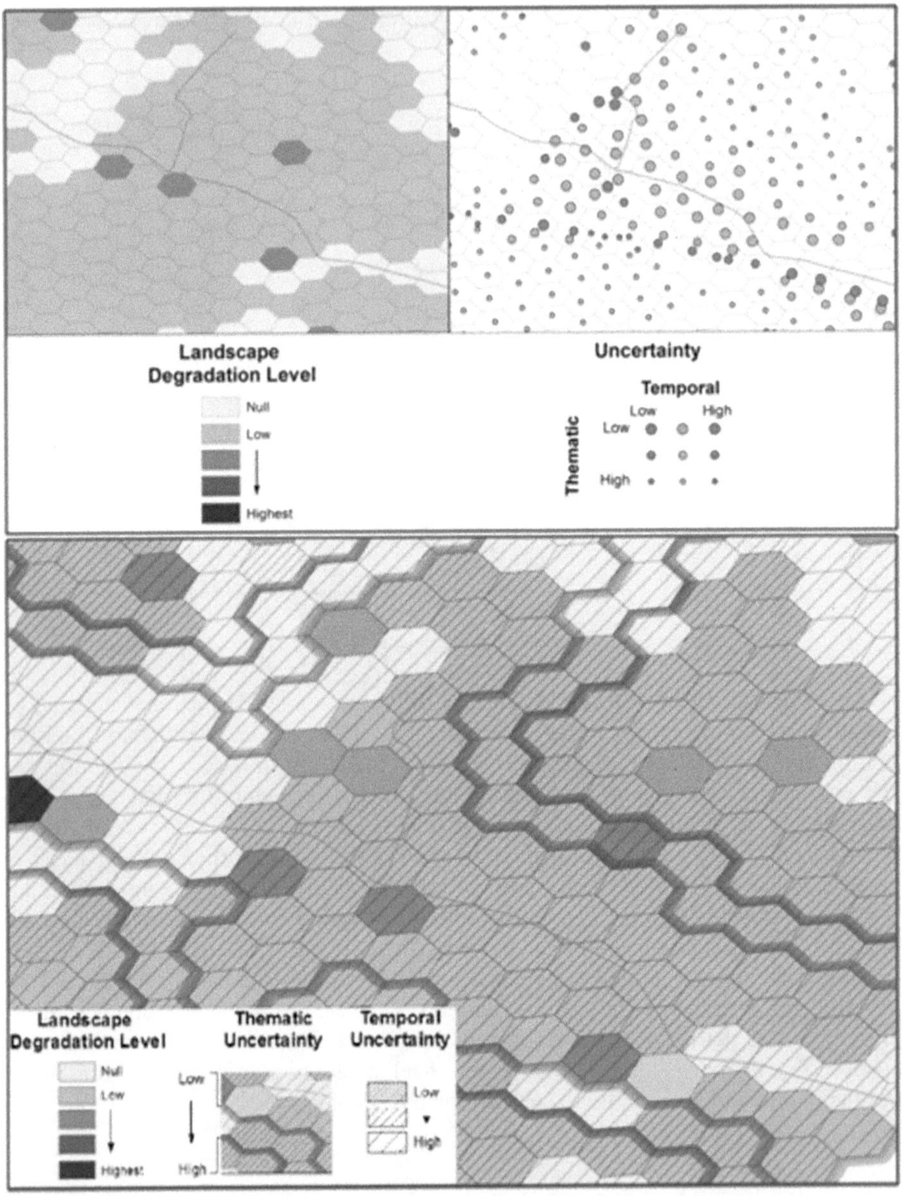

Fig. 5. Comparatives (upper) and combined (lower) uncertainty maps (illustration)

References

1. Demek, J., Embleton, C. (eds.): Guide to Medium – Scale Geomor-phological Mapping. IGU Commission on Geomorphological Survey and Mapping. Brno (1978)

2. Dvorsky, J., Snasel, V., Vozenilek, V.: Map Similarity Testing Using Matrix Decomposition. In: Badr, Y.K., Caballe, S., Xhafa, F., Abraham, A., Gros, B. (eds.) International Conference on Intelligent Networking and Collaborative Systems, pp. 290–295 (2009)
3. Edwards, L.D., Nelson, E.S.: Visualizing data certainty: A case study using graduated circles maps. Cartographic Perspectives 38, 19–36 (2001)
4. Feiden, K., Kruse, F., Řezník, T., Kubíček, P., Schentz, H., Eberhardt, E., Baritz, R.: Best Practice Network GS SOIL Promoting Access to European, Interoperable and INSPIRE Compliant Soil Information. In: Hřebíček, J., Schimak, G., Denzer, R. (eds.) ISESS 2011. IFIP AICT, vol. 359, pp. 226–234. Springer, Heidelberg (2011)
5. Goodchild, M.: Visualizing fizzy map. In: Hearnshaw, H. (ed.) Visualization in Geo-Graphical Information Systems, pp. 158–167. John Willey & Sons, London (1994)
6. Gurnell, A., et al.: Urban rivers: Hydrology, geomorphology, ecology and opportunities for change. Geography Compass 1(5), 1118–1137 (2007)
7. Jakubínský, J.: The influence of watercourses regulations on runoff processes and the development of floods (a case study of the Košátecký stream), Diploma Thesis (2011)
8. Jakubínský, J., Báčová, R.: Environmental Values: the Dunajovický and the Košátecký Stream Catchments, Czech Republic. Journal of Maps 9(3), 9 p. (2013), doi:10.1080/17445647.2013.829409
9. Kubíček, P., Šašinka, Č.: Thematic Uncertainty Visualization Usability – Comparison of Basic Methods. Annals of GIS 17, 20 (2011)
10. Leitner, M., Buttenfield, B.P.: Guidelines for the display of attribute uncertainty. Cartography and Geographic Information Science 27, 3–14 (2000)
11. MacEachren, A., et al.: Visualizing uncertain information. Cartographic Perspectives 13(3), 10–19 (1992)
12. MacEachren, A., et al.: Visualising geospatial information uncertainty. What we know and what we need to know. Cartography and Geographic Information Science 32, 139–160 (2005)
13. Pang, A., et al.: Visualizing uncertainty in geo-spatial data. In: Proceedings of the Workshop on the Intersections between Geospatial Information and Information Technology, pp. 1–14. National Academies Committee of the Computer Science and Tele-communications Board, Washington, D. C (1997)
14. Pechanec, V., Janíková, V., Brus, J., Kiliánová, H.: Typological data in the process of landscape potential identification with using GIS. Moravian Geographical Reports 17, 12–24 (2009)
15. Shi, W.: Principles of modelling uncertainties in spatial data and spatial analyses. Taylor and Francis, London (2010)
16. Staněk, K., Friedmannová, L., Kubíček, P., Konečný, M.: Selected issues of cartographic communication optimization for emergency centers. International Journal of Digital Earth 3, 316–339 (2010)
17. Tockner, K., Bunn, S.E., Gordon, C., Naiman, R.J., Quinn, G.P., Stanford, J.: Flood plains: critically threatened ecosystems. In: Polunin, N. (ed.) Aquatic systems, trends and global prospects, pp. 45–61. Cambridge University Press (2008)
18. Tucek, P., Paszto, V., Vozenilek, V.: Regular use of entropy for studying dissimilar geographical phenomena. Geografie-Sbornik CGS 114, 117–129 (2009)
19. Walsh, C.H.J., et al.: The urban stream syndrome: current knowledge and the search for a cure. J. N. Am. Benthol. Soc. 24, 706–723 (2005)
20. Zhang, J., Goodchild, M.: Uncertainty in Geographical Information. Taylor and Francis, London (2002)

The Use of Simulations for the Need of Environment Protection

Jiří Barta

University of Defence, Kounicova 65, 662 10 Brno, Czech Republic
jiri.barta@unob.cz

Abstract. The environmental protection is currently a very hot topic. The article deals with the perspectives and possibilities for deployment of "new technologies" or already known technologies but in new ways in order to protect selected environment. In the first part, the paper is focused on education and technologies used in the educational process at the University of Defence, at the Department of Civil Protection. The second part outlines the possibilities of using this technology to protect specialized workplaces, which is also applicable to environmental protection.

Keywords: Simulator, Environment, Information Technology, Study Programme, Specialized Workplaces.

1 Introduction

We live in a time which can be characterized as a continuous development of technologies. These technologies should provide easy access to information, and especially to their most effective utilization. The computer support of management takes part in many different spheres of human activity. At present, there are many available systems and technologies, which enable the formulation of crisis plans, collaboration between the state and private subjects, monitoring of infrastructure and other necessary services for crisis managers in various organizations [3].

2 Problem Formulation

Information systems and technologies can be used in both public administration and the private sector. They can support successful solutions of crisis situations. One of the areas, where these technologies are used, is education.

Higher education institutions form the highest level of the Czech education. They offer accredited degree programmes at three levels: bachelor´s, master´s and doctoral degree programmes. Apart from public and private higher education institutions there are state institutions in the Czech Republic. The University of Defence is one of the state institutions, which is under the responsibility of the Ministry of Defence.

J. Hřebíček et al. (Eds.): ISESS 2013, IFIP AICT 413, pp. 294–301, 2013.

There is not only an educational process but also research activities both in defence technology for the Army of the Czech Republic as well as in the field of information technology and crisis management system. The focus is on systems that support analyzing, planning and solving emergency and crisis situations. One of these research projects is dealing with interoperability specialized centers of crisis management. This project uses the outputs from modeling programs (see section 3.1) and implements these outputs to the process of crisis management and environmental protection.

3 Education at the University of Defence

This university educates and prepares military professionals and other experts. The scope of university work is defined with focuses of education, formation and training of these professionals as required by the Army of the Czech Republic and research projects in favour of the Ministry of Defence and Public Administrations [8].

The University of Defence is the university-type higher education institution, which offers all types of degree programmes (bachelor´s, master´s and doctoral) and carries out associated research, developmental and other creative activities.

3.1 Professional and Educational Aspects

The University of Defence consists of three faculties: the Faculty of Economics and Management, the Faculty of Military Technology and the Faculty of Military Health Sciences. The academic year 2001/2002 was the first year for the study module "Population Protection" that started at the Faculty of Economics and Management and which is guaranteed by the Department of Population Protection. This department focuses on the area of population protection, crisis management and civil and military cooperation. Furthermore, the department solves the problems connected with military and non-military emergencies. The department emphasizes issues associated with humanitarian aid, environmental security, major accident prevention, information systems and risk analysis.

The Department of Population Protection cooperates with many institutions, e. g. the Population Protection Institute, The Ministry of Interior of the Czech Republic, the Nuclear Research Institute Rez plc., the Transport Research Centre and other universities and institutions not only in the Czech Republic.

Students in both bachelor´s and master´s degree study programmes are informed about actual problems of population protection. They pass theoretical subjects (e.g. Economics, Computer Science, Mathematics, Law: Czech National Security and Defence), applied subjects (e.g. Economic-Mathematic Methods, Accounting, Enterprise Theory) and profile subjects (e.g. Crisis Management, Natural Disasters and Industrial Accidents, Population Protection, Environment Protection, Dangerous Substances, Classification and Modelling of Risks, Public Administration Information System etc.). The department is also aimed at doctoral degree study programme and guarantees the programme Forces and Civil Protection [8].

3.2 Aims and Scientific Activities

The main purpose of this programme is to prepare highly-qualified and specialized experts for basic and applied research, research workers, education and practice of civilian protection in accordance with requirements and needs of the NATO [5].

Scientific activity includes the following areas:

- crisis management in civil protection during non-military crisis situation,
- logistic support of non-military crisis situation, especially at the field of solution of material standards for civil protection at crisis situation and coordination of human aid with utilization of non-profit-making organizations,
- raising of environmental safety at the level of advanced armies of NATO,
- dangerous substances detection at industrial accidents and natural disasters,
- optimization of information and software support for prevention and solution of consequences of extraordinary events and for simulation of these processes,
- interoperability at the crisis management.

It is possible to say that information technologies are used almost in all scientific activities.

4 Information Technologies Used at the Department of Civil Protection

At the Department of Population Protection students have an opportunity to meet with the issue of information technologies in the subject "Information Systems of Crisis Management I and II". In this subject, they can get information not only about IT but also information, which support making up the information systems of public and state administrations.

The aim of this subject is to educate students for principal work with chosen information systems that are closely linked with crisis management. Students learn how to use computer programmes for modelling possible impacts and consequences of industrial accidents with the dangerous substances release.

Information systems are products made in order to support effective decision making of crisis managers. For the solution of the scheduled objectives, especially Geographical Information Systems are used [4]. The overview of the most important information systems that are intended for education at the Department of Population Protection is in the table 1.

Table 1. Information Systems at the Department of Population Protection [1, 2, 3, 5]

Area	Information system	Description
Civilian Emergency Planning	ARGIS	Information system for Planning of Civil Resources, which is important for self-solution of extraordinary events in the Czech Republic

Table 1. *(continued)*

Monitoring	MONIS	This system monitors critical infrastructure.
	PREMIS	This system, which is sponsored by the department of informatics of the Prague municipality, provides public information about the quality of the air of Prague. These systems were made especially for Czech market and monitor not only air but also road practicability, river level etc.
Modelling	TerEx (Terrorist Expert)	The tool for immediate forecast of explosions and chemical release hazards.
	WAVE	Program for modelling of the dam destruction.
	SYMOS´97	The tool for dispersion modelling of emissions from sources of pollution [2].
Planning	EMOFF (Emergency Office)	It is technology to help in emergency planning and management and business continuity. The system integrates software tool for analysis, planning and emergency situation support.
Simulation of Emergency Event	ESIM (Emergency Management Simulator)	Simulator for Crisis Management Training and Exercising. It is made for obtaining of skilled staff and preparedness to deal with various situations.
Analysis	RISKAN (Risk calculator for Managers)	It is used for organization security of governmental and public sector as well as strategic companies from telecommunications, utility and financial sectors.

The aim of this paper is to introduce the product mentioned above that is used for education of students in the Department of Population Protection. It is a methodology for model calculations of the transport and dispersion of pollution in the atmosphere SYMOS'97 (System modeling of stationary sources). This program is also used in the research activities in the field of crisis management and implements its outputs to the process of protecting the environment.

4.1 Symos´97 Software

Methodology / SYMOS'97 software allows:

- calculation of air pollution gaseous substances and dust from point, line and area sources,
- calculation of the contamination from a number of sources,
- to identify the characteristics of pollution in dense geometrical network of reference points,
- to consider the statistical distribution of wind speed and direction relative to the stability classes layers of the atmosphere.

Symos´97 is the tool for quick forecasting of consequences and impacts of chemical materials leaks on the environment (fig. 1). Thanks to this program, it is possible to calculate and visualize the emission leaks and to model the consequences that resulted from contamination of the environment. The model includes connectivity to geographic information system for direct results display in the map. The software can be easily used by non-professionals [3].

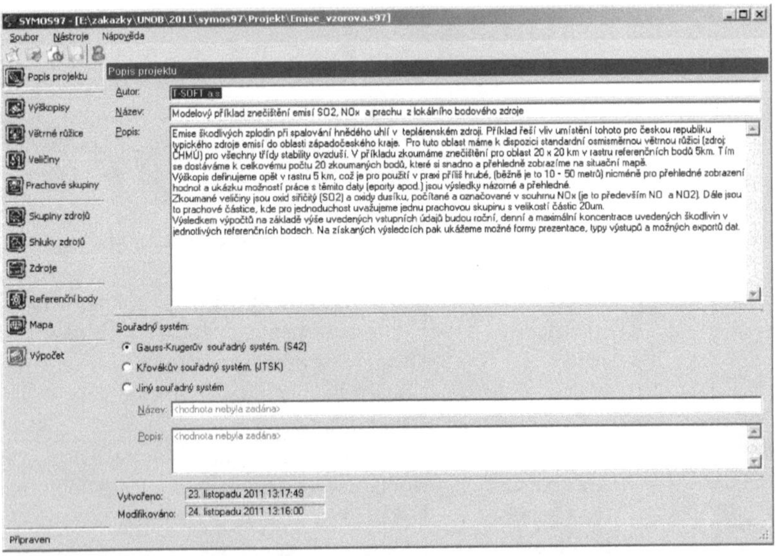

Fig. 1. Symos´97 – start of simulation [2]

Symos´97 calculation results are arranged in a very easy, comprehensive and unambiguous format, which helps to make quick decisions. This effect is achieved by minimizing output items to important ones and by displaying results on the map, as it is in the fig. 2 & 3.

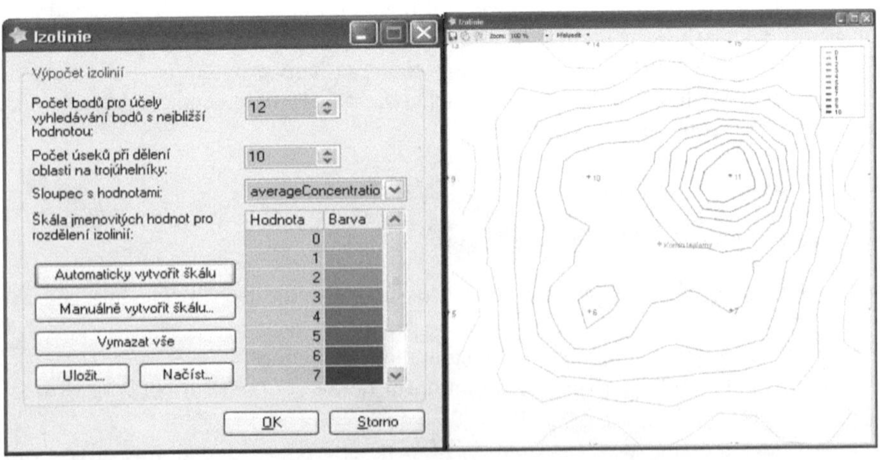

Fig. 2. Concentration and Isolines [2]

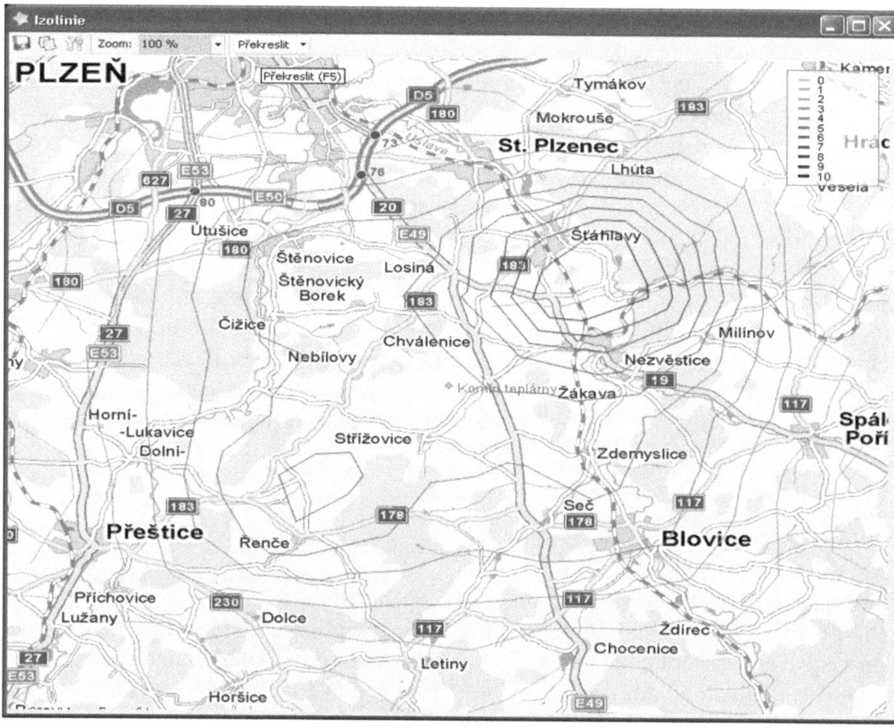

Fig. 3. Isolines on the map [2]

This software is used at distinct projects in terms of its research activities in the field of security, communication, civil-military cooperation and other. One of these projects aims for a novel perspectives and possibilities of "new approaches" to education and training of selected members of management. This research project of the University of Defence is called Interoperability Workplaces to Support Teaching of Security Management in a Computer Network. The primary aim of the research project is to create the new system and technology for education, training and exchange of relevant information with managers of crisis management. The secondary aim of the research project is to use this new system to protect the environment.

To deal with the question of using common security, communication and information technologies and other computer-aided technologies, it is necessary to apply the principle of COTS (Commercial Off The Shelf) as much as possible. That means maximal utilization of commercial products, technologies and services to create a specific system or technology [7].

5 Specialized Workplaces

The research project Interoperability Workplaces to Support uses existing structures and systems of communication. The key idea is to connect existing heterogeneous information systems and other relevant organizations in the area of management.

These individual management entities will be able to exchange information and experiences. They can use selected simulation programs (such as Symos´97) and information systems of other entities connected to the network. Therefore, this will enable the sharing of information and practical experience across the whole spectrum of management and crisis management [4, 6].

The system “Workplaces to support teaching of security management in a computer network” is implemented at interface with communication among different systems of crisis management entities. This system can be implemented among expert workplaces, and among the various computers. The system is used in the existing Internet network for communication. Of course, it uses a secure connection via encryption algorithms. The method primarily applies the basic tools for Windows and network settings. The method allows the sharing of computers, the computing space and simulation programs for the education and training of other entities registered in this system [1]. The open system architecture also allows the use of supercomputers in the VSB-Technical University of Ostrava.

6 Testing

During testing, it is necessary to take into account the fact that throughput communication network is very variable over time. This complicates the stability of information system and the communication between specialized workplaces must be modified to the actual situation. All the results were recorded and the restrictive conditions for using this technology in real situations were defined. The result was a test environment illustrating the real usage of the information and communication system for specialized workplaces.

Developing of a network standard for a facility and connecting interface design is a precondition for building an effective communication expert system, which will satisfy requirements for the needs of management and executive units of crisis management. This system is designed as an open network to which the verification can be accessed by other users of the environment of crisis management and various experts on the issue of terrorist acts, accidents and natural disasters [3].

7 Conclusion

Finally, it is necessary to say that all software products represent the function of crisis manager helper in his decision making. All computer programs are only “lifeless” and “thoughtless” products. The crisis manager is principal, especially his knowledge, abilities and acquirements, and many times his extemporization too.

At the same time, it should be noted that this solution is a compromise in many aspects and it does not fully meet needs of all stakeholders. In addition, relations between the lines of preparation for solution of civilian crisis management and lines of preparation and training for crisis management staff have to be considered carefully [6].

The resulting situation can be recognized as an essential qualitative advance, but it has to be developed henceforth. The “Workplace expert of crisis management” belongs among the priority issues and it needs to be contemplated from the national security point of view, as well as the crisis management viewpoint.

The first steps to resolve such issues meet consistently all requirements resulting from the Conception of population protection. It is important to detect system uncertainties and gaps, when an issue emerges, and to formulate all requirements on education and training in this area to be further developed.

References

1. Barta, J.: Project ADAPTIV and its Uses for Protection of the Selected Objects Infrastructure. TRANSACTIONS of the VŠB - Technical University of Ostrava: Safety Engineering Series I, 1-4 (2011)
2. Bubník, J., Keder, J., Macoun, J.: SYMOS 1997, Systém modelování stacionárních zdrojů. Praha (1998)
3. Heretik, J., Barta, J.: Počítačové modelování možných havárií s únikem nebezpečných látek. In: Interoperabilita managementu ochrany obyvatelstva (2006)
4. Ludík, T., Ráček, J.: Process methodology for emergency management. In: Hřebíček, J., Schimak, G., Denzer, R. (eds.) ISESS 2011. IFIP AICT, vol. 359, pp. 302–309. Springer, Heidelberg (2011)
5. Urbánek, J.F., et al.: Scénáře adaptivní kamufláže. Tribun EU, Brno (2012)
6. Urbánek, J.F., et al.: Crisis Scenarios. University of Defence, Brno (2013)
7. Urbánek, J.F., Průcha, J.: A Development of Wireless Interoper-mobile Application for Outdoor Operation Management. In: Proceedings 8th Int. Conf. on Electronics, hardware, wireless and optical communications, EHAC 2009, pp. 57–64. WSEAS Press (2009)
8. Statute of the Faculty Economics and Management. University of Defence. Brno (2005)

Protection of Information and Communication Systems

Jiří Barta, Veronika Sadovská, Albert Srník, and Jiří F. Urbánek

University of Defence, Kounicova 65, 622 10 Brno, Czech Republic
{jiri.barta,veronika.sadovska,albert.srnik,
jiri.urbanek}@unob.cz

Abstract. Awareness, education and protection of organizational information secure environments are currently highly serious and relevant topics. The paper deals with the perspectives and possibilities of "new approaches" to the protection of information and communication systems and nets as a part of Czech Republic critical infrastructure. It means that selected technologies are used, considering new perspectives to interoperable management training and education. The first part of this paper is focused on the current state of organization management information systems in Czech Republic. The second part concentrates on the possibilities of new technologies application and computer-aided tools, protecting information & communication systems and other assets.

Keywords: Interoperability, Security environment, Assets, Communication and information systems, Safety.

1 Introduction

From the distant past to the present, the people have to satisfy their subsistence needs that closely relate to their existence and to a preservation of their culture and assets. The human society was changing all the time, especially at the scientific and technological platforms. The progress has brought more advantageous living conditions. Along with the benefits, the negative events, such as natural disasters and man-made industrial accidents of great extent, have emerged. Fearsome disruptive events arise from scientific and technical progress. The hazards of such an anthropogenic extraordinary events cause deaths and damage of the people, assets and living environment. These inflictions are increased by the development of new sources for meeting a demand and innovations of production technologies capacity. Infrastructure, designed for meeting a demand is mostly endangered by industrial accidents and environmental disasters. Moreover, it can be distinguished among epidemics, pandemics, epiphyte and epizootics. Besides such events, the humankind is also threatened by itself in connection with wars, genocides, international terrorism and organized crime. These phenomena are the major reasons, why assets intended for population needs satisfying are critically vulnerable.

Assets represent everything that any organization or organization perceives as valuable and it can be eliminated by activated threats, i.e. perils. In business terms, it is considered not only a property but also the employers with their abilities and skills. In respect of this fact, it is accepted as "firm assets" or "company assets". Another

J. Hřebíček et al. (Eds.): ISESS 2013, IFIP AICT 413, pp. 302–310, 2013.

meaning of the term "assets" appears in area of information protection. The "information system assets" can be understood as hardware, software, information system documentation, classified and other critical information that are stored in the information system [1], [14].

2 Needs for Change in the Understanding of Threats

Infrastructure protection is a topical issue nowadays. The risk of various extremist and terrorist actions is very serious in the present time. We believe, this danger is rather underestimated due to the inclusion of Czech Republic into the group with low terrorist attack risk [10]. Despite of above mentioned, terrorist threats became a much discussed subject in professional community. Here, particular areas, technologies or states that could be the next terrorist targets, are controversial and often fuzzy.

To ensure the defence against these attacks on time, various instruments of security, crisis, and emergency management are used. Their basic parts consist from the activities, covered by contingency and emergency planning, which are based on the analysis and assessment of risks and threats. It is an expert analysis of the required goals, which identifies the appropriate ways of goals achievement, and person-oriented leadership that supports the creative work and mutual collaboration [10].

Preventive measures, protecting information infrastructure elements are the most effective but they require the availability of necessary technologies, economic resources and personal qualities. In developed states, instruments for security management are used that help to ensure the high security of communication and information infrastructure elements.

The technology is a key factor in protecting information infrastructure elements [8]. It is not necessary to invent and develop new technologies for the protection and security of individual subjects. It is possible, and for economic reasons often necessary, to use common technologies with a slight modifications or in a new use cases. These procedures create opportunities to secure the selected infrastructure objects at relatively low costs.

There are many research institutes dealing with science and research in information technologies. They toy with the question of the use of common security technologies and other security options that use the principle of COTS (Commercial Off The Shelf) as much as possible. That means the maximal utilization of commercial products and services to create a specific system or technology [13].

3 Security Environment of Information Systems

Each organization should assure security of its information system by reason of property and data loss (or theft) possibility. Comprehensive identification of assets is integral part of the protection securing. The identification consists in compiling a list of organization information system. Other part of the identification process is represented by determination of assets vulnerability within the system realized by application of methods intended for data gathering. The aim of this step is to create the register of assets and system vulnerability (weak points) for the reason that the vulnerability may be the potential target of threats [2].

The active and passive prevention is essential for the assets security. The purpose of active prevention (or just prevention) is to reduce the threat (before its activation as the peril) and potential damage of the assets. It includes:

- Threat elimination or reduction and its transfer,
- Asset resistant strengthening, e.g. building guarding.

The passive prevention (or preparedness) is losses reduction after activated threat. The passive prevention includes:

- Information about threat activation,
- Operations during emergency event occurrence,
- Clearance operations, loss reduction and prevention of domino and synergetic effects.

Principally, building up the preparedness results from the vulnerability determination, because the preparedness can be defined as an ability to manage any emergency event or crisis. On the one hand, the preparedness refers to a community preparedness level. On the other hand, it is linked to the preparedness level of rescue system integration. The preparedness should cover the protection of all assets and it manages all consequences of any emergency/disruptive events. Therefore, it means that any organization has to design such organizational structure of all resources in order to ensure efficient mobilization at the appointed time and place.

Although, the preparedness means creation of certain safety measures because it is impossible to be prepared for every threats. The design of preparedness should contain more than simple balance of resources (staff and equipment), which is more frequent than desirable. The preparedness is mainly aimed at two states within the framework of crisis management – contingency and emergency states.

The organization should make its own security policy that should contain base rules for ensuring the security of its assets with the help of organizational, personal and technological measures. The security policy helps to manage difficulties, connected with intense utilization of information systems that are perceived as essential resources for any organization and its performance [11].

The security policy is based on a security survey, processed by informal team, organized from one external specialist and staff members, responsible for particular areas regarding security. The security survey deals with assessment and description of the security situation within the organization. It consists from the identification of organization assets, the assessment of their importance and the detection of security weaknesses. The planning of basic security measures is also integral part of the security survey. The survey scope is determined by organization management, which should reflect the security as a complex issue. The comprehensive processing also enables efficient access and implementation of security policy.

is recognized as a weakness of certain asset. The weakness can be the target of one or more threats. Therefore, it indicates the possible extent of damage to certain asset.

The security policy objectives are focused on information documentation and information system operations. Following objectives have ensured of interoperable:

- Confidentiality – data leak protection,
- Integrity – asset protection against unauthorized and random modification and guarantee of asset accuracy and completeness,
- Accessibility – stable access to assets in accordance with organization objectives,
- Incontestability – clear identification of transaction author or communicating parts,
- Correspondence between legislation and obligations – the legislative, contractual obligations of the organization.

Benefits of security policy are:

- Uniformity in information management,
- Quick recovery after incident,
- Adoption of efficient measures – future prevention of errors.

4 Asset Evaluation on Security Environments

The value of the information system is recognized as the basic attribute that is derived from objectively quantified price of the system and/or from subjectively evaluated importance of the system by certain organization on certain security environments. One possible option of asset evaluation is to assess potential impacts of undesirable incidents on the organization. Its interests may be directly or indirectly affected by data leak, modification, inaccessibility and/or damage on the asset. Such incidents may consequently result in loss of income or profit, market share, or image and reputation. The organization has to count with all these factors within the framework of asset evaluation.

The information value is often relative in connection with the evaluation process and the evaluator point of view. The typical example of this problem is linked with information which may be perceived as valuable or not, e.g. impact of harmful substances on living organisms, or current economic score of bond issuers. The assessment of the asset value may be based upon the level of impact, measured in money units and associated with the asset damage or loss.

In the asset evaluation, the following aspects have to be taken into account at various security environments:

- Purchase price (or costs),
- Asset importance for organization existence and activities,
- Damage abatement costs,
- Speed of damage control,
- Profit characteristics,
- Others, e.g. potential asset profits.

In the asset evaluation is desirable to divide it into appropriate groups, according to their common features. Assets may be classified as computer assets, security assets, commercial assets and services. It is also necessary to assign owner to each asset. The owner is an individual who is responsible for asset development, production, maintenance, utilization and security. This owner can be viewed as the real holder with all property rights [2, 3].

The determination of the asset value in connection with the impact assessment, which may result from the loss of the asset confidentiality, availability or integrity, it includes selection of scale and evaluative criteria. The scale may be expressed in money units and/or qualitative values. The qualitative values are perceived as numerical or verbal values (from low to critical value). It is practical to use colour resolution and order as you can see in table 1.

Table 1. Example of qualitative evaluation of assets [2]

1	No impact on the organization
2	Insignificant impact
3	Difficulties or financial loss
4	Serious difficulties or financial loss
5	Existential difficulties

The principle of the asset (information system) value determination is based on costs derived from impact of damage to the organization in connection with confidentiality, availability or integrity losses at pertinent security environments. The evaluation may have different results in respect of the viewpoint used during the evaluation. Therefore, the asset may have different values. The final value can be determined by maximal value, by using calculation of average value or using sum of individual values. It is important to use the same method for the evaluation of each asset. The most frequent method is represented by the average calculation. Individual viewpoints are evaluated with the help of the scale illustrated at the table 1. and the final value is calculated. Final asset values are used for risk analysis and calculation of expenses set for the asset protection [12].

The software can be used in the asset evaluation. The organization may choose from specialized software or spread sheets (e.g. LibreOffice or OpenOffice). The organization has to define the scale and evaluative criteria in order to carry out the asset evaluation. This scale may be qualitative or quantitative.

During the asset evaluation, possible threats and impacts on the organization are assessed in respect of the possibility that certain threat will be activated. The threat can potentially trigger undesirable incident, which may result in damage to the system or the organization and its assets. Threats may be natural or anthropogenic and they may be intended or random. The damage may be classified as temporary or permanent.

Many different threats potentially lead to undesirable damage of the assets. It is necessary to identify such threats and assess their likelihood. Therefore, there is enough space for realization of the risk analysis. The term “risk” means a probability that given threat will occurred and cause direct or indirect damage of the organization.

5 Risk Analysis

Risk analysis is defined as a process of likelihood assessing of security failure that may result from threat occurrence, assets vulnerability and accumulated impacts on particular assets. The basic step is to determine adequate security level according of the asset value, which includes cost and risk analysis. The purpose of the analysis is to detect weaknesses of assets and to determine the risk level for each vulnerable asset and threat. Risks have to be reduced to acceptable level. The residual risks arisen from inefficient risk minimization have to be adopted. There are several methods for risk analysis implementation, e.g. informal procedures with the use of individuals' knowledge and experience and/or structured procedures. Risks can be also evaluated with the help of scale with defined points according to risk level or by using your own method or appropriate software. A list of risks is the output from this procedure [2-3].

The risk analysis should lead to selection of appropriate security measures. After the identification and assessment of risks and revision of existing and planned security measures, there are so-called risk residues. Risk residues can be accepted in the case of their low likelihood or if the organization has financial limits in connection with measures availability, or they are not acceptable. If they are not acceptable, the risk analysis has to be performed again, but there is a possibility of adopting such measures which are expensive or useless [2-3].

6 Information System Protection

In ensuring organization security, the key task is protection of the people, information and other properties on relevant security environments. These elements belong to the fundamental assets of the organization. The protection level is influenced by availability of money, people, time and other factors. The protection level is also influenced by organization willingness to accept certain risk level. There are many threats, which endanger the organization information systems. Therefore, it is necessary these threats to identify [5].

Data stored in information systems may be intentionally or unintentionally misused or threaten by a nature. Information system protection is understood as protection from viruses, hackers, spam, irresponsible users and natural disasters.

According of the legal act about information protection [15], the protection of information or communication systems is defined as a system of measures for ensuring confidentiality, integrity and availability of classified information stored in information system. This definition implies that information security consists in many rules and mechanisms for ensuring data confidentiality, integrity and availability. Therefore, information security is extremely complex issue.

Data confidentiality represents prevention of unauthorized access to the system. Data integrity security signifies prevention of unauthorized data modification. Requirement of data availability lies in ensuring access to the system in the right time [6]. Basic components of information system security are illustrated at the Fig. 1.

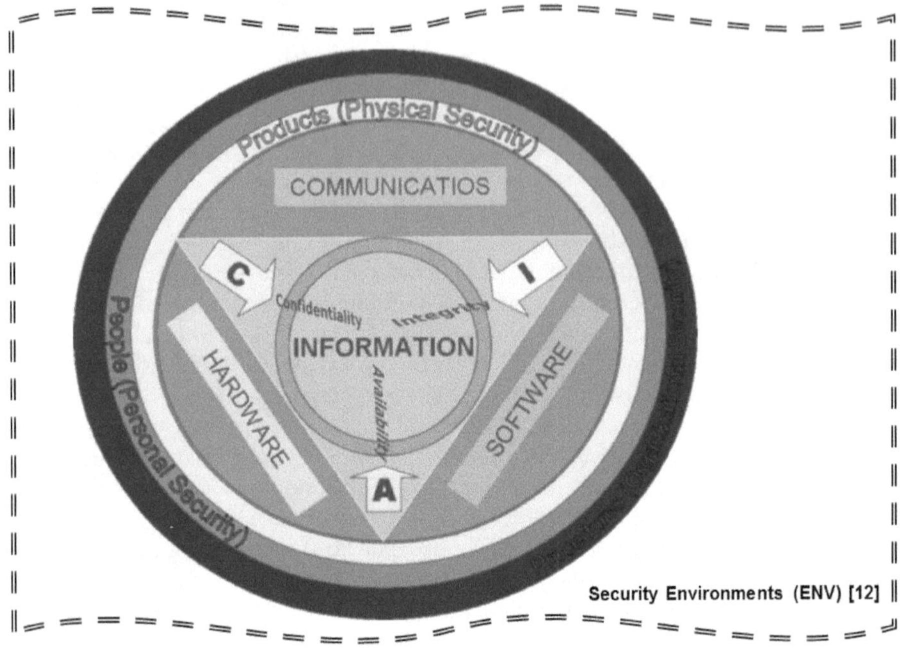

Fig. 1. Information Security Components [4, 12]

In recent years, the progress of ICT has brought certain need for bigger amount of collecting data within information systems and databases. It has resulted in a possibility of information systems destabilization. Today, we can hack into any information system, disturb any military operation, and connect people ideologically all over the world with help of any electronic device [6]. Threats endangering information system security may have fatal consequences, not only as unconstitutional impacts on privacy or personal integrity, but also as violation of national security. Therefore, importance of information system protection reposed on government authorities and private sector is significant [9].

Protection of assets owned by organizations is very important in view of business plans. Entrepreneurs act in more complex legal environment, where they may be exposed to unexpected turns and attacks, which threaten their assets. If their business is spread abroad, then it is necessary to apply foreign law, the situation is even more complicated and risks are higher. Planning the structure aimed at asset protection, that is able to eliminate or reduce adverse impacts when emergency event occurs, is needed [5].

7 Process Approach to Asset Protection

The aim of business process modelling is to create model of processes in order to understand all activities, relations among different activities and all assets and their roles within a process. Every process has several basic characteristics. In asset protection, especially crisis asset protection, the creation of appropriate set of activities is emphasized in order to preserve process potential even during emergency event occurrence. The continuity process principle focuses on problem mentioned above [12].

Process management focuses on integration of activities within the organization for more rapid and flexible cooperation [7]. It optimizes activities of all processes from the first supplier to the end-customer, and it delegates responsibilities for any process. Performance measurement of individual activities enables easier identification of specific problems. It addresses problem causes. Therefore, it has significant value because in the savings of the money and other resources, which can be, on the contrary, spend for elimination of impacts as consequence of activities disharmony. The process management supports teamwork, brings higher flexibility, enabling better sharing of knowledge and maintains communication during whole process.

8 Conclusion

The majority of the organizations do not focus on information systems protection during times of prosperity. Organizations accumulate certain amount of assets and information, which may be threaten in case of extraordinary event occurrence and they may lose their activities and stable profit. The organizations should make their own security policy that should contain base rules for ensuring the security of its information systems with the help of organizational, personal and technological measures. Therefore, it is important to prevent such events and select appropriate measures in order to protect organization assets.

References

1. Allen, R.: The PENGUIN English Dictionary. Penguin Books (2003)
2. Božek, F., Urban, R.: Management rizika. Univerzita obrany, Brno (2008)
3. Božek, F., Ješonková, L., Dvořák, J., Božek, A.: General Procedure of Risk Management. Ekonomika a management 3, 15–24 (2012)
4. Dardick, G.: Cyber Forensics Assurance. In: The Proceedings of the 8th Australian Digital Forensics Conference, pp. 57–64. Cowan University, Perth Western Australia (2010)
5. Halibozek, E.P., Kovacich, G.L.: Securing corporate assets: Protecting a business against cloak-and-dagger tactics requires a manager's eye for efficiency. Industrial Engineer: IE 35, 39–43 (2003)
6. Knopová, M.: Bezpečnost dat v informačních systémech. Ikaros 15 (2011), `http://www.ikaros.cz/node/6946`
7. Ludík, T., Navrátil, J., Langerová, A.: Process Oriented Architecture for Emergency Scenarios in the Czech Republic. World Academy of Science, Engineering and Technology 1, 2342–2351 (2011)
8. Ludík, T., Ráček, J.: Process methodology for emergency management. In: Hřebíček, J., Schimak, G., Denzer, R. (eds.) ISESS 2011. IFIP AICT, vol. 359, pp. 302–309. Springer, Heidelberg (2011)
9. Mitnick, K.D., Simon, W.L.: The art of deception: controlling the human element of security. Wiley, Indianapolis (2002)
10. Proházková, D. a kol.: Bezpečnost a krizové řízení. Police history, Praha (2006)

11. Urbánek, J.F., et al.: Scénáře adaptivní kamufláže. Tribun EU, Brno (2012)
12. Urbánek, J.F., et al.: Crisis Scenarios. Univerzity of Defence, Brno (2013)
13. Urbánek, J.F., Průcha, J.: A Development of Wireless Interoper-mobile Application for Outdoor Operation Management. In: 8th Int. Conf. on Electronics, Hardware, Wireless and Optical Communications, EHAC 2009, pp. 57–64. WSEAS Press (2009)
14. Czech Republic Legislation Decree. 523/2005 Coll., on the security of information and communication systems..In Czech Republic Statute Book
15. Czech Republic Act No. 412/2005 Coll., on the protection of Classified Information and Security. In Czech Republic Statute Book

Crime Open Data Aggregation and Management for the Design of Safer Spaces in Urban Environments

Antonios Bonatsos, Lee Middleton, Panos Melas, and Zoheir Sabeur

University of Southampton IT Innovation Centre,
Faculty of Physical Sciences and Engineering, United Kingdom
{zas,ljm,pm}@it-innovation.soton.ac.uk

Abstract. This paper describes the major research and development activities which have been achieved so far since the launch of the DESURBS project (www.desurbs.eu) in 2011. The project focuses on the development of a Decision-Support System Portal (DSSP) which integrates information, data and software modules representing city assets, hazards and processing models that simulate exposures to risks and potential compromise to safety and security. The use of the DSSP will aid the design of safer and more resilient urban spaces. Specifically, it provides security related scenarios with contextual information to support various types of users who specialise in urban spatial design and planning. The DSSP is a web enabled system which is also adapted to mobile devices usage. It is supported with geographic maps and visualised aggregated data from a number of heterogeneous sources. A responsive web design which adapts to the resolution of smart mobile devices has also been achieved. That is, low powered mobiles can still provide map oriented data in a responsive fashion, while using multiple platforms (Android and iOS currently). The first DSSP prototype employs the United Kingdom crime statistics feed of year 2012 and analyses crime trends in 13 English Cities (including Greater London) which are distributed into four major-regions. The DSSP displays raw crime data via a marker on a map, while they are aggregated under specific crime type threads and visualised as "heat maps". The specific visualisations are aligned to the various administrative regions such as neighbourhoods, catchments and postcodes. It also allows users to explore historical crime trends for a region over time, where crime statistics are contrasted. The scalability of the DSSP was also tested under increasingly large datasets and numbers of users, with tested loads on the map server and the main Django user application. The difference in speed between the mobile and desktop interfaces for a defined set of tasks using the application shall also be performed and presented in the near future.

Keywords: Smart Cities, urban safety, urban crime, data visualisation, decision-support, web-enablement, mobile applications.

J. Hřebíček et al. (Eds.): ISESS 2013, IFIP AICT 413, pp. 311–320, 2013.

1 Introduction

The traditional view of cities is that they bring together "hard" infrastructure to facilitate the lives, activities, security and communication between large and permanent settlement of people in urban zones. This view is becoming less valid since the emergence of digital and mobile technologies in the last few decades and the most recent ICT driven Future Internet concepts of "Smart cities" [5], [2]. For example, security in the smart cities context involves now the management of public spaces with the use of the internet to access to geo-information, overlaid with processed data and information for decision-support [4, 6]. From this technology perspective alone, a key driver which will aid smart cities to be managed at localised levels involves the deployment and use of large scale sensor networks for monitoring and recording crime incidents and beyond [1]. With the pervasive nature of smart mobile devices and their capability to be adopted as sensor platforms [3], sensors can be considered for exploitation as key Future Internet enablers in smart cities [7]. In particular, the generated big sensor data from mobile devices shall become the prime Citizens observatories for aiding local authorities achieve detailed and real-time measurement of crimes at catchment levels in cities. Furthermore, the new generation of decision-support systems for managing crimes in smart cities shall require access to data generated by citizens together with generic big data processing and analytics tools. These shall provide advanced real-time crime situation awareness at localised scales for citizens. The following sections of this paper provide the pertinent aspects of the new generation decision-support system design for monitoring crimes in cities. This is performed in a top down approach to provide the status of the current prototype system development.

2 Decision-Support System Portal (DSSP) Design

The DSSP is a web enabled information system environment which gives access to integrated data and information for the management of hazards and potential crimes incidents in cities. It allows users perform scenarios for the design of safer spaces. The DSSP is split into two main parts, the server side and the user or client side.

Fig. 1 illustrates a high level view of the overall system. The server-side is comprised of 3 distinct layers. The top layer concerns the various data sources. The middle layer is the analytics layer which incorporates local cache for data processing. Finally, the lower layer specialises in data visualisation. The client-side at the user end, is simple and relies on an HTML5 capable browser. The User Interface (UI) is platform agnostic, and user-friendly for tablets and mobile devices. In addition, a mobile application has been developed in order to facilitate end users to ingest data (reports, surveys) into the system.

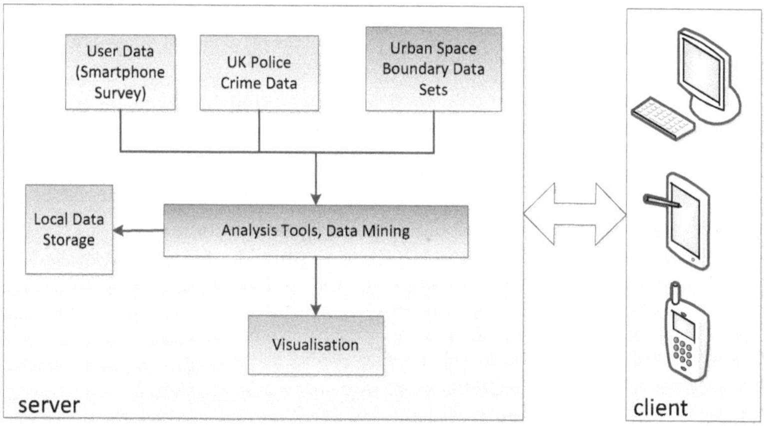

Fig. 1. System Top Level Design for the DSSP

2.1 Urban Space Boundary Sets

An important feature in DSSP is how the abstract idea of an urban space is defined. Having defined an urban space then data analysis on crime or on safety perception from other data resources can be easily applied. Urban spaces in DSSP are defined as boundaries in the form of geospatial data polygons. These polygons are used as our base boundary layer which allow us to filter and group the crime incidents and on top of which we do all the visualisations and analysis. Although a city can be defined in geographical terms by a single, but complex, polygon, smaller divisions of a city such as boroughs, districts, neighbourhoods, should also be defined in similar terms i.e. geospatial data polygons. These additional polygons will form the basis of the different zoom levels in our analysis tools. The transitions between different and with existing zoom levels should be consistent. For instance, polygons at the same zoom level should not overlap with neighbouring polygons. Similarly peripheral polygon boundaries should match higher zoom level polygon boundaries. For our analysis we examine 13 UK cities in which we define three distinct resolution levels. Data from the UK Census[1] and UK Ordnance Survey[2] was used to produce the required zoom level boundary sets. The different levels employed in the application are now defined.

- **Cities:** a boundary dataset containing polygons for each one of our cities. The census Districts boundary set is used for the council level.
- **Neighbourhoods:** This level offers a more in-depth view on each city's crime rates. The census Middle Layer Super Output Arias (MSOAs) boundary data set was used. MSOA polygons are population-wise similar (~7400), this provides a level of normalisation for our data analysis.

---

[1] http://www.ons.gov.uk/ons/guide-method/census/census-2001/data-and-products/output-geography/output-areas/index.html

[2] http://www.ordnancesurvey.co.uk/oswebsite/products/boundary-line/index.html

- **Postcodes:** The highest level of resolution. It allows a very detailed view of the crime rates in every part of the city.

Although boundary polygons for council and neighbourhood levels match, postcode boundary polygons do not match with either neighbourhood or council boundary polygons. This means that a postcode polygon can intersect two or more neighbourhood polygons. For this reason work was performed to make the different polygons consistent. The data sets are extracted from ESRI shapefiles and converted to use latitude and longitude coordinates within the WGS84 datum. The OS data for instance uses its own grid coordinate scheme as illustrated in Fig. 2a. Additionally, all necessary associations between neighbourhoods, councils and convex hulls were also created. An example of the city and neighbourhood boundaries is illustrated in Fig. 2b.

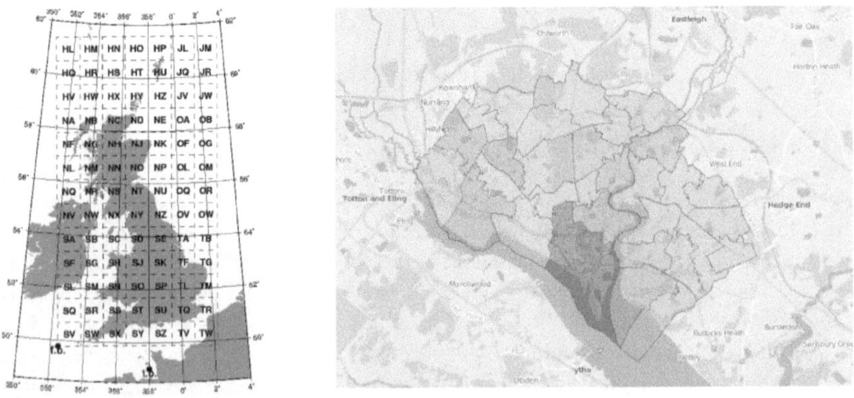

(a) (b)

Fig. 2. (a) Ordinance Survey UK National Grid (b) The city and neighbourhood boundaries for Southampton

2.2 DSSP Data Sources

Having defined urban spaces in terms of boundary data with clear polygon sets, and zoom levels for all 13 UK cities, the next step is to aggregate data related to those cities. In addition as the data sources all have different coordinate systems we need to project them all onto the same system. We employed the WGS84 coordinate system for all our work.

Crime Data

The UK Police Crime database is based on data about reported crime incidents. In order to protect the privacy of individuals who are involved in the incidents, the police provide approximate and not completely accurate coordinates. For example crimes are not included unless they are from a street with 8 or more postal addresses. The raw data is publicly available via a police API which provides information on the type of crime, location (in Easting and Northing), and the date of the crime.

The DSSP system periodically checks for updates and downloads newly released crime data to keep its database up to date. UK crime data from January 2011 to April

2012 period were initially gathered and processed and cached locally. Gathered data is pre-processed to use a common grid reference system. Reported coordinates for each crime incident are provided in the form of Eastings and Northings 2-D Cartesian coordinates using the Transverse Mercator projection and the OSGB36 datum. The DSSP converts crime incidents coordinates into longitude and latitude within the WGS84 datum. Finally crime data are filtered and stored. We use a modified ray casting algorihtm to decide if a crime lies within a specific polygon at a particular zoom level.

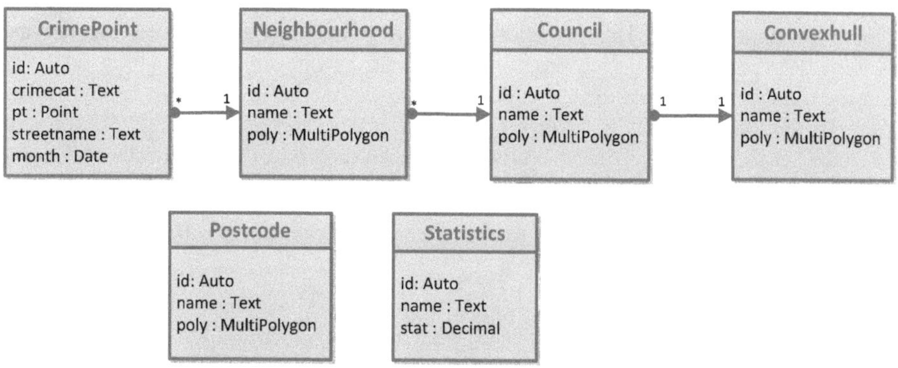

Fig. 3. Crime database schema

Fig. 3 shows the schema for the database. Notice that each crime is associated with exactly one neighbourhood, each neighbourhood is associated with exactly one council, and each council has exactly one convex hull. These associations allow us to group the data into areas, so that they can be easily retrieved later, by minimising the number of spatial queries. This is necessary as while our database provides fast spatial queries they are still slower than simple lookup queries. Postcodes are not associated with the other geographical entities, because their boundaries are not matching. Currently we only use the postcode information for visualisation so this is not a problem.

Participative Users Data

The system allows individual user to submit their own reports about their personal perception of safety in urban spaces. Perception of safety can have the form of a report or a survey that can describe an incident or opinion from the individual's point of view. Individuals therefore can submit reports directly to the DSSP portal over a browser or via the portal API. The API is REST based and is fully self-describing and web browseable. This simplifies client side development. Submitted surveys and reports are stored in system user data database. A user data report includes name of report, description, submission date (including time), geospatial information, a user identifier (which is hashed to preserve privacy), and optional attachments such as images.

2.3 Analysis Tools and Data Mining

The DSSP provides all the necessary tools to analyse and harvest information from the different data sources. Processed data is stored in a local database. Analysis of data can take the form of normalisation, processing, and conversion between coordinate systems. Upon data ingest the data is filtered and aligned to the boundary polygons for use in creation of the crime data reports. For each crime type we also generate a heat-map. The data is interpolated to create a continuous function over the polygon. The heat-maps are generated at neighbourhood and postcode boundaries. Data mining takes place on user reported surveys to correlate with actual incidents from other data sources. This provides a measure of truthfulness to the user generated content. For example a police reported crime can be chosen, as the actual incident, and the data mining operations, e.g. correlation, can be applied on user reported data pool. In this way an incident can be enriched with user submitted set of data, and presented in the form of a timeline arrangement as shown in Fig. 4. Similarly the perception of safety for a particular location at a particular time can be detected, measured, and analysed.

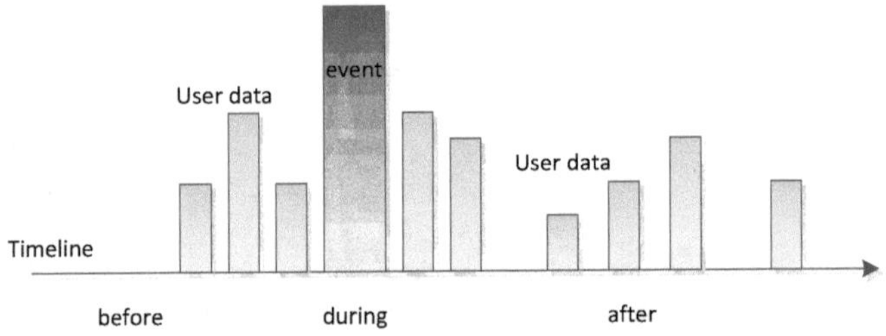

Fig. 4. Timeline of an incident

2.4 Visualisation

The DSSP provides a rich set of data visualisations including interactive maps with different levels of information for different resolution levels. Also, area statistics with pie and line charts and comparisons and correlations between two different areas of interest. Finally, it also provides heatmaps for each crime type and crime point projections. The DSSP mapping service is comprised of the following parts, the base map data, the javascript interaction library, and the map server. Since the Google Map API could not be used due to its licence constrains, the base map data is based on OpenStreetMap which is free and not bound by usage limits. For the Javascript interaction library we used Leaflet which provides all the necessary functionality. Additionally it is lightweight and designed to be mobile friendly.The map server is based on Tilestache server, a python-based tile server that can serve and cache map tiles based on geographic data. Tilestache takes requests in the form of tile URLs such as:

```
http://{address}/{layer name}/{zoom}/{column}/{row}.{extension}
```

This request describes the address of a map tile using the Google XYZ tile addressing format. The format defines that tiles are referenced by their zoom level and X and Y values. It also takes a layer name as parameter. This allows many layers to be created each illustrated a different set of pertinent data. The DSSP currently has the base OpenStreetMap layer and 11 additional layers for different types of heat-maps. The charts are created with jqPlot which is a javascript based plotting library. It plots data transmitted from the server via JSON objects. The basis for the user interface is Twitter Bootstrap. This employs a 960 pixel grid approach which allows a large degree in versatility in layout and design. The consequence of this is the responsive design which allows the DSSP to support a number of different devices, both mobile and desktop, via a simple code base.

2.5 Client Applications

The DSSP offers its client functionality over a web interface, the UI is HTML5 based and therefore platform agnostic. Additionally we make our interface repoonsive (adapt to client device resolution). Special considerations ware taken into account in order to provide a UI that is convenient for tablets and mobile devices too. As an example an interface based on mouse clicking items to bring up information is not directly useable on a touch based device. Smartphones with their unique characteristics such as wide spread adoption, permanently carried, always on, interactiveness; provide a powerful complementary and disruptive mass media. DSSP considers and targets smartphones as the major source of the User Data, has developed a customised mobile application to submit reports and surveys. DInR is a simple mobile app, which is under development at the moment, in order to facilitate individuals submitting reports into the DSSP directly using their mobile phones. Currently DInR is running on Android 4.1 phones or tablets and provides most of required report fields. DInR can successfully create and submit reports to the current version of DSSP system. DInR can provide automatically 3/5 data fields in the user report, consequently smartphone users can submit to DSSP user data with a minimal effort.

3 DSSP Prototype

3.1 Functionality

The current implementation of the system prototype runs on two distinct machines. The first machine hosts the main application while the second on the Tileserver. This separation of machine allows for independent optimisation and deployment of the two main components. The main application runs on a machine with an 2.83 GHz CPU and 4GB of RAM. The Tileserver has an 2.83 GHz CPU, 4GB of RAM, and 2 TB of fast storage. Currently a demonstration version of the system application is hosted externally[3]. The application allows the user to respond to a number of questions of the raw crime data. It can also perform them at a variety of zoom levels. These levels include Cities, Neighbourhoods, and Postal Codes which are in accord with the standard Ordinance Survey (OS) classifications in the UK. An example of the browse mode is shown in Figure5(a).. The top part of the figure is a GIS view that shows a zoomed version of the region of interest (South of England) with a blue polygon hig-

[3] http://desurbs.it-innovation.soton.ac.uk/police

hlighting the city under investigation. The lower part of the figure shows a series of graphs concerning crime data analytics. The pie chart illustrates the relative proportion of the specific crime types. The line chart shows the historical rates of crime types, where the specific crimes illustrated in the graph are selectable. Further, a colour coded matrix shows the cross-correlations of the different crime types in the City of interest. This allows the user to cluster crimes of various types in the investigation. As illustrated in Figure5(b) "Drugs" related crimes statistically correlate well with "Criminal Damage and Arson" based crimes in the same City. This feature in the DSSP enables crime knowledge experts to demonstrate crime status in specific areas of a City with greater emphasis to a collective audience of decision-makers in urban design of safer spaces.

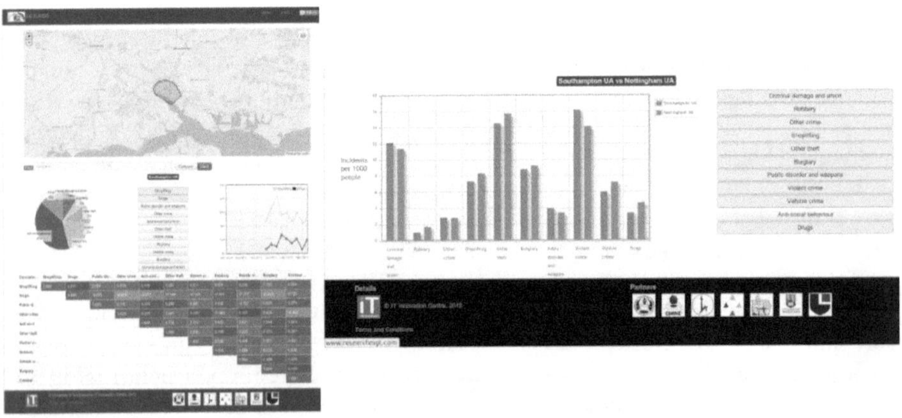

(a) (b)

Fig. 5. (a) Illustration of the application showing crime data for a specific city (b) Comparison of two cities

The comparison between various cities crime trends can also be processed by the DSSP as illustrated in **Chyba! Nenalezen zdroj odkazů.b**. The bar chart shows the number of incidents of a specifically selected crime type in Cities. The specific crime types are also user selected. Furthermore, the crime data is normalised in accord with the respective population densities in each city. With this feature, user queries about the relative numbers of crimes in different urban regions of a given country can be investigated. In Fig. 6 a Heat-Map is visualised and aligned to the OS administrative boundaries. In addition to the above, the locations of all specific crime types in a City can be added as new layers above the Heat-Map for further analysis and comparative studies. These are visualised with representative points that can be removed by the user if needed. The visualisation of the crimes provides the respective crime-specific locations in a city of interest together with the intuitive discovery of urban areas that are more exposed to specific crimes than others.

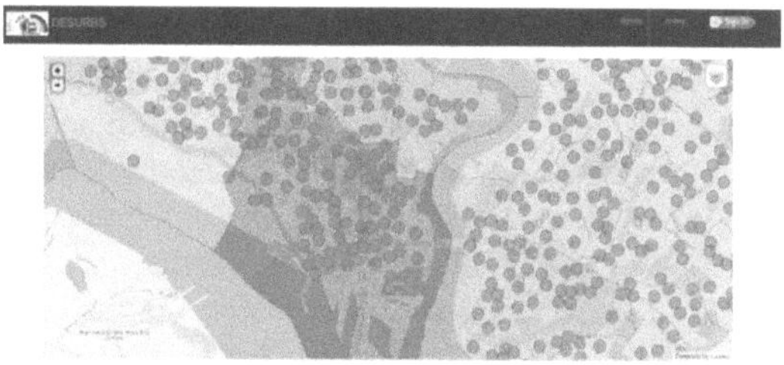

Fig. 6. Heatmap based on the crime data with specific crime points overlaid

3.2 Optimisation

The DSSP was built to provide both functionality and performance on multiple platforms. Many optimisations carried out in order to improve the system performance and responsiveness where the goal was to present similar user experiences even with low powered mobile platforms. The general strategy to meet this goal is to perform most of the processing work on the server and send to the client only the minimal amount of data required. The layers for the boundary datasets at different resolutions are chosen carefully to match their borders. The vertices of the polygons are simplified wherever possible to increase performance at different zoom factors. The UK crime data is filtered, trimmed, and stored locally for the regions of interest only. In order to enforce consistency between the boundary datasets data normalisation at different resolutions was performed. The use of a unified coordinate scheme also makes visualisation and analysis simpler. Geospatial features of PostgreSQL are also not used unless there is a processing benefit from it. The mapping server (Tilestache) is lightweight and scalable and has the benefit of caching data. In order to reduce network traffic lightweight javascript libraries are used. Prior computation and caching takes place on the DSSP server extensively. Specifically, pre-calculation is used to compute spatial features. For example the Southampton convex hull was reduced from 2834 vertices to 53 points. When the visualisation is computationally expensive the imagery is cached on the tileserver. Statistical analysis of the data is also performed offline and displayed via a single lookup in a database. Finally, the clustering of the crime points is also performed offline at different zoom levels.

4 Conclusions and Future Work

An early DSSP prototype which specialises in the provision of integrated data and information for designing safer spaces in urban agglomerations has been described in this paper. The new system is enabled under the Future Internet context of 21st century smart cities; where citizens participate in providing collective observations on potential hazards, incidents and their perceptions of safety around their local areas. The supporting hardware infrastructure of the DSSP consists of a number of special purpose machines with specific roles. These include the front-end application server and the tile server for data and information mapping. A set of optimisations have been performed in order to make the system work efficiently, particularly on low-powered mobile clients. The DSSP is currently innovative for potentially collecting citizen's perception of safety in local urban zones and contrasting it with the existing UK Police crime data. Nevertheless, the system prototype has still a "work in progress" status considering the number of visualisation features and functionalities which shall be added to it prior to the end of summer 2013. For example, the current tile-server which serves identical sized tiles for different devices will be optimised and improved to perform more efficiently across all type of platforms. Additional enhancements shall be made to the mobile interface too. Specifically, the device dimensions will be used to inform the layout of the user interface. The mobile survey tool will also be expanded with more generic features in order to apply to wider participatory observations under the smart cities context. This is pertinent to making the system useful for a variety of smart cities oriented applications.

References

1. Filipponi, L., Vitaletti, A., Landi, G., Memeo, V., Laura, G., Pucci, P.: Smart City: An Event Driven Architecture for Monitoring Public Spaces with Heterogeneous Sensors. In: Fourth International Conference on Sensor Technologies and Applications (SENSORCOMM), pp. 281–286 (2010)
2. Gibson, D.V., Kozmetsky, G., Smilor, R.W.: The Technopolis Phenomenon: Smart Cities, Fast Systems, Global Networks. Rowman & Littlefield, New York (1992)
3. Lane, N.D., Eisenman, S.B., Musolesi, M., Miluzzo, E., Campbell, A.T.: Urban Sensing Systems: Opportunistic or Participatory? In: Proceedings of the 9th Workshop on Mobile Computing Systems and Applications, pp. 11–16 (2008)
4. Nam, T., Pardo, T.A.: Smart City as Urban Innovation: Focusing on Management, Policy, and Context. In: Proceedings of the 5th International Conference on Theory and Practice of Electronic Governance, pp. 185–194 (2011)
5. Sabeur, Z.: Future Internet Smart Cities User Behaviour Driven Innovation Platforms. Presented at the 9th Edition of the Future Internet Assembly, Aalborg, Denmark (2012)
6. Schaffers, H., Komninos, N., Pallot, M., Trousse, B.: Smart Cities and the Future Internet: Towards Cooperation Frameworks for Open Innovation. In: The Future Internet, pp. 431–446 (2011)
7. Walravens, N.: The City as a Platform A Case-based Exploration of Mobile Service Platform Types in the Context of the City. In: International Conference on Intelligence in Next Generation Networks, pp. 283–288 (2011)

Multi-scale Robust Modelling of Landslide Susceptibility: Regional Rapid Assessment and Catchment Robust Fuzzy Ensemble

Claudio Bosco[1], Daniele de Rigo[2,3], Tom Dijkstra[4], Graham Sander[1], and Janusz Wasowski[5]

[1] Loughborough University, Department of Civil and Building Engineering Loughborough LE11 3TU, United Kingdom
c.bosco@lboro.ac.uk
[2] European Commission, Joint Research Centre, Institute for Environment and Sustainability, via E. Fermi 2749, I-21027 Ispra (VA), Italy
daniele.de-rigo@jrc.ec.europa.eu
[3] Politecnico di Milano, Dipartimento di Elettronica e Informazione, via Ponzio 34/5, I-20133 Milano, Italy
[4] British Geological Survey, Environmental Science Centre Keyworth NG12 5GG, United Kingdom
[5] National Research Council (CNR), Bari, Italy

Abstract. Landslide susceptibility assessment is a fundamental component of effective landslide prevention. One of the main challenges in landslides forecasting is the assessment of spatial distribution of landslide susceptibility. Despite the many different approaches, landslide susceptibility assessment still remains a challenge. A semi-quantitative method is proposed combining heuristic, deterministic and probabilistic approaches for a robust catchment scale assessment. A fuzzy ensemble model has been exploited for aggregating an array of different susceptibility zonation maps. Each susceptibility zonation has been obtained by applying heterogeneous statistical techniques as logistic regression (LR), relative distance similarity (RDS), artificial neural network (ANN) and two different landslide susceptibility techniques based on the infinite slope stability model. The sequence of data-transformation models has been enhanced following the semantic array programming paradigm. The ensemble has been applied to a study area in Italy. This catchment scale methodology may be exploited for analysing the potential impact of landscape disturbances. At regional scale, a qualitative approach is also proposed as a rapid assessment technique – suitable for application in real-time operations such as wildfire emergency management.

Keywords: Landslide susceptibility, Modelling, Ensemble, Semantic Array Programming.

1 Introduction

One of the main challenges in modelling landslides is related to the assessment of their spatial probability of occurrence. Estimating landslide spatial probability

J. Hřebíček et al. (Eds.): ISESS 2013, IFIP AICT 413, pp. 321–335, 2013.

may be supported by many different analytical approaches: heuristic, deterministic and statistical [1]. A subset of those approaches may be described as data-transformation models (D-TM) which propagate input data layers and parameters so as to estimate landslide susceptibility as a spatially explicit derived data layer.

Statistical landslide susceptibility methods are based on establishing relationships among measurable variables whose combination is empirically found to correlate with observed landslide occurrences. Theoretical insights guide the selection and analysis of landslide triggering factors (geological, hydrogeological and geomorphological) to determine the most appropriate input – the best suited set of predictors or covariates [2] – to use for statistically reconstructing landslide susceptibility. Domain expertise is therefore essential for supplementing semi-automated computational methods with meaningful semantic heuristics. Estimations are then made of their relative contribution in slope failure and areas are classified as having different hazard or susceptibility degree [3–5]. Multivariate statistical methods are commonly used methodologies for these analyses. This kind of analysis is based upon the presence or not of stability phenomena within the classified areas [6]. The main problem using this approach is the high sensitivity of the results to the input data and the difficulty in deriving the probability of occurrence from the susceptibility [5].

In deterministic approaches, the landslide susceptibility is determined by carrying out a slope stability analysis. It results in computing the distribution of a factor of safety characterizing the study area. The deterministic approaches should be able, in theory, to provide more reliable results (especially where no field measurements are available on landslides and stable areas) but require detailed datasets describing the triggering factors. Many different models, which are usually composed of coupled dynamic hydrological and slope stability models, have been developed by several authors [7–9].

The evaluation of slope stability conditions in a landscape veneer can be performed by considering the local equilibria of forces along pre-determined, shallow slip surfaces representative of translational slide mechanisms. These equilibria can be expressed as a local factor of safety (FS), a ratio between the resisting and driving forces acting on a slip surface. If this FS reaches values lower than unity, the slope will fail.

$$FS = \frac{\text{resisting forces}}{\text{driving forces}} = \frac{\text{shear strength}}{\text{shear stress}} = \frac{S}{\tau}. \tag{1}$$

In this equation S represents the shear strength mobilised along the slip surface and can be represented, in its simplest form, as:

$$S = c + (\sigma - u)\tan\phi, \tag{2}$$

In this equation c represents the cohesive resistances in the soil (these resistance may be enhanced by the presence of roots and other elements), σ represents the total stress and u represents the pore pressure, both acting on the slip surface and affecting the amount of frictional resistance ϕ that can be mobilised.

The driving force is represented by τ, the slope parallel component of gravity acting on the soil mass above the slip surface

$$\tau = \gamma Z \cos\beta \sin\beta, \tag{3}$$

where γ is the bulk unit weight and Z is the depth to the shear surface.

Statistical methods have become popular in recent years [10], including those employed by the British Geological Survey for their GEOSURE product that incorporates an assessment of landslide potential based on an expert elicitation process [11]. One of the main problems in applying these methods is related to changing terrain conditions after a mass movement. Environmental parameters such as slope, land cover or soil depth can be totally different following landslide events. Furthermore, temporal changes in triggering factors in these dynamic landscapes are difficult to evaluate using statistical methods. Deterministic methods may provide more physically based results, but the current limitations due to data quality and availability seriously hampers our ability to simulate landslide processes over relevant spatial and temporal scales [10,12,13]. Some of the soil properties required by deterministic methods, such as soil cohesion or friction angle, are difficult to measure over large areas and show a high spatial variability. Therefore, currently these models still need a high degree of simplification [12]. Because of this limit on present approaches, research is urgently needed for improving landslide susceptiblity assessment methodologies.

Landslide susceptibility is also closely linked to many geomorphological and environmental aspects whose perturbation may be driven by natural or anthropogenic changes. This may even affect the landscape/regional scale along with challengingly complex networks of relationships among natural resources and ecosystem dynamics [14–16], also involving climate change and land use alterations [17]. The transdisciplinary analysis of the involved uncertainty and complexity is addressed within the broad domain of integrated natural resources modelling and management (INRMM) [18]. A new approach based on a robust modelling ensemble at catchment scale is here proposed for improving landslide susceptibility estimation. A preliminary regional-scale rapid assessment is also introduced as a possible support in real-time operations such as wildfire emergency management [19,20].

2 Methods

2.1 Study Area

The study area is located in the Puglia region of southern Italy. The study site (Fig. 1) covers almost 16 km^2 in the municipal territory of Rocchetta Sant'Antonio which is situated in the southern part of the Daunia Appennines. The Daunia Appennines are highly susceptible to landsliding [21,22] and over 20% of the Rocchetta Sant'Antonio area has experienced landslide activity [23].

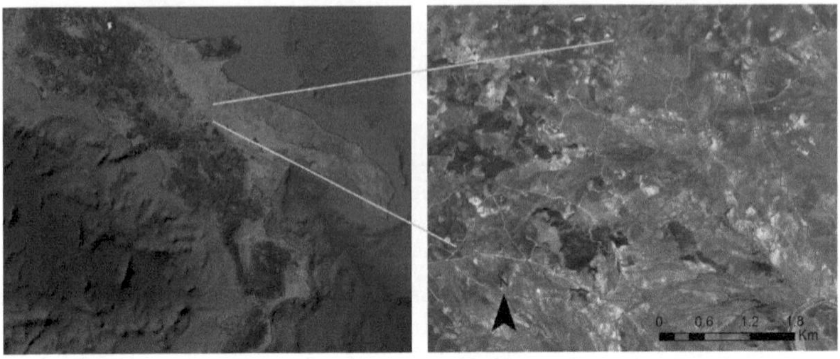

Fig. 1. The study area (Rocchetta Sant'Antonio, Italy). Google Earth, ©2013 Google.

This area has seen a large number of landslides affecting the local economy and has been studied for some years resulting in a database of information (including topography and Digital Elevation Model (DEM), site investigation and geotechnical test data, soil distribution and land use maps) [24–26].

2.2 Applied Techniques

Within the study area, precipitation is the main triggering factor for landslide occurrence. In order to improve the spatial prediction of landslides where water is the triggering factor, a combined total of five different deterministic and statistical models have been applied. In order to enhance the determination of landslide susceptibility of the study area, a new method based on an ensemble approach has been used for aggregating the modelling results. The ensemble approach is a reproducible D-TM applied to the results of the array of models and is based on relative-distance similarity (RDS), a flexible technique in INRMM problems [27–29]. The application of an ensemble approach, especially in data-poor regions, could potentially reduce the uncertainty and mitigate local poor performance associated with individual models, by excluding outlier estimations.

Uncertainty may affect these models from the inaccuracy of required inputs data layers X and parameters θ to the approximation of their reconstruction (e.g. by means of other specialised D-TMs) when not directly accessible as available datasets. Uncertainty may also be exacerbated by modelling simplifications or overcomplication. In the latter case, a perhaps theoretically accurate approach might sometimes result in a poorly performing D-TM implementation due to site-specific information gaps and possible impacts of site complexity where multiple conceptual mechanisms coexist as landslide drivers. Furthermore, nontrivial computational models may be affected by *software uncertainty* [18, 30, 31], namely the distance from the theoretical mathematical formulation and the actual model implementation in one or more artificial programming languages.

$Y = f^*(X) = f(\theta^*, X)$	Theoretic D-TM: instance of a model family f by means of selected parameters θ^*
$Y = f^\zeta(\theta^{*,\zeta_\theta}, X^{\zeta_X})$	Real D-TM with uncertainty affecting input data, parameter calibration and the model itself (software uncertainty [18,30,31])
$Y = \bigcup_{i=1}^{n} ::\left\| f_i^\zeta(\theta^{*,\zeta_\theta}, X^{\zeta_X}) \right\|_{::}^{sem_i}$	Mitigating uncertainty by ensembling an array of $i \in \{1, \cdots n\}$ heterogeneous models (design diversity [18,30]) with semantic checks sem_i (semantic array programming [31–33])

(4)

In order for the uncertainty to be mitigated, a robust fuzzy ensemble model is proposed to aggregate an array of different susceptibility zonation maps. Each susceptibility zonation has been obtained by applying heterogeneous models, so as to increase design diversity [18,30]. The technique is designed to scale to different arrays of models. Each model is adapted to fit the ensemble array by wrapping its interface so as to behave as a semantically enhanced module. Semantic checks further mitigate inconsitencies between input data, parameters and outputs, following the paradigm of semantic array programming (SemAP) [31–33]. SemAP complements the compactness of array programming notation with an effort towards the most concise generalisation of autonomous tasks as modules which are subject to array-based semantic checks. Each model is considered as a semantically-enhanced module of the ensemble. A few straightforward semantic checks sem_i are exemplified in the following with the notation *::sem::* with hyperlink to the corresponding online description.

Exploiting the available landslide maps and the available environmental information (DEM, land cover and lithological map, geomorphology) two deterministic models based on the infinite slope equation (SINMAP [34,35] and a simple slope stability model inspired by van Beek's PROBSTAB model [7]), as well as three statistical models based on Artificial Neural Network (ANN), Logistic Regression (LR) and Relative Distance Similarity (RDS), were calibrated and validated. The theoretical basis of SINMAP involves a mechanistic infinite slope stability model linked with a topographically based steady-state hydrology model. The theory at the basis of the model applies to translational slides where fluctuating pore pressures form the dominant trigger factor. The landslide susceptibility distribution is governed within the model by calculating slope and specific catchment areas starting from a DEM.The model parameters are allowed to be flexible, following a uniform distribution between an upper and lower limit. The parameters may be calibrated using geographic calibration regions based on lithological, land cover or soil characteristics [34,35].

The implemented slope stability model (hereafter named TransSlide) is based on a translational slope stability function that calculates the factor of safety at a potential shear plane based on variations of groundwater level and volumetric moisture content. TransSlide is based on eqs. 1, 2 and 3 where:

$$\sigma = Z\gamma \cos^2 \beta \tag{5a}$$

$$u = \gamma_w WL \cos^2 \beta \tag{5b}$$

(γ_w is the density of water (kNm^-3) and WL is the groundwater height above the shear surface). The static inputs include soil shear strength (cohesion and friction) and an additional root cohesion function representing land use. The model functions in a raster-based environment and calculates the local instability on the balance between resisting and driving forces. The paucity of available data in the test region meant that the full functionality of TransSlide could not be mobilised.

Logistic regression analysis, introduced by Cox [36] in 1958, is one of the more commonly used statistical methods in earth science [37]. However, it does not predict presence or absence of landslides, but the probability of occurrence [2]. The predicted values, ranging from 0 to 1, representing the landslides hazard can be expressed as:

$$P(y) = \frac{1}{1+e^{-y}}, \tag{6}$$

where P is the probability of landslide occurrence. It describes the relationship between a dependent variable (y) (assuming a distribution between presence [1] and absence [0]) and n explanatory variables (the covariates C_c^1, C_c^2, ..., C_c^n in the spatial cell c) where:

$$y = b_0 + b_1 C_c^1 + b_2 C_c^2 + .. + b_n C_c^n \tag{7}$$

and b_0 is the intercept of the model with b_1, b_2, ..., b_n the regression coefficients.

The maximum-likelihood method is used for estimating the coefficients of the logistic multiple regression model. Because of the non-linearity between independent variables and probability, parameters estimation requires the application of an iterative algorithm [12].

Among the most recent approaches in landslide spatial susceptibility evaluation there are artificial neural network models. An ANN can be considered as a "computational mechanism able to acquire, represent and compute a mapping from one multivariate space of information to another, given a set of data representing that mapping" [38]. A neural network may be able to learn and describe more complex relationships among input covariates than logistic regression could. Actually, a logistic regression model might be seen as a simple special case of neural network. An ANN architecture is characterised by one or more processing elements (nodes or neurons) which propagate weighted inputs received by other nodes [39]. The learning process comes from adjusting the weights between neurons analysing the error between the predicted and target output. The output of a neural network, after the training, is a model that starting from an input dataset is capable of predicting a target value [40].

The RDS is based on the relative distance between two values C_1^j and C_2^j of th j-th nonnegative covariate. The relative distance is a dimensionless number between 0 (maximum dissimilarity) and 1 (maximum similarity) and is simply the ratio between the minimum and the maximum value:

$$\frac{\min(C_1^j, C_2^j)}{\max(C_1^j, C_2^j)}. \tag{8}$$

The RDS index of a given multi-dimensional point c with respect to a set A of reference points (either stable areas or unstable ones) involves the relative distance among the pairs $\{C_c^j, C_\alpha^j\}$ for each $\alpha \in A$ and each dimension j of the N^C covariates.

2.3 Susceptibility Forecasts

2.3.1 Data and Explanatory Variables

For running the statistical models a set of calibration parameters was selected from international literature [37, 39, 41]. Seven parameters, commonly assumed as directly or indirectly related to landslide occurrence [37, 39], have been used for calculating the ANN, LR and RDS models. The Topographic Wetness Index (TWI) [42], the slope angle and aspect, the profile and plan curvature have been calculated starting from a digital elevation model with a resolution of 5 meters. The remaining models covariates are the land cover and the DEM itself.

The input parameters for running the deterministic models (root and soil cohesion, internal friction angle, bulk density of the soil, groundwater height, soil depth and the effective recharge rate), have been calculated using different methods. When actual measured data did not exist, estimated values from the literature were applied, if available, otherwise physically sensible values based on our judgment were used.

The recharge rate used in SINMAP (50 mm/d) was derived comparing the available climatological data (Rocchetta SantAntonio and Rocchetta scalo meteorological stations) and approaches from literature [43]. The transmissivity rate required by the model (m^2/hr) was calculated using the equation

$$T = K \cdot b, \tag{9}$$

where K(m/hr) is the saturated hydraulic conductivity and b (m) is the soil depth [43]. The hydraulic conductivity as the soil cohesion and the internal friction angle (used both in SINMAP and TransSlide) were derived from soil texture data using typical soil properties values from literature [44, 45].

The general soil characteristics of the catchment were derived by comparing the lithological map of the study area [23] and field data.

2.3.2 Calibration of the Models

The three statistical models were executed using both the same training set and the same set of covariates. The training set consists of 2091 points. The selected areas were subject to **::binary::**[1] classification as stable (0) or not stable (1)

[1] `http://mastrave.org/doc/mtv_m/check_is#SAP_binary`

Table 1. Mean Absolute Error (MAE) and Root Mean Square Error (RMSE) related to the singles and combined applied models. U and S refer on MAE and RMSE calculated respectively on unstable (U) and stable (S) areas only.

	RDS	ANN	LR	SINMAP	TransSl.	median	ENSEMB.
MAE	0.003	0.44	0.37	0.45	0.51	0.35	0.001
MAE U.	0.002	0.42	0.36	0.61	0.68	0.45	0
MAE S.	0.003	0.45	0.38	0.3	0.34	0.25	0.001
RMSE	0.02	0.47	0.43	0.54	0.58	0.4	0.019
RMSE U.	0.01	0.45	0.4	0.65	0.7	0.47	0
RMSE S.	0.03	0.48	0.46	0.4	0.42	0.32	0.026

on the basis of a dataset presented by [23]. The calibration points represent the 0.6% of the catchment area. These have been selected applying the RDS ensemble technique for analysing the relative distance between the model's covariates within the catchment. Applying this method, for selecting the training and validation points, it is possible to minimize the presence in the training set of areas having similar characterisics.

The multicollinearity of the dataset of **::nonnegative::**[2] **::finite::**[3] predictors was also analysed in order to avoid high correlation between different predictors. This was done for preventing the possibility that small changes in the data can cause an erratic change of the coefficient estimates.

For estimating the best architecture of the ANN the sum of squared errors (SSE) was determined for every combination of weight decay and number of neurons calculated by the ANN model during the network training. The final architecture of the ANN comes from the selection of weight decay and number of nodes minimizing the SSE. The calculation of SSE was then used for evaluating the fitting performance of the models, that is the ability of the model to reproduce the calibration set.

The catchment was split into areas based upon vegetation or lithologic data and calibration of the deterministic models was performed using data from the literature. Due to the paucity of high resolution data for calculating the parameters required by SINMAP and TransSlide only a rough calibration was possible. The low resolution of the available data does not allow the modelling of local variations within the catchment. The applied methodology includes a visual calibration based on expert judgment for adjusting the model's parameters in accordance with landslide observations. The adjustments were necessary for capturing a higher proportion of landslides in areas having a low stability index.

2.3.3 Validation of Model Performance

The main objective of this method is to estimate the fitting performance of a model with an independent set of data. Using cross-validation it is possible to estimate how accurately the models perform. An initial set of data was parti-

[2] `http://mastrave.org/doc/mtv_m/check_is#SAP_nonnegative`

[3] `http://mastrave.org/doc/mtv_m/check_is#SAP_finite`

tioned into two subsets, with the calibration carried out on a training subset (2091 pixels) and the validation performed on the testing subset (682 pixels). For analysing the performance error, mean absolute error (MAE) and root mean square error (RMSE) of the models against validation data were calculated.

Heterogeneous quantities provided as model outputs need to be harmonised in order for the models' performance to be comparable. The **::nonnegative::** `output` of the two deterministic models has been remapped to the corresponding **::possibility::**[4] values $\in [0\ 1]$ by means of Piecewise Cubic Hermite Interpolating Polynomials (PCHIP) with the codelet (MATLAB language): `pchip(` $[0\ 0.5\ 1\ 1.25\ 1.5\ 10\ \infty]$`,` $[1\ 0.8\ 0.6\ 0.5\ 0.4\ 0\ 0]$`, output )`.

The results of validation are summarized in table 1. The table shows a better performance of the statistical methods when compared with deterministic approaches. Between the applied models, the ensemble and RDS give the lowest errors. Although the high error rate of some model, the simple models' median is the next best result. Its application as a staightforward unsupervised ensemble might prove useful even where no additional information is available (black box output data).

3 Results and Discussion

The results of the models are shown in Fig. 2. The classification criteria used for measuring the landslide susceptibility fluctuate in a range between 0 (stable conditions) and 1 (unstable conditions). The adopted classification scheme is as follows: 0-0.4 (stable), 0.4-0.6 (area of model uncertainty) and 0.6-1 (unstable). The better performance of the statistical approaches when compared to deterministic techniques could be linked with both the weakness of data necessary for estimating some of the deterministic model's parameters and the use of spatial data having a sub-optimal resolution.

The proposed ensemble, being a supervised method, slightly improves the best model in the array of outputs (in the worst case, the enseble would have been equal to the best model in the array). The high performance showed by the RDS approach could be linked with the criterion used for the selection of the training and testing set of data. The possible presence of bias in using a similar technique for selecting the data and calculating the landslide susceptibility need to be further investigated. Because the quality of spatial landslides forecast is largely dependent on the quality of the available datasets, the good performance of the combined model broadens the possibility of applying a quantitative assessment in data poor regions. One of the main difficulties resides in determining how to combine the models for application to a specific area. In landslide susceptibility assessment this can be considered as an unresolved problem [46–48] in [41]. In other disciplines, for example meteorology, a long experience in combining multiple model forecasts already exist [49]. Exploiting ensemble analysis to obtain the most probable forecast is a common practice [41].

[4] `http://mastrave.org/doc/mtv_m/check_is#SAP_possibility`

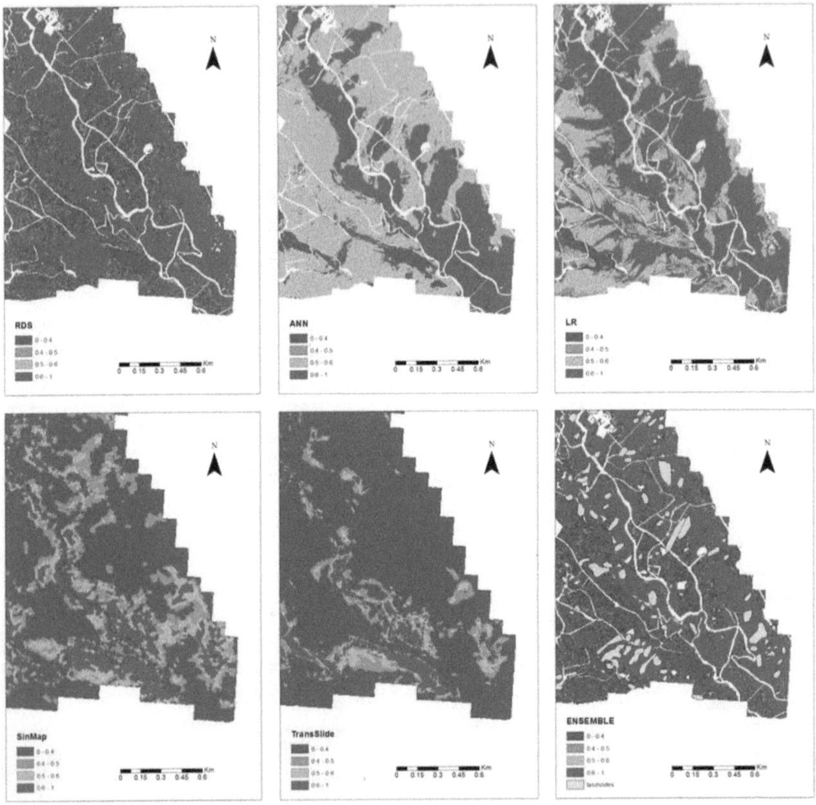

Fig. 2. Landslide susceptibility maps of the south-eastern section of the study catchment. The top three diagrams represent the statistical models outputs, the lower left two maps represent the deterministic models. The lower right diagram represents the output of the ensemble modelling.

4 Post-fire Regional Scale Rapid Assessment

European landscapes are affected by many disturbances [50,51]. Among them, wildfires have a major impact especially in southern Europe (Mediterranean region) [52]. Besides their direct environmental impact, wildfires can result in future secondary effects such as shallow landslides and debris flows [53,54].

Applying landslide susceptibility models at a regional scale, in order for the slope stability changes to be rapidly assessed in burnt areas, is a challenging task. Rapid assessment of landslide susceptibility requires fast computational modelling estimations. The availability – for a given spatial extent involving high wildfire risk – of known areas where landslides occurred and of areas of stability cannot be expected. Therefore, empirical models might not be best suited for this kind of analysis. Complex physically based models are also generally inapplicable. However, less data demanding physically based models might provide pre-

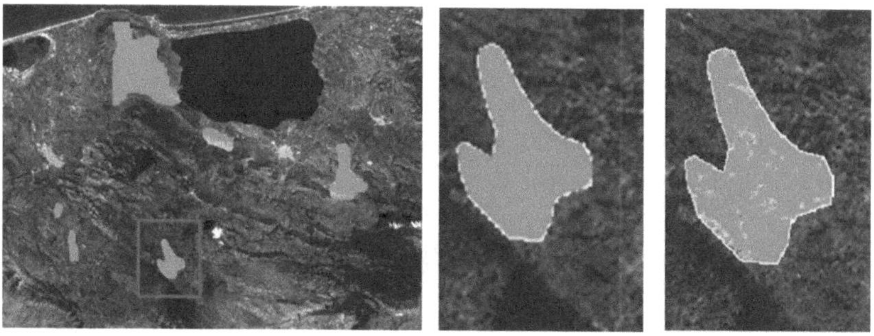

Fig. 3. Pre and Post-fire shallow landslide susceptibility in Northern Puglia (Italy). Ground layer: Google Earth, ©2013 Google.

liminary assessments of landslide susceptibility. Here, an area of approximately 600 km^2 was analysed calculating the changes in pre and post-fire landslide susceptibility. Six large fires were selected from the European Forest Fires Information System (EFFIS). An increased soil water content and a reduced root cohesion are the main factors inducing shallow sliding after fires [55, 56]. The SINMAP model was applied setting the model parameters for pre and post-fire condition on the basis of literature data [57,58]. Land cover, aggregate stability and saturated hydraulic conductivity, i.e. the post-fire parameters most susceptible to change significantly [57, 59], were calibrated into the models.

Comparing the pre and post-fire modelling results within the area shown in Fig. 3, indicates that almost 10% of the burnt area changes from stable to unstable conditions. The application of the SINMAP model in post-fire landslide susceptibility analysis has to be considered as a first attempt for applying these techniques at a regional scale. Future research will be carried out to extend the multi-model ensemble architecture from catchment to regional scale, overcoming the possible unavailability of observations on stable and unstable areas.

5 Conclusions

Five different techniques for modelling shallow landslide susceptibility were applied to a catchment located in Southern Italy (Rocchetta Sant'Antonio, FG). The stability maps were obtained applying two deterministic models (SINMAP and TransSlide) and three statistical methods (LR, RDS and ANN). A final map was prepared combining the output of the applied models using a robust fuzzy ensemble model. The performance of the models was evaluated against a landslide inventory of the year 2006. The good results of the ensemble model, when compared with the single techniques, make this method suitable to be applied in data poor regions with a lack of calibration and validation data. Because of the uncertainty in selecting a single suitable method for modelling spatial landslide susceptibility in areas characterized by data weakness, the applied ensemble method can potentially result in a less uncertain zonation. Alhough these preliminary results are promising, further research is required before this method can be used to communicate the findings with relevant authorities. The landslide susceptibility maps calculated applying the statistical methods were obtained using three different scripts [60] implemented using MATLAB and R languages respectively in GNU Octave and GNU R free software along with the modelling library Mastrave which implements the semantic array programming (SemAP) paradigm. The script applied for calculating the ANN is based on the work of Rossi [41]. Also the scripts used for selecting the training and testing points and for calculating the combined model [60] were written following the SemAP paradigm in MATLAB language, within GNU Octave as computing environment.

Acknowledgements. This paper is published with the permission of the Executive Director of the British Geological Survey and the support of the Maieutike Research Initiative.

References

1. van Westen, C.J., Rengers, N., Terlien, M.T.J., Soeters, R.: Prediction of the occurrence of slope instability phenomena through GIS-based hazard zonation. Geologische Rundschau 86, 404–414 (1997)
2. Brenning, A.: Spatial prediction models for landslide hazards: review, comparison and evaluation. Nat. Hazards Earth System Sci. 5(6), 853–862 (2005)
3. Aleotti, P., Chowdhury, R.: Landslide hazard assessment: summary review and new perspectives. Bulletin of Eng. Geology and the Env. 58, 21–44 (1999)
4. Suzen, M.L., Doyuran, V.: A comparison of the GIS based landslide susceptibility assessment methods: multivariate versus bivariate. Environmental Geology 45, 665–679 (2004)
5. Acharya, G.: Analysing the Interactions between Water-induced Soil Erosion and Shallow Landslides. Phd Thesis. University of Canterbury (2011)
6. Van Westen, C.J.: The modelling of landslide hazards using GIS. Surveys in Geophysics 21, 241–255 (2000)
7. Van Beek, L.P.H.: Assessment of the Influence of Change in Landslide and Climate on Landslide in Mediterranean Environment. PhD Thesis, Utrecht University, 363p. (2002)
8. Chen, H., Lee, C.F.: A dynamic model for rainfall-induced landslides on natural slopes. Geomorphology 51(4), 269–288 (2003)
9. Van Westen, C.J.: Geo-information tools for landslide risk assessment an overview of recent developments. In: Proceed. of the 9th Int. Symposium on Landslides, Evaluation and Stabilization, Rio de Janeiro, pp. 39–56 (2004)
10. Van Westen, C.J., Van Asch, T.W.J., Soeters, R.: Landslide hazard and risk zonation; why is it still so difficult. Bulletin of Engineering geology and the Environment 65(2), 167–184 (2006)
11. Foster, C., Harrison, M., Reeves, H.J.: Standard and methods of hazard assessment for mass-movements in Great Britain. Journal for Torrent and Avalanche Control 166 (2011)
12. Dai, C., Lee, C.F.: Landslide characteristics and slope instability modeling using GIS Lantau Island, Hong Kong. Geomorphology 42, 213–238 (2002)

13. Dijkstra, T.A., Dixon, N.: Climate change and slope stability in the UK: challenges and approaches. Quarterly Journal of Engineering Geology and Hydrogeology 43(4), 371–385 (2010)
14. Lee, C.F., Ye, H., Yeung, M.R., Shan, X., Chen, G.: AIGIS-based methodology for natural terrain landslide susceptibility mapping in hong kong. Episodes 24(3), 150–158 (2001)
15. Karsli, F., Atasoy, M., Yalcin, A., Reis, S., Demir, O., Gokceoglu, C.: Effects of land-use changes on landslides in a landslide-prone area (ardesen, rize, NE turkey). Environ. Monit. Assess. 156(1-4), 241–255 (2009)
16. de Rigo, D., Bosco, C.: Architecture of a Pan-European Framework for Integrated Soil Water Erosion Assessment. In: Hřebíček, J., Schimak, G., Denzer, R. (eds.) ISESS 2011. IFIP AICT, vol. 359, pp. 310–318. Springer, Heidelberg (2011)
17. Winter, M.G., Dixon, N., Wasowski, J., Dijkstra, T.A.: Introduction to land-use and climate change impacts on landslides. Quarterly Journal of Engineering Geology and Hydrogeology 43(4), 367–370 (2010)
18. de Rigo, D.: Behind the horizon of reproducible integrated environmental modelling at European scale: ethics and practice of scientific knowledge freedom. *F1000 Research*, submitted (exp. 2013)
19. de Rigo, D., Rodriguez-Aseretto, D., Bosco, C., Di Leo, M., San-Miguel-Ayanz, J.: An Architecture for Adaptive Robust Modelling of Wildfire Behaviour under Deep Uncertainty. In: Hřebíček, J., Schimak, G., Kubásek, M., Rizzoli, A. (eds.) ISESS 2013. IFIP AICT, vol. 413, pp. 367–380. Springer, Heidelberg (2013)
20. Di Leo, M., de Rigo, D., Rodriguez-Aseretto, D., Bosco, C., Petroliagkis, T., Camia, A., San-Miguel-Ayanz, J.: Dynamic Data Driven Ensemble for Wildfire Behaviour Assessment: A Case Study. In: Hřebíček, J., Schimak, G., Kubásek, M., Rizzoli, A. (eds.) ISESS 2013, vol. 413, pp. 11–22. Springer, Heidelberg (2013)
21. Iovine, G., Parise, M., Crescenzi, E.: Analisi della franosita nel settore centrale dell Appennino Dauno. Mem. soc. Geol. Ita. 51, 633–641 (1996)
22. Magliulo, P., Di Lisio, A., Russo, F., Zelano, A.: Geomorphology and landslide susceptibility assessment using GIS and bivariate statistics: A case study in southern Italy. Natural Hazards 47(3), 411–435 (2008)
23. Wasowski, J., Lamanna, C., Gigante, G., Casarano, D.: High resolution satellite imagery analysis for inferring surface-subsurface water relationship in unstable slopes. Remote sensing of Environment 124, 135–148 (2012)
24. Wasowski, J., Casarano, D., Lamanna, C.: Is the current landslide activity in the Daunia region (Italy) controlled by climate or land use change. In: Proceedings of the International Conference on Landslides and Climate Change, Ventor, Isle of Wight, UK, vol. 124, pp. 41–49. Taylor and Francis, Balkema, London, Rotterdam (2007)
25. Mossa, S., Capolongo, D., Pennetta, L., Wasowski, J.: A GIS-based assessment of landsliding in the Daunia Apennines, southern Italy. In: Graniczny, M., Czarnogorska, M., et al. (eds.) Proceedings of the International Conference Mass Movement Hazard in Various Environments, Krakow, Poland, vol. 20, pp. 86–91. Polish Geological Institute, Warsaw (2005)
26. Wasowski, J., Lamanna, C., Casarano, D.: Influence of land-use change and precipitation patterns on landslide activity in the Daunia Appennines. Quarterly Journal of Engineering Geology and Hydrogeology 43, 387–401 (2010)
27. Bosco, C., de Rigo, D., Dewitte, O., Poesen, J., Panagos, P.: Modelling Soil Erosion at European Scale. Towards Harmonization and Reproducibility (in prep.)

28. de Rigo, D., Guariso, G.: Rewarding Open Science: A Collaborative Review System for Semantically-Enhanced Free Software and Environmental Data Modelling. IFIP AICT (submitted)
29. de Rigo, D., Barredo, J.I., Busetto, L., Caudullo, G., San-Miguel-Ayanz, J.: Continental-Scale Living Forest Biomass and Carbon Stock: a Robust Fuzzy Ensemble of IPCC Tier 1 Maps for Europe. In: Hřebíček, J., Schimak, G., Kubásek, M., Rizzoli, A. (eds.) ISESS 2013. IFIP AICT, vol. 413, pp. 271–284. Springer, Heidelberg (2013)
30. de Rigo, D.: Software Uncertainty in Integrated Environmental Modelling: the role of Semantics and Open Science. Geophys. Res. Abstr. 15, 13292+ (2013)
31. de Rigo, D., Corti, P., Caudullo, G., McInerney, D., Di Leo, M., San Miguel-Ayanz, J.: Toward Open Science at the European Scale: Geospatial Semantic Array Programming for Integrated Environmental Modelling. Geophys. Res. Abstr. 15, 13245+ (2013)
32. de Rigo, D.: Semantic array programming for environmental modelling: Application of the mastrave library. In: International Environmental Modelling and Software Society (iEMSs) 2012 International Congress on Environmental Modelling and Software. Managing Resources of a Limited Planet: Pathways and Visions under Uncertainty, Sixth Biennial Meeting, pp. 1167–1176 (2012)
33. de Rigo, D.: Semantic Array Programming with Mastrave - Introduction to Semantic Computational Modelling. The Mastrave Project (2012)
34. Pack, R.T., Tarboton, D.G., Goodwin, C.N.: The SINMAP Approach to Terrain Stability Mapping. In: 8th Congress of the International Association of Engineering Geology, Vancouver, British Columbia, Canada (1998)
35. Pack, R.T., Tarboton, D.G., Goodwin, C.N., Prasad, A.: SINMAP 2. A Stability Index Approach to Terrain Stability Hazard Mapping, technical description and users guide for version 2.0. Utah State University (2005)
36. Cox, D.R.: The Regression Analysis of Binary Sequences. Journal of the Royal Statistical Society. Series B (Methodological) 20, 215–242 (1958)
37. Yilmaz, I.: Landslide susceptibility mapping using frequency ratio, logistic regression, artificial neural networks and their comparison: a case study from Kat landslides (Tokat Turkey). Comput. Geosci. 35(6), 1125–1138 (2009)
38. Garrett, J.: Where and why artificial neural networks are applicable in civil engineering. J. Comput. Civil Eng. 8, 129–130 (1994)
39. Pradhan, B., Lee, S.: Landslide risk analysis using artificial neural network model focusing on different training sites. International Journal of Physicical Sciences 4(1), 1–15 (2009)
40. Lee, S., Ryu, J.H., Kim, I.S.: Landslide susceptibility analysis and its verifica tion using likelihood ratio, logistic regression, and artificial neural network models: case study of Youngin, Korea. Landslide 4, 327–338 (2007)
41. Rossi, M., Guzzetti, F., Reichenbach, P., Mondini, A.C., Peruccacci, S.: Optimal landslide susceptibility zonation based on multiple forecast. Geomorphology 114, 129–142 (2010)
42. Beven, K.J., Kirkby, M.J.: A physically based variable contributing area model of basin hydrology. Hydrological Sciences Bulletin 24, 43–69 (1979)
43. Witt, A.C.: Using a GIS (Geographic Information System) to Model Slope Instability and Debris Flow Hazards in the French Broad River Watershed, p. 165. Carolina State University, North Carolina (2005)
44. Morgan, R.P.C., Quinton, J.N., Smith, R.E., Govers, G., Poesen, J.W.A., et al.: The European Soil Erosion Model (EUROSEM): Documentation and User Guide. Silsoe College, Cranfield University (1998)

45. Horn, R., Fleige, H.: A method for assessing the impact of load on mechanical stability and on physical properties of soils. Soil. Till. Res. 73(1-2), 89–99 (2003)
46. Guzzetti, F., Cardinali, M., Reichenbach, P., Carrara, A.: Comparing landslide maps: a case study in the upper Tiber River Basin, central Italy. Environmental Management 25, 247–363 (2000)
47. Guzzetti, F.: Landslide Hazard and Risk Assessment. Ph.D. Thesis, Mathematisch-Naturwissenschaftlichen Fakultät der Rheinischen Friedrich-Wilhelms-Universität, University of Bonn, Germany, p. 389 (2006)
48. Guzzetti, F., Reichenbach, P., Ardizzone, F., Cardinali, M., Galli, M.: Estimating the quality of landslide susceptibility models. Geomorphology 81, 166–184 (2006)
49. Collins, M., Knight, S.: Ensembles and probabilities: a new era in the prediction of climate change. Philosophic. Transact. of the Royal Soc. A, 1471–2962 (2007)
50. Bengtsson, J., Nilsson, S.G., Franc, A., Menozzi, P.: Biodiversity, disturbances, ecosystem function and management of European forests. Forest Ecology and Management 132(1), 39–50 (2000)
51. Seidl, R., Fernandes, P.M., Fonseca, T.F., Gillet, F., Jnsson, A.M., Merganiov, K., et al.: Modelling natural disturbances in forest ecosystems: a review. Ecological Modelling 222(4), 903–924 (2011)
52. Rodriguez-Aseretto, D., de Rigo, D., Di Leo, M., Cortes, A., San-Miguel-Ayanz, J.: A data-driven model for large wildfire behaviour prediction in Europe. Procedia Computer Science 18, 1861–1870 (2013)
53. Shakesby, R.A., Doerr, S.H.: Wildfire as a hydrological and geomorphological agent. Earth Sci. Rev. 74, 269–307 (2006)
54. Cannon, S.H., Gartner, J.E.: Wildfire-related debris flow from a hazards perspective. In: Debris-flow Hazards and Related Phenomena, pp. 363–385. Springer, Heidelberg (2005)
55. Swanston, D.N.: Principal soil movement processes influenced by roadbuilding, logging and fire. In: Proceedings of a Symposium: Forest Land Uses and Stream Environment, pp. 29–40. Oregon State University, Corvallis (1971)
56. Swanson, F.J.: Fire and Geomorphic Processes. In: Gen. Tech. Rep. USDA For. Serv. WO-26. pp. 401–420, Washington DC (1981)
57. Neary, D.G., Klopatek, C.C., DeBano, L.F., Ffolliott, P.F.: Fire effects on belowground sustainability: A review and synthesis. For. Ecol. Manage. 122, 51–71 (1999)
58. Blesius, L., Weirich, F.: The use of high-resolution satellite imagery for deriving geotechnical parameters applied to landslide susceptibility. In: ISPRS Hannover Workshop (2009)
59. Humphreys, F.R., Craig, F.G.: Effects of fire on soil chemical,structural and hydrological properties. In: Gill, A.M., et al. (eds.) Fire and the Australian Biota, Canberra. Aust. Acad. Sci, ch. 8, pp. 177–200 (1981)
60. de Rigo, D. and Bosco, C.: A multi model robust ensembling architecture for estimating landslide susceptibility. Free Software and Semantic Array Programming Research (in prep., 2013)

RainPortal – A Web Portal for Providing Climate Change Related Precipitation Data Using SUDPLAN Services

Peter Kutschera[1], Jonas Olsson[2], and Lars Gidhagen[2]

[1] Austrian Institute of Technology – AIT, Austria
Peter.Kutschera@ait.ac.at
[2] Swedish Meteorological and Hydrological Institute – SMHI, Sweden
{Jonas.Olsson,Lars.Gidhagen}@smhi.se

Abstract. RainPortal (http://sudplan.ait.ac.at/RainPortalServer) is a web portal that has three main targets: 1) Provide access to scientifically sound data related to some key climate change aspects in Europe, 2) Provide concrete data about climate change related local precipitation changes by using climate model output and 3) Easy usage and incorporation of the provided information into existing planning workflows.

Keywords: climate change, mitigation, precipitation, IDF, SUDPLAN, Delta Change.

1 SUDPLAN

The SUDPLAN (Sustainable Urban Development Planner for Climate Change Adaptation) project ([2] and http://sudplan.eu), which ran from 2010 to 2012, produced a set of services that provide climate change related information, including services to carry out projections of historical local data provided by a user. These projections to the future up to the year 2100 can be done using different assumptions about expected climate changes. These services are based on OGC standards and are accomplished by a decision support system [7] as a frontend. This frontend, called "Scenario Management System" (SMS), targets the whole planning process for cities, including management of resources. While this works well and brings a lot of climate change services to the planner's desk it also adds the requirement to get familiar with a complex tool.

2 Calculation and Usage of IDF Data

IDF is the acronym for Intensity, Duration, and Frequency. Based on measured or calculated high-resolution precipitation data over at least 30 years IDF curves are created using different standardized statistical methods [1]. These curves describe, for a given recurrence frequency (often also named return period) and rain event duration, the maximum precipitation intensity to be expected.

J. Hřebíček et al. (Eds.): ISESS 2013, IFIP AICT 413, pp. 336–344, 2013.

The intensity can be expressed in mm rain per event duration or in mm per time step – SUDPLAN services uses mm/h. A typical IDF curve is shown in Fig. 1, featuring for different frequencies (1, 10, 100 years, color coded) the relation between duration (up to 10 hours, to the right) and intensity (up to 100 mm per hour, to the top). As an example, this curve shows that for a 2h rainfall period the strongest intensity to expect every year is about 10 mm/h, i.e., in total 20 mm during the event. But over a 10 year period the maximum intensity to expect is 20 mm/h, and in a 100 year period even more than 30 mm/h has to be expected.

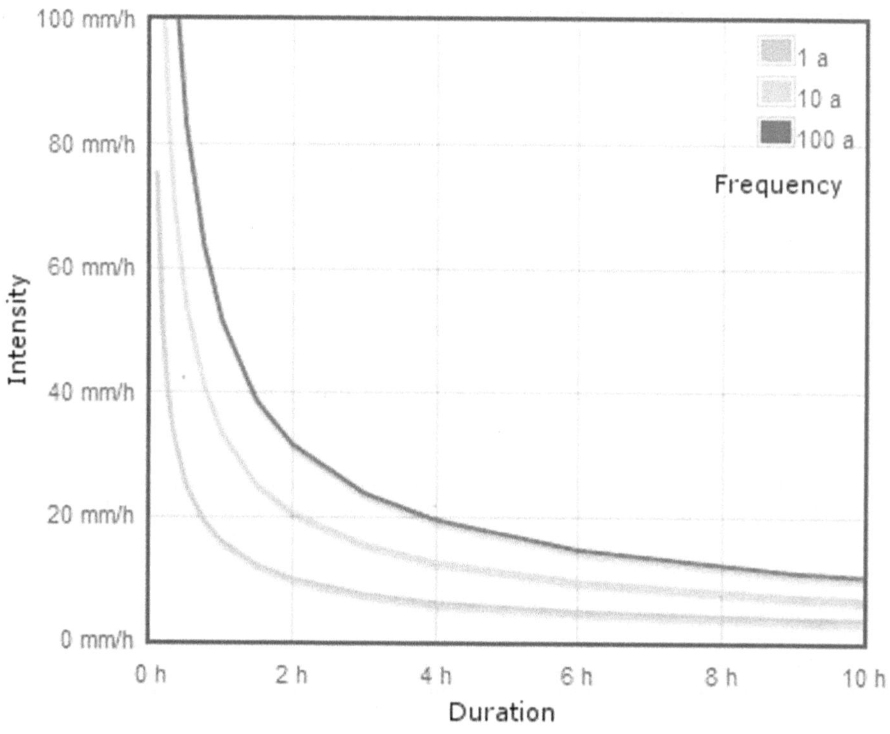

Fig. 1. Typical IDF curves

The most prominent usage of these curves is in the dimensioning of infrastructure and buildings for drainage of rain water from a given area. One of the SUDPLAN pilots ([5, 6]) considered problems related with surface water runoff in the city of Wuppertal.

In general, recommended IDF curves are available in most European countries. In Austria we are in the lucky position that one of our ministries (Ministry of Life; http://lebensministerium.at) provides this information free of charge (http://ehyd. gv.at/).

3 Why Take Climate Change into Account?

The actual planning process is based on the knowledge of past conditions. So in the case of planning taking IDF data into account, these actual data are based on statistics about rain conditions over the last 30 years. From this database a sound statement can be given about what type of rain events to expect.

Thus the planning for buildings and infrastructure that should last some 10 up to 100 years is based on historical data. This leads to the question if these historical statistics will still be valid in the future. How much will climate change influence and invalidate the IDF data used? But this question cannot be answered on a global level; local conditions need to be taken into account.

4 IDF Changes from Different Climate Projections

Future precipitation from global (GCM) and regional climate models (RCM) indicates a large variability. This suggests combining outputs from various models to get a more "robust" estimate of what the future precipitation will be [9, 10]. Currently the SUDPLAN rainfall downscaling is performed with an ensemble of four future projections from one RCM representing different GCMs and IPCC emission scenarios [10]. Updates with more projections are expected during 2013.

Existing results from RCMs do not accurately reproduce short-term intensity variations (intense short-lived rainfall events) nor small-scale local variations related to, e.g., topography. The Delta Change method used for further downscaling in SUDPLAN is a way to overcome these limitations and provide expected changes of present rainfall characteristics for a specific future 30-year period (up to period 2071-2100). In the Delta Change method, historically observed rainfall or IDF data is modified to reflect expected future conditions, based on statistical analyses of climate model results. For details about the approach used in SUDPLAN, see further [3].

SUDPLAN's services thus require local historical data to be projected to the future. The usage of local data provides actual local conditions to the result. The influence of future climate change is highly uncertain as there are a lot of different climate models and global emission scenarios available [9, 10]. Today, no one can tell which of them provides the best approximation - or if the future conditions will be outside all the bandwidth of all model results. This is the reason why SUDPLAN services provide results for an increasing list (Table 1) of emission scenarios and global as well as regional climate model combinations. These combinations are referenced as "climate projections" and are described in detail on the SUDPLAN website [8]. Table 2 describes the main characteristics of the used emission scenarios.

Using all these climate projections will result in a broad range of possible results. This range gives the user an impression of the bandwidth of conditions to expect in the future. Subsequently, it is still up to the planner to decide based on his knowledge and experience – but now this is based on scientifically sound estimations of expected future conditions.

Table 1. Actual available combination of emission scenarios and climate models

IPCC scenario	Global Model	Regional Model
A1B	ECHAM5	RCA3
A2	ECHAM5	RCA3
A1B	HADCM3	RCA3
A1B	CCSM3_A1B	RCA3

Table 2. Main assumptions of the used IPCC emission scenarios

IPCC scenario	Main assumptions
A1B	• Rapid economic growth. • A global population that reaches 9 billion in 2050 and then gradually declines. • The quick spread of new and efficient technologies. • A convergent world - income and way of life converge between regions. Extensive social and cultural interactions worldwide. • A balanced emphasis on all energy sources.
A2	• A world of independently operating, self-reliant nations. • Continuously increasing population. • Regionally oriented economic development

5 RainPortal

In contrast to the complex GUI ([7]) developed in SUDPLAN, RainPortal targets only the projection of precipitation data with the explicit goal of providing a simple user interface. The idea is that a planner already has his tools to do his job in terms of, e.g., hydrologic or hydraulic modeling. Therefore, the RainPortal only provides additional model input information in a format that can easily be used in the established planning process. The first version of the RainPortal allows only the projection of IDF (Intensity-Duration-Frequency) data.

To use the RainPortal, historical IDF data are needed as a table. Different formats can be used. The 3-column format contains the columns *duration* (in minutes), *frequency* (in years) and *intensity* (either in mm per event duration or in mm per hour) in this order (Fig. 2).

	A	B	C
1	Duration [minutes]	Frequency [years]	Intensity [mm]
2	5	1	6,30
3	10	1	8,40
4	15	1	9,90
5	5	2	8,40
6	10	2	11,10

Fig. 2. IDF data in 3 column format

Another format is one column per frequency. The units to use are the same as before (Fig. 3).

As RainPortal was developed in Austria there exists also the option to use the data format as downloaded from the Austrian ministry.

	A	B	C	D	E
1		1	2	3	5
2	5	6,3	8,4	9,5	11,1
3	10	8,4	11,1	12,8	14,9
4	15	9,9	13,2	15,2	17,7
5	20	11	14,7	16,9	23,4
6	30	12,7	17	19,6	22,7

Fig. 3. IDF data in one column per frequency format

IDF data are always based on a rain time series of a specific location and are therefore only representative for specific coordinates. These coordinates need to be provided in the WGS84 reference system.

The input data need to represent at least a historical period of 30-years. As the data from the whole time span are used, the year in the center of this period is taken as time reference. This year has also to be specified.

The first step using the RainPortal is loading the IDF curves into the browser (Fig. 4).

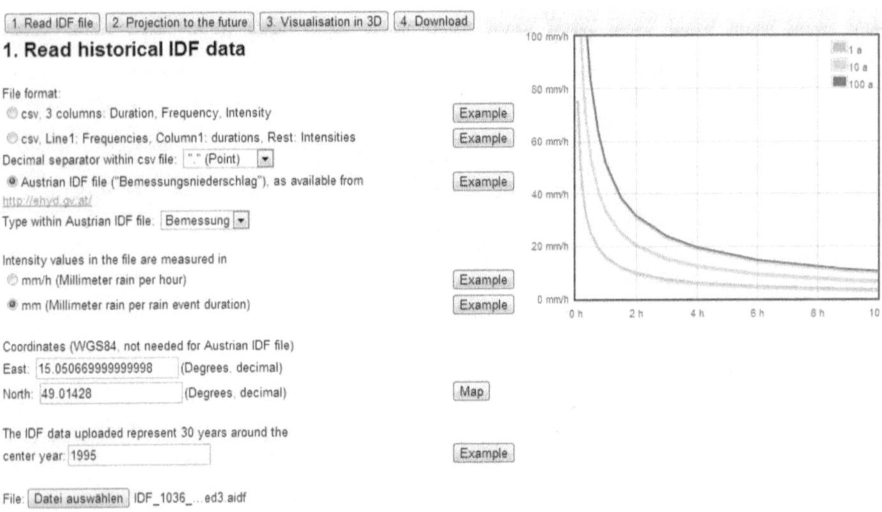

Fig. 4. Loading local IDF curves

The loaded data are shown on the web page as 2D graph and as a table immediately allowing checking.

The second step is the specification of the target period, this is the period used to project the historical data to. Again the center year of a 30-year period has to be specified. The service can perform calculations between years 2011 and 2100, which means a center year in the interval 2026 to 2085.

The service can provide results for different climate projections. But for practical use the option to calculate the minimum and maximum expected values over all climate projections would be most useful.

Pressing the "Start" button will send the input data to the server, triggering the calculations on SHMIs server (Fig. 5), which will take some minutes.

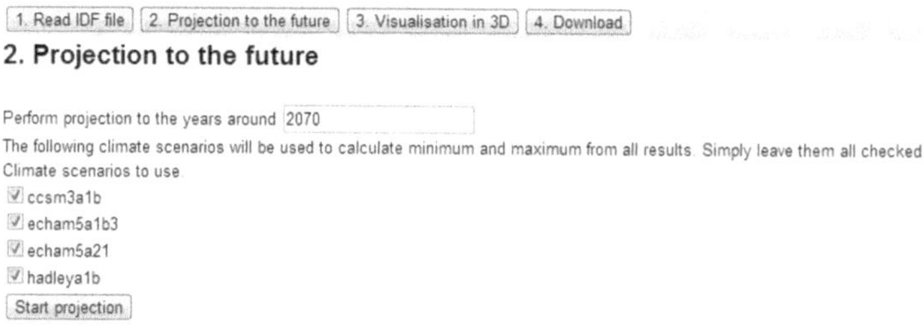

Fig. 5. Requesting projection to the future

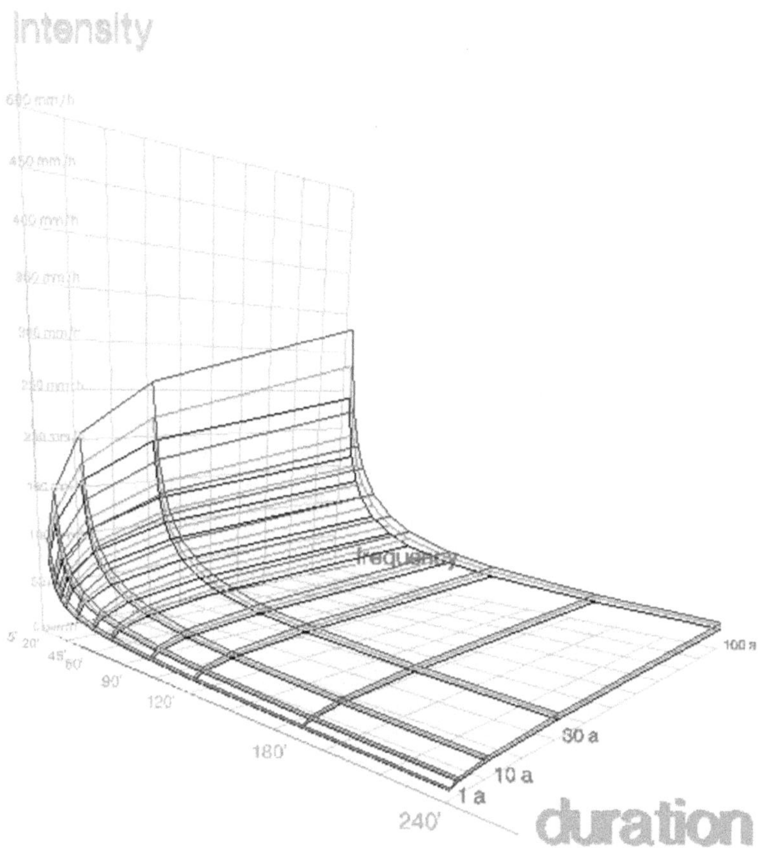

Fig. 6. 3D visualization of results

In the third step, the historical data as well as the projections can be examined in a 3D graph (Fig. 6) that can be moved, rotated and zoomed interactively to visualize different aspects. The graph compares historical data (black) with the minimal (green) and maximal (red) values from all calculated climate projections. This example shows a general increase in the intensities to expect.

The same values are available as table (Fig. 7). Additionally, the result for each single climate projection is available. It should be mentioned that it is not clear in advance which climate projection will deliver the maximum values. This is even true within one IDF table, where sometimes one projection delivers the highest intensities for lower frequencies, and another projection for the higher frequencies.

Duration (Minutes)	Scenario	Frequencies (Years)				
		1	3	10	30	100
		Intensities (mm/h)				
5	historic, 1995	75.60	114.0	157.2	195.6	240.0
	ccsm3a1b, 2070	85.10	128.6	179.1	228.9	306.8
	minimum	83.70	126.4	175.1	220.6	282.2
	maximum	98.60	148.8	205.9	258.9	329.2
	hadleya1b, 2070	98.60	148.8	205.9	258.9	329.2
	echam5a21, 2070	83.70	126.4	175.1	220.6	282.2
	echam5a1b3, 2070	93.60	141.1	194.4	240.9	291.3
10	historic, 1995	50.40	76.80	106.2	132.6	162.6
	ccsm3a1b, 2070	56.70	86.70	121.0	155.1	207.8
	minimum	55.80	85.20	118.3	149.5	191.2
	maximum	65.70	100.2	139.1	175.5	223.0
	hadleya1b, 2070	65.70	100.2	139.1	175.5	223.0
	echam5a21, 2070	55.80	85.20	118.3	149.5	191.2
	echam5a1b3, 2070	62.40	95.10	131.3	163.3	197.4
15	historic, 1995	39.60	60.80	84.00	105.6	129.2

Fig. 7. Results

The fourth and last step is to download the result. This can be done in different formats to ease the integration of the results into the established planning process.

6 Privacy and Data Security

RainPortal is not interested in storing any data about the user. Therefore, the easiest option would be to have the service and all results available to every user without the need for any login. However, this would imply that the results cannot be assigned to a specific user and the uploaded data and results will be visible to everyone. This could raise a problem, as the user might not have the right to publish the data.

Because of this a user management is needed in RainPortal, allowing each user to see only their results. The input data and results will be available to the user for visualization and download for some days and will be deleted afterwards to avoid resource consumption on the server. RainPortal will only keep some statistics and comments provided by the users.

7 Concluding Remarks

We, the developers of RainPortal, have two goals in mind: 1) The provision of climate change related data to the community and 2) We want to learn about who is this community, how do they use the data, what are their needs and how can we make RainPortal and SUDPLAN as a whole an even better product. So we are asking every user for feedback.

Depending on the success of the RainPortal, future versions might also allow projections of rain timeseries. In contrast to IDF curves, this provides information about changes in overall precipitation volume and changed rain patterns between the yearly seasons.

A key issue is to make the user fully aware of the uncertainties and limitations in the estimated future rain intensities. Local short-term rainfall is notoriously difficult to simulate by climate models, due to the high variability in time and space. In addition, IDF calculations require extreme value analyses that add further uncertainty to the estimated future changes. The final results thus must be considered as very approximate indications of what is likely to happen in the future; which is the probable sign of the change (increase or decrease) and which is the probable magnitude of the change (small or large). In RainPortal, it is strongly recommended to evaluate all available projections to see the range of possible outcomes and on that basis make a qualitative interpretation of the probable change. By combining this with a careful "climate sensitivity analysis" of the system under study (e.g., by identifying rain thresholds at which system response dramatically changes), we hope to provide a useful and sound basis for climate change impact assessment.

References

1. DWA-A 531: Arbeitsblatt DWA-A 531, Starkregen in Abhängigkeit von Wiederkehrzeit und Dauerregen, DWA (2012)
2. Gidhagen, L., Denzer, R., Schlobinski, S., Michel, F., Kutschera, P., Havlik, D.: Sustainable urban Development Planner for Climate Change Adaptation (SUDPLAN). In: Proceedings of ENVIP 2010 Workshop at EnviroInfo 2010, October 6-8. CEUR-WS, vol. 679 (2010) ISSN 1613-0073, urn:nbn:de:0074-679-9

3. Olsson, J., Gidhagen, L., Gamerith, V., Gruber, G., Hoppe, H., Kutschera, P.: Downscaling of Short-Term Precipitation from Regional Climate Models for Sustainable Urban Planning. Sustainability 4(5), http://www.mdpi.com/2071-1050/4/5/866
4. Kutschera, P., Olsson, J., Havlik, D., Gruber, G.: Future IDF curves for regional planning in Europe - a SUDPLAN result. Geophysical Research Abstracts, EGU General Assembly (2012), http://meetingorganizer.copernicus.org/EGU2012/EGU2012-9483.pdf
5. Michel, F., Steffen, D., Schlobinski, S., Sander, S.: Considering the Impact of Future Climate Change on the Resilience of a City – Surface Run-Off due to Heavy Storm Events in the City of Wuppertal. In: RealCORP 2012 – Proceedings 17th International Conference on Urban Planning and Regional Development in the Information Society (2012)
6. Sander, S., Hoppe, H., Schlobinski, S.: Integrating climate change in the urban planning process – A case study. In: Hřebíček, J., Schimak, G., Denzer, R. (eds.) ISESS 2011. IFIP AICT, vol. 359, pp. 631–640. Springer, Heidelberg (2011)
7. Schlobinski, S., Denzer, R., Frysinger, S., Güttler, R., Hell, T.: Vision and requirements of scenario-driven environmental decision support systems supporting automation for end users. In: Hřebíček, J., Schimak, G., Denzer, R. (eds.) ISESS 2011. IFIP AICT, vol. 359, pp. 51–63. Springer, Heidelberg (2011)
8. Climate scenario descriptions on the SUDPLAN website, http://www.sudplan.eu/About-SUDPLAN/Pan-European-input-data/Climate-scenarios
9. Willems, P., Olsson, J., Arnbjerg-Nielsen, K., Beecham, S., Pathirana, A., Bulow Gregersen, I., Madsen, H., Nguyen, V.-T.-V.: Impacts of Climate Change on Rainfall Extremes and Urban Drainage Systems, 238 p. (2012) ISBN: 9781780401256
10. Solomon, S., Qin, D., Manning, M., Chen, Z., Marquis, M., Averyt, K.B., Tignor, M., Miller, H.L. (eds.): IPCC, 2007: Climate Change 2007: The Physical Science Basis. Contribution of Working Group I to the Fourth Assessment Report of the Intergovernmental Panel on Climate Change. Cambridge University Press, Cambridge (2007)

Crowdtasking – A New Concept for Volunteer Management in Disaster Relief

Georg Neubauer[1], Andrea Nowak[1], Bettina Jager[1], Christian Kloyber[3], Christian Flachberger[2], Gerry Foitik[3], and Gerald Schimak[1]

[1] AIT Austrian Institute of Technology GmbH
{Georg.Neubauer,Andrea.Nowak,Bettina.Jager, Gerald.Schimak}@ait.ac.at
[2] Frequentis AG
[3] Austrian Red Cross

Abstract. Based on governmental institutions and strong volunteer organizations, Austria provides a comprehensive and well developed emergency response system. As an important factor for the maintenance of the current quality standard of Austria's protection and emergency system, the further engagement of voluntariness has to be ensured and enhanced in the light of the ongoing societal change. On the one hand, the involvement of new media provides opportunities to expand the organizations' service portfolio to create a broader participation for citizens' engagement; on the other hand, long lasting and formal memberships are often a challenge for the current dynamic lifestyle. To face this situation, the involvement of new media services for volunteer management in order to enable new modes of voluntary binding is a promising strategically effort. A new process called "crowdtasking" dedicated to the improvement of volunteer management applying new media is discussed; new processes of volunteer management are presented by exemplary initiatives of humanitarian non-governmental organizations, such as the Austrian Red Cross.

Keywords: crisis and disaster management, volunteer management, humanitarian non-governmental organizations, new media, crowdtasking.

1 Introduction

A major concern of current crisis and disaster management approaches is to enhance the resilience of all involved actors such as authorities, emergency services and mainly the affected population in the various phases of the disaster life cycle. As defined by the International Federation of Red Cross and Red Crescent Societies a resilient society is characterized through the abilities to anticipate, manage and analyze their own risks and the development of new procedures to prevent crisis or reduce the damage caused by crisis and/or disasters [1]. In Austria, the National Crisis and Disaster Protection Management (SKKM) is guided by the principles of self-protection on local areas and subsidiarity on higher levels of public authority [2]. Based on governmental institutions and strong volunteer organizations, Austria provides a comprehensive emergency response system. As an important factor for the maintenance of the current quality

J. Hřebíček et al. (Eds.): ISESS 2013, IFIP AICT 413, pp. 345–356, 2013.

standard of Austria's civil protection and emergency response system, the further engagement of voluntariness has to be ensured and promoted in the light of the debate on the societal change. On the one hand, societal change accounts for changing volunteering patterns: long lasting and formal memberships nowadays appear as hardly compatible to the more individualistic and flexible lifestyles taking place particularly in urban environments. On the other hand, it provides opportunities for new and advanced forms of citizen's participation by developing a multifaceted range of voluntary engagement. In the order of preparedness measures, the recruitment of volunteers to mobilize them in the case of a hazardous event is a significant factor to match the requirements of crisis and disaster management. With the term readiness, which describes the ability of an affected society to quickly and appropriately respond when required, the time-critical dimension of crisis and disasters might be expressed very well [3]. Therefore, current disaster risk reduction strategies aim to foster the resilience of the public while at the same time relieving emergency and humanitarian organizations from an uncoordinated influx of information or spontaneous volunteers. As one strategically effort, complementary offers of voluntary engagement is envisaged by humanitarian non-governmental organizations, e.g. the Red Cross, to create loose forms of organization and providing a low-threshold participation offers for potential volunteers. A promising approach to enable such new modes of voluntary binding is the use of new media services for recruiting and mobilizing volunteers.

2 Traditional Voluntary Engagement

In Austria, voluntarism can be regarded as a main pillar of the social system, with volunteers providing services in manifold areas such as education, sports, culture, politics or disaster relief – the latter being the main scope of our paper. While the concept of voluntarism is often bound to national traditions and differing cultural contexts a generic aspect of voluntarism is that the actions provided by volunteers are undertaken of their own choice and motivation without concern for financial gain, and they benefit the individual volunteer, communities and society as a whole [4]. For this paper we refer to broad definitions of voluntarism which address both formal and informal forms of volunteering regardless of their duration (full-time / part-time) or their locality (at home / abroad). While formal voluntary engagement is provided within an institutionalized context (organization, association or other institution), informal forms of volunteering are triggered by individual engagement beyond the context of an organization and are often referred to as "neighborly help". For many years dedicated disaster relief actors in Austria, such as fire brigade or Red Cross have relied on formal forms of volunteering. Thus the typical Austrian volunteer is between the age of 30 and 50, married, Austrian citizen, employed or self-employed and volunteering within an organization [5]. Especially in the younger parts of society as well as on the upper scope (over 50) there is still great and partly un-addressed potential. With societal change mainly younger generations are less willing to bind themselves to one single organization for a lifetime, but more eager to help in a flexible and task based way, where they can see the results of their doing in their immediate area of life. A phenomenon of informal volunteering particularly specific for disaster relief organizations are convergent volunteers: They are individuals or groups that spontaneously offer their help in the aftermath of a disaster or an emergency and may arrive unsolicited at the scene. Convergent volunteers have not been specially recruited and are not affiliated to the existing emergency management system. They provide service without additional or specific compensation [6]. Just as younger parts of Austrian society seek for more flexible forms of voluntary engagement, convergent volunteers also want to temporarily provide their help in case of an emergency without affiliating to traditional volunteering response organizations. For the response system on the other side, they are still hard to address and pose a major challenge. While not all European response systems may rely on volunteers and must subsequently address societal change, convergent volunteering is a more universal pattern, analogous to more altruistic behaviors like neighborly help, why its relevance is not confined to single countries but is at least a European one [7].

3 Existing Solutions

There are various examples of existing solutions for volunteer management; mainly in the Anglo-American area various forms of voluntary engagement have been already established. Different approaches to mobilize volunteers by non-conventional procedures will be discussed below to identify starting-points for the Austrian use case. In 2007, an initiative called Team Österreich[1] was established in Austria. Those people willing to become a member of the team have to undergo an extensive online registration process - having an e-mail address and a cell phone is compulsory. After creating an initial profile they have to fill in their personal data, their knowledge and skills and must state what kind of contributions they wish to make (e.g. physical work, administrative help, and contributions only in case of disaster). After a training course they become an active member in case-specific events. By April 2013 about 30.000 people registered in this database and became members of the Team Österreich. This form of coordinating and mobilizing volunteers has several positive effects: The number of people potentially willing to help (potential convergent volunteers), their skills and their location are known to emergency managers. Following the principle of neighborly help, local people respond to local disasters. The knowledge and skills of the members become accessible and usable – as long as they are specified in the database. Team Österreich follows a "bring your own skills"- principle with members bringing in their individual skills and knowledge and only getting a basic training on alarming procedure and deployment. If an event occurs, the alerting and communication process is initiated by an alerting-SMS and an e-mail with detailed information about the emergency mission. Members themselves decide whether they want to be on the mission by replying either with "yes" or "no". Their answers are automatically processed and positive requests are immediately seen by Red Cross coordinators. For their time on mission, members have insurance coverage like regular Red Cross volunteers and stand under commandment and supervision of an emergency service officer. Furthermore people are only alarmed when they are really needed – conversely when there is need they will be solicited according to their profile and proximity to the scene. In this way Team Österreich minimizes the convergence to a disaster site,

[1] `http://apps.teamoesterreich.at/`

while transforming the potential of voluntary work into a low-threshold structure of coordination and mobilization for those people willing to help only in case of emergency but not wanting to become regularly involved in a voluntary response organization. The projects Team CP[2] and SAVE.regions[3], both launched by the Red Cross are aimed at enhancing the societal resilience. The project Team CP places emphasis on strengthening the voluntary engagement by IT-supported coordination and mobilization of volunteers. The approach of SAVE.regions focuses on transferring knowledge between the emergency aid organizations in Southern Moravia (Czech Republic) and Lower Austria (Austria). The Czech platform TEAM MORAWA[4] provides a service for volunteers to registry themselves for delivering aid in future events. Training courses in the area of emergency assistance are offered to registered volunteers. If an event occurs, registered volunteers will be notified by SMS to deliver low-threshold engagement, e.g. filling sandbags, providing management report. Similarly, the approach of Team MV[5] in Northern Germany, an initiative of the German Red Cross and the Norddeutscher Rundfunk (NDR) works on the principle of an online platform to register potentially volunteers. The focus lies on supporting the emergency staff by low-threshold assistance. Team Morawa and Team MV were launched in the aftermath of the initiative Team Österreich and were coming into practice through co-financed projects funded by the European Union. The Swiss platform Benevol Jobs[6] allocates jobs or single tasks for engaging volunteers. Particularly the service includes location based tasks in the non-profit area aimed at strengthening the capacity of citizen's self-help. The citizen's mobilization at the pre-crisis level appears as a main concern to cause a mutual benefit for the community and emergency services. A process-driven approach including coordination activities, e.g. communication with volunteers is not taken into account. ZiviCloud[7], a web-application providing services for creating and completing tasks for civil society projects enables a flexible temporary management by online and micro volunteering. It supports every user in creating tasks or engaging for jobs, there is no restriction relates to the individuals' legal status. The UN Disaster Relief Report 2.0 asserts that there is the willingness of volunteers to support emergency services in the case of crisis or disasters. During the Libyan conflict in 2011, it was observed, how non trained volunteers had taken on tasks, e.g. translating contents or visualizing of important information from the crisis area on the Libya Crisis Map, via mobile phones and internet [8], [9]. The platform United Nations Volunteers[8] enabled the recruitment and mobilizing of the majority of the volunteers. In case of crisis or disasters, volunteers receive the opportunity to participate in a tender of several UN organizations. The initiative CrisisCommons[9], organizes training camps for volunteers who are comfortable with technology to support crisis management activities through programming of web sites or blogs,

[2] `http://www.roteskreuz.at/site/team-civil-protection/home/`
[3] `http://www.roteskreuz.at/site/saveregions`
[4] `http://www.teammorava.cz/en/who-is-team-morava`
[5] `http://www.team-mv.info/`
[6] `https://www.benevol-jobs.ch/`
[7] `http://zivicloud.roteskreuz.at/`
[8] `http://unv.org`
[9] `http://crisiscommons.org`

preparing campaigns and developing apps. The platform VolunteerMatch[10] is an American non-profit organization aimed at strengthening communities by bringing together non-profit organizations and volunteers. For volunteers this platform is free to use, while several modes of accounts are available for organizations. Inspired by the concept of virtual volunteering, this organization is focusing disaster-related volunteering by providing tools for corporate volunteer engagement. Subsequent on Hurricane Katrina in 2005, a relationship with the American Red Cross was established. As one innovative product, VolunteerMatch introduced a live map, which displays search activity in its network in real-time [10]. As argued in a Danish project to develop and test a mobile application to engage people in volunteering, the system architectures of many existing approaches are not appropriate to meet the challenge of national voluntarism. Nations have their own volunteering structure, and this particularity is provided by the slogan "every nation extra". In the Danish's approach the focus is set on an open access, this means that both organizations as well as individuals can create new tasks or request, as long as they are registered in the system. The application "can't be controlled by a single cause or organization", and the responsibility for the processing of the tasks is not limited to humanitarian organizations [11].

4 The Need for New Approaches

The described developments and the limitations of existing solutions stress the importance of searching for new, alternative and lasting ways on how to involve people in a response, and more general in civil protection, by developing new flexible and task based offers. Relief organizations dependent on volunteers are exposed to various influences such as globalization, rapid technologic transition, continuous societal change or even the uncoordinated influx of convergent volunteers on-site. One of their major challenges is adapting their structures and offers to these (changing) realities in order to meet the needs of the beneficiaries. In this sense, the described decline of formal engagement in various European countries doesn't mean necessary the end of voluntarism; instead people are becoming more selective regarding the voluntary work. They are willing to spend their time with and are at the same time more critic regarding considerable expenditure of time due to lack of adequate opportunities. The result from a micro-census survey, conducted in 2006, demonstrates the strength of willingness for voluntarism: more than 58% of interviewed people without any voluntary engagement stated, the cause for their non-engagement, was that they haven't been asked for participating [12]. In Austria the tendency for informal and flexible forms of voluntary work is clearly increasing, particularly within younger groups and more in urban than in rural areas where formal membership (i.e. voluntary fire services) is still widely common among young (male) Austrians [14]. Thus the organizations within this field need to adapt themselves to these changing needs of people and provide additional and more flexible offers for people to contribute within disaster relief. Notwithstanding the future informal voluntary engagement in disaster relief should be regarded as an additional offer to the public and support to the existing disaster management system, since informal volunteers cannot provide the same type

[10] `http://www.volunteermatch.org/`

of activities as dedicated, continuously trained volunteers which are part of the professional response system. One novel approach within disaster management for meeting the above needs of volunteers and addressing the changing realities might be the combination of new media with informal ways of voluntary engagement. In this sense disaster response organizations could seize the opportunities given by new media to bi-directionally communicate with and coordinate a large crowd with little personnel resources and offering people a more task based, time and location independent means of assisting the response.

5 A New Approach to Classify Different Types of Volunteers

As already discussed in chapter "Traditionally voluntary engagement", voluntary work can be classified into formal and informal voluntary services. Formal voluntary services are understood as activities, which are embedded in the framework of organizations, institutions and incorporated associations. In contrast, informal voluntary engagement is given on basis of a personal initiative without an institutional embedment, e.g. neighborly help [15]. Our observations show that conventional types of voluntary work, like permanently engagements in a preferred organization over decades, don't harmonize with recent draws on life in urbanized societies even more. Since some years, more flexible types of voluntary engagement have been requested. As a consequence, emergency organizations have been faced with new challenges, for instance, to redefine formal and informal voluntary engagement [14]. Therefore, new target groups for volunteers' management with different characteristics have to be addressed with new methods. An investigation of the population's willingness to participate voluntarily led to the following classification as illustrated by Fig. 1 below.

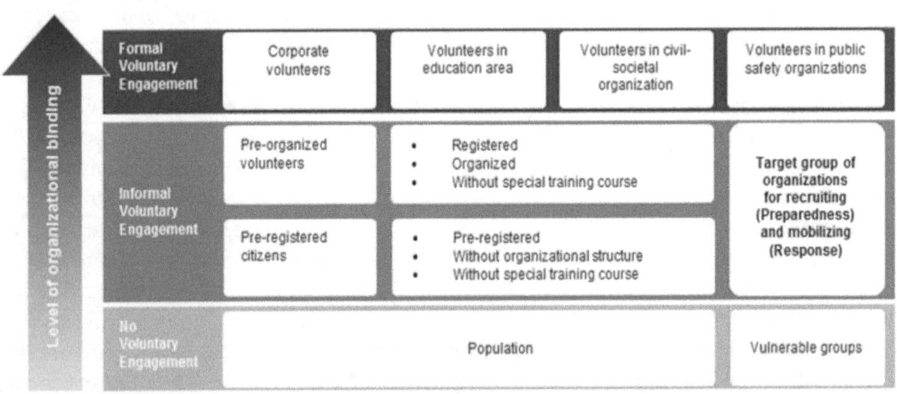

Fig. 1. Classification of target groups for disaster prevention, coordination and mobilization of volunteers

The types of volunteers shown in Fig. 1 differ in regard of their degree of binding to an organization. Pre-registered citizens are characterized by their general willingness to participate with focus on a less-formal organization structure and no special training for crisis response. They would be willing to register themselves on a platform, but they don't want to become member of an organization. Pre-organized volunteers are belonging to another category of voluntary group. Although they are not specifically trained and are dedicated for case-specific assistance like the pre-registered citizens, they seem less reluctant for joining an organization. With the introduction of these two analytic categories we can better assess and address the needs of different target groups of informal volunteers, which were formally subsumed under broader categories such as "spontaneous volunteers". The new concept of crowdtasking described in chapter "The concept of crowdtasking" is addressing both types of informal voluntary engagement (pre-registered citizens and pre-organized volunteers).

6 The Concept of Crowdtasking

Depending on the selected volunteer management approach the definition of requirements or demands on volunteers can be rather generic, the way of executing tasks can be left to volunteers. This approach has the advantage to provide everybody from the crowd the opportunity to participate in relief activities but the disadvantage that aspects such as qualification, reliability and credibility of the volunteering group are a priori unknown. This can lead to several problems such as unknown reliability, but also a hardly predictable number of volunteers available in case of incidents making resource planning challenging. Exact definition of tasks accompanied by a registration process may lead to the opposite situation, i.e. smaller number of available volunteers but in average higher degrees of qualification and reliability of the volunteers. In this context a registration process may allow better prediction of the degree of execution of tasks. Moreover, the degree of freedom for volunteers executing tasks, including the way tasks are specified and the workflow management of tasks is implemented, is also open to the selected strategy of volunteer management. In the frame of this paper we focus on the management of a priori selected volunteers that are managed by processes with low degree of freedoms in process execution, i.e. pre-registered citizens according to Fig. 1. We call this approach Crowdtasking and define it as a comprehensive concept, composed of structures, processes and tools with the goal (1) to build up and maintain an informal community of pre-defined and informal volunteers, (2) to mobilize them when needed, (3) to control their activity and (4) to collect data generated by the mobilized volunteers for enhancing situation awareness of the disaster relief effort as a whole. The following paragraphs describe this concept step by step.

1. Volunteers Community Building and Maintenance

The management of the network of volunteers is a basic and continuous process. Therefore, a specific role is dedicated to the continuous execution of this process. We call this role the "volunteer's community manager". As a result of the volunteer's community building and maintenance, a network of registered potential volunteers shall be available, which can be addressed for mobilization. In order to achieve this, the following tasks have to be performed:

- **Community building** – the volunteers community manager promotes the idea and motivates persons to register themselves as potential volunteers, efforts which proved to be the most effective when accompanied by different forms of media campaigning [7]. By registering, they become members of an informal online-community without the need to join an established response organization and undergo specific training.
- **Registration** – volunteers use a web portal to register to the community by providing specific information on their skills and their availability. During the registration phase specific tools allowing the volunteers to perform their tasks are provided, e.g. downloading of a crowdtasking app for their smartphones.
- **Data maintenance** – the data of the registered volunteers are to be stored, updated and secured.

In the course of the community building process, requirements and needs of involved actors play an important role. Crisis managers and emergency services have specific requirements on the type and quality of services they may need from the volunteers. Volunteers themselves do have individual personal needs and motivations. In the context of these needs and requirements, basic information from the volunteers, such as their individual qualifications, limitations, or objections to perform certain kind of tasks can help to improve the quality of the entire process.

2. Mobilization

The goal of this process is the mobilization of selected volunteers for the execution of a specific task within the context of the whole crisis- and disaster relief effort. Again, this process is controlled by a specific role. We call this role the "crowdtask manager". The process consists of the following tasks:

- **Crowdtask set-up** – the crowdtask manager defines and describes the specific task to be performed by the volunteers. This crowdtask will be a single effort, which can be executed within a limited amount of time. Examples are: Taking of two pictures with the smartphone out of the window plus answering a set of short and simple questions after an earthquake. Crowdtasking requires definition and description of the tasks in a concise way including a simple and structured form for reporting. Reporting may involve sending of a picture or ticking of a few selection boxes within a multiple choice system. As part of the preparation phase of crisis management, crowdtasks can be configured and stored in a database for later use.
- **Selection of volunteers** – the crowdtask manager selects a group of volunteers from the database according to the area and the type of task to be performed.
- **Initiation of crowdtask** – the pre-configured or newly defined crowdtask is activated and sent to the selected volunteers after release by the authorized official. The addressed volunteers receive the crowdtask e.g. on their smartphone.

3. Crowdtask Execution and Control of the Activity

The process not only includes the activities by the volunteers, but also the interaction with the crowdtasking system and the crowdtask manager behind.

- **Execution of the tasks** – this implies reception and execution of tasks by the addressed volunteers, workflow management of single or multiple tasks and provision of reports. The whole process can be supported by using specific user-interfaces provided by dedicated crowdtasking apps on the volunteer's smart-phones.
- **Compensation** – Volunteers may receive compensation. "Compensation" in this context is not referring to monetary aspects since this would contravene the concept of voluntarism, but more to aspects directly related to the motivation behind the engagement of the volunteers. The compensation may include specific "insider" information on the development of the crisis-situation (e.g. location based rain forecast), guidance for personal behavior or measures to be better prepared for an emergency, etc.
- **Information gathering** – all the information gathered from volunteers is collected, stored and processed. The processing of the information has the purpose to derive crisis management relevant knowledge out of the obtained information. This process needs also validation of the obtained information such as plausibility checks.

4. Reporting and Visualization

A specific crowdtasking action shall provide its impact not only in the field, but also for the decision making of the emergency management organizations and crisis managers by enhancing their situation awareness. The goal of the reporting process therefore comprises the generation of intelligence from the field out of the data provided by the volunteers. Situation reports are generated and may be forwarded to other organizations involved into the disaster relief effort e.g. as contribution to a common operational picture. This process is composed of two important tasks:

- **Reporting** – Generation of specific reports and visualizations based on the data received from the volunteers
- **Distribution** – Distribution of reports and visualizations via open interfaces to other organizations.

Functional Blocks of a Crowdtasking Tool

In order to implement the process described above a system consisting of several functional blocks as shown in Fig. 2 is proposed.

The starting point is a web portal for registration and supervision of volunteers. This portal allows promotion of the approach, communication between existing and potential volunteers and finally downloading of services such as apps. Data obtained via the portal needs to be maintained taking into account privacy and security aspects. Data managed in the maintenance tool have to be stored in an adequate deposit. The core component is the crowdtasking management module dedicated to allow setting up of tasks, selection of volunteers by adequate filter tools, task distribution, storing of results and communication with volunteers in case of incidents. Apart from the crowdtasking management module the mobile application is another major

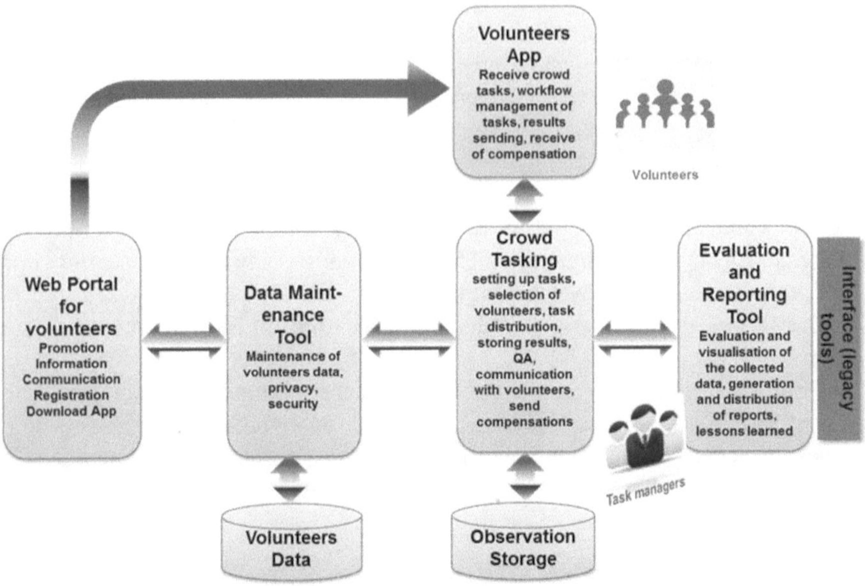

Fig. 2. Structural concept of an advanced volunteer management system

component providing the volunteers the possibility to receive tasks, to ensure an adequate workflow management, to document the execution of these tasks and finally to establish a communication process with the crisis managers including the possibility to document different types of observations such as text, photos or videos. An imperative pre-requirement the mobile application needs is a well-designed graphical user interface, allowing a distinct workflow of the tasks making the possibility of misunderstandings very unlikely. Finally, there is need to evaluate the obtained information in order to extract knowledge out of it. This encompasses the verification of the information by checking the plausibility, inspecting the credibility of the information provider, filtering, e.g. in terms of relevance, and finally the aggregation of the information. Reporting and storing in a data repository are other requirements to be fulfilled by a crowdtasking system.

7 Discussion and Conclusions

The application of the new concept of crowdtasking in the crisis and disaster management is accompanied by several challenges. First of all, there is a need to take into account the acceptance of a future system by the users, i.e. both the volunteers and the volunteer's community managers. The processes have to be designed in a way ensuring that tasks are in line with volunteer's capabilities on one hand and do not require activities incompatible with individual ethic values of volunteers on the other, e.g. collection of sensible private information,. Moreover, readiness to use a future system in case of incidents has to be ensured. For this purpose adequate user interfaces have to be provided and the availability of the service needs to be ensured. In addition, there is need to ensure social inclusion for volunteers from multiple social groups in order to avoid exclusion effects due to parameters such as age, social class or gender. Temporarily and loose forms of voluntary engagement require legal certainty both for volunteers and crisis managing organizations. A major question is the liability in case of accidents of volunteers. Currently, a statutory accident insurance is established to volunteers of one of the Austrian emergency service organizations, when the performed tasks were related either to training, exercise or an operation. In these cases, they are regarded as occupational accidents and are regulated by Austrian law § 176 Abs 1 lit. 7a ASVG [17]. In other areas of voluntarism there is no statutory insurance in the case of an accident, but organizations, which engage volunteers, have the opportunity to insure volunteers privately, wherever necessary. An additional important question is the liability in case of mistakes, e.g. provision of false critical information. Owed to the employee-like relationship between humanitarian non-profit organizations and professional volunteers, the obligation to compensate in the case of volunteer's fault during a mission, is regulated as specified in the employees liability act [19]. The formulation of this law precludes the application to loose commitments without a regular and appointed task area. In this case other alternatives have to be explored to ensure an optimal protection of informal volunteers (see Fig. 1). On the legal side, organizations might need to insure informal volunteers privately, i.e. by procuring an all-in insurance. To avoid many risks a priori, it would be reasonable to follow a strategically approach by defining a limited area of assisting as well as riskless activities for those volunteers being not protected by insurance. Furthermore, equality between different social groups has to be ensured. This encompasses several aspects, e.g. developing special offers to female volunteers and the creation of adequate recruitment procedures by considering gender-specific preferences of voluntary engagement. For single mothers, elderly people or people with disabilities, an engagement will not fail due to lack of willingness, but rather due to lack of opportunities caused by multiple constraints such as lack of time or physical limitations. Another very important aspect is to ensure privacy protection. Both critical data of volunteers as well as potential victims from an affected population have to be treated confidentially. A future volunteer management system has to ensure therefore that requirements of data protection laws are fulfilled and, if necessary, requests of volunteers beyond the legal requirements are also implemented in order to ensure the acceptance of the system. In order to test the presented crowdtasking concept the authors intend to implement a demonstrator designed to manage volunteer's contribution for the Team Österreich. Evaluation and analysis of user-specific requirements shall lead to an application focused concept. Validation efforts lead to an improved process concept for the mentioned demonstrator. Based on the outcome of the evaluation of the above mentioned demonstrator, the application and adaption of the developed crowdtasking concept for crisis and disaster management organizations with a similar structure, e.g. the volunteer fire brigade, will be examined. The individual structures and procedures of different organizations with voluntary engagement will require specifically adapted concepts and solutions.

References

1. International Federation of Red Cross and Red Crescent Societies: The road to resilience – Bridging relief and development for a more sustainable future, Geneva (2012), http://www.ifrc.org/PageFiles/96178/1224500-Road%20to%20resilience-EN-LowRes%20(2).pdf
2. Jachs, S.: Einführung in das Katastrophenmanagement. Tradition, Hamburg (2011)
3. UNISDR: UNISDR Terminology on Disaster Risk Reduction. (2009), http://www.inpe.br/crs/geodesastres/conteudo/artigos/UNISDR_Terminology_on_Disaster_Risk_Reduction_2009.pdf
4. COUNCIL DECISION of 27 November 2009 on the European Year of Voluntary Activities Promoting Active Citizenship (2010/37/EC), L 17/43 (2011), http://ec.europa.eu/citizenship/pdf/20091127_council_decision_en.pdf
5. Statistik Austria: Struktur und Volumen der Freiwilligenarbeit in Österreich (2008)
6. Governor's Office of Emergency Services California: They will come. Post-Disaster Volunteers and Local Governments (2001)
7. Austrian Red Cross: Teaming Up For Civil Protection. Final Report. (2011), http://ec.europa.eu/echo/civil_protection/civil/prote/pdfdocs/TeamCP_ATRedCross_final_report.pdf
8. Crowley, J., Chan, J.: Disaster Relief 2.0: The Future of Information Sharing in Humanitarian Crisis (2011)
9. Meier, P.: Do "Liberation Technologies", change the Balance of Power between repressive States and Civil Society, Dissertation (2011), http://irevolution.files.wordpress.com/2011/11/meier-dissertation-final.pdf
10. VolunteerMatch, http://en.wikipedia.org/wiki/VolunteerMatch
11. Christensen, C. G.: Social Media Volunteering Application (2013), http://www2.imm.dtu.dk/pubdb/views/edoc_download.php/6558/pdf/imm6558.pdf
12. Bundesministerium für Soziales und Konsumentenschutz Hrsg.: Struktur und Volumen der Freiwilligenarbeit in Österreich. Wien, p. 3 (2008)
13. http://www.bmask.gv.at/cms/site/attachments/0/1/0/CH2174/CMS1218022135039/presseunterlage_freiwilligenarbeit_21.1.08[1].pdf
14. Austrian Federal Ministry of Labor, Social Affairs and Consumer Protection (ed.) Volunteering in Austria. 1st Report on Volunteering, Vienna 2009 (summary available) (2009) (in German)
15. CISESK - Certification of Civil Services non formal and informal Skills Transfer: National Report on Voluntary Work in Austria (2009)
16. http://www.ciseskproject.eu/docs/2.2_cisesk_at_desk_research.pdf
17. Legislation on the Social Security Act
18. http://www.ris.bka.gv.at/Dokumente/Bundesnormen/NOR40146957/NOR40146957.pdf
19. Andreaus, F.: Rechtliche Grundlagen des Österreichischen Rettungswesens. Dissertation, Wien (2009)
20. http://othes.univie.ac.at/5527/1/2009-01-27_9547713.pdf

GIS-Based Route Planning for HAZMAT Transportation

Alexander Preda[1], Mauno Rönkkö[2], Stefan Pickl[1], and Mikko Kolehmainen[2]

[1] Faculty of Computer Sciences, University of Federal Armed Forces Munich
{alexander.preda,stefan.pickl}@unibw.de
[2] Department of Environmental Science, University of Eastern Finland,
P.O. Box 1627, 70211 Kuopio, Finland
{mauno.ronkko,mikko.kolehmainen}@uef.fi

Abstract. In this article we discuss the route planning of Hazardous Materials (HAZMAT). For HAZMAT transportation, safety and security is of utmost importance. Because of this, the route planning needs to consider potential risks and costs to find an optimal route for the transportation. It has been shown that the route planning is an NP hard problem. In this article, we approach the problem from a novel point of the view. As the main contribution, we present in this article a GIS-based route planing method that has many complementary advantages to graph-based route planning methods. We also illustrate the use of the method with a simple case study.

Keywords: Hazardous Materials, HAZMAT Transportation, Route Planning, GIS.

1 Introduction

Today's living standards and living comfort would not be possible without the use of HAZMAT in industry. For instance, medicine, paints, cleaning agents, and fuels require the use of HAZMAT for their production. Furthermore, the production of such goods creates hazardous waste. Consequently, there is an increasing need for methods to handle ever more complex HAZMAT transportation problems.

Graph-based algorithms are typically used for planning HAZMAT transportation routes. The problem, however, is shown to be NP-hard, whereby one can only hope for good enough solutions. Moreover, route planning involves risk assessment which is based on estimates at best because of the underlying complexities as well as ethical and political considerations.

Because of this evolving nature of the route planning, there is a need to compute alternative settings with varying valuations and factors. This, however, is not well supported by graph-based algorithms that typically require full computations if some of the factors change. To overcome this problem, and as the main contribution, we propose here a GIS-based route planning method. It supports

J. Hřebíček et al. (Eds.): ISESS 2013, IFIP AICT 413, pp. 357–366, 2013.

integration of partial valuations, and thereby it also supports computations of alternative scenarios with varying valuations.

We shall start by discussing HAZMAT transportation and risks in Section 2. In Section 3, we introduce the route planning problem. In Section 4, we present the proposed GIS-based route planning method. In Section 5, we illustrate the use of the method with a simple case study. And finally, in Section 6, follows the conclusion.

2 About HAZMAT Transportation and Risks

The research on HAZMAT transportation began already in the 1990's. A more recent overview about HAZMAT transportation can be found, for instance, in [1]. The need for HAZMAT transportation is a direct consequence of our modern society, where the amount goods produced by using hazardous materials has increased. Consequently, HAZMAT transportations have increased especially in urban areas. Partially because of this, the HAZMAT Transportation process is planned and carried out by considering certain national and international standard procedures, like the "UN Recommendations on the Transport of Dangerous Goods" [2].

In HAZMAT transportation, there is always a risk involved. According to [3], risk in the context of HAZMAT transportation is a measure of the probability and severity of harm to an exposed receptor due to potential undesired events. In this sense, risk can be assessed either qualitatively or quantitatively. In a qualitative assessment, the goal is to identify possible scenarios. In a quantitative assessment, consequences are modeled and made explicit, so that the exposed receptors can be identified. All of this information is then used to calculate numerical risk evaluations.

An alternative approach to HAZMAT transportation risk assessment is given by Current and Ratick [4]. They consider the location of facilities coping with HAZMAT in the assessment of risks and costs. Consequently, their definition for risk is based on the amount of transported waste, and not on the losses when an incident exposes transported HAZMAT to the environment. The analysis by Current and Ratick proves that minimizing a risk is highly sensitive to its definition. In particular, minimizing one part of the risk can degenerate the other parts. It could even turn out that there is no stable solution for minimizing a risk for a given objective. As an example, Current and Ratick define "minimizing the risk per population density" as the objective 1, and "minimizing the total transportation risk" as the objective 2. When focusing on the objective 1, one can easily find solutions that increase transportation costs disproportionally. In contrast, when focusing on the objective 2, one can find a stable solution. Still, minimizing the costs for the objective 2 does increase the costs for the objective 1. All in all, the analysis by Current and Ratick suggests that the costs are relatively strongly correlated with the risk.

3 The Route Planning Problem

In the planning of HAZMAT transportation routes, both local authorities and carriers need to interact. In effect, this means that the route planning problem is a two player game, as understood in game theory. In such a game, the local authority is interested in minimizing the risk of transportation, whereas the carrier is oriented, as pointed out by [1] and [5], to minimize the costs. In this process, diffuse information increases the difficulty for the local authority to determine a fix route. To date, the usual approach taken by the local authority is to design the whole transportation network and to allocate it to the carrier. In this way, the maximal transportation risk is to be kept below a certain threshold, while giving the carrier a choice to select the actual route within the network.

There are many route planning algorithms for the HAZMAT transportation. In particular, they involve explicitly risk assessment for the transportation, thus considering one or both of the following [6]: an incident probability during the transport and the scale of the impact on the population.

Route planning algorithms are typically based on graph theory. For instance, Dijkstra's Shortest Path algorithm [7] is one of the most famous algorithms that is used. In graph-based algorithms a directed graph models the transportation network. The usual coding is that the nodes represent crossroads or connection points, while the arcs represent streets or links in the transportation network. The weights for the arcs represent the "cost" for passing that link. The cost is an evaluation of, for instance, money, time, and risk factors.

As another example of a graph-based algorithm, Erkut and Alp formulate in [8] a method for designing a HAZMAT transportation network based on minimizing the total transportation risk. In the algorithm, heuristics are used for the cost optimization. The optimization is performed in two phases. Firstly, the optimization is done by the local authority to minimize the risks. Secondly, the carrier optimizes transportation costs by selecting the most convenient route. This process minimizes the probability that the route selected by the carrier increases the risk more than what is expected by the local authority. As in [4], Erkut and Alp distinguish between minimizing the total risk on the route and minimizing the selective risk on a certain road.

In later works [9], Erkut and Gzara propose a method for the HAZMAT transportation network design problem (HTNDP). The method yields stable solutions which are close to the optimal solutions. Their proposal is a bi-level model with an heuristic solution method, because the HTNDP may be ill-posed. In the center of their analysis is the trade-off between costs and risks.

Amaldi et al. [10] developed an improved mixed integer linear program with a bi-level inter-program based on the results of [9] and [8] for solving the HTNDP, and to guarantee the stability of the solution. Although Amaldi et al. deal with the HTNDP, the article concentrates on computational complexity. The conclusion is that the HTNDP with the option to forbid the use of certain arcs is NP-hard even when only a single commodity should be transported. Amaldi et al., however, point out that the HTNDP is often reduced to the shortest path problem extended with a heuristic. The heuristic consists of an iterative elimination of allowed roads until the carrier is no longer able to use a different route than the one with the minimal risk as expected by the local authority. As emphasized in [4], this elimination method may not only increase the costs, but also the lower bound of the minimal risk.

4 GIS-Based Route Planning Method

There are many sources of information that needs to be considered for route planning. The data is either readily available as GIS data, or it can be converted into GIS data. Also, all the information is typically available for the local authority doing the route planning. For instance, information about buildings, occupants, and potential risk groups are typically maintained by city officials for a given city. Depending on the city, such city records may also indicate increased probability of accidents for specific crossroads and junctions. City records also maintain information about the central infrastructure, including water supplies, electricity, and maintenance resources. As for the actual roads, there are many services maintaining digital road data.

In practice, the local authority needs to consider several alternatives and variant scenarios. These "what if" cases provide a valuable insight into the robustness and vulnerability of the transportation network, which is critical when defining a valuation of risks for individual cases. This kind of analysis is not well supported by graph-based approaches, as changing some parameters leads to re-computation of the entire network. As discussed earlier, such a task may be computationally expensive and time consuming. To support the exploration, we shall now present an alternative GIS-based route planning method.

The overall workflow of the method is depicted in Figure 1. In the method, the local authority encodes all individual risk influence factors as separate map layers. Consequently, road maps and residential areas are captured as separate layers. There are also disjoint layers for each neuralgic points, such as hospitals, schools, super markets, and bridges. The layers may also contain some strategic factors, such as temporal traffic congestion. External and diffuse knowledge are encoded as fuzzy rules. Such rules consider, for instance, the risk of an explosion or a chemical dispersion with respect to the location of the incident. Fuzzy rules are then used reproduce a map representation of the diffuse knowledge. All the maps are then summed together cell-wise, to produce a cumulative risk map. The cumulative risk map is then analyzed for transportation routes having the minimal cumulated risk with respect to a transport list.

The heart of the method is the algorithm that takes as an input a map, where each cell indicates the summed risk of all the factors at that location, and produces as an output a list of transportation routes with minimal cumulated risk. For such routes, the total risk is the sum of the risk values of all the individual cells in the route.

The algorithm was implemented as an application of the Distance Transform Path Planning algorithm suggested by Zelinsky et al. [11] for robotics. In short, the algorithm initializes the map with obstacle comfort weights (here summed

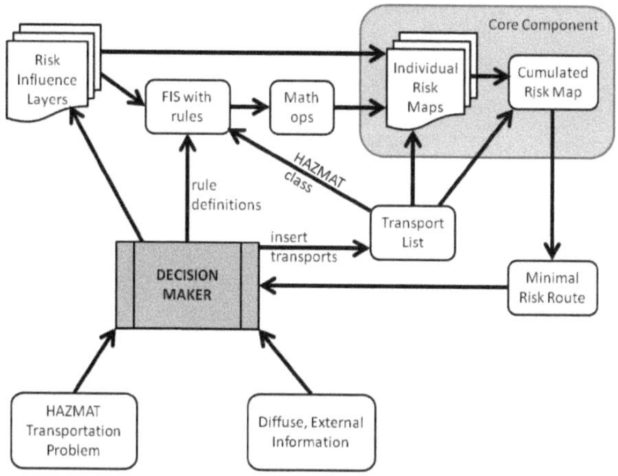

Fig. 1. The workflow of the GIS-based route planning method

prepare r as the risk map
prepare c as the initial cumulated risk map
repeat
    compute m, such that for each cell
        $m_{i,j} = \min(c_{i-1,j}, c_{i+1,j}, c_{i,j-1}, c_{j,j+1})$
    recompute $c_{i,j}$ as $\min(r_{i,j}+m_{i,j}, c_{i,j})$
until c does not change

Fig. 2. Risk cumulation propagation algorithm

risk values). It then computes a distance transform on the map with respect to a fixed starting and ending location. The distance transform is practically a function that returns the cumulated comfort weights (here summed risk values) along a chosen path. The algorithm then minimizes the transform function, yielding a set of paths that have the minimal cumulated comfort weights (here summed risk values).

Our implementation of the algorithm by Zelenski et al. [11] has two phases. In the first phase, the risk value cumulation is propagated from the fixed starting location throughout the entire map. The propagation of risk values was implemented as an iterative update, where each cell is updated as the sum of its value and the minimal risk value of its neighboring cells. For the iteration to work, the cumulated risk map was initialized with large values except for the starting point which is initialized with value 0. The propagation algorithm is presented in Figure 2, and the iteration is illustrated in Figure 3. In the second phase, the routes with minimal cumulated risks are computed for a fixed ending point by using the steepest path of decent as in [11].

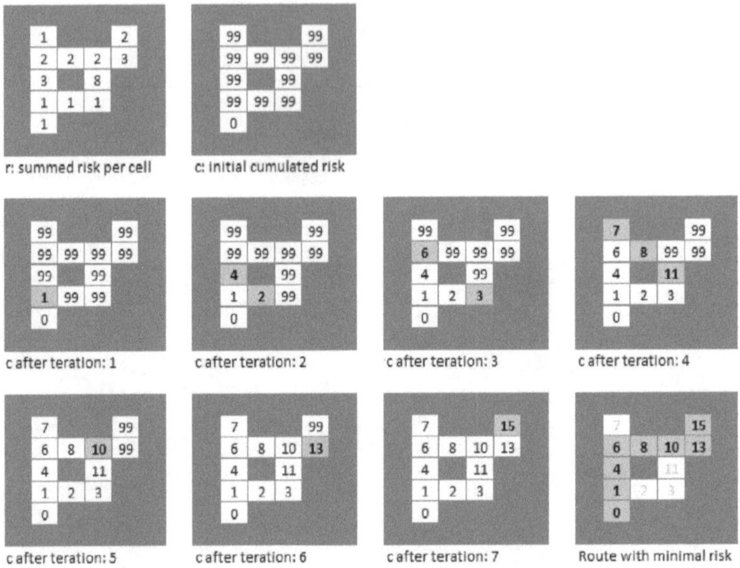

Fig. 3. Illustration of risk cumulation propagation; the two top tables show the initial matrices and the rightmost table on last row shows the obtained route with the minimal cumulated risk

5 Case Study

The purpose of this case study is to illustrate how the GIS-based route planning method works. Therefore, we consider here only a simple street map, as shown in Figure 4. In the street map, the risk level of each street cell is initialized to zero.

As the first test, we consider only one layer, which is the population layer. It is created and initialized with a random population density. Thus, the risk cumulation is computed only with respect to the population layer. The starting point is set to be the street cell at the bottom left corner and the ending point is set to be the street cell at the top right corner. The risk cumulation for the population layer is shown in Figure 5. Note that, as the map shows the cumulated risk, the highest risk value is actually found at the ending point from where it decreases towards the starting point. After running the route planning algorithm, the route with the minimal cumulated risk is found, as plotted in Figure 5. The found route goes from the starting point to the ending point as straight as possible. The found route avoids the bottom right corner which has higher risk values.

In the second test, we create a layer with two neuralgic points having diffuse risk values. The layer is associated with transportation of class-1 HAZMAT (explosives), and the neuralgic points represent buildings that could be severely damaged by such materials. The associated risk values, however, do not localized

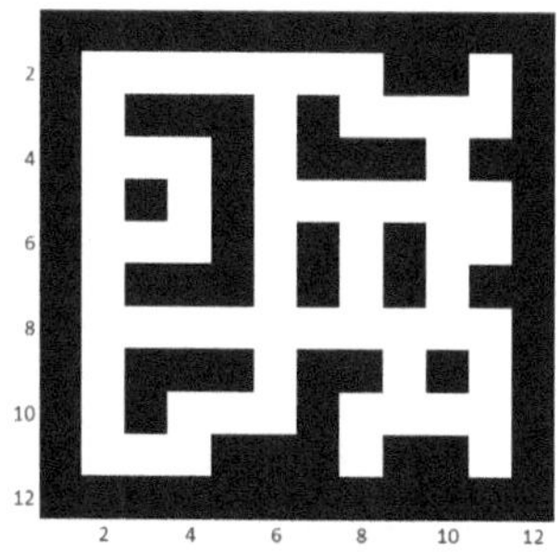

Fig. 4. Street map of the case study

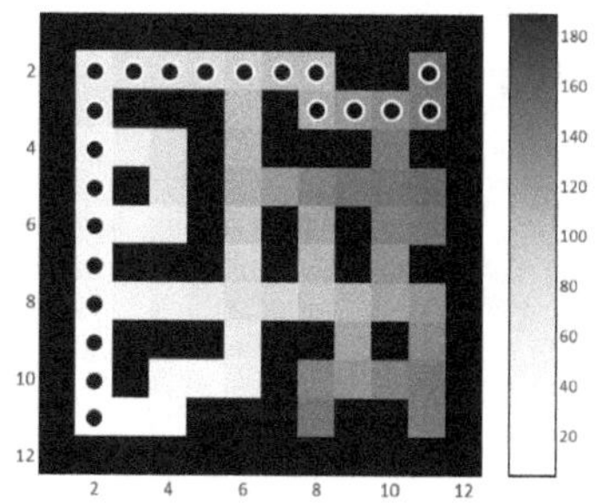

Fig. 5. A population layer showing the cumulated risk and the plotted route with the minimal risk cumulation

just to these buildings, but also to their surroundings. Therefore, the actual risk value varies depending on the distance between the incident location and the buildings. Because of this, fuzzy logic is used to construct a risk layer that captures the described diffuse knowledge. In the construction, the membership functions for the input variable "distances", as depicted in Figure 6, are used. For instance, if the distance between the incident location and a building is about 400 meters, it is mostly considered to be "intolerable"; however, as indicated in Figure 6, such a distance could also fall into classes "close" and "tolerable" depending on the case. Similarly to the variable "distances", the membership functions for the variable "risk values" are constructed, as shown in Figure 6. By applying these membership functions, we can compute how the risk values spread on the map. The result of such a computation is shown in Figure 7 for a neuralgic point located at coordinates (5,7) having the maximal risk value of 10.

By using the technique described above, we can construct a risk layer for transportation of class-1 HAZMAT with two neuralgic points located at the coordinates (3,5) and (10,9) with highest risk values 50 and 40, respectively, As in the first test, we consider in this second only one layer, which is the layer for class-1 HAZMAT. The starting and ending points are kept as in the first

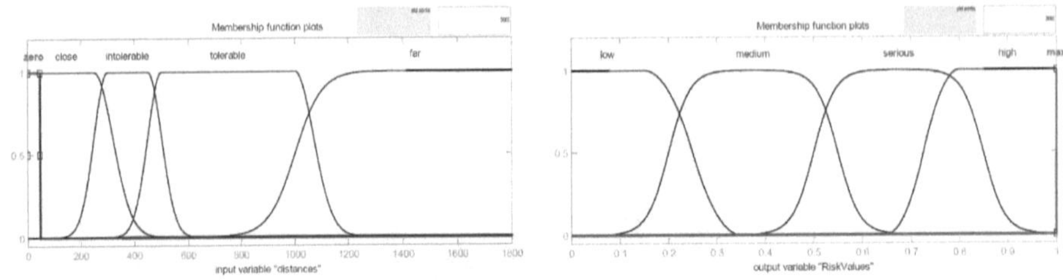

Fig. 6. Fuzzy membership functions

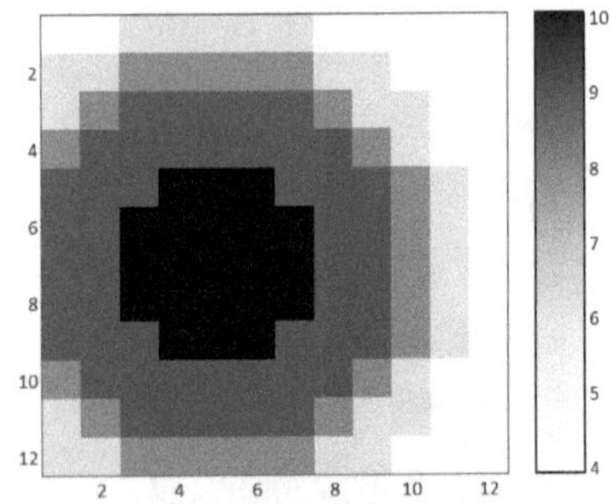

Fig. 7. Risk values computed using a fuzzy realization of the diffuse knowledge

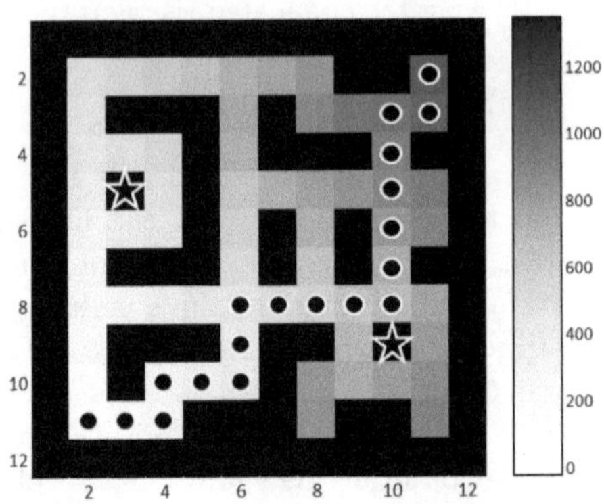

Fig. 8. The risk cumulation map for two neuralgic points and the plotted route with the minimal risk cumulation. The neuralgic points are indicated with stars.

test. After computing the risk cumulation for the starting and ending points, a cumulated risk map is obtained. It is shown in Figure 8. Note that, as the map shows the cumulated risk, the risk value is at highest at the ending point and not around the neuralgic points. For the map, the route planning algorithm yields now a different route than in the first case. The route is plotted in Figure 8. As expected, the route goes along the neuralgic point having the smallest risk value.

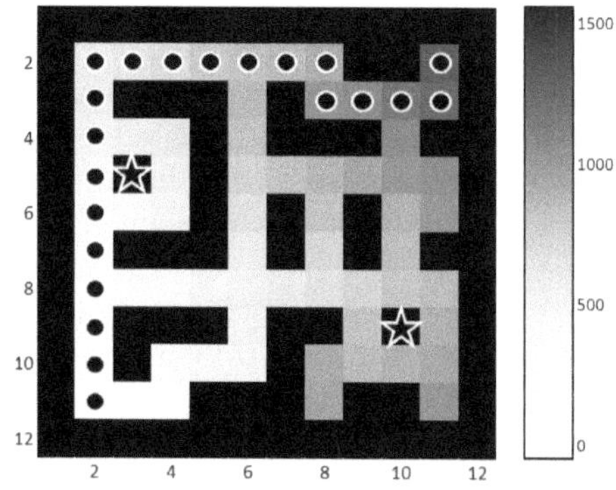

Fig. 9. Risk cumulation with combined layers and the plotted route with the minimal risk cumulation. The neuralgic points are indicated with stars.

As the last test, we combine the two previously introduced layers by summing them up. Thus, there is no need for recomputations, as the new map reuses the earlier computation results. The resulting combined map is shown in Figure 9. This time, when the route with the minimal cumulated risk is computed, it takes the same route as for the first test. The randomly generated risk layer for the population density, although having much smaller values than the layer for the neuralgic points, is sufficient to redirect the route away from the neuralgic point with the smallest risk value.

6 Conclusion

In this article we discussed route planning for HAZMAT transportation. It is an NP hard problem, and as the main contribution, we presented here a GIS-based route planning method that is complementary to graph-based route planning methods. At the heart of the method is the route planning algorithm that we implemented as an application of the Distance Transform Path Planning algorithm suggested by Zelinsky et al. [11] for robotics.

The proposed method exhibits several advantages that are critical for the decision maker. Firstly, each risk factor is encoded as a separate layer, whereby each risk factor and its evaluation can be considered in isolation. Secondly, the decision maker can provide several alternative valuations for the same risk factor, to analyze the overall effect of that specific risk. Thirdly, the method merges for consideration any subset of the risk layers that the decision maker is interested in. This supports the "what if" analysis that is of utmost importance for the decision maker. Fourthly, the risk assessment computation is split into a preprocessing phase and into a path analysis phase, whereby the computational results for an individual risk assessment can be reused. This, in turn, lowers the overall computational costs. Lastly, as the computation is performed on maps, it fully supports the distribution of the computation among parallel processes. Thus, the computational complexity can be tackled simply by introducing more parallel processes working on disjoint portions of the same maps.

The research presented in this article is at early stages. The method is currently implemented by using MATLAB. For repeated use in varying scenarios, it has to be wrapped as a computational service. This requires some consideration, as such a service also requires dataset management along with sufficient data security measures.

As for the future research, we also plan to study the use of the presented method for analyzing not only HAZMAT transportation, but also correlation between traffic infrastructure and occurrences of traffic accidents.

Acknowledgements. Mauno Rönkkö and Mikko Kolehmainen would also like to thank European Regional Development Fund and TEKES for support to continue research in cooperation with Amphion Loudspeakers Oy, Genelec Oy, Mediatuotanto Mindcom Oy, Observis Oy, and Defence Forces Technical Research Centre.

References

1. Erkut, E., Tjandra, S.A., Verter, V.: Hazardous Materials Transportation. In: Barnhart, C., Laporte, G. (eds.) Transportation, vol. 14, pp. 539–621. Elsevier (2007)
2. United Nations: Recommendations on the Transport of Dangerous Goods. United Nations (2007)
3. Erkut, E.: Special issue: Hazardous materials transportation - Introduction. Computers and Operations Research 34(5), 1241–1242 (2007)
4. Current, J., Ratick, S.: A Model to assess Risk, Equity and Efficiency in Facility Location and Transportation of Hazardous Materials. Location Science 3(3), 187–201 (1995)
5. Gumus, A.T.: Evaluation of hazardous waste transportation firms by using a two step fuzzy-AHP and TOPSIS methodology. International Journal of Expert Systems with Applications 36, 4067–4074 (2009)
6. Kara, B.Y.: Accurate calculation of hazardous materials transport risks. Operations Research Letters 31, 285–292 (2003)
7. Dijkstra, E.W.: A Note on Two Problems in Connexion with Graphs. Numerische Mathematik 1, 269–271 (1959)
8. Erkut, E., Alp, O.: Designing a road network for hazardous materials shipments. Computers and Operations Research 34, 1389–1405 (2007)
9. Erkut, E., Gzara, F.: Solving the HAZMAT transport network design problem. Computers and Operations Research 35, 2234–2247 (2008)
10. Amaldi, E., Bruglieri, M., Fortz, B.: On the Hazmat Transport Network Design Problem. In: Pahl, J., Reiners, T., Voß, S. (eds.) INOC 2011. LNCS, vol. 6701, pp. 327–338. Springer, Heidelberg (2011)
11. Zelinsky, A., Jarvis, R.A., Byrne, J.C., Yuta, S.: Planning paths of complete coverage of an un structured environment by a mobile robot. In: The Proceedings of International Conference on Advanced Robotics, pp. 533–538 (1993)

An Architecture for Adaptive Robust Modelling of Wildfire Behaviour under Deep Uncertainty

Daniele de Rigo[1,2], Dario Rodriguez-Aseretto[1], Claudio Bosco[3], Margherita Di Leo[1], and Jesús San-Miguel-Ayanz[1]

[1] European Commission, Joint Research Centre, Institute for Environment and Sustainability, via E. Fermi 2749, I-21027 Ispra (VA), Italy
[2] Politecnico di Milano, Dipartimento di Elettronica e Informazione, via Ponzio 34/5, I-20133 Milano, Italy
[3] Loughborough University, Department of Civil and Building Engineering Loughborough LE11 3TU, United Kingdom

Abstract. Wildfires in Europe – especially in the Mediterranean region – are one of the major treats at landscape scale. While their immediate impact ranges from endangering human life to the destruction of economic assets, other damages exceed the spatio-temporal scale of a fire event. Wildfires involving forest resources are associated with intense carbon emissions and alteration of surrounding ecosystems. The induced land cover degradation has also a potential role in exacerbating soil erosion and shallow landslides. A component of the complexity in assessing fire impacts resides in the difference between uncontrolled wildfires and those for which a control strategy is applied. Robust modelling of wildfire behaviour requires dynamic simulations under an array of multiple fuel models, meteorological disturbances and control strategies for mitigating fire damages. Uncertainty is associated to meteorological forecast and fuel model estimation. Software uncertainty also derives from the data-transformation models needed for predicting the wildfire behaviour and its consequences. The complex and dynamic interactions of these factors define a context of deep uncertainty. Here an architecture for adaptive and robust modelling of wildfire behaviour is proposed, following the semantic array programming paradigm. The mathematical conceptualisation focuses on the dynamic exploitation of updated meteorological information and the design flexibility in adapting to the heterogeneous European conditions. Also, the modelling architecture proposes a multi-criteria approach for assessing the potential impact with qualitative rapid assessment methods and more accurate a-posteriori assessment.

Keywords: Wildfire Behaviour, Deep Uncertainty, Integrated Natural Resources Modelling and Management, Semantic Array Programming.

1 Introduction

Landscape and ecosystem dynamics show intense interactions with disturbances [1–4]. The complexity and uncertainty associated with these interactions – along with the severity and variety of the involved impacts – urge robust, holistic and

J. Hřebíček et al. (Eds.): ISESS 2013, IFIP AICT 413, pp. 367–380, 2013.

transparent approaches [5]. In Europe [6], wildfires are one of the major treats at landscape scale. Wildfires have a particular impact in the Mediterranean Europe [7–9] while affecting the aerosol load even in Northern part of the continent [10,11]. Whilst their immediate impact ranges from endangering human life to the destruction of economic assets, other damages exceed the spatio-temporal scale of a fire event [9]. This generates a complex network of transdisciplinary relationships. Such connections need to be properly taken into account in order for meaningful wildfire behaviour modelling to be able to support assessment and management at the science-policy interface.

Here, a general modelling conceptualisation of the interactions among wildfire behaviour, related transdisciplinary impacts, fire management and control strategies is proposed. The special focus on the many sources of uncertainty [5,12] leads to a robust semantically-enhanced modelling architecture based on the paradigm of Semantic Array Programming (SemAP) [13–15]. Arrays of wildfire behaviour models, dynamic information forecasts (i.e. meteorology), static parameterisation (i.e. spatial distribution of fuel) and transdisciplinary impact assessment models (e.g. [17–21]) are considered within a robust ensemble-based architecture.

The concept of Dynamic Data Driven Application Systems (DDDAS) [22, 23] is exploited along with a Partial Open Loop Feedback Control (POLFC) approach [24] for mitigating the overall costs associated with fire events by means of an adaptive use of timely available information. Preliminary prototypes of the proposed architecture have been discussed in [9,25]. The whole architecture is made possible by essentially relying on free software [5,16,26,27] for ensuring semantically enhanced design diversity [12].

1.1 Transdisciplinary Wildfire Behaviour: Continental Scale INRMM

Wildfires constitute a direct risk for human life and health [28] while being associated with intense carbon emissions and alteration of surrounding ecosystems [29,30], from habitat connectivity [31,32] to biodiversity [33]. The induced land cover degradation [34] has also a potential role in exacerbating soil erosion [18,19] and shallow landslides [17] while altering the precipitation-runoff relationship [35] along with flood regimes [36,37]. Off-site effects [38,39] such as fire-induced sediment yield in downstream rivers may cause losses in water storage and affect water quality [40,41] and long-range transport of aerosol may induce transboundary effects [10,11].

Other elements have a role in perturbing large-scale patterns of interactions. Climate change [42] modifies the spatial distribution and habitat suitability of several plant species [5,43]. Along with pest outbreaks, this may alter vegetation response to fire (e.g. changes in deadwood, litter, undergrowth, average moisture) [44]. This context of uncertainty [45] and multiplicity of transdisciplinary assessment criteria [46] suggests integrated natural resources modelling and management (INRMM) [47] as an appropriate frame.

The INRMM approach may result even more appropriate when wide and heterogeneous spatial extents are considered, for which a broad variety of environments is expected. This is the case of the continental scale, here addressed for Europe. Besides and complementing national scale systems, fire information in Europe can rely on the European Forest Fire Information System (EFFIS) which operates over 30 countries in the European and Mediterranean regions, covering the full cycle of forest fire management [6]. Ongoing research [9, 25] investigates integrated fire behaviour modelling with the help of high resolution meteorological data (e.g. [48, 49]) within a modelling paradigm which exploits geospatial and computational tools [15] for supporting robustness and semantic transparency [5, 12]. This work is set within this exploratory context.

2 Wildfire Behaviour and Management of Fire-Events

Wildfire behaviour intensely interacts with the anthropogenic strategies for controlling fire. The complete modelling methodology is presented in section 3. Here, the anthropogenic control for mitigating the fire evolution is focused. Fire control includes broad options for fire suppression or retardation and fuel reduction, which constitute the set $\mathcal{U}^{\mathbf{u}}$ of controls $\mathbf{u}$. (see eq. 2). Many different computational models have long [50] supported wildfire behaviour analysis [51]. The actions for containing the damages of a given fire event may be split in the sequence of controls $\mathbf{u}_t, \mathbf{u}_{t+\Delta t}, \cdots$ which are applied before, during and after the fire lifespan $\mathcal{U}^t = [t_{\text{start}}, t_{\text{end}}]$, where t_{end} itself could be influenced by the controls. For fire event controls, this sequence may be viewed as instantiating a management policy $u(\cdot)$ for the particular event conditions (i.e. the fire state $\mathbf{x}$):

$$\mathbf{u}_t, \mathbf{u}_{t+\Delta t}, \cdots \quad \text{such that} \quad \forall \tau \in \mathcal{U}^t, \; \mathbf{u}_\tau = u(\mathbf{x}_\tau) \tag{1}$$

The temporal distance $\Delta t(t, \mathbf{x}_t)$ between subsequent controls could vary depending on the state $\mathbf{x}_t$ and the exogenous information available at time t. For example, sudden meteorological changes – or approaching highly populated areas – may suggest the need for intensifying the frequency $1/\Delta t$ of controls. Control policies may either be provided to the wildfire behaviour model as external input or they may be generated within the model. The related control sequences may be concisely expressed as $\mathbf{U}^t_{t_1, t_2} = \{\mathbf{u}^t_{t_1}, \mathbf{u}^t_{t_1+\Delta t}, \cdots \mathbf{u}^t_{t_2}\} \in \mathcal{U}^{\mathbf{u}}_{t_1, t_2}$ when they are known at time t and refer to the interval $[t_1, t_2]$.

Wildfire Behaviour Simulation with External Control Scenarios. A planned control sequence $\mathbf{U}^t_{t, t_{\text{end}}} = \{\mathbf{u}^t_t, \mathbf{u}^t_{t+\Delta t}, \cdots\} \in \mathcal{U}^{\mathbf{u}}_{t, t_{\text{end}}}$ might be passed as model input in t without requiring the model to internally generate it. Here, a series of wildfire control actions is decided elsewhere and notified to the model. The planned sequence is also suitable to be periodically revised as $\mathbf{U}^{t+\Delta t}_{t+\Delta t, t_{\text{end}}}$ by exploiting updated information available at time $t + \Delta t$. A posteriori analysis of the already implemented control sequence $\mathbf{U}^{\text{hist}, t_{\text{end}}}_{t_{\text{begin}}, t_{\text{end}}}$ – or alternative sequences $\mathbf{U}^{\text{alt}\, i, t_{\text{end}}}_{t_{\text{begin}}, t_{\text{end}}}$ which could have been implemented instead – is also possible.

Generating Fire-Event Control Strategies. Instead of externally provided control scenarios, wildfire impacts might be assessed for internally generated fire management options $\mathcal{U}_{t,t_{\mathrm{end}}}^{\mathbf{u}}$ so as to explore the huge space of possible alternatives. Approximate optimisation with machine learning and heuristic methods can help to generate reasonable sub optimal candidates. Information updates during an event may suggest iterative re-optimisation for adapting the suggested control strategy to dynamically support the firefighting and emergency operations. At a broader scale, the approach might serve for simulating possible future fires and contributing to a better spatial allocation of fire crews and equipment.

3 Modelling Methodology

3.1 Dynamic Parameterisation and DDDAS

Fuel, meteorology and topography are known as the *fire environment triangle* [52]. Fire behaviour is also shaped by temporal changes of some of these factors. Therefore, time has been proposed as an explicit variable inside fire models into the *square of fire factors* [53,54]. Dynamic Data Driven Application Systems (DDDAS) [23] programmatically exploit data-driven feedbacks for updating the information used as model input (e.g. adaptive estimates of required information via specialised data transformation models, D-TM [12,13,15]). This way, each predictive step is associated with a calibration phase for properly handling information updates on the evolving fire system.

3.2 Uncertainty and Complexity: Semantic Array Programming

The use of updated information (e.g. real-time) for adapting the management of dynamic systems is the essence of many dynamic/adaptive control strategies [55,56]. Some strategies allow varying uncertainties – even related to data-driven information – to be explicitely modelled. Disturbances routinely affect environmental systems. They may show nonlinear and complex patterns of relationship outside and within the system's spatial extent. For example, meteorological forecasting is perturbed by regional (short-term) or global (mid-term) climatic conditions as well as by seasonal and interannual variations (long-term).

Adaptive control is frequently exploited for natural resources management [57]. Water resources management widely uses stochastic dynamic programming (SDP) for optimal seasonal allocation of water within reservoirs (as dams or lakes) [58]

The growing complexity in forest resources management also increasingly benefits from SDP applications (e.g. fire risk management [59,60]).

Unfortunately, the so called *curse of dimensionality* has a strong impact on SDP applicability, since SDP demand for computing resources exponentially grows with the number of state variables. Accurate descriptions of complex systems often imply wider dimensionality of the state vector $\mathbf{x}$. Several strategies [56,61] support natural resources applications in mitigating the intractability of complex problems, by e.g. approximating the SDP with modified algorithms [24,62,63].

The dynamics of a fire event may be described as the D-TM evolution of the system's state $\mathbf{x}$ (the probability for each spatial cell to be a burnt area) in the discrete time interval $\Delta t(t, \mathbf{x}_t)$ with respect to the disturbances $\xi_{t+\Delta t(t,\mathbf{x}_t)}$ (the spatial field of wind speed and direction, humidity, rainfall), the anthropogenic control $\mathbf{u}_t$ and the system characteristics (slope, aspect, fuel), parameterised as θ_t in the following equation system which extends the one presented in [9]:

$$\mathbf{x}_{t+\Delta t(t,\mathbf{x}_t)} = ::\left| f\left(\theta_t, \mathbf{x}_t, \mathbf{u}_t, \xi^{\tau}_{t+\Delta t(t,\mathbf{x}_t)}\right) \right|^{sem}_{::} \tag{2}$$

where:

$$\begin{array}{llll} t & \in \mathcal{U}^t & \mathbf{x}_t = \mathcal{U}^{\mathbf{x}}_t & \mathbf{u}_t \in \mathcal{U}^{\mathbf{u}}_t(\mathbf{x}_t) \\ \xi^{\tau}_{t+\Delta t(t,\mathbf{x}_t)} & \sim \phi(\,\cdot\,|\, I_\tau) & \in \mathcal{U}^{\xi}_t(\mathbf{x}_t, \mathbf{u}_t, I_\tau),\ \tau \in \mathcal{U}^t & \theta_t = \theta(\mathbf{x}_t, \mathbf{u}_t) \end{array} \tag{3}$$

which enables modular modelling of legal obligations by just redefining $\mathcal{U}^{\mathbf{u}}_t(\cdot)$. The D-TM module $f(\cdot)$ is subject to the semantic checks *sem* as pre-, post-conditions and invariants on inputs, outputs and the D-TM itself [12, 64]:

$$y = \ ::\!| f(\theta, x, u, \xi)\, |^{sem}_{::} \ \Leftrightarrow \begin{cases} y = f(\theta, x, u, \xi) \\ \Box\, sem(y, f, \theta, x, u, \xi) \end{cases} \tag{4}$$

The modal/deontic logic operator $\Box sem$ means: "it ought to be that *sem*", where *sem* is a set of valid array-based semantic constraints (within the *semantic array programming* paradigm [13, 14]). A few more obvious constraints are exemplified in the next section as active links ***::constraint::***). The notation for time refers to the instant in which a given quantity will be known without uncertainty. The disturbance vector $\xi^{\tau}_{t+\Delta t(t,\mathbf{x}_t)}$ may be described in terms of a pdf $\phi(\,\cdot\,|\, I_\tau)$ – generally a function of the state and control at time t, also dependent from the available data-driven information I_τ. This notation expresses the uncertainty associated with the disturbances forecast. Depending on the typology of information and on the time τ in which the information is available, the uncertainty of the predicted disturbances will typically be highly variable.

The controls could either be external inputs – in which case the system's evolution can only be simulated – or be internally generated for supporting decision-making, either in real-time or as off-line assessment. The control problem in this case implements a Partial Open Loop Feedback Control (POLFC) approach [24] for minimising the overall costs associated with the fire event, from the time $t \in \mathcal{U}^t$ onwards:

$$u^t(\cdot) = \underset{\mathbf{u} \in \mathcal{U}^{\mathbf{u}}_{t,\, t_{\text{end}}}}{\arg\min} \left[\mathcal{C}^{1,t} \mathcal{C}^{2,t} \cdots \mathcal{C}^{i,t} \cdots \mathcal{C}^{n,t}\right] \tag{5}$$

where each cost $\mathcal{C}^{i,t}$ is linked to an impact assessment criterion (section 4) and is described as a composition of step-wise costs c^i_τ which are estimated with the data-driven information I_t available at time t for the future time $\tau \in \{\tau_0, \tau_1, \cdots \tau_K\} \equiv \{t, t+\Delta t(t, \tilde{\mathbf{x}}_t) \cdots t_{\text{end}}\}$:

$$\begin{aligned} \mathcal{C}^{i,t} &= \underset{\xi^t_{\tau_1},\cdots\xi^t_{\tau_k},\cdots,\xi^t_{t_\text{end}}}{\mathrm{E}} \left[\left(\sum_{\tau=t\equiv\tau_0}^{\tau_{K-1}} c^i_\tau(\tilde{\mathbf{x}}_\tau, \mathbf{u}_\tau, \xi^t_{\tau+\Delta t(\tau,\tilde{\mathbf{x}}_\tau)}) \right) + c^i_{t_\text{end}}(\tilde{\mathbf{x}}_{t_\text{end}}) \right] \\ \tau_k &= \tau_0 + \sum_{\lambda=0}^{k-1} \Delta t(\tau_\lambda, \tilde{\mathbf{x}}_{\tau_\lambda}), \qquad \tau_0 \equiv t, \qquad \tau_K \equiv t_\text{end} \end{aligned} \tag{6}$$

with E a statistical operator and $\tilde{\mathbf{x}}_\tau$ estimated by iterating eq. 2 from the known state $\mathbf{x}_t$ and information I_t. The time associated with each subscript k of τ_k is a function of the sequence $\tilde{\mathbf{x}}_{\tau_0} \cdots \tilde{\mathbf{x}}_{\tau_{k-1}}$. This sequence also depends on a particular disturbance realisation $\xi^t_{\tau_1} \cdots \xi^t_{\tau_k}$ among the possible ones whose distribution is shaped at time t.

3.3 Design Diversity: The State Transition Function

An ensemble approach is proposed for mitigating the software uncertainty [5,12] associated with the state transition function (eq. 2). Although two particular free software simulators are described, the proposed method is general and applicable to other simulators.

FireSim is the first selected simulator. It implements the library FireLib [65] and the Rothermel model [66] by means of a cellular-automata simulator. FireSim is a deterministic parametric and discrete event type simulator structured as a sequential set of D-TMs with a pipeline structure of four stages.

The second selected simulator is implemented in the free software GRASS GIS [67]. The model, based on the least cost path algorithm and simulating elliptically anisotropic spread, is implemented in the function *r.spread* [68], which takes as input the rate of spread (ROS) generated by the function *r.ros* [69]. This latter computes the ROS following the Rothermel model [66] and is based on the Fortran code by [70]. The direction of the maximum ROS is the vector sum of the forward ROS in wind direction and that in upslope direction. The obtained raster map layers serve as inputs for *r.spread*.

The two instances of the state transition function (eq. 2) are deterministic. Within each run, the state $\mathbf{x}_{t,c}$ of a given cell c at time t is converted to be a **::binary::**[1] flag recording whether c is burnt. Ensembling multiple runs of the instances allows an uncertainty analysis to be performed for different scenarios (ensemble inputs might be more general: valid runs are expected to provide **::probability::**[2] values $\in [0,1]$). The weighted-quantile analysis of the ensemble requires all output **::matrix::**[3] layers to have the **::same_size::**[4] for generating a **::sortable::**[5] 3-dimensional array **::3-array::**[6].

[1] http://mastrave.org/doc/mtv_m/check_is#SAP_binary
[2] http://mastrave.org/doc/mtv_m/check_is#SAP_probability
[3] http://mastrave.org/doc/mtv_m/check_is#SAP_matrix
[4] http://mastrave.org/doc/mtv_m/check_is#SAP_same_size
[5] http://mastrave.org/doc/mtv_m/check_is#SAP_sortable
[6] http://mastrave.org/doc/mtv_m/check_is#SAP_3-array

4 Assessing Wildfire Impacts

Wildfires involving forest resources are associated with intense carbon emissions (CE) and alteration of surrounding ecosystems. Forest fires can affect many different soil properties (chemical, mineralogical and biological) [71]. Soil properties, jointly with vegetation cover and hydrology are among the main factors influencing the relation fires - geomorphic processes.

High-severity wildfire, can induce soil water repellency and reduce infiltration capacity [71,72]. Soil hydrophobicity together with the reduced ground cover has been postulated as one of the main causes of soil erosion (SE) and runoff increasing in after-burn areas [73,74]. The reduced ground cover is also responsible of a reduced root strength affecting the soil cohesion and of a reduced leaf area evapotranspiration affecting the soil water level [74]. This change in hydrological condition is the main triggering factor for landslide initiation. Soil erosion and landslide susceptibility (LS) can then be considerably increased by wildfires.

The impact of landslide susceptibility $\mathcal{C}^{LS,t}$ and soil erosion $\mathcal{C}^{SE,t}$ (eq. 5) are mostly determined by their post-fire component $c^i_{t_{\text{end}}}(\tilde{\mathbf{x}}_{t_{\text{end}}})$, $i \in \{LS, ES\}$ (eq. 6) while the impact of wildfire carbon emissions is substantially determined by the step-wise costs $c^{CE}_t(\tilde{\mathbf{x}}_t, \mathbf{u}_t, \xi^t_{t+\Delta t(t,\tilde{\mathbf{x}}_t)})$ and shaped not only by the initial spatial distribution of carbon stock [20] but also by the pattern of fire intensity.

A fast computational modelling is required for the rapid assessment of carbon emissions, soil erosion and landslide susceptibility in burnt areas. Physically based models can in principle be more suitable for being applied in areas presenting complex characteristics without any possibility for model calibration. The gap between the number and accuracy of the required input parameters is a limit, therefore complex physically based models are generally inapplicable. Empirical models and less data demanding conceptual models might provide a robust preliminary assessments of post-fire soil erosion and landslide susceptibility. The landslide susceptibility has been calculated by applying a deterministic model (SINMAP) [75,76] based on the infinite slope equation. A modified version of the Revised Universal Soil Loss Equation (e-RUSLE) [18,19] has been applied for evaluating pre- and post-fire soil erosion.

Landslide Susceptibility. In SINMAP a topographically based steady-state hydrological model is linked with the infinite slope model for evaluating stability conditions. The model applies to shallow translational landslides triggered by water. The slope stability analysis is governed, within the model, by calculating slope and specific catchment areas starting from a Digital elevation model. The input parameters of SINMAP follow a uniform distribution between an upper and lower limit. Because of the flexibility of its input parameters SINMAP can be considered as a suitable model for a rapid slope stability assessment. Further studies are still ongoing for overcoming the limits of the SINMAP approach. A more accurate a-posteriori analysis is proposed in [17] where a new fuzzy ensemble is applied for aggregating an array of different landslide susceptibility models.

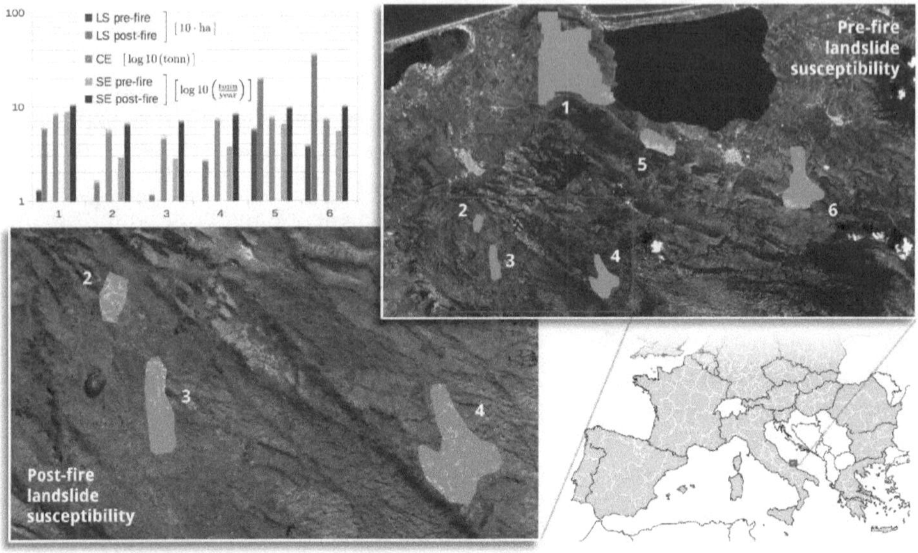

Fig. 1. Rapid assessment of fire carbon emission (CE), pre- and post-fire shallow landslide susceptibility (LS) and water soil erosion (SE) in Northern Puglia (Italy)

Soil Erosion. The e-RUSLE is based on an architecture designed for easing the integration of erosion-related natural resources models. The semantic array programming paradigm [13, 14] and the computational reproducibility [19, 77] are at the basis of the applied modelling architecture. Its flexibility, low data demanding and architecture make the model as one of the best option for rapid post-fire analysis. The e-RUSLE has been applied, for a rapid assessment, using only public available datasets [18, 78]. For a more accurate a-posteriori assessment, remote sensing techniques might be applied. The changes in vegetation cover could be better analysed by estimating vegetation indexes as NDVI or fA-PAR [79]. Furthermore the soil erosion models based on the USLE provide long term erosion averages and have been applied in many different environments and climatic conditions in numerous countries [80–82]. A rapid assessment for the interval of uncertainty associated with the post-fire soil erosion is provided by the DerBos nondimensionalised equation 7. It describes a RUSLE based lower- and upper-bound for the expected soil erosion:

$$\frac{E_c^{burnt}}{E_c^{pot}} = \begin{cases} \max\left(\min\left(1, C_c^{\alpha}\frac{KP_c^{burnt}}{KP_c^{pre}}\right), \frac{E_c^{pre}}{E_c^{pot}}\right) & \text{max. estimation} \\ \min\left(1, \max\left(C_c^{\beta}\frac{KP_c^{burnt}}{KP_c^{pre}}, \frac{E_c^{pre}}{E_c^{pot}}\right), C_c^{\gamma}\frac{KP_c^{burnt}}{KP_c^{pre}}\right) & \text{min. estimation} \end{cases} \quad (7)$$

where subscripts refer to a given spatial cell c and the variables are defined as:

E_c^{pre}	:	pre-fire soil erosion	K_c^{pre}	:	pre-fire K factor
E_c^{pot}	:	potential soil erosion	K_c^{burnt}	:	post-fire K factor
E_c^{burnt}	:	post-fire soil erosion	P_c^{pre}	:	pre-fire P factor
$C_c^{\alpha}, C_c^{\beta}, C_c^{\gamma}$	:	post-fire C factors	P_c^{burnt}	:	post-fire P factor
KP_c^{pre}	:	$K_c^{pre} \cdot P_c^{pre}$	KP_c^{burnt}	:	$K_c^{burnt} \cdot P_c^{burnt}$

C_c^{α},C_c^{β},C_c^{γ} are dimensionless factors representing a **::proportion::**[7] $\in [0\ 1]$ and their values must be ordered in descending order (**::sorted-descend::**[8]) so as to have $C_c^{\alpha} \geq C_c^{\beta} \geq C_c^{\gamma}$. C_c^{α} represents the maximum degradation of the cover factor due to the wildfire. C_c^{β} and C_c^{γ} describe the C-factor minimum expected degradation. While C_c^{β} generally worsens E_c^{burnt} versus E_c^{pre} where $\frac{E_c^{pre}}{E_c^{pot}}$ is low, C_c^{γ} may even cause the post-fire erosion to be less than the pre-fire one in already high-exposed areas in which E_c^{pre} was close to the potential erosion (A substantial ash deposition covering the soil after a fire can reduce the splash detachment and the surface runoff [83]). The ratios $\frac{E_c^{pre}}{E_c^{pot}}$ and $\frac{E_c^{burnt}}{E_c^{pot}}$ are also **::proportion::**s $\in [0\ 1]$.

Carbon Emission. Wildfire carbon emission in a given spatial cell c depends on the overall amount of carbon (total fuel load) and on the local characteristics of fuel (e.g. dead and live fuel, woody and herbaceous one, typical diameters of these fuels) [84]. Also, each component of the a given fuel class is associated with a specific burning efficiency and emission factor for estimating CO_2 releases in atmosphere which may vary for flaming or smoldering fire [84]. The availability of spatially distributed estimations of carbon stock is essential so as to avoid basing CE assessments on rough approximations with fixed conversion factors for each fuel class. The recent pan-European map of living forest carbon stock at IPCC Tier 1 level [20] has been used for integrating the method of [84] with a better spatialisation of CE rapid assessment.

5 Conclusions

A flexible architecture for adaptive and robust modelling of wildfire behaviour is proposed, following the *semantic array programming* paradigm [13,14]. It relies on an array-based and semantically enhanced application of the *dynamic data driven application systems* [22,23] for addressing dynamic simulations under an array of multiple fuel models, meteorological disturbances and control strategies for mitigating fire damages.

The mathematical conceptualisation focuses on the dynamic exploitation of updated meteorological information and the design flexibility in adapting to the heterogeneous European conditions. Uncertainty is associated to meteorological forecast and fuel model estimation. Software uncertainty [5,12] is also considered

[7] `http://mastrave.org/doc/mtv_m/check_is#SAP_proportion`
[8] `http://mastrave.org/doc/mtv_m/check_is#SAP_sorted-descend`

in the data-transformation models needed for predicting the wildfire behaviour and its consequences. The complex and dynamic interactions of these factors define a context of deep uncertainty [5] whose proposed mitigation is based on robust ensembling.

The modelling architecture proposes a multi-criteria approach for assessing potential wildfire impacts with qualitative rapid assessment methods and more accurate a-posteriori assessment. The rapid assessment of soil erosion, landslide susceptibility and carbon emissions is discussed and exemplified.

References

1. Bengtsson, J., Nilsson, S.G., Franc, A., Menozzi, P.: Biodiversity, disturbances, ecosystem function and management of European forests. For. Ecol. Manage. 132(1), 39–50 (2000)
2. Turner, M.G.: Landscape ecology: What is the state of the science? Annu. Rev. Ecol. Evol. Syst. 36(1), 319–344 (2005)
3. Turner, M.G.: Disturbance and landscape dynamics in a changing world. Ecology 91(10), 2833–2849 (2010)
4. Seidl, R., Fernandes, P.M., Fonseca, T.F., Gillet, F., Jönsson, A.M., et al.: Modelling natural disturbances in forest ecosystems: a review. Ecol. Model. 222(4), 903–924 (2011)
5. de Rigo, D.: Behind the horizon of reproducible integrated environmental modelling at European scale: ethics and practice of scientific knowledge freedom. F1000 Research (submitted 2013)
6. San-Miguel-Ayanz, J., Schulte, E., Schmuck, G., Camia, A., Strobl, P., et al.: Comprehensive Monitoring of Wildfires in Europe: The European Forest Fire Information System (EFFIS). In: Tiefenbacher, J. (ed.) Approaches to Managing Disaster - Assessing Hazards, Emergencies and Disaster Impacts, ch. 5. InTech (2012)
7. Díaz-Delgado, R., Lloret, F., Pons, X., Terradas, J.: Satellite evidence of decreasing resilience in Mediterranean plant communities after recurrent wildfires. Ecology 83(8), 2293–2303 (2002)
8. Pausas, J.G., Llovet, J., Rodrigo, A., Vallejo, R.: Are wildfires a disaster in the Mediterranean basin? – a review. Int. J. Wildland Fire 17(6), 713+ (2008)
9. Rodriguez-Aseretto, D., de Rigo, D., Di Leo, M., Cortés, A., San-Miguel-Ayanz, J.: A data-driven model for large wildfire behaviour prediction in Europe. Procedia Computer Science 18, 1861–1870 (2013)
10. Barnaba, F., Angelini, F., Curci, G., Gobbi, G.P.: An important fingerprint of wildfires on the European aerosol load. Atmos. Chem. Phys. 11(20), 10487–10501 (2011)
11. Saarikoski, S., Hillamo, R.: Wildfires as a source of aerosol particles transported to the northern european regions. In: The Handbook of Environmental Chemistry, pp. 1–21. Springer, Heidelberg (2012)
12. de Rigo, D.: Software Uncertainty in Integrated Environmental Modelling: the role of Semantics and Open Science. Geophys. Res. Abstr. 15, 13292+ (2013)
13. de Rigo, D.: Semantic Array Programming for Environmental Modelling: Application of the Mastrave Library. In: Int. Congress on Environmental Modelling and Software. Managing Resources of a Limited Plant, Pathways and Visions under Uncertainty, Sixth Biennial Meeting, pp. 1167–1176 (2012)

14. de Rigo, D.: Semantic array programming with Mastrave - introduction to semantic computational modelling (2012)
15. de Rigo, D., Corti, P., Caudullo, G., McInerney, D., Di Leo, M., San Miguel-Ayanz, J.: Toward Open Science at the European Scale: Geospatial Semantic Array Programming for Integrated Environmental Modelling. Geophys. Res. Abstr. 15, 13245+ (2013)
16. de Rigo, D., Guariso, G.: Rewarding Open Science: A Collaborative Review System for Semantically-Enhanced Free Software and Environmental Data Modelling (in prep., 2013)
17. Bosco, C., de Rigo, D., Dijkstra, T., Sander, G., Wasowski, J.: Multi-Scale Robust Modelling of Landslide Susceptibility: Regional Rapid Assessment and Catchment Robust Fuzzy Ensemble. In: Hřebíček, J., Schimak, G., Kubásek, M., Rizzoli, A. (eds.) ISESS 2013. IFIP AICT, vol. 413, pp. 321–335. Springer, Heidelberg (2013)
18. Bosco, C., de Rigo, D., Dewitte, O., Poesen, J., Panagos, P.: Modelling Soil Erosion at European Scale. Towards Harmonization and Reproducibility (in prep.)
19. de Rigo, D., Bosco, C.: Architecture of a pan-european framework for integrated soil water erosion assessment. In: Hřebíček, J., Schimak, G., Denzer, R. (eds.) ISESS 2011. IFIP AICT, vol. 359, pp. 310–318. Springer, Heidelberg (2011)
20. de Rigo, D., Barredo, J.I., Busetto, L., Caudullo, G., San-Miguel-Ayanz, J.: Continental-Scale Living Forest Biomass and Carbon Stock: a Robust Fuzzy Ensemble of IPCC Tier 1 Maps for Europe. In: Hřebíček, J., Schimak, G., Kubásek, M., Rizzoli, A. (eds.) ISESS 2013. IFIP AICT, vol. 413, pp. 271–284. Springer, Heidelberg (2013)
21. Barredo, J.I., San-Miguel-Ayanz, J., Caudullo, G., Busetto, L.: A European map of living forest biomass and carbon stock. In: Reference Report by the Joint Research Centre of the European Commission. EUR – Scientific and Technical Research, vol. 25730 (2012)
22. Darema, F.: Dynamic Data Driven Applications Systems: A New Paradigm for Application Simulations and Measurements. In: Bubak, M., van Albada, G.D., Sloot, P.M.A., Dongarra, J. (eds.) ICCS 2004. LNCS, vol. 3038, pp. 662–669. Springer, Heidelberg (2004)
23. Denham, M., Cortés, A., Margalef, T., Luque, E.: Applying a Dynamic Data Driven Genetic Algorithm to Improve Forest Fire Spread Prediction. In: Bubak, M., van Albada, G.D., Dongarra, J., Sloot, P.M.A. (eds.) ICCS 2008, Part III. LNCS, vol. 5103, pp. 36–45. Springer, Heidelberg (2008)
24. Castelletti, A., de Rigo, D., Tepsich, L., Soncini-Sessa, R., Weber, E.: On-Line design of water reservoir policies based on inflow prediction. IFAC-PapersOnLine 17, 14540–14545 (2008)
25. Di Leo, M., de Rigo, D., Rodriguez-Aseretto, D., Bosco, C., Petroliagkis, T., Camia, A., San-Miguel-Ayanz, J.: Dynamic Data Driven Ensemble for Wildfire Behaviour Assessment: A Case Study. In: Hřebíček, J., Schimak, G., Kubásek, M., Rizzoli, A. (eds.) ISESS 2013. IFIP AICT, vol. 413, pp. 11–22. Springer, Heidelberg (2013)
26. Stallman, R.M.: Viewpoint: Why "open source" misses the point of free software. Commun. ACM 52(6), 31–33 (2009)
27. Rodriguez Aseretto, D., Di Leo, M., de Rigo, D., Corti, P., McInerney, D., et al.: Free and Open Source Software underpinning the European Forest Data Centre. Geophys. Res. Abstr. 15, 12101+ (2013)
28. Richardson, L.A., Champ, P.A., Loomis, J.B.: The hidden cost of wildfires: Economic valuation of health effects of wildfire smoke exposure in southern California. J. For. Econ. 18(1), 14–35 (2012)

29. Herrando, S., Brotons, L., Llacuna, S.: Does fire increase the spatial heterogeneity of bird communities in Mediterranean landscapes? Ibis 145(2), 307–317 (2003)
30. Torre, I., Díaz, M.: Small mammal abundance in Mediterranean post-fire habitats: a role for predators? Acta Oecologica 25(3), 137–142 (2004)
31. Estreguil, C., Caudullo, G., de Rigo, D., Whitmore, C., San-Miguel-Ayanz, J.: Reporting on European forest fragmentation: Standardized indices and web map services. IEEE Earthzine 5(2), 384031+ (2012); 2nd quarter theme: Forest Resource Information
32. Estreguil, C., Caudullo, G., de Rigo, D., San-Miguel-Ayanz, J.: Forest landscape in Europe: pattern, fragmentation and connectivity. EUR – Scientific and Technical Research 25717(JRC 77295) (2013)
33. Moreira, F., Russo, D.: Modelling the impact of agricultural abandonment and wildfires on vertebrate diversity in Mediterranean Europe. Landsc. Ecol. 22(10), 1461–1476 (2007)
34. Di Piazza, G.V., Di Stefano, C., Ferro, V.: Modelling the effects of a bushfire on erosion in a mediterranean basin. Hydrol. Sci. J. 52(6), 1253–1270 (2007)
35. Moody, J.A., Martin, D.A., Haire, S.L., Kinner, D.A.: Linking runoff response to burn severity after a wildfire. Hydrol. Proces. 22(13), 2063–2074 (2008)
36. Candela, A., Aronica, G., Santoro, M.: Effects of forest fires on flood frequency curves in a Mediterranean catchment. J. Hydrol. Sci. 50(2), 193–206 (2005)
37. Rulli, M.C., Rosso, R.: Hydrologic response of upland catchments to wildfires. Adv. Water Resour. 30(10), 2072–2086 (2007)
38. Ice, G.G., Neary, D.G., Adams, P.W.: Effects of wildfire on soils and watershed processes. J. Forestry 102(6), 16–20 (2004)
39. Shakesby, R., Doerr, S.: Wildfire as a hydrological and geomorphological agent. Earth Sci. Rev. 74(3-4), 269–307 (2006)
40. Smith, H.G., Sheridan, G.J., Lane, P.N.J., Nyman, P., Haydon, S.: Wildfire effects on water quality in forest catchments: a review with implications for water supply. J. Hydrol. 396(1-2), 170–192 (2011)
41. Smith, H.G., Blake, W.H., Owens, P.N.: Discriminating fine sediment sources and the application of sediment tracers in burned catchments: a review. Hydrol. Proces. 27(6), 943–958 (2013)
42. Krawchuk, M.A., Moritz, M.A., Parisien, M.-A., Van Dorn, J., Hayhoe, K.: Global pyrogeography: the current and future distribution of wildfire. PLoS ONE 4(4), e5102+ (2009)
43. Pautasso, M., Dehnen-Schmutz, K., Holdenrieder, O., Pietravalle, S., Salama, N., et al.: Plant health and global change some implications for landscape management. Biol. Rev. 85(4), 729–755 (2010)
44. Nijhuis, M.: Forest fires: Burn out. Nature 489(7416), 352–354 (2012)
45. Hou, Y., Burkhard, B., Müller, F.: Uncertainties in landscape analysis and ecosystem service assessment. J. Env. Manag. (2013)
46. Spangenberg, J.H., Settele, J.: Precisely incorrect? Monetising the value of ecosystem services. Ecol. Complex. 7(3), 327–337 (2010)
47. de Rigo, D.: Integrated Natural Resources Modelling and Management: minimal redefinition of a known challenge for environmental modelling. Excerpt from the Call for a Shared Research Agenda Toward Scientific Knowledge Freedom, Maieutike Research Initiative (2012)
48. Hollingsworth, A., Engelen, R., Textor, C., Benedetti, A., Boucher, O., et al.: Toward a monitoring and forecasting system for atmospheric composition. Bull. Amer. Meteor. Soc. 89(8), 1147–1164 (2008)

49. Bouttier, F.: The Météo-France NWP system: description, recent changes and plans. CNRM (2010)
50. Perry, G.L.W.: Current approaches to modelling the spread of wildland fire: Areview. Prog. Phys. Geogr. 22(2), 222–245 (1998)
51. Andrews, P., Finney, M., Fischetti, M.: Predicting wildfires. Sci. American 297(2), 46–55 (2007)
52. Countryman, C.M.: The concept of fire environment. Fire Manag. Today 64(1), 49–52 (2004)
53. Viegas, D.X.: Parametric study of an eruptive fire behaviour model. Int. J. Wildland Fire 15(2), 169+ (2006)
54. Rodriguez Aseretto, R., Cortés, A., Margalef, T., Luque, E.: An Adaptive System for Forest Fire Behavior Prediction. In: 11th IEEE Int. Conf. on Computational Science and Engineering, CSE 2008, pp. 275–282 (2008)
55. Bellman, R., Kalaba, R.: On adaptive control processes. IRE Transactions on Autom. Control 4(2), 1–9 (1959)
56. Wang, F.-Y., Zhang, H., Liu, D.: Adaptive Dynamic Programming: An Introduction. IEEE Computational Intell. Mag. 4(2), 39–47 (2009)
57. Lara, M., Doyen, L.: Sustainable Management of Natural Resources: Mathematical Models and Methods. Springer (2008)
58. Rani, D., Moreira, M.: Simulation-Optimization Modeling: A Survey and Potential Application in Reservoir Systems Operation. Water Resour. Manag. 24(6), 1107–1138 (2010)
59. Ferreira, L., Constantino, M.F., Borges, J.G., Garcia-Gonzalo, J.: A Stochastic Dynamic Programming Approach to Optimize Short-Rotation Coppice Systems Management Scheduling: An Application to Eucalypt Plantations under Wildfire Risk in Portugal. Forest Sci. 58(4), 353–365 (2012)
60. Minas, J.P., Hearne, J.W., Handmer, J.W.: A review of operations research methods applicable to wildfire management. Int. J. Wildland Fire 21(3), 189+ (2012)
61. Lee, J.M., Lee, J.H.: Approximate dynamic programming strategies and their applicability for process control: a review and future directions. Int. J. Control Autom. Syst. 2(3), 263–278 (2004)
62. de Rigo, D., Rizzoli, A.E., Soncini-Sessa, R., Weber, E., Zenesi, P.: Neuro-dynamic programming for the efficient management of reservoir networks. In: Proceedings of MODSIM 2001, Int. Congress on Modelling and Simulation Held in Canberra, Australia, vol. 4, pp. 1949–1954 (2001) ISBN: 0-867405252
63. de Rigo, D., Castelletti, A., Rizzoli, A.E., Soncini-Sessa, R., Weber, E.: A selective improvement technique for fastening neuro-dynamic programming in water resources network management. IFAC-PapersOnLine 16, 7–12 (2005)
64. de Rigo, D.: Applying semantic constraints to array programming: the module "check_is" of the Mastrave modelling library. Mastrave Project Technical Report (2011)
65. Bevins, C.D.: FireLib user manual and technical reference. Systems for Environmental Management (1996)
66. Rothermel, R.: How to predict the spread and intensity of forest and range fires. US Department of Agriculture, Forest Service, Intermountain Forest and Range Experiment Station, Ogden (1983)
67. Neteler, M., Bowman, M.H., Landa, M., Metz, M.: GRASS GIS: A multi-purpose open source GIS. Environmental Modelling & Software 31, 124–130 (2012)
68. Xu, J., Lathrop, R.G.: GRASS GIS manual: r.spread. In: GRASS GIS 6.4.3svn Reference Manual (2006)

69. Xu, J.: GRASS GIS manual: r.ros. In: GRASS GIS 6.4.3svn Reference Manual (2009)
70. Andrews, P., LLoyd, P.: Fire modeling and information system technology. Int. J. Wildland Fire 10(4), 343–352 (2001)
71. Certini, G.: Effects of fire on properties of forest soils: A review. Oecologia 143, 1–10 (2005)
72. DeBano, L.F.: The role of fire and soil heating on water repellency in wildland environments: A review. J. Hydrol. 231-232, 195–206 (2000)
73. Letey, J.: Causes and consequences of fire-induced soil water repellency. Hydrol. Processes 15, 2867–2875 (2001)
74. Swanson, F.J.: Fire and Geomorphic Processes. In: Gen. Tech. Rep. USDA For. Serv. WO-26. Washington DC. pp. 401–420 (1981)
75. Pack, R.T., Tarboton, D.G., Goodwin, C.N.: The SINMAP Approach to Terrain Stability Mapping. In: 8th Congress of the International Association of Engineering Geology, Vancouver, British Columbia, Canada (1998)
76. Pack, R.T., Tarboton, D.G., Goodwin, C.N., Prasad, A.: SINMAP 2. A Stability Index Approach to Terrain Stability Hazard Mapping, technical description and users guide for version 2.0. Utah State University (2005)
77. Bosco, C., de Rigo, D., Dewitte, O., Montanarella, L.: Towards the reproducibility in soil erosion modeling: a new Pan-European soil erosion map. In: Wageningen conference on applied soil science: 'Soil Science in a Changing World, p. 209+ (2011)
78. Panagos, P., Jones, A., Bosco, C., Senthil Kumar, P.S.: European digital archive on soil maps (EuDASM): preserving important soil data for public free access. Int. J. of Digital Earth 4(5), 434–443 (2011)
79. Myneni, R.B., Williams, D.L.: On the relationship between FAPAR and NDVI. Remote Sensing of Environment 49(3), 200–211 (1994)
80. Kinnell, P.I.A.: Event soil loss, runoff and the Universal Soil Loss Equation family of models: A review. Journal of Hydrology 385, 384–397 (2010)
81. Angima, S.D., Stott, D.E., O'Neill, M.K., Ong, C.K., Weesies, G.: Soil erosion prediction using RUSLE for central Kenyan highland conditions. Agriculture Ecosystems and Environment 97, 295–308 (2003)
82. Bosco, C., Rusco, E., Montanarella, L., Panagos, P.: Soil erosion in the Alpine area: risk assessment and climate change. Studi Trent. Sci. Nat. 85, 117–123 (2009)
83. Cerdà, A., Doerr, S.H.: The effect of ash and needle cover on surface runoff and erosion in the immediate post-fire period. Catena 74, 256–263 (2008)
84. Barbosa, P., Camia, A., Kucera, J., Libertà, G., Palumbo, I., San-Miguel-Ayanz, J., Schmuck, G.: Assessment of Forest Fire Impacts and Emissions in the European Union Based on the European Forest Fire Information System. Developments in Environmental Science 8, 197–208 (2008)

TiTiMaKe: A Distributed Service Architecture for Security Applications

Mauno Rönkkö[1], Markus Stocker[1], Markku Huttunen[2], Ville Kotovirta[2], and Mikko Kolehmainen[1]

[1] Department of Environmental Science, University of Eastern Finland, P.O. Box 1627, 70211 Kuopio, Finland
{mauno.ronkko,markus.stocker,mikko.kolehmainen}@uef.fi
[2] VTT Technical Research Centre of Finland, P.O. Box 1000, 02044 VTT, Finland
{markku.huttunen,ville.kotovirta}@vtt.fi

Abstract. In this article, we discuss integration of distributed information services, to support the implementation of applications for fields of safety and security. Although there are standard interfacing technologies, such as SOAP and REST, there is very little architectural support for the development of personalized and case specific applications, where the same data sources and computational services are integrated in different ways. As the main contribution, we present here a distributed service architecture, called TiTiMaKe, that supports the development of scenario based applications. We also illustrate TiTiMaKe by describing its use in applications in fields of surveillance and emergency rescue.

Keywords: Distributed Systems, Internet Computing, Service Networks, Safety and Security.

1 Introduction

In this article, we discuss information system solutions to crisis prevention and incident management. These phases appear in various models for crisis management, such as in the crisis management chain by Carter [1] and in the relational model by Jaques [2]. In crisis prevention and incident management, enhanced perception by using state-of-the-art sensory technology coupled with computational methods can significantly improve situational awareness [3], that is, the comprehension and projection capabilities for the actors involved in the action. Also, through specifically tailored systems architecture, the enhanced communication capabilities can significantly improve the shared situational picture, leading to faster and more accurate responses.

Applications for crisis prevention and incident management are classified as safety critical systems. According to Knight [4] "safety-critical systems are those systems whose failure could result in loss of life, significant property damage, or damage to the environment". Safety-critical systems require significant effort

J. Hřebíček et al. (Eds.): ISESS 2013, IFIP AICT 413, pp. 381–390, 2013.

in specification, architecture design, implementation verification, and processing. In particular, reliability, efficiency, and resilience are key factors for the design. In this respect, Internet computing and cloud computing offer solutions for improving not only robustness and performance, but also extensibility and adaptation.

When developing service-based applications, however, there is virtually no support for features like adaptation, personalization, and automated reasoning at runtime. The same applies, if there is a need for data manipulation and transformation on-the-fly. Currently, the support is found at the technical level of integration. For instance, modern programming environments do support code generation for service integration, especially for standard interfacing techniques like SOAP and REST. Not surprisingly, some work on identifying integration patterns [5] has been done recently.

In this article, as the main contribution, we present an extensible, distributed service architecture, called TiTiMaKe, that supports the development of scenario based safety and security applications. The TiTiMaKe architecture is based on core logic design, where design patterns [6], such as factory, bridge, facade, chain-of-responsibility, mediator, observer, thread pool, and messaging patterns, are applied. The architecture not only supports the integration of new data sources and computational services, but also their concurrent use in multiple scenarios. Concurrency issues are resolved with an internal event queue and attachable event listeners running in multiple threads. Because of this, the architecture supports also the reuse of components throughout different scenario implementations. This is also how the TiTiMaKe architecture differs from some existing architectures for safety and security applications, such as UHHA [7]. In this article, we also illustrate the use of the TiTiMaKe architecture in two scenarios that use the same sensor network as the data source, but integrate it with very different kinds of computational services. Both scenarios were implemented and tested for functioning.

The rest of the article is organized as follows. We start by briefly describing the used information services in Section 2. In Section 3, we describe the TiTiMaKe architecture. In Section 4, we illustrate the use of TiTiMaKe in implementing two security applications. Finally, in Section 5, follows the conclusion.

2 Distributed Service Network

The service network considered in this article consists of multiple, heterogeneous services. These are the ToMoVaKe sensor network, Vehicle identification service, Dispersion modeling service, Reachability analysis service, and Spatio-temporal population modeling. We shall now briefly introduce each of these services.

The ToMoVaKe sensor network operated by Lappeenranta University of Technology, LUT Savo Laboratory of Green Chemistry, comprises of three zones: two control centers in two cities (Mikkeli and Kuopio) and an exercise field at the training site of Emergency Services College (PEO). There are many kinds of sensors at the field connected to the sensor network. The sensors include, among others, vibration sensors, acoustic sensors, weather stations, surveillance camera systems, and chemical detectors. The network supports versatile communication technologies, including wireless communications protocols, fiber optics, and Zigbee. The ToMoVaKe sensor network provides both an interface for accessing raw measurement data as well as a user interface framework for applications.

The University of Eastern Finland (UEF) and the VTT Technical Research Centre of Finland (VTT) have developed a non-intrusive vehicle classification service [8] based on vibration measurements. The service is accessed on-line and it uses trained neural networks to classify vehicles based on preprocessed vibration measurement data. The service runs in real-time. It can be trained for different environmental conditions, including annual weather changes, temporal climate changes, and sustained changes in ambient noise.

Aalto University (Aalto) has developed an on-line reachability analysis service. It computes the reachability analysis for both on-road and off-road vehicles. Therefore, it uses not only road network data, but also forest and landscape vegetation data. It computes a reachability map based on location data and vehicle type classification. The map is provided as raster data.

The Finnish Meteorological Institute (FMI) has developed a on-line chemical dispersion modeling service, called ESCAPE [9,10]. It computes the dispersion forecast based on weather data, type of the chemical, and other properties, such as emission location and altitude. The computed dispersion map contains also estimated concentration values. The map is provided as vector data.

Aalto University (Aalto) has also developed an on-line spatio-temporal population modeling service [11]. The service computes an estimation of inhabitants in a region at a given time of a day. The computation is, thus, dynamic taking into account variations in population dynamics during different times of the year and day. The estimation is based on building and road data. It differentiates between building types (e.g. office or residential). The service provides population estimation in XML format based on given location parameters.

3 TiTiMaKe Architecture

The TiTiMaKe architecture is depicted in Figure 1. It consists of the TiTiMaKe server, ToMoVaKe system, FMI server, Aalto server, and PEO sensor network. ToMoVaKe stores the measurements. It also provides messaging and query services that are used by the TiTiMaKe server to fetch the measurements. The TiTiMake server contains the core logic, which is responsible for running the application workflow. The core logic is used to detect anomalies in the measurements and to send alarms to ToMoVaKe through the Sensor ML interface. The core logic is also used to initiate relevant computations on FMI and Aalto servers over HTTP interfaces. The TiTiMaKe server relays computational results to ToMoVaKe for visualization. This logical decomposition is also depicted in Figure 1.

TiTiMaKe uses design patterns [6], such as factory, bridge, facade, chain-of-responsibility, mediator, observer, thread pool, and messaging. The architecture

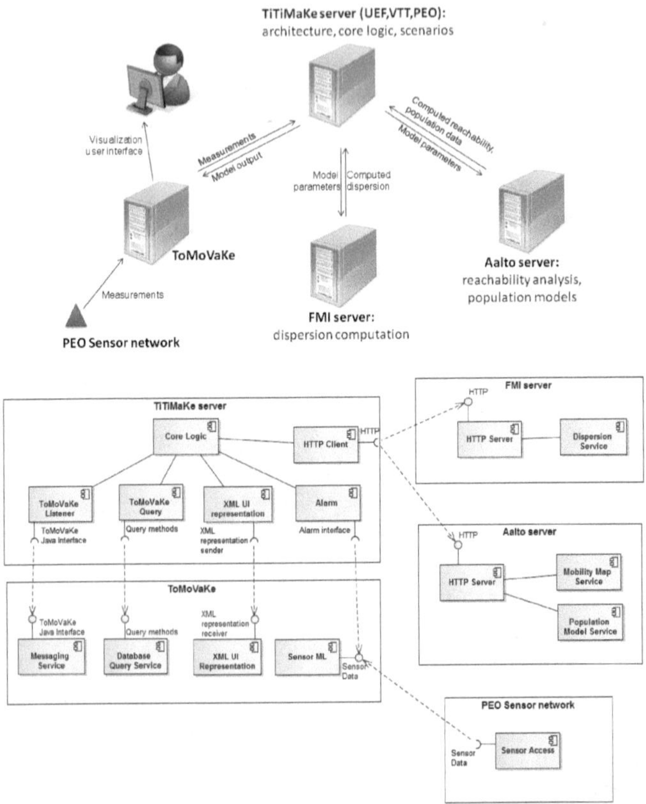

Fig. 1. TiTiMaKe architecture and logical decomposition

supports the integration of new data sources and computational services, as well as their concurrent use in multiple scenarios. Concurrency issues are resolved with an internal event queue and attachable event listeners running in multiple threads. Because of this, the architecture supports also the reuse of components throughout different scenario implementations.

Figures 2 and 3 represent the domain view and the dynamic view of the TiTiMaKe core logic, respectively. The core logic has monitors, each of which executes a single scenario. A monitor has an associated listener that implements the scenario by orchestrating calls to distributed modeling services, integrating their results, and communicating with ToMoVaKe to present the modeling results. Listeners use detectors to detect certain events that are produced by sensors. For instance, a sensor could be collecting classified vehicle observations to a queue accessed by a detector. If the observations in the queue correspond to messages of interest, the detector forwards the observations to the listener. Note that the implementation of listeners, detectors, and sensors are specific to each scenario.

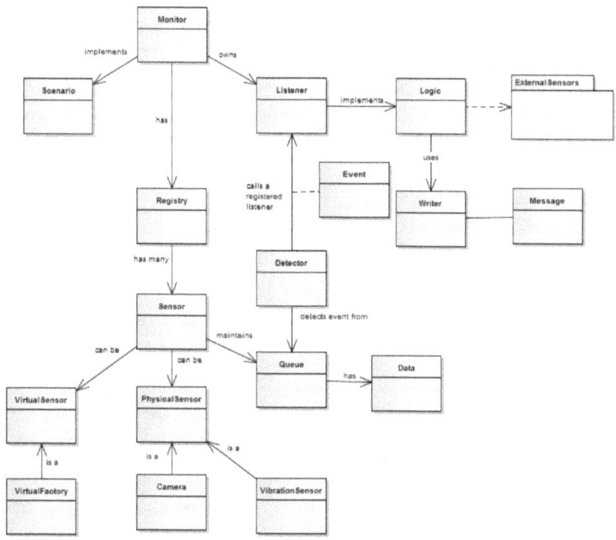

Fig. 2. Domain view of the TiTiMaKe core logic

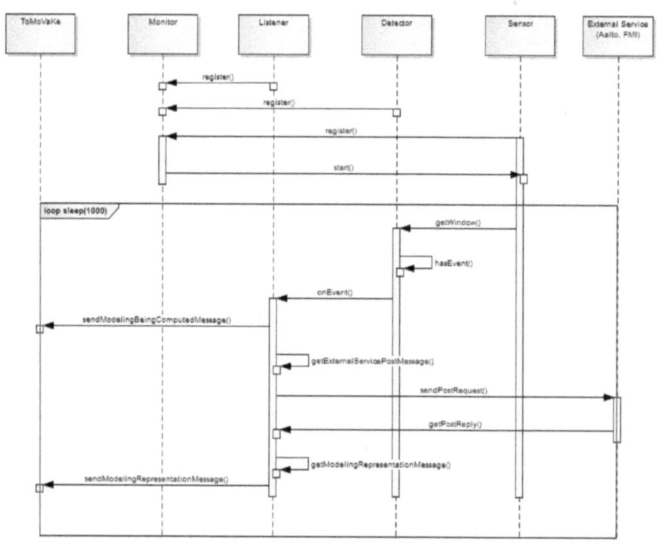

Fig. 3. Dynamic view of the TiTiMaKe core logic

4 Case Studies

We present now two scenarios that were implemented using the TiTiMaKe architecture: “Unauthorized driving on a private road”, and “Fire at an industrial site with chemicals”.

4.1 Unauthorized Driving on a Private Road

The scenario starts with vibration sensors detecting a vehicle on a private road. The system alerts about this by flashing a red icon on top of map on screen. At the same time, the system initiates a reachability computation, to find out how far the vehicle could drive within the next few minutes. Once the reachability computation is completed, the results are shown on the map. The person on duty can also bring up pictures of the vehicle as taken by the surveillance camera. The pictures are fetched when the person on duty clicks the flashing red icon. Based on the images of the vehicle, and the reachability results, the person on duty can either identify the vehicle or raise an alarm, if vehicle cannot be identified. Figure 4 depicts this scenario. The implementation of the scenario is depicted in Figure 5 as a sequence diagram. The communication logic of the sequence diagram was implemented as sensor, detector, and listener components in the TiTiMaKe architecture.

4.2 Fire at an Industrial Site with Chemicals

This is a simulated scenario. The scenario starts when a fire in a factory causes an automatic fire alarm. The system fetches data about the factory and informs the person on duty and firemen about a danger of a chemical hazard. Upon an arrival, the firemen inform the control center about toxic smoke, whereby the system informs about potential toxins and visualizes a computed estimate of dispersion. Based on this, the control center starts to plan the need for additional resources and procedures. Based on a continuous on-site measurement activity, the system alerts the control center about how toxic the smoke is and computes using weather data an estimated dispersion and concentration on the map. Next, the system fetches population data within the concentration area and computes an estimate for the number of affected people and for potential risk groups. Then, the system produces an evacuation and alert message for the interest groups. The scenario is depicted in Figure 6.

As this scenario included externalities, such as an industrial site, chemical containers, and firemen, the implementation of the scenario was limited to its core functionality. The implementation is depicted in Figure 7 as a sequence diagram. In particular, the implementation focused on the interaction between the TiTiMaKe server and the computational services, namely dispersion modeling computed by the FMI server and dynamic population modeling computed by the Aalto server. The logic of the sequence diagram in Figure 7 was implemented as sensor, detector, and listener components in the TiTiMaKe architecture.

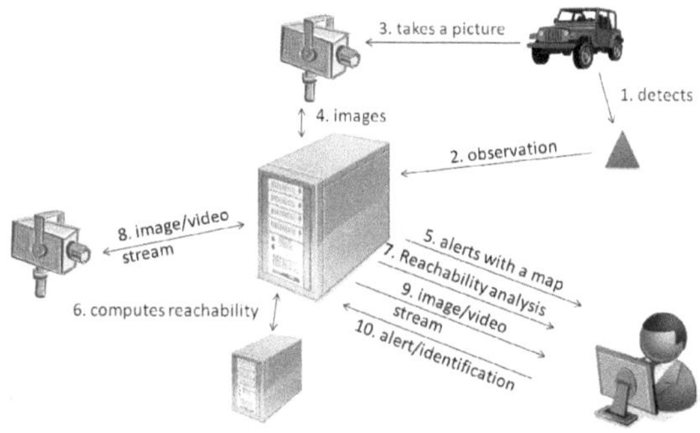

Fig. 4. Unauthorized driving on a private road

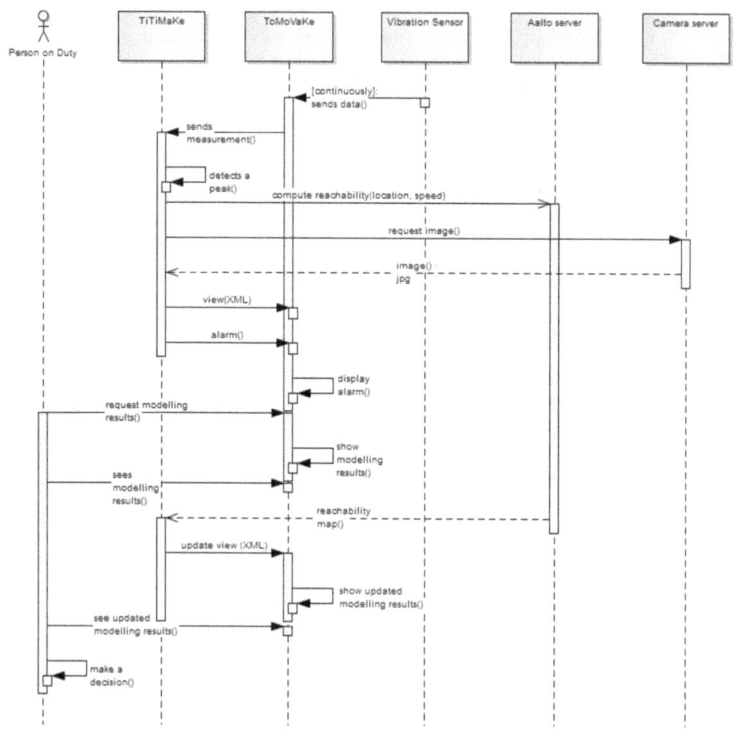

Fig. 5. Implementation of the scenario "Unauthorized driving on a private road" depicted as a sequence diagram

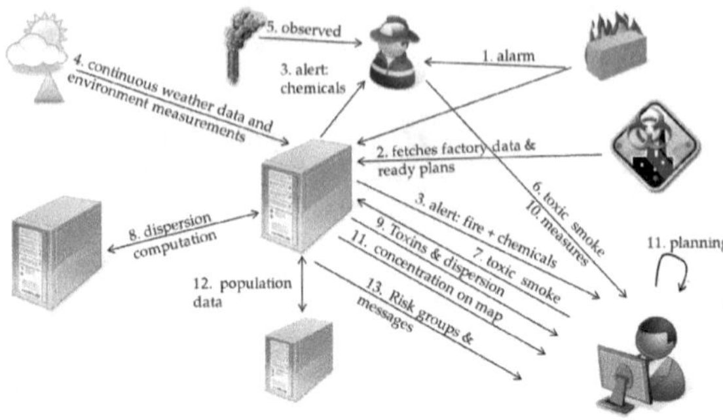

Fig. 6. Fire at an industrial site with chemicals

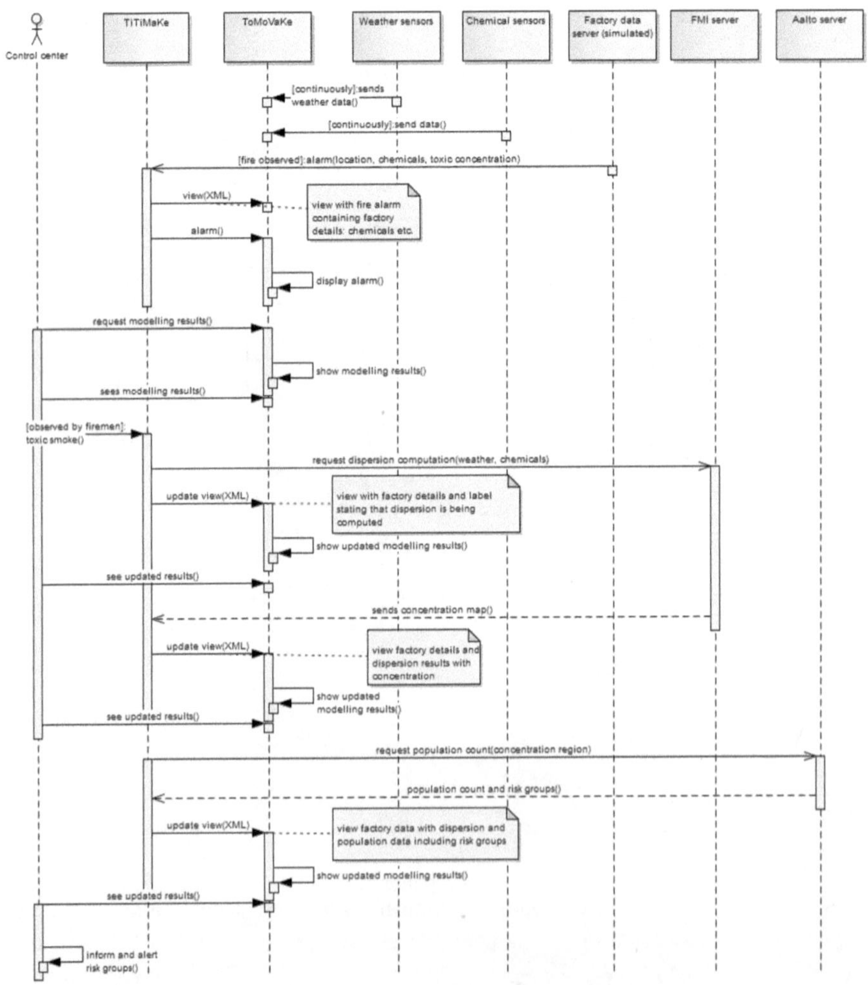

Fig. 7. Implementation of the scenario "Fire at an industrial site with chemicals" depicted as a sequence diagram

5 Conclusion

In this article, we discussed the integration of distributed information services to support the implementation of applications for the fields of safety and security. To support the development of such applications, and as the main contribution, we presented here the TiTiMaKe architecture. We used the architecture to implement two scenario based applications: "Unauthorized driving on a private road", and "Fire at an industrial site with chemicals".

The implementation of both scenarios confirmed that the TiTiMaKe architecture does indeed support the development of scenario based applications by using a distributed service network. The tests, however, revealed how sensitive such a system is to defects, if robustness and resilience are not considered in the communication logic. To overcome such defects, central services of the system need to be duplicated or communication with minimal information has to be guaranteed some other way, for instance by providing alternative communication media. The TiTiMaKe architecture supports both of these measures.

One clear topic for future research is to integrate more services to the TiTiMaKe architecture and thereby to implement different kinds of scenario based applications. In this way, the core logic of the architecture becomes more versatile, thus supporting the development of more advanced safety-critical applications.

Another topic for future development has to do with the centralized aspect of the TiTiMaKe architecture. In current form, the core logic has a central supervisor. Although parts of the logic can be distributed, the supervisory logic must reside in a central server. This raises the issue of resilience, if the central server becomes unaccessible. An alternative is to investigate cooperative decision making by implementing loosely coupled autonomous servers that cooperate by communication. Such an arrangement would facilitate not only better robustness and resilience, but also better performance scaling when the number of concurrent applications increases.

Acknowledgements. We wish to thank the Finnish Funding Agency for Technology and Innovation (TEKES) and the industrial partners (APL Systems Oy, Control Express Finland Oy, Environics Oy, Finnish Defence Forces Research Centre, and Symo Oy) for funding this research. We also wish to thank all our academic partners (Aalto University, Emergency Services College, and Finnish Meteorological Institute) for excellent work and support during the project. Mauno Rönkkö and Mikko Kolehmainen would also like to thank European Regional Development Fund and TEKES for support to continue research in cooperation with Amphion Loudspeakers Oy, Genelec Oy, Mediatuotanto Mindcom Oy, Observis Oy, and Defence Forces Technical Research Centre. Last but not least, we wish to thank Heikki Isotalus and Jarmo Kastinen for technical support.

References

1. Carter, W., Disaster, N.: management: A disaster manager's handbook. Asian Development Bank, Manila (1991)
2. Jaques, T.: Issue management and crisis management: an integrated, non-linear, relational construct. Public Relations Review 33, 147–157 (2007)
3. Salmon, P.M., Stanton, N.A., Jenkins, D.P., Walker, G.H., Young, M.S., Aujla, A.: What really is going on? Review, Critique and Extension of Situation Awareness Theory. In: Harris, D. (ed.) HCII 2007 and EPCE 2007. LNCS (LNAI), vol. 4562, pp. 407–416. Springer, Heidelberg (2007)
4. Knight, J.C.: Safety critical systems: challenges and directions. In: Proceedings of the 24rd International Conference on Software Engineering, ICSE 2002, pp. 547–550 (2002)
5. Hohpe, G., Woolf, B.: Enterprise Integration Patterns: Designing, Building, and Deploying Messaging Solutions. Addison-Wesley (2003)
6. Gamma, E., Helm, R., Johnson, R., John Vlissides, J.: Design Patterns: Elements of Reusable Object-Oriented Software. Addison-Wesley (1994)
7. Molarius, M., Huovila, H., Kotovirta, V., Wessberg, N., Yliaho, J.: UHHA - a framework for emergency management in chemical accidents. In: Proceedings of 7th Symposium on CBRNE Threats. Meeting the Future Challenges, Defence Forces Technical Research Centre, Finland, Publications, vol. 18, pp. 112–116 (2009)
8. Stocker, M., Silvonen, P., Rönkkö, M., Kolehmainen, M.: Detection and classification of vehicles by measurement of road-pavement vibration and by means of supervised learning. Submitted to IEEE Transactions on Intelligent Transportation Systems (2012)
9. Kukkonen, J.: Modelling source terms for the atmospheric dispersion of hazardous substances. Commentationes Physico-Mathematicae 115, Dissertationes No. 34. The Finnish Society of Sciences and Letters, Helsinki (1990)
10. Riikonen, K., Nikmo, J., Kukkonen, J.: A validated assessment tool for consequence analysis of chemical emergencies - ESCAPE for Windows. In: Kujala, E., Laihia, K., Nieminen, K. (eds.) Proceedings of the NBC 2000 Symposium on Nuclear, Biological and Chemical Threats in the 21st Century, pp. 286–287 (2000)
11. Zhang, Z., Sunila, R., Virrantaus, K.: A spatio-temporal population model for alarming, situational picture and warning system. In: Joint International Conference on Theory, Data Handling and Modeling In Geospatial Information Science (2010)

Modelling of Environmental Risk Management under Information Asymmetry

Petr Šauer, Petr Fiala, and Antonín Dvořák

University of Economics, W. Churchilla 4, 130 67 Prague 3, Czech Republic
{sauer,pfiala,advorak}@vse.cz

Abstract. The Environmental risk management is an important component of governmental environmental policies. Alternative mechanisms for achieving cost-effective environmental risk reduction have been discussed in environmental economics. The paper presents a relatively new approach to environmental risk management – a model of negotiation between polluters and authorities under information asymmetry when also economic instruments are applied. A combinatory model that serves computing the first best solution was developed. The CRAB software (CombinatoRial Auction Body Software System) was used for this model. The computed first best solution was compared to the results of small economic laboratory experiments. Students played the role of the subjects in the experiments. The research concluded that under economic pressure in the form of known limitation of financial resources, the experiment results are closer to minimal financial supports. Even in a one-round game, a more cost-effective solution is achieved compared to experiments where such limitation was not introduced.

Keywords: environmental risk, management, pollution, combinatorial auctions, environmental subsidy, economic laboratory experiments.

1 Introduction

Reducing environmental risks in a cost-effective way is an important issue in environmental economics and policy. Three kinds of environmental policy instruments are able to achieve this solution: tradable pollution rights (permits), environmental charges/taxes and financial supports.

This paper presents a relatively new approach to environmental risk management. It is a negotiation between polluters and authorities (mostly government in practice) where economic instruments of environmental policies in the form of financial support are applied and where there is an information asymmetry between the authority and the polluters. The authority distributes financial resources in a one-round auction. The polluters know their individual abatement costs and try to maximize their surplus, which they can get if they apply for and negotiate higher support from the authority than the minimum they need, so their project is efficient for them. They are also allowed to establish coalitions to achieve goals in environmental risk reduction. After

J. Hřebíček et al. (Eds.): ISESS 2013, IFIP AICT 413, pp. 391–402, 2013.

negotiating with the coalition partner(s), they apply for financial support to the coalition project(s) together with their individual projects.

The research questions were as follows: (i) to what extent can the negotiated result be close to the computed optimum solution, (ii) what is the role of the financial pressure on the subjects and pay-offs, and (iii) how do the coalition subjects distribute the "extra cake" from getting higher support than the minimum?

Economic laboratory experiments were used as the method to answer the research questions. It is possible to observe increasing use of economic laboratory experiments in research. Supporters of these experiments argue that the experiments can serve well in fundamental research for (pre)testing theoretical hypotheses and continuing to build theories. In applied research, the experiments serve testing various institutional settings that are considered for introduction in policy practice. For the use of economic laboratory experiments for research purposes and its advantages and disadvantages, see for instance [1-7].

The paper is structured as follows: The general model of combinatorial auctions is presented first. Theoretical optimum results for a case with 10 polluters computed with the use of the CRAB software are compared with experimental results, where financial pressure and pay-offs were not introduced. A new series of laboratory experiments where financial pressure and pay-offs are introduced is described and the results are presented in the last part.

2 Combinatorial Auctions – General Model

Combinatorial auctions serve as a theoretical framework for the approach presented in the paper. Auctions are important market mechanisms for the allocation of goods and services. Auctions have emerged as a particularly interesting tool for negotiations. An auction provides a mechanism for negotiation between buyers and sellers. Combinatorial auctions [8-9] are those auctions in which bidders can place bids on combinations of items, so-called bundles. The advantage of combinatorial auctions is that the bidder can express his preferences more fully. This is particularly important when the items are complements. The auction designer also derives value from combinatorial auctions. Allowing bidders to express preferences more fully often leads to improved economic efficiency and greater auction revenues. It is possible to formulate single-sided combinatorial auctions, forward auctions and reverse auctions. In forward auctions, a single seller sells resources to multiple buyers. In reverse auctions, a single buyer attempts to source resources from multiple suppliers, as is common in procurement.

The problem, called the winner determination problem, has received considerable attention in the literature. The problem is formulated as follows: Given a set of bids in a combinatorial auction, find an allocation of items to bidders that maximize the seller's revenue. Many important ideas are introduced, such as the mathematical programming formulation of the winner determination problem, the connection between the winner determination problem and the set packing problem, as well as the issue of complexity.

Many types of combinatorial auctions can be formulated as mathematical programming problems. Among the different types of combinatorial auctions, we present the reverse auction of indivisible items with one buyer and several sellers. This type of auction is important for the supplier selection problem. This type of auctioning serves as the theoretical basis for our case, where the polluters sell their contributions to environmental risk reduction when investing in pollution reduction projects.

Let us suppose that m potential sellers S1, S2, ..., Sm offer a set R of r items, j = 1, 2, …, r, to one buyer B (see Fig. 1).

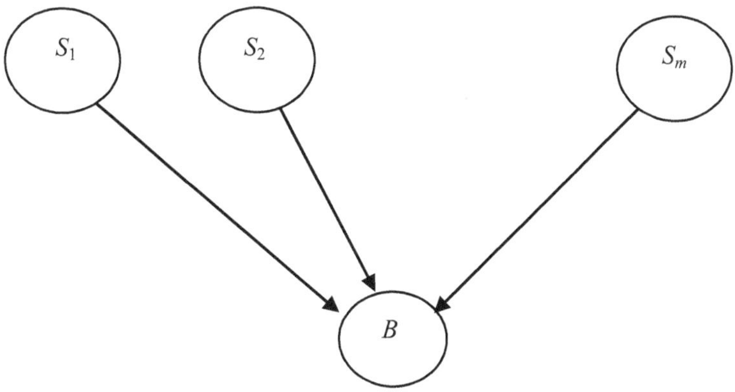

Fig. 1. Reverse auction

A bid b_h made by the seller S_h, $h = 1, 2, \ldots, m$, is defined as

$$b_h = \{C, c_h(C)\}, \tag{1}$$

where

$C \subseteq R$, is a combination of items,

ch(C), is the offered price by seller Sh for the combination of items C.

The objective is to minimize the buyer's cost given the bids made by sellers. Constraints establish that the procurement provides at least a set of all items.

Bivalent variables are introduced for model formulation:

$y_h(C)$ is a bivalent variable specifying whether the combination C is bought from the seller S_h ($y_h(C) = 1$).

The reverse combinatorial auction can be formulated as follows

$$\sum_{h=1}^{m} \sum_{C \subseteq R} c_h(C)\, y_h(C) \rightarrow \min \tag{2}$$

subject to

$$\sum_{h=1}^{m} \sum_{C \subseteq R} y_h(C) \geq 1, \quad \forall j \in R, \tag{3}$$

$$y_h(C) \in \{0, 1\}, \ \forall \ C \subseteq R, \ \ \forall h, \ \ h = 1, 2, \ldots, m. \tag{4}$$

The objective function expresses the cost. The constraints ensure that the procurement provides at least a set of all items.

The CRAB (CombinatoRial Auction Body) software system was proposed [10]. The CRAB is a non-commercial software system for generating, solving, and testing of combinatorial auction problems. The system solves problems using Balas' method or the primal-dual algorithm.

3 Base Experimental Case

An ideal model territory was created for the laboratory experiment. There were 10 polluters in 5 groups specified as A, B, C, D and E. This group also shows where it makes sense (from the economic point of view) to think about coalition projects (see Fig. 2).

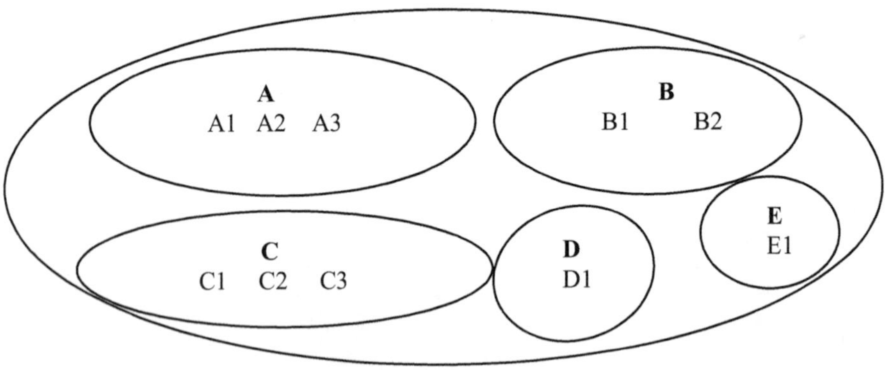

Fig. 2. Polluters in the region

Information about 10 individual projects (A1, A2, A3; B1, B2; C1, C2, C3; D; E1) and for 9 coalition projects (A1-A2, A1-A3, A2-A3, A1-A2-A3; B1-B2; C1-C2, C1-C3, C2-C3, C1-C2-C3) was prepared. The environmental effects were expressed as emission reduction (as an easy-to-understand indicator), but environmental risk reduction could also be applied. The following indicators were prepared for each of the projects:

Pollution reduction dEi, i = 1, 2, …, 19,

Minimum financial support Pmini, i = 1, 2, …, 19, needed from an external source, so the projects are economically efficient for the subject. The minimum financial support was based on the idea of so-called net abatement costs [11].

A table with all the experimental data can be found in [12]. The subjects playing the role of the managers were asked to reduce the pollution by a given amount for each of the groups. They could apply for support for individual projects and/or negotiate within the group for support for coalition projects. Students of the University of Economics played the role of the polluters. Teachers – authors of the paper – played the role of the authority. A cost-effective solution was computed with the CRAB software. The results of 8 experiments (E1 … E8) were compared to this first best solution.

The main goal of these economic laboratory experiments was to test whether and to what extent it is possible to establish coalitions that will solve the problem of pollution reduction with the lowest costs. The experiment was already complex enough, so it would be difficult to stimulate the subjects to reduce their bids. It would be possible via introducing multi-round negotiations. For more details about this experiment, see [12].

Table 1. Comparing computed and experimental results

Computed optimum solution								270
Experiment no.	E1	E2	E3	E4	E5	E6	E7	E8
Negotiated financial support	520	348	470	417	521	370	500	510
Minimum support	340	280	270	310	315	270	300	290

Most of the experiments achieved close to the optimum (first best) solution in terms of the structure of selected projects (i.e. the "right" coalitions were found). It can be seen in Table 1 that without the application of multi-round negotiations or/and a financial pressure on the subjects, the negotiated financial support is very high.

4 Partially Detailed Experiments – Design and Results

Two experiments were prepared. The case for them was derived from the base experiment mentioned in Section 3, including the data where also the first best solution was computed. The first experiment was designed for groups of two subjects. The second experiment was elaborated for groups of three subjects.

4.1 Experiment Design

Experiment Design for Groups of Two Subjects (A and B)

Pollution reduction by individual projects: dPA, dPB
Pollution reduction by a joint project: dPAB = dPA + dPB
Costs of individual projects: CA, CB
Costs of joint (coalition) project: CAB

Experiment Design for Groups of Three Subjects (A, B and C)

Pollution reduction by individual projects: dPA, dPB, dPC
Pollution reduction by joint projects: dPAB = dPA + dPB; dPAC = dPA + dPC; dPBC = dPB + dPC; dPABC = dPA + dPB + dPC
Costs of individual projects: CA, CB, CC
Costs of joint (coalition) projects: CAB, CAC, CBC, CABC

Sixteen students of the University of Economics, Prague (8 in international trade major and 8 in theory of decision making major) and sixteen students of the College of Polytechnics Jihlava (management of tourism major) were subjects in the experiments. They played the roles of managers applying for the financial support from the authority. The role of the authority was played by the teacher (the same person in all the experiments).

The subjects in the role of managers were given the following materials (see attachment for more details):

- The situation (case) description
- Training (non-confidential) data for better understanding the experiment
- Confidential data for the experiment(s)
- Application form for financial support from the authority.

The subject playing the role of the authority had the same materials as the students plus a description of details of the experimental process to keep the same conditions in all the repetitions of the experiments.

The subject playing the role of the authority distributed the materials, made a picture of the situation on the blackboard and explained the case, demonstrated the managers' decision making process on the training data, and collected the filled-in application forms for financial support from the subjects at the end of the experiments. He also informed the subjects that there is one round of application possible, that the financial resources for financial support are limited (i.e., not all applications can be supported) and that the subject will be financially rewarded after the experiments (CZK 30 for showing up + part of the surplus they will negotiate in the experiment).

4.2 Experiment Results for Groups of Two Subjects (Table 2)

Solution for Individual Groups

The first-best solution for groups of two subjects (A, B) is to establish the coalition AB with a total pollution reduction of 70 units and a minimum support of 50 monetary units. All these groups established coalitions with minimum requested support of 72.5 monetary units on average in the University of Economics experiments (71.4 monetary units in the Polytechnics Jihlava experiments). In the base experiment (see Section 3) without pressure, the requested support was 85 monetary units on average.

Solution for the Whole Region

If we presume pollution reduction in the whole region by 60%, the total pollution reduction would be 336 units (0.6*8*70 = 336). By solving the combinatorial reverse auction problem (see Section 2), we get a solution of selected coalitions: 1AB, 4A, 5AB, 6AB, 7AB, 8A; the minimum support would be 323.6 monetary units in the University of Economics experiments (1A, 4AB, 5AB, 6AB, 7A, 8AB); the minimum support would be 316 monetary units in the Polytechnics Jihlava experiments). The solutions are not trivial; they are combinations of one- and two-element coalitions.

Table 2. Results of the experiment for groups of two subjects

Project	Pollution reduction	Univ. of Economics, Prague			Polytechnics Jihlava		
		Required support	Surplus A	Surplus B	Required support	Surplus A	Surplus B
1A	30	30	10	-	25	5	-
1B	40	78	-	18	70	-	10
1AB	70	70*	9	11	70*	10	10
2A	30	35	15	-	50	30	-
2B	40	95	-	35	90	-	30
2AB	70	80*	10	20	80*	15	15
3A	30	30	10	-	30	10	-
3B	40	70	-	10	75	-	15
3AB	70	75*	15	10	75*	15	10
4A	30	27	7	-	32	12	-
4B	40	70	-	10	75	-	15
4AB	70	80*	10	20	72*	10	12
5A	30	23.5	3.5	-	30	10	-
5B	40	80	-	20	68	-	8
5AB	70	70*	8	12	64*	6	8
6A	30	30	10	-	30	10	-
6B	40	80	-	20	75	-	15
6AB	70	70*	10	10	70*	5	15
7A	30	35	15	-	25	5	-
7B	40	65	-	5	90	-	30
7AB	70	65*	7	8	80*	10	20
8A	30	21.6	1.6	-	30	10	-
8B	40	70	-	10	70	-	10
8AB	70	70*	10	10	60*	4	6

Distribution of Surplus

The total surplus in the experiments under pressure is lower than in the experiments without pressure. The distribution of surplus in negotiations between the subjects A and B is 9.9 : 12.6 average in the University of Economics experiments (9.4 : 12 in the Polytechnics Jihlava experiments).

4.3 Experiment Results for Groups of Three Subjects (Table 3)

Solution for Individual Groups

The first-best solution for groups of three subjects (A, B, C) is to establish the big coalition ABC (total pollution reduction of 230 units) with a minimum support of 60 monetary units.

Six groups of three subjects (1-6) established the big coalition with a minimum requested support of 94 on average. Group no. 7 proposed the best solution by establishing the coalition AB (requested support 45) and C (requested support 50), totalling 95 monetary units. The requested support for the big coalition was 99 monetary units.

Table 3. Results of the experiment for groups of three subjects

Project	Pollution Reduction	Required support	Surplus A	Surplus B	Surplus C
1A	90	35	5	-	-
1B	40	65	-	35	-
1C	100	75	-	-	25
1AB	130	50	7	8	-
1AC	190	100	10	-	20
1BC	140	90	-	15	15
1ABC	230	90	10	10	10
2A	90	41	11	-	-
2B	40	45	-	15	-
2C	100	66	-	-	16
2AB	130	55	10	10	-
2AC	190	88	8	-	10
2BC	140	78	-	9	9
2ABC	230	84	8	7	9
3A	90	50	20	-	-
3B	40	40	-	10	-
3C	100	85	-	-	35
3AB	130	55	15	5	-
3AC	190	100	10	-	20
3BC	140	80	-	5	15
3ABC	230	90	5	5	20
4A	90	55	25	-	-
4B	40	40	-	10	-
4C	100	60	-	-	10
4AB	130	50	10	5	-
4AC	190	85	5	-	10
4BC	140	70	-	5	5
4ABC	230	80	5	5	10
5A	90	40	10	-	-
5B	40	35	-	5	-
5C	100	75	-	-	25
5AB	130	45	5	5	-
5AC	190	90	10	-	10
5BC	140	90	-	15	15
5ABC	230	100	14	13	13
6A	90	40	10	-	-
6B	40	45	-	15	-
6C	100	60	-	-	10
6AB	130	55	10	10	-
6AC	190	100	15	-	15
6BC	140	90	-	15	15
6ABC	230	120	20	20	20
7A	90	40	10	-	-
7B	40	40	-	10	-
7C	100	50	-	-	0
7AB	130	45	3	7	-
7AC	190	85	8	-	7
7BC	140	70	-	4	6
7ABC	230	99	13	13	13

In the base experiment (see Section 3) without pressure, the requested support was 133.125 on average, which is almost 50% more.

Solution for the Whole Region
If we presume pollution reduction in the whole region by 60%, the total pollution reduction would be 966 units (0.6*7*230 = 966). By solving the combinatorial reverse auction problem (see Section 2), we get a solution of selected coalitions: 1AB, 2ABC, 4ABC, 5AB, 6AB, 7AB; the minimum support would be 359 monetary units. The solution is not trivial; it is a combination of two- and three-element coalitions.

Distribution of Surplus
The total surplus in the experiments under pressure with three subjects is lower than in the experiments without pressure. The distribution of surplus in negotiations among the subjects A, B and C is 10.7 : 10.4 : 13.6 on average."

5 Concluding Remarks

Modelling of environmental risk management under information asymmetry seems to be a very promising area of research. The aim of this paper was to contribute to a theoretical framework and compare the computed theoretical first-best solution with experimental results. Models of combinatorial auctions were selected as the theoretical framework. A base experimental case and partially detailed experiments were created and used for testing the basic research questions in the paper, which were as follows: (i) to what extent can the negotiated results be close to the computed optimal solution, (ii) what is the role of financial pressure on the subjects and pay-offs, and (iii) how do the coalition subjects distribute the "extra cake" from getting higher support than the minimum?

The experiments showed that the experimental results were close to theoretical results mainly in the structure of the solution, i.e., the structure of selected coalitions to the structure of the first best solution. The required supports are significantly closer to the theoretical minimum supports in the experiments under financial pressure than in the situations without the pressure, i.e., in the situations where all the subjects get the supports and there is no pay-off for the subjects in the experiments. The negotiated distributions of the surplus among the subjects are, on average, fairly balanced. Small differences are caused by the position of the subject in establishing coalitions. An interesting observation is the finding that the results are very similar for groups of students (experimental subjects) from different universities.

The results in the paper inspire us to continue experiments for a richer structure of coalitions with multi-round negotiations under pressure and with the introduction of multi-criteria evaluation of environmental and other effects performed by the authorities. These more general cases promise to bring new interesting findings. The EU Operational Programme Environment (OPE) support for water management infrastructure and the reduction of flood risks is a typical field of application for establishing joint (coalition) projects supported from public funds. We believe that it also makes sense to develop one-round models. This procedure is typical for various support programmes, such as the grants provided from the OPE for reducing industrial pollution and environmental risks.

Acknowledgement. The paper was developed with the support of the Czech Science Foundation, grant no.13-07036S "Modelling of Negotiations in Environmental Policy under Information Asymmetry".

References

1. Davis, D.D., Holt, C.A.: Experimental Economics. Princeton University Press, Princeton (1993)
2. Smith, V.L.: Experimental Economics: Induced Value Theory. The American Economic Review 66(2), 274–279 (1976)
3. Walker, J.: Economic Instruction: Experimental Economics in the Classroom. Journal of Economic Education 18(1), 51–56 (1987)
4. Roth, A.E.: Introduction to Experimental Economics. In: Kagel, J.H., Roth, A.E. (eds.) The Handbook of Experimental Economics, pp. 3–109. Princeton University Press, New Jersey (1995)
5. Sauer, P., Dvorak, A., Lisa, A., Fiala, P.: A Procedure for Negotiating Pollution Reduction under Information Asymmetry. Environmental and Resource Economics 24, 103–119 (2003)
6. Levitt, S.D., List, J.A.: What Do Laboratory Experiments Measuring Social Preferences Reveal about the Real World? The Journal of Economic Perspectives 21(2), 153–174 (2007)
7. Falk, A., Heckman, J.: Lab Experiments are a Major Source of Knowledge in the Social Sciences. CESifo Working Paper Series, No. 2894. Available at SSRN on-line (2009), `http://ssrn.com/abstract=1536400` (assessed March 20, 2013)
8. Cramton, P., Shoham, Y., Steinberg, R. (eds.): Combinatorial Auctions. MIT Press, Cambridge (2006)
9. de Vries, S., Vohra, R.V.: Combinatorial auctions: A survey. INFORMS Journal of Computing 15(1), 284–309 (2003)
10. Fiala, P., Kalčevová, J., Vraný, J.: CRAB—CombinatoRial Auction Body Software System. Journal of Software Engineering and Applications 3(7), 718–722 (2010)
11. Šauer, P., Dvořák, A., Fiala, P.: Negotiation between authority and polluters: A model for the support of decision-making in ecological policy (in Czech). Politická Ekonomie, XLVI/6, 772–787 (1998)
12. Fiala, P., Šauer, P.: Application of Combinatorial Auctions on Allocation of Public Financial Support in the Area of Environmental Protection: Economic Laboratory Experiment (in Czech). Politická Ekonomie, 59(3), 379–392 (2011)

Annex - Text Distributed to the Subjects (Polluter a as an Example; Case of the Groups of Three Subjects)

Case Situation

In the region in which your Firm [A] is located, there are several sources of environmental pollution. It is important to reduce this pollution due to fulfilment of an international agreement in the field of environmental risk reduction. Each of the polluters has four potential projects for reducing the pollution in the region. The individual project is one of the options. Joint (coalition) projects with the other firms [B, C] is the other option. Each of the projects (i.e., individual or coalition ones) offers some pollution reduction (dP), requires some costs (C) and needs some minimum external financial support (Smin). The firms and/or their coalitions apply for this financial support at the authority. The application procedure has one round. In the application forms, you are supposed to fill in:

a) Pollution reduction offered (see dP in the Confidential data …) and

b) Required amount of financial support; its amount is based on your negotiation strategy; you must not request less than the shown minimum amount (Smin in the Confidential data …).

The financial resources for the support are limited. The support will be distributed to the most efficient projects, i.e., not all projects can be supported. If you succeed in getting higher support than the minimum you need (Smin), you get a surplus. The whole surplus belongs to you in the case your individual project is supported. In case a coalition project is supported, you share the surplus with the other firms in the coalition. The shares in the surplus are negotiated with the other firms before the application is submitted.

You will be rewarded for your participation in the experiment. The remuneration will consist of a show-up fee of CZK 30 and the amount derived from the amount of surplus you get.

Training data (non-confidential data for better understanding the experiment by the subjects; case of the group of two subjects):

Data for polluter X	Pollution reduction (dP)	Cost of the project (C)	Minimum financial support (Smin)
Individual solution (project)	20	50	25
Joint solution (project A+B)	40	80	40

Data for polluter Y	Pollution reduction (dP)	Cost of the project (C)	Minimum financial support (Smin)
Individual solution (project)	20	60	30
Joint solution (project A+B)	40	80	40

Confidential data for the subject playing the role of the polluter [A] in the experiment with three subjects for whom it might be efficient to establish a coalition:

	Pollution reduction (dP)	Cost of the project (C)	Minimum financial support (Smin)
Individual solution (project)			
A	90	60	30
Joint solutions			
A+B	130	70	35
A+C	190	140	70
A+B+C	230	120	60

Application form for financial support from government (case of the group of three subjects):

Polluter [A]

Experiment no.: ....Date: ....

Subject (student): ....

	Pollution reduction offered (dP) from Confidential data....	Financial support required Your decision	Your share in the surplus1)
Individual solution (project)			
Joint solutions (projects) with your participation	X	x	X
A+B			
A+C			
A+B+C			

In the case of joint projects, the sum of the individual firms' shares in the surplus has to be equal to the whole surplus of the joint project.

An Information System Supporting WFD Reporting

Thorsten Hell[1], Eckhard Kohlhas[2], Sascha Schlobinski[1],
Ralf Denzer[3], and Reiner Güttler[3]

[1] Cismet GmbH, Altenkesseler Strasse 17, 66117 Saarbrücken, Germany
{thorsten.hell,sascha.schlobinski}@cismet.de
[2] Landesamt für Umwelt, Naturschutz und Geologie Mecklenburg-Vorpommern (LUNG),
Goldberger Straße 12 18273 Güstrow, Germany
eckhard.kohlhas@lung.mv-regierung.de
[3] Environmental Informatics Group (EIG), Goebenstr. 40, 66117 Saarbrücken, Germany
ralf.denzer@enviromatics.org, reiner.guettler@enviromatics.net

Abstract. The EU Water Framework Directive (WFD) defines the quality goals for European water bodies until 2027. Water bodies are managed on very different governmental levels in the member states. Due to the federal system, Germany has a complex structure of responsibilities regarding the management and maintenance of water bodies in Europe.

The paper will present a highly integrated information system to support water body management in the context of the WFD. The system was designed and developed for the German federal state Mecklenburg-Vorpommern. The objective was to provide a state-wide information platform accessible through an environmental information system (EIS) for all activities related to the water body management, including the maintenance of the water body network itself, managing related information regarding statistics, agriculture and environmental protection, as well as water body quality like chemical and biological status.

This EIS integrates disparate data sources from a large number of individual providers through standardized services. Moreover, the EIS provides a homogenous view on very different data regarding their topic, content, structure and spatial reference and supports all WFD reporting related workflows.

Keywords: Water Framework Directive, WFD, water reporting, environmental information system, EIS.

1 Software Support in the Context of the EU WFD

The EU Water Framework Directive [1] (WFD, [1]) requires the EU Member States to provide reports on the biological and chemical status of ground and surface water, to plan and execute programs for improvement and to report about plans, measures and results as well. In Germany these reporting requirements are managed by the federal states and the national government. The reporting is carried out through a formalized

1 http://ec.europa.eu/environment/water/water-framework/

J. Hřebíček et al. (Eds.): ISESS 2013, IFIP AICT 413, pp. 403–413, 2013.

data model in order to harmonize data across Europe. Reported information is available in the Water Information System for Europe (WISE) [5].

In the German state Mecklenburg-Vorpommern, the WFD reporting is carried out in the context of the **FIS Water MV** (Water Information System of Mecklenburg-Vorpommern), which aims at providing an integrated water information base for all water-related management and reporting activities in the state. Besides water information, data from sources like general statistics, agriculture, land use, nature protection and bio diversity etc. are relevant for the assessment of water body quality.

The diverse water-related data and processes (like runoff, drainage, gages, flood forecasts, maintenance plans etc.) require the same baseline and common base geometries. For this reason, the WFD reporting cannot be regarded as an isolated activity that happens once a year.

The FIS Water MV (Fig. 1) is operated by the "Landesamt für Umwelt, Naturschutz und Geologie Mecklenburg-Vorpommern" (**LUNG**), the state agency for environment, nature protection and geology. The different systems are at different stages of realisation.

While the FIS WRRL (WFD Reporting) is productive, other components are under development or only at a conceptual stage. E.g. the FIS for flood risk management is currently under implementation and the information system for water use and wastewater will be upgraded from 2014 on.

All systems operate on catchments and water bodies regardless of political boundaries. They use a common water network with unified numbering and stationing, the same references for coastal zones, the same geometry and referencing for still waters and a common spatial reference system.

Fig. 1. Overview of FIS Water MV

This paper gives an overview of the **IS WRRL** (Information System for WFD Reporting).

2 Users, Requirements and Functional Architecture

The FIS WRRL must support users from several organisations involved in WFD activities that have a variety of different requirements (Fig. 2). The organisatons are:

- the LUNG as the responsible organisation,
- subsidiary water administrations, namely 4 **state agencies for agriculture and environment** who are responsible for the management of 4 territories (shortly called "state agencies" in this paper),
- subsidiary lower water administrations,

- water and soil management associations who are not direct subsidiaries and
- external users like contractors.

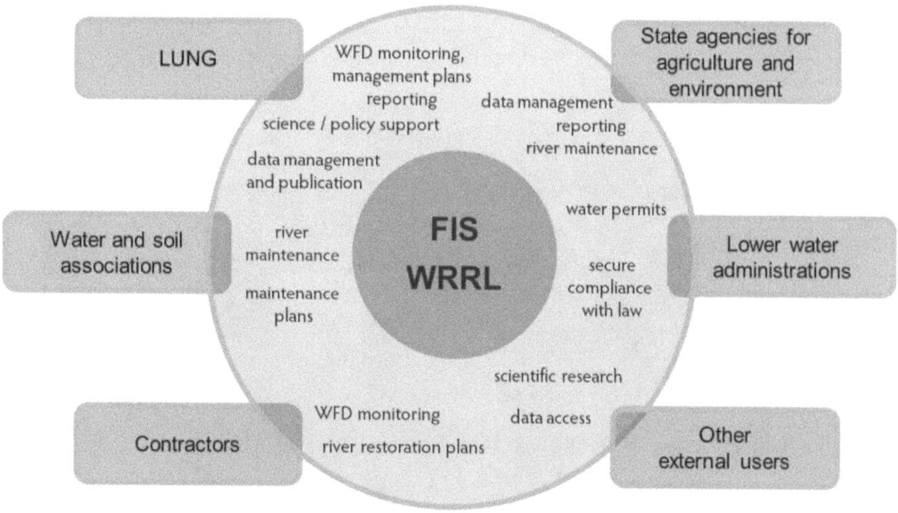

Fig. 2. Users of the FIS WRRL

Some of these users operate on the scale of the whole state, others (like lower water administrations, regions and cities) only within their territories. Also, some users work on information beyond WFD data and may need to connect their concepts and data with data in the WFD reporting system.

The following fundamental goals were defined before the FIS WRRL was developed. To:

- enable ***on-the-fly reporting*** (push-button-reporting) at any time,
- support those workflows in the LUNG which are required for WFD reporting,
- provide ***current and historic*** data on a day-to-day basis,
- hold the WFD-related data in ***a central common data repository*** in the LUNG,
- enable ***decentralized data input and maintenance***, also by external users, while avoiding redundancies
- enable the integration of data ***from other data holders***
- ensure that the solution is ***scalable***,
- ensure that the solution is extendable towards***other components of the FIS Water MV and beyond***,
- use ***open standards*** and to ***avoid licensing costs***.

Some of these requirements lead directly to a ***service-based approach***, based on open source / open standards software. A critical issue was to provide a fine grain security and access control framework as users from different administrative bodies interact with the system. Fig. 3 shows an overview of the functional architecture. At the core is a WFD data repository which serves to store all WFD-related data, including photo documentation and related features. Beyond the data as such, information regarding management strategies, ongoing and past projects, project monitoring and cost is also included in this core repository.

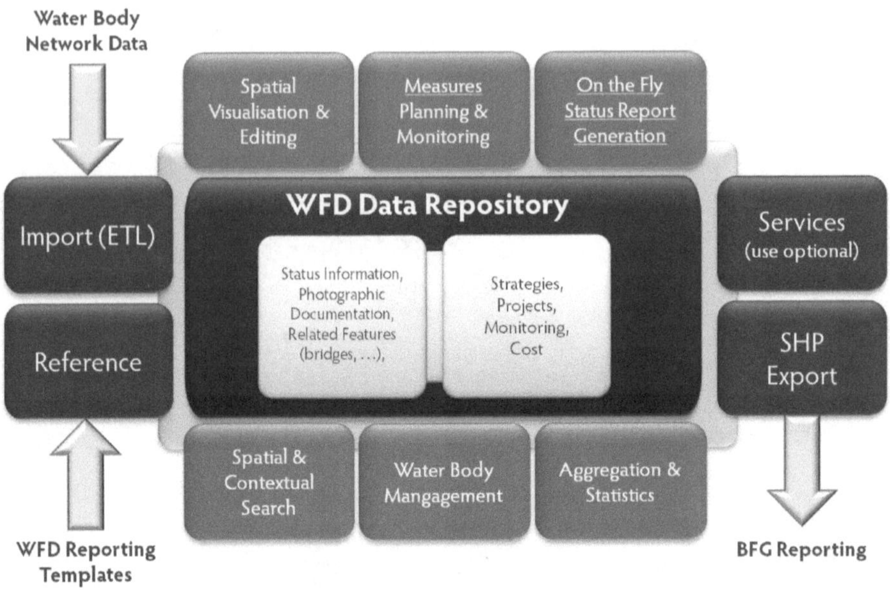

Fig. 3. Overview of functional architecture

The core repository supports users with general functionality such as:

- visualisation and editing functionality (both geography and database content),
- spatial and contextual search and
- aggregation and statistics.

Specific WFD-related functionality includes:

- water body management,
- measures planning and monitoring and
- on-thy-fly report generation.

Water body network data can be imported from external sources. The actual WFD reporting is based on WFD Reporting Templates which are used to configure the reports to the BFG (Bundesanstalt für Gewässerkunde) [2]. The current format for reports to the BFG is the Shape-File format [3]. Additionally the system allows service-based access to reports.

[2] http://www.bafg.de
[3] http://www.esri.com/library/whitepapers/pdfs/shapefile.pdf

3 Software Stack for the Solution

The solution has been implemented with a stack of freely available open source software. The core of the solution is the ***CIDS product suite*** of CISMET GmbH. This product consists of a set of software components, application programming interfaces (APIs), management and development tools, services and applications, with a special focus on interactive solutions which need to integrate geo-spatial systems with databases, sensor networks, document-oriented systems, unstructured information sources and numerical models. CIDS is particularly suited for solutions which have to be built across existing heterogeneous information systems, which may be under control of different organisations.

CIDS uses freely available open source software exclusively, is itself freely available as open source software, distributed as Github [4] projects, and is compliant with all major OGC services. CIDS has been used in numerous projects since 1999 [2]. For recent projects see [3,4].

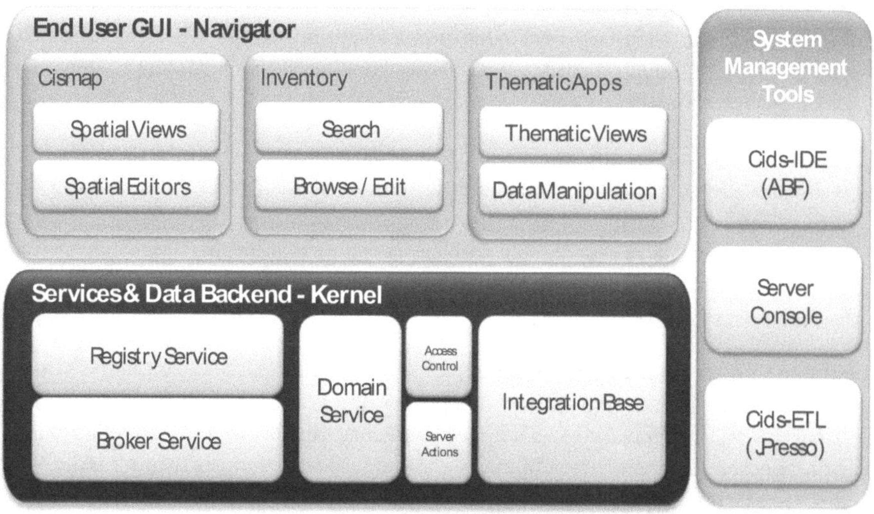

Fig. 4. Overview of CIDS product suite

The ***CIDS Kernel*** consists of one or more ***Integration Base(s),*** which are connected to a network of services. One Integration Base represents one ***CIDS Domain***. Domains are used to organise thematic application areas within larger solutions. Domains in a large municipality can for instance be "traffic", "noise", "water bodies", "land register", "street light inventory" or "waste water". In cross-governmental applications, domains can also be organised as "City of X", "City of Y", "Water Supplier of City of Y", "Region Z". The domain concept includes fine grain control of access rights and access policies even across domains.

[4] https://github.com/cismet

The Integration Base consists of a standard relational Database Management System (DBMS) augmented by meta-information that facilitates integration tasks. It is the basis for a concrete CIDS solution and is able to model arbitrary objects (real-world objects, services, models, geographical features, other information systems, etc.), their attributes (e.g. data types, geographic location) and relationships. The Integration Base also models users, user groups, navigational structures for users, permissions, and many other properties

The Integration Base is protected through ***Access Control*** mechanisms. Moreover, the CIDS Kernel consists of a ***Server Actions*** module that can be used to run procedures on the integration base, for instance for aggregation or statistics.

The concrete data of a CIDS solution can reside in external systems, or within the Integration Base, or both. For each individual attribute, CIDS can be the "leading system" or the "following system", and there are diverse mechanisms for synchronisation and mapping of data structures in both directions (*from* or *to* an external system). As an example, in a waste water taxation application, the user interacts with a "land parcel", which in physical terms consists of geo-data coming from external WMS/WFS services, geo-data stored in the Integration Base itself (individual parcels within the premise, which are only needed in this application), ownership information from a database and documents from a standard document management system. Three types of services constitute a ***CIDS Service Network***:

- A ***Domain Service*** publishes one Integration Base.
- A ***Broker Service*** acts as proxy hiding details of distributed access from client applications.
- A ***Registry Service*** is responsible for service name resolution, rights management, parallelised distributed search as well as core infrastructure functionalities such as service infrastructure monitoring.

A complex system can contain any number of Integration Bases and can be scaled in any direction using an appropriate number of Integration Bases and services. The physical design of this network is adapted to the organisational and computational set-up of a concrete customer. A minimal system consists of one software bundle which publishes one Integration Base, and which integrates no external data sources at all (similar to operating one standard RDBMS server and using a dedicated database client software).

A CIDS Service Network is configured and developed with an interactive development and configuration environment (***CIDS-IDE***), which consists of the configuration and development tool ***ABF*** and the ETL tool ***JPresso*** for synchronisation with external systems. The ***Server Console*** is a management interface to the CIDS services.
As open source software the CIDS platform uses many open source tools to provide its functionality. Among the most prominent are: Netbeans RCP [5] and PostGres/PostGis[6].

---

[5] http://netbeans.org/features/platform

[6] http://www.postgresql.org

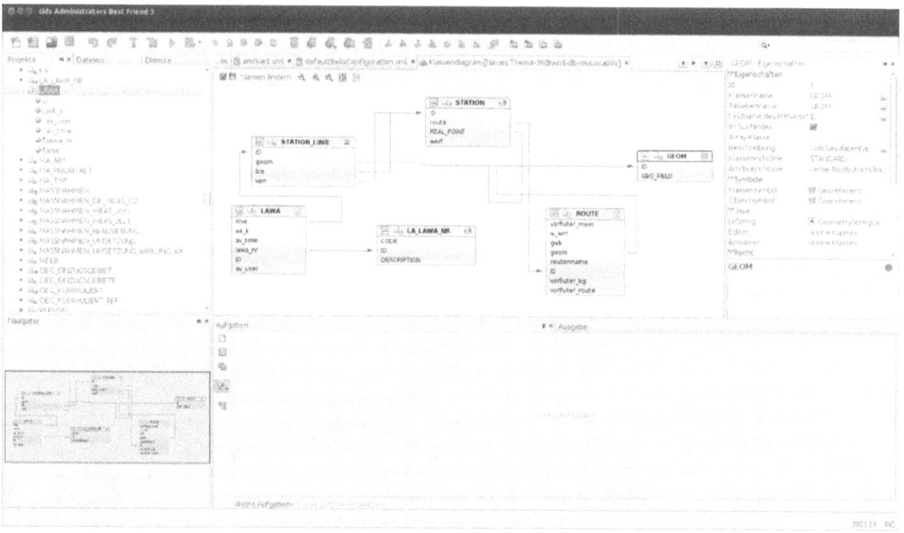

Fig. 5. CIDS ABF

4 End User View of a Highly Integrated Interactive Application

End users interact with the CIDS solution through customized interactive applications. At the core of most developments is the ***CIDS Navigator***. The Navigator offers different metaphors to interact with the system. The Inventory at the left (in Fig.6) is a customized poly-hierarchical view over user interests and serves for "browsing" and "search". Entries in the inventory are often data objects that represent arbitrary data sets ("water bodies" as top level of a water domain), but can also be information like organisational processes ("maintenance", "people", "billing") or interactions with and results of numerical models ("runoff model", "model configuration", "model run"). The Inventory can be customized for user groups and individual users.

In the case of the WFD reporting system described in this paper, the inventory contains the following themes: chemical monitoring sites, biological monitoring sites, rivers, still waters, ground water bodies, coastal water bodies, buildings, projects, drainage pipes, routes, catchments, photo documentation, land use, protected areas, hydrology and several waterway maintenance themes.

For spatial information, the map component ***Cismap*** is provided. This component is a map application specialized to access OGC compliant services like WFS and WMS for information from internal and external geospatial data sets. The individual layers being visualized can be either layers generated by CIDS or external layers as required.

The third way to interact with the system are "data pages", which can be relatively simple views but also full blown applications (***Thematic Apps***), customized to the user's view on the data provided. Data pages can include visualization and interaction with time series data, photo documentation, document views, spatial previews and many more types of information; data pages can be configured to allow for editing of the data contained as required.

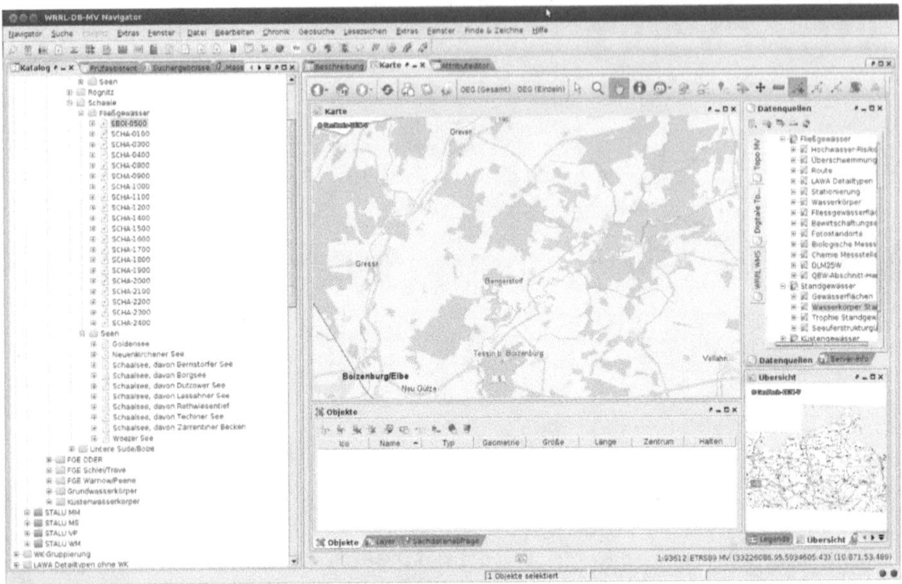

Fig. 6. CIDS Navigator

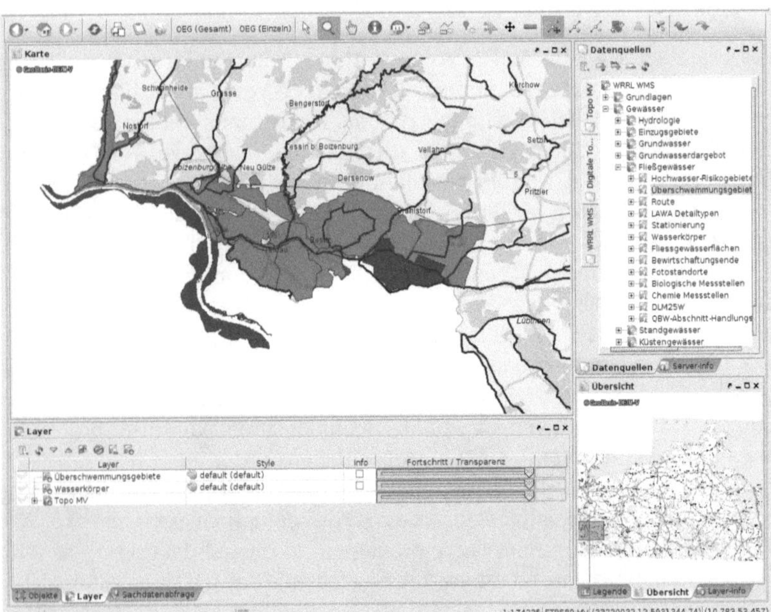

Fig. 7. Geographic interaction with Cismap

An outstanding feature of the Navigator is the fact that users can search, browse and interact seamlessly across these different metaphors. For instance data objects can be dragged on to the map, and polygons on the map can be used to query data themes in the service network, *including all connected external systems*. The software of the *Navigator* includes extension points for interactive editors and renderers which can be included in map views and data pages respectively. There is also an extension point for information aggregation over a set of objects.

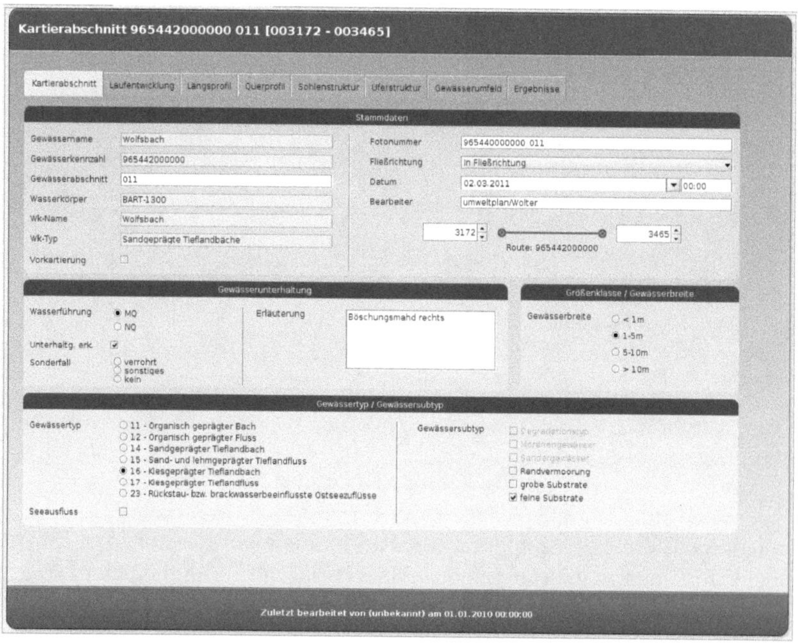

Fig. 8. Thematic view / data editing pages

5 Customization: Linear Referencing

The CIDS system provides a powerful framework to build geo-spatial information systems. However, in every project customizations are necessary particularly at the graphical user interface to support end users in an optimal fashion. An example of such a customization is the ***Linear Referencing*** implemented as part of the FIS WRRL implementation.

In a river network, the so-called ***route*** is an important concept. A route defines one river with a name and a unique number. The river is represented in the system by a polyline "M" which contains measurement information. Every location along a river can be referenced along the route by the distance to the estuary. This reference is called ***linear reference***. Linear references are used throughout waterway management and are a central concept to connect information to a location (which is along a river). Because this concept is so central, it was essential that end users be supported in ***stationing*** (locating) arbitrary objects interactively along a route, and to enable them to create ***station segments*** which are for instance used for waterway maintenance projects.

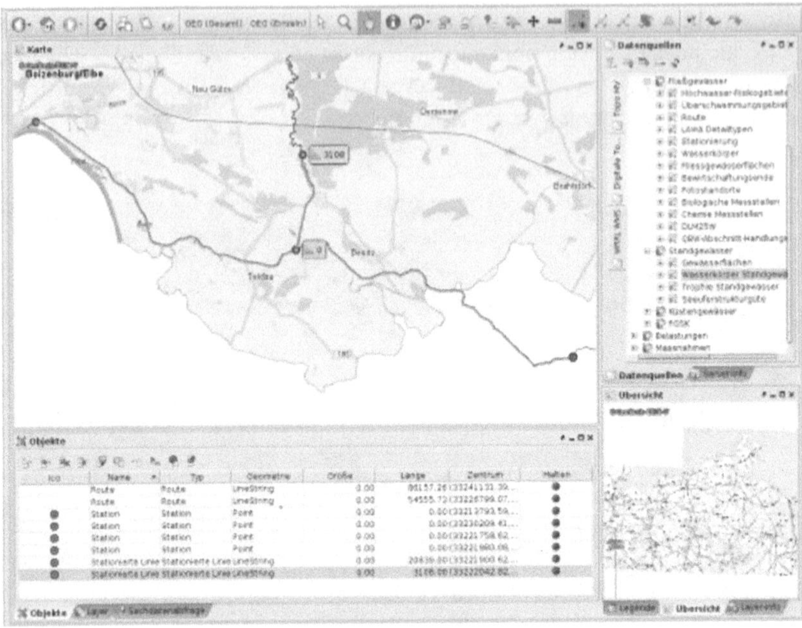

Fig. 9. Linear referencing

Through a custom extension point of CIDS an interactive editor for CISMAP was implemented to support the stationing of objects (Fig. 9). With this, users are enabled to position objects along rivers and to create segments.

6 Assessment and Conclusions

The FIS WRRL has been in operation since 2011 and has supported WFD reporting since. The FIS Water MV is currently being extended with additional systems in order to provide an integrated water information framework for the state. The choice of open software has proven to be important for long-term protection of investments into the water information infrastructure of the state.

With the realisation of the FIS-WRRL, the state Mecklenburg-Vorpommern has made an important step towards co-ordinated water information management. All WFD-related information can now be managed in a coherent way at a central location while data maintenance is performed in a decentralized fashion by the LUNG, the 4 state agencies and other organisations in a federated fashion based on geographical and organisational access rights.

Now the reports required for EU WFD reporting can be generated in an on-the-fly procedure based on actual daily data.

At the same time, the kernel of an integrated water information system has been developed which serves administration, policy and experts in their daily work.

The system has been in productive operation in the state agencies for two years, has operated without problems and has become an important information asset for the daily work of experts. Since one year, water management agencies and soil management agencies use the system. In total, 300 users are registered and daily logins are around 70 (May 2013). The users can access a total of 204 information themes supplied by the water management agencies. These themes are structured according to the needs of the WFD. Users generate approximately 15000 map access interactions each day.

The supported functionalities of the system are also extended. Since 2011 a business process supporting the mapping of morphological features of the waterbodies (FGSK – Fließgewässerstrukturgütekartierung) is integrated. Since then, contractors of the 4 state agencies, typically engineering companies, can directly maintain results of the FGSK mapping from their sites. Both contractors and water management agencies thus operate on a highly actual information base.

Currently (June 2013) a solution for river maintenance planning is in the test installation while a project has been started to build an information management application for the inventories of watercourses, profiles and technical plants maintained by the water and soil associations.

The technical solution is fully compliant with OGC standards [7] and additional information sources complying with these standards can be accessed on-the-fly by end users. Most of these geographical data sources are maps and aerial photos provided by the "Landesamt für innere Verwaltung (LAIV-MV)". Access to mapping services of the neighbouring states Brandenburg, Niedersachsen und Schleswig-Holstein make sure that the view is not restricted at the state borders.

References

1. eur-lex.europa.eu/LexUriServ/LexUriServ.do?uri=OJ:L:2000:327:0001:0072:EN:PDF
2. Güttler, R., Denzer, R., Houy, P.: An EIS Called WuNDa, Environmental Software Systems. Environmental Information and Decision Support, vol. 3, pp. 114–121. Kluwer Academic Publishers (2000)
3. Denzer, R., Torres-Bejarano, F., Hell, T., Frysinger, S., Schlobinski, S., Güttler, R., Ramírez, H.: An Environmental Decision Support System for Water Issues in the Oil Industry. In: Hřebíček, J., Schimak, G., Denzer, R. (eds.) ISESS 2011. IFIP AICT, vol. 359, pp. 208–216. Springer, Heidelberg (2011)
4. Schlobinski, S., Denzer, R., Frysinger, S., Güttler, R., Hell, T.: Vision and Requirements of Scenario-Driven Environmental Decision Support Systems Supporting Automation for End Users. In: Environmental Software Systems. Frameworks of eEnvironment, vol. 9, pp. 51–63. Springer Advances in Information and Communication Technology 359 (2011)
5. Water Information System for Europe (WISE), http://water.europa.eu/

[7] http://www.opengeospatial.org

Web 3D Visualization of Noise Mapping for Extended INSPIRE Buildings Model

Lukáš Herman and Tomáš Řezník

Masaryk University, Faculty of Science, Department of Geography,
Laboratory on Geoinformatics and Cartography (LGC),
Kotlářská 2, 611 37 Brno, Czech Republic
herman.lu@mail.muni.cz, tom@mail.muni.cz

Abstract. This paper deals with cross-domain environmental aspects of noise mapping which is commonly understood as a stand-alone initiative. Scientific core of this research lies in the integration of noise mapping with interoperable data sources, where the European Union's INSPIRE (INfrastructure for SPatial InfoRmation in Europe) Directive is nowadays one of the main.

Principles of noise mapping according to the END (European Noise Directive), OGC (Open Geospatial Consortium) CityGML specification, X3D (Extensible 3D) and application schemas in INSPIRE are depicted at the beginning. Special attention is then paid to the specifics of INSPIRE spatial data theme called buildings, including its extension possibilities. Motivations for 3D application schema of INSPIRE buildings are discussed while only 2D application schema is the legally binding one. On the other hand, 3D application schema is recommended by the INSPIRE Technical Guidelines. We also deal with OGC CityGML specification since it is the basis for the 3D application schema. We took a step forward to publish the results of noise mapping based on INSPIRE application schema through X3D technology on the Web. This enables to disseminate information on noise mapping to wide audience through next generation Web applications.

The paper proposes formalised methodology on X3D-based application development through the XSLT (eXtensible Stylesheet Language Transformations). Noise mapping was taken as a use case. There are discussions on benefits and limitations of X3D visualization as well as differences between the X3D presentation and storage of 3D geospatial data according to INSPIRE application schema on buildings at the end.

Keywords: geographic information, CityGML, X3D, INSPIRE, noise mapping.

1 Introduction

Geographic information has always been crucial for research in a wide variety of sciences within and beyond Geography. On the other hand, contemporary Web search engines are only partly suitable for efficient discovery and presentation of geographic information. The main difference between geographic and other data is the spatiotemporal dimension of geographic data. For that reason it is not possible to search for

J. Hřebíček et al. (Eds.): ISESS 2013, IFIP AICT 413, pp. 414–424, 2013.

geographic data for a certain area using search engines such as Google, Yahoo, Bing, etc. Furthermore, such a limitation of geographic information discovery limits the re-use of geographic information [11].

To deal with this problem, the European Commission started the initiative called INSPIRE (*INfrastructure for SPatial InfoRmation in Europe*). This initiative was transformed in 2007 into the Directive of the European Commission and the Council with designation 2007/2/EC. This Directive was transposed into national legal systems of the EU (European Union) Member states between 2007 and 2009. The Directive itself contains the general concept, while more detailed information may be found in the corresponding Commission Regulations addressing specific issues, as well as in the underlying technical guidelines.

These are, so far, interoperability of spatial data, metadata of spatial data and services, data and service sharing, network services. The directive defines 34 spatial data themes covering a wide range from coordinate reference systems, cadastral parcels, transport networks, hydrography, land cover, orthoimagery, soil, human health and safety, natural risk zones, habitats and biotopes, energy resources, buildings, and many others.

We may also define interoperability on the conceptual and policy levels. INSPIRE Directive is closely related to other European Directives, international standards and standardization activities, etc. We may find examples in the linkages to European Noise Directive, Water Framework Directive, International Organisation for Standardization (ISO) 19100 series of standards for geographic information or implementation specifications of the Open Geospatial Consortium (OGC). This paper analyses the issues of the INSPIRE application schema for spatial data theme Buildings, its connection to the European Noise Directive, presentation and storage of 3-dimensional geographic information created through innovative Web technologies with noise mapping as a use case.

2 Legal and Standardization Framework

2.1 INSPIRE Application Schemas for Buildings

The INSPIRE concept of geographic data follows an object-oriented approach of modelling the entities of the real world. This means that one or more application schemas covering each specific point of view(s) on the domain are created. Each application schema then contains explicit definitions of feature types, their aggregation into classes, attributes of feature types, domains of these attributes, etc. The primary aim is to develop a model which will ensure interoperability and harmonisation within each spatial data topic. Each application schema is offered in a number of ways, such as through a UML (*Unified Modelling Language*) class diagram, feature catalogue, XSD (*eXtensible Stylesheet Document*), and, of course, textual descriptions. Besides the application schemas themselves, the concept of INSPIRE geographic data also includes related aspects. These are definitions of the reference systems (coordinate, vertical as well as temporal), quality of geographic data, metadata and many others. On the other hand, common issues for all spatial data topics are written in the Generic Conceptual Model (GCM) document, which comprises definitions of unique identifiers for geographic data, explicit definition of data types, principles of network application schemas, multi-lingual geographic information, etc.

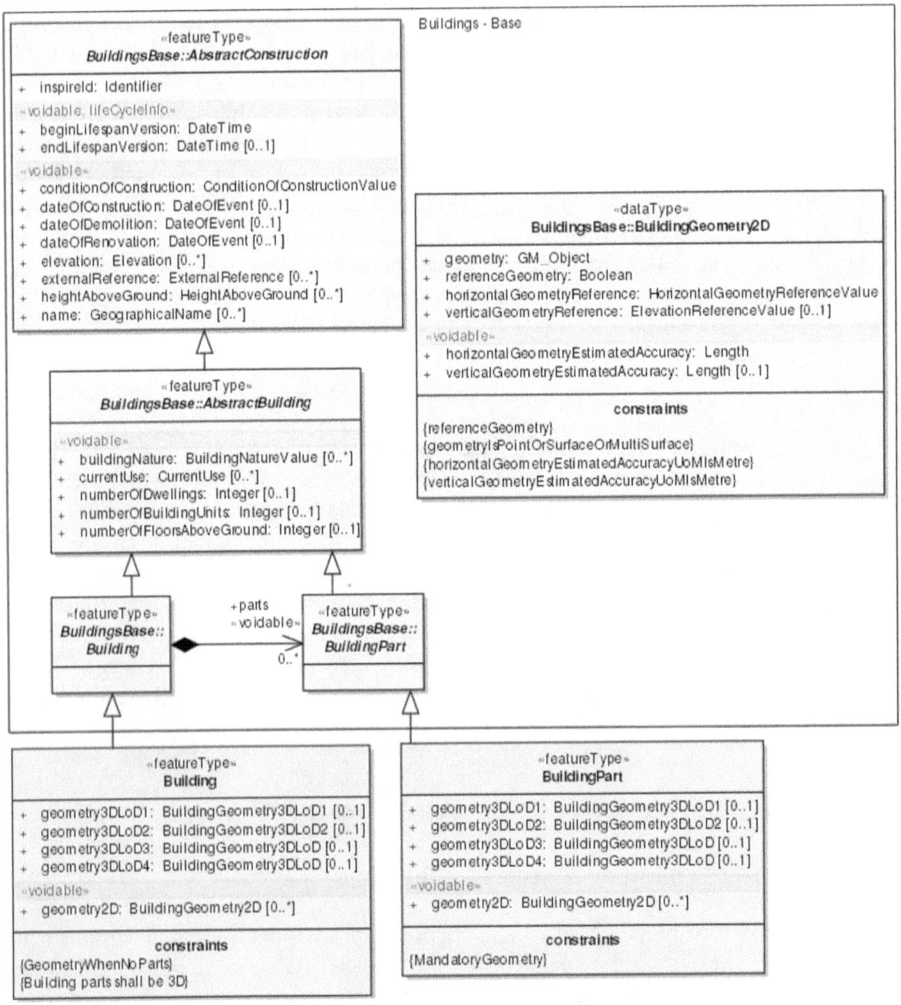

Fig. 1. UML class diagram for the INSPIRE application schema *Buildings3D* according to the INSPIRE Data Specification on Buildings – Draft Technical Guidelines [8].

The INSPIRE application schema for buildings is[1] in its almost final phase, designated as release candidate 3 version. There may be minor changes in this version if the Amendment to the Commission Regulation No 1089/2010 will be modified. The latest version of the INSPIRE application schema for buildings mainly focuses on the physical description of real world entities seen as constructions. A capability of a building to provide services is a subject of other INSPIRE application schemas in

[1] 24 April 2013

another spatial data themes, above all utility and governmental services, production and industrial facilities, agricultural and aquacultural facilities. We may define four application schemas within the *Buildings* spatial data theme:

- **The Buildings2D profile** includes various geometrical representations of buildings as 2D or 2,5D data.
- **The Buildings3D profile** has same semantic content as the Buildings2D profile and allows in addition, the geometric representation of buildings in any of the four levels of detail of CityGML.
- **The BuildingsExtended2D profile** is a semantic extension of Buildings2D profile with additional thematic attributes (material of construction, official area or value, connection to utility networks ...), classes (building units, installations, other constructions) and references to other data (like cadastral data and addresses).
- **The BuildingsExtended3D profile** is an extension of the Buildings3D profile for rich 3D representations at different levels of details. It includes the possibility to represent many building components, such as the building boundaries (wall, roof), the openings (doors – windows) and building interior (rooms, internal installations) and the textures associated with the main feature types. It also contains all the semantic information of extended 2D profile.

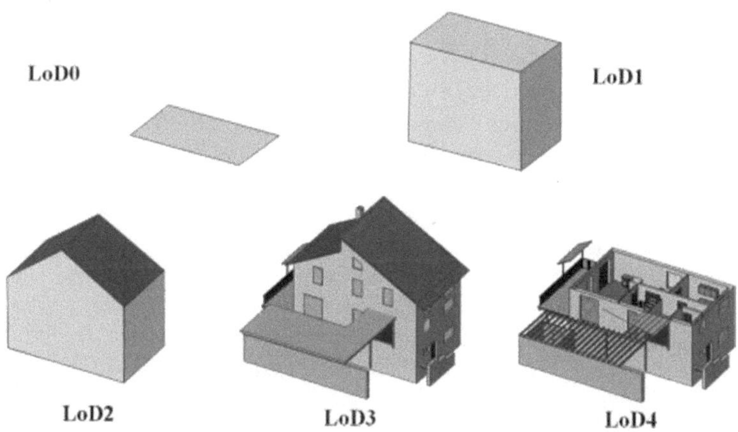

Fig. 2. Levels of detail (LoD) within the OGC CityGML implementation specification according to Data for 3D project, 2013 [3]. Please note, that INSIPRE 3D buildings application schemas incorporates only LoD1 to LoD4, since LoD0 is reserved for the 2D application schemas.

We will hereinafter deal only with the Buildings3D profiles since they are based on the OGC CityGML specification, enable to link to noise data and allow presenting through innovative X3D Web technology. Figure 1 contains the UML class diagram of the 3D application schema for buildings as a common starting point. Data type *BuildingGeometry2D* on the right side of the Figure is originating from the 2D application schemas for buildings. The pink bottom part enables to incorporate any building or even any part of the building in four levels of details (LoD) as depicted in

Figure 2. This application schema also enables to aggregate information from buildings and building parts into two feature types: *AbstractBuilding* and *AbstractConstruction*.

2.2 European Noise Directive

Omnipresent noise enters more and more often and intensively into human subconsciousness as a part of everyday life and only few realize how harmful it is. Long term influence of noise can cause not only hearing impairment, but also many other illnesses. In particular, it has negative psychological effect in the form of stresses, neuroses, etc. Negative effects of noise caused that another Directive of the European Union (EU) was developed.

Directive 2002/49/EC of the European Parliament and of the Council of 25 June 2002 relating to the assessment and management of environmental noise is commonly known as the Environmental Noise Directive and hereinafter is referred to as the 'END'.

The END requires Member States to undertake strategic noise mapping. It also refers to legislation regarding access to the information from strategic noise maps, since it requires "that the strategic noise maps … are made available and disseminated to the public in accordance with relevant Community legislation … and in conformity with Annex IV … to this Directive".

Strategic noise map (and mapping) is formally defined by the END as „a map designed for the global assessment of noise exposure in a given area due to different noise sources or for overall predictions for such an area". For more information on the END please refer to the Directive itself [4] or scientific articles with this topic such as [14].

The Aarhus Convention has established the right of everyone to receive environmental information held by public authorities ("access to environmental information"). The parties to this important Convention, including all European Union (EU) Member States, are required to make the necessary provisions so that public authorities at national, regional or local level ensure that this right is upheld.

Together the END (2002/48/EC), INSPIRE Directive (2007/2/EC), Directive on public access to environmental information (PSI; 2003/4/EC) and Aarhus convention have established two distinct aspects with regard to informing the public; (1) the availability of noise information upon request, and (2) the active and systematic dissemination of noise information to the public. We may identify the following chain in noise mapping: the END defines the noise mapping itself, the INSPIRE Directive spatial representation and discovery mechanisms and the PSI Directive re-use of data and information that was created within the END and INSPIRE.

3 Technological framework

3.1 CityGML

CityGML (*City Geography Markup Language*) is an open data model and XML-based format for sets of 3D urban objects. It is implemented as an application schema for the GML3 (*Geography Markup Language version 3*). Both are extendible international standards for spatial data exchange issued by the OGC (Open Geospatial Consortium). CityGML is used for the representation, storage, and exchange of virtual 3D city models. Generalization hierarchies between thematic classes, aggregations and relations between objects are also included in this application schema. CityGML standard has been used as well as a basis for developing INSPIRE application schemas for spatial data theme *Buildings*.

CityGML provides a standard model and mechanisms for describing 3D objects with respect to their geometry, topology, semantics and appearance; furthermore defines five discrete levels of detail (LoD) as depicted in Figure 2. CityGML provides much more than 3D content for visualization to diverse applications. CityGML provides class definitions, regulations and explanations of the semantics for the important geographic features in virtual 3D city models including digital terrain models, buildings, water bodies, vegetation, transportation objects and city furniture [15]. CityGML has been designed as a universal topographic information model according to [10].

Practical applications often require storing and exchanging data, which do not belong to any of the predefined classes. CityGML provides two different ways of extension for such cases. The first one is a usage of generic features and generic attributes; however this method reduces semantic interoperability. Concept called Application Domain Extension (ADE) is used to eliminate this problem. In contrast with generic objects and attributes, ADE has to be defined within a separate XML Schema definition. CityGML data model extended by an ADE has been used for example for noise mapping in North-Rhine Westphalia in Germany [2].

3.2 X3D

X3D (Extensible 3D) is an XML based 3D graphic format. X3D originates from another 3D graphical format – VRML (*Virtual Reality Modelling Language*). X3D provides all functionality included in the second version of VRML, which is commonly known as VRML97. XML encoding enables better integration of X3D and contemporary Web technologies. X3D supports 3D graphics and programmable shaders, 2D graphics, animations, spatialized audio and video, user interaction and navigation, networking, scripting and user-defined objects. Several profiles are defined for various levels of capabilities (e.g. X3D Core, X3D Interchange, X3D Immersive, which most closely matches VRML97, and X3D Full). X3D was approved as an ISO standard (ISO/IEC 19775) in 2004 [9].

JavaScript library for rendering the X3D scenes directly in a Web browser is called X3DOM. X3DOM is based on direct integration of X3D elements into the HTML (*HyperText Markup Language*) code. The main goal is to achieve X3D scene available for JavaScript DOM (*Document Object Model*), which allows modifying the 3D scene by only adding, removing or changing DOM elements. No specific plug-in is needed. Common JavaScript events, like *onclick*, on 3D objects are supported as well. There is also available an X3DOM runtime API, which provides a proxy object to programmatically read and modify runtime parameters. The API functions are available for resetting views, cycling cameras or changing navigation modes [1]. X3D elements can also be stored as part of external files; their connection is realized via the element *Inline*. Particular X3D elements can be clearly distinguished through their attribute *DEF*.

4 Use Case

Standardization frameworks, INSPIRE application schemas for Buildings and CityGML, are suitable for storing 3D data about buildings and noise pollution. On the other hand, X3D has a bigger potential for the visualization in Web environment. Therefore, it is necessary to deal with the ways of transformation from mentioned data standards into X3D a as first step. XSLT (*Extensible Stylesheet Language Transformations*) language has been used for these purposes.

4.1 Input Data

Technologies described in this paper have been tested on example of a Brno city district called Nový Lískovec, the Czech Republic. CityGML model consisting of building in LoD1, terrain model in LoD0 and noise maps stored as generic objects in LoD1 were used as input data. Geometry of buildings originally comes from stereophotogrammetry and terrain model as developed for the official Czech topographic database called ZABAGED. Noise maps show spatial distribution of equivalent continuous sound pressure level in the night (22:00 – 6:00) and in the daytime (6:00 – 22:00). These maps were calculated on basis of compendious traffic from land transport: road, railway and tram system transport. Model of noise emissions has been calculated according to the situation in year 2004.

4.2 XSLT

XSLT is a language for transforming the XML documents into the differently structured XML documents and/or open data formats. The original document is not changed. However a new document is created – based on the content of an existing one. XSLT uses two types of input data: input datasets and templates (stylesheet). The XSLT processor takes one or more XML input documents together with one or more XSLT stylesheet modules, and processes them to produce an output document. The XSLT stylesheet contains a collection of template rules: instructions and other directives that guide the processor in the production of the output document. The main principle of transformation is in this case simplification of the semantic structure. Both, described and possible inputs (CityGML and INSPIRE application schemas for Buildings), are based on GML3, thus form of both stylesheets is very similar.

```
<xsl:if test="$from > $to">
  <xsl:call-template name="iterate">
    <xsl:with-param name="from" select="(($from) + 1)"/>
    <xsl:with-param name="to" select="$to"/>
    <xsl:with-param name="list" select="concat($list,
    $from, ' ')"/>
```

```
    <xsl:with-param name="points" select="$points"/>
  </xsl:call-template>
</xsl:if>
<xsl:if test="$from = $to">
  <xsl:param name="f_list" select="concat($list, '-1')"/>
  <IndexedFaceSet coordIndex="{$f_list}">
    <Coordinate point="{$points}"/>
  </IndexedFaceSet>
</xsl:if>
```

XSLT source code listing above depicts the list of coordinates (X, Y, Z) which are transferred into the value of the attribute *point*. The attribute *coordIndex* contains sequence indicating the order in which the points are connected. These two key elements define geometry of element *IndexedFaceSet*, which defined the geometry of all volume and surface features.

4.3 Differences between (City)GML and X3D

The most significant difference between X3D and CityGML consists in the method of geometry storage. CityGML as well as all INSPIRE application schemas use a subset of the GML3 geometry models. Geometries of geographic features are in GML3 are represented as objects having an identity and further substructures. GML3 provides classes for geometric primitives (e.g. *Point*), composite geometries (*CompositeSurface*) and geometry aggregates (*MultiSurface*). Volumetric representations are modelled according to the Boundary representation (*BRep*), where each solid is bounded by a closed surface (typically *CompositeSurface*).

Geometry may be defined in X3D as a solid figure (elements *Box*, *Cone*, *Cylinder*, *Sphere*); while this method is not convenient for complex shapes of geographical features. Spatial vector data are represented in Geographical Information Systems (GIS) as points, lines and polygons. We may identify following X3D elements that are suitable for storing these types of spatial data: *PointSet*, *IndexedLineSet* and *IndexedFaceSet*. Raster data can be stored in X3D as *ElevationGrid*.

Definition of materials in CityGML is adopted from X3D, so there is not any issue with their transformation. Definitions of materials are in both cases composed of colour value, transparency and shines. For these purposes, X3D elements *Material* and *ImageTexture* are used. CityGML and X3D also allows linking to texture images. Since appearance is not inevitable part of input CityGML data, their definition may be a part of XSLT template. Appearance is not specified in the 3D profiles of INSPIRE application schemas for *Buildings* spatial data theme.

Semantic data structures are largely removed during the transformations, because X3D does not support storing attributes and semantic classes while enables only a general element (*Group*) for creating taxonomies. On the contrary, data originating from attributes can be stored in external files (e.g. XML or JSON) and connection would be provided through unique identificators (attributes *DEF* or *id*) and selected programming language (e.g. JavaScript). The second type of semantic information connection is their storing in additional attributes. This method describes [16].

4.4 Resulting Web Presentation

Use case Web application consists of HTML page, JavaScript files and X3D files comprising digital terrain model, noise maps and buildings. Saving 3D data as X3D and using JavaScript and X3DOM library allow to implement interactivity such as interactive and non-interactive movement, setting up type of interactive movement, changing rendering mode and switching between noise maps for calculated for day and for night. Further enrichment of the user interface, which mentions [17], is also possible. Using internet technologies like X3DOM provides 3D visualization of spatial data, such as noise pollution, for wide audience.

Fig. 2. Developed use case Web application for noise mapping with INSPIRE-based buildings. Visualization shows noise pollution in Brno city district Nový Lískovec, the Czech Republic

5 Conclusions and Future Work

It may be stated that the European legal frameworks of the END, INSPIRE and PSI directives represent complementary activities in the field of noise mapping. Models developed according to the END can be inserted into the models based on the INSPIRE application schemas and re-used in line with the principles of the PSI directive. The weakness of above mentioned legal frameworks lies in the lack of visualisation rules and best practices for the domain of noise mapping. A developed noise map shall follow strict rules for the data structure; however its visualization does not belong between the primary aspects. The same situation appears also in the documents of the INSPIRE directive: legally binding documents do not contain visualization principles. Visualization rules are formalised through the OGC Styled Layer Descriptor (SLD) in the INSPIRE technical guidelines; however any rule does not relate to a 3D profile.

We have therefore developed and tested X3D-based visualization in comparison to common Web services as depicted, for instance, by [7]. To sum up, we have identified four major advantages of the X3D visualization for noise mapping:

- X3D enables user-friendly 3D visualization based on standardized Web technologies;
- X3D does not require to install any new software or plug-in;
- X3D enables to present noise maps calculated in 2D as 3D visualization;
- X3D enables exploration of barrier effect caused by high buildings blocks and/or other impacts of terrain and buildings.

Future work should aim at issues of noise mapping in 3D. Currently the END requires only the noise map for a level of 4 meters above the terrain, i.e. typically the second floor of a building. The X3D technology allows displaying a noise value at each part of a building. Such technique would be valuable for visualization of areas highly polluted by noise, such as city centres or built-up areas near highways. The first attempts on such noise mapping may be found in [12].

Acknowledgements. This research has been supported by funding from the project of Masaryk University under the grant agreement No. MUNI/A/0902/2012, which is called 'Expression of Global Environmental Change in Component Earth's Spheres'.

References

1. Behr, J., Jung, Y., Drevensek, T., Aderhold, A.: Dynamic and Interactive Aspects of X3DOM. In: Proceedings of the 16th International Conference on 3D Web Technology (2011)
2. Czerwinski, A., Sandmann, S., Stőcker-Meier, E., Plűmer, L.: Sustainable SDI for EU noise mapping in NRW – best practice for INSPIRE. International Journal of Spatial Infrastructures Research 2, 90–111 (2007)
3. Data for 3D project, http://www.agency9.com/products/city-planner/data-for-3d-project/
4. Directive 2002/49/EC of the European Parliament and of the Council of 25 June 2002 relating to the assessment and management of environmental noise - Declaration by the Commission in the Conciliation Committee on the Directive relating to the assessment and management of environmental noise, http://eur-lex.europa.eu/LexUriServ/LexUriServ.do?uri=CELEX:32002L0049:EN:HTML
5. Directive 2003/4/EC of the European Parliament and of the Council of 28 January 2003 on public access to environmental information and repealing Council Directive 90/313/EEC, http://eur-lex.europa.eu/LexUriServ/LexUriServ.do?uri=OJ:L:2003:041:0026:0032:EN:PDF (accessed April 25, 2013)

6. Directive 2007/2/EC of the European Parliament and of the Council of 14 March 2007 establishing an Infrastructure for Spatial Information in the European Community (INSPIRE), http://eur-lex.europa.eu/LexUriServ/LexUriServ.do?uri=OJ:L:2007:108:0001:0014:en:PDF
7. Horák, J., Ardielli, J., Růžička, J.: Performance Testing of Web Map Services. In: Nguyen, N.T., Trawiński, B., Jung, J.J. (eds.) New Challenges for Intelligent Information and Database Systems. SCI, vol. 351, pp. 257–266. Springer, Heidelberg (2011)
8. INSPIRE Data Specification on Buildings – Draft Technical Guidelines, http://inspire.jrc.ec.europa.eu/documents/Data_Specifications/INSPIRE_DataSpecification_BU_v3.0rc3.pdf
9. ISO/IEC 19776:2007 Information technology — Computer graphics and image processing — Extensible 3D (X3D) encodings, http://www.web3d.org/x3d/specifications/ISO-IEC-19776-X3DEncodings-All-Edition-1/
10. Kolbe, T.H.: Representing and exchanging 3D city models with CityGML. In: Lee, J., Zlatanova, S. (eds.) 3D Geo-information Sciences, pp. 15–31. Springer, Heidelberg (2009)
11. Konečný, M.: CARTOGRAPHY: Challenges and Potentials in Virtual Geographic Environments Era. Annals of GIS 17(3), 135–146 (2011)
12. Law, C., Lee, C., Lui, A.S., Yeung, M.K., Lam, K.: Advancement of three-dimensional noise mapping in Hong Kong. Applied Acoustics 72(8), 534–543 (2011)
13. Mao, B., Ban, Y.: Web-based Visualisation of the Generalised 3D City Models using HTML5 and X3DOM (2013), http://jselab.njue.edu.cn/download/conference/Web-based%20Visualisation%20of%20the%20Generalised%203D%20City%20Models%20using%20HTML5%20and%20X3DOM.pdf
14. Murphy, E., King, E.A.: Strategic environmental noise mapping: Methodological issues concerning the implementation of the EU Environmental Noise Directive and their policy implications. Environment International 36(3), 290–298 (2010)
15. OGC City Geography Markup Language (CityGML) Encoding Standard, http://www.opengeospatial.org/standards/citygml
16. Pittarello, F., De Faveri, A.: Semantic Description of 3D Environments: a Proposal Based on Web Standards. In: Proceedings of the 11th International Conference on 3D Web Technology (2006)
17. Staněk, K., Friedmannová, L., Kubíček, P., Konečný, M.: Selected issues of cartographic communication optimization for emergency centers. International Journal of Digital Earth 3(4), 316–339 (2010)

National INSPIRE Geoportal of the Czech Republic

Jan Prášek[1], Jiří Valta[1], and Jiří Hřebíček[2]

[1] Czech Environmental Information Agency, Vršovická 1442/65
100 10 Praha 10, Czech Republic
[2] Masaryk University, Institute of Biostatistics and Analyses
Kotlářská 2, 611 37 Brno, Czech Republic
{jan.prasek,jiri.valta}@cenia.cz, hrebicek@iba.muni.cz

Abstract. Environmental information systems of the Czech Republic´s Ministry of Environment have been developed since 1990. Creation of the National Information System to Collect and Evaluate Information on Environmental Pollution began in 2010. This project has been co-financed by the European Regional Development Fund. The project is split into three parts: (a) Integrated Reporting System (ISPOP); (b) Environmental helpdesk (EnviHELP) and (c) National INSPIRE Geoportal (Geoportal), which were developed in 2010-2013. In this article, Geoportal as national implementation of the INSPIRE Directive in the Czech Republic is presented in more detail including several usage examples.

Keywords: INSPIRE, ISPOP, EnviHELP, National INSPIRE Geoportal, GIS, CENIA.

1 Introduction

Environmental Information Systems (EIS) deal with processing, retrieval and presentation of environmental data and information. In the Czech Republic, European Union (EU) and the World, we are witnessing the creation of large-scale information and communication infrastructures. In the Czech Republic, EIS in public administration have been under development since the beginning of 1990 [1]. The Act No. 2/1969 Coll., on the establishment of ministries and other central state administration bodies of the Czech Republic (Article 4 § 19), provides that the Ministry of Environment (MoE) of the Czech Republic secures and manages development of the *Single Environmental Information System* (JISŽP), including overall monitoring of the entire Czech Republic. The information system (JISŽP) was defined by the MoE and consolidated including all non-operational information systems of MoE, its infrastructure and management systems [1-2].

In 2008, the MoE approved the *Information Strategy of MoE and JSŽIP for the years 2008-2010* which focused on the importance of achieving interoperability within the EIS of MoE, the efficient management of environmental information content and information flow (from data monitoring and creation through their acquisition, validation, evaluation and presentation of highly aggregated information), in order to achieve a high level of electronic services in public administration

J. Hřebíček et al. (Eds.): ISESS 2013, IFIP AICT 413, pp. 425–438, 2013.

(eGovernment) and also to support information provision by the services and to evaluate effectiveness of the policy instruments in environmental protection [2-3].

The basic objective of the *Register of rights and obligations* (RPP) of the *Basic public administration registers* within eGovernment [4] is to facilitate for citizens, companies and other entities the communication with the authorities. The RPP also contains their rights and obligations concerning environmental protection, including MoE agendas in JISŽP. At the same time, public authorities have to ensure a safe, efficient and transparent exchange of accurate and up-to-date environmental data for all the entities [2].

Focusing on a strategic approach related to the development, JISŽP emphasis is on achieving a high level of electronic services (eGovernment) and building up support for information services, to evaluate the effectiveness of policy instruments for environmental protection.

JISŽP Development Council, an advisory board of the minister of environment, is the institutional component to ensure data and access to information. To put this in practice, various application tools (ICT tools) have been gradually created within several individual projects in the respective MoE departments.

Development of the *National Information System to Collect and Evaluate Information on Environmental Pollution* (CISAŽP), which was co-financed by the European Regional Development Fund (ERDF), began in 2010 [6]. CISAŽP supported the implementation of *Shared Environmental Information System* (SEIS) and eEnvironment [7] in the Czech Republic. The main users of the new CISAŽP are the entities subject to environmental laws and public authorities. The aggregated environmental data and other information, however, will have to access to the public. The advantages and benefits of CISAŽP consist in a substantial increase of transparency, consistency and efficient communication between the Ministry of Environment and entrepreneurs, the authorities and the public.

CISAŽP consists of three main parts:

- Integrated reporting system (ISPOP) [8];
- Environmental helpdesk (EnviHELP) [9];
- National INSPIRE Geoportal (Geoportal) [10].

CISAŽP outputs allow computerization of administrative agendas throughout the environmental sector, simplifying considerably the reporting obligations and state administration. All information which enters the CISAŽP system is analyzed, evaluated comprehensively and placed into a broader context, taking into account the economic and social aspects and constraints. The outputs form an important basis and starting point for the formulation of priorities and objectives to protect and improve the environment in the Czech Republic.

The *Czech Environmental Information Agency* (CENIA) is in charge of development and technical operations related to CISAŽP.

The main users of CISAŽP are governmental institutions, professionals and the general public.

Other EIS projects were developed in the Czech Republic: IRZ (Integrated Pollution Register) [11]; CITES (Convention on International Trade in Endangered Species of Wild Fauna and Flora) register [12]; ISOH (Waste management information system) [13]; ISOP (Information system of nature conservancy) [14] and geological IS, e.g. Czech Geological Survey [15].

Formally, these EIS are reflected in a new concept of the *Information Strategy of MoE* [5], the attestation of which took place in November 2012.

In the following text, we briefly introduce the ISPOP and EnviHELP systems and describe the National INSPIRE Geoportal in more detail.

2 Integrated Reporting System (ISPOP)

Environmental legislation of the Czech Republic (in compliance with the European Union (EU) legislation) imposes on economic entities (organizations, companies, enterprises, etc.) the duty to notify public authorities of their activities´ environmental impacts.

These entities are thus becoming the *declarants of environmental records.* Their notification duty is provided for by the relevant environmental legislation of the Czech Republic, which obliges them to keep records and to report. The declarants´

Fig. 1. Basic ISPOP website (Source [8])

duties include submitting of relevant environmental data, information and messages to the respective public authorities and institutions. This mainly includes primary environmental data such as those concerning air pollution, water pollution, waste, soil pollution, waste electrical and electronic equipment, car wrecks etc., which are processed and reported to the CENIA SEIS data centres.

ISPOP makes it possible to incorporate a reception of selected environmental reports (reporting obligations) in electronic format and also to distribute environmental reports to relevant institutions and public authorities including data flows to the SEIS data centres [6]. Operation of ISPOP was launched on 1th January 2010. The Ministry of Environment is the founder of ISPOP and it is also responsible for the forms to be filled in within the single reporting obligations. Technical operation and development of ISPOP is ensured by CENIA.

The ISPOP's main objective is to create a universal data input gateway providing the incoming data files in a specified standardized format. It controls distribution of the target processing systems. ISPOP must handle both the management reports submitted in a form of environmental records and archiving and evaluation of their content (in cooperation with other related systems). The ISPOP architecture was structured to ensure flexibility in terms of expansion to the other input agendas and also to guarantee, inter alia, compliance with updated environmental legislative obligations.

3 Environmental Helpdesk - EnviHELP

EnviHELP provides support in a particular operating state, e.g. administration for incumbents fulfilling the reporting obligations under the Act No. 25/2008 Coll., as well as business communities and the general public in providing environmental information in relation to the Act No. 123/1998 Coll., public access to environmental information and the Aarhus Convention.

Knowledge basis of the Environmental Helpdesk is under preparation at present. It aims at collecting the maximum available information and methodological materials on environmental issues. Currently, it comprises:

- about 500 items of the following types: law, regulations, government regulations, etc.;
- about 1,400 terms related to the environment;
- about 300 life situations;
- 350 agendas arising from legislation;
- 400 most frequently asked questions concerning life situations and MoE agendas;
- 700 inputs and outputs of MoE agendas or life situations and other contexts;
- about 500 objects, especially guidance and reporting support for ISPOP.

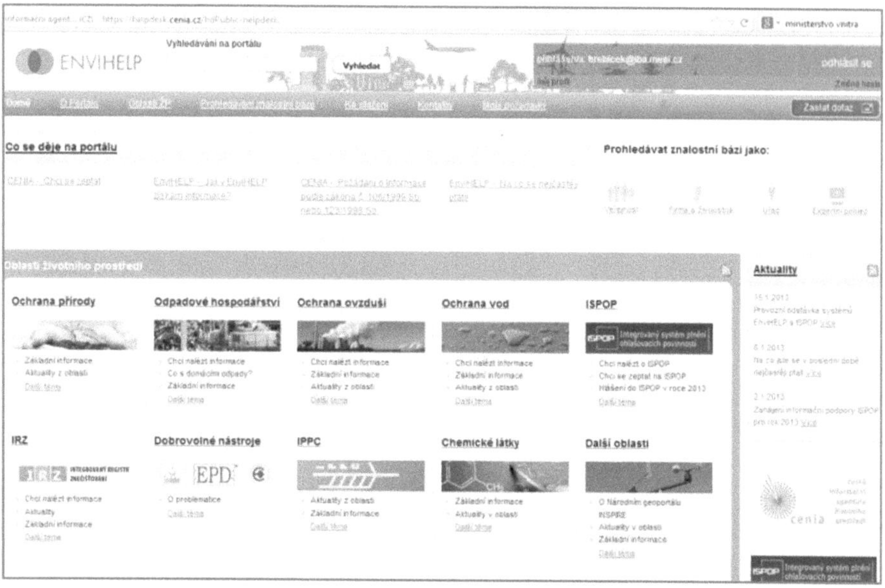

Fig. 2. Environmental helpdesk (EnviHELP) website (Source [9])

4 National INSPIRE Geoportal

At present, the amount of GIS (Geographical Information Systems) users is growing continuously. This technology improves work and decision-making at both national and local government level. The INSPIRE (Infrastructure for Spatial Information in Europe) initiative of the European Commission (EC), described in the Directive No. 2007/2/EC, specified the setup of a spatial data infrastructure within the Community (the INSPIRE Directive). This Directive aims at creating a European legislative framework which is needed to build a European *Spatial Data Infrastructure* (SDI). It lays down general rules for the establishment of a European SDI, in particular to support environmental policies and policies that affect the environment.

INSPIRE employs the following basic principles to organize the European geo-information infrastructure [16]:

- geo-information will be stored, managed and made available on one suitable level;
- geo-information coming from different sources in the European Union can be combined in a consistent manner and exchanged between different users and applications;
- geo-information that has been collected from one public sector level can be exchanged with other public sector levels;
- geo-information will be made available in such a way that large-scale use of the system will not be unnecessarily impeded;

- any available geo-information can easily be looked up, and the suitability and user conditions can easily be checked.

EU Member States are providing discovery and view services according to the INSPIRE Regulation on Network Services through the INSPIRE Geoportal [17] which was launched in November 2011. The main objective of INSPIRE is to make more high-quality and standardized spatial information for the development and implementation of EU policies at all levels of the EU Member States. The INSPIRE Geoportal provides the means to search for spatial data sets and spatial data services, and subject to access restrictions, to view spatial data sets from the EU Member States within the framework of the INSPIRE Directive.

The INSPIRE Directive was transposed to the Czech Republic´s legislation through the amended Act No. 123/1998 Coll., on public access to environmental information, which entered into force on 23 October 2009. On the basis of this Act, the National INSPIRE Geoportal (Figure 3) was established to allow the general public access to spatial data relating to at least one of the topics of the INSPIRE Directive attachments.

Fig. 3. Homepage of the Czech National INSPIRE Geoportal (Source [6])

The construction of Czech National INSPIRE Geoportal (i.e. Geoportal) made available the current and validated spatial data concerning the Czech Republic to improve the design, implementation and monitoring of environmental policies at the national, regional and local levels. The Geoportal project is also designed to enable solutions to life situations by linking to the EnviHELP knowledge base within Georeport generation. Geoportal was implemented during the period 2009-2013 and it aims at providing more transparency, consistency and efficiency to the communication between entrepreneurs, government and the public on one side and the MoE on the other, and also at fostering implementation of eEnvironment in the Czech Republic.

4.1 Technical Solutions to Geoportal

Geoportal is based on the ICT solutions ESRI, ArcGIS10.x Server for Server Enterprise Advanced (4 cores) [18]. The portal solution is built on the open source applications company DxoLogicLiferay Enterprise [19].

Geoportal uses information services to connect CENIA to map servers and other data providers, see Figure 4.

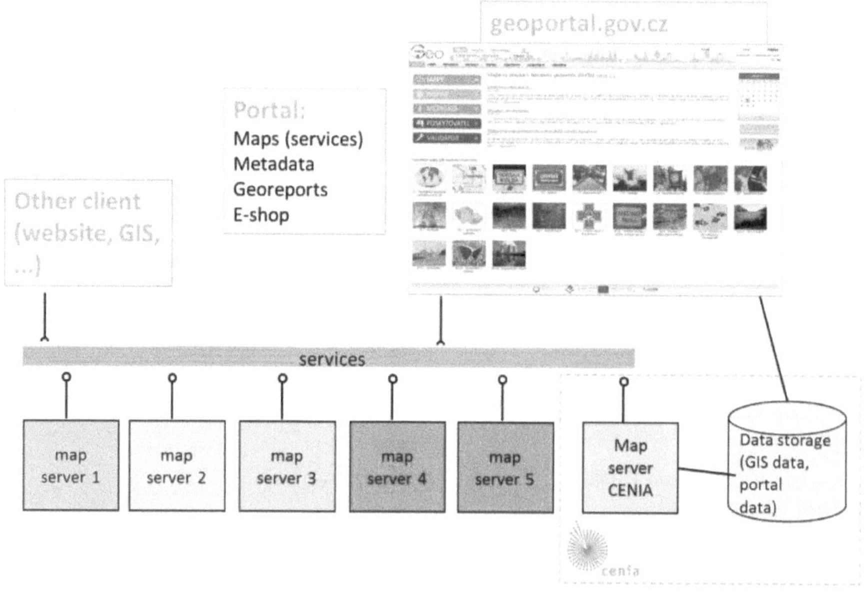

Fig. 4. Geoportal Architecture (Source: authors)

4.2 Geoportal Services

Geoportal services are provided on a subscription basis. For *unregistered users,* only the essential basic tools of the Geoportal are available. An ID-based registration of a Geoportal's user is required for all the other Geoportal activities and services. Registration allows users to store their own map compositions, to download data (via e-shop), to create and store metadata at the metadata server and much more. If providing data, metadata, data monitoring, etc., the users must also include, in addition to the registration, a verification of the data provided, even if the activity is performed by an authorised person, organization or company.

All data included in Geoportal documents contain Metadata; they are used to perform different operations to search for or create the user´s own metadata according to a national or INSPIRE profile using the metadata editor. Metadata can also be imported into the user´s own metadata records but they must subsequently be validated against the supported profiles.

Geoportal also lets users use their own services (catalogue, viewing, ...). These services can be registered within Geoportal. This function is available only to the registered users, and only for services that are equipped with metadata.

The system also allows the *registered users* to make their own data accessible. The data can be uploaded to Geoportal and the administrators can then create a service or can be entrusted with publishing the data.

According to the INSPIRE Directive, each data and service *provider* is annually required to fill in details about their data sets and services. For this purpose, a monitoring tool is being operated. Every provider, from early January to late February, completes the required data for the previous calendar year.

Current map layers published by Geoportal are as follows [9]:

- Cadastral maps:
  - Land Register;
  - definition points of parcels;
  - Land Cadastre.
- Topographic maps of the Czech Office for Surveying, Mapping and Cadastre;
- Main Basis of Geographical Maps of the Czech Republic (ZABAGED ®);
- Digital model area (DMÚ25);
- Orthophoto maps (historical);
- Orthophoto maps (current);
- III. Military Survey (1870s and 1880s, scale 1: 25,000), a new height, contours and dimensions;
- II. Military Survey (1806 - 1869 in Austria-Hungary, scale 1:28 800);
- Military maps (raster);
- Road maps.

In Geoportal menu, Home is made up of a menu containing the favoured map tasks according to INSPIRE categories, see Figure 3.

The Geoportal map window consists of a chart and menu displaying various maps and layers that can be seen, added or removed. The menu is also a guide post to metadata of the individual map layers, see Figure 5.

Maps displayed at the National INSPIRE Geoportal are also available in the form of map services, which can be used in so-called thick clients (software applications) for other purposes. Map services are offered as standard WMS and are also available via Simple Object Access Protocol (SOAP) services of ESRI ArcGIS Server [18]. Both types of services have the same names and data visualized in the same way.

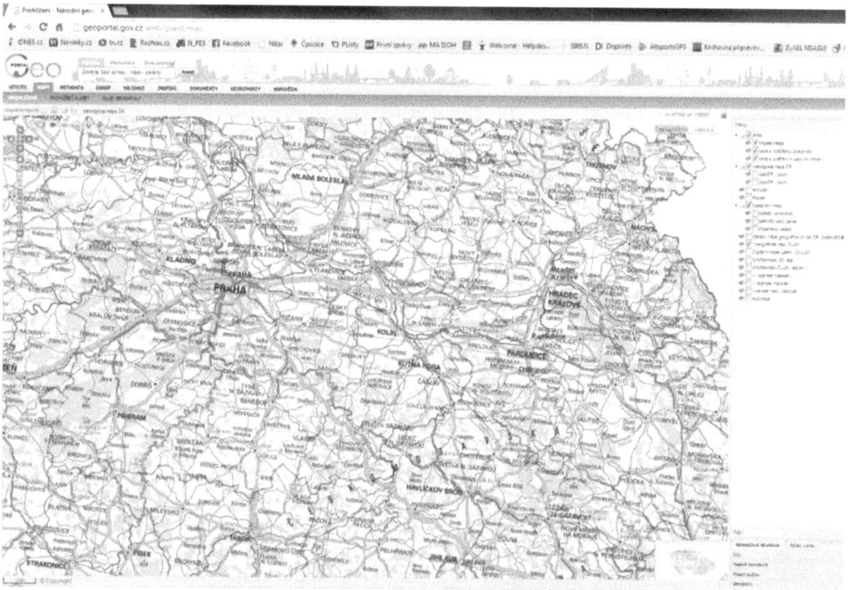

Fig. 5. Map Window (Source: authors)

The individual objects in spatial compositions can obtain a detailed description of their attributes, such as information from air quality monitoring stations, Figure 6.

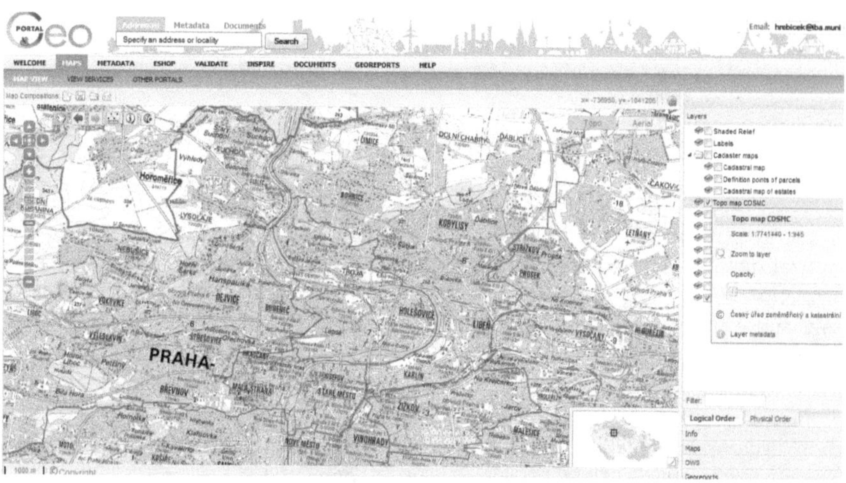

Fig. 6. Detail of Attributes (Source: authors)

5 Map Examples

In this section we present several interesting examples of how to use Geoportal.

5.1 Noise Map of Prague

The noise map of Prague is based on the evaluation of measured noise values in different parts of the metropolitan area. Currently available data are extrapolated from the evaluation of noise indicators Ln (dB) and Ldvn (dB), see Figure 7.

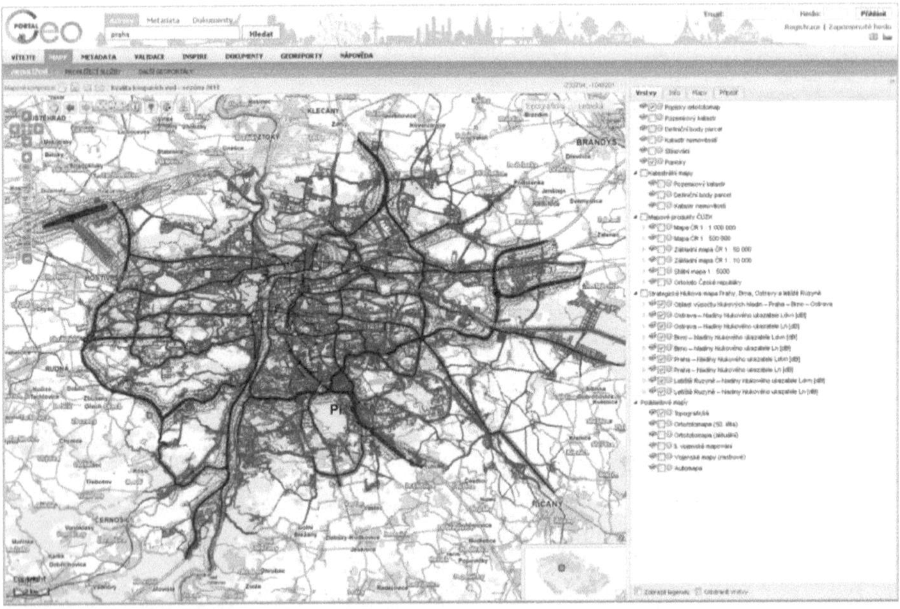

Fig. 7. Noise Map of Prague (Source: authors)

5.2 Geological Map of the Czech Republic

The task of the Czech Geological Survey (CGS) [20], established in 1919, is to provide the national geological services for the Czech Republic. The CGS is the national organization with the statutory responsibility to gather, store and interpret geological information so that the authorities can take appropriate decisions about national economic and environmental issues. It provides the results of systematic regional geological mapping and investigation to all interested persons. It also performs related tasks as required by the MoE on the basis of the Act No. 62/1988, on geological work. This involves the collection, processing and provision of information regarding geological structure, protection and use of mineral and groundwater resources, and also concerning geohazards. The CGS map portal is connected to Geoportal.

The Geological Map of the Czech Republic (Digital Geo-atlas of the Czech Republic - GEOČR 500, see Figure 8) developed by CGS provides 11 maps covering different geological and scientific topics, arising mostly from the original analogue data at the scale 1:500,000 (Geological Map, Radiometric Map, Map of Radon Risks, Map of Mineral Water, Gravimetric Map, Digital Elevation Map, Satellite Map, Map of Land Cover, Topographical Map). Individual maps form a unique GIS, which is also made in the three coordinate systems: S-JTSK, S-42, and geographic coordinates.

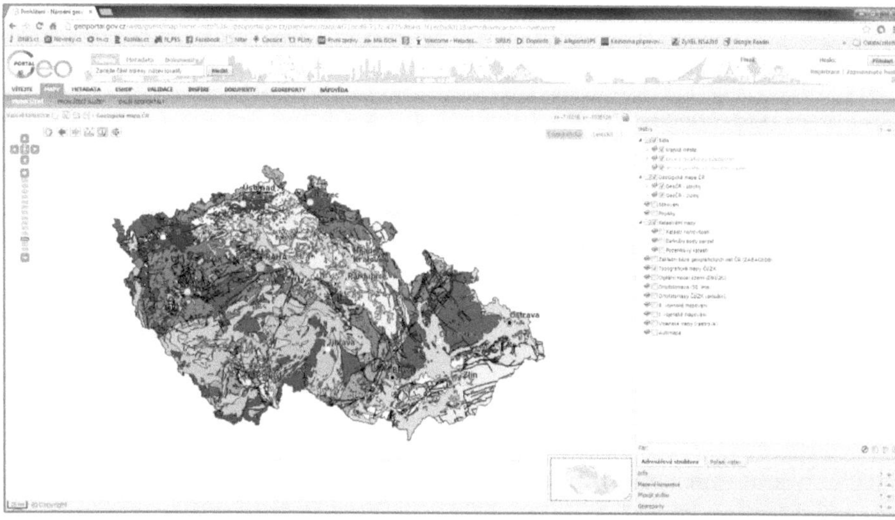

Fig. 8. Geological Map of the Czech Republic (Source: authors)

5.3 Historical and Present-Day Orthophoto Maps

Within the methodological section (Stage 1) of the project National Inventory of Contaminated Sites (NIKM) [21], an orthophoto map of the Czech Republic was created from historical images collected within the first nationwide aerial photographing

Fig. 9. Comparing orthophoto maps of 1952 and 2010 – development of the settlement (Source: authors)

dating back to 1950s. The company Geodis BRNO Ltd[1] has processed the source materials (aerial images provided by the Military Geographic and Hydrometeorological Office (VGHMÚř) in Dobruška[2]) and supplied the historical orthophoto maps.The current orthophoto map is provided via the browser-based services WMTS-ORTOFOTO as a public viewing service related to the data from the orthophoto map of the Czech Republic. In order to optimise the speed, the data are provided in a form of map tiles. The orthophoto data service covers the entire usable range of scaling, i.e. including small scales. The service meets the technical guidance for INSPIRE View Service v. 3.1 [22] and is compliant with the standard OGC WMTS 1.0.0[3]. To compare development of settlement captured in the maps of 1952 and 2010, see Figure 9.

5.4 The 2006 Floods

On the basis of evaluation of the measurement data, it is possible to see the area which the Elbe River flooded on April 3, 2006 (Figure10).

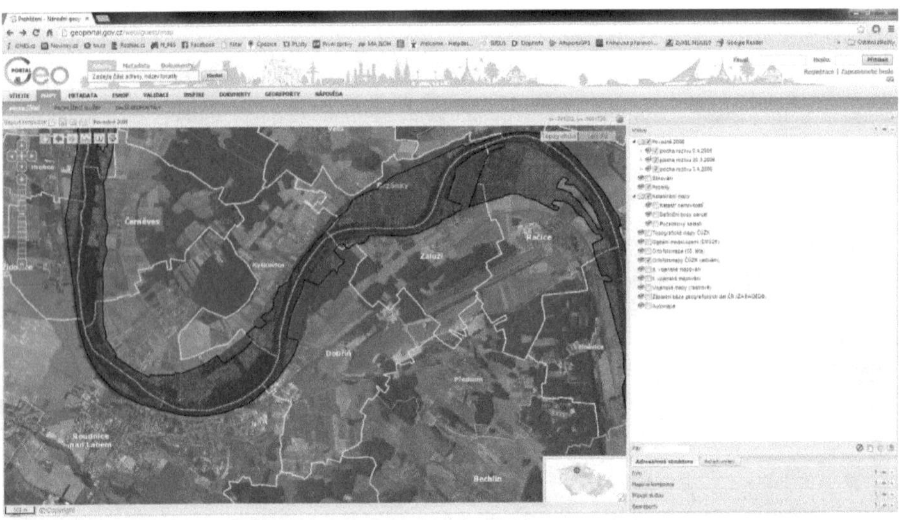

Fig. 10. Area flooded by the Elbe River on April 3, 2006 (Source: authors)

6 Georeports

Georeports represent an application of Geoportal, which offers, in a simple form, interpretation of selected useful information based on data related to a selected location. To create a Georeport output, the users simply choose a Georeport topic and select an area of their interest on the map.

---

[1] http://sluzby.geodis.cz/products/coloured-orthophotomap-of-the-czech-republic?lang=2

[2] http://www.geoservice.army.cz/htm/s_urad.html

[3] http://www.opengeospatial.org/standards/wmts

In a Georeport, the map section with the results of the spatial data analyses of the selected areas is accompanied by text. In the text part, the Georeport contains a final list of relevant authority offices which concern the chosen topic in the locality, and information about possible deadlines, permits, forms and fees is also provided. The resulting, automatically generated Georeport takes the form of a PDF document that can be viewed, saved, and sent to an e-mail address.

E-shop is a further application of Geoportal and it provides access to paid services consisting of:

- Data sharing/providing tool (including licensing).
- Providers themselves decide to whom they wish to provide the data and under what conditions.
- Geoportal output module exports the desired data to the approved applicants.
- RSS feeds also serve without any fee, in the form of a "click" license.

7 Conclusion

Development of CISAŽP project of the MoE began in 2010; it has been co-financed by the European Regional Development Fund during 2010-2013. Environmental information is generated by CISAŽP in its three main components: IS ISPOP, EnviHELP and Geoportal. IS ISPOP is used for electronic reporting of mandatory data from environmental polluters. The environmental helpdesk EnviHELP provides support to users of the subsystems within the project CISAŽP. EnviHELP is designed to serve as the main help desk platform managed by the MoE.

Geoportal provides updated and validated spatial data to improve the design, implementation and monitoring of environmental policies at the national, regional and local levels. It includes ICT tools for receiving and validating spatial data and services from statutory and other providers. It also enables display and publication based on standardized service creation, management and retrieval using metadata, transformation and loading of data and it guarantees payment for the data.

It can also be linked to EnviHELP knowledge base to solve different life situations within generating Georeports. Geoportal receives in average 4,300 visits per day by users who download about 44.6 GB of data.

References

1. Hřebíček, J.: Globale Umweltinformationssysteme in internationalen Masstab. In: Internationales Umweltmanagement, Band II: Umweltmanagementinstrumente und - systeme. Gabler Verlag, Wiesbaden (2003)
2. Hřebíček, J., Kubásek, M.: Environmentální informační systémy (Environmental information systems). Akademické nakladatelství CERM, Brno (2011), http://www.iba.muni.cz/res/file/ucebnice/hrebicek-environmentalni-informacni-systemy.pdf

3. Bízková, R.: eGovernment na Ministerstvu životního prostředí (eGovernment at the Ministry of Environment) (2010), http://vsol.obce.cz/clanek.asp?id=2010209
4. Základní registry veřejné správy (Basic public administration registers), http://www.mvcr.cz/clanek/zakladni-registry-zakladni-registry-verejne-spravy.aspx
5. Informační strategie Ministerstva životního prostředí (Information strategy of the Ministry of Environment), http://www.mzp.cz/cz/informacni_strategie
6. Projekt CISAŽP, http://www.mzp.cz/cz/projekt_cisazp
7. Hřebíček, J., Pillmann, W.: eEnvironment: Reality and Challenges for eEnvironment Implementation in Europe. In: Hřebíček, J., Schimak, G., Denzer, R. (eds.) ISESS 2011. IFIP AICT, vol. 359, pp. 1–14. Springer, Heidelberg (2011)
8. Integrated reporting compliance, http://www.ispop.cz
9. Environmental help desk, http://helpdesk.cenia.cz
10. Geoportal, http://geoportal.gov.cz
11. Integrated register pollution, http://www.irz.cz/
12. CITES, http://www.mzp.cz/en/cites
13. ISOH, http://www1.cenia.cz/www/odpady/isoh
14. ISOP, http://portal.nature.cz/publik_syst/ctihtmlpage.php?what=3&nabidka=hlavni
15. Czech Geological Survey, http://www.geology.cz/extranet-eng
16. INSPIRE standards, http://geostandards.geonovum.nl/index.php/1.3.3_INSPIRE
17. INSPIRE geoportal, http://inspire-geoportal.ec.europa.eu/
18. ArcGIS for Server, http://www.esri.com/software/arcgis/arcgisserver
19. DxoLogicLiferay Enterprise, http://www.doxologic.com/produkty/liferay-portal
20. Czech Geological Survey, http://www.geology.cz/extranet-eng/sgs
21. Suchánek, Z.: NIKM Project. The national inventory of contaminated sites., http://www.cenia.cz/__C12572160037AA0F.nsf/pid/CPRJ7RCKSQQ7/FILE/GEO%20GEOSS%20Seminar%20130212%20CENIA.pdf
22. Technical Guidance for the implementation of INSPIRE View Services, http://inspire.jrc.ec.europa.eu/documents/Network_Services/TechnicalGuidance_ViewServices_v3.0.pdf

INSPIREd Air Quality Reporting

European Air Quality e-Reporting Based on INSPIRE

Katharina Schleidt

Umweltbundesamt GmbH, Vienna, Austria
Katharina.Schleidt@umweltbundesamt.at

Abstract. The new Air Quality Directive (2008/50/EC) foresees the provision of air quality reporting data utilizing the data specifications and services specified by the INSPIRE Directive (2007/2/EC). For this purpose, existing INSPIRE data specifications must be extended as required to support the legal requirements for air quality e-Reporting. Both the legal background for air quality reporting and INSPIRE as well as the process for extending the INSPIRE data models as required for air quality e-Reporting are described in this paper.

1 Introduction

The new Air Quality Directive 2008/50/EC of the European Parliament and of the Council of 21 May 2008 on ambient air quality and cleaner air for Europe (AQD) foresees the provision of air quality reporting data utilizing the data specifications and services specified by the INSPIRE Directive (2007/2/EC). E-Reporting under the AQD is foreseen to begin with the reporting cycle starting in 2014.

For this purpose, the requirements ensuing from the AQD must be analysed and the appropriate classes from the INSPIRE data specifications that can serve as base classes for AQD e-Reporting must be identified. These INSPIRE base classes must then be extended as required to fulfil all requirements ensuing from the AQD. In addition, mandatory attributes stemming from the INSPIRE data specifications must be integrated with the existing reporting requirements from the AQD.

In this paper, we provide both information on the legal background pertaining to the AQD and INSPIRE, as well as detailing the process required for the extension of the existing INSPIRE based classes to the requirements of e-Reporting under the AQD.

2 Background

2.1 Air Quality Directive and Reporting

2008/50/EC on ambient air quality and cleaner air for Europe.
The Directive 2008/50/EC of the European Parliament and of the Council of 21 May 2008 on ambient air quality and cleaner air for Europe entered into force on 11 June

J. Hřebíček et al. (Eds.): ISESS 2013, IFIP AICT 413, pp. 439–450, 2013.

2008. This Directive merges most of existing air quality legislation into a single directive (except for the fourth daughter directive) with no change to existing air quality objectives, repealing the Directives 96/62/EC, 1999/30/EC, 2000/69/EC and 2002/3/EC as well as Decision 97/101/EC.

In the process of defining Directive 2008/50/EC the previous air quality directives where reviewed and streamlined with the following goals:

- Simplify/replace specific reporting obligations on exceedances of information and alert thresholds by a semi-automated exchange based on established near-real-time data exchange mechanism;
- Update and optimize exchange of monitoring data and metadata (current exchange of information under 97/101/EC EoI) to minimize administrative burden by providing data exchange tools; these should enable semi-automated aggregation/cascading and through further Quality Control (QC) provisions and availability of metadata minimize further need for interaction between supplier and user;
- Update and optimize provisions on reporting assessment information (current 2004/461/EC questionnaire) that will enable semi-automated aggregation/cascading of the assessment information regardless of its origin (monitoring, modelling), enabling spatial presentation of assessment and unambiguous links to all other information flows;
- Eliminate any double reporting;
- Reduce administrative burden of Member States (MS), the Commission and European Environment Agency (EEA) by automating tools for checking report consistency etc. (applies to all data-flows);
- Enable faster availability of basic assessment information (provisional monitoring data) at a European scale, even in near real time; information will not be used for compliance purposes, but will enable further and cost-effective public dissemination and provide input to European models supporting national and local assessment and forecasting;
- Replace provisional monthly summer ozone reporting and notifications of alert threshold situations by automated near-real time exchange

For provision of data stipulated by Directive 2008/50/EC, the provisions laid down in Directive 2007/2/EC (INSPIRE) shall be taken into account.

2004/107/EC relating to arsenic, cadmium, mercury, nickel and polycyclic aromatic hydrocarbons in ambient air.
Directive 2004/107/EC is the 4th daughter directive passed under the previous air quality framework directive (96/62/EC) on ambient air quality assessment and management, and the only daughter directive not repealed by 2008/50/EC. The directive sets target values for all the pollutants except mercury, complementing the environmental objectives harmonised in 2008/50/EC.

2011/850/EU Implementing Decision.
As stated in Article 28(2) of Directive 2008/50/EC, the Commission shall determine the additional information to be made available by MS as well as the timescales in

which such information is to be communicated. The Commission shall also identify ways of streamlining the way data are reported and the reciprocal exchange of information and data from networks and individual stations measuring ambient air pollution within the MS.

This task has been realised in the form of the Commission Implementing Decision of 12 December 2011 laying down rules for Directives 2004/107/EC and 2008/50/EC of the European Parliament and of the Council as regards the reciprocal exchange of information and reporting on ambient air quality (IPR).

The following data flows have been defined to cover the requirements of 2008/50/EC:

(B) Information on zones and agglomerations (Article 6);

(C) Information on the assessment regime (Article 7);

(D) Information on the assessment methods (Articles 8 and 9);

(E) Information on primary validated assessment data and primary up-to-date assessment data (Article 10);

(F) Information on generated aggregated data (Article 11);

(G) Information on the attainment of environmental objectives (Article 12);

(H) Information on air quality plans (Article 13);

(I) Information on source apportionment (Article 13);

(J) Information on the scenario for the attainment year (Article 13);

(K) Information on measures (Articles 13 and 14);

The annexes of the IPR also provide information on data processing and encoding, mandatory thresholds such as limit and target values, aggregates and statistics to be provided as well as the units of measurement to be used for the provision of data.

2.2 INSPIRE and AQD Data Models

The INSPIRE Directive (2007/2/EC).
The INSPIRE Directive entered into force on the 15th of May 2007. It recognises that Community policy on the environment must aim at a high level of protection taking into account the diversity of situations in the various regions of the Community. The Infrastructure for Spatial Information in the European Community (INSPIRE) should assist policy-making in relation to policies and activities that may have a direct or indirect impact on the environment.

INSPIRE is based on the infrastructures for spatial information that are created by the MS and that are made compatible with common implementing rules and are supplemented with measures at Community level. These measures should ensure that the infrastructures for spatial information created by the MS are compatible and usable in a Community and trans-boundary context.

The INSPIRE Directive states that infrastructures for spatial information in the MS should be designed to ensure that spatial data are stored, made available and maintained at the most appropriate level; that it is possible to combine spatial data from different sources across the Community in a consistent way and share them between several users and applications. Since the wide diversity of formats and structures in which spatial data are organised and accessed in the Community hampers the efficient formulation, implementation, monitoring and evaluation of Community legislation that directly or indirectly affect the environment, implementing measures should be provided for in order to facilitate the use of spatial data from different sources across the MS. Network services are necessary for sharing spatial data between the various levels of public authority in the Community. Those network services should make it possible to discover, transform, view and download spatial data and to invoke spatial data and e-commerce services.

INSPIRE and Reporting.
While the basic INSPIRE data specifications consciously do not take reporting requirements into account, they are built in such a manner as to allow for extension for reporting purposes. Various environmental reporting obligations were provided as Use Cases to the Thematic Working Groups, and taken into consideration in the definition of the INSPIRE data specifications for relevant themes.

Annex F of the INSPIRE Generic Conceptual Model describes how the INSPIRE data specifications can be extended in order to support other purposes while maintaining the INSPIRE requirements. Thus, data provided using an INSPIRE based extended data specification can be used to fulfil obligations stemming from both INSPIRE as well as a thematic reporting requirement, thus reducing the burden on MS data providers.

However, the inclusion of the explicit reporting concepts directly into the INSPIRE Data Specifications would not be beneficial, as this would increase the legislative burden each time thematic reporting requirements were adjusted. In addition to the necessary modifications of the thematic legislation, changes to the INSPIRE Data Specifications would also be required. Thus, the INSPIRE Data Specifications have been designed in such a manner to provide the core thematic framework, which can then then be extended to also support the additional requirements stemming from reporting requirements.

AQD Data Model.
The AQD e-Reporting data model has been designed based on the INSPIRE data specifications. The UML data models created by the INSPIRE Thematic Working Groups (TWG) have been extended to support all requirements stemming from the IPR; these data models have then been used to generate the AQD e-Reporting schemata. The INSPIRE Annex III data specifications, that are of specific relevance to the provision of air quality data under the AQD, are in the process of finalisation.

At the present time, information on the INSPIRE based AQD e-reporting data model is available only from the EEA Air Quality (AQ) Portal. This information includes UML diagrams, presentations on the data model, a mapping file showing the correspondence between the individual elements of data flows B-G to the data model as well as preliminary XSD schema files.

Further background information is also available from the INSPIRE data specification on Environmental Monitoring Facilities which provides the basis for the AQD e-reporting data model as well as the Guidelines for the use of Observations &

Measurements (O&M) and Sensor Web Enablement-related standards in INSPIRE Annex II and III data specification development. While these documents do not specifically cover the AQD e-reporting data model, they provide valuable background information and insights to this specification.

3 Definition of the AQD e-Reporting Data Model

3.1 Data Model Definition Process

In order to achieve full synergies between INSPIRE and the newly defined AQD e-Reporting, it is necessary to define an INSPIRE based data model fulfilling all requirements ensuing from the AQD. The following steps have been identified as necessary for the extension of INSPIRE Data Specifications to reporting requirements:

- Analyze reporting concepts from legal requirements
- Identify relevant INSPIRE Themes
- Identify relevant INSPIRE Feature Types
- Map reporting concepts to INSPIRE Feature Types
  - Direct mapping to attributes possible
  - Direct mapping to associations possible
  - Additional attributes necessary → derive class from INSPIRE
  - If required, define additional classes
  - INSPIRE requires additional attributes → add to reporting guidelines
- Provide Mapping from reporting concepts to (extended) INSPIRE classes

The air quality reporting concepts have been harmonized within Annex II of the Implementing Provisions (2011/850/EU) of the AQD and described as discrete data flows. Data flows B-G are of relevance to INSPIRE, as the data to be provided under these data flows corresponds directly to INSPIRE Themes. The information to be provided under data flows H-K pertain to air quality Plans and Programmes; while the data models for these data flows are also modelled in accordance with all specifications and requirements from the INSPIRE Directive for simplicity of reporting, there is no direct connection between these data flows and INSPIRE data types. Thus, data flows H-K will not be further discussed in this paper. In addition, in ANNEX II the section (A) Common Data types provides common data types that are referenced and reused by the data flows B-K.

In the table below, a mapping to the INSPIRE Themes identified as relevant to serve as base classes for the implementation of these data flows is provided:

Table 1. Mapping of IPR Data Flows to INSPIRE Themes

AQD e-Reporting Data Flow		INSPIRE Theme
(A)	Common Data types	INSPIRE Base, GML
(B)	Information on zones and agglomerations	III.11. Area management/restriction/regulation zones and reporting units
(C)	Information on the assessment regime	
(D)	Information on the assessment methods	III.7. Environmental monitoring facilities O&M in INSPIRE
(E)	Information on primary validated assessment data and primary up-to-date assessment data	III.7. Environmental monitoring facilities O&M in INSPIRE
(F)	Information on generated aggregated data	III.7. Environmental monitoring facilities O&M in INSPIRE
(G)	Information on the attainment of environmental objectives	I.4. Administrative units
(H)	Information on the air quality plan(s)	
(I)	Quantitative source apportionment	
(J)	Evaluation - Baseline and projection	
(K)	Documentation of measures	

In addition, the following INSPIRE themes are of relevance, as the data specifications either link to elements provided by other institutions under these themes or in turn provide input to them:

I.4. Administrative units

III.5. Human health and safety

III.10. Population distribution — demography

Links to and from the relevant AQD e-Reporting classes must be provided within the data model.

3.2 Extension Examples

The relevant classes from the INSPIRE data specifications must be identified, and the required reporting concepts mapped to these classes. Where necessary, the base classes provided by INSPIRE must be extended with Air Quality reporting specific concepts. This mapping and extension process pertains not only to the spatial information, such as station and zone geometry and metadata; it also covers both primary and aggregated data available for both the station and zone levels, utilizing the Observations and Measurements Standard (ISO 19156) for the provision of measurement data. In other cases, the documentation of reporting relevant concepts must be extended in order to fulfill INSPIRE requirements in addition to those ensuing from the AQD.

In the following sections, we will illustrate this extension process based on the INSPIRE EnvironmentalMonitoringFacility Class and the e-Reporting requirements for air quality stations under IPR data flow D. The diagram below shows the INSPIRE EnvironmentalMonitoringFacility Class as well as an excerpt of the the e-Reporting requirements for air quality stations under IPR data flow D:

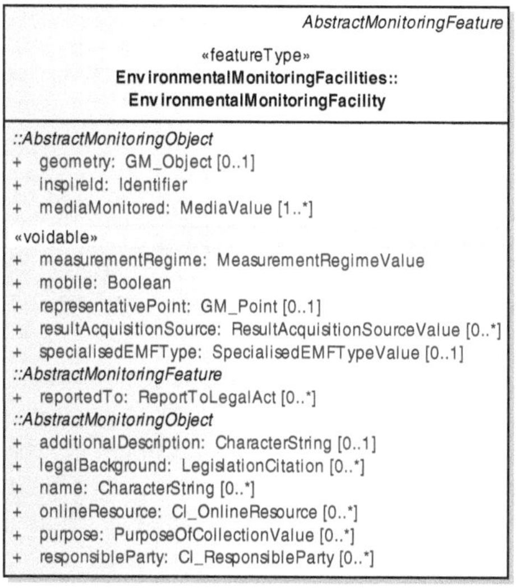

D.5.2	Monitoring station
D.5.2.1	Station Identifier
D.5.2.2	National station code
D.5.2.3	Name of the monitoring station
D.5.2.4	Name of the municipality
D.5.2.5	European station code
D.5.2.6	Station time references
D.5.2.6.1	Station start date
D.5.2.6.2	Station end date
D.5.2.7	Geographical coordinates (geographical longitude and latitude)
D.5.2.7.1	Geographical longitude
D.5.2.7.2	Geographical latitude
D.5.2.7.3	Geodetic altitude
D.5.2.7.4	Reference geodetic coordinate system
D.5.2.8	Meteorological parameters measured
D.5.2.9	Documentation of station information, including Maps and photographs

Fig. 1. INSPIRE Class and IPR Requirements

Mapping AQD Requirements to INSPIRE.

In many cases, the classes provided for the specific INSPIRE Themes already possess the necessary attributes for AQD e-Reporting. In the diagram below one can see how various attributes defined for air quality monitoring stations correspond to requirements stemming from the IPR:

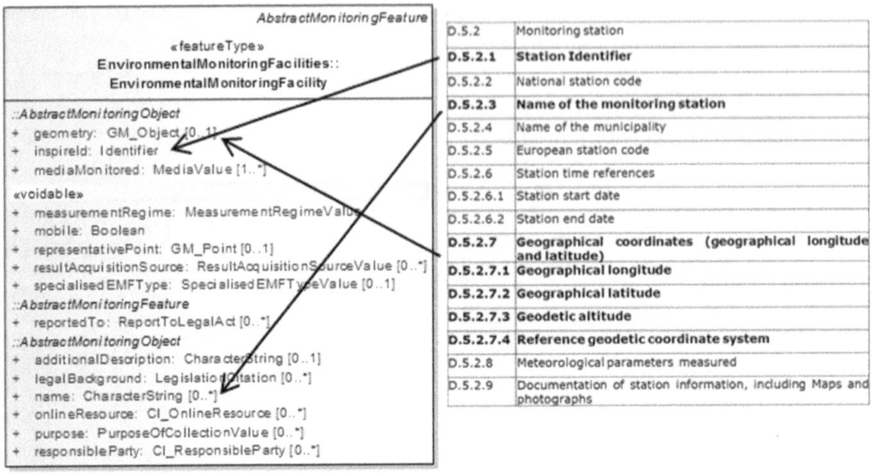

Fig. 2. Mapping of IPR Requirements to INSPIRE Class Attributes

In other cases, the IPR requirements will be mapped to associations between INSPIRE classes. In the example below, we see how the reference to the Air Quality Network a specific Air Quality Monitoring Station belongs to is modeled as the association between the INSPIRE classes EnvironmentalMonitoringFacility and EnvironmentalMonitoringNetwork:

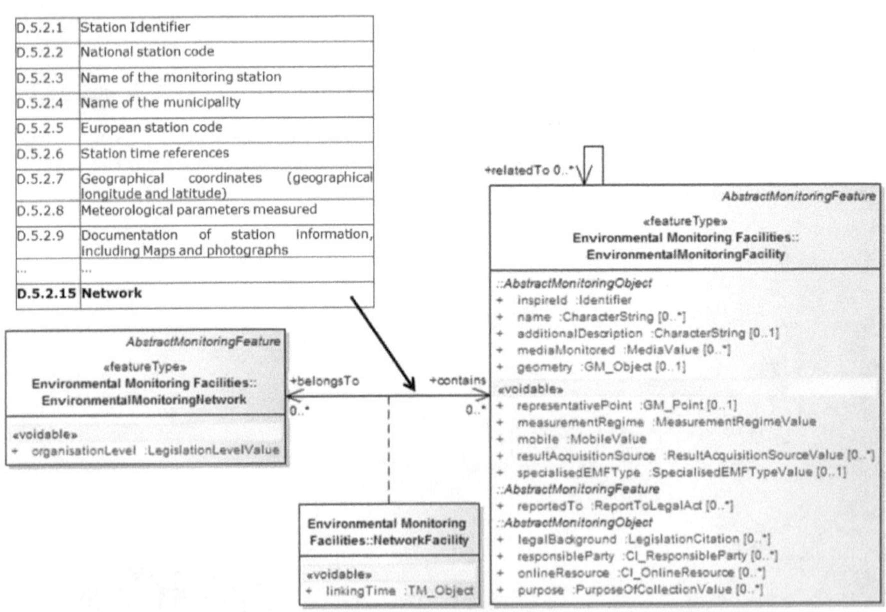

Fig. 3. Mapping of IPR Requirements to INSPIRE Associations

Extending INSPIRE Base Classes.

In certain cases, attributes required for AQD e-Reporting will not be available from the INSPIRE based classes as the INSPIRE classes are defined in a very generic manner and do not encompass all thematic requirements stemming from various environmental domains and usages. In these cases, the additional attributes must be identified, and a new class derived from the basic INSPIRE classes with these additional attributes included. In the example below, the attributes National station code, European station code and Name of the municipality have been identified as additional requirements to the INSPIRE EnvironmentalMonitoringFacility class when using this as a basis for AQD e-Reporting Air Quality Monitoring Stations:

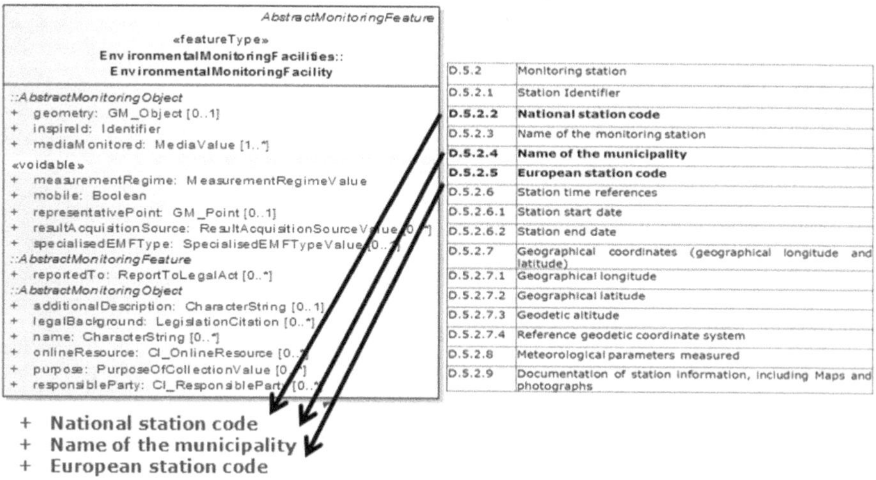

Fig. 4. Extending INSPIRE Classes

A new class AQD_Station must be defined extending the base INSPIRE EnvironmentalMonitoringFacility class as follows:

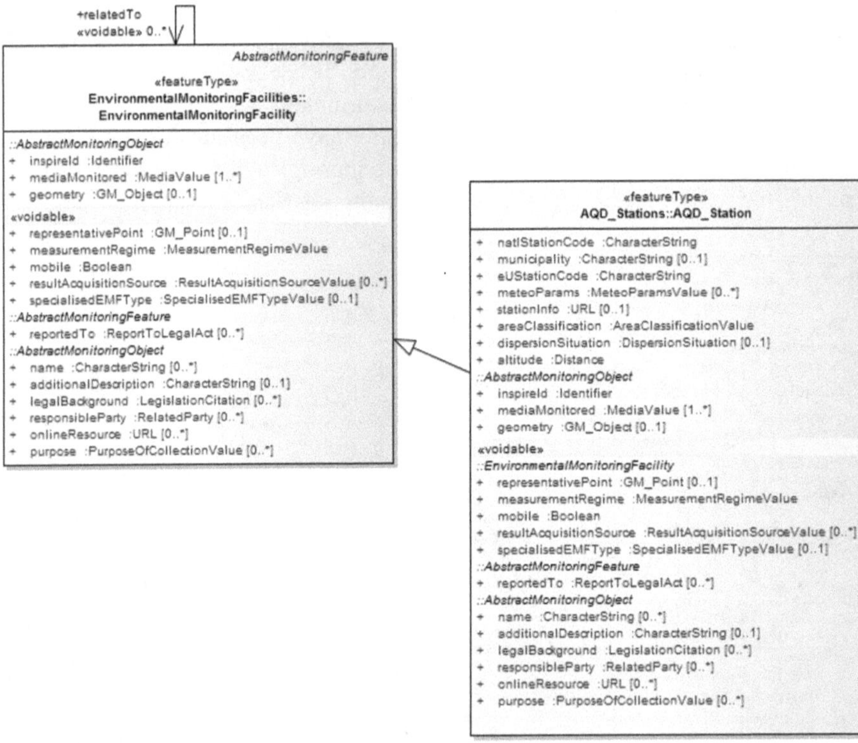

Fig. 5. Deriving an extended INSPIRE Class for AQD e-Reporting

Adding INSPIRE Requirements to Reporting Requirements.
Finally, mandatory attributes stemming from the base INSPIRE data specifications must be identified and added to the reporting requirements. In the example below, the INSPIRE attribute mediaMonitored is appended to the requirements for AQD e-Reporting from the IPR:

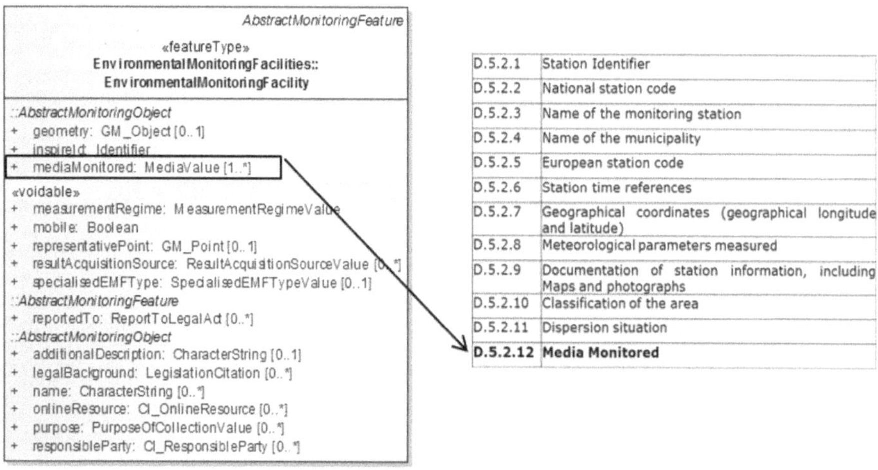

Fig. 6. Extending the AQD Requirements based on INSPIRE

3.3 Mapping from AQD Requirements to e-Reporting Data Model

As a final step in the process of defining an INSPIRE based AQD e-Reporting Data Model, a mapping between the reporting requirements stemming from the IPR to the classes and attributes defined within the data model must be provided. As the air quality responsibles in the MS are familiar with the requirements structure from the IPR, this served as the basis for the mapping tables. For each element in these requirements tables, a mapping to the class and attribute from the AQD e-Reporting Data Model is provided as follows:

Table 2. Mapping from AQD Requirements to Data Model

IPR #	IPR Name	Class	XPATH
D.5.1	Monitoring Station	AQD_Station	/aqd:AQD_SamplingPoint/ef:broader/@xlink:href Linked to /aqd:AQD_Station
D.5.2.1	Station Identifier	AQD_Station	/aqd:AQD_Station/ef:inspireId/base:Identifier
D.5.2.2	National station code	AQD_Station	/aqd:AQD_Station/aqd:natlStationCode
D.5.2.3	Name of the monitoring station	AQD_Station	/aqd:AQD_Station/ef:name
D.5.2.4	Name of the municipality	AQD_Station	/aqd:AQD_Station/aqd:municipality
D.5.2.5	European station code	AQD_Station	/aqd:AQD_Station/aqd:EUStationCode
D.5.2.1	Station Identifier	AQD_Station	/aqd:AQD_Station/ef:inspireId/base:Identifier
D.5.2.6	Station time references	AQD_Station	/aqd:AQD_Station/ef:operationalActivityPeriod/ef:OperationalActivityPeriod

4 Conclusions

Air Quality e-Reporting, as defined in the new Air Quality Directive 2008/50/EC of the European Parliament and of the Council of 21 May 2008 on ambient air quality and cleaner air for Europe stipulates the use of INSPIRE for data provision. Thus, an INSPIRE based data model must be created for this purpose. The base classes provided by the various INSPIRE Themes provide the basic building blocks required by reporting for the AQD e-Reporting. However, the base classes provided by INSPIRE will need to be extended to cover the full breadth of reporting requirements. Conversely, the reporting requirements guidelines will need to be extended to cover all INSPIRE requirements.

Through the use of INSPIRE data specifications as a basis for the AQD e-Reporting, great synergies can be attained. Member States only need to set up one INSPIRE service providing relevant Air Quality Features and Data to cover both their INSPIRE obligations as well as their legal reporting obligations pertaining to Air Quality. Air quality data, including both measurement metadata as well as real-time measurement data flows, will be made widely available in a standardized manner allowing for reuse beyond European Reporting obligations.

Using the HOS Method for Evaluating the Efficiency of Environmental Information Systems

Zuzana Chvátalová and Miloš Koch

Brno University of Technology, Faculty of Business and Management,
Kolejni 4, 612 00 Brno, Czech Republic
{koch,chvatalova}@fbm.vutbr.cz

Abstract. In this paper the authors apply the principles of the HOS method developed at the Faculty of Business and Management at the Brno University of Technology for evaluating the efficiency of information systems including environmental information system (EIS). Selected results of the authors´ research based on monitoring more than four hundred organizations are presented in a case study. The authors also take into account some aspects of the formulation of environmental information of the Czech Republic (supported by, e.g., the Czech Statistical Office). The primary principle of the research is to determine the level of balance of the organization's information system, this being the key prerequisite of its effectiveness. The effectiveness is understood as maximizing the output with respect to minimizing the total cost. It is a way of implementing information and communication technologies and respecting their characteristics as properties of the information system generally and within the limits of sustainability of their development and implementation throughout their life cycle. Organizations respecting these rules support the use of the tools and state-of-the-art environmental informatics, and thus significantly contribute to their sustainable development.

Keywords: Balance, Czech Statistical Office, Efficiency and Effectiveness, Environmental Information Systems, Evaluation, HOS Method and HOS Diagram, Information System, Statistics, Subsystem.

1 Introduction

Since 1990, informatics, information systems (IS) and information and communication technologies (ICT) in the Czech Republic (CZ) have become an important developmental post-revolutionary phenomenon. Since the end of the 1990s when the first concept of state information policy was formed, other information systems in state administration, institutions and business units have been developed and thus CZ informatics has been integrated into world and European structures.

Hřebíček and Kubásek [5] have said about the situation in CZ: "*After 1990 there has been a big expansion of information and communication technologies, which has resulted in increased access to the growing amount of data and information through the Internet and Web services. In the creation and protection of the environment there*

J. Hřebíček et al. (Eds.): ISESS 2013, IFIP AICT 413, pp. 451–463, 2013.

are many projects in the private and public spheres (businesses, institutions, public authorities and non-governmental organizations), which are being used to collect environmental data and information and also for providing information services concerning the environment."

Until approximately the 1980s we may assume that the "*criteria of success*" (efficiency, or rather, effectiveness) of implementing and using an IS were, above all, reaching the pre-determined goal, i.e., a concrete task, for example calculation, graphic output, simulation in a reasonable time-frame, etc. Economic and organizational difficulty, operational literacy, leading the processes, safety or data-accessibility may have been very disadvantageous all the while and not always effective.

Hamilton and Chervany wrote in [4]: "*Evaluations of Management Information Systems (MIS) tend to be subjective and are influenced by the perceptions of system objectives, as well as the experiences with system performance in accomplishing organizational objectives. Consequently, the assessments of MIS effectiveness are often controversial and can be sources of disagreement and conflict between different functional groups involved in MIS implementation – users, MIS development, internal audit, and top management personnel.*"

The Chairwoman of the Czech Statistical Office (CZSO) professor Ritschelová, et al, wrote [13] about public environmental data, information and statistics: "*One of the very important aims of the Czech Statistical Office is to produce, among others, national statistical data on environmental development. During the last two decades one could trace the constantly increasing interest in and attention given to environmental statistics both at the international, national and regional levels.*" The primary environmental data and information are monitored and collected for the CZSO through the organizations' IS. CZSO also devotes considerable attention to the monitoring of many aspects of the development of ICT and IS in CZ.

Given the growing importance of ICT and their enormous impact on the economy, society and the environment from approximately the beginning of the 21th century, the CZSO has been dealing with statistical monitoring, analysis of the spectrum of ICT and the publication of collected information which provide information to government authorities, organizations, business entities and the public. It carries out extensive research - *Information Society, Information Economy, Information Technologies (Information Society in Figures, Use of ICT in the Public Administration Sector, Use of ICT in Households, Use of ICT in Enterprises of the CZ*, etc.), more in [1].

The ICT have changed the availability and use of information everywhere. They contribute to changes in laboratory productivity, patterns of resource use, the life of society and the environment.

The emphasis on the importance of current trends of corporate performance evaluation and corporate reporting in the CZ is discussed in detail in [6], [7], [8]. Corporate reporting requires monitoring (measurement of) the parameters / variables using quantitative and qualitative methods, and advanced mathematical or statistical methods and models as in [9], [10] implemented through efficient ICT tools as in [3].

It is necessary to measure the balance level of the corporate reporting IS – a key assumption of the effectiveness of its complex IS. The effectiveness is understood as maximizing the output with respect to minimizing the total cost connected with the IS. Corporations can benefit from a comprehensive approach to implementing their ICTs. Being aware of the characteristics of the ICTs as also being general properties of the IS will also be beneficial. The organisation's overall sustainable development will impose limits on the IS throughout its life cycle. Organizations that respect these facts support the use of the ICT tools, taking into account state-of-the-art environmental informatics, and thus they significantly contribute to their own sustainable development.

Our outcomes of measuring IS efficiency are based primarily on the practical applications of the HOS method for many organizations. The discussed results obtained through the Zefis portal [11] are visualized using Microsoft Excel office applications and some values in the paper are calculated using the Maple system [9-10].

2 HOS Analysis: Methodology

HOS analysis is a method used by organizations for the purposes of assessment and investment in IS, or for the modification of IS management strategies using the Zefis portal [11]. The HOS method classifies IS and organizations for their managers and allows them to act, if the efficiency of IS is sufficient, or to see whether the IS should be improved, and in what areas (subsystems), more in [12]. The following formal descriptions are analogous to [12].

2.1 HOS Analysis: Principle

The information system *IS* is composed of individual subsystems S_i, i=1,…,8 (also can be called components, parts, elements, areas, etc.), see Table 1.

Table 1. The subsystems in the HOS method (Source: Authors´ elaboration, [11], [12])

SUBSYSTEMS					
Abbreviation	**Level**	**Subsystem**	**Abbreviation**	**Level**	**Subsystem**
HW	L_1	*Hardware*	DW	L_5	*Dataware*
SW	L_2	*Software*	CU	L_6	*Customers*
OW	L_3	*Orgware*	SU	L_7	*Suppliers*
PW	L_4	*Peopleware*	MA	L_8	*Management IS*

- *Subsystem levels* L_i, i=1,…,8 of the information system *IS*, where the level L_i is for the subsystem S_i, (i=1,…,8). Each subsystem is rated on a four-point scale as 1 – *bad*, 2 – *somewhat bad*, 3 – *somewhat good*, 4 – *good.* The assessment of each subsystem can be made by qualified assessment specialists (but this excludes the use of methods for the initial assessment by organizations themselves), or using control questions for each subsystem in a questionnaire, [2], [11], [12].

– *A overall level* L_w of the *IS* (where $1 \leqq m \leqq 8$ for $\forall i, j \in \{1,..,m\}, i \neq j : L_i \neq L_j$) is equal to:

$$L_w = \min\{L_1,..,L_m\}. \tag{1}$$

– The definition of a balanced IS, unbalanced IS or heavily unbalanced IS is given in Table 2.

Table 2. The definition of a balanced / unbalanced / heavily unbalanced IS[1] (Source: Authors´ elaboration, [11])

INFORMATION SYSTEM	CONDITIONS	
Balanced	$[\sum_{i=1}^{8}(L_i - L_w) \leq 3] \wedge [\forall i \in \{1,..,8\} : L_i - L_w \leq 1]$	(2)
Unbalanced	$[\sum_{i=1}^{8}(L_i - L_w) > 3] \wedge [\forall i \in \{1,..,8\} : L_i - L_w \leq 1]$	(3)
Heavily Unbalanced	$\exists i \in \{1,..,8\} : L_i - L_w > 1$	(4)

– *A recommended level* is based on the importance which an organization attaches to the IS.

2.2 Application of HOS Analysis in an Individual Organization

As an example of our research, we have selected a Czech manufacturing organization with 145 employees. This is a case of an enterprise information system of a monitored organization that leads to environmental reporting. The CZSO and the Ministry of the Environment (MoE) of the CZ require obligatory annual reporting (waste, air quality, integrated pollution register, etc.). These subsystems of the enterprise information system (ERP) support these reporting outputs to CZSO and MoE as the organization's environmental information subsystems.

We have used the HOS method for evaluating the organization's IS as follows:

– An assessment of each subsystem of the monitored IS has been made by a qualified assessment specialist and is shown in Table 3.
– In Table 4 we implement all the incremental steps of the HOS method.

[1] *Note*: In (2), (3), (4) for $\forall i \in \{1,..,8\}$: $0 \leq L_i - L_w \leq 3$.

Table 3. The subsystem levels of the monitored IS (Source: Authors´ elaboration, [11])

SUBSYSTEMS	**HW**	**SW**	**OW**	**PW**	**DW**	**CU**	**SU**	**MA**
Level	L_1	L_2	L_3	L_4	L_5	L_6	L_7	L_8
	3	2	4	2	2	2	3	2

Table 4. An evaluation of the IS using the HOS method (Source: Authors´ elaboration, [11])

HOS METHOD			
	HOS DIAGRAM	**CONDITIONS**	**IS**
Subsystems Level	*Light grey irregular octagon boundary*	--	--
Overall Level	*Grey regular octagon boundary*	(1) $L_w = \min\{4,3,2\} = 2$	2
Current Status		(4) $\sum_{i=1}^{8}(L_i - 2_w) = 4 > 3$ $\exists i = 3 : L_3 - 2 = 2 > 1$	*Heavily unbalanced*
Recommended Level	*Bold black regular octagon boundary*	The *recommended level* is based on the importance which an organization attaches to the IS.	3 *Somewhat good*

– In Figure 1 we make use of all the facts obtained for the visualization of the HOS diagram.

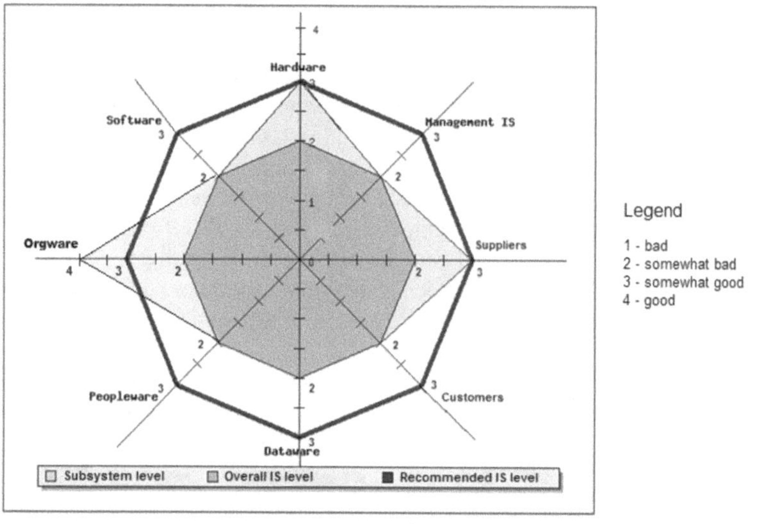

Fig. 1. The HOS diagram corresponding with Table 4 (Source: Authors´ elaboration, [11])

Results and *discussion*: The IS of the monitored organization has an overall level equal to 2. Two subsystems (HW, SU) have levels equal to 3, one subsystem (OW) has a level equal to 4, i.e., differing from the overall level by more than 1 evaluation point. Thus (4) is valid. Therefore the IS is *heavily unbalanced.* Its recommended level is equal to 3 (*somewhat good*).

HW: The level of the technical equipment (hardware products such as HW for environmental data collection and emissions measurement, sensors for identifying the sources of pollution, noise, registration of hazardous waste, energy burden, etc.) of this organization is *somewhat good.*

SU: It is clear that this monitored organization has quite a good level of the sophisticated supply-chain system (i.e., a set of external organizations), i.e., organizations that provide operations and support of the IS on a level that is *somewhat good.*

OW: It is remarkable that this organization was able to develop good quality rules for the operation of information systems, recommended operating procedures, and security rules. Orgware is evaluated as *good.*

Other subsystems are assessed as *somewhat bad.* (For example, the software has a significant disparity between the system and the application subsystems.)

These large disparities of the assessment of the ERP subsystems lead to an overall categorization of "*heavily unbalanced* IS."

Due to the outputs of the HOS methods and the recommendation of an expert to evaluate the level as *somewhat good*, we can say that we are dealing with an IS here that has the potential to improve its quality.

3 Case Study: Selected Aggregated Research Results

3.1 The Zefis Portal as a Research Source

The Zefis portal [11] offers organizations (free of charge, anonymously) the advantage of evaluating the efficiency and balance of the IS using the HOS method (by means of questionnaires). A registered user can carry out surveys and get results immediately. The results can be compared with references in the Zefis databases. The languages of this portal are Czech and English. Many of its users are from Slovakia.

The HOS method was initially designed for the three subsystems (hardware, orgware, software), and the user database had contained more than 2,000 users. In 2010, the method was modified to include the eight abovementioned areas (subsystems), see third paragraph. At present, our research of IS efficiency deals with and analyses the most current facts and data through the Zefis database.

Currently, the HOS method has been tested on a sample of 425 organizations from the Czech and Slovak Republics in the period from 2010 up until now (as at March 4, 2013).

Although our results reflect a stricter classification, we will respect the general category (in particular, supported by the activities of CZSO) in this paper.

- For the purposes of comparison with the EU, the CZSO regularly presents statistics on the CZ information society in the following areas: *ICT Infrastructure*; *Households*; *Individuals*; *Enterprises*; *Public Authorities*; *Schools*; *Health Establishments*. *Note*: we have merged Households with Individuals into the category

Others. The representation of the abovementioned groups in our survey sample is presented in Figure 2.

- We will also respect the categorization of organizations according to the number of staff as given by the CZSO: 10 – 49; 50 – 249; 250 and more. *Note*: We will include the category of number of staff: 1 – 9. The representation of the mentioned groups in our survey sample is presented in Figure 3.
- *Note*: For the purposes of greater of clarity in this paper, we have rounded to the next integer of percentage.

Comment: Based on Figure 2, we can say that the IS evaluation is mainly interesting for enterprises, namely (on a deeper analysis): commercial and manufacturing, with services and financial services. In terms of their number of staff, enterprises are represented quite equally (see Figure 3). It is interesting that even micro-enterprises are represented comparably.

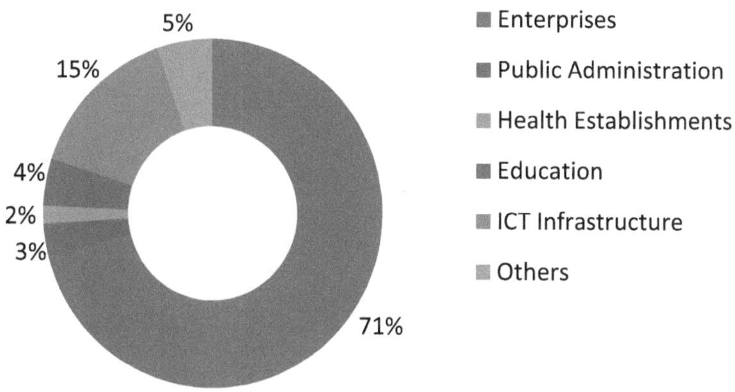

Fig. 2. The information society in our survey – the representation in % (Source: Authors´ elaboration according to [11] in Microsoft Excel)

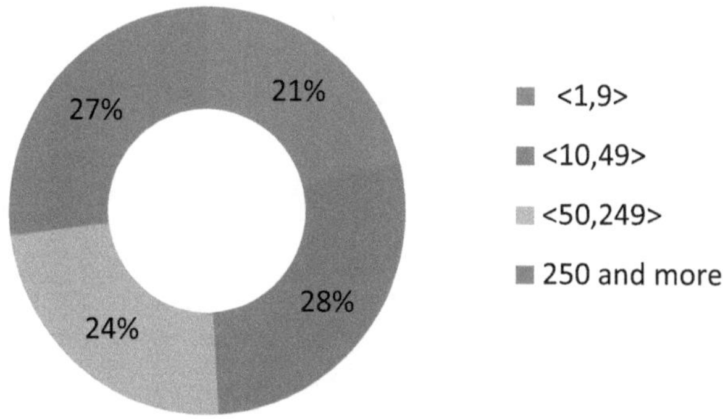

Fig. 3. Numbers of staff in % (Source: Authors´ elaboration according to [11] in Microsoft Excel)

3.2 Aggregated Average Research Results

Here we present the selected *aggregated average results* of the 425 monitored organizations. For the purpose of IS evaluation using HOS analysis, we have employed the Zefis portal (in the reporting period stated above, see 3.1). *The average subsystem levels* are represented by a light grey irregular octagon boundary and *the average overall level* is represented by a grey regular octagon boundary, see Figure 4.

Tables 5 and 6 present the aggregate values (*average*) of subsystem levels, and the overall and recommended levels (plotted on the individual axes), as well as the corresponding simple statistical diagnostics of the obtained results (*variances, standard deviations*).

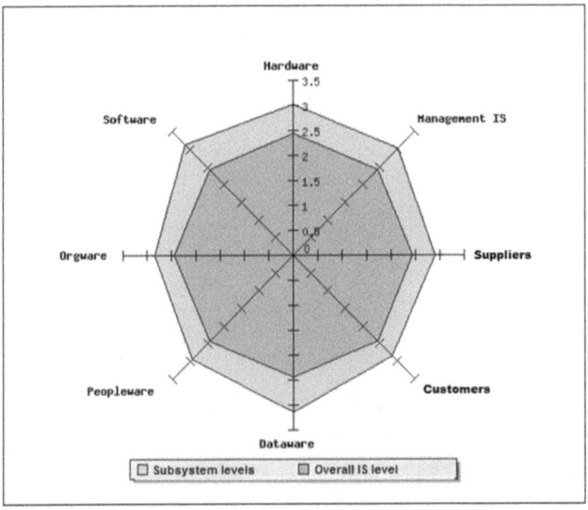

Fig. 4. The aggregated average results – HOS diagram (Source: Authors´ elaboration, [11])

Table 5. The aggregate levels of the individual subsystems and the simple statistical diagnostics (Source: Authors´ elaboration in the Maple system according to [11])

SUBSYSTEM	LEVEL	AVERAGE	VARIANCE	STANDARD DEVIATION
HW	L_1	3.02	0.285	0.533
SW	L_2	3.15	0.334	0.578
OW	L_3	2.84	0.566	0.752
PW	L_4	2.94	0.401	0.633
DW	L_5	3.15	0.371	0.609
CU	L_6	2.84	0.399	0.632
SU	L_7	2.90	0.292	0.540
MA	L_8	3.04	0.463	0.680

Table 6. The aggregated overall and recommended levels of the subsystems and the simple statistical diagnostics (Source: Authors´ elaboration in the Maple system according to [11])

LEVEL IS	AVERAGE	VARIANCE	STANDARD DEVIATION
Overall	2.44	0.359	0.599
Recommended	3.58	0.309	0.556

Discussion: Let us summarize the results for the sample of the monitored organizations. From Figure 4, Table 5 and Table 6, by using the results analysis of the axes, we can see that:

- On any axis the aggregate levels do not show significant differences, i.e., each subsystem of the monitored information systems can be assessed as sufficiently *harmonized* (*balanced*);
- A regular dark grey octagon (smaller) presents the aggregate overall level of the monitored information systems (that is given by the perception of the importance of the information system in the organization) and its level is close to the value of 2.5, the whole status of the IS can be identified between *somewhat good* and *somewhat bad*;
- An irregular light grey octagon (larger) indicates the level of the individual subsystems of the monitored information systems, and these levels move around the value of 3, i.e., *somewhat good*;
- The monitored information systems show the worst results (imbalance) in the subsystems of: OW – *Orgware*, MA – *Management IS* and CU – *Customers* (users).

3.3 Aggregated Results as Related to Subsystems of the Information Society

We can see a representation of the balanced / unbalanced / heavily unbalanced 425 monitored information systems, as related to the areas of the information society, see Figure 5.

Discussion: Based on the outputs we have identified by means the Zefis portal and the HOS method, we believe that:

- It is basically logical that in the category of the *ICT Infrastructure* there be a dominance of balanced IS. The large number of heavily unbalanced information systems seems astonishing.
- The proportion of balance / imbalance / heavy imbalance in the category of *Education* is similar. This not so good situation may be the result of frequent changes in this category.
- An interesting negative aspect is the comparable ratio of all three types of balanced / unbalanced / heavily unbalanced information systems in the category of *Public Administration*. These organizations have a unified environmental management system at their disposal – a result of the progression towards e-government in the CZ. The resulting proportions (balanced / unbalanced / heavily unbalanced IS) of organizations involved in our research are very unfavourable for this category. It is necessary to further analyse the reasons for this in detail.

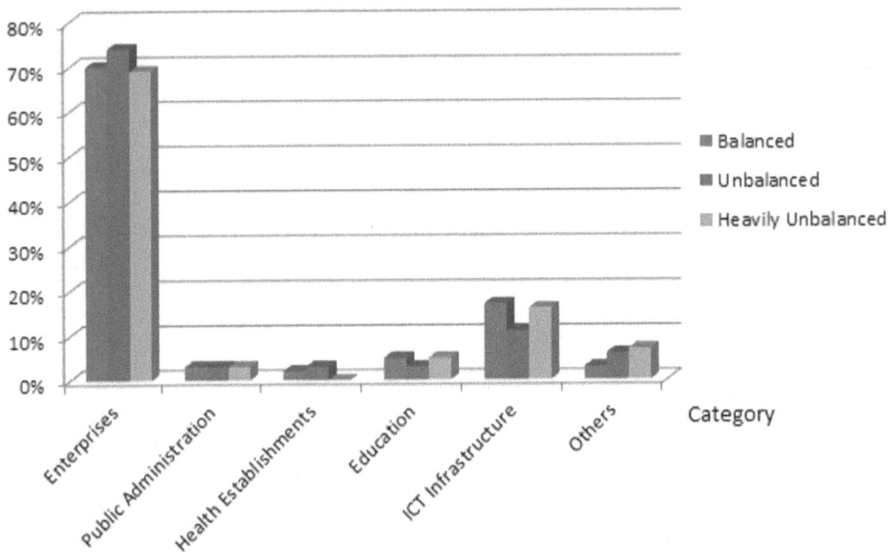

Fig. 5. IS evaluation as related to areas of the information society (Source: Authors´ elaboration according to [11] in Microsoft Excel)

– The situation in the category of *Enterprises* is cause for concern. It can be estimated that many enterprises do not utilize their information systems effectively. It appears that the management of many enterprises place other values above an appropriate information system. Their activities can, however, have a significant impact on the environment.
– A partly good tendency (this is only due to the fact that the number of heavily unbalanced information systems is quite small) appears in the category of *Health Establishments.* We believe that this reason may be applied, as the number of unbalanced information systems is large enough.
– The situation in the category of *Other*s which consists mainly of *Households* and *Individuals* can be attributed to poor professionalism. This may explain the rather negative proportions of balance / imbalance / heavy imbalance.

3.4 Aggregated Results Regarding Number of Staff

We can see a representation of the balanced / unbalanced / heavily unbalanced state of the 425 monitored information systems, as classified by the size of the organization, see Figure 6 (for each number of staff category separately) and Figure 7 (for the whole sample of monitored organizations).

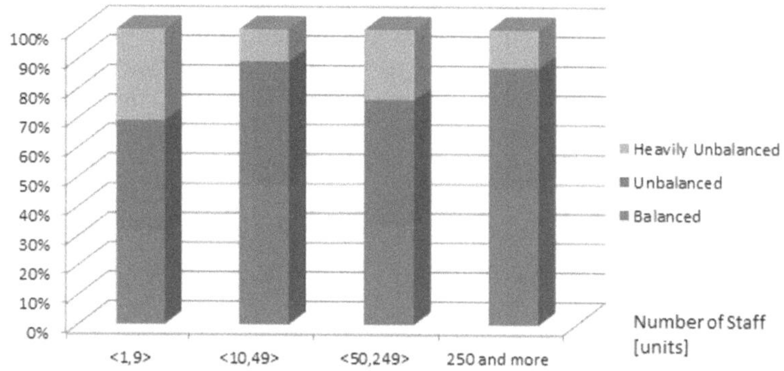

Fig. 6. IS evaluation by number of staff, for each organization separately (Source: Authors´ elaboration according to [11] in Microsoft Excel)

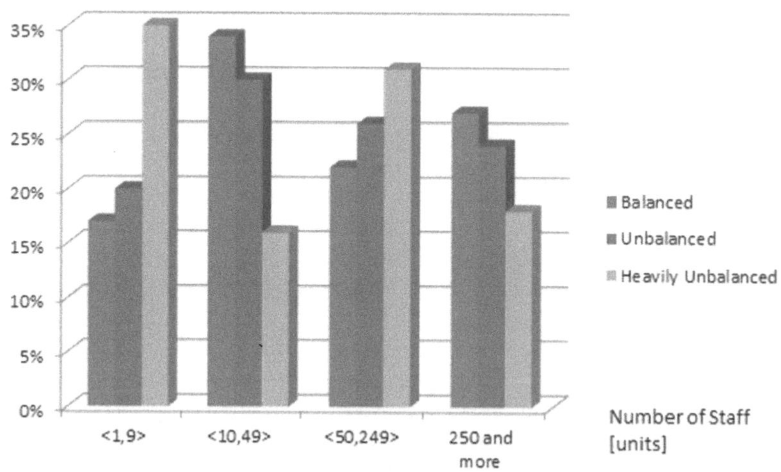

Fig. 7. IS evaluation by number of staff (Source: Authors´ elaboration according to [11] in Microsoft Excel)

Discussion: Based on the outputs that we have identified by using the Zefis portal and the HOS method, we believe that:

- In the case of large organizations employing 250 and more people, the situation is good enough (except for very large organizations with over 1,000 employees, as it has turned out by using other, more detailed investigation methods). These organizations are probably beginning to manage their information systems. However, the scope of the systems is not so large, it can therefore still be rationally managed in a better way, and it can be adapted to changing conditions in a flexible manner. *Note*: Probably, for very large organizations (with more than 1,000 employees) we must begin to take into account the 'heaviness' of large systems that cannot be flexibly changed and readily adapted to individual requirements.

– The explanation of this phenomenon requires further investigation. However if the obtained results are true, it shows that small organizations with 11 – 49 employees have already surpassed the amateur approach to the use information systems (which we can see in the case of micro-organizations). They operate on a relatively satisfactory level. This is due, among other factors, to a relatively small scale of information systems of small organizations. This allows them to rationally manage and use information systems without complicated methodologies, such as ITIL or COBIT. This may be a positive trend in terms of environmental protection.

4 Conclusion

The Zefis portal is designed for the evaluation of the efficiency of organizations' information systems. It uses the HOS method to provide an inclusion in one of the classes: balanced / unbalanced / heavily unbalanced IS. A balanced system supports the effectiveness of the IS. This classification is based on the evaluation of the eight subsystems of IS. The result is visualized in the HOS diagram, which also includes the recommended level, and the overall level of the IS. The survey captures the aggregate outputs relative to the monitored organizations. Our outcomes for the set of 425 monitored organizations are as follows:

The best results have been achieved by small organizations with 11 – 49 employees. From this perspective, it also seems that organizations with 11 – 49 employees have great potential in terms of impact on the environment. They usually operate in the local area, but they can help minimize the environmental burden on a national, and also on a global scale. Due to their size, they can work very effectively, and have an interest in the introduction of environmental supports, environmental information systems, environmental reporting, etc. The worst ratio of balanced and unbalanced systems is expected in the case of micro-small-organizations with a staff of less than 10. Medium-sized organizations (with the number of employees from 50 to 249) report worse results than large organizations (with the number of employees starting at 250). *Note*: In performing a deeper analysis, which is not the subject of this post, we have found, nevertheless, that very large organizations with 1,000 and more employees have worse results than organizations not exceeding 1,000 employees.

Obviously, the category of *ICT Infrastructure* has reached the best results in terms of balance. Negative findings in the category of *Public Administration* have obtained an approximately equal representation in all the three classes of balanced / unbalanced / heavily unbalanced IS. These organizations have a unified environmental management system at their disposal – as a part of the progression towards e-government in the CZ. It is relatively surprising that in the category of *Enterprises* it appears that many enterprises do not utilize their IS adequately and effectively. Currently, they should be supporting any activities (if they have not done so yet) leading to environmental protection, reporting (at least for the purpose of making a good name for the organization, for more competitiveness, a better market position, the creation of economic value, corporate performance, etc.). The situation is not very satisfactory in the categories of *Education* and *Health Establishments*. In these categories, we would expect a dominant balance that would support IS efficiency in the environmental field. The *Others* category is probably not very interesting as far as the efficiency of the organizations' IS is concerned.

Acknowledgments. This paper is supported by The Czech Science Foundation. Name of the Project: Construction of Methods for Multifactor Assessment of Company Complex Performance in Selected Sectors. Reg. Nr. P403/11/2085.

References

1. Czech Statistical Office, `http://www.czso.cz/`
2. Dovrtěl, J.: Selected Aspects of the Effectiveness of Information Systems (Dissertation in Czech). BUT FBM, Brno (2004)
3. Gander, W., Hřebíček, J.: Solving Problems in Scientific Computing Using Maple and Matlab, 4th edn. Springer, Berlin (2004)
4. Hamilton, S., Chervany, N.L.: Evaluating Information Systems Effectiveness - Part II: Comparing Evaluator Viewpoints 5, 79–86 (1981)
5. Hřebíček, J., Kubásek, M.: Environmental Information Systems, 1st edn. CERM, Brno (2011) (in Czech)
6. Hilty, L., Page, B., Hřebíček, J.: Environmental Informatics. Environmental Modelling & Software 21, 1517–1518 (2006)
7. Hodinka, M., Štencl, M., Hřebíček, J., Trenz, O.: Current Trends of Corporate Performance Reporting Tools and Methodology Design of Multifactor Measurement of Company Overall Performance. Acta Univ. Agric. et Silvic. Mendel. Brun. 60, 85–90 (2012)
8. Hřebíček, J., Pillmann, W.: eEnvironment: Reality and Challenges for eEnvironment Implementation in Europe. In: Hřebíček, J., Schimak, G., Denzer, R. (eds.) ISESS 2011. IFIP AICT, vol. 359, pp. 1–14. Springer, Heidelberg (2011)
9. Chvátalová, Z., Hřebíček, J.: Modeling and Simulation Utility Functions with Maple. Mendel Journal Series 1, 552–557 (2012)
10. Chvátalová, Z., Hřebíček, J.: Computational Finance and Finance Economics with Maple. International Journal of Mathematical Models and Methods in Applied Sciences 7, 541–550 (2013)
11. Koch, M.: ZEFIS: Online Assessment for your Information System. Brno (2013), `http://www.zefis.cz/`
12. Koch, M., Chvátalová, Z.: Information System Efficiency as an Attribute in Environmental Information Systems. In: Hřebíček, J., Schimak, G., Kubásek, M., Rizzoli, A. (eds.) ISESS 2013. IFIP AICT, vol. 413, pp. 31–43. Springer, Heidelberg (2013)
13. Ritschelová, I., Sidorov, E., Hřebíček, J.: Environmental Statistics at the CZSO: Past Development and Future Trends. In: Hřebíček, J., Schimak, G., Denzer, R. (eds.) ISESS 2011. IFIP AICT, vol. 359, pp. 33–46. Springer, Heidelberg (2011)

How to Build Integrated Climate Change Enabled EDSS

Ralf Denzer [1], Sascha Schlobinski [2], Lars Gidhagen [3], and Thorsten Hell [2]

[1] Environmental Informatics Group (EIG)
ralf.denzer@enviromatics.org
[2] Cismet GmbH, Altenkesseler Strasse 17, 66115 Saarbrücken, Germany
{sascha.schlobinski,thorsten.hell}@cismet.de
[3] Swedish Meteorological and Hydrological Institute, Folkborgsvägen 1, 60176 Norrköping, Sweden
lars.gidhagen@smhi.se

Abstract. Results from global and regional climate scenario modelling predict significant changes in temperature and precipitation during the rest of the century. Episodes with extremely high temperatures and more intense rainfall will occur more often. Climate change will affect the urban environment and should be taken into account in long term and sustainable urban planning for adaptation to the new climate conditions. To consider the effects of climate change in today's planning, tools are needed to support the development of climate change enabled applications. These tools have to support several aspects of application development in the context of EDSS: a) How can the results from climate change models be integrated in today's applications? b) How to access climate change information? c) How to compare local effects of climate change scenarios? d) How to gain access to downscaling functionality required for local problems? e) How to feed local models with climate scenario data. Or in other words: How to build climate change enabled applications.

Within the FP7 project SUDPLAN, which integrates environmental modelling and software expertise, a system of standardised services and end user applications has been developed. The system delivers long term projections of environmental data for different aspects of local modelling (air quality, hydrological conditions and intense rainfall) based on different climate scenarios. The services are interactive and require input of local observations to improve downscaled projections. This is necessary to support wide applicability as the resolution requirements of local models can be very different. All services are provided through standardized service interfaces. On top of these services, the so-called Scenario Management System (SMS) provides end user components for scenario management, visualisation, model integration as well as wizards to support the relevant workflows and the interaction with common downscaling services. The SMS is based on the geo-spatial application suite CIDS which allows to build decision support systems (DSS) with a focus on using climate change projections in the workflow.

Keywords: Environmental Decision Support System, EDSS, 3D-visualisation, Climate Change.

J. Hřebíček et al. (Eds.): ISESS 2013, IFIP AICT 413, pp. 464–471, 2013.

1 Introduction

Climate Change (CC) is expected to have influence on the general environmental conditions of today and the future. As of today existing national action plans like the "Adaptation Action Plan of the German Strategy for Adaptation to Climate Change" [1]indicate that CC is recognized as an unavoidable reality. The potential impact of CC on society requires decision makers to consider direct or indirect effects of CC in many planning and decision making processes. Thus, CC is important to consider in EDSS that work on a larger temporal scale.

Available Climate change information from Global or Regional Climate Models (GCM's, RCM's) comes with a number of properties that need to be dealt with when using climate change projections in EDSS's.

In essence, climate change data consists of projections of meteorological variables like temperature and precipitation. To utilize this information in other environmental contexts the meteorological variables need to be mapped to environmental variables of the specific decision context (e.g. by a model). Climate change data is usually available on large scales. GCM's provide grids with a grid size of hundreds of km's, RCM's grids with a grid size of tens of km's. For many applications these grids need to be adapted to the higher resolution scale needed by local EDSS.

Another aspect is that climate change models produce climate change projections (climate scenarios) based on assumptions e.g. about future emissions. There is a number of established Climate Change Models available which produce different projections [1,2]. For each CC model different projections are available that might be relevant for the integration into an EDSS. In addition, CC models come with uncertainties that need to be somehow addressed if to be used in an EDSS.

How to utilize CC information for instance in the context of urban planning is subject to on-going research [3,4]. Issues to be dealt with include more than pure CC data related problems – particularly how to couple CC information with local experiments and how to visualise CC related phenomena.

This paper presents a discussion of which functionality is required in Climate Change enabled software. Concepts and possible realisations are based on the results of the FP7 project SUDPLAN [5], which integrates environmental modelling and software expertise, a system of standardised services and end user applications.

The SUDPLAN Scenario Management System (SMS) delivers long term projections of CC data (air quality, hydrological conditions and intense rainfall) based on different climate scenarios, which can in turn be used for different aspects of local modelling and decision support. The example of the SUDPLAN software is used to demonstrate how CC can be integrated in EDSS, from providing access to adequate climate change information, to the tools to link this information, and finally to the individual decision making problem.

---

[1] `http://www.bmu.de/fileadmin/bmu-import/files/pdfs/allgemein/application/pdf/aktionsplan_anpassung_klimawandel_en_bf.pdf`

2 Requirements for Climate Change Enabled EDSS

CC enabled software needs to provide access to appropriate climate change data. Appropriate in this context means CC information tailored to decision making process of the EDSS. The tailoring to be done includes:

- the temporal scale
- the spatial scale
- the environmental variables
- a sufficient number of climate change scenarios (as a means to deal with uncertainties of CC models)

Therefore EDSS using CC data need support to discover, explore, filter and select CC data to drive a specific local decision scenario. When dealing with CC a comparison of several relevant CC driven local scenarios is also needed, to estimate the potential impact of CC on the local problem. For this reason Scenario Management functionality, data comparison features and advanced simulation play an important role.

Another aspect is that CC enabled software needs to support the coupling of climate change data and the local decision problem. In the context of running a local simulation model to support the decision making process this means to transform CC data according to the requirements of a local model regarding formats, frequencies or anything that requires pre-processing of CC data before it can be used as input data (e.g. as boundary conditions) for a local simulation model. Moreover, flexible conversion, import and export mechanisms enable the analyst to use arbitrary tools to visualise, analyse, aggregate and correlate results of CC driven local scenarios.

3 Approach

Since 2010, the SUDPLAN project has designed and implemented a software framework for use of urban planners who need to assess the environmental impacts of storm water flooding, hydrological conditions and air quality in urban environments, under present and future conditions taking into account climate change scenarios. The primary target users for the SUDPLAN products are scientific users and city planners as well as environmental service providers [6]. The SUDPLAN solution provides two key elements of climate change aware urban planning:

1. Downscaled climate change projections of environmental variables based on ensembles of CC scenarios
2. Functionalities needed to consider and evaluate the effect of Climate Change in local planning processes, including climate data integration support facilities, model control, scenario management and advanced visualization capabilities.

SUDPLAN provides the means to downscale rainfall, hydrological and air quality data, taking climate change into consideration (Fig. 1). To obtain urban scale climate change data, end users provide local data which the downscaling services use to generate downscaled versions of the CC data of the needed local resolution. This data can then be used as input for "CC-enabled" local models. This infuses the effect of climate change into the runs of local models. Planners can use their conventional local models to produce local model experiments (local scenarios) and climate change is considered through the CC-enabled input data and boundary conditions [7].

Different CC scenarios can be used as input to local planning scenarios, to compare the uncertainty of the climate change projections. Intervals of possible climate change effects can be identified and thus adaptation strategies can be considered in the planning process.

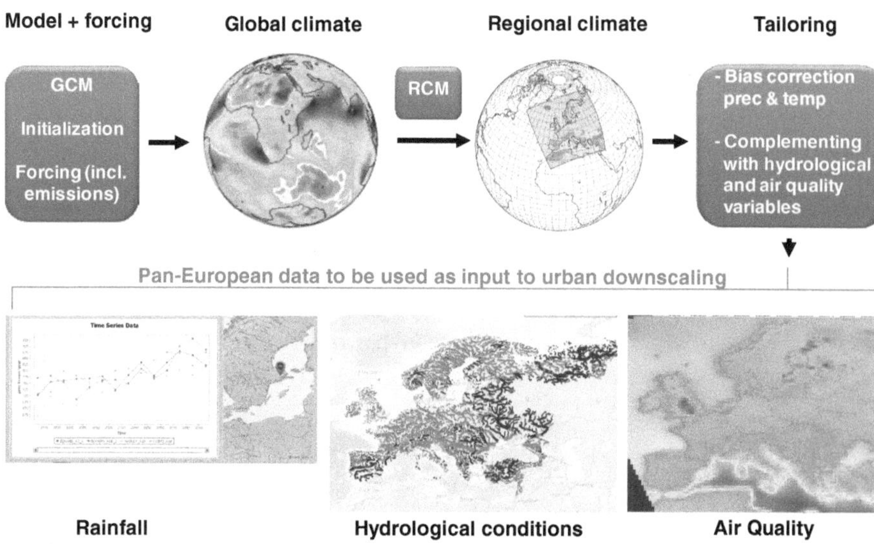

Fig. 1. CC data downscaling

The applicability of this approach to a large variety of decision processes is supported by using prominent open standard service interfaces. The transfer of time series or gridded data is through OGC Sensor Observation Services (SOS) and the model execution takes place through the OGC Sensor Planning Services (SPS) [8]. The use of these two standard interfaces also allows external users to access CC data in a well-defined manner and to incorporate CC data into applications.

4 Elements of CC Enabled EDSS

In the SUDPLAN framework the requirements discussed above are addressed by software components. Access to climate change data and downscaling functionality is provided through highly interactive GUI components. Wizards support the use of downscaling functionality and local model runs.

As a first step, end users gain access to CC data, including comparison and data analysis features for gridded and time series data from different CC scenarios (Fig. 2).

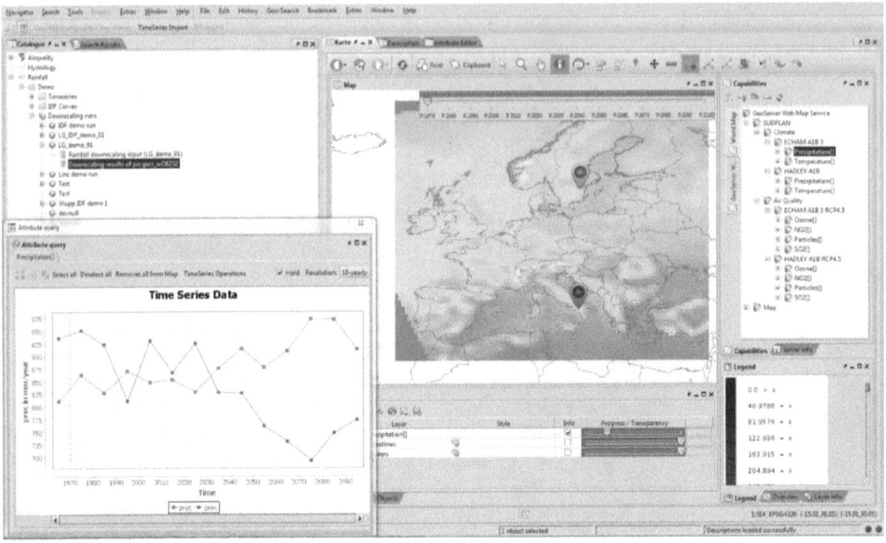

Fig. 2. Step 1: CC data access and visualization

In step 2, users can "upload" local data which serves to initiate the downscaling process. These processes are supported through wizards which guide users through the work flow.

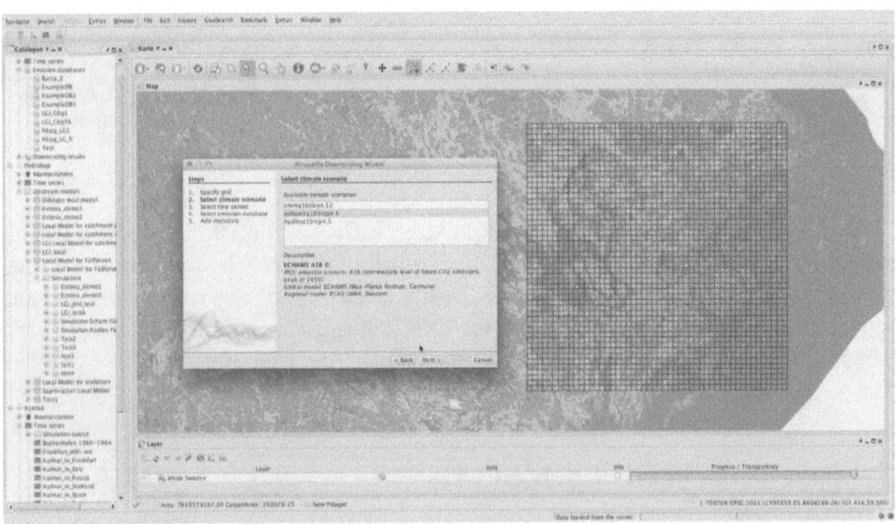

Fig. 3. Step 2: Downscaling wizard(example of air quality downscaling)

In step 3, users can couple the downscaled CC data to local models. This is where the actual "CC infusion" into the local decision process takes place, as the local models are run with the CC projections to produce local scenarios that are CC-enabled. This work flow is also supported through wizards.

In step 4, the exploration of multiple local scenarios, the management and visualisation of all related artefacts (model runs, model results,...) is handled by so-called model control components. These components enable users to run multiple variations and to compare results. Fig. 4 for instance shows the visualisation of results of a local sewer overflow model (SWMM, see [9]) which has been driven by downscaled rainfall projections.

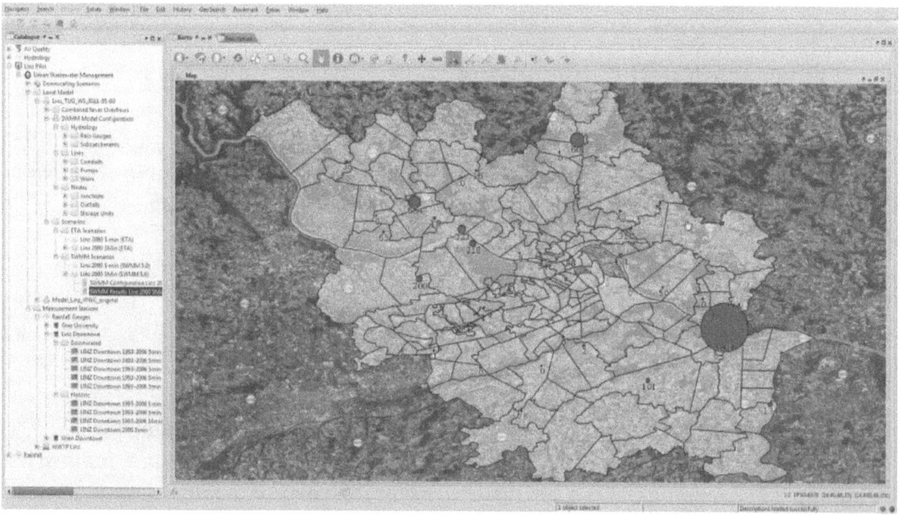

Fig. 4. Step 4: Comparison / visualization of local scenarios(example of a local sewer overflow scenario)

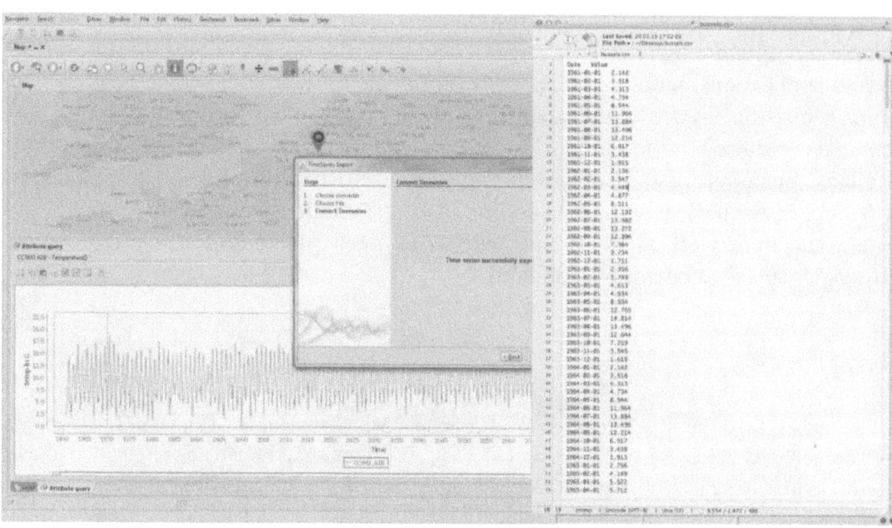

Fig. 5. Step 5: time series export

Finally in step 5, import, export and conversion functions supporting prominent formats in the respective domains, which enable users to utilize results in other tools like data analysis, statistics and specific post-processing tools. As an example **Fig. 5** shows the export of time series data to a csv file.

In order to produce material for decision makers that do not use an EDSS directly but rather are supported by an expert, the publication of results like the generation of maps and reports is part of the framework functionality.

5 Discussion and Conclusions

Dealing with CC data in EDSS means facing certain challenges. This paper identifies requirements and shows options for possible solutions as part of flexible framework to develop CC-enabled EDSS. The framework has been developed and validated in the context of the FP7 SUDPLAN project.

Currently, the development of CC-enabled software in general and EDSS in particular plays an insignificant role in integrated real world applications. However, this might change over time when the effects of climate change become more important for local planning.

Acknowledgements. SUDPLAN is a Collaborative Project (contract number 247708) co-funded by the Information Society and Media DG of the European Commission within the RTD activities of the Thematic Priority Information Society Technologies.

Software Baseline. The SUDPLAN solution has been implemented with a stack of freely available open source software. The core of the solution is the ***CIDS product suite*** of CISMET GmbH. This product consists of a set of software components, application programming interfaces (APIs), management and development tools, services and applications, with a special focus on interactive solutions which need to integrate geo-spatial systems with databases, sensor networks, document-oriented systems, unstructured information sources and numerical models. CIDS uses freely available open source software exclusively, is itself freely available as open source software, distributed as github [2] projects, and is compliant with all major OGC services. The framework is freely available as an open source software project under http://sudplan.cismet.de/download.html.

References

1. Roeckner, E., Arpe, K., Bengtsson, L., Christoph, M., Claussen, M., Dümenil, L., Esch, M., Giorgetta, M., Schlese, U., Schulzweida, U.: The atmospheric general circulation model ECHAM4: model description and simulation of present-day climate. Report No. 218. Max-Planck-Institut für Meteorologie, Hamburg (1996)

[2] https://github.com/cismet

2. Gordon, C., Cooper, C., Senior, C.A., Banks, H., Gregory, J.M., Johns, T.C., Mitchell, J.F.B., Wood, R.A.: The simulation of SST, sea ice extents and ocean heat transports in a version of the Hadley Centre coupled model without flux adjustments. Climate Dynamics 16, 147–168 (2000)
3. Schlobinski, S., Denzer, R., Frysinger, S., Güttler, R., Hell, T.: Vision and requirements of scenario-driven environmental decision support systems supporting automation for end users. In: Hřebíček, J., Schimak, G., Denzer, R. (eds.) ISESS 2011. IFIP AICT, vol. 359, pp. 51–63. Springer, Heidelberg (2011)
4. `http://www.grabs-eu.org/`
5. `http://www.sudplan.eu/`
6. SUDPLAN Final Report, http://sudplan.eu/polopoly_fs/1.30418!SUDPLAN_final.pdf
7. Schlobinski, S., Gidhagen, L., Olsson, J., Frysinger, S., Denzer, R., Kutschera, P.: Integration of Climate Change Effects in Local Models and Urban Planning Processes, In: International Congress on Environmental Modelling and Software, pp. 826–834 (2012), `http://www.iemss.org/society/index.php/iemss-2012-proceedings`
8. `http://www.opengeospatial.org`
9. Camhy, D., Gamerith, V., Steffelbauer, D., Muschalla, D., Gruber, G.: Scientific Data Management with Open Source Tools – An Urban Drainage Example. In: Proceedings IWA/IAHR 9th International Conference on Urban Drainage Modelling, Belgrade (2012)

Qualified Safety Management Ensures Sustainable Human Life Quality

Dana Procházková

Czech Technical University in Prague
Faculty of Transportation Sciences
Konviktská 20,110 00 Prague 1, Czech Republic
prochazkova@fd.cvut.cz

Abstract. The quality of both, the safety management and the ways of realisation of management tasks, are fundamental importance for ensuring the human security and sustainable development. The capability of correct application of risk management process is important. For assembling the qualified data for the FOCUS, the EU project, it was proposed the tool that leans on method of application of case study at decision-making. The tool has four main parts and it acts for obtaining the qualified groundwork for territory safety management and it can be easy adapted for needs of objects, partial safeties as it is the labour safety and health, and organisations. The paper shows aims of individual steps and linking the partial tools from management domain.

Keywords: safety management, FOCUS, sustainability, security.

1 Introduction

Present goal of humans is their security and sustainable development. In agreement with the EU and UN proclamations, the professional knowledge and experience, there is necessary for human society conservation and sustainable development to create the safe territory, safe community, safe state, safe Europe and safe world. The safe space undoubtedly includes the quality environment, on which the human lives are strongly dependent. For investigation purposes, the safe space is represented by safe open dynamically variable system that we denote in agreement with the UN report from 1994 [1] as the Human System. The environment is also the system and from the general view we can say that the environmental system is understood as biosphere elements which include the atmosphere, the hydrosphere, the lithosphere and the ecosphere, the set of interactions among these elements that are caused by links and flows among elements [2]. In detail, both systems are system of systems (SoS), i.e. several overlapping systems [3].

Safety of each system is disturbed by disasters, i.e. internal or external phenomena that lead or can lead to damages, harms and losses on system assets and on system oneself. The sources of some disasters are also inherently connected with human lives and activities [3]. It means that both systems under account (the environmental system and the human system) are affected by both, the processes, actions and phenomena

J. Hřebíček et al. (Eds.): ISESS 2013, IFIP AICT 413, pp. 472–488, 2013.

that are under way in human society, environment, planet system, galaxy and other higher systems, and the human management acts.

The basic terms of each system safety management [3] are:

Security is a state of system at which the occurrence of harm or loss on system assets has an acceptable probability (it is almost sure that harm and loss do not origin). To this there is also belonged a certain stability of system in time and space, i.e. a sustainable development in time and space.

Safety is a set of human measures and activities for ensuring the security and sustainable development of system and its assets (protected interests).

Secure system is a system that is protected against to internal and external disasters.

Safe system is a system that is protected against to internal and external disasters and it does not threaten its vicinity.

Safe environment is an environment which is secure and it can sustainable develop.

Safe human system represented by ***safe territory*** including the human society, is a human system, the assets of which are public assets (Figure 1) being in security and with potential of sustainable development. Similarly ***safe organisation*** is the organisation, the assets of which are in security and they can sustainable develop; the organisation assets are public assets and specific assets directly connected with organisation existence [3].

Human system safety management is the management of human system directed to human system safety, the product of which is security and sustainable development of all public assets denoted in Figure 1 [3].

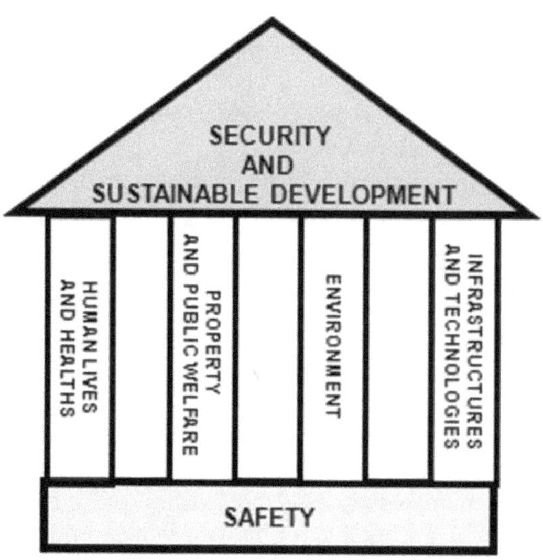

Fig. 1. Human system safety management process model and its public assets (Source: [3])

By analogy, ***the environmental system safety management*** denoted in main professional sources (www.epa.gov.ems/, www.inem.org) only as an environmental management is realised by an environmental management system (EMS). It refers to the management of an organization's environmental programs in a comprehensive, systematic, planned and documented manner. It includes the organisational structure, planning and resources for developing, implementing and maintaining policy for environmental protection. According to rules given in materials presented in www.epa.gov.ems/, www.inem.org) it:

- serves as a tool to improve environmental performance,
- provides a systematic way of managing an subject's environmental affairs,
- is the aspect of the subject's overall management structure that addresses immediate and long-term impacts of its products, services and processes on the environment,
- gives order and consistency for organizations to address environmental concerns through the allocation of resources, assignment of responsibility and on-going evaluation of practices, procedures and processes,
- focuses on continual improvement of the system.

It means that it deals with an organisation or territory safety management with the asset "environment".

Enthusiasm that was in developed countries in the middle of last century, when humans believed that the human kind would have power over disasters (wind, rain, earthquake etc.), was replaced step by step during the time by respect to the Earth's Planet System. At present the human actions and management are implemented by a pragmatic approach based on evaluation of credible and relevant data and the humans take into account that their knowledge and capabilities are:

- too small to prevent the origination of disasters that are the manifestation of the Earth's Planet System development,
- adequate to mitigate the impacts of disasters that are the manifestation of the Earth's Planet System development,
- sufficient to prevent the origination of disasters that are connected with the human actions and with development of human society (so called man-made).

Generally, there is known that every time when we want to control some phenomenon or to avert it, we must know its cause, size, repeat and nature of impact effects on assets. The disaster sizes, namely extreme ones have basic importance for a system safety. From them it is unreeled the created protection system, i.e. the system of measures and activities for averting or mitigating the disasters and / or their impacts [3].

There is necessary to take into account that disasters from the viewpoint of the Planet or environment development might be inventive changes supporting these system states and that human with his / her wishes and management goes against them. If it is reasonable and profitable for humans in near or distant future, we cannot estimate because our fittings in this domain are very poor and such question has been only recently appeared in a professional domain. Therefore, there is necessary to use the precaution principle in each system management. This principle is inherently included in safety management that denotes the strategic management ensuring the identification, diagnose, enforcement and implementation [4]. According to present knowledge the processes, induced the disasters in both considered systems, are given and characterize in [3]. At the disaster occurrence there are originated chains of undesirable phenomena (impacts, consequences) of external and internal character, primary and secondary, which affect negatively system assets with different intensities and in different time moments. The substantial role plays the local vulnerability and pertinent faults in a human behaviour or a management on all levels. The disasters are the cause of emergency situations, the severity of which substantially increases if cascade impacts occur, see data in [3]. Regarding to present knowledge and experience each system safety management has to arrange to:

- precede disasters if possible, e.g. in a case of natural disasters it is impossible,
- eliminate the causes of severe disaster impacts or at least to reduce their occurrence frequency,
- mitigate unacceptable disaster impacts by preventive measures, preparedness, optimal defeating the disaster impacts and by them defeating the induced critical situations (i.e. really the reducing the emergency duration to acceptable amount),
- ensure renovation and start of further development of system considered.

It holds for both, the human system safety management [3] and for environmental management (www.epa.gov.ems/, www.inem.org). From the present knowledge viewpoint there is necessary to ensure in order that each system management directed to safety might be proactive, strategic and might consider facts, findings, experiences and their correct evaluation. The reactive management is only admissible on operative level when emergency or critical situations have been occurred, i.e. in times when there is necessary to solve problems immediately and when no time for deep analysis and assessments that are challenging for data, methods and processing time [5]. Therefore, in practice we realise third level safety management, Figure 2.

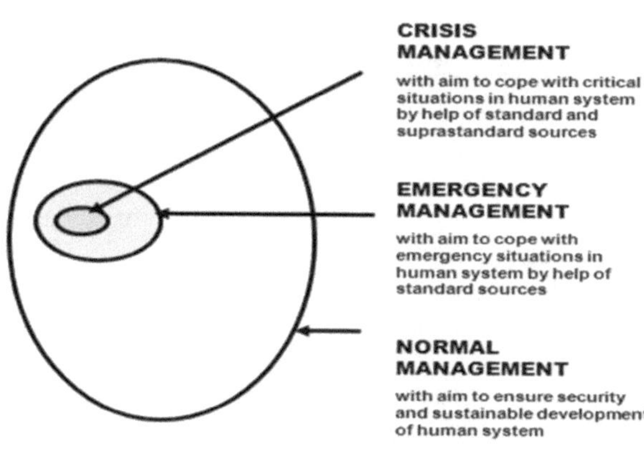

Fig. 2. Levels of safety management (Source: [3])

The paper deals with the relation between human system safety management and environmental management (in the sense of above given fact it goes on environmental system safety management but in the most professional sources the word "safety" is not given, and therefore, we respect it).

2 Risk and Safety Management Principles

Strategy of management for ensuring the security and sustainable development of managed subject consists in negotiation with risks [6-7]. According to present possibilities of human society we apply several ways of negotiation with risk:

- part of risk is reduced, i.e. by preventive measures the risk realisation is averted,
- part of risk is mitigated, i.e. by preventive measures and by preparedness (warning systems and another measures of emergency and crisis management) there are reduced or averted non-acceptable impacts,
- part of risk is re-insured,
- part of risk for which response and renovation will be necessary there are prepared resources for response and renovation,
- part of risk for which there is only prepared contingency plan, i.e. this part of risk is non-controllable, too expensive or low frequent.

To this it is joined the distribution of risk defeating among all stakeholders [6]. The distribution in the good governance is performed according to a rule that all stakeholders have responsibility for risk defeat and that the defeat of real risk is assigned to a subject, the preparedness of which is the best.

For classical / standard risk management and for risk governance in a subject it is necessary:

1. To understand the process of disaster origination and the conditions, under which the process goes on.
2. To know the sites in which the disaster can originate and its probable physical and other characteristics.
3. To identify the hazard that disaster means for a given subject according to stipulated rules.
4. To determine the impacts of disaster with size equals to a hazard on assets.
5. To eliminate the unacceptable disaster impacts in cases in which it is possible with acceptable expenses.
6. For residual impacts to calculate by help of predictive models their occurrence probability, respecting the fact that there are also considered possible failures of preventive measures.
7. To calculate possible harms of assets in the considered subject with regard to assets, that are really in the subject considered and by help of occurrence probability to determine the risk size.
8. To identify and to realize the mitigation measures with regard to considered assets in the way that they may be ALARP (so low as reasonable achievable).
9. To prove that all measures for averting and mitigating the risks were performed.
10. The acceptable risk can be achieved by hazard reduction of disasters, which are only connected with human activities, and above all by reduction of vulnerability of a given subject that is an object of risk assessment.

To ensure the subject security the subject assets governance has to be established on the safety management that is proactive and it is based on project and process approaches that are concentrated to good handling with risks that are inherent to a given subject. If with risks are not dealt correctly, so it is impossible to reach successfully targets, and therefore, the feasibility of procedures is reviewed in advance [3]. The relevance of risk roles is caused by a fact that the costs of project / process target implementation and the all successfulness depend on a risk distribution. Therefore, it is necessary that each project may hold a special structure, risk distribution and financing that correspond to its character.

The risks have different sources, i.e. they depend on the disasters, local vulnerabilities, methods of defeating and on response management, they originate on side of all participated stakeholders. To reach comprehension and following risk reduction it is necessary to perform its analysis that consists in the following phases: risk identification; risk assessment; risk allocation including the risk and the risk assignation to participated subjects; risk treatment; and continuous monitoring and in a case of need the application of corrective measures.

The risk management model directs the subject to a continuous proactive risk management that leans on the promt identification, analysis, countermeasure planning, monitoring and governance of risks. Each risk passes through these steps at least once, and often several times [3], [7-9]. In the first step there is determined the risk source, the character of possible failure, operational and commercial connections. In the second one there is determined the probability and impact (for calculation and mutual risk comparison). In the third one there are defined countermeasures leading to the risk elimination, risk transfer to somebody else, negotiation with risk or its impact. In the fourth one there is obtained information on changes of individual risk elements in time. In the fifth one there are performed planning actions as reactions to appurtenant changes.

Outputs from risk management process for need of assets governance are the following:

1. Risk assessment document – including the all information on appurtenant risk.
2. Top risks list - including the list of selected risks, the solution of which has the highest demands on sources and time.
3. Retired risk list – serving as historical reference for future decision-making.

The safety management in comparison with the classic risk management uses the set of optimal measures against to all possible disasters respecting the physical nature of disasters possible in a given subject and it includes the precaution principle in concept promoted by the European Union at present [4]. Therefore, on the basis of present knowledge analysis, performed in work [3], [6], there is necessary for ensuring the security and sustainable subject development to change the standard risk management to the risk governance profiting the safety, called safety management. It is necessary to:

- establish synergic relations among the risks, vulnerability and safety,
- model the process of decision-making the public administration with regard to risks and uncertainties (to perform support decision-making systems),
- specify legal conditions and protected measures,
- improve activities of institutions (institutional changes).

Everyday facts and analyses of behaviour of both followed systems show that risks are going reality and that during the time new risks emerge. Therefore, it is necessary to live according to the concept of live with risks [10].

Reduction of each risk is connected with a cost increase, with a lack of knowledge, technical means etc. Therefore, in practice we search for boundary to which it is reasonable the risk reduction in order that expended costs may be reasonable. This level of risk reduction (certain optimality) is mostly a subject of top management and of political decision-making, at which there are used the present scientific and technical findings and considered the economic, social and other conditions.

The basic turn of a subject management with regard to required targets is not possible to reach by individual partial measures, but only by a complex approach. To ensuring the security and sustainable development of subject there is necessary to use the co-ordinated and intentional approach. It enables step by step and in agreement with their importance and urgency to solve set of tasks in all domains and parts and by this to reach the required subject state [3].

Both followed management systems fulfil demands above put on safety management system - see to [3] and www.epa.gov.ems/, www.inem.org.

3 Real Preventive and Mitigation Measures and Activities

As it was above said at the disaster occurrence there are originated chains of undesirable phenomena (impacts, consequences) of external and internal character, primary and secondary, which affect negatively the subject under consideration in different intensities and in different time moments. The substantial role plays the local vulnerability of subject and pertinent faults in a human behaviour or management on all levels.

With regard to the historical development there are a lot of preventive and mitigation measures that are applied into practice by legal rules, technical standards and norms and public instructions. These ensure the protection against to design basis disasters. In the case of beyond design disasters this protection does not exist and severe harms, damages and losses are caused not only by direct impacts on system safety but also through interdependencies arranged by couplings created in system by links and flows of a permanent or random character. These secondary impacts mostly affect the system assets and the system security strongly and lengthy [11].

For both considered management types the land-use planning plays a big role because it ensures a basic prevention against disasters of all kinds; i.e. natural, technological, environmental, social and those caused by interdependencies in the critical infrastructure, including the terrorist attacks. In the frame of safety management there are performed measures in the land-use planning, designing, building and operation of objects and infrastructures. The measures are technical, legal, organisational, economical etc. [11]. For this management type there is necessary:

1. To consider all disasters that can occur in the subject under account (so called "All Hazard Approach" [12]).
2. At possible disasters there is necessary to take into account hazards of the 10th, 100th and may be more year disasters.

3. To carry out measures for vulnerability (and risks) reduction against disasters that can have unacceptable impacts on the subject assets.
4. To carry out mitigation measures against unacceptable impacts on the assets, the occurrence of which cannot be prevented.
5. To concentrate to critical assets, critical functions and critical activities which create the base for fulfilment of management aims in the subject.
6. From the economical viewpoint it is necessary to implement only measures suitable for a given subject and effective not only in the limit time interval but in reasonable time period.

For realisation in practice there is necessary:

- formation (generation) of professional background for the decision making by the research and science support,
- application of suitable management structure that will ensure the rational and qualified planning the measures,
- building up and training the executive forces,
- specialist training and systematic population education,
- legal regulations, norms and standards,
- inspections and other check mechanisms including the QA system,
- special management types for emergency and critical cases.

From this it follows that a great role at creation of safe space there is played by level of management. As it was said above the subject management has to be proactive, come out from sophisticated grounds, be tailored to real conditions and understandable to all subjects to which it is addressed. That it is necessary in order that the subject top management might recognise that the safety is not something in advance given, but that it must be created by conscious, directed and linked system measures and interventions, which from the theory viewpoint means to carry out management of safety [3], [13]. If we consider the emergency situation scale with degrees 0 – 5 [3], the subject safety management ensures:

- each human is capable to put under control the emergency situations of the 1 - 2 categories due to his/her education, training and preparation,
- subject top management has the emergency management system for successful defeating the emergency situations of the 2 - 4 categories,
- subject top management has the crisis management system for successful defeating the emergency situations of the 5 category (i.e. the crisis).

In the frame of mentioned management systems, the public administration and considered subject top managements and their administrative sections:

- build executive forces (fire-fighters, security guards, technical services, cyber safeguard etc.), that are prepared and trained for putting under the control the emergency situations,

- form financial and material resources in order to they may put under the control all possible emergency situations,
- form the temporary systems which ensure the executive forces support under critical situations.

For ensuring the subject stability and its sustainable development there is necessary to ensure the continuity, i.e. the survival of emergency and critical till crisis situations [14]. To this aim there is necessary to perform minimally the risk management and optimally the safety management. The risk management is a planning, organisation, allocation of work tasks and check-up of resources of organisation so that there might be reduced losses, damages, harms, injuries or deaths caused by various disasters. The safety management consists in a planning, organisation, allocation of work tasks and check-up of sources of organisation with target to reach requested safety level. Its targets are higher than the risk management goals. It includes the principals that are important for sustainable development [3], i.e.:

- adequate expenses principle (i.e. the spent expenses on measures must be optimal - it must not happen wasting),
- integration principle (i.e. the activities in domain of one protected interest must also take into account the impacts on other protected interests – it goes on finding the mutual balance),
- preliminary cautions (precaution) principle (i.e. there is necessary to perform the prevention and mitigation measures also then, when there is a very low probability that undesirable severe impact will occur, i.e. it does not rely on size values obtained by common statistical methods, but on the results of methods based on extreme values or marginal estimate),
- prevention principle (i.e. at a decision making there is taking into account the experience that in a time scale comparable to the human life they are generally preventive measures feasible and financially less exigent than corrective measures, and quite it pays when we take into account the needs for the next generation existence),
- synergy principle (i.e. in case that there are possible several variants, so there might be preferred such variant, that is an optimal one from the view of strategy and accessible resources – there is recommended the proactive management application),
- subsidiary (decentralisation) principle (i.e. the problems might be solved on the lowest possible level – there are recommended methodologies for the regional level of public administration),
- public space cultivation principle (i.e. there is assumed the participation of the state authority, public administration, their executive and other forces, research and professional bases, legal and physical persons and citizens),
- partnership principle (the relations among the protected interests are equivalent),
- openness and transparency principle (groundwork, decisions and their justification are accessible for public),

- exactness principle (i.e. there is came out from scientifically well-founded, measurable and checked up concept tasks and measures, which are short-term, medium-term or long-term character),
- well-judged receiving the ecological and other liabilities principle (i.e. there is enforced that ecological and other liabilities cannot be accepted without responsible evaluation of their impacts on further protected interests).

Both followed management systems fulfil demands put on safety management system - see to [3] and www.epa.gov.ems/, www.inem.org.

4 Groundwork for Management

A clear model of the safety management system (SMS) has been established based on current knowledge [3], [6], [13], Figure 3. It is necessary to note that this SMS model applies to systems with not very complex structure and with clearly defined relationships and flows among the elements of the system. Even here, however, considering the diversity of systems that are object of management, it is necessary to elaborate each particular SMS in accordance with a concept that respects the individual structure and specifics of a system, by which we replace the object that we want to manage. By such concept we also determine risks that we follow and the way of their consideration, i.e., if process of decision making is based on evaluation of results of partial, integrated or integral risks. It has to be emphasized that only the integral ones include the cross sectional risks that are associated with internal dependencies among interconnected assets of system or among interconnected individual systems in a case of system of systems (SoS - System of Systems) [6].

According to a current knowledge, the real systems like the human system, critical infrastructure, environment, human society, human body etc. are complex systems that consist of several systems of different nature and different locations that are interconnected and aimed to ensure certain operations and activities in a specific place and time and in a certain quality. It is necessary to note that there is a fundamental difference between the complex system which we have been monitored and managed for several decades, and the SoS, which we have systematically examined in recent times [6].

A complex system is a set of interconnected systems which have closed architecture. This model nowadays represents foundation for management of power [15], gas, heat, and water, supply systems etc. According to the nomenclature of engineering disciplines focused on safety [4], the aim of such systems is to ensure system security, whereas no attention is paid to the system surroundings or to neighbouring systems. Therefore, e.g. in a case of black-out of electricity supply system, the electricity distribution management system only deals with technical problems associated with the restoration of electricity supply and it does not address to issues of impact of such power failure impacts on lives of people, on running businesses, etc., i.e. on public assets [15].

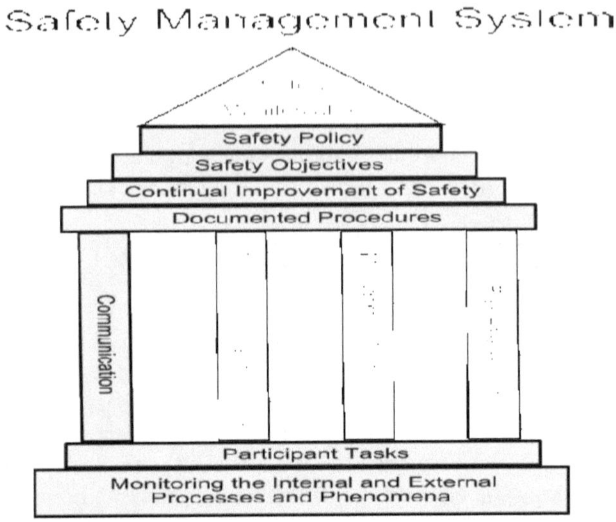

Fig. 3. A general model of the safety management system of real objects

On the other hand, the current SoS concept is understood as a set of interconnected systems which have open architecture, i.e., different elements are connected as long as they fulfil the conditions of interoperability and user requirements [16]. With regard to the above-mentioned public assets, the aim of SoS safety management is to ensure not only the safe SoS as such, but also its safe surroundings, i.e. its goal is not just a system security, but also requirement that the system does not threaten its surroundings, i.e. it goes on the integral safety management [3], [16]. In other words, the SoS safety management system takes into account requirements of surroundings, e.g. such as demands of electricity users etc. From the above it is clear that the concept of SMS for SoS is different from those for complex systems now commonly used in a practice.

Qualified SMS for the SoS safety management are currently being looked for. They have to deal with internal dependencies among systems and appropriately resolve conflicts among systems of different nature. Conflict resolution with aim to ensure the SoS safety in the real environment means finding a consensus among aims of individual systems and among ways of their reach; priority target for SoS safety is coexistence of partial systems [17].

The biggest challenge for SoS safety management systems are the identification, understanding and appropriate management of cross-cutting risks which cause or can cause different cascades of failures of SoS functionality, which deplete the SoS assets. For their discovery and strategic management it is necessary to analyse internal dependencies among SoS individual systems, across entire SoS and between the SoS and its surrounding. Systems of systems have specific properties such as nonlinearity, different steady states (attractors), catastrophic behaviour, chaotic behaviour, etc., which are the cause of cross-cutting risks that disturb both, the followed SoS security and the SoS vicinity security [6]. In order to ensure the safe SoS and safe SoS vicinity, we have to be able to cope with the cross-cutting risks, i.e. to identify and appropriately manage them.

We have to realize that for SoS risk management it is valid basic knowledge of risk management domain [6], i.e. it is not enough to know the risk size, but it is necessary to know its specific causes, their localisation in a controlled system and the particular vulnerabilities of assets in a given allocation. For identification, analysis and assessment of risks, there are many classical methods, tools and techniques [6]. Unknown there are the tools, methods and techniques for identifying, analysing and managing the cross-cutting risks.

With regard to the Czech national project results [11] there is necessary for needs of safety management of a given subject to ensure qualified replies to the following questions:

1. What disasters can occur in a given subject and what impacts have they?
2. Where disasters can occur and how their impacts are spread in a given subject?
3. Under what conditions can disasters occur in a given subject and what conditions can cause escalation of their impacts?
4. How often can disasters occur in a given subject?
5. From what disaster sizes have disasters in a given subject unacceptable impacts, that caused losses, harm and damages on assets?
6. What maximum sizes could reach disasters in a given subject?
7. What assets damages can be caused by maximum possible disaster on specified credibility level in a given subject and what are its impacts on a given subject?
8. What is possible to do in a given subject against unacceptable disaster impacts on section of land-use planning, design, construction and operation of civil and technological objects and infrastructure, and may be in other domains as are the monitoring, inspection, education etc. with the aim to prevent the occurrence of disasters if possible or at least to prevent or to mitigate unacceptable impacts by preventive measures, preparedness, fit response to a disaster and by renovation, at which there has to be respected the losses prevention and targets of sustainable development?
9. What are necessary measures against real disasters in a given subject in the technical, organisational, financial, social, legal, education and training domains?
10. What unacceptable and residual risks (i.e. undesirable impacts with probability occurrence superior to a limit stipulated) with regard to possible disasters in a given subject will stay, when there are fulfilled rational measures that public administration can ensure in the technical, organisational, financial, social, legal, education and training domains?
11. How does perform the response to disaster with an aim to stabilize the subject state and to start renovation.
12. How does perform the renovation of subject and its assets with an aim rationally to use resources, forces and means for the prohibition of further losses, the upgrade of resistance against possible disasters and for the start of further subject development with all items on which it is dependent?
13. What is suitable the form of management and of a given subject renovation and its assets performance after disaster and how is it possible to realise it?
14. How does create the financial / monetary reserve for a rational renovation of a given subject and of its assets after disaster?

Then, for the subject safety management there is necessary with regard to hazard size and local vulnerability to divide known disasters into the following groups:

- disasters which cannot have impacts on a given subject,
- disasters which have only acceptable impacts on a given subject, i.e. ***the relevant disasters***,
- disasters which have on a given subject only such impacts that might be put under the control at performance of preventive and mitigation measures, i.e. ***the specific disasters***,
- disasters which have on a given subject inadmissible impacts, and hence there is necessary to carry out the principal preventive measures the technical, organisational, legal and educational domains, i.e. ***the critical disasters*** which cause or can cause crisis situations.

The human system safety management of territory is based on these data [3], the environmental management system is so in-depth only in some cases - see www.epa.gov.ems/, www.inem.org (the environment impact assessment and strategic environment assessment are not so detailed).

5 Procedure for Ensuring the Safety, Continuity and Sustainable Development of Subject

With regard to results in works [3], [5], [7], [10-11], [14] there is necessary for the subject safety ensuring to consider and to evaluate all disasters that have or can have unacceptable impacts on the subject, i.e. above all to determine disasters that have unacceptable risks which are impossible to avert by measures performed in advance and to split up them on:

- specific, which may put under control by qualified and in-depth preparedness of response to these disasters using the standard forces, measures and resources,
- critical, which may put under control by standard and beyond standard forces, measures and resources including the limitation of the rights and freedoms of staff and citizens.

After this sorting there is necessary to specify and to apply measures for averting the disasters or their unacceptable impacts there, where it is possible, and there, where it is impossible to prepare in the frame of preparedness to response the measures for mitigation of unacceptable impacts.

The human system safety management of territory respects these facts [3] and the environmental management system goes on so in-depth only in some cases - see www.epa.gov.ems/, www.inem.org.

6 Planning for Safety Management Needs

Because there is a reality that it is impossible to avert some of unacceptable disaster impacts on assets being important and mostly requiring the investments, and because it is impossible to ward off by preventive measures, because they either do not exist or are not accessible financially, technologically or personally, therefore, ***there is necessary to perform emergency and crisis planning*** in a given subject.

The planning is consciously regulation of development. It is conscious activity of a subject management that consists in the selection and presuming the aims, tasks, variants and ways, which condition the achievement of these aims. The most important feature of planning is the selection of aim. The planning is not the make-up of hierarchical commands file, which might be unthinkingly fulfilled, it is a creative activity, which stipulates the real aim and determines the optimal way for its achievement. The planning is a fundamental section of each management. For achievement of the long-term aims there is used the strategic planning and for achievement the short-term aims objectives the operative planning; both have their particularities, which pre-determinate the selection of methods and ways.

The human, however, wants and has to do either preceding the critical situations or put under a control the emergency, critical and crisis situations, and therefore, he / she has to apply a higher attention to the equation "insufficient awareness and insufficient understanding to crisis = insufficient preparedness, which means bad planning". Planning the measures, based on an objective evaluation have to be carried out during the land-use planning, designing, construction and operation, i.e. in the EU it is divided into ex ante, during and ex post.

In the planning domain there is necessary for determination, specification and realisation of preventive and mitigation measures in a case of every relevant disaster to know the impact sizes and their distribution on the area. This groundwork must be prepared by pertinent research and scientific institutions, because they have data and needed knowledge to their interpretation. The role of decision making groundwork lastingly increase with the time, because also during the intervention (response to disaster) there are the disaster characteristics that influence the commander decision making, e.g. the rate of disaster start (sudden or slow beginning), warning, preparation time, size of danger, risks for participants of intervention, casualties, assessment, number of members of executive forces for intervention, stage of disaster development (beginning, period of secondary impacts etc.), main risks, used forces, number of commanders, required decision (routine, known, complex or unknown), sufficiency or insufficiency of material for intervention, knowledge of site, time of intervention, space location (one or more sites).

In the planning domain there is not still uniformity - most often there are used:

- security plan, the aim of which is to arrange the subject needs and development,
- emergency plan, i.e. the set of response plans to emergency situations of the 3 – 4 categories for foreseeable emergency situations,
- continuity plan,
- crisis plan, i.e. the set of response plans for putting under control the critical situations,
- contingency plan, i.e. the response plan for unforeseen situation.

The planning in the subject, based on the stipulation of possible impacts and on costs, which a given subject will pay for failure, ***has to be particularly taken to this assets and property that mostly require investments at renovation*** [3].

The human system safety management of territory respects these facts [3] and the environmental management system respects these facts only in some cases - see www.epa.gov.ems/, www.inem.org (no obligatory realistic response plans to emergency and critical situations).

7 Comparison of Human System Safety Management and Environmental Management

From the above given results there is clear that both followed management types, i.e. the human system safety management and the environmental management in the sense of environment safety management are based on advanced risk management and that the human system safety management has more tools supporting the safety (all hazard approach, procedure for disaster assessment, emergency plans, crisis plans). Logic comparison of basic items shows that considered management types differ by:

1. Assets, i.e.:

   – environmental management has one asset "environment" and in advanced concept of the EU three pillars (environment, social, economic) [4],
   – human system safety management has several public assets and environment is only one of them (Figure 1).

2. Aims, i.e.:

   – recent environmental management is aimed to secure environment with sustainable development enabling the human existence; the first is environment needs and the second human needs (www.epa.gov/ems; www.inem.org), which does not enable to solve human needs in the sense of Maslow pyramid [18],
   – human system safety management is aimed to secure human system with sustainable development and it also takes care on environment but not as the main priority but as one of basic system assets, i.e. it has a chance to solve human needs in the sense of Maslow pyramid [18].

Regarding these facts we can conclude that for humans the human system safety management is more acceptable because it has a chance to fulfil human needs in the sense of Maslow pyramid [18]. But the present problem is that we have not enough knowledge on interdependences among the human system assets that are sources of cross-section risks, on criteria and limits for negotiation with all types of risks existing in the human system. The aimed research is very necessary.

8 Conclusion

The security situation in the world, territory and organisation has been changing with the time, and therefore, there must be systematically built the safety culture, which taking into account actual piece of knowledge and experience. The safety culture promotion into practice requires both, the aimed management and the broad participation of all staff of a subject under consideration with emphasising that the top management has the biggest responsibility. It understandably leads to the assignment of higher priority to planning and safety management as well as to higher demands to the understanding level of all participants.

Facts given above show that for effective management of all risks to both subjects considered, the environment and the territory represented by human system, there is necessary to apply in the subject under account the sophisticated safety management. This represents integral, complex and system tool ensuring the subject safety and enabling its development. It acts in dynamically uncertain surroundings and systematically ensures the stability (prevention of crisis) and forms conditions for the next development. It leans on the monitoring, situation diagnostics, sophisticated analyses and processes appreciation, prediction of further process development, preventive and mitigation measures application before activities start and in their course, preparations of measures for correction of undesirable processes development or for putting under a control the undesirable processes development and for ensuring the renovation etc. All partial tools are one another linked and sequenced by a way that leads to a synergic effect [19].

The differences between the followed management types are in the priority aims: for environmental management the aim is to ensure secure environment with sustainable development enabling the human existence; and human system safety management is aimed to humans, however, it also respects the other assets among which is also the environment. If we stipulate that humans build the system for them, so we must agree that human system safety management inherently includes the environmental management by way that is more acceptable for humans. It means that we could not create too many management types in practice but to create only one which will build world for humans on a correct base.

Acknowledgement. The author thanks for support to the Czech Technical University in Prague, the EU – project CITI-SENSE, ENV.2012.6.5-1, grant No308524.

References

1. UN: Human Development Report. New York (1994), `http://www.un.org`
2. Raven, P.H., Berg, L.R., Johnson, G.B.: Environment. Saunders College Pub. (1998)
3. Procházková, D.: Strategic Management of Safety of Territory and Organisation. ČVUT, Praha (2011)
4. EU: European Commission Adopts Communication on Precautionary Principle. Brussels (2000), `http://europa. eu.int/rapid/pressReleasesAction.do?reference=IP/00/96&format=ML&aged=0&language=EN&guiLanguage=en`
5. A Guide to the Project Management Body of Knowledge. US Project Management Institute, Washington (2004)
6. Procházková, D.: Risk Analysis and Risk Management. ČVUT, Praha (2011)
7. OECD: Assessing Societal Risks and Vulnerabilities. OECD Studies in Risk Management, Denmark, Paris (2006)

8. EU: Risk Assessment and Mapping Guidelines for Disaster Management. Working paper SEC(2010) 1626. Brussels (2010).
9. ISO: Draft International Standard ISO/DIS 31000, Risk management – Principles and guidelines on implementation (2008)
10. US ISDR: World Conference on Disaster Reduction Declaration. Kobe (2005), `http://www.un.org`
11. Procházková, D.: Methodology for Estimation of Costs for Renovation of Property in Territories Affected by Natural or Other Disaster. SPBI SPEKTRUM XI Ostrava (2007)
12. FEMA: Guide for All-Hazard Emergency Operations Planning. State and Local Guide (SLG) 101. FEMA, Washinton (1996)
13. OECD: Guidance on Safety Performance Indicators. Guidance for Industry, Public Authorities and Communities for developing SPI Programmes related to Chemical Accident Prevention, Preparedness and Response. OECD, Paris (2002)
14. Disaster Advisors Inc.: Bussiness Continuity Planning (2003), `http://www.disasteradvisors.com/disasterplanning.htm`
15. Procházková, D.: Real Problems of Critical Infrastructure Threatening the Region safety. In: Ale, Papazoglou, Zio (eds.) Reliability, Risk and Safety, pp. 2387–2394. Taylor & Francis Group, London (2010)
16. Procházková, D.: Critical Infrastructure Safety. ČVUT, Praha (2012)
17. Bossel, H.: Systeme, Dynamik, Simulation – Modellbildung, Analyse und Simulation komplexer Systeme. Books on Demand, Norderstedt/Germany (2004), `http://www.libri.de`
18. Maslow, A.H.: Motivation and Personality. Haper, New York (1954)
19. EU: FOCUS, `http://www.focusproject.eu`

Novel Approaches for Web-Based Access to Climate Change Adaptation Information – MEDIATION Adaptation Platform and ci:grasp-2

Markus Wrobel[1], Alexander Bisaro[2], Dominik Reusser[1], and Jürgen P. Kropp[1]

[1] Potsdam Institute for Climate Impact Research (PIK), Potsdam, Germany
{wrobel,reusser,kropp}@pik-potsdam.de
[2] Global Climate Forum, Berlin, Germany
sandy.bisaro@globalclimateforum.org

Abstract. This paper presents two novel web-based applications for disseminating climate change adaptation related information. (i) The **MEDIATION Adaptation Platform**, one of the core outputs of the European FP7 research project MEDIATION (Methodology for Effective Decision-making on Impacts and AdaptaTION), offers a set of decision trees that can be browsed graphically to navigate over adaptation challenges and available types of methods for addressing them. This framework is interlinked both with a Toolbox and a set of case study descriptions. (ii) **ci:grasp-2**, the successor of the BMU-funded Climate Impacts: Global and Regional Adaptation Support Platform (ci:grasp), provides access to a growing pool of global and regional information on climate stimuli and impacts, as well as on adaptation projects. The paper identifies central user tasks for both applications and presents interaction metaphors that have been chosen to support users in fulfilling these tasks.

Keywords: Climate change adaptation, web-based information systems, user interfaces, HCI.

1 Introduction

Efficient communication of the scientific understanding of anthropogenic climate change is a prerequisite to induce the necessary shifts towards mitigation and adaptation. Against the background that such communication faces a multitude of challenges (e.g., [1-2]), web-based information systems hold some promise to facilitate the necessary dissemination of scientific information and to lower access barriers. However, designing effective information systems in the context of climate change is not a straightforward exercise due to two complimentary sets of challenges.

On the one hand, the information to be taken into account is of high complexity, ranging from climate change impacts and vulnerability to available adaptation options. In addition, existing case studies on climate change impacts, vulnerability and adaptation use different terminology and focus on different aspects of adaptation [3], which further impedes information usage. E.g., considerable work has been done to systematically characterize adaptation to climate change (e.g., [4]), including

J. Hřebíček et al. (Eds.): ISESS 2013, IFIP AICT 413, pp. 489–499, 2013.

definition of frameworks for assessing adaptation (e.g., [5-6]), as well as towards systematic analyses (e.g., [7]) or ranking of available adaptation options (e.g., [8]). However, besides some agreement on the broad categories we are noticing the absence of a common classification scheme for adaptation options. The variety of approaches to classify adaptation originates from the inherent complexity of the field and is further complicated by the specifics of the various sectors that need to adapt to climate change. The lack of a consistently used classification scheme is not only reflected in the literature, but also in the variety of web based applications aiming to provide access to adaptation related information.

On the other hand, designing interactive climate change information supply systems faces various user interface (UI) challenges, ranging from general UI challenges (e.g., [9-10]) over challenges of web-based applications (e.g., [11-12]) to challenges related to the use of interactive maps. The latter are increasingly used to depict climate change related information. However, web-based interactive maps in general come with own usability challenges (e.g., [13]), and using maps for climate (change) communication is as well not always straightforward (e.g., [14-15]). As today there are no established standards on how to design interactive web-based applications for complex climate change and adaptation related information [16], and current web-based climate change information applications are not free of usability problems [17].

In the remainder of this paper, we present central user tasks and interaction metaphors for the MEDIATION Adaptation Platform and for ci:grasp-2, two web-based applications to be launched in 2013 that each tackle different aspects of the complex overall adaptation information supply challenge. Section 2 briefly introduces the background of both applications. Section 3 then presents central user tasks and interaction metaphors for both applications, and Section 4 concludes the paper with a brief discussion of open challenges.

2 Central User Tasks and Interaction Metaphors for the MEDIATION Adaptation Platform

An increasing number of methods and tools from a diverse field of disciplines is being developed to provide answers to adaptation related questions. This emerging set, however, is complex and thus not easily accessible in a systematic manner. It is the aim of the web-based MEDIATION Adaptation Platform[1] for climate change adaptation methods and tools to aid users in disentangling it. To this end the MEDIATION Adaptation Platform builds on other core outputs of the MEDIATION project: a diagnostic framework for problem-oriented adaptation research, bringing together insights generated from different perspectives on adaptation into one coherent framework [18], and a Toolbox. In addition, a diverse set of case studies is made available to serve as illustrative examples of identifying and selecting salient methods and tools in adaptation research. The diagnostic framework as well as the MEDIATION Adaptation Platform provide *generic* methodological guidance applicable across sectors, regions and hazards for addressing vulnerability, impacts and

[1] http://www.mediation-project.eu/platform/

adaptation and do *not* aim at providing sector or regionally specific information. The intended users are scientists, policy advisors or practitioners with a technical or scientific background; the platform is not built for untrained decision-makers or the general public.

The diagnostic framework is structured along five stages: (i) assessing vulnerability (impacts & capacity); (ii) identifying adaptation measures; (iii) appraising adaptation options; (iv) implementing adaptation actions; and (v) monitoring and evaluation. It defines a set of decision trees based on empirical and theoretical criteria to support analysts in identifying salient adaptation challenges and choosing appropriate methods to address them. The aim of the MEDIATION Adaptation Platform is to enable interactive access to this framework and efficiently interlink it with Toolbox and case study information; an intuitive and flexible navigation over the decision trees is thus one of the application's core user tasks.

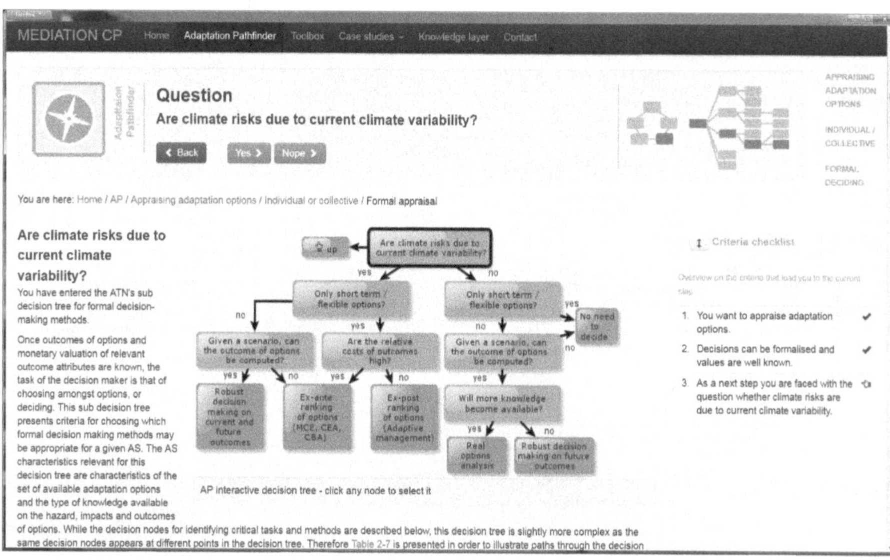

Fig. 1. Adaptation Pathfinder: Screenshot showing the sub decision tree for formal decision-making methods

To enable access to the diagnostic framework, the application provides the *Adaptation Pathfinder*. All decision trees are made available to the user in form of interactive graphic representations (Fig. 1). Due to spatial limitations and to reduce complexity, the trees are separated into and provided as set of sub-trees. By clicking a tree node, the user either (i) views available information related to the specific methodological choice or approach represented by this node, or (ii) navigates up or down in the overall decision tree structure.

To support user orientation, a combination of feedback mechanisms is applied. A popover indicating a node's content can be obtained by moving the mouse over the graphical representation of any node. A dynamically updated criteria checklist (Fig. 1, right hand) is displayed to permanently offer an overview on the choices that have led to the currently selected node. An additional overview diagram at the top right of the page provides further orientation aid by depicting the overall framework structure and indicating the currently selected section within this structure (Fig. 1, top right). To accommodate for different interaction preferences, the Adaptation Pathfinder offers further redundant navigation mechanisms: the trees can be alternatively browsed using answer-buttons available above the graphical representation of the trees (Fig. 1 top); in addition, shortcut lists to select a specific approach are offered. Importantly, the Adaptation Pathfinder does not prescribe any sequential access order, rather, the user can select any of the 5 stages as entry point and navigate subsequently to any sub-decision tree to focus on content of interest.

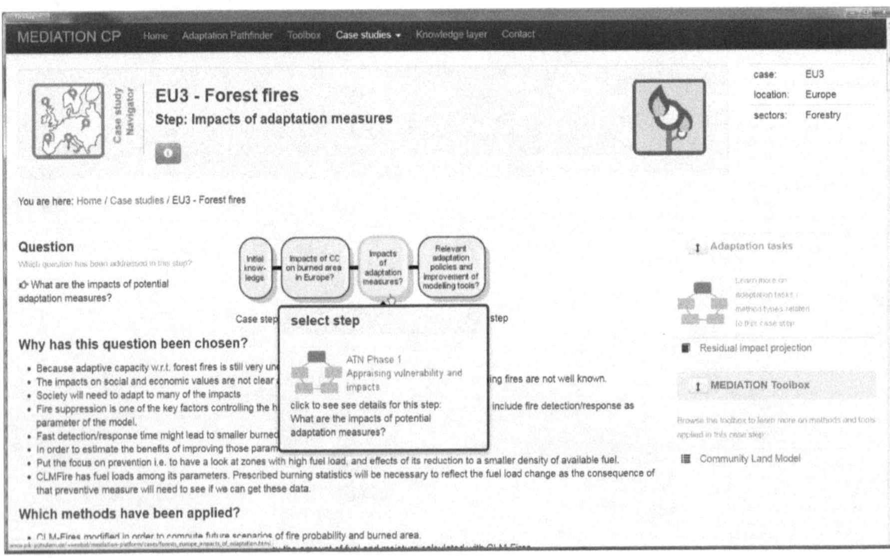

Fig. 2. Case Study Navigator showing a MEDIATION case study [19] on the effect of climate change on forest fires in Europe (screenshot)

Within the MEDIATION project, a diverse set of case studies has been conducted and informed the development of the diagnostic framework. Based on this, the MEDIATION Adaptation Platform applies a novel concept in providing access to case study information. All case studies have been broken down into a set of steps according to the diagnostic framework, with each step describing the respective question addressed, the method applied, and the results achieved.

To enable the user to flexibly browse these steps, the Case Study Navigator has been developed. Utilizing a somewhat similar interaction metaphor as applied in the Adaptation Pathfinder, interactive graphical representations are used to both depict the step structure of a case study and to allow selecting any step by clicking on its graphical representation (see Fig. 2).

A third central user task relates to integrated navigation over the three main sections of the MEDIATION Adaptation Platform (Adaptation Pathfinder, Case Study Navigator, Toolbox). To this end, while browsing through any of these three main sections, the user needs to be effectively supported in identifying related content within the other two. E.g., having selected a specific approach in the Adaptation Pathfinder, the user should be able to easily identify those case study steps that have applied this approach, as well as Toolbox entries related to this approach, and *vice versa*.

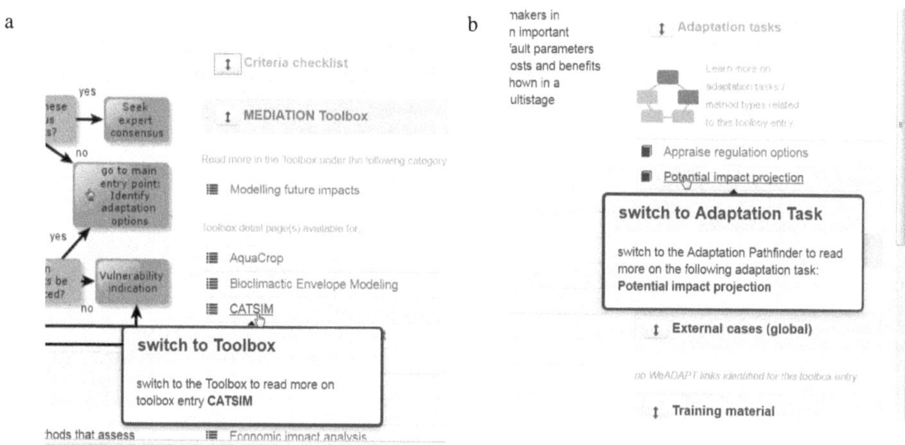

Fig. 3. Interlinking of and switching between main content sections using link boxes: (a) Adaptation Pathfinder with link box for accessing toolbox entries (b) Toolbox page with link box to related nodes in the Adaptation Pathfinder (screenshot details)

To support the user in fulfilling this task, the platform interlinks its three main sections logically in form of a triangle. To this end a tagging scheme is applied that interlinks the content elements of each main section with respective content elements of the other two. Link boxes are then derived and displayed in each main section to enable navigation to the related content elements of the respective other two main sections (Fig. 3 depicts an example for interlinking Adaptation Pathfinder and Toolbox entries). Further link boxes point to external content, enabling the user to additionally access related information as well as training material in the weADAPT[2] platform.

3 Central User Tasks and Interaction Metaphors for ci:grasp-2

The web-based Climate Impacts: Global and Regional Adaptation Support Platform (ci:grasp)[3] has been initiated by the Potsdam Institute for Climate Impact Research (PIK) and the German Agency for International Cooperation (GIZ) in a project funded by the Federal Ministry for the Environment, Nature Conservation and

[2] http://weadapt.org/

[3] http://www.pik-potsdam.de/cigrasp-2/

Nuclear Safety[4]. ci:grasp performs as a climate information service and aims at providing sound knowledge on current and projected climate stimuli and climate impacts, as well as on adaptation options, and to integrate both global and regional information on climate change. It aims to support decision makers in developing and emerging countries to prioritise adaptation needs, and to plan and implement appropriate adaptation measures. ci:grasp-2 is the successor of this application and provides both extended content and enhanced functionalities. Core features include content organisation based on cause-effect-relations (impact chains), support for interactive visual representations of multidimensional spatio-temporal data, and a module specifically addressing the representation of city related research results.

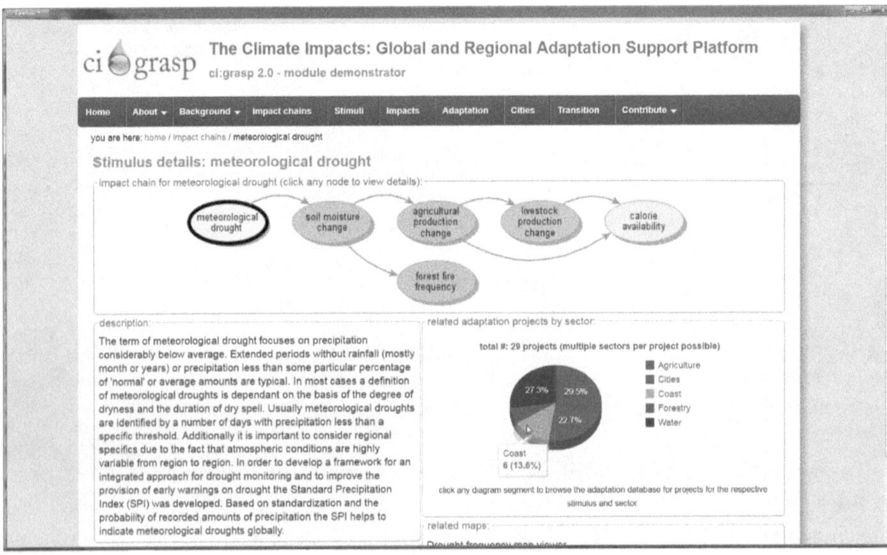

Fig. 4. Example of the interactive graphical representation of an impact chain for meteorological drought in ci:grasp-2 (screenshot)

ci:grasp-2 provides the user with information on a diverse set on climate change stimuli and impact as well as on adaptation projects. A central user task thus is efficiently gaining an overview over this complex pool of content, in order to subsequently explore it. To aid the user in the first task, the application is structured based on impact chains, i.e. simplified cause-effect-relations demonstrating how a given climate stimulus propagates through a system of interest via direct and indirect impacts it entails. cigrasp-2 offers interactive graphical representations of impact chains as means to navigate over the content in a structured manner (Fig. 4). Colour coded nodes depict a climate stimulus together with its potential impacts, as well as potentially affected livelihood conditions. By clicking the graphical representation of any node in this graphical representation the user gains immediate access to respective descriptions as well as to an overview on available content, covering respective

---

[4] Bundesministerium für Umwelt, Naturschutz und Reaktorsicherheit (BMU).

stimulus or impact maps as well as related entries from a database of adaptation projects.

To provide a vivid overview on the latter, an interactive pie chart diagram displays the sectoral distribution of available adaptation projects related to a currently selected stimulus or impact (Fig. 4, bottom right). By clicking the diagram segments the user obtains a sortable result table providing a more detailed overview on the currently selected projects by title, location, approach and scale. This overview can then be used to access the detailed descriptions for every project, as well as adjusted by applying different filter criteria. Exploration of the available map content is supported by the impact chains in a similar manner. For each stimulus or impact selected, an overview over available related maps is provided; any of these entries can then be selected to be displayed in ci:grasp-2's interactive map viewer. Here, the user can subsequently zoom and pan to areas of interest as well as select different maps by applying the map viewer's filter functionality.

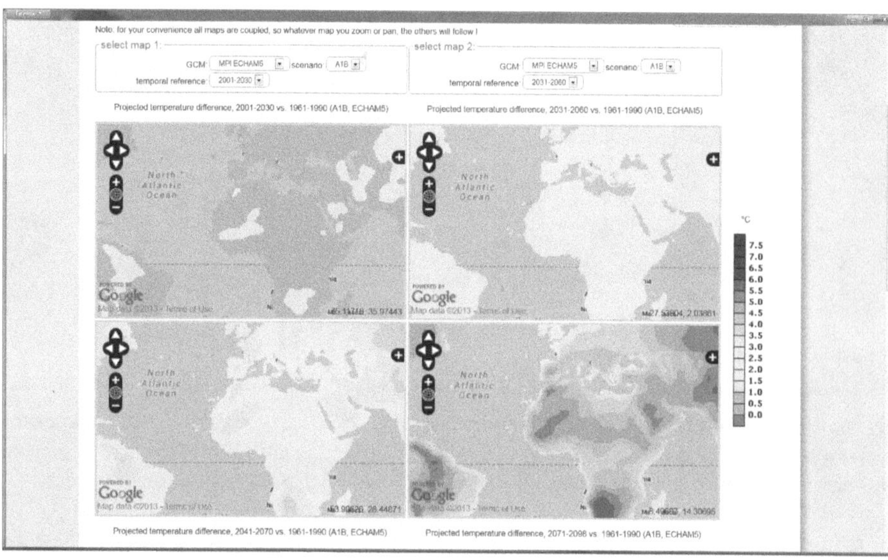

Fig. 5. Multi-map viewer showing temperature difference projections for four time slices for SRES A1B generated by ECHAM5 (screenshot detail)

De facto standards for interactive selecting of map content to be subsequently displayed on interactive maps can as today not be observed across different climate change information applications [16]. To support the user in selecting maps from a diverse pool of content, ci:grasp-2 utilizes a hierarchical multi-level set of dynamically updated UI controls. Any selection in a control of a higher level (e.g., the spatial reference) updates the selection options made available in the subsequent dependent control (e.g., the variable to be displayed), which in turn updates the next level (e.g., GCM), etc. This mechanism ensures that only meaningful attribute combinations can be selected by the user and prevents ending up with 'empty' filter results. The concept is applied both for a central map viewer providing access to ci:grasp-2's overall map pool, and for a set of further map viewers tailored to support accessing multidimensional information. The latter viewers combine a set of adjacent interactive maps in tabular form such that the user can jointly display characteristics of various user selectable dimensions (e.g., different GCMs, scenarios, or temporal references; Fig. 5). The maps are coupled such that any zooming or panning interaction with one of the maps directly triggers all remaining maps to adjust accordingly, so that defining an identical spatial reference for the complete set of maps is greatly facilitated. ci:grasp-2 uses this multi-map approach to represent, e.g., climate projections for AR4 and AR5 scenarios and impact projections from the ISI-MIP project[5].

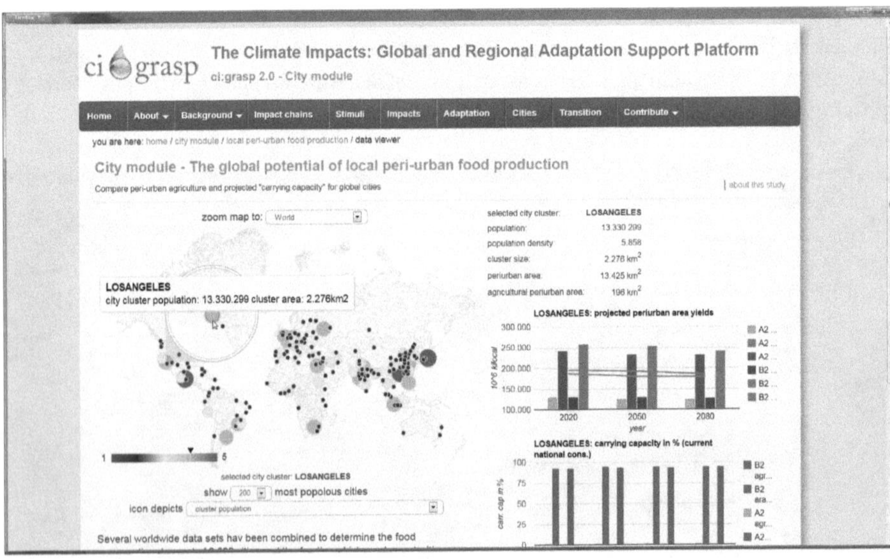

Fig. 6. Example of a tailored content viewer from ci:grasp-2's city module: Data explorer for results on the global potential of local peri-urban food production [20], allowing to compare peri-urban agriculture and projected carrying capacity for a global set of cities (screenshot)

A further core user task for ci:grasp-2 relates to flexible and intuitive exploration of a growing and necessary diverse set of research results and datasets on a variety of themes and scales. Themes to be made available comprise the transitions of dietary patterns and associated GHG emissions [21], a set of city related studies, e.g. on urban livelihood vulnerability under rapid coastal city growth [22], on city heat island effects [23], or the influence of adaptation measures during extreme heat events [24], as well as results related to adaptation to sea-level rise of the agricultural sector in Indonesia [25], a human development framework for CO_2 reductions [26], Climate impacts on livelihood conditions [27] or an integrated transferable climate change vulnerability assessment [28]. To accommodate this high degree of content diversity, ci:grasp-2 refrains from a one-size-fits-all content exploration approach (which would have been desirable for the sake of simplicity, but would have come with the down-

[5] Inter-Sectoral Impact Model Intercomparison Project; http://www.isi-mip.org/

side of restricted filter, representation, and comparison options for individual content subsets). Instead, the application incorporates an extensible set of tailored content viewers to display specific thematic subsets, e.g., to explore content subsections of a city module (see Fig. 6 for an example). The viewers are built using common UI APIs for interactive maps and diagrams to provide common interaction metaphors. Where possible, consistent interaction metaphors are used across different viewers to increase usability and facilitate transfer of training.

The concept of embedding an extensible set of tailored viewers allows for flexible integration of available results from climate change and adaptation research projects in an ongoing manner. Furthermore, already existing explorative tools developed within other projects can be included as well; an example for the latter is CATHY, a web-based Climate Assessment Tool for Hyderabad [29].

4 Discussion and Conclusions

We have identified central user tasks as well as described the respective UI metaphors of two novel climate change adaptation related web-based information supply applications. It shows that such applications need to address a range of different user tasks, ranging from browsing a set of decision trees to identify adaptation related tasks and methods (MEDIATION Adaptation Platform), to navigation over a complex set of stimulus, impact, and adaptation information and the exploration of multi-dimensional spatio-temporal datasets (ci:grasp-2). A universal challenge in designing appropriate interactive systems to fulfil such user tasks is the recurrent need for balancing between interface flexibility and complexity.

Extending classical UI components by appropriate interactive graphical representations can aid the user in fulfilling such tasks. However, due to as yet missing UI standards in the climate change adaptation information context, careful testing during the development should be applied to ensure that selected interaction metaphors match users' mental models. Repeated interface improvements and adjustments during development based on obtained feedback are required and desirable. E.g., feedback gathered in an earlier user test during developing the MEDIATION Adaptation Platform showed that most users judged a graphical navigation through the decision trees attractive, but some got lost in the overall structure. This led to the inclusion of an additional overview diagram to indicate overall structure and current selected section (Fig. 1, top right). After the diagram was included, it showed that some test users tried to click the nodes in this overview diagram, assuming that it could also be used for navigation. As a consequence, a next iteration included this interaction feature so that the overview diagram now provides additional 'short cuts' to the main stages and the main decision trees of the Adaptation Pathfinder.

A specific challenge in designing appropriate information applications for scientific content originates from the often required co-development of scientific content and information deploy application, due to the established approach of addressing scientific tasks within research projects with a restricted runtime. These temporal restrictions and hence the need for advancing both content and application in parallel can lead to initial *chicken-and-egg* situations between UI experts and domain experts until the required mutual understanding is achieved. A further challenge in this context relates to sustaining and updating the content of information supply applications after the respective research projects' lifetime.

Appropriate interactive web-based applications have the potential to greatly facilitate access to current knowledge on climate change and adaptation. However, further systematic research is needed to better understand which interaction approaches work best in this context, as well as the development of rigorous approaches to measure and evaluate the effectiveness of the envisaged information transfer from the scientific supply side to the demand side.

Acknowledgements. This work has been supported by the project MEDIATION (Methodology for Effective Decision-making on Impacts and AdaptaTION) (FP7, grant agreement n°244012) and the project Methodeninventar zur Klimaanpassung (PN 10.9062.0).

References

1. Moser, S.: Communicating climate change: history, challenges, process and future directions. Wiley Interdisciplinary Reviews: Climate Change 1, 31–53 (2010)
2. Patt, A.: Communicating uncertainty to policy makers. In: Uncertainties in Environmental Modelling and Consequences for Policy Making. Springer, Heidelberg (2009)
3. Hofmann, M.E., Hinkel, J., Wrobel, M.: Classifying knowledge on climate change impacts, adaptation, and vulnerability in Europe for informing adaptation research and decision-making: a conceptual meta-analysis. Global Environmental Change 21, 1106–1116 (2011)
4. Carter, T., et al.: IPCC Technical Guidelines for Assessing Climate Change Impacts and Adaptations (1994)
5. Smit, B., Burton, I., Klein, R.J.T., Wandel, J.: An Anatomy of Adaptation to Climate Change and Variability. Climatic Change 45, 223–251 (2000)
6. Füssel, H.M.: Adaptation planning for climate change: concepts, assessment approaches, and key lessons. Sustainability Science 2, 265–275 (2007)
7. Berrang-Ford, L., Ford, J.D., Paterson, J.: Are we adapting to climate change? Global Environmental Change 21, 25–33 (2011)
8. Barr, R., Fankhauser, S., Hamilton, K.: Adaptation investments: a resource allocation framework. Mitig. Adapt. Strateg. Glob. Change 15, 843–858 (2010)
9. Shneiderman, B.: Designing the user interface: strategies for effective human-computer interaction. Addison-Wesley, Reading (1998)
10. Norman, D.: The psychology of everyday things. Basic books (1988)
11. Krug, S.: Don't make me think: A common sense approach to web usability. New Riders (2006)
12. Petrie, H., Power, C.: What do users really care about? In: Proceedings of the 2012 ACM Annual Conference on Human Factors in Computing Systems, CHI 2012, p. 2107. ACM Press, New York (2012)
13. Nivala, A.-M., Brewster, S., Sarjakoski, T.L.: Usability Evaluation of Web Mapping Sites. The Cartographic Journal 45(2), 129–138 (2008)

14. McKendry, J.E., Machlis, G.E.: Cartographic design and the quality of climate change maps. Climatic Change 95(1-2), 219–230 (2009)
15. Ishikawa, T., Barnston, A.G., Kastens, K.A., Louchouarn, P., Ropelewski, C.F.: Climate Forecast Maps as a Communication and Decision-Support Tool: An Empirical Test with Prospective Policy Makers. Cartography and Geographic Information Science 32(1), 3–16 (2005)
16. Wrobel, M., Costa, L., Lissner, T., Moneo Laín, M., Weiß, T., Kropp, J.P.: A review of user interface conventions in web applications for climate change information. In: International Congress on Environmental Modelling and Software, Ottawa (2010)
17. Wrobel, M., Costa, L.: Web based applications for climate change related information: a usability test. In: Seppelt, R., Voinov, A., Lange, S., Bankamp, D. (eds.) International Congress on Environmental Modelling and Software. Managing Resources of a Limited Planet, Sixth Biennial Meeting, Leipzig, Germany (2012)
18. Hinkel, J., Bisaro, A.: A diagnostic framework for problem-oriented adap-tation research. Regional Environmental Change, Special Issue: Approaches for Problem-oriented Adaptation Research (in preparation)
19. Migliavacca, M., Dosio, A., Kloster, S., Ward, D.S., Camia, A., Houborg, R., Houston Durrant, T., Khabarov, N., Krasovskii, A.A., San Miguel-Ayanz, J., Cescatti, A.: Modeling burned area in Europe with the Community Land Model, J. Geophys. Res. Biogeosci. 118, 265–279 (2013), doi:10.1002/jgrg.20026
20. Kriewald, S., García Cantú Ros, A., Sterzel, T., Pradhan, P., Kropp, J.P.: Feeding cities in the post-carbon age: can peri-urban agriculture reclaim a place? (in preparation)
21. Pradhan, P., Reusser, D.E., Kropp, J.P.: Embodied greenhouse gas emissions in diets. PloS One (accepted, 2013)
22. Sterzel, T., Lüdeke, M.K.B., Walther, C., Kok, M.J.T., Lucas, P., Janssen, P., Kropp, J.P.: A global typology of urban livelihood vulnerability under rapid coastal urbanization (in preparation)
23. Zhou, B., Rybski, D., Kropp, J.P.: The statistics of urban heat island intensity (in preparation)
24. Schubert, S., Grossman-Clarke, S.: The Influence of Green Areas and Roof Al-bedos on Air Temperatures during Extreme Heat Events in Berlin, Germany, Meteorologische Zeitschrift (in press)
25. Förster, H., Sterzel, T., Pape, C.A., Moneo-Laín, M., Niemeyer, I., Boer, R., Kropp, J.P.: Sea-level rise in Indonesia: On adaptation priorities in the agricultural sector. Regional Environmental Change (2011), doi: 10.1007/s10113-011-0226-9
26. Costa, L., Rybski, D., Kropp, J.P.: A Human Development Framework for CO2 Reductions. PLoS ONE 6(12), e29262 (2011, 2012), `http://www.plosone.org/article/info%3Adoi%2F10.1371%2Fjournal.pone.0029262`, doi:10.1371/journal.pone.0029262
27. Lissner, T.K., Reusser, D.E., Lakes, T., Kropp, J.P.: Measuring livelihood conditions in the context of climate change and sustainability (in preparation)
28. Holsten, A., Kropp, J.P.: An integrated and transferable climate change vulnerability assessment for regional application. Natural Hazards 64, 1977–1999 (2012), doi:10.1007/s11069-012-0147-z
29. Kit, O., Lüdeke, M.K.B., Reckien, D.: Defining the bull's eye: satellite imagery-assisted slum population assessment in Hyderabad/India. In: Urban Geography 2013 (2013), `http://www.tandfonline.com/doi/full/10.1080/02723638.2013.778665`

Integration of Macro and Micro Sustainability Issues: The Need for 'Engineering Accounting'

Joanne L. Tingey-Holyoak, Roger L. Burritt, and John D. Pisaniello

Centre for Accounting, Governance and Sustainability
University of South Australia, Adelaide, South Australia 5000, Australia
{joanne.tingey,roger.burritt,john.pisaniello}@unisa.edu.au

Abstract. The paper argues for the integration of engineering and accounting professionals through a transdisciplinary approach to on-farm water issues associated with farm dams. An integrated macro and micro solution is put forward in the context of poor farmer behaviour as spillways are blocked to pursue individual aims at the expense of other stakeholders, including downstream communities and the environment. Integrated engineering accounting is proposed as a solution through transdisciplinary thinking.

Keywords: integrated engineering accounting, transdisciplinarity, farm dams, spillway blocking.

1 Introduction

As accountancy professional bodies worldwide move to calls to expand the boundaries of the accounting discipline to incorporate environmental sustainability measures, there is evidence to suggest that there has been difficulty in integrating aspects of the environment into practice [8]. In the main this is because of the intricacies of the engineering aspects involved in accounting for elements of the environment. Ngwakee [8] suggests, in the context of sustainable development, accounting and engineering are like uncongenial twins (see earlier analogy between accounting and economics [2]) and should be recognised through the notion of 'engineering accounting'. Engineering accounting is portrayed to be a specialized field of accounting designed to handle the complex technical and socio-political accounting problems that climate change and carbon trading present, and which brings together macro (economy wide) and micro (company based) notions of accounting [21]. But this narrow setting is generalizable to all settings where engineering and accounting contribute to a transdisciplinary solution.

Accounting for carbon, water, biodiversity, electricity usage and other environmental sustainability measures all require engineering knowledge and without this, the accounting profession's involvement in the critical organisation-environment boundary is limited. Despite acknowledgement of the transdisciplinary nature of sustainability accounting, it is apparent that continuous reliance on engineers reduces the power of accountants to evolve in line with the changing demands of macro and micro stakeholders such as society, international agencies, clients of firms and professional bodies.

J. Hřebíček et al. (Eds.): ISESS 2013, IFIP AICT 413, pp. 500–507, 2013.

'Engineering accounting' would embody all of the aspects of any recognised profession, education and training for technical proficiency, independence of perspective, support for public interest as well as recognition of the importance of member needs and a system to measure, evaluate, record, interpret and report organizational sustainability information but not just for audit purposes as suggested by Ngwakwe [8]. The purpose is much broader and can be mainstreamed in traditional and environmental accounting practices providing sustainability information to all types of managers – production, product, divisional, marketing, environmental and top management [3]. The accounting and engineering professions face an opportunity to capture their presently unrecognised relevance in a joint initiative mainstreaming an integrated set of sustainable development activities common to the currently separate, siloed professions – accounting and engineering.

The behaviours of actors over water, a critical environmental aspect of sustainability and the suggested future of global tensions to rival those over oil [9] can be accounted for by a combination of engineering and accounting expertise and is considered here to illustrate the way in which 'engineering accounting' can help to understand the reasons for and the behaviours of, for example, farmers desperate to secure water for their farming activities. Engineering accountants would stand together as joint participants in the engineering and accounting professions. The code of ethics for professional accountants APES110 prescribes that accountants safeguard against threats to their ability to perform duties with the appropriate degree of professional competence by consulting with independent experts or obtaining additional training themselves [1]. This was evidenced with the recent release of the first Australian Water Accounting Standard [19] which raises interesting issues of the reliance on experts, specifically engineers, to provide the numbers to be used by accountants. The development of and training for 'engineering accounting' would reduce this reliance thereby creating specialised accountants equipped to measure, record and interpret organisational environmental sustainability information. Such accountants would be educated at the tertiary level and by the professional firms through specially designed university courses and professional development as well as certification designed to recognise skills acquired.

The paper addresses the bringing together of a team of accountants and engineers to study the possibility of specialised engineering accounting for one of the world's largest surface fresh water resources, farm dams. The research problem being addressed is how macro and micro sustainability solutions can be encouraged through engineering accounting using the example of the urgent need for solutions to the risks from inequitable water sharing and safety threats.

2 Accounting for Sustainability – Farm Dams

It is critically important to measure and account for fresh water resources, especially in areas with increasing frequency of not only droughts but also flash flooding [4], [18]. Not accounting for fresh water resources has far reaching consequences at the micro level for businesses and industries, and increasingly at the macro level for society, the economy and aquatic and land ecosystems [10]. For agriculture, dams are the

Fig. 1. An under-designed spillway[1]. SH = Max. Spillway Height, SW = Max. Spillway Width

[1] The spillway (located in South Australia) size is clearly too small for this 100 ML impoundment. The spillway is also badly maintained as the owner allows physical obstructions, including a walkway bridge and sandbags, to be present across the spillway restricting its potential capacity. Cost-effective spillway safety technology enables the existing dimensions of the spillway (i.e. Spillway Height, SH and Width, SW) to be reviewed for adequacy by 'non engineers' due to its simplicity. It can also be used in design mode to readily determine what spillway size is needed (i.e. by how much SH and/or SW must be increased) to ensure adequate flood safety for the dam whilst being able to accommodate both the farmer's minimum storage capacity requirements and the lowest cost earthworks. At the same time the technology is in line with modern engineering best practice [13].

"lifeblood" of most farming businesses as they provide stocks of essential water supply for irrigation and other farming activities. However, a problem exists with a lack of accounting for farm dams and their safety in that thousands of dam structures have failed and many more pose significant safety threats [11-14]. In China the Shimantan and Banquia dams failed in 1975 because of the cumulative failure of 60 smaller upstream dams, resulting in the deaths of 230,000 people [20]; in the United States, the 5m Evans and Lockwood dams, which held only 89ML and 39ML of water respectively, both collapsed in a cascade manner in 1989, devastating communities and killing 2 people [5]; and, in Quang Binh Province, Vietnam in 2010, 85,000 homes and businesses were flooded and at least 16 people died after water rose more than one metre over a large public dam following heavy flooding in the catchment which contained many small dams [16]. Dam failure for a farming business can mean: no longer being able to water valuable crops, the subsequent loss of those crops, serious inconvenience to farming activities, substantial cost to repair or replace the dam, and possible liability for downstream consequences [7], [14].

Furthermore in times of drought, farmers have an incentive to store water using inequitable and unsustainable tactics such as the blocking of spillways (see Figure 1) [14]. Thus, whilst farm dams can reduce farming risks associated with water shortages, during periods of low rainfall in the catchment, they also reduce water available for downstream users and the environment [15], especially when unfairly managed in this way. Hence, management of a farm dam to maintain its performance, structural integrity and safety are vital for the sustainability of farming business [14]. But, accounting for the structure and its stocks and flows is also important for the safety and sustainability of the downstream community and environment.

Unfortunately, many countries and regions do not have an integrated water accounting system for dams [14] that can effectively capture the linkages and interactions within or between macro and micro economic production and consumption, fresh water resource depletion and socio- and hydro-ecosystem degradation [6]. By bringing together engineers and accountants the research is finding that sustainability accounting provides a tool for identifying and making transparent the costs to business, the community and policy-makers associated with dam failure and inequitable water sharing.

3 Accounting Issues

Lessons from the past indicate that key issue for accounting for sustainable farm dams is the institutionally fragmented nature of actors who are operating, managing, regulating and overseeing the accounting for sustainable dam management [25] (Figure 2). These actors are often pursuing individual management goals because of sub-optimal local decision making rather than global optimal decisions for flood risks to communities in any catchment [22]. Changing climate conditions and forecasts of increases in extreme weather events mean that the narrow focus of actors will only increase [26]. However, before moves can be made to overcome institutionally divided approaches, a clear understanding of the accounting issues for sustainable farm dam management is required.

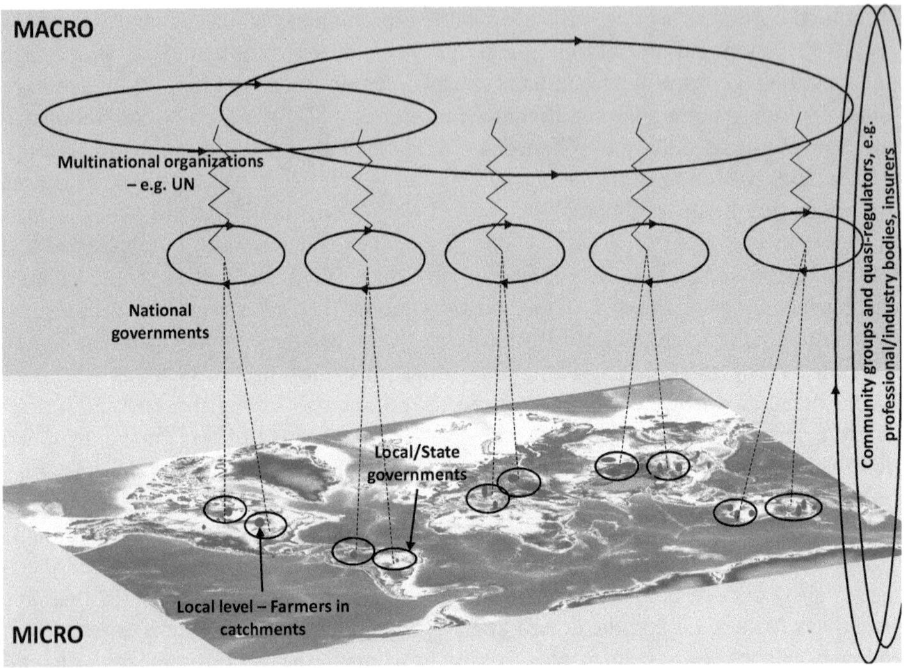

Fig. 2. Fragmented nature of institutional actors involved in accounting for sustainable farm dams

3.1 Micro Accounting Issues

Engineering Accounting for Direct, Indirect and Hidden Costs

The cost of sustainable farm dam management includes conventional direct and indirect costs such as equipment for maintaining a dam, or monitoring equipment to measure dam capacity. However there are also four types of costs which potentially remain hidden because neither engineering nor accounting systems separately identify them. Upfront costs are incurred prior to the operation of a dam and include costs related to the choice of site, site preparation, permits, engineering and installation. Regulatory costs of dam operations involve notification, reporting, monitoring and testing, modelling studies, remediation, planning, training, inspections, insurance, assurance or audit, pollution control, spill or failure response, taxes and fees. There are also voluntary costs of operations linked with community outreach and stakeholder engagement, monitoring and testing, training, assurance, reporting, insurance, recycling, habitat protection, landscaping, and financial support to nongovernment organisations. Back end costs relate to site survey, the decommissioning of dams, and post-closure care.

Contingent Liabilities

There are also non-certain contingent costs that can be associated with probabilistic calculations (expected value, variance, etc.). These contingent liabilities include the costs of remedying and compensating for dam failure and related fines and penalties for non-compliance, remediation for property damage or damage to natural resources, personal injury, legal expenses, relationships with downstream communities and relationships with regulators.

3.2 Macro Accounting Issues

Social and Environmental Costs

Image and relationship costs associated with dam management are tangible but the benefits that result from trying to affect the perceptions of different stakeholders such as downstream communities, non-government organisations and regulators are intangible. Reporting and assurance costs provide one example, community relations activities another, and costs of downstream protection activities such as developing an emergency action plan or tree planting a third. However, environmental costs are mostly ignored, with all but the most basic costs being brought to light because of the limited ability of accountants to undertake 'engineering accounting'.

3.3 State of the Art in Engineering Accounting for Sustainable Water Storage

Accounting for sustainable farm dam management can be possible, when complemented by the appropriate institutional response [25]. The technological engineering accounting tool developed by [13] based on best practice catchment analysis, calibration and modelling, flood hydrology and reservoir hydraulics, all in line with [23] provides an example of a possible mechanism to create a sustainability account of farm dams in a region. The accounting tool allows for measurement of the hydraulic response of any size of reservoir and spillway(s) relative to the hydrological flood response of the selected rural catchment in an 'hydrologically homogenous' region for the full range of design storm events up to the Probable Maximum Precipitation (i.e. 1 in 10,000,000 years Annual Exceedence Probability (AEP)). This accounting tool has been developed in South Australia, Victoria and New South Wales [24] in order to promote consistency and uniform standards in the sustainability of farm dams and practical application involves using a regionalised relationship based on simple hydrological/hydraulic variables, for predicting reservoir flood capability – known as Dam Crest Flood (DCF) (see [13]) and is cost-effective and simple to use. The tool would enable diverse actors to access reliable measures for monitoring and overseeing the sustainability of farm dams in a region.

4 Conclusion

This paper addresses the bringing together of a team of accountants and engineers to study the possibility of specialised engineering accounting for one of the world's largest surface fresh water resources, farm dams. By addressing the research problem of how macro and micro sustainability solutions can be encouraged through engineering accounting, the paper draws awareness to the challenge for researchers and professionals to traverse and span multiple disciplines for sustainability solutions, especially when faced with ever increasing weather related disasters. However, what is also highlighted is the need for an urgent future research agenda that includes gathering evidence from key macro and micro stakeholders through government, community and professional body perception, attitude and behaviour surveys.

Perhaps this is a timely opportunity to draw from the lessons of the past, such as those experienced by the accounting profession when faced with the challenge of fraud, who developed the specialised forensic accounting stream, or other professions coping with changes in the environment such as economists who created the ecological economics concentration. In addition, public and investor confidence in the amount of water traded, consumed, recovered and managed for environmental and other public benefit outcomes will be improved in conjunction with understanding of how fresh water resources are sourced, managed, shared and used. Importantly, new opportunities are presented: for the marketing of transdisciplinary professional services, for professional development, for educators of professional services graduates, for water resource managers and engineers requiring the basis of financial accounting for effective water accounting, and for revisiting the mechanism critical to addressing complex problems, integrated multi-level, multi-purpose macro and micro systems at a time when economics itself has been slipping from the main professional interests contained in engineering and accounting.

References

1. Accounting Professional and Ethical Standards Board (APESB): Compiled APES 110 Code of Ethics for Professional Accountants, APESB, Canberra (2011), `http://www.apesb.org.au/uploads/july-2012-bm/attachment-5a-pgs-1-4-47-48-133-135-of-compiled-apes-110-code-of-ethics-for-professional-accountants-may-2012-marked-up.pdf`
2. Boulding, K.E.: Economics and Accounting: The Uncongenial Twins. In: Baxter, W.T., Davisson, S. (eds.) Studies in Accounting Theory, pp. 44–55. Richard D. Irwin, Inc., Illinois (1962)
3. Burritt, R., Hahn, T., Schaltegger, S.: Towards a Comprehensive Framework for Environmental Management Accounting. Links Between Business Actors and Environmental Management Accounting Tools. Australian Accounting Review 12, 39–50 (2002)
4. CSIRO and ABOM: Technical Report, 2007. Climate Change in Australia, CSIRO and The Australian Bureau of Meteorology, Canberra (2007)
5. Graham, W.J.: A Procedure for Estimating Loss of Life Caused by Dam Failure, Office, Report DSO-99-06, United States Dept of Interior, Bureau of Reclamation, Dam Safety, Denver (1999)
6. Guan, D., Hubacek, K.: A New and Integrated Hydro-Economic Accounting and Analytical Framework for Water Resources: A Case Study for North China. Journal of Environmental Management 88(4), 1300–1313 (2008)
7. Lewis, B.: Farm Dams: Planning, Construction and Maintenance. Land Links Press, Victoria (2002)
8. Ngwakwe, C.C.: Rethinking the Accounting Stance on Sustainable Development. Sustainable Development 20, 28–41 (2012)
9. Karlsson-Vinkhuyzen, S.I., Jollands, N., Staudt, L.: Global Governance for Sustainable Energy: The Contribution of a Global Public Goods Approach. Ecological Economics 83, 11–18 (2012)
10. Kingsford, R.T.: Ecological Impacts of Dams, Water Diversions and River Management on Floodplain Wetlands in Australia. Austral Ecology 25, 109–127 (2000)

11. Pisaniello, J.D.: Attitudes and Policy Responses to Australian Farm Dam Safety Threats: Comparative Lessons for Water Resources Managers. International Journal of Water Resources Development 26(3), 381–402 (2010)
12. Pisaniello, J.D., Burritt, R.L., Tingey-Holyoak, J.: Dam Safety Management for Sustainable Farming Businesses and Catchments. Agricultural Water Management 98, 507–515 (2011)
13. Pisaniello, J.D., Tingey-Holyoak, J., Burritt, R.L.: Appropriate Small Dam Management for Minimising Catchment-wide Safety Threats: International Benchmarked Guidelines and Demonstrative Cases Studies from Australia. Water Resources Research 48 (2012)
14. Pisaniello, J.D., Tingey-Holyoak, J., Burritt, R.L.: Dual-Extreme Cumulative Impacts and Threats In Agricultural Catchments: The Need for Effective Integrated Policy. Agricultural Water Management 118, 103–112 (2013)
15. Schreider, S.Y., Jakeman, A.J., Letcher, R.A., Nathan, R.J., Neal, B.P., Beavis, S.G.: Detecting Changes in Streamflow Response to Changes in Non-Climatic Catchment Conditions: Farm Dam Development in the Murray–Darling Basin, Australia. Journal of Hydrology 262(1-4), 84–98 (2002)
16. Thanh Nien News: Central flood exacerbated by dam failure. Thanh Nien News (May 10, 2010), `http://www.thanhniennews.com/2010/Pages/20101005103915.aspx` (viewed November 21, 2012)
17. Tingey-Holyoak, J., Burritt, R.: Transdisciplinary Nature of Accounting: A Pathway Towards the Sustainable Future of the Profession. In: Evans, E., Burritt, R., Guthrie, J. (eds.) Emerging Pathways for the Next Generation of Accountants Institute of Chartered Accountants in Australia and Centre for Accounting Governance and Sustainability, Sydney and Adelaide (2012)
18. UN IPCC: Summary for policy makers. In: Qin, S.D., Manning, M., Chen, Z., Marquis, M., Avery, K.B., Tignor, M., Miller, H. L. (eds.), Climate Change 2007: The Physical Science Basis. Contribution of Working Group 1 to the Fourth Assessment Report of the United Nations Intergovernmental Panel on Climate Change. Solomon. Cambridge University Press, Cambridge (2007)
19. Water Accounting Standards Board (WASB): Australian Water Accounting Standard 1, WASB, Canberra (2012), `http://www.bom.gov.au/water/standards/wasb/wasbawas.shtml`
20. Yi, S., Dai, Q.: The World's Most Catastrophic Dam Failures: The Collapse Of The Banqiao And Shimantan Dams. In: Thibodeau, J.G., Williams, P.B. (eds.) The River Dragon Has Come! The Three Gorges Dam and the Fate of China's Yangtze River and Its People, ch. 3. M. E. Sharpe, Armonk (1998)
21. Yu, S.C.L.: Microaccounting and Macroaccounting. The Accounting Review 41(1), 8–20 (1966)
22. De Bruijne, M., Van Eeten, M.: Systems That Should Have Failed: Critical Infrastructure Protection in an Institutionally Fragmented Environment. Journal of Contingencies and Crisis Management 15(1), 18–29 (2007)
23. IEAust: Australian Rainfall and Runoff - A Guide to Flood Estimation, 1 & 2. Institution of Engineers, Australia, Barton, ACT (1999)
24. Pisaniello, J.D., McKay, J.: A Tool to Aid Emergency Managers and Communities in Appraising Private Dam Safety and Policy. Journal of Disaster Studies, Policy and Management 31(2), 176–200 (2007)
25. Tingey-Holyoak, J., Pisaniello, J.D., Linnenluecke, M.: Resilience Policy for Flood Risk Transferred from Private to Public Dams: Insurance and Accounting Issues. Journal of the Asia-Pacific Centre for Environmental Accountability 18(4), 222–250 (2012)

Benefits of Environmental Management Systems in the Czech Republic

Jaroslava Hyršlová[1], Miroslav Hájek[2], and Gabriela Rajdlová[2]

[1] University of Economics and Management, Prague, Czech Republic
jaroslava.hyrslova@vsem.cz
[2] Czech University of Life Sciences Prague, Czech Republic
hajekmir@volny.cz, g.rajdlova@seznam.cz

Abstract. Environmental Management Systems (EMS) are important voluntary tools used by the business sphere and other organisations in the world, as well as in the Czech Republic, for the management of the impacts of their activities, products and services on the environment. The article presents EMS as a tool for the management of the environmental aspects and impacts of organisations, depicts the development of their implementation into the practice deployed by organisations around the world and in the Czech Republic in the last 15 years and highlights the systems' benefits for the society and for the organisations themselves.

Keywords: Environment, environmental performance, environmental management, environmental management systems.

1 Introduction

A company's responsible approach to environment requires many changes. These have to include not only specific rectifying measures, usually of technical and/or technological nature, but primarily the entire management system in the company. Wider proactive corporate attempts at the integration of environment protection and sustainable development aspects into management systems were one of the most important factors of the environmental politics in the early 1990s. EMS have been formed and gained wide support from the corporate sphere. The aim of this article is to present EMS as a tool for the management of environmental aspects and impacts, to show the development of the implementation of EMS over the last 15 years both globally and in the Czech Republic, and most importantly to highlight the benefits of the system for companies.

2 Implementation of EMS in the World and in CR

EMS are being implemented around the world and in the Czech Republic both according to ISO 14 001 and according to EMAS.

Figure 1 depicts development of the no. of certified organisations in the world and in Europe. As of the end of 2011, a total of 267,457 organisations held EMS

J. Hřebíček et al. (Eds.): ISESS 2013, IFIP AICT 413, pp. 508–516, 2013.

certificates compliant with ISO 14 001; this is 19x more than at the end of 1999. The highest increase in the number of certified organisations was registered in 2009 (up 34,400, i.e. 18 % on the total number of certified organisations registered as of December 31, 2008). In Europe, the certificate was held by a total of 106,700 organisations as of the end of 2011 (in 50 countries). The highest increase in the number of certified organisations in Europe was registered in 2010 (13,889 organisations). Europe is thus the no. 2 continent in the number of certificate holders.

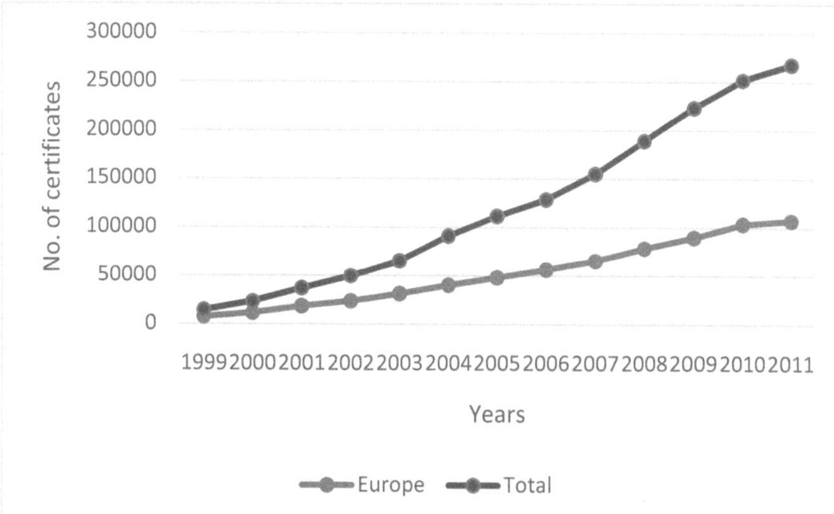

Fig. 1. Development of the no. of certified organisations in the world and in Europe in 1999-2011 (Source: [1])

In CR, the certificate was held by a total of 4,451 organisations as of the end of 2011. CR placed first among EU member states with comparable population. The highest increase in the no. of certified organisations was seen in 2010 (up 1,945, or 41 %).

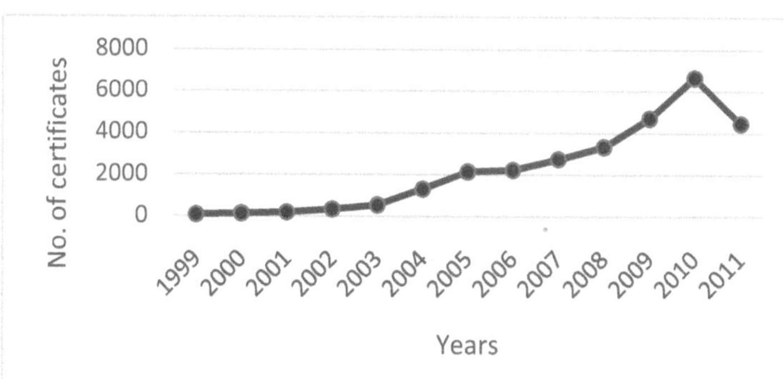

Fig. 2. Development of the no. of certified organisations in the Czech Republic in 1999 – 2011 (Source: [1])

Massive growth was registered also in 2009 (the no. of certified organisations increased by 1,366, i.e. 41 %) and in 2005 (up 834 organisations, i.e. 165 %). Figure 2 depicts the development of the no. of certified organisations in the CR.

There were totally 4,581 organisations of various sizes and activity fields registered with the EMAS on the territory of the EU as of June 1, 2013. Italy (1,652 registrations) has the highest number of organisations registered with the EMAS, followed by Germany (1,283 registrations) and Spain (1,286 registrations) [2]. Figure 3 depicts development of the no. of registered organisations in the CR.

The previous text indicates that the number of certified and registered organisations has been growing significantly since 1999. The organisations clearly prefer to use systems built according to the internationally valid norm, ISO 14 001. The no. of certified organisations in the world is growing faster than in Europe. The Czech Republic is among the countries with a high number of certified organisations.

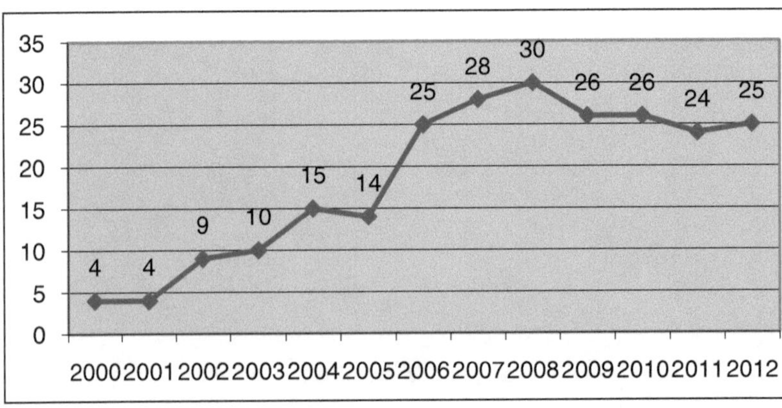

Fig. 3. Development of the no. of registered organisations in the Czech Republic in 2000 – 2012 (Source: [3])

3 Benefits of the Implementation of EMS in the Czech Republic

The most recent extensive survey on the benefits of environmental management systems for the business sphere in the Czech Republic was carried out by the Czech Environmental Management Centre and the Czech Environmental Information Agency in 2005[1]. The survey used written questioning to find out primary information. Totally 1,265 organisations were polled, mostly companies that had EMS implemented at the time of the research. Filled questionnaires were returned by 224 companies (return rate of 17.7%); out of this total, 222 organizations had implemented EMS. The reviewed

[1] The survey was carried out within the solution of research and development project no. VaV – 1C/4/13/04 (Application of Environmental Accounting on Macroeconomic and Microeconomic Levels). The project was financed by the Ministry of Environment. The co-author (J. Hyršlová) was the project's co-researcher.

sample consisted primarily of large and medium companies[2] with turnover exceeding CZK 100 mln p.a. Most strongly represented were organizations from the manufacturing industry (52 %) and construction industry (25 %). The reviewed companies had the system implemented predominantly acc. to ISO 14001 (of the total of 222 companies, 91 % only held certificates acc. to ISO 14001, 6 % held ISO 14001 certificates and simultaneously were registered with the EMAS program, while 3 % had the system implemented, but not certified or registered). The importance of management systems for respondents was proven by the fact that 97 % of the respondents had implemented also the ISO 9001 quality management system. Companies from both manufacturing industry and construction industry were registered in the EMAS program. The registered organizations ranged from large, medium and small enterprises to micro-companies.

The research resulted into the following basic findings [4]:

- The representatives of 92 % of companies (from the total reviewed set) have confirmed that the organisation's management is dealing with environment protection issues. The share of respondents confirming that management is paying attention to the environmental performance regularly (the environmental performance is monitored and rated in regular intervals) grew with the companies' size. Employees were involved in the corporate environmental performance improvement process in all companies.
- 76 % of companies carried out in the year preceding the survey (2004) measures within EMS that led to significant reduction of the consequences of corporate activities, products and/or services for the environment. This means that the companies were improving their environmental performance. According to the respondents, the implemented measures brought benefits also in other areas. 16 % of the respondents have noted social benefits of these measures (primarily the benefit of improved working environment). Almost 14 % of respondents have confirmed the economic benefits of the implemented environment protection measures, 13 % of respondents perceived positive response from the public, business partners or public administration.

The representatives of companies of individual size categories (see Table 1) and individual sectors perceived the benefits of EMS differently.

The research has confirmed that companies viewed EMS as a tool for improving their environmental performance; large and medium enterprises and companies from the manufacturing industry perceived also the system's economic benefits. Half of the respondents pointed to the system's benefits for safety and health protection at work and improvement of working environment. Smaller companies that implemented EMS have not perceived the aforementioned benefits so unambiguously. They tended to point towards the importance of EMS for the improvement of relations with the public, business partners and state authorities (this benefit was most frequently stated also by representatives of the construction industry) and its benefit for higher knowledge potential.

---

[2] Classification of companies: micro-company 0 – 9 employees; small company 10 – 49 employees; medium company 50 – 249 employees; large company 249+ employees.

Table 1. EMS' benefits (acc. to company size) (Source: [4])

Type of benefit	Relative rate (% of respondents from the total no. of respondents in the given category who have confirmed the benefit)				
	Reviewed set (n=222)	Micro-firms	Small firms	Medium firms	Large firms
Improved environmental performance	74	45	62	72	88
Economic benefits	59	27	55	58	68
Improved relations	54	73	55	49	52
Benefits in social area	51	55	51	57	45
Increased knowledge	41	73	36	40	38
Provision of information	27	55	26	30	19

Another survey aimed to identify EMS' benefits for organisations was carried out in the Czech Republic in 2009. The survey used questionnaires and interviews. 3Totally 35 organisations that were registered with the EMAS program at the time of the research were polled. At the time when the research was carried out, mainly large and medium enterprises were registered with the EMAS program (63 % of the companies fell into the category of large enterprises, 23 % into the category of medium enterprises); no micro-company was registered with EMAS. 63 % of the registered companies were generating annual turnovers of over CZK 1 bln at the time of research. The no. of registered organisations increased most significantly in 2005 and 2006; the Czech Republic has thus received the European EMAS award for high increase in the no. of registered organisations. The increase in the no. of registered organisations may be attributed primarily to the changes to the building act; registration with the EMAS program became one of the mandatory criteria for participation in tenders. Totally 31 out of 35 polled organisations participated in the survey (return rate of 88.6 %).

The research brought the following basic results [5]:

- 28 respondents (90 %) were using an environmental management system that was simultaneously certified acc. to ISO 14001.
- More than half of the respondents have stated continuing the implementation of management systems into practice, efforts to boost the company's competitiveness and improvement of environmental performance as the main reason for registering the system.
- The systems' benefits, as perceived by organisations participating in the research, are stated in Table 2.

[3] The survey was carried out by the Institute of Chemical Technology, Prague. The article's authors use collected data and evaluates them in the context of this article's goal.

Table 2. Systems' benefits (Source: [5]

Systems' benefits	Reviewed set (n=31 organisations)	
	Absolute rate	Relative rate
Improvement of environmental performance (reduction of negative impacts)	20	64 %
Improved image of organisation	17	55 %
Compliance with legislative norms on environment protection	10	32 %
Management system improvement (links to ISO 9000, ISO 14001)	9	29 %
Increased competitiveness	3	10 %
Improved supplier-customer relations	3	10 %
Reduced costs of raw materials and energies	3	10 %
Better negotiation with banks and insurers	2	6 %

More than half of the respondents viewed the system's contribution to the improvement of the organisation's environmental performance and improvement of image as the main benefits. According to the opinion of one third of respondents, the system contributes towards compliance with legislative norms on environment protection and generally improves the quality of the organisation's management system.

- 8 respondents (26 %) described the system as an important tool for environment protection; 17 respondents (55 %) regarded the system as a formal instrument that brought order into the organisation. 6 respondents (19 %) have stated that EMAS failed to meet their expectations in all sectors that motivated them to implement the system. Two respondents have confirmed that their organisation does not plan to maintain EMAS for this reason.

Following up on the research carried out in 2009, the corresponding survey was held also in 2013. At the time of the research, 24 organisations were registered in the EMAS program at the time of research. Large companies dominated among registered organisations (12 organisations, i.e. 50 %), followed by medium companies (9 organisations, i.e. 37.5 %); no micro-company has registered with the EMAS program yet. Two municipal authorities and one regional authority have registered the system. All registered organisations (except for municipalities and the regional authority) generated turnover of over CZK 1 bln last year. Half of the registered organisations are doing business in construction industry. All organisations were asked to take part in a survey based on questionnaire enquiries. Filled questionnaires were returned by 21 organisations (return rate of 87.5 %). The survey resulted into the following findings:

- 13 respondents (62 %) have confirmed that the company's environmental performance influences the success of their business also under current conditions and in the current business environment. 4 respondents (19 %) have stated that the environmental performance is not unambiguously directly linked to the success of their business.
- All organisations participating in the research (with the exception of the regional authority) have the environmental management system certified also acc. to the ISO 14 001 norm. The evaluated companies have first implemented the system in compliance with the requirements of the ISO 14 001 norm and then gained the EMAS registration (the evaluated companies gained the first certificates already in 1998, the highest no. of certificates was awarded in 2005; the companies gained first registrations in 2004, the highest increase in the no. of registered companies was reported in 2005 and 2006).
- The respondents have confirmed the following main reasons for registering with the EMAS program: efforts at raising (or maintaining) the company's competitiveness (52 % of respondents) and improvement of image and raising of the organisation's credibility (43 % of respondents).
- The research focused primarily on the systems' main benefits. The overall results are summarized in Table 3.

Table 3. Systems' benefits

Systems' benefits	Reviewed set (n=21 organisations)	
	Absolute rate	Relative rate
Improvement of environ. performance	16	76 %
Compliance with legislative norms	9	43 %
Increased competitiveness	8	38 %
Improved image of organisation	5	24 %
Better preparedness for accidents	5	24 %
Creating environmental awareness in employees	5	24 %
Overview of operating documentation	4	19 %
Improved communication	4	19 %
Reduced costs of raw materials and energies	3	14 %
Improved supplier-customer relations	2	9 %
Better intra-company organisation and management	2	9 %

The respondents have stated the improvement of their corporate environmental performance (76 %) as the main benefit of the system. According to the respondents' opinion, the system also contributes to the compliance with legislative norms on environment protection and the company's competitiveness. 3 respondents have described the system as a solely formal matter; they think that the system has no direct link to the company's environmental performance.

- 8 respondents (38 %) view EMS as an important tool for environment protection; 9 respondents (43 %) perceive the system as a formal instrument contributing to order in the organisation. 12 respondents (57 %) think that the systems' benefits exceeds the implementation and maintenance costs. All respondents have confirmed that they plan to continue maintaining and recertifying the systems acc. to ISO 14 001. 3 respondents are analysing whether they will register the system with EMAS again. The main reason for not registering the system with the EMAS program for them is the high registration costs.

4 Conclusions

The benefits of EMS may generally be perceived on environmental, economic, social and organisational level (this classification of benefits is stated e.g. in [6]).

In early 21st century the companies doing business in the territory of the Czech Republic that had implemented the environmental management system perceived the system as an instrument for the management of the environmental impacts of corporate activities, products and services. Some of the companies (predominantly large and medium enterprises and firms from the manufacturing industry) were also aware of the systems' economic benefits. Part of the companies from the manufacturing industry pointed also to the systems' benefits for safety and health protection at work and improvement of the working environment. The companies clearly prefer to implement the systems in line with the requirements of ISO 14 001 and want to maintain the implemented and certified systems in the long run.

From mid-1990s onwards, some companies, primarily large and medium enterprises, began to register their environmental management systems also with the EMAS program. Registration with the program was a logical extension for the system that complies with the requirements of the ISO 14001 norm. The registration process was joined primarily by large companies from the construction industry, as registration with the EMAS program was an important step towards remaining competitive in public tenders.

The reasons for registration with the EMAS program in the monitored group changed over the monitored period (2009-2013). In 2009, the companies stated continuing with the implementation of management systems and efforts at raising (or maintaining) competitiveness as the main reason. The foreseen benefits for the company's environmental performance (environmental benefits) dominated in the overall listing of reasons. In 2013, the companies accent primarily the economic reasons for registration with the EMAS program – efforts at maintaining or boosting competitiveness.

In 2009, the monitored group of companies perceived the improvement of the organisation's environmental performance and image as the main benefits of registration with the EMAS program. Environmental benefits dominated the overall list of benefits for companies; the organisations were also aware of the impact of the registration on their higher credibility. The survey held in 2013 confirmed the systems' environmental benefits, i.e. their importance for the management and improvement of the environmental performance and compliance with legislative norms on environment protection. Companies continue to perceive the systems' economic benefits, i.e. their contribution towards the maintenance and/or boosting of competitiveness. To a lesser extent, the enterprises note also social benefits (creating environmental awareness in employees) and organisational benefits (overview of operating norms and documentations).

The total no. of organisations registering with the EMAS program has dropped in the monitored period (2009-2013). Future will show whether new organisations will be interested in entering the registration process and whether existing organisations will be interested in renewing their registrations.

To achieve all benefits of EMS, it is necessary to ensure that EMS is not just a formal tool; environmental aspects have to be implemented into the entire corporate management system and reported to stakeholders (see e.g. [7]). The goal is to secure the achievement of corporate goals for economic performance, environmental performance as well as social aspects of doing business.

References

1. The ISO Survey of Management System Standard Certifications (1999-2011), http://www.iso.org
2. EMAS Register, http://www.emas-register.eu
3. Ministry of the Environment of the Czech Republic, http://www.mzp.cz
4. Hyršlová, J., Mísařová, P.: Udržitelný rozvoj a systémy environmentálního managementu v ČR. Planeta, XV (2), 3–6 (2007)
5. Dobrý, J.: Analýza významu a postavení EMAS v České republice a Evropské unii jako nástroje ochrany životního prostředí. VŠCHT, Praha (2010)
6. Kreuz, J., Vojáček, O.: Firma a životní prostředí. Oeconomica, Praha (2007)
7. Hodinka, M., Štencl, M., Hřebíček, J., Trenz, O.: Current Trends of Corporate Performance Reporting Tools and Methodology Design of Multifactor Measurement of Company Overall Performance. Acta Univ.Agric. et Silvic. 60(2), 85–90 (2012)

The Comparison of Selected Risk Management Methods for Project Management

Vladěna Obrová and Lenka Smolíková

Brno University of Technology, Faculty of Business and Management,
Kolejni 4, 612 00 Brno, Czech Republic
{obrova,smolikova}@fbm.vutbr.cz

Abstract. Project management is a set of validated and described procedures that comprehensively solve the implementation and management of defined activities that relate to a specific project. In the Czech Republic, the issue of risk management in projects often neglected and began to be more used to the ESF projects where is the risk management required. There are used most often for risk analysis 3 methods - sensitivity analysis, RIPRAN method, scoring method. The following article focuses on risk management methods that are not used in common practice immediately concern to project risk management. The benefit of this paper is the connection between methods used in project management and the methods of risk analysis. The result is an overview of the main advantages and disadvantages of these methods and the possibility of their use in different phases of the project, taking into account the time and cost effectiveness.

Keywords: project management, risk management, ETA, HRA, FMEA, comparison.

1 Introduction

The purpose of project management is to ensure effective management of these activities so as to bring the expected results and benefits in the shortest time. Project management is thus the application of knowledge, skills, activities, tools and techniques to project so that the project will satisfy the requirements imposed on it and achieve their goals. Risk management is also an integral part of project management, which should be incorporated into.

Risks are associated with all the activities that may occur in the individual life cycle of the project. The paper compares existing methods and their applicability and appropriateness in the management of the project, taking into account the time and cost effectiveness of each method.

The project includes already defined analytical methods that can be further used in risk management. One of these methods is the method of network analysis (CPM, PERT, or other), which is used to observe the total duration of the project and the creation of a critical path. To identify activities for the method of network analysis, there is used the method WBS (Work Breakdown Structure). This method is used for

J. Hřebíček et al. (Eds.): ISESS 2013, IFIP AICT 413, pp. 517–524, 2013.

a detailed segmentation of each activity in the project so that it can be assigned responsibility for these activities. In further analysis of risk, it is appropriate to focus on these critical activities, which may result in a delay that would result in delay of the project as a whole. It is also suitable to use the responsibility matrix to assign and display of responsibilities between team members in projects, processes or parts thereof. There are more variations of the matrixes; the most common are RACI matrix and its extension RASCI. These 3 methods above are used as a basis for risk analysis.

Among the analyzed selected risk management methods belong ETA and FMEA. ETA method is based on event tree analysis - ETA (Event Tree Analysis). ETA is used as a causal analysis technique, which is used to evaluate the process and the possible events leading to the accident. Another method is to analyze the possible defects and their consequences - FMEA (Failure Mode and Effect Analysis). It is an analytical technique, which aims to identify potential sites of defects or faults in systems. The method FMEA is a systematic identification of all possible defects in the product or process and its implications, identifying steps to prevent, reduce or limit the causes of these defects and document the entire process.

The paper deals with risk management issues in project management. In project management, risk management is a systematic process that aims to identify and manage risk, in order to act on its appearance (eliminating, minimizing or controlling it), by implementing systems and procedures to identify, analyze, evaluate and address the risks inherent to any project. [1] [2]

Risk management is a structured approach for the identification, assessment, and prioritization of risks followed by planning of resources to minimize, monitor, and control the probability and impact of undesirable events. [3]

Risk management is an essential part of every project; no project is free from risks. At any stage of a life cycle, a project is plagued with various risks due to the complex and dynamic nature.[4]

Risk management must contribute to define the different project objectives, improve project control, increase the chances of project success, improve communication between project participants and facilitate decision-making and prioritise actions. [5] [6] [7]

Despite this fact, risk management is often left aside in project management practice, or is not given due attention. Therefore, risk management can help project managers to anticipate delays that cause projects not to be delivered on time. [8]

The project team may encounter risks throughout the whole life cycle of the project, i.e. in all of its phases.

Smith described principles and guidelines for effective risk management and emphasized the importance of active risk management for accelerating projects and improving their success rates. [9] Raz et al. performed an empirical study and reported that risk management practice is more applicable for higher-risk projects and appears to be related to project success.[10]

The paper focuses on the comparison of three methods: ETA, HRA, and FMEA; all three are methods that are commonly used for risk management in a company, but have not been incorporated into the field of project management. The first chapter describes and explains these methods, afterwards they are compared and their advantages and disadvantages are reviewed.

2 Theoretical Description of ETA, HRA and FMEA Methods

2.1 ETA

The first selected method is ETA (Event Tree Analysis). It is a causal analysis technique that is used to evaluate the process and its events leading to a possible accident. ETA method does not take into account the causes of an undesirable event, but is assessing further development of the events, thus providing an overview of the probability of possible resulting events. The principle of this method is to observe the events, their sequence and relationships between them, which not only lead to a malfunction, but also to a failure of the system. Within the framework of the method, each activity is analyzed and its sequences are graphically displayed using the event tree model, where these activities can be further evaluated quantitatively. A logical diagram describes the logical development of the scenario, starting from the initiating event to possible serious consequences. It is an inductive, systematic approach developing the initiating event through successive, logical steps (possible sequences). The results of ETA analysis are different accident scenarios. The analysis may be performed by one analyst, but it is preferable to employ more analysts. Analysts can use the results for recommendations on reducing the likelihood and/or consequences of potential failures. The method is used for identification and analysis of the system, project and process weaknesses. The result is a set of recommendations to reduce the likelihood of an accident and to mitigate its consequences. [11]

2.2 HRA

Another method used in identifying risks within the company is HRA (Human resource analysis). This method is used to identify potential risks resulting from human factor that is present in any activity occurring not only in the company, but also in the project team. It is therefore appropriate to analyze the human component in project planning and its implementation to allow not only for current use of the results of the analysis, but also for their future application, thus avoiding the risks associated with human factor failure. It is a systematic evaluation of factors affecting the work of team members including technicians, maintenance engineers and other employees of the company. The aim is to identify potential human errors, their causes and consequences. HRA explores the relationships between certain key areas. The interviews focus on the following areas:

- communication and information;
- duties;
- internal environment;
- education;
- system of rewards;
- work conditions;
- company management;
- total satisfaction with the work.

The method is usually used in conjunction with other methods that allow quantification of results. If we apply the method using event trees or fault trees, it is possible to graphically illustrate the sequence of human failure and its impact on the result of the activity, or project. The method should be employed by two analysts as a minimum that must be familiar with the interviewing technique; it is a very demanding method requiring experience on the part of the analyst in order to interpret the interviews in a proper manner.[12]

2.3 FMEA

Still another method is FMEA (Failure Mode and Effect Analysis). Again, it is an analysis technique that is used to identify areas where a failure or another malfunction of the system might occur. The principle of this method is based on the quantification of the frequency of failures, their severity and ease of their detection. Due to its versatility, it can be used in a number of areas, in particular in the area of risk management and quality management, or security management. The method creates a table of causes of failures and their impacts on the system, company or project. FMEA identifies simple failures that can significantly contribute to an accident, but cannot be included in an exhaustive list of failures. It is easily applicable when a process change or modification takes place, or if a new process is introduced. It can be conducted by one analyst, but should be reviewed by another one. It also includes an estimate of the worst case consequences. It is usually documented in a tabular form including recommendations for improving safety. It is a preventive method which allows for timely identification of possible failures, errors or defects that may affect the operation of the system or the resulting quality or safety. This method recognizes risks at an early stage of planning, thus reducing the level of risk, and, therefore, it is saving time and providing a possibility to invest the time into other areas. The method requires a lot of experience of the team with the analyzed system. An extended version of this method called FMECA (Failure Mode and Effect Critical Analysis) is also used. [13][14]

3 Risk Management Process in Projects

Risk management in a company should be implemented through the following phases: risk identification, risk analysis, risk assessment, risk treatment and risk monitoring. [15] [16]

3.1 Risk Identification

In this phase, hazards that may threaten the project are identified; we are trying to list and describe these hazards as accurately as possible. It is unfeasible to compile a complete list of all risks threatening the project; it is necessary to identify significant risks that can adversely affect the success of the project. At this stage it is advisable to use one of the creativity techniques, such as brainstorming or the Crawford Slip method.

3.2 Risk Analysis

This phase is based on the compiled list of risks, and includes an effort to evaluate the risks: the likelihood of occurrence of a certain hazard, and to estimate the adverse impact on the project (incurred financial damages). Risk analysis can be quantitative (the value of probability and the value of loss using a straight numeric value is determined), or qualitative (used to determine the probability and the loss of verbal value). Finally, the risk value is calculated, again using quantitative and qualitative approach.

3.3 Risk Assessment

The purpose of this step is to make a decision which risks should be treated, which are to be omitted, or, conversely, which are unacceptable. It can be based on the Pareto principle, i.e. the 80/20 rule. Thus, 20% of major risks should be treated very carefully; even using the majority of funds for risk treatment, and the remaining funds should be kept as a reserve.

3.4 Risk Treatment

If the risk value is determined and we have decided to treat the risk in a way, we should consider how we will respond to the risk. The aim of this phase is to reduce the total value of all risks to such a level so that the project is feasible with high probability. A higher value of risk should be treated with appropriate measures reducing the risk value. The following includes typical measures that are most commonly used:

- provide insurance against the adverse event;
- eliminate the risk by finding an alternate solution that does not contain a hazardous event;
- create a reserve;
- create a backup Plan B.

3.5 Risk Monitoring

Once a risk analysis has been completed, and we continue in implementing the project, it is necessary to constantly monitor all risks. Risk monitoring is the most frequent issue on the agenda of project meetings.

The phases of risk identification, risk analysis, risk assessment and risk treatment are carried out within the pre-project (or planning) stage, while risk monitoring is done within the project stage (which is also referred to as the implementation phase); the post-project stage includes evaluation of the whole risk management process in the project.

4 Comparison of Methods

The following tables show an overview of the main advantages and disadvantages of using the methods in project management, as well as a comparison of their time-consuming nature and costs, and last but not least, the phase of the project in which it is appropriate to use the methods, and types of projects for which they are best suited.

Table 1. Advantages and Disadvantages

Method	Advantages	Disadvantages
ETA	1.The ability to take into account the time-sequence, dependency or domino effect factors; 2.Well-arranged overview of consecutive actions; 3.Possibility to continue working and quantify the results of the analysis.	1.An event tree must be built for each event.; 2.Conditional probabilities dependent on previous events in the path → omission of the dependence or its incorrect interpretation.
FMEA	1.Objective evaluation of the design, including functional requirements and alternative suggestions; 2.Increased probability that a failure and its consequences, if any, will be identified in the early stage of the project design or its development; 3.Improvement in the quality of manufactured products, services or processes 4.Development of a list of potential failures (defects) that are listed according to the value of their impact on the project; 5.Identification of additional information to support the development, design and planning; 6.Significant psychological effect, strengthening the involvement of employees; 7.Improving communication between the employees and different departments of the organization.	1.Complexity; 2.Toilsomeness of the task of mapping large amounts of information about the system or project, such as technology, function, operation, range, etc.; 3.Human error.
HRA	1.Elimination of human errors prior to the beginning of the project or during its implementation 2.An overview of the skills and knowledge of team members during the project or when building the team; 3.Measures against discontent of team members and their erroneous performance caused by a lack of communication between the management and other industries.	1.Time required for conducting the interviews with all team members 2.In the case of a bad choice of the analyst, a human error may occur even after completing the analysis; 3.Reluctance of employees to cooperate during the interviews, or intentionally biased information provided about themselves, leading to misinterpretation of the outcome of the interview.

Table 2. Time and Costs

Method	Time	Costs
ETA	1. If the necessary software is purchased, the method requires minimum efforts; 2. If a detailed analysis is made, it is very challenging to incorporate each event into the tree of subsequent events.	1. The cost of software purchase ; 2. Training an operator of the program, and engaging an employee or employees who will develop the analysis.
FMEA	1. Highly time-consuming in the case of systems or projects that are composed of many components and activities, and include numerous functions; 2. The first application of the method in the company or in the project team to a complex system is also time-consuming.	1. The costs of FMEA analysis are minimal and are counterbalanced by the certainty that everything has been done for a smooth implementation of product, process, system, or project; 2. In the case of the first application, increased costs due to staff workload.
HRA	1. Dependent on the length of interviews in direct proportion to the number of analysts performing the analysis.	1. The cost of own employee training who will implement the method as an analyst; 2. Hiring an external company to perform the analysis; 3. Incorporation of the interviews into the working hours, and the necessity to reimburse the employees or team members for this time as hours worked.

Table 3. Project Phase and Project Type

Method	Project Phase	Project Type
ETA	1. Pre-project phase: risk identification, risk analysis.	1. All types of projects: introduction of a new product, service, innovation, software development.
FMEA	1. Pre-project phase: risk identification, risk analysis.	1. Industrial area: introduction of new products and systems, innovation.
HRA	1. Pre-project phase: risk identification.	1. Without any limitations.

The main disadvantage of the above methods is that they only deal with risk identification, and not with risk analysis and identification of countermeasures that should eliminate the risk, or reduce its likelihood or impact on the project.

5 Conclusion

The paper describes the risk management methods that are not commonly used in project management practice. The risk management process in the project is also described, including the various stages of the project. The contribution of the paper is the comparison of ETA, FMEA and HRA methods, and presentation of their use in the context of project management. Advantages and disadvantages of each method are outlined, as well as the time and costs associated with it. The paper further shows the phases and types of projects in which it is suitable to use the methods. It is definitely advisable to use the above methods in project management, but their great disadvantage must not be left aside, i.e. the fact that they do not propose any countermeasures against the risks that are identified. In their application, it is therefore appropriate to focus on the choice of countermeasures, or risk management strategies in compliance with the whole project in its various stages.

References

1. Conroy, G., Soltan, H.: ConSERV, a project specific risk management concept. International Journal of Project Management 16, 353–366 (1998)
2. Raz, T., Michael, E.: Use and benefits of tools for project risk management. International Journal of Project Management 19, 9–17 (2001)
3. Smith, P.G., Merritt, G.M.: Proactive Risk Management: Controlling Uncertainty in Product Development. Productivity Press, New York (2002)
4. Zhao, X., Chen, Q.: Reconsidering Baron and Kenny: myths and truths about mediation analysis. Journal of Consumer Research 37, 197–206 (2010)
5. AFNOR: Norma FD X 50-117 Management des Risques d'un Project (2003)
6. Courtot, H.: La gestion des risiques dans les projects. Edition Economica, Paris (1998)
7. Courtot, H.: Quelques renseignements liés à la mise en oeuvre d'une démarche de gestion des risques dans les projets. La Cible-A.F.I.T.E.P. 74 (1998)
8. Grant, K.P., Cashman, W.M., Christensen, D.S.: Delivering projects on time. Research & Technology Management 49, 52–58 (2006)
9. Smith, P.: Managing risk as product development schedules shrink. Research Technology Management 42, 25–32 (1999)
10. Raz, T., Shenhar, A.J., Dvir, D.: Risk management, project success, and technological uncertainty. R&D Management 32, 101–109 (2002)
11. Korecký, M., Trkovský, V.: Management rizik projektů. Grada Publishing, Praha (2011)
12. Tichý, M.: Ovládání rizika: analýza a management. C.H. Beck, Praha (2006)
13. Nenadal, K.: Moderní systémy řízení jakosti. Quality Management. Management Press, Praha (1998)
14. Zeman, M.: Diplomová práce. Univerzita Tomáše Bati ve Zlíně, Fakulta technologická, Ústav výrobního inženýrství, Zlín (2010)
15. Doležal, J., Mácha, P., Lacko, B.A.: Kolektiv: Projektový management podle IPMA. Grada Publishing, Praha (2012)
16. Smejkal, V., Rais, K.: Řízení rizik ve firmách a jiných organizacích. Grada, Praha (2009)

A Framework for Monitoring and Evaluation of Learning Processes

Lucie Pekárková, Patrícia Eibenová, and Tomáš Pitner

Masaryk University, Faculty of Informatics
Laboratory of Software Architectures and Information Systems
Botanická 68a, 602 00 Brno, Czech Republic
{pekarkova,eibenova,pitner}@fi.muni.cz

Abstract. The paper will describe the features and show advantages of process orientation in all phases of university course design and implementation applied in an introductory course on environmental science at Masaryk University, Faculty of Informatics in Brno, Czech Republic. It will demonstrate the principles of process-oriented learning design and its support in the open-source based Medusy project. Primarily, it will show the integrated approach to learning process design, tooling, runtime environment, and learning process pattern repository.

Keywords: process-oriented teaching and learning, Medusy project, course on introduction to environmental science.

1 Processes in Learning

The contemporary e-learning tools very successfully manage the classic paradigm of teaching, i.e. relatively uni-directional flow of information from teacher to students and subsequent assessment of students. Due to the massive arrival of web technologies and the Web 2.0 phenomena, new modern tools can provide students' feedback, their mutual collaboration or cooperation on content of the lectures (wiki, personal blog, chat etc.). The environmental education is not an exception [10]. The research [7] suggests that well prepared e-learning courses can be even more effective than face-to-face lectures and in most cases reaches the same results as the conventional form of teaching.

The introduction of roles, definition of rules, procedures, repeatable processes and some principles of process management can contribute to a comprehensive learning environment. That also brings advantages such as refactoring and optimization of processes, monitoring and formalization of educational principles and procedures.

Effective use of processes goes along with proper methodology for their management. That is why we focused on principles of process management (Business Process Management – BPM). Even though processes and process management come from business we can find them in non-business domain. In those cases is the process management focused mainly on the technological aspect and the

J. Hřebíček et al. (Eds.): ISESS 2013, IFIP AICT 413, pp. 525–532, 2013.

management part is largely influenced by target domain. That is also a case of BPM in e-learning. citeleyking

We can say that the learning itself is a process. The owner of the process is mainly a teacher. The main (learning) processes are created by activities and events that form the content or goal of the lecture. In connection with time they create the learning plan. Finding such learning processes is a subject of empirical research (or a pedagogical problem). The support processes participate on creation and management of the learning environment and as well as in the classical case provide the additional functionality for main processes.

In the business world, the BPM implementation leads to one goal – cost reduction and profit improvement. In the e-learning domain the goal could be higher efficiency and quality of education. Implementation and automation of processes in e-learning should project as a combination of direct (students have clearly defined tasks and duties including deadlines; easier information sharing and a knowledge base; opportunity to use sophisticated services and utilities, especially for technology-oriented courses) and non-direct (reducing number of administrative tasks, their automation and teachers' team cooperation support leads to even distribution of responsibilities and competences and the teacher has more time to prepare the content; instant overview of the situation status, completed and uncompleted tasks; possibility of continuous evaluation of the effectiveness and further learning process improvement) effects.

2 Process Modeling

As we agreed, creating the learning processes is largely a pedagogical problem - every course or lecture requires different approach. The use of the educational patterns can in many ways simplify creation and optimization of learning processes and at the same time improve the quality of learning. Formalization of each pattern via processes could help during the subsequent discussion or repeated use. [3] In our perception the educational pattern is a small reusable sequence of steps that are included in one blended learning activity. Patterns should be used as building blocks for more complex blended courses. [1] A pattern should define general structure of the activity and should be customizable. One of our aims is to build community based repository of such patterns. Patterns should be built on best practices in blended courses and each pattern should define standard structure of activity that can be extended according to specific needs of other (re-)users. In context of learning activities it should serve as a guideline, how to use certain educational pattern.

BPMN 2.0 was chosen because this modeling language has to be capable of immediate execution after modeling in contrast to BPMN 1.1.

3 Medusy Project

The Medusy project is in an early phase of development such as analysis and design. We have done one functional prototype of process based e-learning application, tool for learning process modeling, pattern repository and now we are in next iteration where we try to integrate monitoring and evaluation tools into the project. A prototype of the engine was designed which enabled to analyze possibilities of process management and patterns utilization in e-learning/blended learning.

3.1 Process Engine

The main technology used in the process engine is *Activity BPM* based on BPMN 2.0. It is an evolution of *jBPM* – an environment working on the *Process Virtual Machine* (PVM) principle. The PVM is an architecture design pattern for process engines. The structure of the process definition is separated from the implementation of the different activity types. And the runtime data structure for process executions is based on a pointer to an activity in the process graph. Activity types and hence whole process languages become pluggable on that single core engine. Main characteristics of this environment are stability, high performance, easy customization, transaction support and asynchronous task support. The resulting architecture is directly derived from used technologies. The main principle is to access the Activity runtime environment from within *Ruby on Rails* (RoR) application.

3.2 Process Modeling

For better user experience an open-source modeling tool was created based on the *Oryx* editor. The Oryx editor was enhanced with modeling of learning patterns for advanced users that are familiar with BPMN 2.0 and with modeling of learning processes in simple notation by parametrizing the pre-modeled patters for casual users. After the learning process is modeled, it can be exported to XML format that is executable in the Activity engine prototype.

4 Learning Patterns

The concept of using patterns was for the first time established in architecture. Christopher Alexander captured and described applied expert design advices as architectural design paradigms. Similar principles were used several years later in the software engineering for describing distinguish and frequent design issues in object-oriented software systems. Years later, pedagogical domain was explored as well. In this context, patterns are used to capture best practices from the educational area. It is crucial that patterns are able to solve one issue appearing in various contexts at a time. [2]Another more adjusted perspective is not searching for best practices, however to locate more various solutions. Final selection of suitable pattern is supposed to be a teachers preference. [3]

4.1 Patterns for an Environmental Course

Within the project, there are two environmental-related courses taught at Faculty of informatics that were innovated by adding new e-learning and interactive elements. Enhanced process models of the courses were created.Innovation of environmental courses as well as the whole project stands on a principle of using *learning patterns.* Creating a course model tends to be as an illustration how the models can be created in the future. In this paper, we present a model of the Environmentalistics course, taught by one of the authors at Faculty of informatics, Masaryk University Brno.

4.2 The Course on Environmentalistics

Environmentalistics is a basic environmental course taught at Masaryk University, Faculty of informatics. The course is an introduction to the topic of environmental science with focus at global environmental problems.

The education is organized into interactive lectures, followed by a colloquial exam. During the course, two essays are assigned. This is very common pattern in education – lectures, homework and an oral colloquial exam.

Innovation requirements of this course should fit into these set rules, therefore *no increased demands* on students or teacher are required. Suitable option is to integrate e-learning elements into the course, thus create a blended-learning course. An advantage of this concept is that students can prepare before coming to lectures and devote more time to extended topics than repeating basic facts. Lectures demand active participation. The Environmental information systems course extends this basic course, see [5]. It is devoted to managing, monitoring a evaluating of data in environmental information systems and likewise design of these very systems.

4.3 Principles of Modeling Learning Processes

In innovation of both courses, learning patterns were used through current pattern database. Due to this method, any teacher would enhance own courses by using a tool which can provide required patterns – educational objects. Educators are expected to make their choice and compile a whole course.

The syllabus of the Environmentalistics were described in informal specification – how is the course organized and running, conditions of evaluation and so forth. The very first step, just before the very modeling is creation of a structured list. This list has a clear start and termination and every further step in the list is a part of a process. According to complexity of steps on the list, it is possible to moderate an abstraction scale of the modeling tool. For instance, a user insists on placing a team project into the course, however he does not have accurate conception. There exist several options how to organize a team project in a course and the user does not have to be familiar with all of them. Therefore there is a possibility of choosing present patterns from the repository. Eventually the whole course is created by compiling single elements together.

Despite of this light description from the users point of view, there is a BPMN model behind every pattern. Compiled courses have to be functional and make sense likewise from the technical point of view, which is not easily controllable. Therefore automatization is only partial and the created model should be manually validated.

4.4 Demonstration of Enhanced Formal Course Model

Final compilation of the complex course model is a pedagogical matter. A teacher decides which learning object should take place in the final course and how to connect them. In this paper, we present the enhanced concept of final colloquium. Colloquium is a method of examination, where small group of student discuss learned knowledge with the teacher. Each student also defends elaborated essay.

The concept of final colloquial exam is slightly enhanced in a new model, see Fig. 1. Students are divided into groups of three. Each student is supposed to elaborate his/her own final essay, submit it, afterwards read essays of two colleagues and write a report which will be discussed.

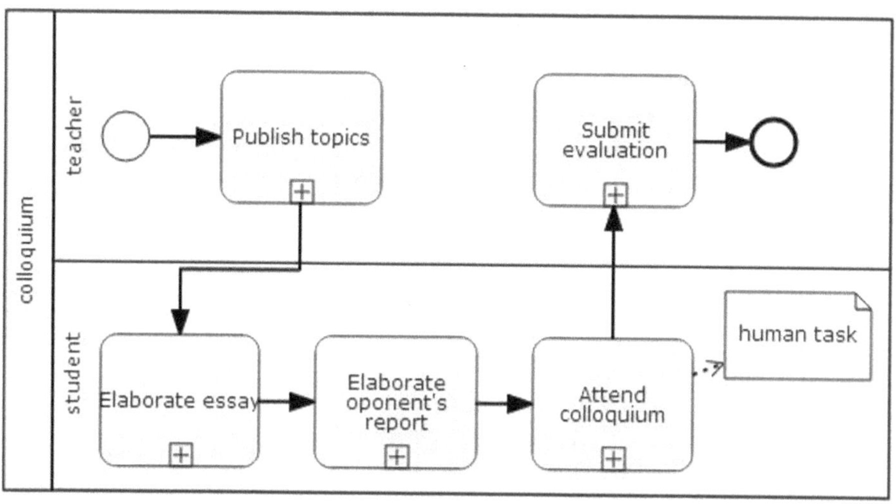

Fig. 1. "Colloquial exam" process

Model of a colloquium is a standard BPMN 2.0 diagram and consists mostly of collapsed subprocesses. Each subprocess represents a set of activities within a business process. A subprocess contains another whole business process with its start and termination and sequence of tasks or another subprocesses. For instance, Fig. 2 shows that subprocess *Publish topics* contains another process.

For every learning object in the course, there is a business model. Particular subprocesses have a precise order and together they compose a whole course.

5 Pattern Repository

The next step within the Medusy project is creation of a *learning patterns repository*. The repository is a module in the framework maintaining a database of the

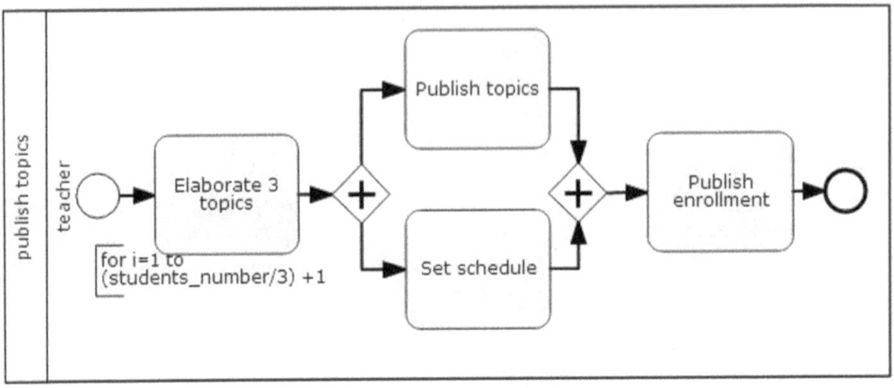

Fig. 2. "Publish topics" subprocess

most used patterns in the educational domain. By module we mean a standalone application connectable to the other part of the system such as modeling tool and execution engine.

The proposed repository deals with several other challenges that define this solution. It is possible to achieve proposed attributes by suitable merge of current modules used in the project. At the moment, the database is deeply vast and unorganized, however forms a core component to the pattern repository. Demands of the proposed repository are as follows:

- interconnection with the modeling tool and execution engine;
- structuring patterns according function, popularity, subject domain or the scale of abstraction;
- simple and friendly user interface;
- adjustment of the pattern by using the BPMN;
- documentation and precise pattern description;
- validation of formal correctness of a pattern.

5.1 Codification vs. Personalization Approach

Another important matter is a scale of customization and whether to provide maximal scale with the possibility of creating custom models or offer fixed steady patterns, see [4].

The codification approach is focused on using the patterns repeatedly – use one single pattern over and over again is the main goal. In the beginning, there is a concern on creating an exquisite database of patterns. User chooses a pattern and sticks with the solution that should fit perfectly. There is a minimal need for optimization considering the high input at the beginning. By contrast the personalization approach enables representing the model thoroughly. The personalization encourages innovation, however, it exceeds our pattern definition, see [9].

5.2 Optimization of the Modeling Tool for Users

We also assume two kinds of users, an inexperienced one, who has not any experiences with modeling in BPMN 2.0 notation. This user has an opportunity to use predefined patterns and assemble them into a new course. Another option is to choose from predefined whole courses assuming there is no need for added adjustments. Second kind of user, an experienced one, has already some knowledge about BPMN 2.0 notation and knows how to model simple processes. In this case, it is possible to create entirely new course according to special demands. This option enables maximal scale of customization. However, the preferable attitude dedicated to the majority of users, is submitting prepared patterns.

5.3 Integration of the Repository into the Framework

Besides the current database of patterns as an essence of the pro-posed pattern repository, there are other modules in the framework. The repository will be a core component to the modeling tool. A teacher creates a model of a course with the aid of the modeling tool and simultaneously uses existing patterns from the repository. The current modeling tool is based on Oryx editor so it is essential that the repository will inherit applied technologies and solutions. A prepared and validated model of a course is an executable model as well. Models are executed in the functional prototype of process engine. There are no compatible issues due to the fact we use BPMN 2.0. Executable XML file is a requirement for execution of the model in the process engine.

6 Conclusion and Future Work

The project Medusy tries to leverage best practices in person-centered technology-enhanced learning. The project introduces a process-oriented approach to course design, development and education flow in general. The designed framework provides background for modeling, execution, monitoring and evaluation of education processes in academic and commercial environment. Process models of two environmental courses at Masaryk University, Faculty of Informatics have been created. The models are described in up-to-date BPMN 2.0 notation and executable in the process engine. All the models were created in systems modeling tool and compiled from the learning patterns from the project repository. Obtained data and findings will be analyzed and the output will be used in the upcoming module for monitoring and evaluation of educational processes.

Acknowledgement. This work was supported by the European Unions territorial cooperation program European Territorial Co-Operation Austria-Czech Republic 2007-2013 of the under the EFRE grant M00171, project iCom (*Constructive International Communication in the Context of ICT*).

References

1. Avgeriou, P., Papasalouros, A., Retalis, S., Skordalakis, M.: Towards a pattern language for learning management systems. Educational Technology & Society 6(2), 11–24 (2003)
2. Bergin, J., et al. (eds.): Pedagogical Patterns: Advice for Educators. Joseph Bergin Software Tools (2012), http://www.pedagogicalpatterns.org
3. Derntl, M., Motschnig-Pitrik, R.: The role of structure, patterns, and people in blended learning. The Internet and Higher Education 8(2), 111–130 (2005)
4. Hansen, M.T., Nohria, N., Tierney, T.: Whats your Strategy for Managing Knowledge? In: Harvard Business Review, pp. 106–116 (March, April 1999)
5. Hřebíček, J., Kubásek, M.: Environmentální informační systémy (Environmental information systems). CERM, Brno (2011), http://www.iba.muni.cz/res/file/ucebnice/hrebicek-environmentalni-informacni-systemy.pdf
6. Leyking, K., Chikova, P., Loos, P.: Competency and Process-Driven e-Learning – a Model-Based Approach. The Electronic Journal of e-Learning 5(3), 183–194 (2007)
7. Means, B., et al.: Evaluation of Evidence-Based Practices in Online Learning: A Meta-Analysis and Review of Online Learning Studies, U.S. Department of Education (2010), http://www2.ed.gov/rschstat/eval/tech/evidence-based-practices/finalreport.pdf
8. Ministr, J., Stevko, M.: Human Resources Requirements for Professional Management of ITSCM Process. In: Doucek, P., Chroust, G., Oskrdal, V. (eds.) IDIMT-2010: Information Technology – Human Values, Innovation And Economy, 18th Interdisciplinary Information Management Talks, Jindrichuv Hradec, Czech Republic, Schriftenreihe Informatik, September 08-10, vol. 32. Springer, Heidelberg (2010)
9. Reimer, U., Heck, U., Streit, S.: Collaboration-Oriented Knowledge Management Using Interaction Patterns. In: Yamaguchi, T. (ed.) PAKM 2008. LNCS (LNAI), vol. 5345, pp. 26–37. Springer, Heidelberg (2008)
10. Talaš, J., Gregar, T., Pitner, T.: Semantic Wiki in Environmental Project Management. In: Hřebíček, J., Schimak, G., Denzer, R. (eds.) ISESS 2011. IFIP AICT, vol. 359, pp. 437–444. Springer, Heidelberg (2011)

Classifying Environmental Monitoring Systems

Mauno Rönkkö[1], Ville Kotovirta[2], and Mikko Kolehmainen[1]

[1] Department of Environmental Science, University of Eastern Finland,
PO Box 1627, 70211 Kuopio, Finland
{mauno.ronkko,mikko.kolehmainen}@uef.fi
[2] VTT Technical research Centre of Finland,
PO Box 1000, 02044 VTT, Finland
ville.kotovirta@vtt.fi

Abstract. In this article the diversity of environmental monitoring systems is studied. The number of such systems is steadily increasing each year, as systems are tailored to specific, growing needs of authorities, corporate users and citizens. Because of this, it becomes harder to compare systems and their functionality. Systems that appear to have the same functionality may turn out to be tailored for different application domains. Likewise, a chosen system may later on turn out to have insufficient support for connectivity and interoperability, although it provides the best support for core functionality requirements. To make sense of the ever growing diversity, and as the main contribution, a method for classification and analysis is proposed. The method is generic to environmental monitoring systems. The use of the method is also illustrated. The classification results yield even for a limited number of systems relevant clusters that help in identifying critical properties for further inspection.

Keywords: Environmental information systems, Environmental monitoring, Systems Architecture, Systems Analysis, Systems Classification.

1 Introduction

There are currently many factors driving the design and development of both national and international environmental information systems. One of the most notable of these factors is the green house effect. Reducing energy consumption, cutting greenhouse gas emissions and eliminating energy wastage are of global interest. Another significant factor has to do with the vulnerability of modern society in case of natural disasters. This has also been highlighted by recent extreme weather phenomena exhibited around the globe.

Not surprisingly, as pointed out by Messer et al. [1] "High-resolution, continuous, accurate monitoring of the environment is of great importance for many applications from weather forecasting to pollution regulation." Environmental monitoring has, thus, become an attractive topic for state-of-the-art research. For instance, monitoring of the Fukushima incident is still relevant for impact assessment as discussed by Cyranoski et al. [2]. Similarly, continuous monitoring of factories and power plants is critical as they may cause long term effects [3].

J. Hřebíček et al. (Eds.): ISESS 2013, IFIP AICT 413, pp. 533–542, 2013.

As the number of new environmental monitoring systems grows, it becomes harder to understand the similarities and differences between the systems. In particular, it becomes harder to understand what the differences mean in terms of future applicability and interoperability. Without this critical information, there is a temptation to develop a new system rather than to adapt and use an existing one; thus, making the overall problem worse.

To make sense of the ever growing number of environmental monitoring systems, a method for the classification is proposed in this article. The method identifies three strategic viewpoints: application domain, functionality, and architecture. The proposed method is used in this article to classify a selected set of environmental monitoring systems. Even for a limited case, the results identify some relevant and interesting clusters. Clustering reveals clearly, which properties are shared by the systems and which properties differentiate a specific system from the rest of the systems within a cluster.

This article is organized as follows. In Section 2 we present the proposed classification method. In Section 3 we describe the environmental monitoring systems considered for classification, and present the classification results. Finally in Section 4 follows the conclusion.

2 Proposed Classification Method

There are three strategic viewpoints that are necessary for classification: application domain, functionality, and architecture. The application domain viewpoint describes not only the purpose and need for the use of the system, but also the domain of interest. The functionality viewpoint describes what the system does to satisfy the needs for its use. The architecture viewpoint describes how the system is implemented to provide its functionality. These viewpoints are intentionally orthogonal, to support the analysis process. However, even for these three viewpoints, there are many properties of interest. Therefore, the analysis is limited here to specific properties per viewpoint, as described next.

Application domain viewpoint emphasizes the purpose and the need for the system, including the domain of interest. From this viewpoint there are four kinds of properties of interest:

1. *Phenomenon:* What is the monitored environmental phenomenon?
2. *Interest:* Why the end-user is interested in the phenomenon?
3. *Task:* What is the system used for?
4. *Benefit:* How do the end-users benefit from using the system?

Functionality viewpoint emphasizes the activities of the system that satisfy the end-user's need. From this viewpoint, there are six properties of interest:

1. *Operation:* What operation is performed by the system?
2. *Interaction:* How is the operation initiated and carried out?
3. *Performance:* What kind of performance is expected from the operation?
4. *Reliability:* How dependable the operation is expected to be?

5. *Methods:* What methods are used to perform the operation?
6. *Maintenance:* What kind of maintenance is expected?

Architecture viewpoint emphasizes the details of the system that realize the activities. From this viewpoint, there are four properties of interest:

1. *Implementation:* How is the system implemented?
2. *Data:* What kind of data is used by the system?
3. *Capacity:* What kind of performance can the system provide?
4. *Connectivity:* How is the system connected to other systems?

The actual analysis is performed by cross-tabulation over the properties for each viewpoint, where each cell indicates if the considered system supports the combined properties. This results in three tables as shown in Figures 1, 3, and 5. Consequently, the tables provide a qualitative analysis, a comparison over the considered systems. The tables show three different kinds of clusters: properties that are unique to some system, properties that are shared among a few systems, and properties that are common to most of the systems. Properties that are unique to a system indicate the competitive edge of the system. Properties that are shared among a few systems indicate a higher level feature that is somehow central to those systems. Lastly, properties that are common to most of the systems indicate compulsory requirements. Thus, meaningful clusters are found by finding clusters of properties that are shared by only a few systems.

3 Results

In this section, environmental monitoring systems that were classified are introduced first shortly. A set of systems were chosen to represent different kinds of environmental monitoring systems. The classification results and discussion are presented at the end of this section.

3.1 Environmental Monitoring Systems to Be Classified

Measuring and Forecasting. For this class of systems, NOAA Integrated Ocean Observing Systems (IOOS) was considered. As presented by Harlan et al. [4], the NOAA IOOS uses a high frequency radar network for measuring ocean surface currents in United States. It consists of more than 100 high frequency radars. As pointed out by Harlan et al., the radar network constitutes the largest set of ocean current observations in existence.

Integrated Service. For this class of systems, a weather service application integrating Weather Underground and INTAMAP [5] was considered, later called INTAMAP WU. As presented by Williams et al. [6], the application provides quality control for weather monitoring over a collective network of privately owned weather stations. It provides an OGC standard Observations & Measurements (O&M) interface and uses UncertML [7] for describing uncertainties for the Weather Underground data.

Household Monitoring. For this class of systems, a system for measuring and monitoring indoor air quality and energy efficiency, AsTEKa, by Skön et al.

[8] was considered. It consists of a sensor network for measurements and data capture, a database for data storage, and an application server for monitoring services and support for external user interface built on top of a web browser. The sensor network gathers data for instance about humidity, temperature, and carbon dioxide, as well as data about water and electricity consumption.

Environmental Safety and Security. For this class of systems, a distributed service network for security applications, TiTiMaKe [9], was considered. According to Rönkkö et al., TiTiMaKe integrates heterogeneous services, including a dedicated sensor network, vehicle detection and classification service, reachability computation and analysis service, chemical dispersion computation and modeling service, and a spatio-temporal population modeling service. TiTiMaKe implements safety and security applications for traffic monitoring, and estimating affected people and risk groups in case of a chemical accident.

Navigation and Planning. For this class of systems, Icebreaker Plot (IBPlott) [10] was considered. IBPlott is a system for icebreaker navigation and assistance planning. It delivers near real-time ice, weather, and hydrological data to users making routing decision in dynamic environmental conditions. The system takes satellite images, weather observations and forecasts, water level observations and forecasts, ice charts and ice model forecasts, as well as ship traffic information as an input.

Participatory Sensing. For this class of systems, EnviObserver [11] was considered. EnviObserver uses people as sensors in collecting observations about the environment. Users use a specialized mobile application to report environmental phenomena and the observations. Time of the observations and the location of the users are stored in the server for later analyses. The observations collected can be viewed graphically near real-time, and alerts can be generated.

3.2 Classification Results for Application Domain

The cross-tabulation over application domain properties are presented in Figure 2. There are four distinct clusters of systems. The clusters are health and comfort, sea and water, improved security, and improved efficiency.

Health and Comfort. INTAMAP WU, AsTEKa, and EnviObserver forms the first cluster. The shared properties have to do with measuring, learning, and analysis related to comfort, health, and well-being for citizens and households. There is also a strong emphasis on delivering personalized information to citizens.

Sea and Water. NOOA IOOS HFR, IBPlott, and EnviObserver form the second cluster. This cluster targets the services for authorities for monitoring and decision making. These systems share strictly the interest of water, sea or lake, as the application domain.

Improved Security. OOA IOOS HFR, TiTiMaKe, IBPlott, and EnviObserver form the third cluster. This cluster subsumes the cluster for sea and water; however, these systems share strongly the common interest for improved security by

	Application Domain	Interest							Task					Benefit						
		Authorities	Researchers	Rescue workers	Industry	City council	Citizens	Households	Measuring	Monitoring	Learning	Analysis	Decision-making	Understanding	Health and well-being	Comfort	Saving costs	Improved efficiency	Improved security	Income
Phenomena	Ocean	A, E	A	A	E	A	F		A, F	E,F	A,E,F	A,E,F	A,E,F	A, E, F	F	F	A, E	A, E	A, E	E
	Weather	A	A,B	A		A	B	B	A,B	E	A,B	A,B,E	E	A,B		B	A	A	A	
	Air quality						F		F	F	F	F	F	F	F	F				
	Indoor air		C					C	C	C	C	C	C	C	C	C	C	C		
	Water consumption		C					C	C	C		C	C				C	C		
	Energy consumption		C					C	C	C	C	C	C	C			C	C		
	Vibration		D						D	D	D	D		D					D	
	Noise		D						D	D	D	D		D					D	
	Chemical dispersion		D						D	D	D	D		D					D	
Interest	Authorities									E, F	E	E	A,E	E			A	A,E	A, E	
	Researchers								A,B,C, D, F	D	A,B,D, F	A,B,C, D,F		A,B,C, D, F	C				D	
	Rescue workers												A				A	A	A	
	Industry				Redundant					E	E	E	E	E			E	E	E	E
	City council											A	A				A	A	A	
	Citizens								B			B	B	F	F	B, F	F		F	
	Households								B,C	C	C	B,C	B,C	C	C	B,C	C	C		
Task	Measuring													A,C,D, F	F	B,C, F	C, F	C, F	A,D, F	
	Monitoring													E,F	F	F	C,E	C,E	D,E	E
	Learning				Redundant						Redundant			A,B,C, D,E,F	C,F	F	C,E	C,E	D,E	E
	Analysis													A,C,D,E ,F	C,F	B,C,F	A,C,E	A,C,E	A,D,E	E
	Decision-making													E,F	F	F	A,C,E	A,C,E	A,E	E

Fig. 1. Cross-tabulation of the properties for application domain. Labeling: A=NOOA IOOS, B=INTAMAP WU, C=AsTEKa, D=TiTiMaKe, E=IBPlott, F=EnviObserver

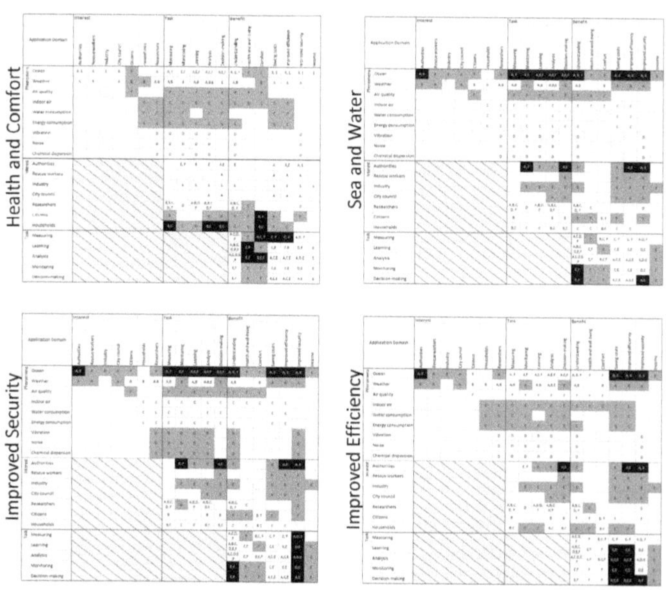

Fig. 2. Clusters found for the application domain viewpoint, Figure 1; the shared properties are indicated with black fill and the disjoint properties with gray fill

targeting the services for authorities. There is also a shared interest for understanding the underlying phenomena.

Improved Efficiency. OOA IOOS HFR, AsTEKa, and IBPlott form the last cluster. These systems show a strong emphasis on saving costs. Interestingly the tasks involve not only monitoring and decision making, but also learning and analysis.

3.3 Classification Results for Functionality

The cross-tabulation over functionality properties are presented in Figure 4. In the figure, properties unique to the clustered systems are indicated with yellow, and shared properties are indicated with red. There are three distinct clusters of systems. The clusters are local-monitoring, decision making, and robust together with reliable.

Local Monitoring. AsTEKa and TiTiMaKe form the first cluster. Both systems perform monitoring of present data with model-based methods. These systems do, however, differ in application domain, which is reflected in the operational properties.

Decision Making. NOOA IOOS HFR, INTAMAP WU and TiTiMaKe form the second cluster. In this cluster, shared operations are: producing new data, forecasting, extracting knowledge, and delayed response. The systems differ on data flow and locality of measurements. Also, NOOA IOOS HFR and TiTiMaKe have maintenance that INTAMAP WU does not really have.

Robust and Reliable. NOOA IOOS HFR, IBPlott, and EnviObser form the third cluster. The cluster is not cohesive, as only IBPlott and EnviObserver have a subscription service for real-time phenomena monitoring. Moreover, NOOA IOOS HFR and IBPlott have uncertainty management, but only NOOA IOOS HFR and EnviObserver use statistical methods.

3.4 Classification Results for Architecture

The cross-tabulation over architecture properties are presented in Figure 6. In the figure, properties unique to the clustered systems are indicated with yellow, and shared properties are indicated with red. There are four distinct clusters of systems. The clusters are open systems, closed systems, large data flow, and externally dependent.

Open Systems. INTAMAP WU and EnviObserver form the first cluster. These systems have loosely coupled, international components and sensors with open interfaces. These two systems differ only by use of programming languages and use of standards.

Closed Systems. NOOA IOOS HFR, AsTEKa, and TiTiMaKe form the second cluster. For these systems, use of standards and locality (or nationality) is

	Functionality	Interaction			Performance							Dependency			Methods		Maintenance				
		Poll	Subscribe	Reactive	Real-time	Delayed response	Local	National	International	MB dataflow	GB dataflow	Safety-critical	Robust	Uncertainty managed	Statistical	Model-based	Calibration	Software revision	Configuration	Self-monitor	Repair
Operation	Relay	A,B,C, D, F	E,F		A,B,C, D,E,F		C	A,D	B,E,F	B,C,D,E ,F	A	E	A, E		E					E	
	Measure	A,B,C, D, F	F		A,B,D, F	C	C,D	A	B,F	B,C,D,F	A			A,F	A,F		A,D	A,D	F		A,D
	Store	A,B,C, D,E,F	E,F		A,B,D, E,F	C	C,D	A	B,E,F	B,C,D,E ,F	A		A		E				A,B,C, D,F	E	A
	Transform	D, F	E,F		E,F	D	D		E,F	DE,F		E	E			D				E	
	Produce new	B,D				B,D	D	B		B,D				B	B	D					
	Forecast	B,D				B,D	D	B		B,D				B	B,D	D		D			
	Extract knowledge	A,D				A,D	D	A		D	A					A,D					
	Present	B,D,E,F	E,F	C	E,F	B,C,D	C,D	B	E,F	B,C,D,E ,F		E	E		E	C,D		D	F		
Interaction	Poll				A,B,C, D, E, F	B,C,D,E	C,D	A,B,D	B,E,F	B,C,D,E ,F		E	A,E	F							
	Subscribe	Redundant			E,F				E,F	E,F		E	E		Not applicable		Not applicable				
	Reactive																				
Performance	Real-time											E	A,E	A	A,F			E,F	F	E	E
	Delayed response													B	B	A,C,D	A,D	A,D	A,B,C, D		A,D
	Local															C,D	D	D	C,D		D
	National	Redundant			Redundant								A	A,B	A,B	A	A	A	A		A
	International											E	E					E,F	B,F	E	E
	MB dataflow											E	E		B	C,D	Not applicable				
	GB dataflow												A		A	A					
Dependency	Safety-critical																				
	Robust	Redundant			Redundant							Redundant					Not applicable				
	Uncertainty managed														A,F						
Methods	Statistical	Redundant			Redundant							Redundant			Redundant			A			
	Model-based																		D		

Fig. 3. Cross-tabulation of the properties for functionality. Labeling: A=NOOA IOOS, B=INTAMAP WU, C=AsTEKa, D=TiTiMaKe, E=IBPlott, F=EnviObserver.

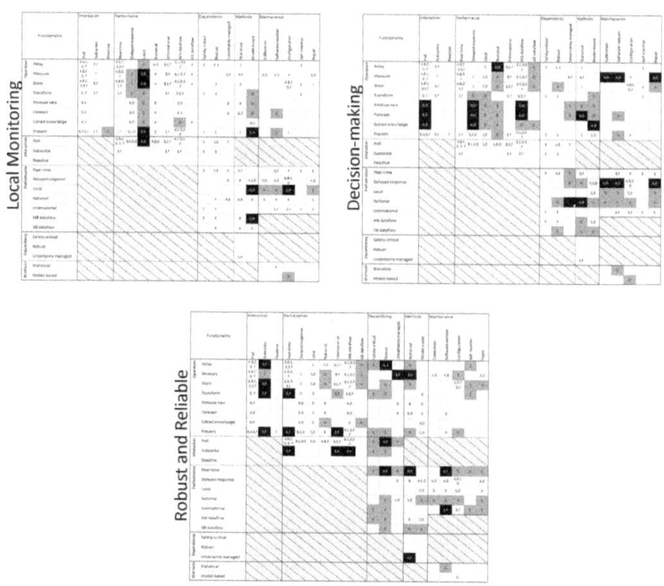

Fig. 4. Clusters found for the functionality viewpoint, Figure 3; the shared properties are indicated with black fill and the disjoint properties with gray fill

		Data										Capacity						Connectivity		
	Architecture	Local	National	International	Real-time	Historical	Forecast	Tens of variables	Hundreds of variables	Quality controlled	External sources	GB storage	TB storage	MB processing	GB processing	Tens of sensors	Hudreds of sensors	Required exteral services	Provided external services	Open interfaces
Implementation	Loosely coupled		F	B, E	E,F	B,E,F	E	A,B, E,F		A	B, E, F	B,F	E	B, F	A, E	A	B,F	D, E, F	A,B,E,F	B, F
	Multiple technologies	D	A, F	B, E	A,D,E,F	B,E,F	B,D,E	A,B,D, E,F				B,D,F	E	B,D,F	A,E	D	B,F	D,E,F	A,B,D, E,F	B,F
	Use of standards	C,D	A	B	F	F		A,C,D			B	C ,D	A	C,D		A,C,D		D	B,C,D	B
	Multiple languages		A	B, E	A,D,E,F	B,E,F	D,E	A, E			B	E		B	A,E		B	E	A,B,E,F	B
Data	Local											C,D		C,D		C,D			C,D	
	National												A	D	A	A		D	A,D	
	International											B,F	E	B,F	E		B,F	E,F	B,E,F	B,F
	Real-time											E,F	A	E,F	A		A,F	E,F	E,F	F
	Historical						Redundant					B,C,D,F	E	B,C,D,F	E	C,D	B,F	A,B,C, D,E,F	A,B,C, D,E,F	F
	Forecast											B,D,E		B,D	E			E	B,D	
	Tens of variables											B,C,D,E ,F	A	B,C,D,F	A,E	A,C,D	B,F	D,E,F	A,B,C, D,E,F	B,F
	Hundreds of variables																			
	Quality controlled															A				
	External sources											B,E,F		B,F	E		B,F	D,E,F	B,E,F	B,F
Capacity	GB storage																	E,F	B,C,D,E ,F	B,F
	TB storage																		A	
	MB processing																	D,F	B,C,D,F	B,F
	GB processing						Redundant								Redundant			E	A,E	
	Tens of sensors																		C,D	
	Hundreds of sensors																		B,F	B,F

Fig. 5. Cross-tabulation of the properties for architecture. Labeling: A=NOOA IOOS, B=INTAMAP WU, C=AsTEKa, D=TiTiMaKe, E=IBPlott, F=EnviObserver.

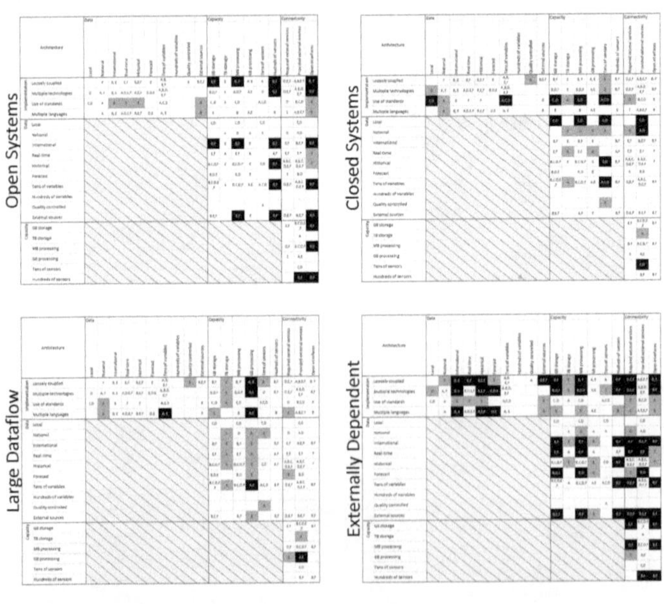

Fig. 6. Clusters found for the architecture viewpoint, Figure 5; the shared properties are indicated with black fill and the disjoint properties with gray fill

strongly expressed. In addition, these systems merely provide external interfaces for specific purposes.

Large Data Flow. OOA IOOS HFR, and IBPlott form the third cluster. Interestingly, these systems share partially the same interest domain, ocean. However, as shown in the cluster, these systems have almost nothing else in common than the use of large data flows.

Externally Dependent. INTAMAP WU, TiTiMaKe, IBPlott, and EnviObserver form the last cluster. These systems have loosely coupled components and they also use multiple technologies and programming languages. The cluster is actually divided into to sub-clusters, where IBPlott and EnviObserver focus more on the real-time aspects, and the other two systems focus more on the forecasting aspects.

3.5 Discussion

The cross-tabulation revealed interesting clusters for the selected systems. For application domain, a cluster for health and comfort emerged, emphasizing delivery of personalized information. Interestingly, these systems operated at very different domains, which is reflected as unique properties in the cross-tabulation. In the cluster for improved security, the systems shared a strong emphasis for targeting the services for authorities.

The cross-tabulation for functionality revealed a cluster for decision making, where the systems shared the property of delayed response. The system in this cluster differed mostly on having maintenance or not.

The cross-tabulation for architecture revealed a cluster for open systems consisting, emphasizing particularly the use of open interfaces. The architectural similarity in this class could be explained by the fact that the systems are targeted for use by citizens. In the cluster for externally dependent systems, the cross-tabulation clearly reveals how the systems are built for different purposes. In this cluster, the combination of systems sharing some specific property varies substantially from property to property.

4 Conclusion

The classification method proposed here, produces relevant information about the considered systems, where clusters highlight interests shared by the systems and, at the same time, indicate how such systems differ from each other. Thus, The method is an excellent tool, when choosing a system from among several candidates. It is also a good tool for mapping competition, and strengthening an existing system, for instance, by emphasizing differentiating features.

One interesting topic for future research is to apply the classification method to understand qualitative properties of environmental monitoring systems, such as interoperability, reliability, robustness, and resilience. For such investigation, however, it would be necessary to involve the system developers, as required details tend not to be publicly available. For instance, interoperability seems to be a term easily used for any systems offering some external interfaces. However, as the clustering result in this article with respect to functionality viewpoint indicates, external dependability is not an unambiguous measure.

Acknowledgements. We wish to thank the Finnish Funding Agency for Technology and Innovation (TEKES) for funding this research. This work has been carried out in TEKES funded CLEEN SHOK programme, "Measurement, Monitoring and environmental Assessment" (MMEA). We wish also to thank CLEEN Ltd. and all our partners in MMEA for collaboration.

References

1. Messer, H., Zinevich, A., Alpert, P.: Environmental monitoring by wireless communication networks. Science 312, 713 (2006)
2. Cyranoski, D., Brumfiel, G.: Fukushima impact is still hazy. Nature 477, 139–140 (2011)
3. Gilbert, N.: Drug waste harms fish. Nature 476, 265 (2011)
4. Harlan, J., Terrill, E., Burnett, B.: National high frequency radar network: Update. In: OCEANS 2009, MTS/IEEE Biloxi - Marine Technology for Our Future: Global and Local Challenges. IEEE (2009)
5. Williams, W., Cornford, D., Ingram, B., Bastin, L., Beaumont, T., Pebesma, E., Dubois, G.: Supporting interoperable interpolation: the intamap approach. In: Swayne, D., Hrebicek, J. (eds.) Proceedings of International Symposium on Environmental Software Systems (2007)
6. Williams, M., Cornford, D., Bastin, L., Jones, R., Parker, S.: Automatic processing, quality assurance and serving of real-time weather data. Computers & Geosciences 37, 351–362 (2011)
7. Williams, M., Cornford, D., Bastin, L., Pebesma, E.: Uncertainty markup language (UncertML). OpenGIS Discussion Paper 08-122r2. Open Geospatial Consortium Inc. (2009)
8. Skön, J., Kauhanen, O., Kolehmainen, M.: Energy consumption and air quality monitoring system. In: Proceedings of the 7th International Conference on Intelligent Sensors, Sensor Networks and Information Processing, pp. 163–167 (2011)
9. Rönkkö, M., Kolehmainen, M.: Titimake final report. In: Rönkkö, M., Kolehmainen, M. (eds.) University of Eastern Finland (2011)
10. Berlund, R., Kotovirta, V., Sein, A.: A system for icebreaker navigation and assistance planning using spaceborne sar information in the baltic sea. Canadian Journal of Remote Sensing 33(5), 378–387 (2007)
11. Kotovirta, V., Toivanen, T., Tergujeff, R., Huttunen, M.: Participatory sensing in environmental monitoring experiences. In: The Sixth International Conference on Innovative Mobile and Internet Services in Ubiquitous Computing, IMIS 2012 (2012) (accepted for publication)

Waste Management Statistics in the Czech Republic

Egor Sidorov[1,2], Iva Ritschelová[1,3], and Jiří Hřebíček[4]

[1] Czech Statistical Office, Na padesatem 81,
100 82 Prague 10, Czech Republic
iva.ritschelova@czso.cz
[2] Jan Evangelista Purkyne University, Faculty of Environment
Kralova vysina 3132/7, 400 96 Usti nad Labem, Czech Republic
egor.sidorov@ujep.cz
[3] Technical University of Liberec, Economic Faculty
Voronezska 13, Liberec, Czech Republic
[4] Masaryk University, Institute of Biostatistics and Analyses
Kotlarska 2, 611 37 Brno, Czech Republic
hrebicek@iba.muni.cz

Abstract. History of statistics in the field of waste management in the Czech Republic is relatively long. At present the waste management survey is focused among others at the generation of waste by enterprises and municipalities, methods of waste use and disposal, and other methods of waste management.

The main objecotive of the paper is to introduce the system development and the processing methods. The paper deals not solely with the determination of the set of reporting units and its changes over years, yet also with technology of data processing, system of checks, corrections, and with a description of control relations. Following to the reduction of administrative burden on respondents and reduction of the respondent pool (from 18 thousand to 10 thousand over 5 years), methods for imputations and statistical checks are also presented. The discussion part of the paper is focused at balancing the purpose of waste management data with respective administrative burden related to their production.

Keywords: waste, secondary raw materials, statistics, data processing, Czech Republic.

1 Introduction

As a part of the State Statistical Service of the Czech Republic (CZ) the Czech Statistical Office (CZSO) aims at acquisition of data and consequent production of statistical information on environmental state of the country among others [1]. Environmental policies both at international and national level lead to the stable demand for high quality environmentally related statistics [2], [6].

Waste management statistics belongs among traditional statistical outputs of the CZSO. Forming an important input for production of the core indicators of environmental and resource efficiency waste statistics is highly demanded by the

J. Hřebíček et al. (Eds.): ISESS 2013, IFIP AICT 413, pp. 543–559, 2013.

variety of stakeholders that use them both for design and consequent evaluation of environmental, production, trade and other policies at different levels. As an essential decision supporting tool of resource management activities waste management statistics becomes more and more related to policies in the field of secondary raw materials. These processes make waste statistics a matter of strategic interest at national level.

The following paper maps the main milestones that have influenced its recent development of the official waste statistics in CZ and presents a bottom line of its current state. The special attention is devoted to current processing methods including issues of determination of the set of reporting units, system of checks, corrections, and description of control relations aimed at keeping quality of statistical data and information. Following the political stress at administrative burden reduction the discussion part of the paper is focused at balancing the purpose of waste management data with respective administrative burden related to their production.

2 Institutional Conditions

The primary transformation of the Czech economy in the beginning of the 90's was related to variety of activities aimed at improving the environmental state of the country. Czech environmental situation of that time could be rated as one of the most problematic in the Central Europe. In the first half of the 90's the important changes of environmental protection, especially in the legislative sphere, took place. These activities widely stimulated development of environmental statistics including statistics in the field of waste management.

During that period legislation has consequently covered the variety of environmental domains, state institutions and environmental funding processes. In addition to acts establishing the Ministry for the Environment (1989), Czech Environmental Inspectorate and State Environmental Fund (1991) number of environmental regulations either newly appeared, or were updated. These were often inspired and/or stimulated by the European legislative framework.

The initial legislative act in the field of waste was introduced back in 1991. It was a historically first codified text to provide a legislative framework for waste issues in CZ. In reaction to these processes the CZSO has started regular waste generation survey in 1992.

The initial act on waste was valid till the end of 1997, when it was replaced by another act aimed at solving number of conceptual and practical issues of its predecessor. The 1997 version of act declined the special waste category of waste. All the waste was divided into two categories: hazardous and other waste. On the one hand conditions of hazardous waste operation were restricted. On the other hand they were simplified in case of other waste types.

These developments were also closely connected to active preparation of CZ to entering the EU. During that time the Czech legal regulations transposed several tens of European environmental directives. As a result of these processes the contemporary legal framework of CZ responds to all new legal environmental regulations adopted by the European Community.

At present the waste operation is regulated by the act on waste of 2001. The reason for its adoption was that compatibility of the former national legislative framework with community law was relatively insufficient. The new act introduced number of updated and totally new definitions, principles, institutes and processes. Today the Czech legislative framework in the field of waste is compatible with waste management laws and regulations of the EU[1].

All the above mentioned institutional conditions during this time were actively forming demand for high quality waste statistics. These requirements have also been consequently finding respective reflections in the activities of the CZSO. The CZSO environmental department was established in 1993. Its personnel capacity have been fluctuating around 8-9 employees during the whole period of its existence with 2 full-time persons performing survey in the field of waste management statistics (excluding processing staff.

3 Waste Survey Evolution

CZSO has been performing statistical surveys in the field of waste generation and waste treatment since the 80's. Two stand-alone surveys of waste landfills and waste production were initially performed by the former Federal Statistical Office in 1980 and 1987. However, regular annual waste generation surveys have been performed since 1992 [1]. And through the number of existing critiques (e.g. waste type coverage, highly aggregated outputs, etc.) this survey from 1992 established a good basis for further time series creation.

The statistical waste survey is a part of the Programme of statistical survey for the relevant year. Under Act No 89/1995 Coll., on state statistical service, the reporting unit is obliged to provide all the required data. On the one hand for the unit there exists a legal duty to provide the required data to the Federal Statistical Office, but on the other hand the office provides these subjects with a guarantee of protection of the provided individual information. In practice this means that any data about the production of waste, the means of its elimination etc. may be provided to another subject only in aggregated form. This means in a form where it is not possible to identify the producing unit. The provision of information to the Ministry of the Environment (MoE) and its subordinate organisations (for example the Czech Environmental Inspectorate) is no exception. This fact is important in particular for those units which do have a duty under the waste act to keep records on waste, to report data about production and handling of waste to the district authority, but which for any reasons (mainly due to ignorance) do not fulfil this duty.

---

[1] At present the new version of act on waste is being discussed. The main reason for that is the fact that since 2001 the act on waste has been amended number of times in order to implement several mutually independent EU directives and therefore became relatively complicated and disorganized.

3.1 Statistical Surveys in 1992 - 1997

In spite of the reservations which existed about the survey – the recording of only part of the issue (category of special and dangerous waste), publication of only aggregated data (with regard to protection of individual data) etc., the data generated at the given time constituted a relatively good basis for the creation of time sequences showing the developmental trends in the area of the production and disposal of a certain category of waste.

The attempt to prevent the duplicate counting of waste (when transferring waste between several subjects) lead to the survey being focussed up to 1994 only on ascertaining the method for the disposal and use of waste at the own facilities of the producer. The interest in the identification of the further flow of waste after its transfer to another unit was the reason for expanding the survey (from 1995) to include the identification number of the organisation to which the given waste is transferred. On the basis of a comparison of the receiving organisations with the register of operators of landfills and incinerators it was possible to identify the elimination of the transferred waste for approximately 60% of production.

Companies which on the basis of predominant activities were classified pursuant to Czech "Statistical Classification of Economic Activities" [8] (referred to hereinafter as NACE) in sections 01, 10 to 41, 45, 55, 601, 602, 62, 747, 7481, 851, 852, 90, 9211, and 93 and with 25 or more workers were included in the statistical survey. From the start the statistical survey was performed using the enterprise method, i.e., the data relates to the headquarters of the company.

No significant changes were made in the survey in 1997 either. One of the most significant changes was an expansion in the survey to include the anticipated means for elimination or use of waste at the receiving organisation.

More than 70 of statistical statements were returned, and there is a slight ongoing increase.

3.2 Statistical Survey of Year 1998

1998 was a year of very extensive changes in the area of waste statistics. The statistical survey of production and elimination of waste was adapted as far as possible to the new Waste Act 125/1997 Coll. and its implementing decrees. The main changes in the survey included:

1. switch to new waste catalogue [9];
2. switch to new system for codifying the methods for the disposal and use of waste;
3. inclusion of category "other waste" in survey;
4. expanding set of respondents.

On the basis of discussions with respondents about these changes, and on the basis of the testing of the statement, it was seen that changes No 1 and 2 caused the greatest problems. The new system for codifying the methods of disposal and use of waste in particular is considered relatively complicated.

From the aspect of the statistical survey there occurred significant problems in subjects where at the time systematic attention had been paid to waste management and they had the records adjusted on the basis of the new rules. But in addition to these subjects there existed hundreds and thousands of smaller subjects for whom it was very difficult to understand the new regulations and fill in the statistical statement.

The switch to the new waste catalogue [9] was a very difficult matter not only for the respondents themselves, but also for those processing the survey. From the aspect of simplifying this switch a conversion key between the old and new catalogue was created. This conversion bridge made it possible to preserve and prolong the timelines concerning the production and disposal of special and dangerous waste which at the time were of relatively high quality. It unified and simplified the conversions for the respondents themselves, and last but not least it at least in part preserved the possibility of checking information provided by the respondent.

The set of respondents was the same as in the previous period, i.e., companies which on the basis of predominant activities were classified pursuant to "Statistical Classification of Economic Activities" (referred to hereinafter as NACE) in sections 01, 10 to 41, 45, 55, 601, 602, 62, 747, 7481, 851, 852, 90, 9211, and 93 and with 25 or more workers were included.

3.3 Statistical Survey 1999-2012

In general terms it can be stated that in view of the above specified set of respondents it is evident that it was not possible to interpret the gained results as the overall volume of produced waste in CZ, but as its decisive part.

Up until 1998 no calculations of production which were common in EU countries were implemented. In 1999 for the first time a rough estimate was made for the total production of dangerous waste based on the methodology of the European Statistical Office. The calculations were based on the industry classification of waste producers, their size (according to the number of workers) and volume of their production.[2]

In the period 1999-2001 there were no significant changes in the organisation of the survey, but in 2002 a special section was introduced for ascertaining municipal waste from towns which the regular annual reports of this indicator to Eurostat followed on from [6].

---

[2] The analysis of production of waste according to size of enterprise (number of workers), according to production of enterprises and according to volume of annual waste production is relatively interesting and, from the aspect of imputations, essential. For example in 1996 enterprises with up to 50 workers contributed only 2% to the production of special and dangerous waste, but in the set of respondents it was virtually 28% of the surveyed enterprises. The analysis of the set of respondents from the aspect of volume of their production of waste showed that more than 95.3% of the ascertained volume is generated by less than 3% of respondents producing more than 6000 tons of special and dangerous waste per year. 74 % of the enterprises surveyed produce less than 150 tons per year, and 12% produce between 150 and 500 tons per year.

After five years, in connection with attempts to simplify the survey, in 2007 a unified form for towns and the corporate sector was introduced.

In 2008 a selected set of respondents was switched to the new classification CZ-NACE Rev. 2 [10], which was prepared by the Czech Statistical Office in compliance with Regulation (EC) No 1893/2006 of the European Parliament and of the Council of 20 December 2006.

The surveyed data is classified according to groups of the official waste catalogue among others [11]. The survey covers data on waste generated by enterprises (broken down by hazardous and other waste), waste generated by municipalities (including separately collected categories), waste management (recovery, disposal, etc.).

Another significant change was the change in the technology for selection of respondents for the corporate sphere in 2009, when a so-called "rotation" survey model was introduced (for more details on this model see below).

So far the last change which occurred in 2011 was the introduction of a survey into secondary raw materials [4]. This change occurred as a result of the amendment of Act No 154/2010 Coll., the Waste Act No 185/2001 Coll.

4 Technology for Processing of Statistical Survey

The actual technical processing of the filled in and collected statements is a highly complicated matter. For more than twenty years the Czech Statistical Office has been using the so-called ProjektMan software. This software is intended for the support and automation of project work on statistical statements. It allows the interactive drafting of sections, the assembly of statements using these sections, the assembly of parts of statements and supra-statements and the creation of projects above these statements and supra-statements.

The information administered by the ProjektMan system can then be used to assemble tasks and check runs in the DataMan system for data acquisition and the EPV electronic reporting system. ProjektMan can be run on PCs using the operating system MS-Windows. The minimum system requirements for ProjektMan include a Pentium class processor with frequency of at least 1GHz and 256 MB RAM.

The technology for processing the survey is described in the so-called technical project [5]. The technical project includes, amongst other things, the following information:

- Timetable for preparation and processing of entire survey;
- Coordination of all involved offices of the CZSO;
- Description of primary input data;
- Description of processing technology;
- Description of checks, corrections, autocorrections and derived indicators;
- Definitions of outputs.

With a view to current version of the statistics act respondents have an option of filling-out the survey both in electronic as well as in paper form. Electronic survey tools include editable .pdf forms and electronic survey software that is available free of charge online on www.czso.cz.

4.1 Data Collection and Processing

Data collection and survey processing is performed in a centralized manner with common procedures of checks, corrections and control relations. Data verification takes place right after survey processing. In case of discrepancies evaluation procedures are in place. In case of necessity, discrepancies may be additionally checked with respective respondents. Centralization of these procedures enables significant reduction of input data verification time. The necessary condition is full survey completion by the unit.

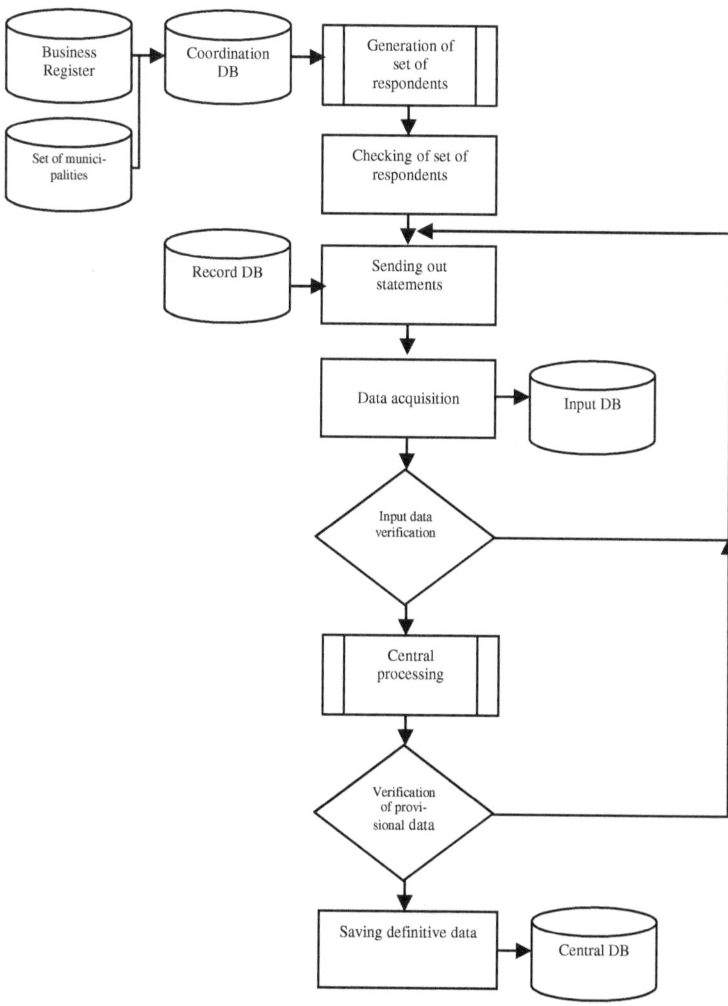

Fig. 1. Principal Survey Processing Scheme [Source: own construction]

The generated set of respondents passes via the coordination database and is also made accessible to the environmental statistics department for checking and agreement (T[3]-10 months), which is then followed by the sending out of electronic forms (or paper forms, if the respondent has no data box) (T-9 months), data acquisition (T-7 months), its input verification and uploading (T-4 months). After the end of gathering the data is sent to the central processing department, where so-called provisional data is generated (T-3 months). This is then subject to further checks, but not now from the aspect of the individual respondents but from the aspect of the entire set.

Any fluctuations and inconsistencies in waste data are traced back to the level of an individual respondent, with whom a verification of the reported data is initiated via the statistical processing department, or the data is immediately amended. In this way definitive data is gained, which is locked and saved to the central database (T-1 month), which is a data source for all publication outputs and public presentations of data for national and international needs.

4.2 Data Verification

In the course of processing of the statements there is verification, amongst other things, of the correctness of the data given by respondents via several tens of check links. After the discovery of any doubts the unit is contacted by workers of the statistical office, and they attempt to resolve the problem together. It is necessary to state that after many years of experiences with this survey the communication with respondents is constantly improving, as is their willingness to cooperate.

In the context of the data processing logical methods for the analysis and checking of all output indicators from various aspects are developed. The following 5 groups of methods for logical checks and links of indicators can be distinguished in the context of the individual statements:

- Relationship to identification of respondent unit;
- Relationship to classifications and code lists;
- Relationship to completeness of filling in of questionnaires;
- Empirical checks.

With regard to the identification of the relevant respondent unit it is possible to name logical checks focussing on the completeness and accuracy of specification of respondent identification data from the aspect of the formal requirements (for example, number of business ID No[4] symbols) and material accuracy (for example, presence of business ID No in Business Register of CZ [12] and/or presence of business ID No in selected set).

Logical checks in relation to classifications and code lists are performed from various aspect (for example according to the classification CZ-NACE [10], according to code lists of types of waste according to the Waste Catalogue [11], according to the waste category or, for example, the method of handling [13]). Logical controls

[3] T denotes September 30 of given year.

[4] ID No is the unique 8 number of identification of business unit in the Czech Republic.

focus primarily on the generation of impermissible codes in compliance with the classifications and code lists.

From the aspect of completeness of filling in of questionnaires, checks are performed focussing on the completeness of filling in of the individual lines (for example, if on a given line a certain item is filled in, the other relevant items in the line must be filled in). In addition many checks are based on the balance principle (for example, whether the amount of waste recorded as transferred waste is equal to the amount of waste recorded as received waste, whether the sum of the individual categories is equal to the value of the category of higher degree). The check sums represent a highly significant tool for checking the completeness of correctness of the filling in of the statement. Incorrect check sums are given including the relevant differences.

Empirical checks focus primarily on the logical interconnectedness and justifiability of the statements' individual elements in relation to the legal standards, methodological instructions and experience with collection of data from the preceding periods. For example, in view of the fact that the import of waste to CZ with the aim of dumping is forbidden, for the given type of waste according to original the handling code characteristic for storage was given. One example of a check from the methodical aspect is the linkage of definitions of waste and secondary raw material. If production of the recorded waste is identical with the amount of recorded production of secondary raw materials, this is probably an error, because in view of the methodology of the statement it is not possible to record the material flow in both sections of the statement. Many checks are associated with the fact that a certain means of handling is only possible in the case of certain waste codes. In addition, for example, in the case of a town there is an expectation that the part of the statement associated with municipal waste will be filled in. Empirical experience also makes possible a check on the presence of extreme values (for example, volume of municipal waste per inhabitant). Frequent errors include not giving the name and address of the firm taking the waste.

The most basic checks include a comparison of the specified production and method of waste disposal with the production and disposal declared in the preceding years. And this comparison helps to identify many errors which the respondents make. This involves in particular incorrect recording or recording of significantly lower or exactly the same production as the preceding period.

On the basis of experience, the following errors constitute other frequent errors in the area of the statistical survey:

1. mixing up units of measurement (for example, kilograms confused with tons and vice versa);
2. incorrectly selected physical indicators (instead of weigh units, volume units or items are given);
3. impossibility of making estimate (if the respondent itself is not capable of making a qualified estimate of weight, the workers of the statistical office help him with the estimate on the basis of previous experience; this involves, for example, the weight of strip lights, rechargeable batteries, content of sumps, sludge from riverbed maintenance, sedimentary water from sludge settling ponds etc.);

4. incorrectly selected financial indicators (for example, costs for waste removal are given differently with or without VAT).

In addition to the calculation of absolute values the analyses also contain an evaluation of the relevant indicators (so-called, provisional indexes). For these purposes in the phase of verification of provisional data so-called check tables are generated in the specified structure mainly allowing a comparison of the data recorded by individual respondents over time. The check focuses on units displaying a significant year-on-year increase (index >150) or dropped (index <50) in the investigated indicators. Attention is also paid to identical year-on-year data.

5 Generation of Basic Set

The set of survey respondents then consist of both legal persons and trading natural persons and towns. In compliance with Regulation (EC) No 2150/2002 on waste statistics the method of selective survey is used in order to minimize the burden on respondents [13].

Economic subjects are selected from a set of all possible respondents from the Business Register [11]. The Business Register plays the role of integrated statistical instrument which is regularly updated and in which the basic characteristics of the subjects allowing stratification and parametric selections are kept. All subjects in the Business Register are unambiguously designated by the identification number of the legal or natural person.

5.1 Criteria for Generation of Basic Set

The basic set of the CZSO survey in the area of enterprises in a general case includes economic subjects classified in the sectors S11, S13 and S15 of CZ NACE [10]. The concrete criteria for the generation of the basic set are as follows [5]: enterprises in context of CZ NACE codes 01-36, 451, 452, 47, 491-494, 55, 56, 59, 611-612, 9601-9603 with number of employees greater than 25; CZ NACE codes 41-43, 742, 812, 813 with number of employees greater than 20; CZ NACE code 4675 with number of employees greater than 5; CZ NACE codes 37, 38, 39, 4677 with zero number of employees and CZ NACE codes 86, 871-873 with number of employees greater than 50.

For the purposes of ensuring the necessary level of quality of the statistical outputs, some selected enterprises (so-called core enterprises) are also surveyed regardless of the number of employees. This involves about 300 firms which as the most significant waste generators with almost three quarters of the total amount of all produced waste in the country.

In order to reduce reporting burden the so-called "rotation model" approach to respondent selection has been used since 2009. It consists in regular surveying of the most significant business entities (according to generated waste, industry type, or pursuant to the number of employees). The remaining "rotation" part of respondents is inquired once every three years with the rest of the data being mathematically imputed. This approach enabled reducing of respondent number by almost a half.

Core and newly formed enterprises are selected for the selection set from the basic set. The remaining firms from the basic set are divided up into three thirds, of which only one third gets into the selection set for each year of the survey.

Sets of respondents also go through a so-called coordination database. This is a tool developed under the conditions of the CZSO making it possible to include the given economic subject in the current year in the minimum necessary number of statistical surveys (naturally with regard to its importance).

In the case of municipalities there is a systematic selection of the pre-designated number of units with randomly selected start stratified according to regions at the level of NUTS2 [14] and number of inhabitants in the municipality [5]:

1. up to 200 inhabitants;
2. 200-999;
3. 1 000 – 9 999;
4. 10 000-99 999;
5. above 100 000.

In context of the given rotation cycle municipalities selected for survey in the preceding year and belonging to strata 1.-3. have been excluded. Using this additional selection the change of sets is ensured, and in the given year the selected units are then assigned a new order of change, and the existing order of change of the remaining units in the selection set is reduced by 1. It should be noted that in the case of strata 4. and 5. there is a blanket survey.

For the purposes of a reduction in the administrative burden of municipalities, mathematical-statistical imputation methods are used as part of the survey of municipal waste production. For the purposes of imputation at least 5 filled statements are required in each stratum. In the case of stratum 2. and 3., 80% response is required, for stratum 4. it is 90%, and for municipalities in excess of 100 000 inhabitants a 100% response is required. It must be noted that the accuracy of the calculated results at the level of NUTS2 [14] even in the case of a 100% response the incidence of extreme values may not wholly satisfactory in view of great spread of calculated data [5].

5.2 Brief Characteristics of Computation Procedure

The scope of the basic set from the *enterprise* sphere is considered final *without the need for further calculations.* The data gained for *municipalities* is then *calculated for the entire Czech Republic.*

With regard to municipal waste, for the municipalities the calculations are made on the basic set of all municipalities of CZ and this is *only* on the basis of values filled in for the relevant year. The calculation is performed in the following manner: the acquired data for the individual variables is multiplied by the designated weightings, and the weightings are the same for all variables.

The calculation of weightings is performed in compliance with the separation of municipalities into regions according to size groups (strata). Weightings are estimated as the share of all inhabitants permanently living in the strata out of the number of inhabitants for the municipality with a filled in statement. Three values of weightings are distinguished:

1. weighting =1 for external units[5] and in a general case municipalities with a number of inhabitants in excess of 100 000,
2. non-zero weighting (>1) for other municipalities with filled in statement,
3. zero weighting for units not included in selection.

In compliance with the technical project [5], if in the given stratum extreme values are identified with the weighting =1, the weighting of the other non-extreme municipalities in the same stratum is increased in such a way that the sum of weightings multiplied by the number of inhabitants for the municipalities to whom a statement has been sent out in the given year in the given stratum is equal to the total number of inhabitants in the given strata.

As has been stated above, for the calculation of weightings for the individual to be possible and reliable, there is a requirement that in each stratum there should be at least 5 non-extreme municipalities with filled in statement. If the given conditions are not met in some stratum, a calculation of weightings is conducted for the larger strata (2 size groups of municipalities are merged into one). But this procedure reduces the accuracy of the calculation. As a priority the minimum number of filled in statements is ensured by sending reminders to respondents [5].

6 Balancing Users Needs and Reporting Burden Optimization

The official purpose of the survey proclaimed by the CZSO as an institution producing European statistics is the meeting of the duties arising from Regulation (EC) No 2150/2002 of the European Parliament and of the Council of 25 November 2002 on waste statistics, Directive 98/2008 of the European Parliament and of the Council of 19 November 2008 on waste and repealing certain Directives, for the purposes of the European Union (EU) and international organisations [15-17], in particular the Organisation for Economic Cooperation and Development (OECD) [18], and to give information to the public .

Among the major international user categories one can name e.g. Eurostat and its data users representing all EU member states and other countries. For more than a decade the CZSO has been transferring waste related data to Eurostat. During 1995-2004 the national data reporting was performed according to the so-called Eurostat/OECD Joint Questionnaire on Waste. After Regulation (EC) No 2150/2002 of the European Parliament and of the Council of 25 November 2002 on waste

[5] A municipality where the share of the value of the reference indictor to the number of permanently resident inhabitants is more than 20 times higher than the average share of the value of the reference indicator to 1 inhabitant in the given stratum is considered to a an extreme unit.

statistics, the national administrator of which is the Czech Statistical Office, Eurostat is sent data about waste in the format designated by Commission Regulation No 782/2005. In compliance with the given regulation the data which was originally classified according to the Waste Catalogue has been transformed into the required classification EWC–STAT (into the so-called statistical nomenclature of waste focussing on the main substance contained in waste) and in the following structure [13]:

- production of waste in classification pursuant to EWC-STAT [13] and CZ-NACE Rev. 2 [10] in classification pursuant to category of waste "dangerous" and "other" waste, level NUTS0 [14];
- handling of waste in classification pursuant to EWC-STAT and pursuant to selected codes of handling R and D, level NUTS0;
- numbers of devices for use and disposal of waste and their capacity, level NUTS2.

With regard to data about the production and handling of municipal waste, this has been sent to Eurostat every year since 1995 for the purposes of evaluating the Structural indicator on Municipal waste [19]. The methodology for the collection of data complies with the instructions in for the OECD and Eurostat questionnaire about municipal waste. The respondent units are the municipalities which ensure that waste produced by subjects integrated in the municipal system of waste collection, such as schools, government offices and small businessmen, is also included in the survey[6].

Among the main data users directly using data on waste are e.g. such national institutions as Ministry for the Environment and its affiliate Czech Environment Information Agency (CENIA), Ministry of Industry and Trade (MIT), variety of research and development institutions and, of course, media. The calculation of indicators for material flows in cooperation with Charles University in Prague (Environmental Center) is just one well-known means of use of the outputs from the statistical survey [20], [21].

The data about secondary raw materials is intended directly for the needs of the concrete institutional user in the shape of the Ministry of Industry and Trade as the basis for strategic and analytical document for the area of use of secondary raw materials pursuant to government resolution No 1398/2009.

Monitoring of secondary raw materials represents the most significant change of the survey of the last years. This change reflects the latest political and economic conditions when producers are highly motivated to transform waste (i.e. negative production) into secondary raw materials and to actively use them as production inputs.

Issues of raw materials safety are paid an exceptional attention at the European level at present. Secondary raw materials are looked at as a strategically important input into production process. Under these circumstances one can expect that extended waste (and secondary raw materials) production survey would be consequently improved and developed in future according to various stakeholder needs.

---

6 The statistical data of the CZSO does not include municipal waste produced by legal persons and natural persons authorized to trade who are not included in the municipality's municipal waste management system.

The flexible nature of the statistical survey [20], which is implemented on the basis of an annually approve programme of statistical survey, is not *a priori* bound by complicated processes for the passing legal regulations and allows the CZSO to react flexibly to the requirements of statistical users. For instance, in CZ the waste related data is also being currently provided from an administrative source by the Ministry for the Environment through Integrated system of reporting compliance at CENIA. However, since its actions are limited by the waste-related legislation, and novelization processes involving number of stakeholders are difficult and time consuming, the Ministry for the Environment has no power to follow the new material flow trends so operatively. This also highlights flexibility of national statistical office in reflecting current socio-economic reality.

The collected data is a part of number of regular publications such as Statistical Yearbook of CZ, Generation, Recovery and Disposal of Waste in CZ [20], Statistical Environmental Yearbook of CZ [22], and Report on the Environment of CZ [23] among others. The last two are joint outcomes of CZSO and the Ministry for the Environment.

One should expect, however, that the range of possible data users is far more extensive. Since the CZSO publishes the majority of data on its web pages, significant amount of regular data users remain anonymous, since they obtain them indirectly via internet.

High-quality and easily accessible information is the key to the correct and effective decision-making and implementation of environmental policy and waste management instruments. Data to evaluate the status of waste management come largely from public resources available [21-23]. Any additional information are used for comparison or detailed descriptions of phenomena that are not tracked in CZSO public statistics. They are based on functional, up-to-date and credible information resources of CZSO and MoE, CENIA, MIT describing the status of waste management in CZ, operated within a defined legal and information environment of CZSO.

The CZSO is preparing data and information outputs for the needs of the MoE, other ministries and the Government of CZ. In addition, it satisfies the reporting requirements of Eurostat and this data and information are used by the European Commission, DG Environment and the European Parliament. Quality of the information obtained from statistical surveys is ensured by internal standards of CZSO compatible with Eurostat.

Range of possible waste statistics outputs is closely related to methodological peculiarities of their production. One should mention that the range of possible waste statistics uses is objectively limited by legislative requirements on the one hand and financial perspective both on the side of respondents and CZSO on the other.

The development in the scope of the survey or the number of respondents over the period of the existence of the statement is characterised by the overview given in Table No 1.

Table 1. Waste Production Survey Respondent Base [Source: CZSO]

Year	1993	1995	1998	2002	2004	2006	2008	2010	2012
No. of respondents	5820	10816	14461	17741	19085	17875	17363	8466	10369

As one can see, the change in the technology of selecting respondents implemented in 2009, when the so-called "rotation model" was introduced, was a truly significant change. This resulted in the administrative burden being almost halved.

The aim of the state statistical service is to produce high quality and consistent data on development trends in a given area. According to their nature statistical data is only valid for producing aggregate outputs and according to law cannot under any circumstances be used for administrative or control activities. The aforementioned technology for the collection of data in combination with the methods for imputation for the entire population guarantees the year-on-year comparability of data and allows the use of the gained data to designate trends of the monitored indicators without the results being influenced by the number of included units. Moreover the design of the statement and use of the line of check links significantly prevents duplicity in the entry of the statement in the processing system. And the linkage with the register of economic units plays an irreplaceable role for gaining statistical data with all the attributes of quality.

With respect to above mentioned properties waste statistics information can be used both for designing environmental policies at the national level, as well as its consequent assessment. Waste production dynamics is capable of indicating critical fields where interventions of the state planner are necessary. The same data may be consequently used for assessing efficiency of respective environmental policy measures based on number of benchmarks available in comparable time series.

Waste statistics data produced by the CZSO based on EU-wide common methodological basis can be applied as an environmental management tool also at international level [15-18]. For instance, they can be used for benchmarking of country performance and achievement of different policy goals in comparison with other member states.

Concerning information needs of the regional policies one can state that this is one of the weaknesses of the statistical survey on waste. Regional policy is primarily aimed at programs at the level of administrative regions' falling under respective executive bodies. One also cannot exclude user requirements for data about of non-administrative, i.e. functional regions. With respect to survey methodology one can find several objective shortcomings that disable production of regional statistics (e.g. [24] in Czech).

From the legislative point of view survey methodology fits the requirements of relevant EU regulations. As it was mentioned above European legislation requires data reporting at the CR (i.e. NUTS0) level, without any need for data regional decomposition. This is directly related to the objective needs of the European executive power data needs. It is also an outcome of active administrative burden reduction policy both at the EU and national levels. Given additional saving measures that are being made by the government the waste survey is designed in a way to satisfy only mandatory legal EU requirements. These features are directly related to a limited extent of respondents and application of rotation and imputation techniques that enable production of required quality outputs at the national level, and not at regional level though.

Another reason for regional data absence is the applied method of data collection. According to methodology indicators are surveyed based on the so-called enterprise method that allocates respondent data according to headquarters geography and not according to respective local units. In this respect regionalization may be accompanied by significant distortions of waste production geography (e.g. establishment with local production unit in one of the regions and headquarters located in the capital city may indicate e.g. production of waste in the capital city where no pollution production actually takes place). Such input data cannot be used for production of quality regional waste statistics. Only municipal waste data is relevant at the regional level, since one can directly allocate municipalities to the specific region.

In order to produce data that would be valid at the regional level the current enterprise method should be replaced. E.g. companies could fill-out as many surveys as many regions they operate in. Number of respondents in this case would remain the same, however total administrative burden would increase. This approach is however related to the possible double counting issue related to common practices of transporting waste materials between the local units.

7 Conclusion

Waste management data collection technology combined with imputation methods enables inter-temporal comparability of data in EU and its use for setting of development trends of the Czech Republic in compliance with EU. Waste statistics outputs provide the whole range of institutional users a valuable information base for environmental policy design and its consequent assessment both at national, as well as international levels.

At present the statistical survey method is primarily based on requirements of EU legal acts in the field of waste statistics. The limited survey extent is also highly influenced by administrative burden reduction and cost saving policies of the government. These are objective factors that disable production of outputs at the regional level. Given the institutional demand and more favourable economic conditions the flexible nature of the statistical survey, however, doesn't exclude the possibility of adaptation of survey design to ever-changing needs of decision-making sphere.

Acknowledgements. This paper is a result of the project of Technology Agency of the Czech Republic No. TD010130 Regionalization of economic performance indicators in relation to environmental quality. The support is gratefully acknowledged.

References

1. Ritschelova, I.: Determinantion of Waste Production in CR. Eurostat - Work session on environmental statistics, Lysebu – Norway, Working Paper No. 13 (1996)
2. Ritschelová, I., Sidorov, E., Hřebíček, J.: Environmental Statistics at the CZSO: Past Development and Future Trends. In: Hřebíček, J., Schimak, G., Denzer, R. (eds.) ISESS 2011. IFIP AICT, vol. 359, pp. 33–46. Springer, Heidelberg (2011)
3. Ritschelova, I., Sidorov, E., Farsky, M.: Impact of Waste Deposition Fees on Enterprises in the Czech Republic. E&M Ekonomie a Management, 2 (2010)
4. CZSO. Generation, Recovery and Disposal of Waste 2011. CZSO, Prague (2012)

5. CZSO. Technical Project of Collection, Processing and Data Presentation in CZSO Odpady 5-01 for 2012 (Technický projekt sběru, zpracování a prezentace dat v resortu CSÚ. Odpady 5-01 za rok 2012). Prague, CZSO (2012)
6. EUROSTAT. Quality in the European statistical system – the way forward. Luxembourg: Office for Official Publications of the European Communities, (2002), http://epp.eurostat.ec.europa.eu/portal/page/portal/quality/documents/ESS_QUALITY_RECOMMENDATIONS_2002_EN_0_1.pdf
7. CZSO. Statistical surveys of the CZSO. CZSO, Prague (2012), http://www.czso.cz/eng/redakce.nsf/i/statistical_surveys_of_the_czso
8. CZSO, http://www.czso.cz/csu/klasifik.nsf/i/odvetvova_klasifikace_ekonomickych_cinnosti_%28okec%29
9. Decree of the Ministry of Environment, issuing waste catalogue and provides other lists of waste (Waste Catalogue) 337/1997 Coll., Prague, Ministry of Environment (1998)
10. CZ-NACE, http://www.czso.cz/eng/redakce.nsf/i/classifications
11. List of waste, http://eur-lex.europa.eu/smartapi/cgi/sga_doc?smartapi!celexplus!prod!CELEXnumdoc&numdoc=32000D0532&lg=en
12. Business Register, http://www.czso.cz/eng/redakce.nsf/i/business_register
13. European Waste Classification for Statistics, version 4, http://ec.europa.eu/eurostat/ramon/other_documents/ewc_stat_4/index.cfm?TargetUrl=DSP_EWC_STAT_4
14. CZ NUTS, http://notes.czso.cz/csu/klasifik.nsf/i/klasifikace_uzemnich_statistickych_jednotek_%28cz_nuts%29
15. EIONET, http://scp.eionet.europa.eu/facts/wastebase
16. DG Environment, http://ec.europa.eu/environment/waste/publications/
17. EEA, http://www.eea.europa.eu/themes/waste
18. OECD, http://www.oecd.org/env/waste/
19. EUROSTAT. Sustainable development indicator on municipal waste. How to deliver data, http://epp.eurostat.ec.europa.eu/portal/page/portal/waste/documents/Guidance_document_on_Municipal_Waste_2011.pdf
20. CZSO. Statistical Yearbook of the Czech Republic 2012. CZSO, Prague (2013), http://www.czso.cz/csu/2012edicniplan.nsf/engkapitola/0001-12-eng_r_2012-0300
21. Kovanda, J., Weinzettel, J., Hák, T.: Material Flow Indicators in the Czech Republic in Light of the Accession to the European Union. Journal of Industrial Ecology 14, 650–665 (2010)
22. Statistical environmental yearbook of the Czech Republic 2011. CENIA, Prague (2012), http://www.cenia.cz/rocenka2011/index.htm
23. CENIA, Report on the Environment of the Czech Republic (2011), http://www1.cenia.cz/www/sites/default/files/Report_on_the_Environment_of_the_Czech_Republic_2011_pro%20web%20a%20USB.pdf
24. Statistical survey South Moravia region (Statistická ročenka Jihomoravského kraje (2012), http://www.czso.cz/csu/2012edicniplan.nsf/krajkapitola/641011-12-r_2012-03

Methodology and Information System for Evaluating Environmental Protection Expenditure Efficiency at the Local Level

Jana Soukopová and Eduard Bakoš

Masaryk University, Faculty of Economics and Administration,
Department of Public Economics, Lipová 41a, 602 00 Brno, Czech Republic
soukopova@econ.muni.cz, e_bakos@email.cz

Abstract. The paper presents a methodology and information system for evaluating the efficiency of current municipal environmental protection expenditures. The methodology and information system were approved by the Ministry of the Environment of the Czech Republic as a voluntary tool for municipal officials.. The proposed methodological procedure for evaluating municipal environmental protection expenditures is based on a weighted assessment of multiple criteria. The procedure gives municipalities an instrument for assessing expenditure efficiency and addresses the three pillars of sustainable development: economic development, social development, and environmental protection. The methodology and information system can be used by other countries and municipalities to evaluate the efficiency of public spending at the local level.

Keywords: Efficiency, environmental protection expenditure, information system, methodology, local level, voluntary tool.

1 Introduction

Defining and measuring *efficiency* in the process of using resources and transforming them into outputs and outcomes seems to be one of the biggest issues of contemporary economic theory. In 1957, Farrell asked how to measure efficiency and stated its importance for economic policy makers: 'It is important to know how far a given industry can be expected to increase its output by simply increasing its efficiency, without absorbing further resources' [6]. Efficiency evaluation and the technology for evaluating efficiency have greatly improved and advanced over the past several decades. However, it remains a conceptual challenge in relation to public expenditures. According to a group of authors at the European Commission [9] this issue is further complicated by the fact that public sector outcomes are often off-market, lacking relevant data, and thus cannot be quantified.

In practice, the ability to effectively manage public expenditures lies in the applicability of the manuals and practices that originated in international organizations. A number of tools have been created for public expenditure management. The handbook

J. Hřebíček et al. (Eds.): ISESS 2013, IFIP AICT 413, pp. 560–570, 2013.

by [1] represents an example of a means to effectively manage public expenditure in general. Contemporary society prioritizes the human relationship to nature. Therefore, some handbooks and guides have been created to manage the environment and especially environmental protection costs and expenditures based on experience with the OECD methodology [2-3], [10], [13-15], which is primarily focused on the economic tools of environmental protection (specifically on public expenditure in this area), as well as on the recommendations of the United Nations Organization and World Bank [4].

The aim of the paper is to introduce a new methodological background developing for municipal environmental protection expenditure efficiency evaluating, taking into account three pillars of sustainable development (economic, social and environmental) and efficiency in the sense of 3E methodology. The methodology for the evaluating environmental protection expenditure efficiency at the local level (hereinafter referred to as 'the methodology') was created by the authors of this paper [16] and approved as a voluntary tool by the Ministry of the Environment of the Czech Republic. Next goal is to present the developed simple information system for implementing the methodology [18].

2 Methodology and Information System

Environmental protection expenditures (EPE) are all the spending on all activities aimed at both preventing and protecting the environment [5]. One of the key criteria is that environmental protection is the primary objective of these activities. Activities that positively affect the environment but do not have protection of the environment as their primary aim are not included. EPE are classified according to funding sources, types of expenditure, and areas of environmental protection. In terms of funding sources, the European Union (EU) statistics divide EPE [5] into the public sector, industrial sector, and environmental specialist manufacturers and producers of environmental services for the private and public sector. EPE are divided into capital (investment) expenditures and current (non-investment) expenditures.

The methodology was designed for the evaluation of current public EPE only. The Classification of Environmental Protection Activities (CEPA 2000) is most frequently used to determine EPE. CEPA 2000 divides EPE into nine areas of environmental protection (EP) [5]:

1. protection of ambient air and climate;
2. wastewater management;
3. waste management;
4. protection and remediation of soil, groundwater and surface water;
5. noise and vibration abatement (excluding external safety);
6. protection of biodiversity and landscapes;
7. protection against radiation;
8. research and development;
9. other environmental protection activities.

The methodology was designed for each area of environmental protection separately. Area 8, research and development, is not included because the municipalities have no such expenditures. The methodology philosophy is:

- *Use the concept of environmental protection, including economic and social aspects* – The methodology is based primarily on the three pillars of sustainable development. Each pillar is evaluated separately. The overall rating is a summary of these three pillars.
- *Use existing methodologies and analysis* – Good practices and satisfactory evaluation indicators of existing methodologies, procedures, and methods were incorporated in the methodology if possible and adapted to the conditions of Czech municipalities.
- *Data availability* – The authors attempt to define indicators for the evaluation that are accessible for regional authorities.
- *Weighted multi-criteria evaluation.*
- *Simplicity and complexity of output* – The methodology result was proposed as an index facilitating the interpretation, publication, and communication of evaluation results.
- *Quality of legislation* – Use of the methodology is strictly limited by the legislation of the country that wants to put this methodology into practice.

2.1 Procedure for Assessing Efficiency

The methodology evaluates current municipal EPE in terms of the "3Es" (economy, effectiveness, efficiency). The suggested assessment process is divided into two main levels: basic and general. The basic assessment is based on a municipal environmental management evaluation and a principle of appropriate budget planning. The general assessment is used for each EPE and proceeds in three (parallel) parts that correspond to the three pillars of sustainable development.

The basic principle of the methodology is structured and easy-to-survey tables, which must be completed in steps. The methodology uses both qualitative and quantitative methods of assessment. The qualitative methods include a simple questionnaire for each EPE. The quantitative methods incorporate weighted multi-criteria analysis techniques. The methodology includes closed and open questions.

Figure 1 shows a simplified schema of the procedure for assessing the effectiveness and design of the methodology as an algorithm.

Basic Assessment. To keep the evaluation simple, it was appropriate to use existing and available data. Expenditures on selected budget items were divided into eight areas of environmental protection. The areas of EP defined by CEPA 2000 (see above) were changed by the Ministry of the Environment of the Czech Republic[1].

---

[1] Two areas - Noise and vibration abatement protection of biodiversity and landscapes and Protection against radiation were united. New area - Administration in environmental protection was formed by dividing from the area - Other environmental protection activities.

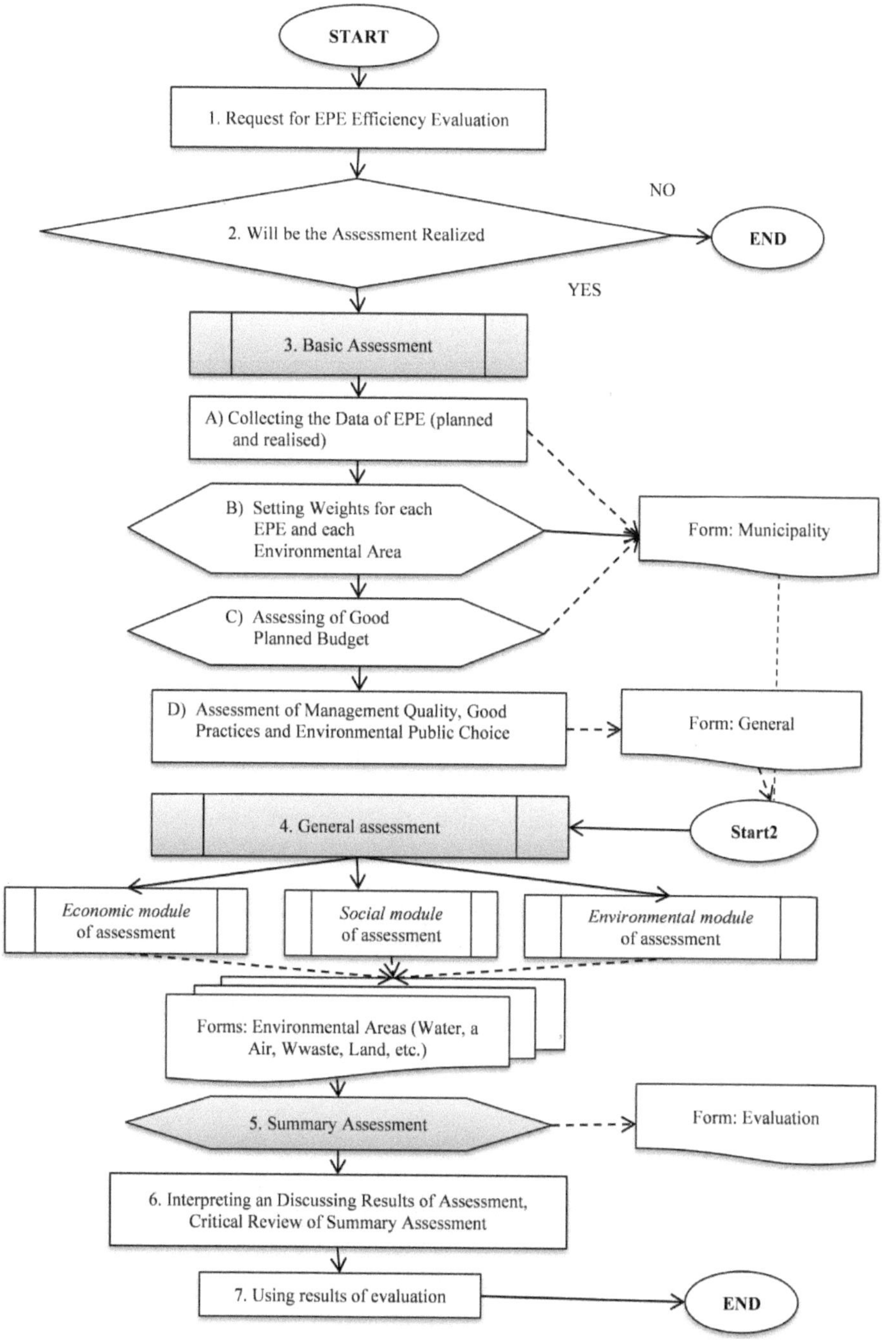

Fig. 1. Scheme of the methodology (Source: authors)

Environmental protection expenditures defined according to [2] (items) C_{io} (i=1…N, o=1…M), N = 49, M = 8, are used for a basic assessment, which the first function is setting weights of each EPE and each area of EP. The weights of each EPE are set as follows:

$$w_{iO} = \frac{C_{iO}}{\sum_{i,j=1}^{n,8} C_{iO}} \tag{1}$$

where:

w_{iO} is the weight of the i-th EPE in the O-area of EP (i = 1…n, O = 1… 8);
C_{iO} is the i-th EPE in the O-area of EP;
n is the number of implemented EPE of the municipality.

The second function of the basic assessment is the evaluation of the budget planning. The difference between the budgeted and actual EPE is analyzed, and the results are entered into the general assessment.

The third function of the basic assessment is the evaluation of management quality, good practices and supporting the elected bodies of the municipality/city. The results of this evaluation are entered into the general assessment.

General Assessment. When considering efficiency, the general assessment is based on multi-criteria evaluation of efficiency anchored in 3 basic pillars of sustainable development (economic, social and environmental).The evaluation of both economic and social pillars of sustainable development is consistent for all areas of the environment. Questions for the economic pillar are still further divided into the following areas: legality, effectiveness, economy, and efficiency. Questions for the social pillar detect how environmental protection policies create space for the participation of stakeholders, enhance the quality of life, improve of working conditions, and more. Questions for the environmental aspects of the assessment are different in each area of EP and include municipal priorities in the area 7 – Administration in EP.

To simplify the methodology, answers are set and assessed from 0 to 3 points:

- 3 – fully satisfactory;
- 2 – rather satisfactory;
- 1 – rather unsatisfactory;
- 0 – unsatisfactory.

The methodology describes the point system in more detail [16]. The set of questions for each area and each pillar can reach up to 100 points. These points are weighted. The weights were subsequently determined by experts:

- economic pillar – weight 0.35;
- social pillar – weight 0.25 and
- environmental pillar – weight 0.30 and budget planning – weight 0.10.

The point assessment of each EPE is as follows:

$$EC_{iO}=Ec_{iO}w_{Ec}+S_{iO}w_S+En_{iO}w_{En}+GPB_{iO}w_{GPB} \quad (2)$$

where

EC_{iO} is the point evaluation of i-th EPE in the O-area of EP (0-100 points),
Ec_{iO} is the point evaluation of the economic pillar (0-100 points),
S_{iO} is the point evaluation of the social pillar (0-100 points),
En_{iO} is the point evaluation of the environmental pillar (0-100 points),
GPB_{iO} is the point evaluation of the actual budget (0-100 points),
w_{Ec} is the weight of the economic pillar,
w_S is the weight of the social pillar,
w_{En} is the weight of the environmental pillar,
w_{GBP} is the weight of budget planning, where: $1=w_{Ec}+w_S+w_{En}+w_{GPB}$

The overall rating is a sum of the weighted sums of the all three pillars of sustainable development and budget planning:

$$E=\sum_{i=1}^{n} EC_{iO}w_{iO} , \quad (3)$$

where

E is the efficiency of the sum of EPE
w_{iO} is the weight of the i-th EPE in the O-area of EP (i = 1,...,n O = 1,..., 8), see above (1)
GPB_{iO} is the point evaluation of the actual budget (0-100 points).

The full result of the methodology was presented in a rating which contains results of each pillar of sustainable development and assessment of actual budget.

2.2 Information System for Measurement of EPE Efficiency

The above text shortly introduced developed methodology for evaluating environmental protection expenditure efficiency at the local level. We used properties of the MS Excel and developed information system (IS EPE) [21] for the integration and simple interconnection of above basic and general assessments into all areas of EP and into summary assessment of EPE efficiency [18]. We used our experience with development of similar information system for waste management expenditure [17], [22-24].

This implemented IS EPE enables the simple option of the chosen set of the input EPEs through the single MS Excel sheet – *List of municipality*, which is interconnected with further MS Excel sheets (IS EPE modules), where are implemented assessment of three pillars of sustainable development:

1. the module *Municipality* of socio-demographic variables and planned and realised EPE variables C_{iO} of the municipality/city;

2. the module *General* with closed questions relating to the tools of environmental management, good practices, supporting the elected bodies of the municipality/city;
3. the modules *Water / Air/ Waste/ Land/ Biodiversity/ Noise and Radiation / Administration / Others* of environmental protection areas (8). These modules include closed questions to assess the effectiveness of the three pillars of sustainable development;
4. the module *Evaluation* with dynamically calculated results of evaluation. This module contains an evaluation of each EPE, each EP, a sum of EPE, each pillar and budget planning. Based on communication with local authorities, these indexes of efficiency are presented as a percentage of 100% efficiency;
5. the module *Setting* of main communication interface of the methodology, where are setting points for the answers and weights for pillars of sustainable development.

Users of this IS EPE set up planned and realized measurement of EPE. The answer of open questions in the module *Municipality* of IS EPE corresponds to the instructions on the closed question in the module *General* and depending on the allocation of expenditure in the modules of individual areas *Water, Air, Waste, Land*, etc.. Based on its answers IS EPE will evaluate the effectiveness of EPE, EP, all pillars and budget planning of which is shown in the module *Evaluation.*

The IS EPE is designed as a simple tool for evaluating the effectiveness of current EPE and is freely accessible to all local authorities and other organizations interested in evaluating the environmental effectiveness of their current expenditure points for the answers and weights for pillars of sustainable development through its web page [21]. The use of IS EPE is possible without the acquisition of licenses.

Results of IS EPE. In Table 1 are presented the results of the applied methodology and the IS EPE in Prague, the capital of the Czech Republic in year 2010. Since 2005, the total environmental protection expenditures of Prague have been over 2 billion CZK yearly (in 2010 more than 2.6 billion CZK). Table 1 and Figure 2 present the results of evaluation from the module *Evaluation.*

Table 1. Evaluation of EPE of the capital Prague in 2010 (Source: authors)

Set of ratings	Performance
Efficiency of the economic pillar	71,36%
Efficiency of the social pillar	79,02%
Efficiency of the environmental/ecological pillar	96,52%
Budget planning	99,82%
Efficiency of all environmental protection expenditures	86,68%

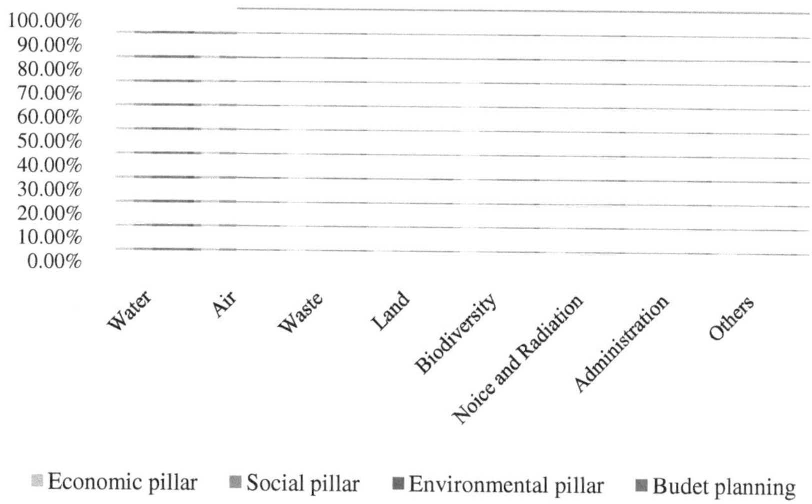

Fig. 2. Efficiency of all pillars of sustainable development and all areas of EP (Source: authors)

The results assessing the effectiveness of each pillar in the relation to each EP provide information within a given area is most efficiently, or how many p are expenditures to which this section or effective compared to other sections. The weights for the individual pillars were determined by an expert, but the methodology makes it possible to set the weights for each pillar in accordance with the priorities of the municipality. Alternatively, the methodology can be based on an evaluation of the sum of all areas and the priorities of the municipality as defined in an assessment.

3 Discussion

It is obvious that the methodology provides a wide variety of information unincorporated in other well-known methodologies [2-4], [10], [12-14]. The results present information on:

1. the efficiency of each EPE and each area of EP;
2. an assessment of each pillar of sustainable development and budget planning;
3. a basis for planning, decision making, negotiating, monitoring, and possibly also taking preventive measures;
4. the possibility of monitoring efficiency in real time;
5. the reduction of the likelihood of incorrect or inefficient allocation of resources.

The following factors could present problems:

- lack of appropriate data;
- false interpretation of results, even with proper data;

- the methodology does not address technical efficiency;
- the methodology depends on the management of public spending.

It is necessary to add that the decision-making about EPE with the support of IS EPE also depends on political decisions, which can significantly influence public expenditures regardless of the results of any evaluation [8], [12], [20]. Even the best methodology does not address this problem, but a good methodology could contribute to general awareness for possible approaches to enhance the current situation. The criterion 'actual budget' is included in the methodology assessment. The inclusion of this criterion means that municipal officers have the opportunity to adjust the environmental priorities of the municipality and then check whether these priorities are met. The authors are aware that the implementation of this methodology can be complicated, but the case study shows that implementation is possible.

4 Conclusion

The Czech government does not currently use indicators for evaluating various areas/ activities at the local (municipal) level. Most municipalities do not work systematically with local environmental indicators, including economic and social indicators. Statistical indicators are used only in isolation and unsystematically. The presented methodology and information system were designed in response to the absence of a simple methodology and information technology (IT) tools for the needs of local authorities.

The developed methodology and IS EPE [21] were also created because is not a national level or globally unified conceptual tool for EPE evaluation in the Czech Republic. Although the OECD methodologies [10], [13-15] exist, there are based generally for evaluation economics instruments. OECD methodologies also emphasize and draw attention to the institutional environment over an assessment of economic instruments. The concept of our proposed methodology and its IT implementation is based on a different philosophy than linking the environmental, economic, and social pillars of environmental protection. Methodology and IS EPE use practical experience from previous "good practice" methodology [7], [19] and "good practices" [14] for the management of public expenditure on environmental protection and developed information system for waste management expenditure [17], [23-24].

The methodology results show information for the evaluation of economic, social, and environmental efficiency and budget planning for individual spending areas as well as protection of the total expenditure on environmental protection and is supported by IS EPE. This makes it possible to get an overview of allocated resources EPE in relation to the level of total expenditure in all areas and activities of environmental protection in municipalities of the Czech Republic. Results from IS EPE provide information about all four values which is graphically displayed for all the different areas of environmental protection.

The evaluation of effectiveness and efficiency of public spending with IS EPE should support decision-making in the political process. It provides information on the extent to which the environmental objectives and other objectives of an implemented municipal environmental policy have been achieved, and with which economic, environmental, and social impacts on all stakeholder were concerned. The results from using IS EPE should include, among other things, the opportunity to compare public spending in relation to other municipalities.

The developed methodology could inspire other Member States of EU and their municipalities to evaluate the effectiveness of public spending at the local level.

Acknowledgement. The paper was elaborated within the project of the Czech Science Foundation no. P403/12/0366 "Identification and evaluation of region specific factors determining outcomes of reforms based on NPM – the case of CEE".

References

1. Allen, R., Tommasi, D.: Managing public expenditure. A Reference Book for Transition Countries. OECD (2001)
2. Barde, J.P.: Economic instruments in environmental policy: Lessons from the OECD experience and their relevance to developing economies (No. 92). OECD (1994)
3. Burns, A., Yoo, K.Y.: Improving the efficiency and sustainability of public expenditure in the Czech Republic. (No. 328). OECD Publishing, Paris (2002)
4. EIG: Annual Report Results and Performance of the World Bank Group (2010), `http://www.worldbank.org/ieg/rap2010/index.html`
5. Eurostat: Classification of Environmental Protection Activities and Expenditure (CEPA 2000) with Explanatory Notes, submitted to the United Nations Expert Group on International Economic and Social Classifications – meeting June 18–20, New York (2001)
6. Farrell, M.J.: The measurement of productive efficiency. J. R. Stat. Soc. Ser A-G 120, 253–290 (1957)
7. Jílková, J. et al.: Efektivnost veřejných výdajů na ochranu životního prostředí. Závěrečná zpráva (Effectiveness of public expenditure on environmental protection. Final Report). VaV/320/3/03. MŽP, Praha (2004)
8. Malý, I.: Veřejné statky a veřejně poskytované statky (Public goods and public services). Polit ekon 6, 861–868 (1998)
9. Mandl, U., Dierx, A., Ilzkovitz, F.: The effectiveness and efficiency of public spending (No. 301). Directorate General Economic and Monetary Affairs, European Commission (2008)
10. Mickwitz, P.: A Framework for Evaluating Environmental Policy Instruments Context and Key Concepts. Evaluation 9, 415–436 (2003)
11. Vyhláška č., M.: 323/2002 Sb., o rozpočtové skladbě (Notice No. 323/2002 Coll., on the budget structure). Praha (2002), `http://www.mfcr.cz/cps/rde/xchg/mfcr/xsl/vyhlasky_32273.htm`
12. Nemec, J., Ochrana, F., Šumpíková, M.: Czech and Slovak Lessons for Public Administration Performance Evaluation, Management and Finance. Ekon cas 4, 353–369 (2008)
13. OECD: Evaluating the Efficiency and Effectiveness of Economic Instruments in Environmental Policy. OECD Publishing, Paris (1997)
14. OECD: Good Practices of Public Environmental Expenditure Management in Transition Economies. OECD Publishing, Paris (2003)

15. Peszko, G.: Good Practices of Public Environmental Expenditure Management in Transition Economies (2002), http://www.oecd.org/environment/outreach/2080960.pdf
16. Soukopová, J., Bakoš, E., Neshybová, J., Hřebíček, J.: Hodnocení efektivnosti výdajů obcí na životní prostředí (Evaluation of effectiveness of municipal spending on the environment) (2010), http://amr.webnode.cz/metodika-hodnoceni-vydaju-obci/metodika-hodnoceni-efektivnosti-vydaju-obci-na-zp/
17. Soukopová, J., Hřebíček, J.: Model of cost and price relationships for municipal waste management of the Czech Republic. Acta Universitatis Agriculturae et Silviculturae Mendelianae Brunensis 59, 371–378 (2011)
18. Soukopová, J., Neshybová, J.: Software pro metodiku hodnocení efektivnosti výdajů obcí na ŽP (Software of methodology for evaluating the effectiveness of municipal spending on the environment) (2010), http://amr.webnode.cz/metodika-hodnoceni-vydaju-obci/software-pro-hodnoceni/
19. Šauer, P., Kreuz, J., Hadrabová, A., Dvořák, A.: Assessment of Environmental Policy Implementation: Two Case Studies from the Czech Republic. Pol. J. Environ. Stud. 21, 1383–1391 (2012)
20. Špaček, D.: Apparent Issues of Self-evaluation in Public Administration: The Case of Common Assessment Framework Implementation. International Journal of Knowledge, Culture and Change Management 10, 45–59 (2010)
21. IS EPE, http://amr.webnode.cz/metodika-hodnoceni-vydaju-obci/software-pro-hodnoceni/
22. Hejč, M., Hřebíček, J.: Primary Environmental Data Quality Model: Proposal of a Prototype of Model Concept. In: Proceedings of the iEMSs Fourth Biennial Meeting: International Congress on Environmental Modelling and Software (iEMSs 2008), pp. 83–90. iEMS, Barcelona (2008)
23. Soukopová, J., Kalina, J.: Mathematical model of economics of municipal waste management. In: Proceedings of the 30th International Conference Mathematical Methods in Economics, pp. 823–829. Silesian University in Opava, Karviná (2012)
24. Hřebíček, J., Kalina, J., Soukopová, J.: Integrated economic model of waste management: Case study for South Moravia region. Acta Universitatis Agriculturae et Silviculturae Mendelianae Brunensis 61, 913–918 (2013)

Modelling and Simulation Support of EMS Processes

Jan Ministr

VSB – Technical University Ostrava, Faculty of Economics,
Sokolská 33, 701 21, Ostrava, Czech Republic
jan.ministr@vsb.cz

Abstract. In Environmental Management System (EMS) area there are inherently included received decisions which entail a great risk. Reducing this risk can be achieved by using dynamic simulation models of EMS processes, such as emergency plans. The dynamic model of problem area within the EMS allows predicting sufficiently the actual results of the proposed decision and supports creativity management. This paper describes the possibilities of modelling tools Witness in modelling and simulation of the EMS processes.

Keywords: EMS, Environmental Management System, Model, Process Modelling, Process Owner, Simulation.

1 Introduction

The Environmental Management System (EMS) processes belong usually to "hard" processes that require a very fast response time. The modelling and simulation tools of these processes represent the important support of decision making in the area of EMS. The usage of these tools increases the quality of decisions and optimizes the costs incurred to realize a decision [2]. The basic premise of this approach is the implementation of process management in the organization and creative approach of the process owners who are responsible for the effective progress of the processes [1].

Computer simulation of the EMS processes in the form of dynamic models represents a robust technique that allows solving many complicated problems in this area. There are two basic approaches to design changes in the EMS processes without using of simulation. The first approach uses the experience of managers and experts and to implement their proposed process modifications. However, this approach does not always improve the process because it is experience-based only. The second approach gradually try running some variant of the EMS process changes in practice and then choose the most suitable solution. The disadvantage of this approach lies in the need of implementing each alternative of process in practice, increasing cost and time requirements [6].

If the process owner wants to avoid the above problems the solution for optimizing EMS processes is to use dynamic simulation required process. First, there is a need to create a simulation model of the process and then choose the best fit solution based on experiments with the simulation model [9], [10] using the dynamic simulation. Simulation is also an important tool in the design of new, previously non-existent

J. Hřebíček et al. (Eds.): ISESS 2013, IFIP AICT 413, pp. 571–578, 2013.

processes, which allows verifying the behaviour of proposed processes and thus avoiding errors at the design stage of the process [4]. The entire simulation process is shown in Figure 1. The great advantage of this method of decision making is the possibility of visualization both the running of processes and their monitored indicators.

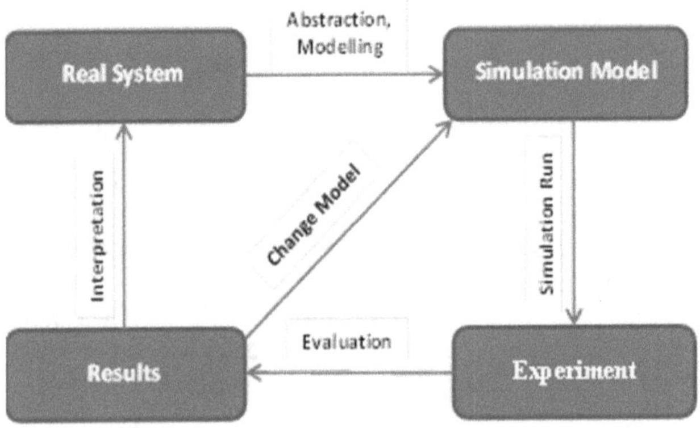

Fig. 1. Process of dynamic simulation (Source: [6])

2 Process Modelling Support by Witness Software

There are many tools for process modelling in practice but only a few are suitable for EMS dynamic process modelling. There are presented author's results based on own experience with modelling of EMS processes in the FirstStep, ARIS and Witness environment. The tool Witness appears to be the most appropriate for its universal properties.

The British software Witness Software from company Lanner Group Ltd. is intended to simulate the discrete event systems. The version "Manufacturing Performance Edition" is primarily used in practice which has the advantage of wide applicability in the simulation of logistics, production and service systems. Using of this version of Witness tool provides next possibilities for creation of EMS dynamic process models and simulation [5]:

- **Risk reduction** in decision making (accurately predicting how proposed changes to a facility or process will work).
- **Proposing the optimal solution** to decision makers (enables scenarios to be compared and the best option for the EMS to be graphically reported).
- **Generating an operational asset** (simulations deployed as robust applications become an asset in operational decision making).
- **Supporting of investment decisions** (provides evidence to identify profitable ideas and avoid costly mistakes).

2.1 Witness Developing Structure of Software

The structure of Witness developing environment consists of three basic modules which enable the efficient creation of simulation models of the EMS processes as it is shown in Figure 2:

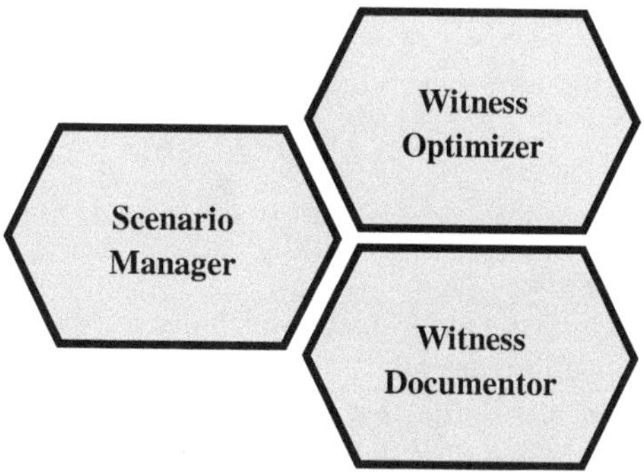

Fig. 2. Structure of witness modules (Source: own)

- **Scenario Manager** that offers a highly flexible framework for experimentation and analysis including visual interactive simulation which allows:
  - Graphical creation of simulation model.
  - Dynamic display of the simulated EMS process model.
  - User interaction with the running model (the simulation halt, requirement input information from user).
- **Witness Optimizer** finds the best solutions for simulation model by the use of the latest sophisticated mathematical techniques and offers an easy-to-use interface and the presentation of optimal results in a selection of useful and innovative tables and graphs.
- **Witness Documentor** allows creation of a simple formatted version of the information included in a WITNESS model.

2.2 Basic Building Blogs of Simulation Model of EMS Processes

The Witness tool provides standard modelling elements which can be used to create a simulation model of the EMS process. These elements are grouped according to their properties into the following groups:

- **Discrete elements** which are main for economic and administrative processes modelling *(Entities, Queues, Activities, Resources).*
- **Physical Handling Elements** *(Conveyors, Paths, Tracks, Vehicles).*
- **Power & free elements** which allow modelling the specialized form of transport systems (*Network, Section, Station and Carrier).*
- **Continuous processing elements** which are used to simulate models where continuous flow is a factor (*Fluids, Processors, Tanks and Pipes).*

- **Logical Elements and Modules** which represent the data and reporting aspects of the model (*Attributes, Variables, Files, Distributions, Functions, Part Files, Shifts and Modules*).
- **Graphical & Reporting Elements** *(Pie Charts, Time series, Histograms and Reports.)*
- **Manipulating Elements** *(Rules, Expressions and Actions).*

Created models of simulation can be stored in various formats which are shortly described in table 1.

Table 1. Storage formats of Witness simulation models (Source: own)

Format	Characteristic
.MOD	Standard model save format. It includes all the definition of the model in a fast loading format.
.SIM	Model and status file that saves the model at a particular point in a model run. When this file is loaded again it is possible to view run statistics and even to run the model on from that saved point in time.
.LST	Library file. This is a text definition of a model file. It provides a way of looking at a model file in a standard editor such as Notepad or Wordpad.
.DES	A designer elements file. This file format saves a single tab of a set of designer elements to a file. This provides easy portability of designer elements between models.
.MDL	A module file. This saves part of a model (or all if a single module) to a file for re-use in another model.
.WXM	A WITNESS model in XML format.

2.3 The Process of Creating the Simulation Model of EMS Project

The Witness GUI (Graphical User Interface) provides users an intuitive way to create a simulation model of the EMS process that can be divided into several phases as it is shown in Figure 3.

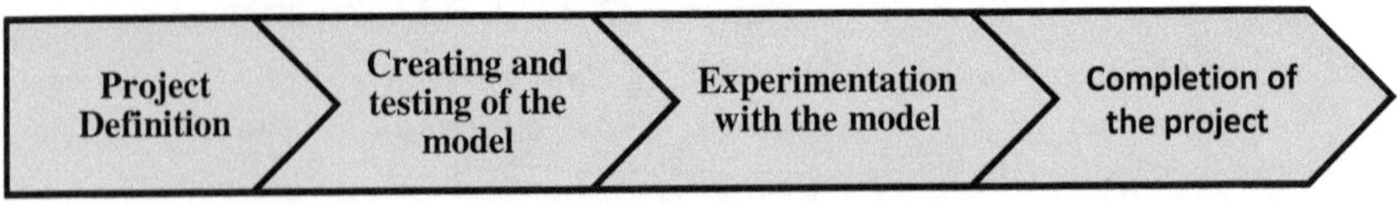

Fig. 3. The process phases of project simulation (Source: own)

The Project definition is the first phase of project simulation that consists of three steps which determine the success of the whole project, these steps are:

- **Establishing objectives** of the project of modelling and simulation of the EMS process.

- **Deciding the scope and the level** of the detail in the simulation model. It is important to limit the scope of the model as far as possible. At the beginning of the model-building process, small additions to the model lead to large increases in its accuracy. However, as the model becomes more detailed each subsequent addition adds less to the model's accuracy. This is particularly true when you are trying to model human behaviour, which is often inconsistent [5]. The graph below shows a typical relationship between the model's level of detail and the model's accuracy.

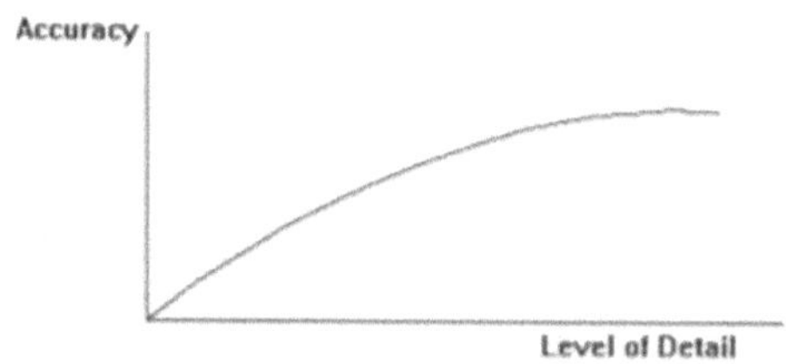

Fig. 4. Effect of adding detail to a simulation model (source: [5])

- **Collecting data** for creating the simulation model where elected information is likely to fall into one of three categories:
  - .*Available* - data is readily available and it is in an appropriate format that the model can use immediately.
  - *Not available but collectable* - data is either in an incorrect format or it has not been collected before. You might need to perform a small work study in order to collect this type of data (for example, timing certain processes manually).
  - *Neither available nor collectable* - data is not currently available and it is not easily collectable (for example, for a model of a new process of the EMS). If the data is neither available nor collectable you must use estimates. You can obtain useful estimates by:
    - *Using manufacturer's data* - Machine or special tools often include information (for example, reliability data) in their promotional literature and machine specifications.
    - *Sensitivity analysis* - involves replacing an unknown parameter (for example, a machine cycle time) with a high value and a low value in turn and comparing the results of the entire simulation.

The second phase of Creating and testing of the model consists of four steps:

- **Structuring the simulation model** identifies the most difficult areas for the simulation model building. It has usually a form of sketch of the situation to be modelled (inputs, outputs, key activities, etc.).
- **Building the simulation model** where elements of the simulation model are created (defining, displaying and detailing them), then linking the elements together with rules of its behaviour. In this step, more complex logic of the model is built by using actions of active elements.

- **Verification of the simulation model** ensures that the content of the model is consistent with expectation of the process owner.
- **Validation of the simulation model** investigates the accuracy of the model compared with the real world.

Experimentation with the simulation model investigates the behaviour of the simulation model in a number of what-if scenarios. It is relatively alone phase of the project whose steps are the simulation runs. Successful experimentation involves using a warm-up period or starting conditions, deciding on a suitable run-length, and running the model with more than one random number stream.

Completion of the project is final phase of the simulation project that consists of next three steps:

- **Generation of the reports and documentation** of simulation where Witness offers standard types of reports (Time series, Histogram, Pie etc.). Documentation makes the simulation model easier to understand if it needs to examine the simulation model at a later date [7], [8]. The WITNESS Documentor module allows creating a report of the names, notes and other property fields within a model.
- **Presentation of results** depends on the size of the simulation project and the culture of user's organization [3]. An animated model provides an effective communication tool to support decision making [6].
- **Implementation of model results** depends on the developer's ability to convince the process owner that the project results should become the part of the decision-making process and that the model is not used to simply justify a past decision. Actions resulting from simulation project should be implemented in the real world.

3 Application of Dynamic Simulation of EMS Process

Witness environment has been used by the author for modelling the activities of Call Centre of Fire and Rescue Corps which coordinates the deployment of fire and rescue techniques. Created simulation model was a part of the research plan no. MSM0021622418 supported by the Czech Ministry of Education, Youth and Sports and the research project no. SP/4H4/168/07 supported by the Czech Ministry of Environment. The application of dynamic simulation in Witness environment pursued two main objectives:

- The first objective was optimization of the activities of Call Centre (respectively department of dispatching) and setting the optimal number of dispatchers.
- The second objective was creation of the data collection for optimization of the deployment of fire and rescue techniques in the terms of arrival time, increase of operational capability and efficiency of Fire and Rescue Corps in the monitored district. On a base of collected data, a number of what-if scenarios have been created and they were used to create the quality emergency plans.

3.1 Potential Benefits of Simulation in Witness Environment

This created simulation model presents a detailed process model where each individual carried out activity of employees can be seen on the screen as shows the figure 5. Each stage of the process is presented to show the input queue. The entities are processed and the employees (dispatchers) are used to perform the required tasks.

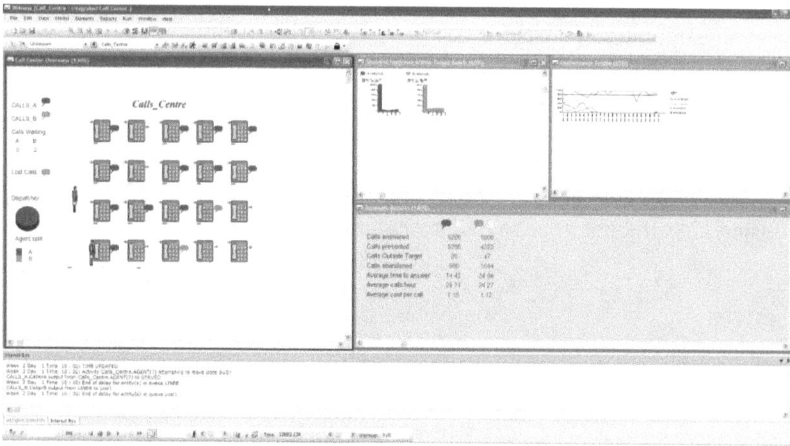

Fig. 5. The simulation model of Call Centre (source: own)

One of the important properties of the Witness environment is the possibility to quickly generate a view of the model using 3D graphics with a choice of the point of view on the model as shown in Figure 6. This property makes easier communication with the process owner who is responsible for its running and maintenance.

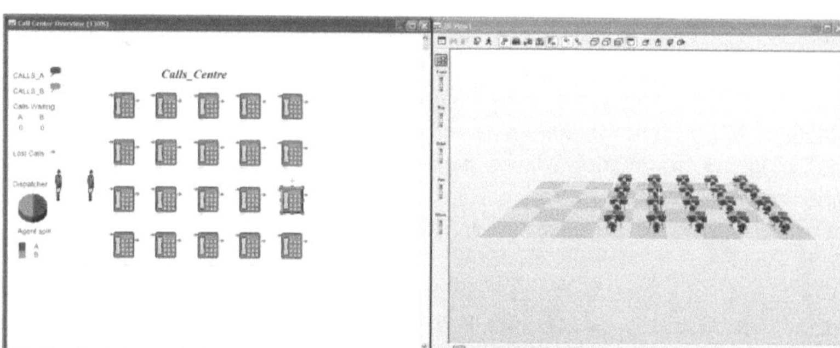

Fig. 6. Using the 3D presentation (source: own)

Next important facility is the possibility of using the Witness Optimizer which offers to user to search for the best answers in the wide range of possibilities and usually finds these within a small percentage of the possible number of simulation runs. Implemented changes of simulation results helped to achieve the objectives of the project as it is published in the report of project no. SP/4H4/168/07.

4 Conclusion

Using the dynamic simulation models of the EMS processes provides useful possibilities of decision making for EMS process owners in critical situations, such as emergency plans.

Using of modelling tools, such as Witness, which were primarily designed to simulate production - logistics processes also appears to be very effective in the case of a dynamic simulation of the EMS processes. Simulation models of processes in both areas generally require a very short response time of the system. This fact opens the possibility of using simulation models and their simulation results for the partial test, the feasibility of emergency plans and behaviour of the EMS systems in crisis conditions on a wider scale.

In the case of an emergency, correct implementation of the simulation results can save considerable amount of economic resources and it can possibly save lives.

References

1. Becker, J., Algermissen, L., Falk, T.: Modernizing Processes in Public Administrations: Process Management in the Age of e-Government and New Public Management (BPM Competence). Springer, New York (2012)
2. Doucek, P.: Dynamic modelling of the software development process. Cybernetics and Systems 27(4), 403–410 (1996)
3. Doucek, P.: Human Resources in ITC – ITC Effect on GDP. In: IDIMT-2010: Information Technology - Human Values, Innovation and Economy 18th Interdisciplinary Information Management Talks. Book Series: Schriftenreihe Informatik, vol. 32, pp. 97–106. Trauner Verlag, Linz (2010)
4. Greasley, A.: Simulation Modelling for Business. Asgate, Burlington (2004)
5. Lanner Group. Witness - Service and proces performance Edition (2010)
6. Manlig, F.: Computer Simulation of Discrete Events. MM Industrial Spectrum 10, 34–35 (1999)
7. Ministr, J., et al.: The IT Service Continuity Management Principles Implementation by Method A2. In: IDIMT- 2009 Systems and Humans – A Complex Relationship – 17th Interdisciplinary Information Management Talks Proceedings. Schriftenreihe Informatik, vol. 29, pp. 131–139. Traunr Verlag, Linz (2009)
8. Ministr, J., Števko, M.: Human resources Requrements for Professional Management of ITSCM Processes. In: IDIMT-2010: Information Technology - Human Values, Innovation and Economy 18th Interdisciplinary Information Management Talks. Schriftenreihe Informatik, vol. 32, pp. 57–64. Trauner Verlag, Linz (2010)
9. Ráček, J., et al.: Integration of eGovernment Services of the Czech Ministry of Environment. In: 6th International Eastern Europe e|Gov Days: Result and Trends. Österreichische Computer Gesellshaft, Linz (2008)
10. Štefánik, M., Hřebíček, J., Ráček, J.: Interoperability in E-Government Services for Environment Protection. In: Eastern European e-Gov Days: Best Practice and Innovation, pp. 19–22. Austrian Computer Society, Wien (2007)

Green Computing Practices as a Part of the Way to the Sustainable Development

Zuzana Bezáková

University of SS. Cyril and Methodius, Faculty of Mass Media Communication, J. Herdu 2, 917 01 Trnava, Slovak Republic
zuzana.bezak@gmail.com

Abstract. Sustainable development is now often mentioned in connection with the further direction of our planet, our economy and the overall human consumption. That is the reason why the companies come with various types of innovation. The use of information technology is now an essential part of our life, our business, our marketing. This area of using computers and their information technology equipment requires a new - greener approach. This post reflects the need for sustainable computing and different approaches to its implementation to the business. We deal with the categorization of green computing performance metrics. One factor of green computing is also the customer and what he wants. Focus on customers' requirements forced the companies to look to the markets and customers as the main drivers of their business in the area of sustainable computing. The contribution also includes the results of the research of Slovak companies approach to sustainable computing in the Slovak business environment.

Keywords: sustainable development, sustainable computing, green computing, innovation.

1 Innovation and Sustainable Development

The concept of innovation evokes in many groundbreaking technological inventions that are the result of long-term activities and research work of scientists. However, innovation can occur in any business, and there is an effort to success in the market, be faster than its competitors, and also to offer products that satisfy customers. Innovation need not be oriented only to products but also to services, or processes, or marketing practices [1]. Creation and management of innovation processes are included with all its principles in the full range of corporate governance [2].

Much attention has recently been devoted to innovation as a means of achieving a radical improvement of the system of corporate environmental practices and performance. Eco-innovation is perceived as key concept that provides direction and vision for tracking overall societal changes necessary to achieve sustainable development [3].

We can harness the power of IT to solve environmental problems but also try to make our IT systems and their use greener. Green IT, also known as green computing is the study and practice of designing, manufacturing and use of computers, servers,

J. Hřebíček et al. (Eds.): ISESS 2013, IFIP AICT 413, pp. 579–587, 2013.

monitors, printers, storage devices, and networking and communications systems effectively and efficiently, with no or minimal impact on the environment. Green IT is used as a support for other environmental activities and help to create green awareness. Thus, green IT includes hardware, software, tools, strategies and practices that improve and promote environmental sustainability.

Green IT benefits the environment by improving energy efficiency, reducing greenhouse gas emissions, using less harmful materials, the re-use and recycling is also important. Thus, green IT includes the dimension of environmental sustainability, economics energy efficiency and total cost of ownership, which includes the cost of disposal and recycling. Increased awareness of the harmful effects of greenhouse gas emissions, new stringent environmental legislation, concerns about procedures for handling electronic waste and concern about corporate image are the driving force for businesses and individuals to become green [4].

Currently available sources talking about sustainability present three broad environmental objectives leading to sustainability:

1. Eco-efficiency - the supply of goods and services at competitive prices to satisfy human needs and bring quality of life, while progressively reducing ecological impacts and resource intensity throughout the life cycle in accordance with the earth's carrying capacity
2. Eco-equity - means equality between people and generations, and equal rights of all people to natural resources
3. Eco-effectiveness - considered the best solution to environmental problems, but requires a shift in thinking and transforming business models. People need to stop environmental degradation and depletion of non-renewable resources [5].

The most significant impacts on the company's ability to behave sustainably include:

- social impact - this is a measurement of the impact of the enterprise in terms of the social contract and the impact of interest groups
- environmental impact - the effect of business activities on the geophysical environment
- organizational culture - involves the relationship between the company and internal stakeholders, particularly employees, and all aspects of the relationship
- finance - are evaluated in terms of adequacy of compensation for assuming the risk.

It is necessary to find a balance between the influences that will ensure not only sustainability, but also sustainable development. The sustainable development is the new view for business, it is a change of the established practices. Addressing sustainability in the business area brings number of sub-issues.

The basic features are:

- sustainable production
- sustainable consumption
- sustainable marketing
- sustainable profit [6].

2 Green Computing

Green computing can be included in the field of sustainable consumption, in terms of access of business entity to sustainable development. It can be defined as the use of computer resources efficiently in practice. It is important that the initiative of green computing was systemic in nature and was designed to address increasingly sophisticated problems. Green computing refers to the requirements of environmental protection and energy saving together with the reduction of operational costs in today's increasingly competitive world. But before it comes to implementation, it is also important to deal with what kind of energy and performance gains can be achieved through it. Therefore, analysis of what we have today and what we have to do is essential and important for achieving the benefits of green computing [7].

Governments of various countries are increasingly pushing for the adoption of green computing and sustainable IT investment and sustainable practices. Due to the immediate impact on business value, it is likely that green computing will continue to focus some time to reduce costs while improving the efficiency of data centers and desktops [8].

IT can be a key factor in greening even other sectors and also the initiator of positive initiatives towards environmental sustainability. Positive change in IT should be reflected in the change of environment. Several studies show that IT contributes only about 3% of greenhouse gas emissions to the total quantity, this means that the vast majority of emissions come from non-IT resources.

Green IT can help create a better environment through these tools, technologies and systems:

- Software tools for analysis, modeling and simulation of environmental impacts and risk management environment
- Platform for eco-management, emissions trading and ethical investment
- Tools for auditing and reporting energy consumption and savings and for monitoring greenhouse gas emissions
- Environmental knowledge management systems, designed for the collection and transmission of environmental knowledge, support systems, environmental ontology
- Environmental engineering systems, including geographic information systems
- Urban planning tools and systems environment
- Technologies and standards for interoperable environment monitoring networks, intelligent sensor networks
- Integration and optimization of existing environmental monitoring networks, easy to plug-in new sensors, Sensors cooperation and networking
- Tools and systems to optimize business processes in organizations.

To reduce their carbon footprint, organizations can dematerialize some of their products and activities, using IT. Dematerialization refers to the transformation of the physical goods Information goods represented in digital form, so-called "Turning atoms to bits" [4].

3 Sustainable Approach to Hardware and Software

3.1 Green PCs, Laptops and Servers

Every company can significantly reduce electricity consumption by making small changes in the way the computer is used. Energy cost savings of one computer does not seem to be an important item, but combined savings of hundreds of computers in the enterprise is already considerable and is not negligible. PC power consumption can be reduced by adopting several measures:

- Enabling power management features
- Turn off the system when not in use
- Use a screen saver
- The use of thin client computers.

These measures will not become a reality without the sincere willingness of users and their active participation. Wise companies tend to adopt innovative environmental strategies, create value and build competitive advantage through innovation in the field of computing [4].

3.2 Green Data Centers

Green data centers (DC) are those in which the computer system, air conditioning, electrical and mechanical systems, which cover data centers are designed and operated for maximum energy efficiency, low carbon footprint and minimal environmental impacts. Operators who want to optimize the energy efficiency of datacenters, have many options to choose from, but usually cannot know in advance exactly identify potential savings before investing. Also, it is very difficult to determine in advance the actual risk of the new approach in a real environment. It is therefore important that the operator can evaluate DC savings, but also the risks of several options before making a choice. For this purpose there are various simulation systems, such as The EoDD (Energy Optimized Data Center Designer) [9].

Performance metrics for green computing data centers can be divided into two categories: basic metrics and advanced metrics.

Basic metrics provide a basic picture of the data center. These include:

- Metrics of greenhouse gas
- Metrics of humidity
- Thermal metrics
- Metrics of Power.

Advanced metrics provide a deeper look at the data centers, and include:

- Multiple indicators
- Total cost of ownership.[10]

3.3 Green Cloud Computing

Cloud computing delivers infrastructure, platform, and software (applications) as services that are available to consumers in the pay-as-you-go model. In the industry, these services are referred to as Infrastructure as a Service (IaaS), Platform as a Service (PaaS) and software as a service (SaaS). Many providers of computing services, including Google, Microsoft, Yahoo, and IBM quickly deploy data centers in different locations around the world, to be able to deliver cloud computing services [11]. In the cloud computing world, IT capabilities are delivered on the fly and on-demand through the Internet when the need arises, instead of drawing from desktop computers. Many design and architectural patterns are emerging around cloud computing that make it difficult to fit everything into a perfect definition [12].

3.4 Green Data Storage

Data and information storage requirements are growing constantly. Storage systems in data centers consume large amounts of energy for running and cooling. For example, in a data center, storage systems consume 24% to 40% of total energy consumption of IT and are the largest centers of power. In addition to the requirements for energy-efficient servers, there is currently the focus on the greening of data storage. Energy management of data storage can be divided into two basic areas: hard-disk energy management and system-level energy management among the various main storage devices, which include hard disk drives, magnetic tape drives and solid state drives, consume more power than tape and SSD. The main techniques for dealing with the energy of individual disks are transitioning state, caching and dynamic rotations per minute (DRPM). System-level energy management is about system-wide management across different or multiple components or devices. Energy management techniques for redundant array of independent disks (RAID) systems, tiered storages, virtualized storage and cloud storage are in this category.

Green data storage is a priority for many European storage professionals, and some of them have begun to implement measures to cut energy usage. Green storage considerations are important when buying storage hardware. Many users would pay extra money for more energy-efficient hardware [13].

3.5 Green Software

Green software is the result of innovation process towards sustainability. Innovation is the successful introduction of new product, service, or process. Software innovation can be seen as a process leading to:

- development of a novel aspect, feature or application of an existing software product or process, or
- introduction of a new software product, service or process or an improvement in the previous generation of the software product or process, and
- entry to an existing market or the creation of a new market [14].

Innovated software is an important element in determining the total energy consumption and computational efficiency. Software is a key element in improving environmental sustainability. Green software is software that is environmentally friendly, and especially helps in improving the environment. Green software can be divided into four categories:

- software that is environmentally friendly, uses less energy to run
- embedded software that helps other things being green
- sustainability reporting software, or carbon management software (CMS)
- software for understanding climate change, assess its impact [4].

3.6 Green Networks and Communication

Networking and communication play more important role than ever before, and facilitate the transfer and sharing of data. It allows us to communicate and share information, trade, learn and socialize online, so that we can more effectively use our working time, and also can do many other things. Green Network shows ways to minimize the impact of networks on the environment through energy-efficient network technologies, protocols and products, and minimize the use of resources. Procedures of green networks include the following:

- the use of newer, more energy-efficient techniques, technologies and products
- update of older devices and replacing them with newer, greener network devices
- the use of intelligent systems, user management and energy conservation through networks to improve energy efficiency
- the use of remote management and video conferencing instead of traveling [4].

4 Sustainable Computing in the Slovak Business Environment

We have made the research dedicated to the point of view of Slovak business entities to sustainable computing. The aim of our research was to find out what are the attitudes of Slovak companies to sustainability but mainly to sustainable computing /sustainable use of PC/. We also examined whether there are specific activities in this area, to which the companies are dedicated. The results should give us an idea of the current state, so we can propose solutions to improve the routing of these companies towards sustainable computing and thereby protect the environment. Also, we wonder whether sustainable computing is an important factor from the point of view of the competitiveness in Slovak business environment and what is the position of the customer in this process.

The survey included 45 Slovak IT companies, advertising agencies and web hosting companies that deal with domain registration, web hosting, online marketing, network management, network solutions, etc.

We focused on small /80%/ and medium enterprises /20%/. The reason was, that [15] in Slovakia recently grows importance of small companies. Their advantage is that they mostly focus on one segment, which can adapt well their marketing mix.

79% of respondents consider their business as sustainable, but this behavior is rather linked with activities such as recycling paper, waste separation, use of electronic invoices and so. We also found that most companies /80%/, which indicate their behavior as sustainable, use the devices - PC, network solutions that are friendly to the environment.

Survey results indicate that these efforts are related to the saving of electricity, which is beneficial to society, but of course it has a positive impact on the environment.

Eco attribute of the device for which they decide in purchase are in favor of ecologic devices only in 27%. Another respondents do not consider this fact as important /33%/, while others argue that the price is mainly important when buying a particular device /33%/.

As mentioned in the article, the energy consumption can be reduced through simple procedures. Most companies /93%/ use at least one procedure to reduce electricity consumption, including the most frequently used such as screen saver, standby PC when it is not in use, or even a complete system shutdown. Just in one case they intentionally use the thin client computers or laptops that have lower power consumption compared to PC.

The status of the customer in the process of sustainable computing is very important. The term "Green Company" is not seen as a big advantage /73%/ in the Slovak business environment.

Customers are not actively interested in environmental aspects of solutions /87%/, Slovak companies meet this interest very rarely /13%/.

Websites are important for the presentation, and also in attracting new customers. The trend is that larger companies are dedicated to their websites and presentation of sustainable behavior, social responsibility, thus helping to improve the environment. Even small and medium-sized companies in our survey indicate in some cases their green activities on websites /27%/, but for many of them it is not at all important to give this information to the customer /47%/. Some companies plan to publish such information on the website /13%/. Of course, companies that do not perform any environmental activities, cannot advertise them /13.3%/.

5 Conclusions

Sustainable IT services are essential for the success in business. They not only focus on data center efficiency and on minimizing carbon emissions. They are aimed directly at the long-term importance of IT for the organization, its customers and for the whole society. Therefore sustainable computing refers to everything what the organization needs to do in order to ensure that IT services will provide excellent value to achieve a strong market position and ability to secure its survival. It is important to align IT with the business strategy to achieve market-leading business value, customer value and societal value. This will ensure the viability of IT within the organization. We define sustainable social services in total value terms as the total amount that is available to the company. Systematic integration and alignment of IT services is important in terms of creating greater social value [8]. Companies should know wishes and needs of customers. It is important to have enough information about them and then use them in their business [16].

The results of the survey show that Slovak companies are trying to implement green approach in the use of hardware and sustainable computing /sustainable use of computers/. Companies themselves do not consider this fact as a competitive advantage in the Slovak business environment. They try to use devices that are environmentally friendly, or trying to use simple procedures to reduce electricity. Also, quite a lot make extensive use of other technical equipment, but also networking solutions that are environmentally friendly.

Based on the survey results, we can conclude that the customer is not an important factor in sustainable computing in Slovakia. The customer is not very interested in greener solutions which are probably the reason why there is even a small degree of companies' promotion of their green activities on the site.

An interesting solution and recommendation is:

- Draw customers into the green activities of the company, gradually acquaint with the importance of sustainable behavior.
- The use of organic labeling and explaining the importance of the environmental behavior importance may lead to an overall improvement of the environment.
- More frequent use of advertising "slogans" with the sustainable theme can help you to become more and more to receive this concept to the attention of the customer
- To offer organic solutions and services and IT solutions, offer sustainable computing initially only as an added value.

These measures can help in the Slovak business environment to build a solid foundation for sustainable use of computers, strengthen the importance of this concept for companies, but especially the need to build sustainable behavior of the customer.

Acknowledgements. This article has been elaborated within the project VEGA 1/0900/12 - "Improving innovation performance and innovation entities through a system of open innovation with support of integrated marketing communication".

References

1. Zaušková, A., Domová, J.: Inovačná schopnosť a inovačná výkonnosť podnikateľských subjektov. UCM Trnava, Trnava (2012)
2. Zaušková, A., Miklenčičová, R.: Tvorivosť v inovačnom procese. In: Trvalo udržateľný rozvoj regiónov a podnikov pomocou inovatívnych a zelených technológií: zborník príspevkov z medzinárodnej vedeckej konferencie uskutočnenej v rámci projektu ITMS 22410420018, 204-213. UCM Trnava, Trnava (2012)
3. Sustainable Manufacturing and Eco-Innovation, http://www.oecd.org
4. Murugesan, S., Gangadharan, G.R.: Harnessing Green It. Willey. IEEE (2012)

5. Information systems and environmentally sustainable development: Energy informatics and new directions for the IS community, http://terrypress.terry.uga.edu
6. Šmida, Ľ., Hrdinová, G., Sakál, P.: Predpoklady budovania spoločensky zodpovedného podnikania ako súčasti udržateľnej spoločnosti v zmysle konceptu udržateľného rozvoja. In: Transfer inovácií 21/201, pp.198–203. Košice (2011)
7. Green computing: Practice of Efficient and Eco-Friendly Computing Resources, http://www.sersc.org
8. Sustainable IT services: Assessing the impact of green computing practices, http://www.sis.pitt.edu
9. Capabilities and Limitations of Direct Free Cooling in Data Centers, http://www.fhnw.ch
10. Review of performance metrics for green data centers: a taxonomy study, http://sameekhan.org
11. Energy-aware resource allocation heuristics for efficient management of data centers for Cloud computing, http://www.sciencedirect.com
12. Green Computing, A.: Dual Technology for Cloud Computing and HPC, http://www.computer.org
13. Green data storage technology survey: Green storage a priority for European storage professionals, http://www.computerweekly.com
14. Lippoldt, D., Stryszowski, P.: Innovation in the Software Sector, pp. 10, 185 p. OECD (2009)
15. Čábyová, Ľ.: Benchmarkingové porovnávanie v oblasti marketingovej komunikácie. FMK UCM Trnava, Trnava (2009)
16. Matúš, J., Čábyová, Ľ., Ďurková, K.: Marketing – základy a nástroje. FMK UCM Trnava, Trnava (2008)

The Plan4business Approach to Transfer Open Data into Real Estate Businesses

Jan Ježek[1], Tomáš Mildorf[1], Karel Charvát Jr.[2], and Karel Charvát[3]

[1] University of West Bohemia, Pilsen, Czech Republic
{jezekjan,mildorf}@kma.zcu.cz
[2] University of Economics, Prague, Czech Republic
xchak07@vse.cz
[3] Help Service - Remote Sensing, Benešov, Czech Republic
kch@hsrs.cz

Abstract. Spatial planning data including urban, regional, spatial or zoning plans are not aggregated so far. Creating time series or comparative analyses on these data sets is not yet possible. The EU funded project Plan4business develops a service platform that can serve to users as a full catalogue of spatial planning data linked with other data sources such as statistics, OpenStreetMap, Urban Atlas and Corine Land Cover that are published as Open Data. The Plan4business platform will offer to clients not only the data itself in an integrated, harmonised and thus ready-to-use form, but also rich analysis and visualisation services via an API and an interactive web frontend. The users include mainly citizens, local authorities and real estate agencies. This paper introduces the problems of data integration and selected technical components of the Plan4business platform supporting data reuse and analysis.

Keywords: plan4business, open data, spatial planning data, land use, integration, harmonisation, analyses, real estate businesses.

1 Introduction

In the last decade, many changes have happened with regard to data provision and data management in both public and commercial sectors. The PSI Directive (Directive 2003/98/EC on the re-use of public sector information) adopted in 2003 triggered a lot of debates about how public information should be handled, shared and priced.

Information, as a meaningful interpretation of data and relations between them [7], plays a crucial role in decision making mechanisms in different domains, such as environment or spatial planning, and on various levels, from local to global. The underlying data for decision making processes represent the most expensive part of information systems. It is therefore important to ensure that:

- data are collected once and maintained at the level where this can be done most effectively;
- it is possible to combine data from different sources and share them between many users and applications;

J. Hřebíček et al. (Eds.): ISESS 2013, IFIP AICT 413, pp. 588–596, 2013.

- data are collected at one level of government and shared between all levels;
- data needed for good governance are available on conditions that are not restricting their extensive use;
- it is easy to discover which data are available, to evaluate their fitness for purpose and to know which conditions apply for their use. [2]

The above mentioned rules are the key principles of the INSPIRE (Infrastructure for Spatial Information in the European Community) initiative aiming to create a European Spatial Data Infrastructure (ESDI). This initiative is supported by the legal framework of the INSPIRE Directive that came into force in May 2007.

Because of different historical developments in countries, heterogeneous data coming from various providers are not easy to combine, share and analyse. The reuse of data is therefore limited. Spatial data, their integration and analysis are the main topics of this paper. The authors present the technological issues of a service platform for data and information reuse that is being developed within the frame of the FP7 project Plan4business (04/2012 – 03/2014).

The next chapter gives and insight into the problem of data interoperability and examples of initiatives aiming to data sharing in an interoperable manner.

2 Interoperability of Spatial Planning Data

Spatial data, sometimes referred to as geographic data, geodata or geospatial data, are defined by INSPIRE as "data with a direct or indirect reference to a specific location or geographic area." [1] The possibility to locate objects and phenomena is an opportunity to interpret, analyse and visualise data in an understandable way familiar to humans. Spatial data create a reference layer capable of linking other non-spatial data, such as statistical data. The relations between objects and phenomena can be then interpreted and analysed in the context of spatial information.

As it was already mentioned, data coming from various countries and data providers are heterogeneous, unless they comply with semantic and technical specifications, and legal and organisational frameworks agreed between all parties involved. The application of common specifications leads to certain level of data interoperability.

The term interoperability is defined by the International Organisation for Standardization (ISO) as the "capability to communicate, execute programs, or transfer data among various functional units in a manner that requires the user to have little or no knowledge of the unique characteristics of those units." [3] As stated by David Schell, founder and chairman of the Open Geospatial Consortium (OGC), "Interoperability seems to be about the integration of information. What it's really about is the coordination of organizational behaviour." [6]

Examples of particular issues categorised according to the European Interoperability Framework are showed in Figure 1. Heterogeneity of spatial data was well documented by the European Plan4all project that collected data specifications from most of the European countries. The Plan4all results show the complexity of the situation.

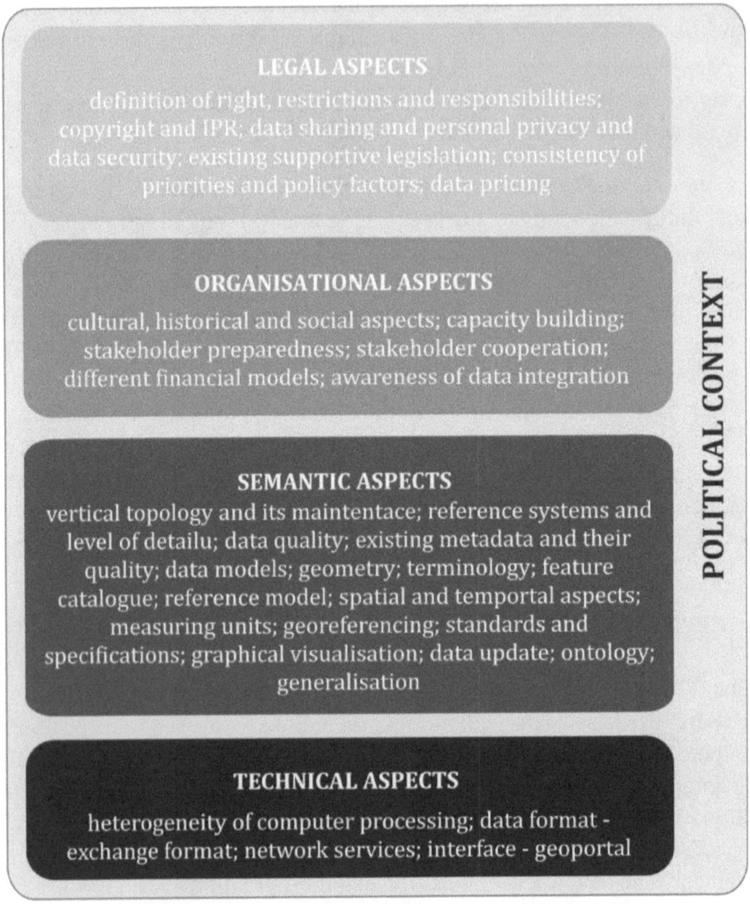

Fig. 1. Aspects of data interoperability [5]

There are several initiatives in Europe aiming to bridge the differences between datasets from various sources by providing legal and organisational frameworks as well as semantic and technical specifications. The already mentioned INSPIRE Directive is the most important initiative as far as spatial data are concerned. Other initiatives include the Shared Environmental Information System (SEIS), Copernicus (previously known as Global Monitoring for Environment and Security - GMES) or the Global Earth Observation System of Systems (GEOSS).

The Plan4business project aims to provide a technical solution for data integration based on the INSPIRE principles, OGC specifications and ISO (International Organization for Standardization) standards of the 19100 series Geographic Information.

Plan4business focuses on integration of spatial planning data from various sources. Spatial planning data represent a valuable source of information that could be used not only for spatial planning activities but also as a reference layer for e.g. real estate agencies, providing tools for data analysis across a certain geographical area where differences between administrative units are diminished. Currently, it is difficult to use results of spatial planning for any other purposes than for printing or simple publishing by the authorities that create them. Creating time series, comparative analyses or data mining on these data sets is not yet possible. However, such capabilities are needed by researchers, spatial planners and professionals from the real estate world.

3 Architecture of the Plan4business Service Platform

3.1 General Architecture

Plan4business develops a platform that can serve to users as a full catalogue of planning data such as transport infrastructure, regional plans, urban plans and zoning plans in relation with other Open Data including statistics, OpenStreetMap, Corine Land Cover or Urban Atlas. The Plan4business platform offers to clients not only the data itself in an integrated, harmonised and thus ready-to-use form, but also offers rich analysis and visualisation services via an API and an interactive web frontend.

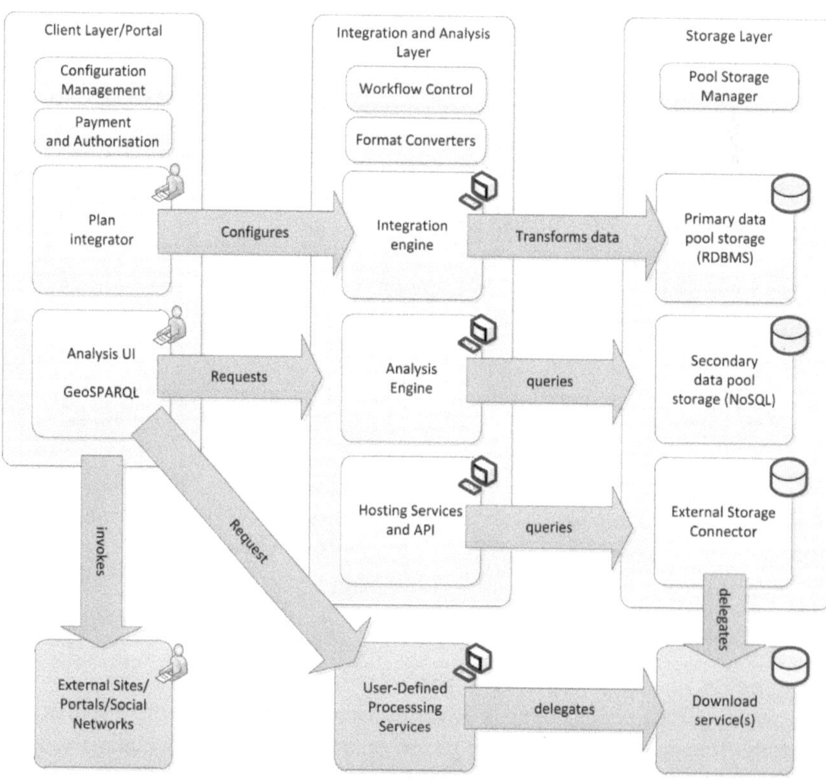

Fig. 2. General architecture of the Plan4business platform [8]

The Plan4business platform consists of several different technical components grouped into three layers. The first layer contains human-machine interfaces for planning data management, integration and conversion as well as for accessing the analytical functions of the platform. The second layer provides two groups of processing engines, again for integration and harmonisation on the one hand and for analytical processing on the other hand. In addition, this layer provides the Plan4business application programming interface (API). The final layer is a storage layer, which contains a storage manager and two different data bases, each optimised for different goals. All the three layers, their components and relations are depicted in Figure 2.

The design and implementation of the platform are partly based on components developed in previous projects and by various initiatives. There are three central components of the server side of the Plan4business platform including Integration, Storage and Analysis Engines (see Figure 3). These components are described in more detail in the next sections.

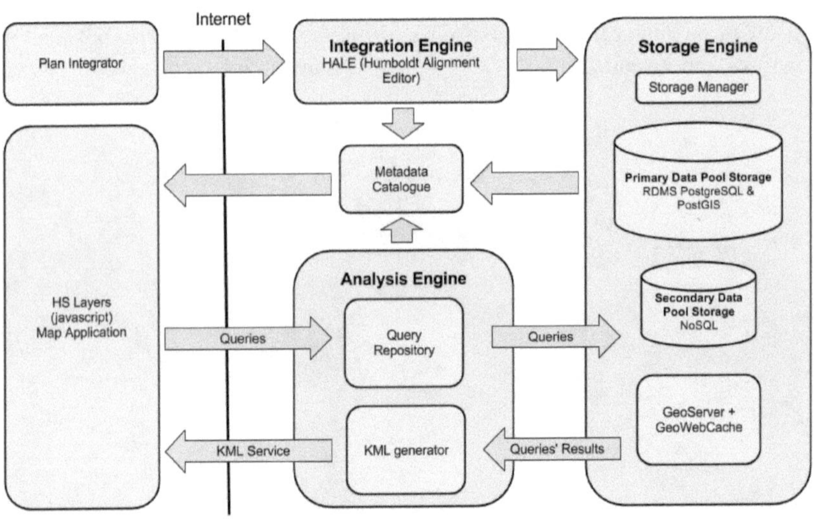

Fig. 3. Main components of the server side of the Plan4business platform [4]

The central part of the components is a metadata catalogue. The catalogue enables to store not only metadata about existing datasets, but also about performed analyses, created map compositions and predefined integration services.

3.2 Integration Engine

The Integration Engine is intended to perform all data transformations necessary to seamlessly integrate heterogeneously structured and externally provided spatial vector data into the Plan4business data pool. Based on the input data and schema alignment provided by the user through the platform interface (Plan Integrator in Figure 3), the

Integration Engine, deployed on the server side of the Plan4business platform, will transforms input data into the schema used by the Storage Engine.

Various conversions and transformation steps such as mapping between different data formats, geometric representations and conceptual schemas will be performed by the Integration Engine. This functionality is being implemented on top of the Humboldt Alignment Editor (HALE), a desktop software that was initiated within the HUMBOLDT project (FP6) and is continuously developed as an open source software.

The Integration Engine will enable several data transformation processes that can be initiated by different users at the same time. Furthermore, the transformed data will not be provided as an output file within the file system but will be directly saved in a database to enable concurrent data retrieval, analysis or extension in a resource-efficient and consistent way. [4]

3.3 Storage Engine

The Storage Engine consists of four subcomponents. The first subcomponent is the Primary Data Pool Storage. This is a hybrid combination of a relational database and a document-oriented storage. This storage provides access to the original (non-harmonised) spatial data sets and to the integrated spatial vector data sets.

The second subcomponent is the Secondary Data Pool Storage which is intended to augment the capabilities that are offered by the Primary Data Pool Storage. While the relational database management system used for the primary storage can be expected to scale well with a growing amount of data and to provide fast data retrieval, the choice of the database management system for the secondary storage needs to be optimised for complex graph-based queries or complex analysis. This can be most likely achieved by using a NoSQL database that breaks with common design decisions of RDBMS.

The third subcomponent of the Storage Engine is the Storage Manager that enables to control the data that are available in the Primary and Secondary Data Pool Storages. The Storage Manager can be used to remove data from the primary data pool or to copy data that is needed for a certain analysis from the Primary Data Pool Storage into the Secondary Data Pool Storage. [4]

The fourth subcomponent supports spatial data publication. The main software solutions include GeoServer, a Java software server that allows users to view and edit geospatial data, and GeoWebCache, a Java web application used to cache map tiles coming from a variety of sources such as OGC Web Map Service (WMS).

3.4 Analysis Engine

The Analysis Engine which is closely related to the Storage Engine ensures processing of analytical queries made by the platform user and retrieval of their results. It enables accessing and processing spatial planning data and their retrieval for visualisation in the map client. The Analysis Engine provides access to all analytical functions of the spatial databases, i.e. the Primary and Secondary Data Pool Storages.

For debugging and testing purposes a web application for performing predefined and user defined queries and displaying query results without any major configuration was implemented (Figure 4).

Fig. 4. A web application for querying the Plan4business database and displaying the results

The results of the analysis are in the KML format that can be visualised in any map client supporting KML such as the Plan4business platform or Google Earth. This feature will ensure interoperability with other solutions and platforms.

4 Czech Pilot Combining Statistical Data

Spatial planning data represent a valuable source of information. In order to fully exploit their potential, it is necessary to combine them with other (non-spatial) data such as socioeconomic or demographic data. The combination is an added value which can be beneficial for end users performing analysis on the Plan4business platform.

Plan4business tested the possibility of inclusion of statistical data from EUROSTAT on European, national and regional levels and from national statistical offices on local level. The spatial coverage of the testing included the territory of the Czech Republic.

EUROSTAT publishes data up to NUTS[1] 3 level. NUTS 1 level includes major socioeconomic regions, NUTS 2 level consists of basic regions for the application of regional policies and NUTS 3 level contains small regions for specific diagnoses.

[1] Nomenclature of territorial units for statistics.

In order to make the analyses through the Plan4business platform interesting for real estate agencies, investors and other user groups, the information from EUROSTAT was complemented by national data from LAU[2] 1 and LAU 2 levels (formerly NUTS 4 and NUTS 5 levels). LAU 1 level includes tens or hundreds of municipalities which are represented as LAU 2 level.

The structure of data published by national statistical offices is not standardised and differs in countries. The data in the Czech Republic were acquired from the Czech Statistical Office (CZSO). The statistical data of CZSO are not published as Open Data but through a Public Database[3]. The Public Database was developed for presentation (viewing) of statistical data mainly for the public where data are presented as a set of predefined tables that vary depending on the type of statistical data. This makes automatic machine collection and processing of these data rather difficult. In the Czech Republic there are currently 6253 municipalities. Obtaining statistical data for each municipality means downloading and processing 6253 files.

The data from individual tables can be exported to XML files. The problem is that the XML files are not accessible through a URL and the files can't be simply generated in a bulk. Manual downloading of such a number of files is practically impossible. Fortunately, every page containing tables with statistical data is accessible through a unique URL which can be generated using codes of municipalities from the CZSO code list. A script for downloading and processing statistical data was created and for each type of statistics the script was modified according to the structure of the data on the HTML pages. This process is rather time consuming and prevents automation of data acquisition.

5 Conclusions

The development of the Plan4business service platform goes in line with the collection and analysis of user requirements and creation of a business plan that ensures the sustainability of the platform after the end of the project.

The project consortium is currently facing challenging problems of data availability and smooth access to data in a machine readable way. Due to different national legislation and other barriers hindering access to data on the pan-European level, the platform will be tested in selected pilot areas. The first one is the Czech Republic where access to statistical, cadastral, topographic and environmental data enables to test the platform functionality in real world scenarios. As described in this paper, the process of data acquisition shown on the example of statistical data is time consuming and cannot be automated.

The primary data sources for the Czech pilot will be data from public administration and governmental bodies. The data will be harmonised and integrated into the Plan4business platform and a set of services will be offered to the main target groups including real estate agencies, citizens and municipalities.

---

2 Local Administrative Units.

3 `http://vdb.czso.cz/vdbvo/en/uvod.jsp`

Acknowledgments. All the authors are supported by the FP7 project Plan4business (Grant agreement no: 296282). The authors from the University of West Bohemia are additionally supported by the European Regional Development Fund (ERDF), project "NTIS – New Technologies for the Information Society", European Centre of Excellence, CZ.1.05/1.1.00/02.0090.

References

1. European Commission: INSPIRE Glossary, http://inspire-registry.jrc.ec.europa.eu/registers/GLOSSARY
2. INSPIRE: INSPIRE website (2012), http://inspire.ec.europa.eu/
3. International Organization for Standardization: ISO/IEC 2382-1 Information technology – Vocabulary – Part 1: Fundamental terms (1993)
4. Ježek, J. et al.: D5.1 Interim Report on Integration, Analysis and Storage Engines, plan4business consortium (2013)
5. Mildorf, T.: Modelová generalizace pozemkového datového modelu (Model Generalisation of a Land Data Model). University of West Bohemia, Pilsen (2012)
6. Open Geospatial Consortium: Open Geospatial Consortium website, http://www.opengeospatial.org (accessed February 18, 2012)
7. Rapant, P. (ed.): Terminologický výkladový slovník pojmů z oblasti geoinformací. Věstník Úřadu pro veřejné informační systémy, II (2001)
8. Fraunhofer-Gesellschaft zur Foerderung der Angewangten Forschung E. V.: Seventh Framework Programme, Grant Agreement No 296282 plan4business - A service platform for aggregation, processing and analysis of urban and regional planning data, Annex I - Description of Work (2012)

SVOD – System for Visualizing of Oncological Data and Their Semantic Enhancement

Miroslav Kubásek, Jiří Hřebíček, Ladislav Dušek, Jan Mužík, and Jiří Kalina

Institute of Biostatistics and Analyses, Masaryk University,
Kamenice 126/3, 625 00 Brno, Czech Republic
{kubasek,hrebicek,dusek,muzik,kalina}@iba.muni.cz

Abstract. We present an overview of the SVOD web portal (System for Visualizing of Oncological Data) which is focused on population risk analyses related to cancer epidemiology and show its integration with the FP7 project TaToo (Tagging Tool based on a Semantic Discovery Framework). The developed TaToo Tools provide a semantic web solution in order to close the discovery gap that prevents full and easy access to specific web resources. Furhter more we will discuss in detail the used ontological resources, the integration of TaToo Tools into the SVOD portal together with the evaluation of this integration.

Keywords: cancer, epidemiology, ontology, SVOD, TaToo, web portal.

1 Introduction

Cancer epidemiology can be regarded as one of the most important and most frequently analysed topic in the field of human risk assessment. It is not only due to a remarkable public concern about growing population risk, cancer incidence and mortality is evident and a clearly reachable endpoint for risk assessment studies. We can enter this problem from the viewpoint of risk factors as agents initiating carcinogenesis, but epidemiological parameters can retrospectively indicate hazardous impact on population on a large scale. The bioindication from epidemiological data of course requires sufficient data sources. It means representative long-term profiles of incidence and mortality and a very good awareness of the most important risk factors.

In order to analyse epidemiological trends we must be able to distinguish between statistically significant trends and random fluctuations. For risk assessors it would be most important to recognize the environmentally related cases that can be attributed to external factors like pollution of air, drinking water and/or food. And again, the influence of these factors must be filtered on substantial background of the other "natural" risk factors like the age structure of the population, genetic factors or the frequently omitted life style. Apart from these complicated circumstances we can concentrate our attention to some cohorts of oncological patients that might probably indicate an external impact – if they increase in incidence and non-randomly in some regional or temporal scales. It means namely the occurrence of less advanced disease stages in age groups that are commonly out of the main risk (for example the breast carcinoma cases in pre-menopausal women younger than 35 – 40 years). In that

J. Hřebíček et al. (Eds.): ISESS 2013, IFIP AICT 413, pp. 597–607, 2013.

context, we can take some information benefit mainly from diagnostic groups that might be related to several external influences (colorectal carcinoma, kidney and lung carcinoma, breast cancer, malignant melanoma, …).

All these analyses require easily available large data sets that are themselves very expensive and typically not directly available, e.g. epidemiological cancer registries. In addition to this, we must aggregate cancer data with demographic data in order to obtain for example age-specific profiles of incidence. That is why there is such a growing interest of many professional groups (health care managers, environmental experts, risk assessors) in accessibility of these data. In our experience however, the demand for data cannot be easily fulfilled by blind databases of primary population data. The analyses are very time-consuming and finally, the outputs might be ambiguous and not assured enough to be communicated to the public. Therefore, we developed a professional web portal SVOD (System for Visualizing of Oncological Data) that offers automatically generated and verified epidemiological analyses on cancer incidence and mortality in Czech Republic [1]. This presentation is aimed to introduce potential users to the technology, architecture and functionality of the portal. The SVOD portal is accessible through the following web address: http://www.svod.cz. The portal user interface can be switched between Czech and English language.

2 System for Visualizing Oncological Data

Creating a web portal about tumour epidemiology in the Czech Republic was primarily motivated by the effort to make this representative and valuable data available to a wide spectrum of users. We anticipate that general epidemiology data about these serious diseases and related population risks should be freely available to everybody in the Czech Republic [2]. Another ambition of this web portal is to provide relevant information about tumour epidemiology in the Czech Republic abroad [3].

The SVOD portal information services can be divided into three sections:

1. *Current news*: regularly updated information about population risk assessment and tumour epidemiology;
2. *Interactive analyses* that allow the user to investigate epidemiological trends of selected oncological diagnoses;
3. *Predefined presentations of important topics* (Authorised information service). These services are available freely to all users. All analyses contain only secured, validated and publishable data of tumour epidemiology, without any privacy data of patients.

The project of creating the web portal SVOD for tumour epidemiology in the Czech Republic is tied to a longstanding development of analytical software for data from the National Oncological Registry (NOR) [4]. It was created in the years 1999-2003 and allows access to all NOR data via a wide range of automated analyses. Although the information and communication technologies (ICT) of SVOD were finalized successfully, there are severe limits with its distribution and availability. The web portal SVOD solves all these problems and provides an effective way of accessing epidemiological analyses by an unlimited number of users.

The SVOD portal processes mainly data from NOR [4] which is managed by the Institute of Health Information and Statistics (UZIS CR)[1]. It offers validated epidemiological data from the years 1977 - 2010. This represents a unique data set (of 1.6 million records) at least in the European region.

Epidemiological trends cannot be produced without relevant demographic data about the examined population. This data was provided from the Czech Statistical Office (CSU)[2] on the basis of general cooperation agreement with the Masaryk Memorial Cancer Institute[3] in Brno and the Masaryk University (MU) in Brno.

The SVOD portal was created by the Institute of Biostatistics and Analyses at the Faculty of Medicine and the Faculty of Science of the Masaryk University in Brno and Masaryk Memorial Cancer Institute in Brno [1-2]. The development of the SVOD portal is vitally supported by the Ministry of Health of the Czech Republic in the context of the National healthcare quality programme. Further development was supported by a research programme of the Masaryk Memorial Cancer Institute (Functional diagnostics of tumours, MZO 00209805) and a research programme of Faculty of Science MU (INCHEMBIOL - RECETOX project, No. 0021622412[4]). These grant projects guarantee a long-term viability of the SVOD portal and ensure regular updates of data and a successive development under supervision of administrators.

Information services of the SVOD portal will be further developed, among others on the basis of user suggestions and requirements. The main goal is to extend the information service in the area of population risk analyses in relation with available environmental data and other external risk factors (like the cooperation with above mentioned INCHEMBIOL project). The current version of the SVOD portal offers only epidemiological data, but NOR database allows even analyses of diagnostics and treatment of oncologic patients and survival analyses - even in relation to the current hospital. All these analyses are prepared only for communication in the Oncological Society[5] and are available in a restricted zone of the portal. It therefore serves as an information source for Czech health management and helps to set up reference standards for healthcare results in oncology [2].

3 Web-Based Analytical Tools

Interactive analyses are one of the core functions of the SVOD portal. Through these functions, you can easily analyze epidemiological trends examined over three decades, stratify and filter cohorts of patients and extract population risk in absolute or age-specific values. Some of these analyses offer benchmarking with respect to the clinical status of the disease. Major epidemiological trends are available in comparison with international data - GLOBOCAN 2008 [5].

---

[1] `http://www.uzis.cz`

[2] `http://www.czso.cz`

[3] `http://www.mou.cz/en/`

[4] `http://www.recetox.muni.cz/inchembiol/index.php?language=en&id=`

[5] `http://www.linkos.cz/`

3.1 Incidence and Mortality Analyses

These analyses calculate time trends of crude incidence, crude mortality and mortality/incidence ratio. Available parameters for calculations are absolute numbers, crude rates (number of cases per 100000 people in population) and age standardized ratios (ASR - European or World standard). User can also specify some filters (constraints) for analyse e.g. sex and age of patients, specify region, time period, stage, TNM and other parameters through the buttons available in the 'Incidence and Mortality' panel (see **Chyba! Nenalezen zdroj odkazů.** - left). These filters are available for all analyses, but some filters are disabled depending on the concrete analyses.

3.2 Regional Overview Analyses

These analyses return a comparison of incidence and mortality in regions of the Czech Republic. Available parameters are the crude rate and the age standardized ratio. The output could be obtained as bar-plots or maps. The area filter for this analyses is disabled (see **Chyba! Nenalezen zdroj odkazů.** - right).

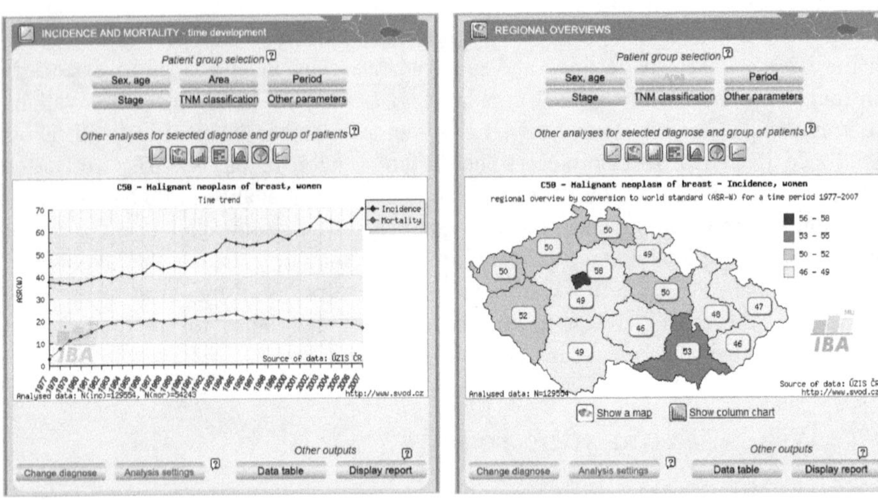

Fig. 1. Example of incidence and mortality and a regional overview analyses. (Source: Authors)

3.3 Time Trends

Time trends describe changes of incidence and mortality with respect to time. Available parameters are growth indexes related to a selected year and between-years changes. Both parameters could be viewed as absolute numbers or as relative proportions. The user can select the reference year for the analyses.

3.4 Clinical Stage Analyses

The Clinical Stage analyses represent time trends of the proportion of patients in specific clinical stages. Available parameters are absolute numbers, % and crude rate of patients in specific clinical stages. Available outputs are time trend bar-plots, time trend line-plots or pie charts of the selected time period. The Stage and TNM classification filters are disabled for this analyses.

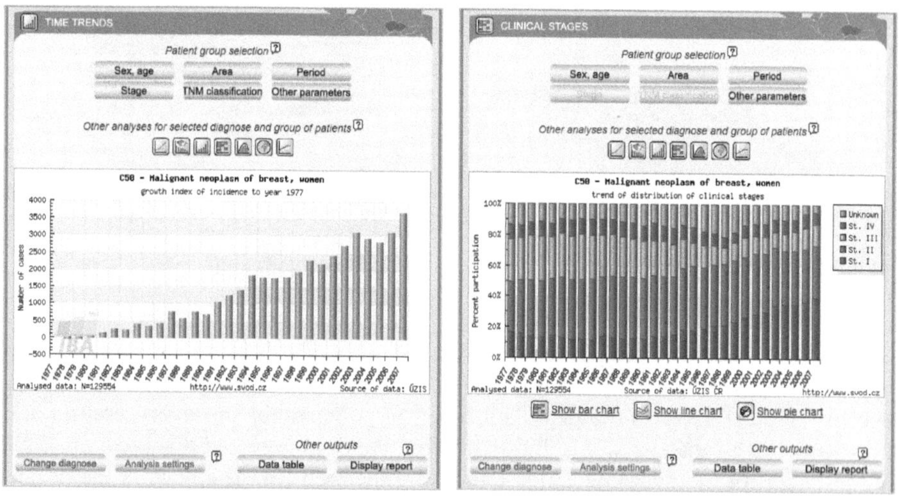

Fig. 2. Example of time trends and clinical stages analyse (Source: authors)

3.5 Age Analyses

These analyses return the age structure of population of patients with a selected diagnose. Available parameters are the absolute numbers of cases in age categories, % of cases in age categories and the age specific rate (number of cases in age category per 100000 people in population cohort of the same age).

3.6 International Data Analyses

These analyses return a comparison of incidence and mortality in the Czech Republic (CZ) with other countries. All these analyses are based on data obtained from the IARC database GLOBOCAN 2008 [5]. Comparative standards: time trend of incidence or mortality in a selected region in comparison with the situation in the whole Czech Republic. Available parameters are the crude rate and the age standardized ratio (ASR - European or World standard).

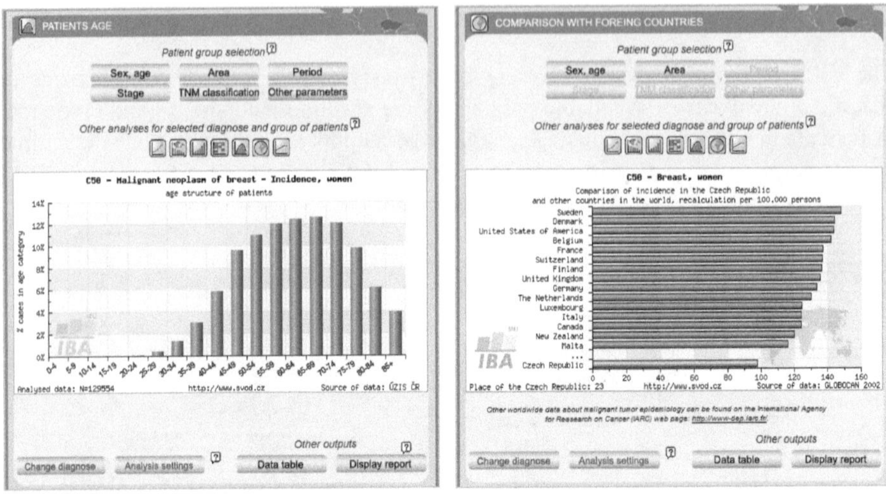

Fig. 3. Example of age analyses and in comparison with foreign countries (Source: authors)

4 Integration with TaToo Tools

The FP7 project TaToo (Tagging Tool based on a Semantic Discovery Framework) aims to set up a semantic web solution to close the discovery gap that prevents a full and easy access to environmental resources on the web [6]. The core of the project will focus on the development of tools allowing third parties to easily discover environmental resources (data and/or services residing on different information nodes) on the web and to add valuable information in the form of semantic annotations to these resources, thus facilitating future usage and discovery, and kicking off a beneficial cycle of information enrichment.

TaToo validates the usability of its developments through the implementation of three different validation scenarios. All three scenarios are embedded in highly complex environmental domains and are therefore mainly addressed to domain expert groups and communities as well as to technically skilled users.

Integration of TaToo Tools into SVOD portal is part of the Masaryk university[6] validation scenario (MU scenario), which is the anthropogenic impact and the influence of global climate change on this impact [7-9]. The purpose of this chapter is to give the overview of implementation of this integration.

4.1 Inventory of Ontologies

First, we briefly describe the knowledge resources related to the cancer and epidemiology domain. These resources vary from ontological resources to non ontological including descriptions of technologies, existing ontologies or concrete vocabularies.

[6] http://www.muni.cz

By analysing the cancer and epidemiology domain we finally proposed the initial version of ontology. For modelling of the ontology we used the Web Ontology Language (OWL) and the NeOn Toolkit[7]. Our proposed ontology contains 351 concepts and was used as the base concept for describing (tagging) resources from the cancer and epidemiology domain .

ICD-10 Ontology

The International Statistical Classification of Diseases and Related Health Problems 10th Revision (ICD-10) [11] is a coding of diseases and signs, symptoms, abnormal findings, complaints, social circumstances and external causes of injury or diseases, as classified by the World Health Organization (WHO)[7]. The code set allows more than 14,400 different codes and permits the tracking of many new diagnoses. Using optional sub-classifications, the codes can be expanded to over 16,000 codes. Using codes that are meant to be reported in a separate data field, the level of detail that is reported by ICD can be further increased, using a simplified multiaxial approach. The International version of ICD should not be confused with national Clinical Modifications of ICD that include frequently much more detail, and sometimes have separate sections for procedures. Work on ICD-10 began in 1983 and was completed in 1992.

Disease Ontology

The mission of the Disease Ontology (DO) is to provide by open source ontology for the integration of biomedical data that is associated with human disease. DO will have a formally correct (in the ontology sense), semantically computable structure. Terms in DO will be well defined, using standard references. These terms will be linked to well-established, well-adopted terminologies that contain disease and disease-related concepts such as SNOMED[8] (we are working with SNOMED to see if we can release SNOMED codes linked via UMLS[9] to the community), ICD-10, MeSH[10], and UMLS. The combination of a semantically computable structure and the external references to these terminologies will enable useful inference between disparate datasets using one or more of these standard terminologies to code disease. The Disease Ontology will be, at the end of this project, a community-driven, community-accepted ontology of diseases for clinical research and medicine inclusive of genetic, environmental and infectious diseases. It will encapsulate, therefore, a comprehensive theory of disease. The design of the disease ontology will enable greater understanding of disease states by placing heritable disorders in the context of other infectious diseases and related diseases. The structure of Disease Ontology and the external references to other terminologies will enable the integration of disparate datasets through the concept of disease.

---

7 `http://www.who.int/en/`

8 `http://www.nlm.nih.gov/research/umls/Snomed/snomed_main.html`

9 `http://www.nlm.nih.gov/research/umls/`

10 `http://en.wikipedia.org/wiki/MeSH`

TNM Ontology

We did not find an overall ontology for the structuring of TNM classifications, but there are few research initiatives in this field. One ontology which can be taken into account for creating an ontology is "An Ontology for Carcinoma Classification for Clinical Bioinformatics " by Anand Kumar et. al.[12] . They outline a formal theory for addressing these issues in a way that inferences drawn upon the ontologies would be helpful in interpreting and inferring on the entities which exist at different anatomical levels of granularity. Their case study is on the colon carcinoma, one of the commonest carcinomas prevalent within the European population.

4.2 Ontology Enhancement of Analytic Tools

The purpose of ontology enhancement of SVOD portal analytic tools is to provide the users of the SVOD portal with the possibility to indirectly use the TaToo functionality for the discovery of similar resources based on analysed objects. The TaToo discovery can be started directly from the web analyses tools. The relevant information needed for the search is already entered via the web interface during the analyses and can be submitted to the TaToo Framework.

In this integration we implemented a "TaToo button" (see Figure 5) into SVOD portal analytic tools and applied the TaToo Discovery Service to retrieve a list of related data resources from the TaToo Repository (see Figure 4). Queries used in calling the Discovery Service are automatically built from the actual settings of the SVOD analyses tool. Communication between TaToo Discovery Service and developed discovery application is based on SOAP.

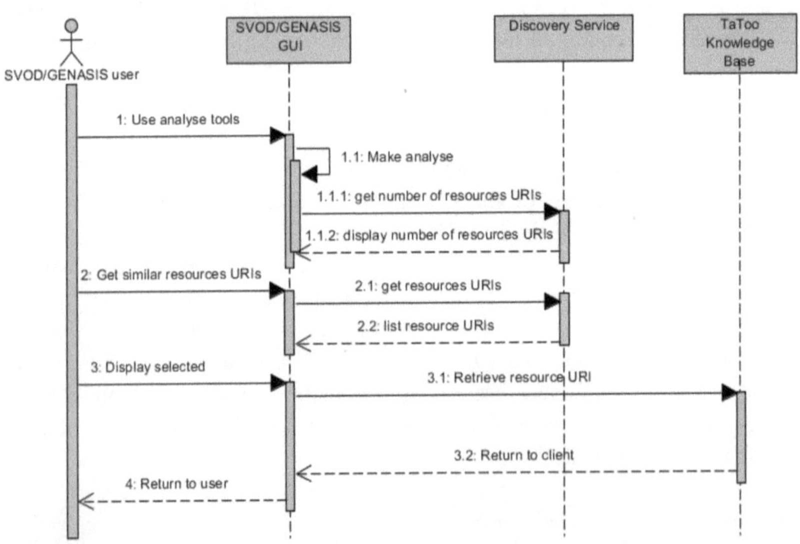

Fig. 4. Sequence diagram of communication between SVOD portal and TaToo services (Source [13])

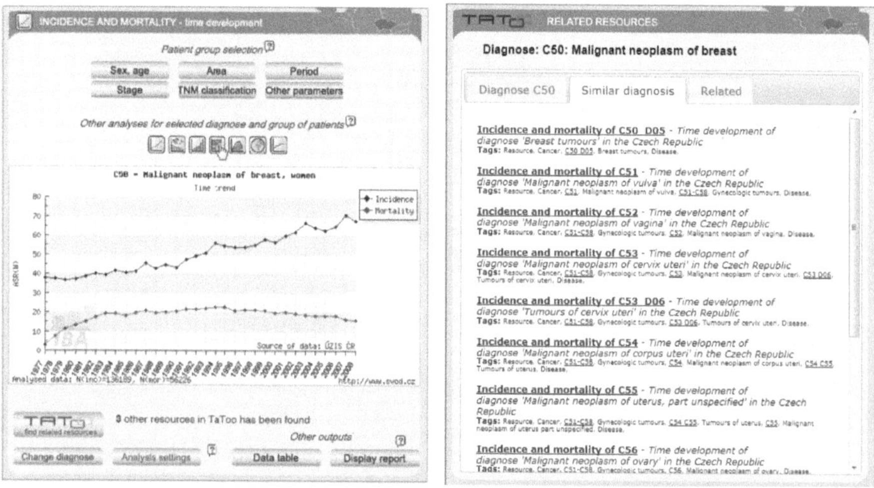

Fig. 5. GUI of integration of TaToo functionality into SVOD portal analytic tool

4.3 Evaluation

The objective of the evaluation of TaToo Tools was to ask external users whether they agree that the objectives of Enhancement of analytic tools have been achieved and how satisfied they are with the provided functionalities.

The evaluation process was organized on three levels depending on the availability of the of the evaluators and additional three workshops. The first group "A" (9 persons) had been invited to three workshops. Two of them were organised during 8th and 9th summer schools of applied informatics in September 2, 2011 and September 7, 2012, where all participants were introduced to the Validation Scenario "Anthropogenic impact and global climate change" and the basic idea of TaToo Project. The discussion and feedback of participants were positive and they obtained public access to the TaToo and SVOD portal. After this introduction in two workshops, we gave all participants of the third workshop, held on November 26, 2012, credentials to access both the TaToo portal and the SVOD portal application which integrated the TaToo Services. We provided them a questionnaire for collecting their feedbacks. This group was given one week's time to get everything tested and sent back thereafter their questionnaires.

The questionnaire has been structured as follows: Two statements regarding relevance and usability could be rated on a five-point scale ranging between 1 ("I agree") and 5 ("I disagree"). The statement regarding relevance is "The function ... is interesting for my work", the statement regarding usability is "The function ... is user-friendly". Furthermore, there was an open question asking for further comments and proposals how the application could be improved.

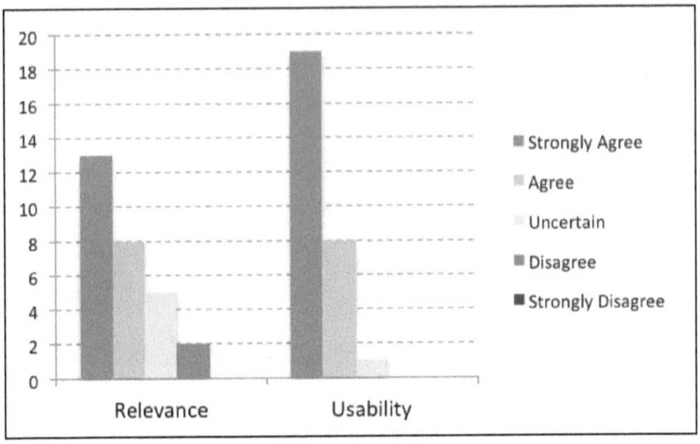

Fig. 6. Assessment of extension of SVOD portal (Source [13])

Overall, the idea of tagging resources through the topics of interest was evaluated very positively by the domain users. Linking TaToo with the other specialized portals was recognised as a good idea, but the problem may be in the fulfillment of the repository. TaToo portal users perceive as a working tool, but for its practical application are sceptical. To administer the repository it would probably be better to use some easier and user-friendly solutions. It would also be appropriate for the end user to hide the ontological background of the TaToo Tools and offer only understandable (e.g. more human readable) concepts.

5 Conclusion

We have introduced the SVOD (System for Visualizing of Oncological Data) portal and described the vision of its integration and ontological enhancement by the FP7 project TaToo. The aim of TaToo is providing a collaborative platform for the semantic enrichment of environmental information resources on the web. The development of domain ontologies in the TaToo Ontology framework is an evolutionary process which will be extended by adding new specific concepts and relationships that are currently not provided. We also gave a short overview of realized functionalities, the software quality evaluation and the usability evaluation of the user interface. A number of domain experts were asked to assess the relevance and usability of SVOD tools and its TaToo extension. We can conclude that the collected feedback was predominantly positive.

Acknowledgments. The research leading to these results has received funding from the European Community's Seventh Framework Programme (FP7/2007-2013) under grant agreement nr. 247893.

References

1. Dušek, L., Mužík, J., Koptíková, J., Brabec, P., Žaloudík, J., Vyzula, R., Kubásek, M.: The national web portal for cancer epidemiology in the Czech Republic. In: 19th International Conference Informatics for Environmental Protection, Enviroinfo 2005, pp. 434–439. Masaryk University Press, Brno (2005)
2. Dušek, L., Hřebíček, J., Kubásek, M., Jarkovský, J., Kalina, J., Baroš, R., Bednářová, Z., Klánová, J., Holoubek, I.: Conceptual model enhancing accessibility of data from cancer–related environmental risk assessment studies. In: Hřebíček, J., Schimak, G., Denzer, R. (eds.) ISESS 2011. IFIP AICT, vol. 359, pp. 461–479. Springer, Heidelberg (2011)
3. Dušek, L., et al.: Czech Cancer Care in Numbers, pp. 2008–2009. Grada Publishing, Praha (2010)
4. Czech National Cancer Registry (2013), `http://www.uzis.cz/ang/ccr/ccrindx.htm`
5. GLOBOCAN 2008: Estimated cancer Incidence, Mortality, Prevalence and Disability-adjusted life years (DALYs) Worldwide in 2008 (2008), `http://globocan.iarc.fr/`
6. Rizzoli, A., Schimak, G., Donatelli, M., Hřebíček, J., Avellino, G., Mon, J.: TaToo: tagging environmental resources on the web by semantic annotations. In: International Congress on Environmental Modelling and Software. Modelling for Environment's Sake, iEMSS 2010, pp. 1192–1199. iEMSS, Ottawa (2010)
7. Hřebíček, J., Dušek, L., Kubásek, M., Jarkovský, J., Brabec, K., Holoubek, I., Kohút, L., Urbánek, J.: Anthropogenic Impact and Global Climate Change. Description of Validation Scenario in TaToo Project. In: Hřebíček, J., Pitner, T., Ministr, J. (eds.) 7. Summer School of Applied Informatics. Indikátory účinnosti EMS podle odvětví, pp. 6–23. Nakladatelství Littera, Brno (2010)
8. Hřebíček, J., Dušek, L., Kubásek, M., Jarkovský, J., Brabec, K., Holoubek, I., Kohút, L., Urbánek, J.: Validation Scenario for Anthropogenic Impact and Global Climate Change for TaToo. In: Proceedings of the Workshop Environmental Information Systems and Services - Infrastructures and Platforms, Aachen. CEUR-WS (2010)
9. Kubásek, M., Hřebíček, J., Kalina, J., Dušek, L., Urbánek, J., Holoubek, I.: Global Environmental Assessment Requires Global Functional Searching Engines: Robust Application of TaToo Tools. In: Hřebíček, J., Schimak, G., Denzer, R. (eds.) ISESS 2011. IFIP AICT, vol. 359, pp. 507–518. Springer, Heidelberg (2011)
10. Gómez-Pérez, A., Motta, E., Suárez-Figueroa, M.C.: NeOn Methodology in a Nutshell (2010), `http://www.neon-project.org/nw/NeOn_Book`
11. World Health Organisation. ICD-10 Classifications of Mental and Behavioural Disorder: Clinical Descriptions and Diagnostic Guidelines. Geneva. World Health Organisation (1992), `http://www.who.int/classifications/icd/en/`
12. Kumar, A., Yip, Y.L., Smith, B., Marwede, D., Novotny, D.: An Ontology for Carcinoma Classification for Clinical Bioinformatics, MIE 2005;116:635-40 (2005)
13. Kubásek, M., Hřebíček, J., et al.: Deliverable D5.3.9 - Final Implementation and Validation Report - Case 3 – V3, TaToo Project, Public Document (2012), `http://www.tatoo-fp7.eu`

Intelligent Facility Management for Sustainability and Risk Management

Adam Kučera and Tomáš Pitner

Masaryk University, Faculty of Informatics
Laboratory of Software Architectures and Information Systems
Botanická 68a, 602 00 Brno, Czech Republic
akucera@mail.muni.cz, tomp@fi.muni.cz

Abstract. Building construction has gone through substantial change with the emerging spread of ICT during last decades. In the field of construction industry, the term intelligent buildings describes facilities equipped with devices and systems that can be remotely controlled and programmed and that are able to communicate and collaborate in order to ensure convenient building environment and effective operation. However, installing devices with such capabilities is only one part of the task of effective facility management and risk management. Facility managers have to be provided with tools that allow them to inspect and analyze gathered building operational data and make decisions to improve building performance. Decision support tools for facility managers usually lack deep integration with building systems. This article aims on technical aspects of providing data from intelligent buildings to the responsible personnel in order to help them to operate the building effectively with emphasis on its energy consumption and risk management.

Keywords: computer-aided facility management, energy management, building management systems, intelligent buildings, business intelligence, monitoring systems, sustainable development, risk management.

1 Introduction

Facility management as a term emerged in the 1970's. *International Facility Management Association* defines facility management as a profession that encompasses multiple disciplines to ensure functionality of the built environment by integrating people, place, process and technology.

The technology part of the facility management have experienced great change as information and communication technologies (ICT) are widely used in various building systems such as building automation, security systems or decision support software. In the construction business, the term intelligent building is used mostly for facilities equipped with remotely controlled and monitored interoperating devices. Other definitions of the term are discussed in [1].

In the opposition to use of the ICT on the low level of ensuring the technical part of building operation such as air conditioning, heating or security in

J. Hřebíček et al. (Eds.): ISESS 2013, IFIP AICT 413, pp. 608–617, 2013.

real time and constantly, ICT helps facility managers to analyze and evaluate the building performance in a long-term perspective using business intelligence and decision support tools, namely CAFM (Computer-Aided Facility Management) software products. CAFM software covers wide range of activities that are not part of organization's core business, but are necessary for its operation – space management, hoteling and reservations, preventive and on demand maintenance, service desk, portfolio management, and accounting. As the main goal of CAFM products is to lower the expenses spent on every day operation, energy management became a regular part of CAFM software products.

Reliable data about organization's facilities (device location, building plans, documentation) can be also used for preparation for the critical sitations (natural disasters, technologcial failures,...) or for direct decision support during disasters and thus increase safety of building users and reduce recovery costs.

By intelligent facility management, we mean utilizing above mentioned ICT technologies (namely capabilities of intelligent building systems and facility management software) to make facility operation inexpensive and more efficient, energy undemanding and thus more sustainable and resistant to failures or disasters. This article proposes selected improvements and extension to the building systems concerning the reliability, accessibility of the data, and cooperation with other systems that do not participate on the operation directly, namely CAFM software and other information systems used in the organization.

2 Integration Technology in Facilities

A wide variety of electronic systems is usually installed in modern buildings – e.g. *HVAC* (heating, ventilation, air condition) regulation and automation (closely connected with environmental protection), security system, access control system, CCTV (Closed-Circuit Television), fire alarm system, lighting control. Integration of all the systems that take part on the building operation is mostly motivated by economical reasons. The ability to control remotely and to monitor all systems in the building from a common user interface reduces the number of staff needed for ensuring the operation as well as demands on expert knowledge of the system operators. A common interface for all the systems also allows them to cooperate (e.g. an air conditioning unit in a particular room cannot be switched on when a window is open – in this use case, the security system provides data from the magnetic sensor on the window frame and passes them to the air conditioning control unit). Rules similar to this one improve the work environment in the buildings and save energy. In this way, purely economical considerations and actions help to make the building operation more sustainable. A building that is able to efficiently decrease its energy consumption is cheaper and more environment-friendly.

The integration of building systems is usually accomplished by establishing a common communication bus that uses one of the integration protocols. Hardware and software vendors support various protocols that can be used on a common bus, e.g. BACnet, LONWorks or KNX/EIB. The bus interconnects operators' workstations and application servers (web user interface, archive database servers, monitoring systems) with the back-end devices such as programmable logic controllers (PLCs), central units of security systems or with protocol gateways and translators. The described infrastructure (workstations, servers, gateways, common bus) is referred to as a *Building Management System* (BMS).

2.1 Building Management at the University Campus Bohunice

The project of the Building Management System that spreads over the whole Masaryk University (BMS MU) and its facilities started together with the construction of the first part of the *University Campus in Brno-Bohunice* (UKB) that was completed in 2007. Since then, the UKB site has grown to 30 buildings controlled by a common BMS. Additionally, other buildings spread through the City of Brno and belonging to various organization units (Institute of Computer Science, Faculty of Economics, University Rectorate/Headquarters, Accommodation and Catering Services) were connected to the system at the occasions of replacing obsolete building equipment with new units. Other projects under construction are planned to be connected to the BMS MU in the next few years. See Figure 1 for schema of the BMS UKB infrastructure.

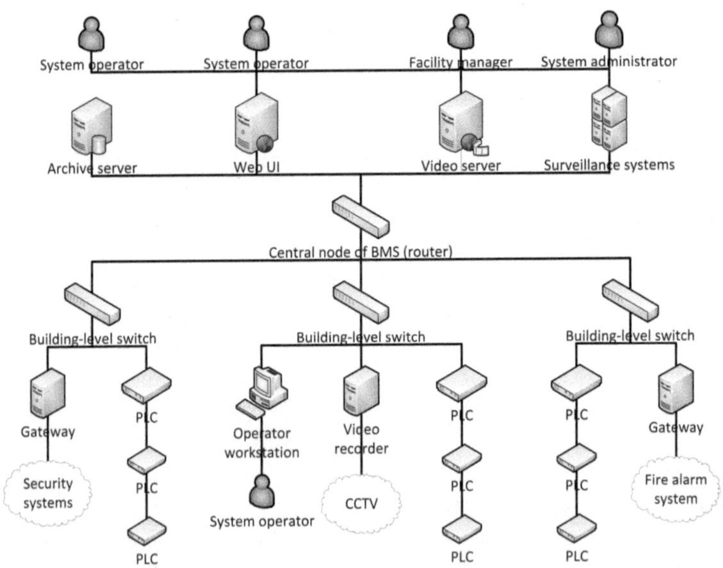

Fig. 1. BMS infrastructure schema (Source: Authors)

The systems integrated in BMS of MU vary on different phases of construction of UKB and other sites. The following systems are integrated under the BMS interface on most of the buildings (bold systems have environmental impact):

1. **HVAC**
2. **Lighting control**
3. **Elevator monitoring**
4. **Energy monitoring**
5. Access control system
6. Security system
7. **Fire alarm system**
8. CCTV (Closed-Circuit Television)
9. UPS monitoring

The core of the system is formed by PLCs and application servers that communicate over *BACnet* protocol standardized by ISO 16484-5. BMS MU today contains over 700 BACnet-capable devices from more 10 vendors including Delta Controls, Saia-Burgess, Honeywell, Johnson Controls, Siemens, and WAGO.

The users are provided with a web interface that allows them to control and monitor all integrated systems. The system contains two archive servers for saving historical data of the building operation.

3 Requirements for an Effective Facility Management

Equipping a building with advanced technologies with the ability to communicate and collaborate on the building operation does not guarantee efficient use of the resources or environment-friendly behavior of the facility. To make the building intelligent in such a way that installed systems are actually helping to optimize expenses and energy consumption of the organization facilities or help to react to the crises (natural disasters, failures), additional tools are required according to our experience gathered during five years of the UKB operation. There is a need for advanced monitoring and evaluation of building performance emerging especially at relatively large sites (dozens of buildings) such as the UKB. Currently available state-of-the-art CAFM systems support analyses based on indirect indicators such as energy consumption (electricity, water, gas) based on data gathered from the invoices from energy suppliers. These techniques usually do not provide enough detail to detect the source of poor operation effectiveness.

Further parts of the paper introduce advanced tools and methods for building performance analysis and evaluation with respect to environmental impact and risk management based on the data gathered directly from the Building Management System devices. These data provide us with a direct insight into the processes and relations among various systems incorporated in building operation and thus allows us to perform a detailed drill-down in order to inspect and resolve possible malfunctions, breakdowns, problems or errors emerging in the systems.

4 Monitoring

During the operation of relatively large sites of the UKB, the need for specialized monitoring tools for the BMS emerged. With increasing number of devices connected to the BMS, constant surveillance of all parts of the system by human operators becomes impossible. Since devices involved in the building operation usually do not support standard methods for network management such as SNMP and ICMP protocols, existing tools must either be adapted for the BMS environment or new tools must be developed. At the UKB, both approaches have been tested and used in the production environment.

An IP traffic flow-based network monitoring tool was extended by a BACnet stack for the analysis of the BACnet traffic , see [2]. A monitoring device (probe) acts as a passive listener on central nodes of the BMS network and monitors most of the network traffic occurring in the network by inspection of Ethernet frames. The data from the probe can be used for detection of sources of suspicious, atypical communication which usually comes from misconfigured devices. An incorrectly communicating device represents serious threat to the stability of the whole network as misconfigured devices tend to use broadcast messages extensively. An early detection of such devices saves time of the building operators which can be spent on more productive tasks such as optimization of building performance and system cooperation which lead to efficient operation.

A specialized monitoring tool for BACnet (named *BMSSentinel*) network is developed at Masaryk University for the purposes of permanent accessibility and functionality surveillance of key elements in the BMS network such as application servers, protocol gateways, specific PLCs ensuring precise climate conditions in labs, or routing devices. BMSSentinel utilizes *SCADAEngine BACnet Server API*[1] for accessing the BACnet network and defines variety of custom algorithms for detecting specific types of faults occurring in the BMS network. BMSSentinel provides users with web user interface and automatic notifications about detected problems. The tool is similar to the *Nagios* infrastructure monitoring tool well known from the field of ordinary ICT. Since Nagios software is used at Masaryk University, BMSSentinel is able to pass detected problems to the Nagios server where they can be processed the same way as other alerts coming from computer network.

Capabilities of BMSSentinel ensure instant reactions to network accidents, errors or failures. Besides that, the system focuses on ensuring data validity and completeness. Data correctness is a fundamental requirement for further analysis of building performance which can lead to operation optimization. Real-time monitoring of system function provides decision support in the case of various failures happening in the system and helps system operators to detect the root cause and correspondingly react to it. An overview of currently used monitoring and traffic analysis tools is provided in Table 1.

5 New Control Mechanisms for the BMS

Building management systems usually base their control decisions (switching devices on and off, opening and closing valves,...) on the data from following inputs:

[1] `http://scadaengine.com/`

Table 1. Monitoring tools, protocols and functionality (Source: Authors)

Tool \ Device	Application server	Gateway	PLC	Switch/router
BMS Sentinel	Functionality	Functionality, Accesibility	Functionality, Accesibility	
Network traffic probe	Communication characteristics	Communication characteristics	Communication characteristics	
Nagios	Accesibility	Accessibility		Functionality, Accessibility

Legend of used protocols:

BACnet	SNMP	ICMP	Ethernet

1. Static regulation algorithms
2. Human operator commands
3. Preset values (requested temperature in the room)
4. Attached sensors (temperature sensors, humidity sensors, pressure sensors, magnetic sensors,...)

Other data sources should be considered, because the effectivity of the BMS is influenced by variety of other factors that cannot be obtained from the above sources.

Other information systems used in the organization often contain data that could be used for more precise setting of the regulation algorithms. The most significant examples of such data sources are the *room reservation system* and the *lecture rooms timetables* in case of an educational organization. Time intervals of the reservations can be used as inputs for the algorithm that automatically changes state of centrally controlled air conditioning units. A basic approach with relatively static schedules set by human operators usually reflect different settings for workdays and weekends or holidays and does not reflect dynamic changes in room utilization during the day. The further step to adjust performance of the HVAC systems according to the actual needs requires additional information about the number of occupants in the room in the given time intervals.

Using weather information is another way of improving HVAC performance. The characteristics and amount of the weather data varies from simple usage of outside temperature (which can easily be measured by standard BMS devices) to utilization of other data such as the intensity of solar radiation which must be obtained by specialized devices.

An extensive overview of various approaches to HVAC systems optimization is provided in [3] and [4].

6 Business Intelligence Tools

A BMS deployed at a large site is able to generate huge amount of data. In case of the UKB, the archive server stores approximately 100 000 new records per day. To obtain useful information about complex processes and events occurring during the building operation, advanced methods of data processing and analysis must be applied [5]. Tools provided by BMS vendors (reporting capabilities, data export tools) are very ineffective (user unfriendly with insufficient speed and limited management options) when dealing with large systems.

Complexity and size of the system makes the task of maintaining constant awareness of the state of all parts of the system difficult. However, overview of all the important processes and states occurring in the whole system during crisis is crucial to succesful risk management.

This section introduces several solutions and methods used at the UKB that simplifies the analysis of building behavior. Figure 2 provides an overview of system connections proposed further.

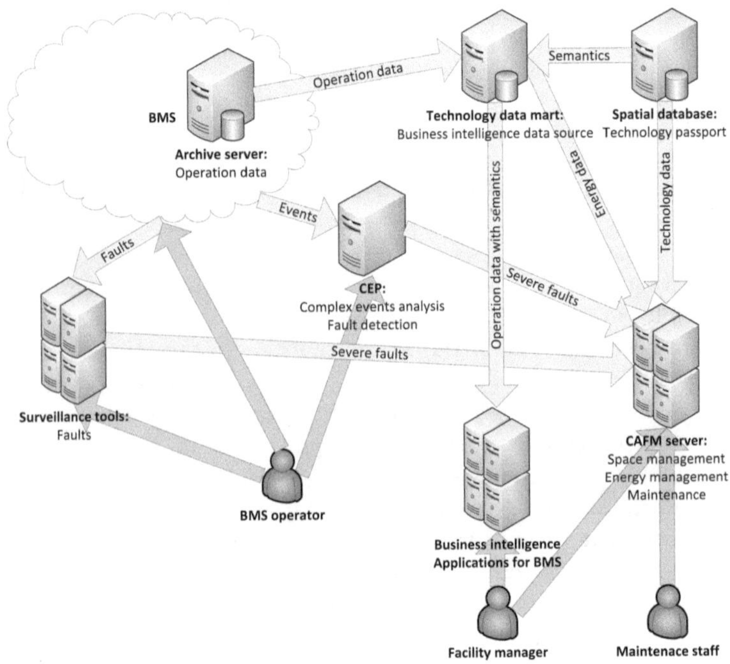

Fig. 2. BMS for Business intelligence use (Source: Authors)

6.1 Adding Semantics – Integration with Geographic Information Systems

Data points (values that are monitored using BMS) do not carry any kind of semantic information about their meaning. They are identified using their network addresses for the purpose of the BMS itself. They have names which are designed to be easily understandable by human operators and thus not suitable for machine processing.

In order to perform advanced data analyses such as aggregations of values based on particular features of data points (e.g. average of all room temperatures in specific building), we need additional source of information about the value represented by a particular data point. For the purposes of Masaryk University, spatial database of the *technology passport* can be employed. The technology passport carries information about devices installed in buildings of Masaryk University. Along with its geographical position, each device has a set of attributes attached – the most important for the purpose of integration with the BMS is the type of device. By extending the relation between a data point in BMS and a device in technology passport by a measured or controlled quantity, the data point gains complete semantic information about its precise location, its source device (temperature sensor, humidity sensor, energy meter,...) of the value and the measured or controlled quantity (room temperature, requested room temperature, relative air humidity, current, voltage,...). The relation of data points and devices in the technology passport can be utilized in business intelligence and decision support tools that are described in the following text.

6.2 Technology Data Mart

Database schema of BMS archive server has several features that limit its use for business intelligence application. Besides missing semantic information, the archive server suffers from low performance when retrieving results of complex queries that work with long time periods. This is caused by optimization of database structure (both schema and physical) for write operations. In this way, archive server is typical example of OLTP database. For this reasons, implementation of separate data warehouse (Technology data mart – TDM) was tested and is planned for deployment into the operation environment.

TDM (implemented using Microsoft SQL Server 2012) contains information about relations between BMS and technology passport as well as optimized indexing schemas, history tracking and advanced features for operations with time interval.

6.3 Complex Event Processing

Complex Event Processing (CEP) is set of tools and techniques allowing us to handle events from large systems in real time. CEP in commonly used in network security, algorithmic stock-trading, fraud detection or business process monitoring for correlating, aggregating and analyzing various events.

For the use at UKB, existing CEP framework (Microsoft StreamInsight) was adapted to support operations based on data point semantics and is able to provide building operators with easily understandable outputs that help to optimize building operation and detect various types of faults. A prototype of the system was implemented and tested in real operation environment. The system is able to perform selections, filtering and computations (average, summa, maximum, minimum) based on the data point semantic information. System runs continuous user-defined queries that are evaluated in sliding time windows. Results of the computations performed by the system can by further analyzed automatically (e.g. detecting faults of a particular device by comparing other devices on the same type) or stored for later visualization that can be examined by users.

6.4 Machine Learning and Data Mining

The data from the BMS can be examined by methods of machine learning (ML) and data mining. Algorithms as decision trees, outlier detection or clustering can be used to discover complex rules that occur in the building behavior. ML analysis, however, suffers from unavailability of data that prevents the algorithms to provide useful results. Many factors that influence the building behavior are not measurable by the BMS (the number of occupants in the room), not measured because of technological or economical limitations (a limited number of sensors that can be attached to the device) or are not saved into the archive server due to the performance reasons.

Despite this limitations, the use of ML and data mining from building management data is a promising research topic. Different approaches to analysis based on ML approaches are presented in [6], [7] and [8].

7 Risk and Crisis Management

The University Campus of Masaryk University is not prone to specific kinds of natural disasters and crises such as floods or earthquakes thanks to its location on the hill in geologically stable area of Central Europe. The main threats concerning natural influences are risk of fire and massive downfalls. However, events such as electricity blackouts or extensive system failures require immediate and corresponding reactions to current situation development.

For such situations, BMS in conjunction with CAFM software are able to provide managers and operators with precise, credible, relevant and up-to-date information. Real-time data about system statuses enriched by semantics obtained from spatial databases can be visualized using the facility management data (building plans) and provide overall view current situation that helps responsible personnel to make correct decisions in the time of crisis, disaster or failure. Cooperation of systems integrated in the BMS allows the facility to automatically react to the situations comprising health and/or safety risks and minimize the impacts of the crisis without the human intervention (i.e. fire safety system is able to control HVAC devices in order to prevent the fire from spreading caused by running ventilation).

8 Conclusion

This paper presents the field of facility management of intelligent buildings and proposes technical requirements for ICT parts of intelligent building systems that must be fulfilled in order to ensure a cost effective, energy efficient and sustainable building operation that is not prone to crises such as system failures or natural disasters. The requirements are based on experience gathered during the first 5 years of operation of the Masaryk University campus in Brno-Bohunice. Next, we propose methods to satisfy defined requirements. Large sites require a complex approach to data analysis and fault detection. Some of the presented solutions are used in the production environment while some of them are planned for future development. The building management data brings a significant potential to improve the building performance in the means of energy efficiency as well as management of risks and crises, but mining useful information from large amounts of data is not a trivial task and suitable methods for data processing must be applied.

Acknowledgement. This work was supported by the European Unions territorial cooperation program European Territorial Co-Operation Austria-Czech Republic 2007-2013 of the under the EFRE grant M00171, project iCom (*Constructive International Communication in the Context of ICT*).

References

1. Wong, J.K.W., Li, H., Wang, S.W.: Intelligent building research: a review. Automation in Construction 14(1), 143–159 (2005)
2. Krejčí, R., Čeleda, P., Dobrovolný, J.: Traffic Measurement and Analysis of Building Automation and Control Networks. In: Sadre, R., Novotný, J., Čeleda, P., Waldburger, M., Stiller, B. (eds.) AIMS 2012. LNCS, vol. 7279, pp. 62–73. Springer, Heidelberg (2012)
3. Prívara, S., Široký, J., Ferkl, L., Cigler, J.: Model predictive control of a building heating system: The first experience. Energy and Buildings 43(2-3), 564–572 (2011)
4. Cigler, J., Prívara, S., Váňa, Z., Žáčeková, E., Ferkl, L.: Optimization of Predicted Mean Vote index within Model Predictive Control framework: Computationally tractable solution. Energy and Buildings 52, 39–49 (2012)
5. Kriksciuniene, D., Pitner, T., Kucera, A., Sakalauskas, V.: Sensor Network Analytics for Intelligent Facility Management. In: Frontiers in Artificial Intelligence and Applications. Intelligent Interactive Multimedia Systems and Services, vol. 254, pp. 212–221. IOS Press (2013)
6. Popelínský, L., Glos, P.: Towards Detection of Anomalies in Building Management Data. In: 15th IBIMA conference on Knowledge Management and Innovation: A Business Competitive Edge Perspective, pp. 664–669. IBIMA Publishing (2010)
7. Seem, J.E.: Pattern recognition algorithm for determining days of the week with similar energy consumption profiles. Energy and Buildings 37(2), 127–139 (2005)
8. Seem, J.E.: Using intelligent data analysis to detect abnormal energy consumption in buildings. Energy and Buildings 39(1), 52–58 (2007)
9. Shen, W., Hao, Q., Xue, Y.: A loosely coupled system integration approach for decision support in facility management and maintenance. In: Automation in Construction, vol. 25, pp. 41–48. Elsevier B.V., Amsterdam (2012)

Visual Analytics in Environmental Research: A Survey on Challenges, Methods and Available Tools

Martin Komenda and Daniel Schwarz

Institute of Biostatistics and Analyses, Masaryk University,
Kamenice 126/3, 625 00 Brno, Czech Republic
{komenda,schwarz}@iba.muni.cz

Abstract. Visual analytics approaches bring an innovative and effective way how to deliver the knowledge from a particular domain to an individual user. With the use of visual analytics methods we can easily discover the unexpected relations and interesting patterns, which are hidden in the huge data warehouses. It builds on the human mind's ability to understand the complex visualization of information. In this paper we introduce the potential usefulness of visual analytics for researchers working in the field of environmental informatics. Current challenges beyond the survey are described here, including the summary of particular well-proven tools and scenarios, which can be applied in many various fields of environmental research.

Keywords: Visual analytics, learning analytics, human cognition, environmental data visualization.

1 Introduction

Visual analytics (VA) is a new interdisciplinary field of study that calls for a more structured scientific approach to understanding the effects of interaction with complex graphical displays on human cognitive processes. Its primary goal is to support the design and evaluation of graphical information systems that better support cognitive processes in areas as diverse as scientific research and emergency management [1]. Technologies based on VA have been moving from research to a widespread use, driven by the increased power of analytical databases and computer graphics hardware. People adopt VA to see and understand complex data, new visual paradigms and hidden insights [2]. Visual analytics techniques allow analysts to look at the data from different perspectives or in various domains and to fulfill diverse analytical tasks. Visual displays and interactive techniques are often combined with computational processing, which, in particular, enables analysis of a larger sets of data than would be possible with purely visual methods [3]. A sufficiently deep understanding of visual analytics guides developers of visualization techniques in creating more effective tools in terms of user requirements, and helps evaluators in designing future studies that are more representative of insights sought by users from their data sets [4].

J. Hřebíček et al. (Eds.): ISESS 2013, IFIP AICT 413, pp. 618–629, 2013.

That knowledge brings an interesting way how to optimize delivering knowledge from huge amount of available data in a particular domain to an individual user. The need for more effective processing of data by users has been identified by several researchers in the Technology Enhanced Learning (TEL) field [5], [6], [7]. With the establishment of TEL, a new research field, called Learning analytics (LA), has emerged in recent years. According to the 1st International Conference on Learning Analytics and Knowledge, LA is defined as measurement, collection, analysis and reporting of data about learners and their contexts for purposes of understanding and optimizing learning processes and the environments in which it occurs [8]. Similarly to VA, LA borrows and synthesizes techniques from different related fields, such as data mining, social network analysis or business intelligence to harness them for converting data into useful information and thereon to motivate actions, like self-reflecting ones previous teaching or learning activities, to foster improved teaching and learning [9]. At present, data mining tends to concentrate on developing new tools for discovering patterns in data. LA – at least as it is currently contrasted with data mining approaches — focuses on the use of intelligent data, learner-produced data, and analytical models to discover information and social connections, and to predict and advise on learning [10], [6]. The emergence of LA as a field in its own right meant that there were now separate groupings focusing on each of the challenges driving analytics research [11]. Regardless of the great enthusiasm that is currently surrounding LA, it also raises substantial questions for research. In addition to technically focused research questions such as the compatibility of educational datasets, or the comparability and adequacy of algorithmic and technological approaches, there remain several "softer" issues and problem areas that influence the acceptance and impact of LA. Among these are questions of data ownership and openness, ethical use and dangers of abuse, and the demand for new key competences to interpret and act on LA results. It is also necessary to mention that any application that analyses student data must comply with privacy regulations, which adds cost and complexity [12]. This means that the implementation of LA in learning processes requires to be carefully crafted in order to be successful and beneficial [13].

2 Human Cognitive Abilities and Limitations

Many researchers concentrate on human cognition and user interaction issues raised by the VA community. The top ten challenges of extreme-scale VA were identified in [14], among them being the problem of the human bottleneck. It is predicted that performance of all major computing components – power, memory, storage, bandwidth, concurrence, etc. – will improve by a factor of 3 to 4,444 by 2018. Human cognitive capability will certainly remain constant. One of the challenges is, therefore, finding alternative ways to compensate for human cognitive weaknesses. Thus, it is necessary to develop new visualization approaches that facilitate the flow of human reasoning. Human cognitive models or human cognitive frameworks, including subprocesses of human thinking such as accommodation, categorization or drawing conclusions will play a key role in that development, as it is shown in [15], [16]. A novel approach how the user interface can be adapted in relation to a current process-task and under consideration of the user's behaviour is described in [17] – a work related to FUPOL project (www.fupol.eu) aimed on social computing and crowd sourcing in e-participation. The research performed in economic decision-making [18] addresses overcoming bounded rationality issues that arise from the cognitive limitations through the application of a VA tool that can aid in complex decision making tasks in real world applications with uncertainty. A more integrative approach to the interplay of science and technology in VA would build new scientific methods for developing and empirically evaluating theories of human-computer "mixed-initiative" cognitive processing [19].

Users may also benefit from a personalized use of Adaptive Learning Environments with underlying context-aware recommender systems for learning [20], which provide tools such as content-based filtering, collaborative filtering, knowledge-based filtering and their hybridizations. A similar example is the content infrastructure of the EU-oriented Open Discovery Space CIP PSP initiative (http://www.opendiscoveryspace.eu) that also attempts to create a very big data infrastructure collecting learning content and usage data around it. Such global learning data infrastructures can help in scaling up recommender systems by allowing them to consume, process and use a rich variety of contextualized usage data streams, and thus enable novel forms of real time intelligence that can only become possible when processing extremely large data volumes.

3 Data Representation and Processing

Interaction with visualizations enables the dialog between user and the visual representation of the underlying data. The interactive manipulation of the data, the visual structure or the visual representation provides the ability to solve various tasks and discover insights. Classifications of interaction describe an abstract view on various visual tasks and may provide a mapping to the human information processing.

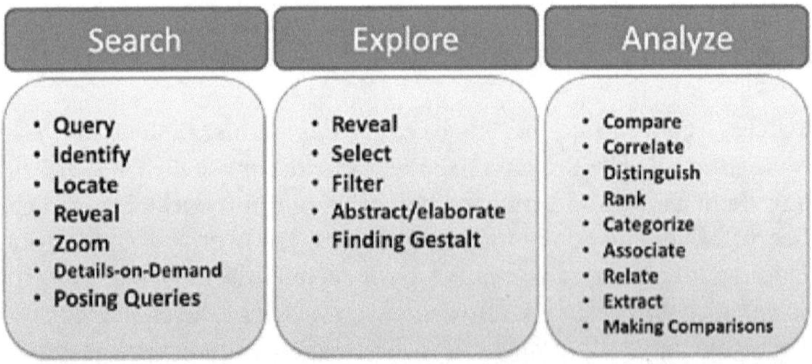

Fig. 1. High-level tasks with assigned subtasks and interactions [21]

Figure 1 illustrates three important high-level tasks and interactions assigned to them as well as subtasks derived from the existing visual task classifications. Various models for differentiating visual variables in context of human information processing are important for an accurate and beneficial adaptation of visualizations [21].

At the core of successful visual analytics systems are computational techniques that transform data into concise, human comprehensible visual representations. Computers store, move and analyze data that, on initial examination, are a simple collection of bits in various primitive data forms such as structured (for example, relational tables, geometry), semi-structured (for example, e-mail, network packets) or unstructured (for example, a collection of text). Many algorithmic approaches are organized around key data characteristics, which are applied to all primitive data forms rather than algorithms that apply to specific data types. Transforming data into an effective visual representation is fundamentally different from transforming incoming raw data. In addition to the raw performance needed to support interactive analytics, characteristics of the data itself affect the transformations and representations for both computation and visualization. Successful data transformation and representation methods combine mathematical, statistical and linguistic analysis with hardware and software techniques to handle various kind of raw data such as massive data, geospatial and temporal data, imperfect data, heterogeneous data, and users-in-the-loop [22]. For example Starlight information graphics [23] provides the classification according to the type of information they portray. The two major classes are non-spatial information graphics such as text and numeric data, and inherently spatial information graphics, which depict information containing intrinsic spatial coordinates, such as geospatial or CAD data. One of Starlight's unique capabilities is its ability to integrate these two types of information so that they may be analyzed together.

4 Current Challenges

Recently scientists have been working on creating sophisticated models of existing domains, to simulate the domain behavior and to extrapolate gained knowledge into the future. They use VA approaches as an integral part in each step of this process, especially for visualization used to verify model setup and to estimate its parameters. High-performance computing simulations of massive amounts of data are then analyzed using visualization software specifically designed for parallel processing of large amounts of structured and unstructured data [24]. Given the challenge of automatic computational analysis of complex data, much recent effort in the analytics community has focused on the design of interactive graphical representations of information that might better support the human ability to perceive and to construct meaningful patterns from data [25]. The use of various virtual environments can provide log files that can be studied with web mining techniques to understand the behaviour of the users. However, statistical metrics are not always easy to interpret for the stakeholders, which suggest the need to develop novel approaches for the easy discovery of usage patterns:

- Combining human domain knowledge with automatic data analysis techniques by providing users with interactive visual interfaces, whereby ‘interactive’ means that users can actively participate in the analytical process.
- Context-aware recommender systems for learning based on VA to allow users for contextualized multiscale exploratory analytical reasoning.

5 Methods

The visual analytics can be applied in many various fields of environmental research. In the following section we briefly introduce the key domains and describe the beneficial use of VA methods.

The utility of VA algorithms was reported among many fields of human scientific activity also in environmental information systems known for their large volumes of multidimensional spatial-temporal data. For illustration, POPs GMP presents analysis of available Stockholm Convention Global Monitoring Plan data on persistent organic pollutant concentrations in core matrices. It provides a set of electronic tools for storage and on-line visualization of data published therein, and proposed a comprehensive IT solution for future data collection campaigns. The visualization provides easily accessible information on performance of monitoring programs in individual countries and/or regions sorted by matrices, time, and compounds [26].

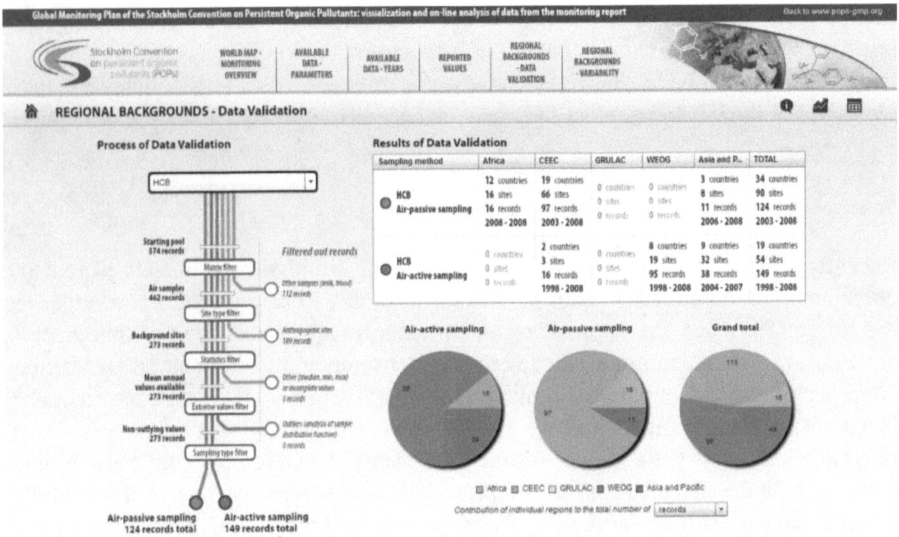

Fig. 2. A screenshot from a statistical summary of reported background atmospheric POPs concentrations [26]

Mike Sips et al. have proposed an approach for detection of interesting patterns in environmental time series: an algorithm to characterize the temporal behavior, visual detection of potentially interesting patterns in a matrix visualization, and interactive

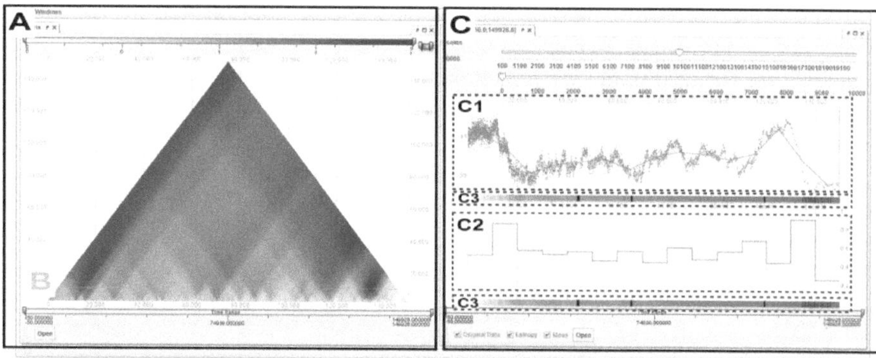

Fig. 3. The visual interface of the prototype which shows the Antarctica time series [27]

exploration of detected patterns. The system computes statistical measures for all possible time scales and starting positions. A specific feature of mentioned approach is its flexibility due to the wide range of statistical measures provided, e.g., mean, variance, entropy. This allows for capturing a large set of temporal patterns to address different analysis tasks and different types of data [27].

Natalia and Gennady Andrienko presented illustrated structured survey of the state of the art in visual analytics concerning the analysis of movement data and have demonstrated how different visual analytics techniques can support our understanding of various aspects of movement.

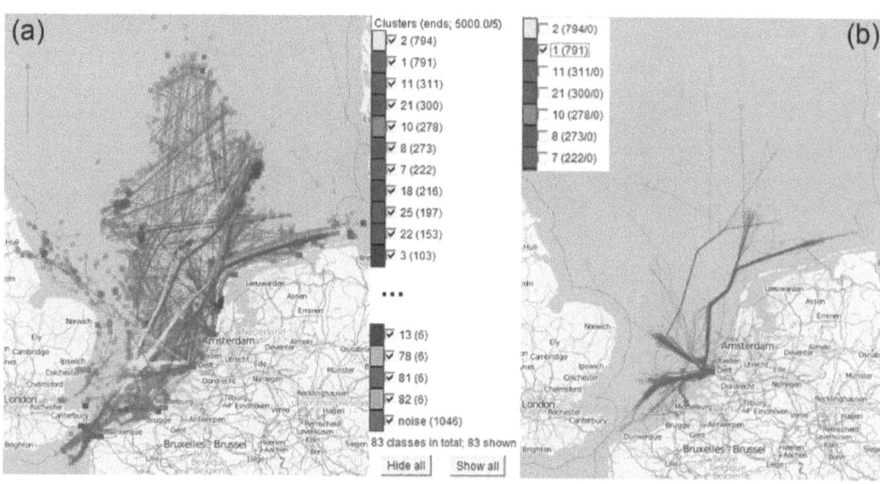

Fig. 4. Interactive progressive clustering of trajectories. (a) The ship trajectories have been clustered according to the destinations. (b) One of the clusters is selected [3].

This hot research topic in visual analytics utilize the legacy of cartography, with its established techniques for representing movements of tribes, armies, explorers, hurricanes, and so on; time geography, with its revolutionary idea of considering space and time as dimensions of a unified continuum (space–time cube) and representation of behaviours of individuals as paths in this continuum; information visualization, with its techniques for user-display interaction supporting exploratory data analysis; and geovisualization, with its interactive maps and associated methods enabling exploration of spatial information [3].

The U.S. Department of Energy's Office of Environmental Management currently supports an effort to understand and predict the fate of nuclear contaminants and their transport in natural and engineered systems.

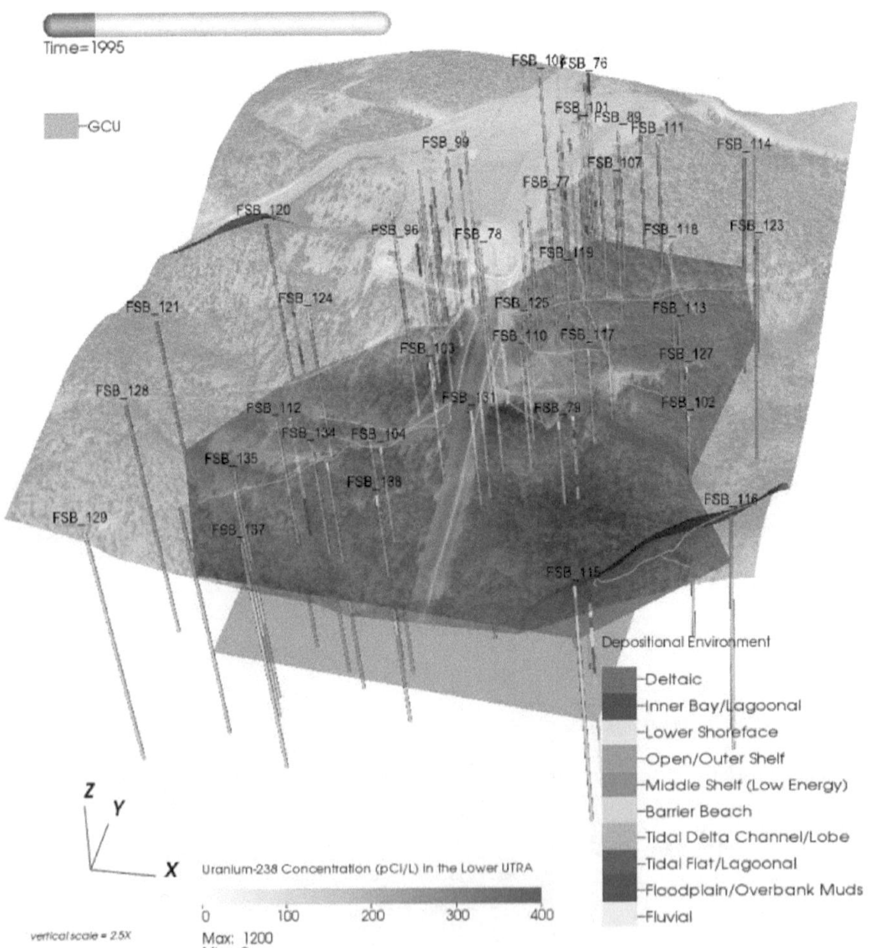

Fig. 5. Three dimensional visualization (computed in VisIt) of Uranium-238 plume migrating through the subsurface of the F-area [24]

Joerg Meyer et al. demonstrates that by bringing together experts from a variety of domains, including geology, hydrology, physics, and computer science, we can develop an integrated system that helps scientists make well informed decisions regarding environmental management. They have developed set of tools that was designed to help determine the extent of soil and water contamination from radioactive isotopes from measured data and to develop models that will be used to predict the future progression of a contaminated site with respect to these contaminants [24].

Watershed management, in its very nature, represents a participatory process, requiring horizontal and vertical collaborations among multiple institutions and stakeholders. Alexander Sun has shown the way of the migration of an Environmental decision support systems module (EDSS) module from the traditional client-server based architecture to a client of cloud-computing services. Google Drive, which is behind the new version of the EDSS module, provides a number of basic visual analytics features that can be used to increase the collaborative decision-making experience while drastically reducing the cost of small-scale EDSS. This approach can fundamentally change the way many decisions are made by allowing collaborations to be set up quickly with less overheads than the client-server approach. The promise of using Google Fusion Table to support collaborative decision-making at only a fractional cost of a full-scale, client-server application was shown. Although only a simple analytical model is described in this study, other models can be easily incorporated into the Cloud to perform more sophisticated analyses [28].

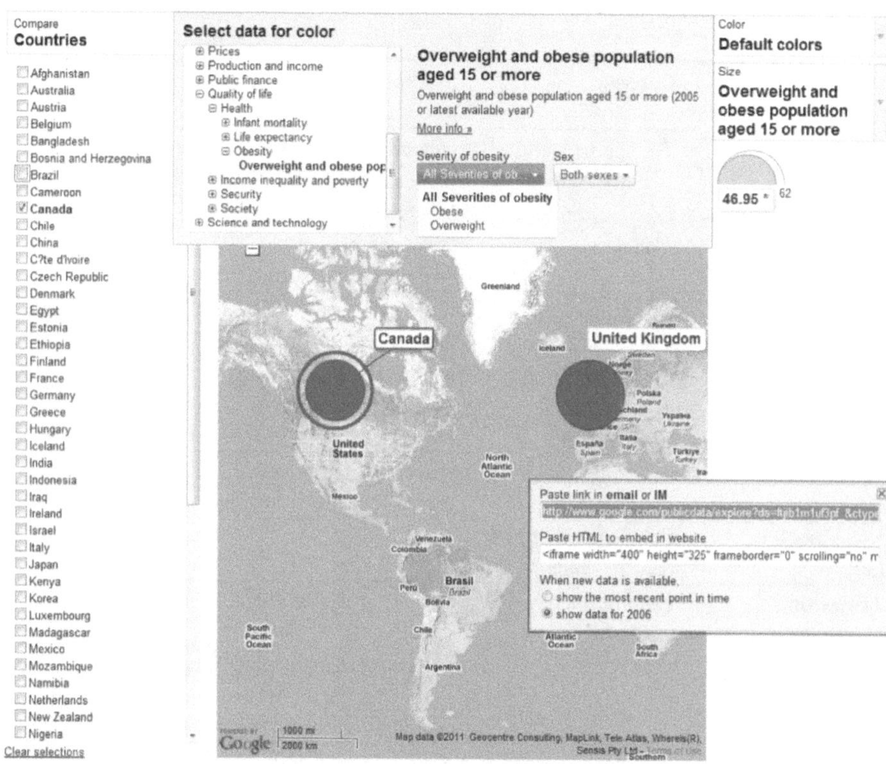

Fig. 6. Screenshot of Google Public Data Explorer showing some of its interface features and options [29]

Geospatial visual analytics, a specialized subtype of visual analytics, is an emerging multidisciplinary area which supports spatio-temporal analytical reasoning and decision-making through interactive visual interfaces (such as maps and other visual artifacts) that are linked to computational methods. Microsoft Pivot enables powerful visual zooming in and out of Web databases and the discovery of new patterns and relations in them that would otherwise be invisible in standard Web browsing of large datasets. It can also be used for geographic classification and offers a fresh way how to visually browse and arrange massive amounts of data and images online and also supports geographic and temporal classifications of datasets featuring geospatial and temporal components [29].

6 Summary on VA Tools and Their Usage

After surveying issues raised by the VA community in the paragraphs above, we present here a summary on particular VA tools which are available for various sorts of use, see Table 1.

Table 1. Selected available VA tools, their main purpose, underlying methods and scenarios of their use

VA Tool [reference]	**Purpose**	**Methods & Techniques**	**Scenarios & Data sets**
POPs GMP [26]	management and visualization of Global Monitoring Plan data	data pre-processing, outlier identification, LOQs replacement, data aggregation, descriptive statistics	persistent organic and pollutant data
Visualization of Multiscale Exploration [27]	detection of patterns in numerical time series from environmental sciences	time scales statistical values computing, pattern visual identification, detected pattern exploration	Earth's climate system and ocean data
Multivariate Trajectories Visualization [30]	customization and versatile exploration using multiple density fields	trajectory aggregation, density field computation	anomalous behaving ships and maritime safety data
Trajectory Contingency Table [31]	exploration of attributes of trajectories	automatic identification system, continuous density model, sampled trajectories computation	density vessel positions data
Visualization via Progressive Clustering [32]	trajectories (dis)similarity assessment	progressive clustering, OPTICS method, trajectory clustering	car GPS-tracking data

VA Tool [reference]	Purpose	Methods & Techniques	Scenarios & Data sets
Animal Ecology Explorer [33]	iterative generation of metadata based on exploratory findings	exploration through enrichment approach, segmentation, clustering	animal tracking data
Visualization Using Graph Centrality Measures [34]	places and explore city structure identification	data pre-processing, trajectory aggregation, graph centrality, visual and algorithmic methods	georeferenced photo captures data
Ship and Weather Information Monitoring [35]	weather parameters visualization along ship routes	gridded binary format processing, parallel coordinates plots, voyage time graphs, geospatial data interaction	ship voyages data
ASCEM Project [24]	decision-making process support in planning of treatment options for contaminated sites	model setup verification, critical parameters estimation, parallel processing analysis, simulation results validation	nuclear contaminant data
EDSS for Watershed Management [28]	effective watershed-scale management	total maximum daily load model, simple analytical modeling, cloud-computing approach	coastal river basin data
Geospatial Visual Analytics [29]	visual zooming and discovery of new patterns and relations	Microsoft Live Labs Pivot technology (descriptive component collection, visual data representation)	mortality dataset

7 Conclusions

This paper has discussed the potential usefulness of visual analytics methods, techniques and particular tools for researchers working in the field of environmental informatics – one of extreme-scale-data areas. Visual analytics tools are helpful in situations where a higher level of user interactivity and involvement is needed to process effectively huge amounts of multidimensional data, understand them and use the acquired knowledge to draw a decision. Challenges beyond the state-of-art in visual analytics were outlined here, including the human bottleneck lying in the limited human cognitive capability. Selected existing and well-proven VA tools were described here together with their main purpose and scenarios of their use. Mental or human cognition modeling as well as adaptive virtual environments with underlying context-aware recommender systems - these are issues which should tease the attention of scientists and developers who need to design visual representation of extreme-scale datasets.

References

1. Fisher, B., Green, T.M., Arias-Hernández, R.: Visual Analytics as a Translational Cognitive Science. Top. Cogn. Sci. 3, 609–625 (2011)
2. Chabot, C.: Demystifying Visual Analytics. Comput. Graph. Appl. Ieee. 29, 84–87 (2009)
3. Andrienko, N., Andrienko, G.: Visual analytics of movement: An overview of methods, tools and procedures. Inf. Vis. 12, 3–24 (2013)
4. Saraiya, P., North, C., Lam, V., Duca, K.A.: An Insight-Based Longitudinal Study of Visual Analytics. Ieee Trans. Vis. Comput. Graph. 12, 1511–1522 (2006)
5. Romero, C., Ventura, S.: Educational data mining: A survey from 1995 to 2005. Expert Syst. Appl. 33, 135–146 (2007)
6. Siemens, G.: What are learning analytics (2010)
7. Duval, E.: Attention please!: learning analytics for visualization and recommendation. In: Proceedings of the 1st International Conference on Learning Analytics and Knowledge, pp. 9–17 (2011)
8. Banff, A.: 1st International Conference on Learning Analytics and Knowledge (2011), `https://tekri.athabascau.ca/analytics/`
9. Dyckhoff, A.L., Zielke, D., Bültmann, M., Chatti, M.A., Schroeder, U.: Design and implementation of a learning analytics toolkit for teachers. J. Educ. Technol. Soc. 15, 58–76 (2012)
10. Bienkowski, M., Feng, M., Means, B.: Enhancing Teaching and Learning Through Educational Data Mining and Learning Analytics: An Issue Brief (2012), `http://www.ed.gov/edblogs/technology/files/2012/03/edm-la-brief.pdf`
11. Ferguson, R.: Learning analytics: drivers, developments and challenges. Int. J. Technol. Enhanc. Learn. 4, 304–317 (2012)
12. Norris, D.M.: 7 Things You Should Know About First-Generation Learning Analytics. Educ. Learn. Initiat. Eli (2011)
13. Greller, W., Drachsler, H.: Translating Learning into Numbers: A Generic Framework for Learning Analytics. Educ. Technol. Soc. 15, 42–57 (2012)
14. Wong, W., Liu, W., Bennamoun, M.: Ontology Learning from Text: A Look Back and into the Future. Acm Comput. Surv. 44 (2012)
15. Green, T.M., Ribarsky, W.: Using a human cognition model in the creation of collaborative knowledge visualizations. SPIE Defense and Security Symposium. p. 69830C–69830C (2008).
16. Green, T.M., Ribarsky, W., Fisher, B.: Visual analytics for complex concepts using a human cognition model. In: IEEE Symposium on Visual Analytics Science and Technology, VAST 2008, pp. 91–98 (2008)
17. Burkhardt, D., Nazemi, K.: Dynamic process support based on users' behavior. In: 15th Int. Conf.on Interact. Collab. Learn., ICL 2012, pp. 1–6 (2012)
18. Savikhin, A., Maciejewski, R., Ebert, D.S.: Applied visual analytics for economic decision-making. In: Ieee Symp. Vis. Anal. Sci. Technol., Vast 2008, pp. 107–114 (2008)

19. Simon, H.A.: The sciences of the artificial. MIT press (1996)
20. Verbert, K., Manouselis, N., Ochoa, X., Wolpers, M., Drachsler, H., Bosnic, I., Duval, E.: Context-Aware Recommender Systems for Learning: A Survey and Future Challenges. IEEE Trans. on Learn. Technol. 5, 318–335 (2012)
21. Nazemi, K., Kohlhammer, J.: Visual Variables in Adaptive Visualizations (2013)
22. Kasik, D.J., Ebert, D., Lebanon, G., Park, H., Pottenger, W.M.: Data transformations and representations for computation and visualization. Inf. Vis. 8, 275–285 (2009)
23. Risch, J.S., Rex, D.B., Dowson, S.T., Walters, T.B., May, R.A., Moon, B.D.: The STARLIGHT information visualization system. In: Proceedings of IEEE Conference on Information Visualization, pp. 42–49 (1997)
24. Meyer, J., Bethel, E.W., Horsman, J.L., Hubbard, S.S., Krishnan, H., Romosan, A., Keating, E.H., Monroe, L., Strelitz, R., Moore, P., Taylor, G., Torkian, B., Johnson, T.C., Gorton, I.: Visual Data Analysis as an Integral Part of Environmental Management. IEEE Trans. Vis. Comput. Graph. 18, 2088–2094 (2012)
25. Thomas, J.J., Cook, K.A.: Illuminating the path: The research and development agenda for visual analytics. IEEE Comput. Soc. (2005)
26. Hůlek, R., Jarkovský, J., Borůvková, J.: Global Monitoring Plan of the Stockholm Convention on Persistent Organic Pollutants: visualization and on-line analysis of data from the monitoring reports (2013)
27. Sips, M., Kothur, P., Unger, A., Hege, H.-C., Dransch, D.: A Visual Analytics Approach to Multiscale Exploration of Environmental Time Series. Vis. Comput. Graph. Ieee Trans. 18, 2899–2907 (2012)
28. Sun, A.: Enabling collaborative decision-making in watershed management using cloud-computing services. Environ. Model. Softw. 41, 93–97 (2013)
29. Boulos, M.K., Viangteeravat, T., Anyanwu, M.N., Nagisetty, V.R., Kuscu, E.: Web GIS in practice IX: a demonstration of geospatial visual analytics using Microsoft Live Labs Pivot technology and WHO mortality data. Int. J. Heal. Geogr. 10, 19 (2011)
30. Scheepens, R., Willems, N., van de Wetering, H., Andrienko, G., Andrienko, N., van Wijk, J.J.: Composite density maps for multivariate trajectories. Vis. Comput. Graph. IEEE Trans. 17, 2518–2527 (2011)
31. Willems, N., Van De Wetering, H., Van Wijk, J.J.: Visualization of vessel movements. Computer Graphics Forum, pp. 959–966 (2009)
32. Rinzivillo, S., Pedreschi, D., Nanni, M., Giannotti, F., Andrienko, N., Andrienko, G.: Visually driven analysis of movement data by progressive clustering. Inf. Vis. 7, 225–239 (2008)
33. Spretke, D., Bak, P., Janetzko, H., Kranstauber, B., Mansmann, F., Davidson, S.: Exploration through enrichment: a visual analytics approach for animal movement. In: Proceedings of the 19th ACM SIGSPATIAL International Conference on Advances in Geographic Information Systems, pp. 421–424 (2011)
34. Vrotsou, K., Andrienko, N., Andrienko, G., Jankowski, P.: Exploring city structure from georeferenced photos using graph centrality measures. In: Gunopulos, D., Hofmann, T., Malerba, D., Vazirgiannis, M. (eds.) ECML PKDD 2011, Part III. LNCS, vol. 6913, pp. 654–657. Springer, Heidelberg (2011)
35. Lundblad, P., Eurenius, O., Heldring, T.: Interactive visualization of weather and ship data. In: 13th International Conference on Information Visualisation, pp. 379–386 (2009)

Information System for Global Sustainability Reporting

Ondřej Popelka, Michal Hodinka, Jiří Hřebíček, and Oldřich Trenz

Department of Informatics, Faculty of Business and Economics, Mendel University, Zemědělská 1, 61300 Brno, Czech Republic
{ondrej.popelka,michal.hodinka,jiri.hrebicek, oldrich.trenz}@mendelu.cz

Abstract. In the article we describe various methodologies considered for company performance assessment – such as Sustainability Assessment of Food and Agriculture systems (SAFA) by the Food and Agriculture Organization (FAO) and other standardized reporting frameworks. We analyse and consider common characteristics of the used performance indicators. We further examine different properties of the defined key performance indicators and additional performance indicators in order to propose a data model for a generic information system. The proposed data model aims to be reusable for different performance assessment methodologies. We introduce several abstractions of the assessment methodologies on the level of performance indicators and report outputs. It is important to allow the organization to use the same data in several different reports and simplify and speed-up the reporting process. The resulting core data model of the information system is described in the article, along with a brief description of the implemented prototype system.

Keywords: Corporate performance, Information system, SAFA, Corporate reporting, Performance evaluation, Performance indicators.

1 Introduction

Current trends promote reporting as one of the means of ensuring sustainable production and preserving the global environment. Many experts and institutions have invested considerable effort into the development of reporting methodologies and frameworks. Special attention has been given to reporting focused on sustainable development and aligned indicators. A considerable effort has also been invested into promoting the concept of corporate responsibility and quantification methods [11].

In our research as part of the GACR No P403/11/2085 project– *Construction of Methods for Multifactor Assessment of Company Complex Performance in Selected Sector*, we concentrate mainly on the agriculture sector that has strong effects on the environment, though other fields of human activity are important too. The research is carried out by teams of Faculty of Business and Management (FBM) of Brno University of Technology (BUT) and Faculty of Business and Economics (FBE) of Mendel University in Brno (MENDELU) since January 2011. The project is being funded by the Czech Science Foundation and solved through the years 2011-2014. The main research goals of this project are specified by its partial research targets [13], [2]:

J. Hřebíček et al. (Eds.): ISESS 2013, IFIP AICT 413, pp. 630–640, 2013.

1. Analyze the current state of corporate performance by means of research of the global information and database sources available.
2. Analyze the current implementations of the economic, environmental, social and governance reporting systems in selected business branches.
3. Assess and categorize existing characteristics of basic corporate performance pillars – economic, environmental, social and governance – in relation to the development of overall corporate performance.
4. Use Key Performance Indicators (KPIs) to identify the roles and importance of environmental, social and corporate governance (ESG) factors used in the overall company performance.
5. Construct quantitative and qualitative methods of multifactor measurement of corporate performance in the chosen economic activities with the use of ICT tools.
6. Apply the developed methods for the measurement of corporate performance into practice, and determine possible further improvements.

Further on in the article we describe the chosen KPIs and the software tools designed to support performance reporting, with special attention to sustainable reporting and corporate responsibility.

Currently, the reporting systems are undergoing a change in usage and interpretation. Starting from a specialized system designed for the use of little data, analysts are developing them further, so that their use be universal. Nearly all organization units need to take part in the reporting process and take advantage of the data reported by other company units. Our goal is to support this change especially in small to middle-sized enterprises. Such companies currently struggle even with the mandatory reporting required by law only to fulfill regulatory demands.

We aim to facilitate the mandatory reporting operations for small and middle-sized enterprises and allow them to transition into a more scalable reporting system. We see this as an important step to a more wide-spread sustainable reporting. This should both enhance strategic decision-making and planning and improve risk management and sustainability of the business. In order to fulfill this goal we propose the development of a specialized information system or of an information system module for company performance reporting [7]. Organizations would be able to use such a system to generate standardized reports and assess their performance. In the implementation details we do not focus on large corporations where an extensive reporting system is already used, though the architecture and system design still applies.

2 Current Reporting Frameworks

Reporting of sustainability and environmental, social and governance (ESG) performance is a crucial step towards a market that rewards the creation of long-term wealth in a just and sustainable society. Sustainability key performance indicators form the backbone of sustainability disclosure that allows for improvement of the issues most tied to the corporation's environmental and social impact and which are most material to a company's financial performance.

The Food and Agriculture Organization of the United Nations is developing the *Sustainability Assessment of Food and Agriculture systems* (SAFA) guidelines to assist in the achievement of fair practices in food and agriculture production and trade on a local and regional level [3]. The SAFA framework is the result of an extensive iterative process. It is built on the cross-comparisons of codes of practice, corporate reporting, various standards, indicators and other technical protocols currently being used by food and agricultural enterprises that implement sustainability assessment [3].

The structure and methodology of the SAFA Guidelines is built upon: ISO 14040:2006, the ISEAL Code of Good Practice [12], the Reference Tools of the Global Social Compliance Programme, the Sustainability Reporting Guidelines [6] and its Food Sector Supplement of the Global Reporting Initiative. The SAFA Guidelines – currently in the testing version – will be revised and finalized in 2013 in order to improve their practicality and applicability.

The guiding vision of SAFA is that food and agriculture systems worldwide are characterized by *environmental integrity, economic resilience, social well-being* and *good governance*. In recent years, there has been some progress in the realization and acknowledgement of sustainable development, which is summarized [9], [14] and [17]. Many stakeholders in the agriculture sector have contributed to this progress by improving agricultural productivity, protecting natural resources and human resources by implementing standards for assessing and improving sustainability across the agricultural sector [5], [9].

Currently the feedback summarized at the *Workshop of SAFA Practitioners and Partners* held in March 2013 in Rome is being evaluated for the purposes of implementation. The draft of finalized version of the guidelines is available since June 2013 [4]. Apart from the methodology strengths, there are also a number of weak spots in the SAFA framework which have been identified during the evaluation process by the test practitioners. These include: unclear language in some parts of the guidance, definitions that are too complex, and problems with measuring the actual performance, etc., all of which is to be addressed in the revised guidelines. The positive elements of the framework are mainly the completeness of the framework and an overall full, complex coverage of various themes [5].

The SAFA guidelines consist of three core sections – *the SAFA framework*, *SAFA implementation* and *Sustainability dimensions*. In the following chapters, we describe a specialized information system, whose role should cover the following operations necessary for sustainable reporting:

- data collecting including data normalization;
- standardized reporting for company stakeholders;
- required regulatory reporting depending on the current law in the current country;
- reporting in standardized interchange XBRL format;
- customized reporting for the purpose of ad-hoc reports;
- evaluating company performance by means of computing standardized performance indicators.

These operations correspond to the last four steps of Section 2 of the SAFA framework implementation (select a tool, collect data, aggregate the results, and perform reporting) (see Fig. 1). Covering the first two steps (setting goals and checking compliance and relevance) seems impractical at present, due to the highly specific nature of the problem for each reporting corporation. Therefore, these steps are not considered a part of the proposed system, although there is some demand to provide as much automation as possible [11].

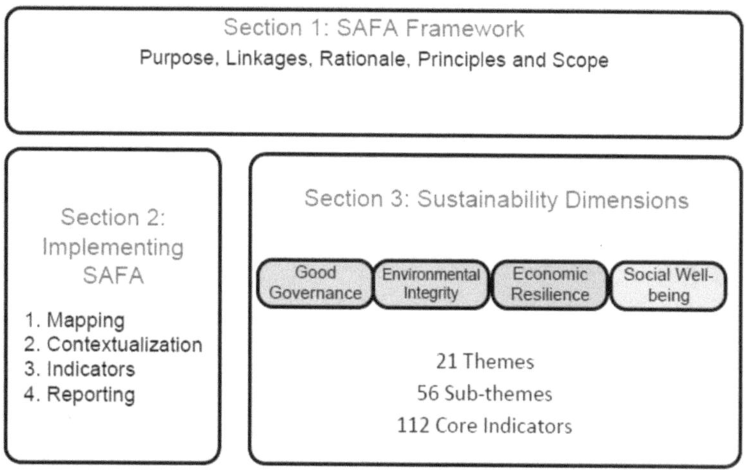

Fig. 1. SAFA Guidelines structure version 2.0 [3]

Among other deficiencies of the current SAFA framework are the challenges which are faced by smallholders, relative to implementing this framework. These include the lack of appropriate user-friendly tools, limited existing data, lack of capacity to obtain the data or of reporting indicators, the values of which are costly to obtain [11]. It was agreed that the measurement of key performance indicators is a must; there may be, however, other types of indicators, the inclusion of which might prove necessary– e.g., practice-based indicators. There is a strong demand for the improvement of reporting tools. The current SAFA reporting tool is presented on an excel sheet; this is found to be insufficient by the participants. Other types of data-collection tools such as questionnaires are being proposed and more automation is required [11].

2.1 Overview of Existing Performance Indicators

Because we concentrate mainly on agricultural organizations, the SAFA methodology is highly important to us. Therefore we closely follow the results of SAFA pilot testing and the feedback obtained from it. The proposals and remarks mentioned in the previous chapters are of great importance and we have considered them in the proposal of the system described further in this article. Also, to allow a greater generality of the information system, we have also considered other performance methodologies and their indicators because these may provide us with other points of view [16].

In June 2013, the GRI released the G4 Sustainability Reporting Guidelines [19] updating the G3.1 Guidelines from 2011 [6] and updating and completed the G3 Guidelines from 2006. The new G4 Sustainability Reporting Guidelines (the Guidelines) [19] offer *Reporting Principles*, *Standard Disclosures* and an *Implementation Manual for the preparation of sustainability reports* by organizations, regardless of their size, sector or location.

There are a number of other alternative reporting methodologies. An example is AEI EU (Agri-Environmental Indicators), whose indicators [1] are not yet generally usable in practical generic reporting. Some of the indicators cannot be computed because of missing data, sometimes the data required are not homogeneous enough [8], [10]. The main reason for the evaluation of these AIE indicators is that it is highly desirable to develop tools, which will be capable of producing required values of these indicators if necessary.

2.2 Key Performance Indicators

Within the solution of the GACR 403 project, we have proposed a minimalistic set of key generic performance indicators [14] applicable to a wide range of enterprises. The latest version of these indicators is listed in Table 1 and they form one set of indicators considered for implementing and validating the information system prototype. Second set of testing indicators are those described in *Methodology for*

Table 1. ESG indicators proposed for manufacturing enterprises

Measurable	Performance Indicator	Scale and (Unit)
Environmental indicators		
Investment	**EN01 – Environmental protection investments**	Total investments to environmental protection / value added [%]
	EN02 – Environmental protection expenditures	Cost of Environmental protection expenditures /value added [%]
Emission	**EN03 – Total air emissions**	Total direct and indirect emissions to air / value added [t / CZK]
	EN04 – Total greenhouse gas emissions	Direct and indirect emissions / value added [t / CZK]
Resource Consumption	**EN05 – Total annual energy consumption**	Direct and indirect energy consumed / value added [MWh / CZK]
	EN06 – Total renewable energy consumption	Energy from renewable resources * 100 / total energy consumption [%]
	EN07 – Consumed materials	Consumption of materials by weight / value added [t / CZK]
	EN08 – Recycled input materials	The share of raw materials from recycled materials, percentage of total input materials [%]
	EN09 – Total annual water consumption	The total volume of water removed / value added. [m^3 / CZK]

Table 1. (*Continued*)

Waste	**EN10 – Total annual waste production**	The total amount of each type of waste generated by operation of the company / value added [t / CZK]
	EN11 – Total annual production of hazardous waste	The total amount of hazardous waste generated by operation of the company / value added [t / CZK]
Social indicators		
Company	**SO01 – Community**	Social investment to local communities * 100 / value added [%]
	SO02 – Contributions to municipalities	Monetary value of projects of municipalities * 100 / value added [%]
Human rights	**HR01 – Discrimination**	Number of discriminatory cases * 100 / number of employees [%]
	HR02 – Equal opportunities	Number of women * 100 / average number of employees [%]
Labor relations	**LA01 – The rate of staff turnover**	Terminated employment relationships * 100 / average number of employees [%]
	LA02 – Expenditure on education and training	Total expenditure on education * 100 / value added [%]
	LA03 – Occupational Diseases	Occupational illness * 100 / average number of employees [%]
	LA04 – Deaths	Number of fatal accidents * 100 / average number of employees [%]
Responsibility for products	**PR01 – Customer Loyalty**	Number of customers at the end of the year – the newcomers during the year * 100 number of customers at the beginning of the year [%]
	PR02 – Marketing Communications	Communications – website, etc. [y/n]
	PR03 – Health and safety of customers	The total monetary value for non-compliance * 100 / value [%]
Corporate Governance (CG) indicators		
Monitoring and reporting	**CG01 – Company Information**	Information about the objectives and strategy of the company [y/n]
The effectiveness of CG	**CG02 – CG Responsibility**	Collective agreement [y/n]
	CG03 – CG Standardization	Publishing of standardized (GRI,CSR) reports [y/n]
	CG04 – Ethical behavior	Code of ethics [y/n]
	CG05 – CG Codex	Code of corporate governance [y/n]

Table 1. (*Continued*)

CG Structure	**CG06 – CG Remuneration**	Amount of remuneration of board of directors and the supervisory board * 100 / Annual labor costs [%]
	CG07 – CG Membership	Number of independent members of the CG * 100 / number of top management members [%]
	CG08 – Equal opportunities	Number of women in the total number of members of CG [%]
Compliance	**CG09 – Compliance with regulatory standards**	Monetary value of fines for noncompliance * 100 / value added [%]
Economic indicators		
Value added	**EE01 – Value added per employee**	value added * 100 / average number of employees in the year [%]
	EE02 – Value added to personnel costs	value added * 100 / payroll costs [%]
Market position	**EE03 – Market Share**	turnover size in manufacturing * 100 / turnover [%]
Efficiency	**EE04 – Profit EAT**	Earnings after taxes – EAT [CZK]
	EE05 – Profit EBT	Earnings before taxes – EBT [CZK]
	EE06 – Profit EBIT	Earnings before interest and taxes – EBIT [CZK]
	EE07 – ROE Performance	ROE = EAT / Equity [CZK]
	EE08 – ROA Performance	Return on assets ROA = EBIT / Total assets [CZK]
	EE09 – ROS Performance	Return on sales ROS = EAT / Total sales [CZK]
	EE10 – ROI Performance	Return on invested capital ROI = EBIT / Total equity [CZK]
	EE11 – ROCE Performance	ROCE = EBIT / Equity + Long-term liabilities [CZK]
	EE12 – Turnover size	Sales of own products and services * 100 / value added [%]
Cash Flow	**EE13 – Operating Cash Flow**	Total cash resources of the company * 100 / Total operating costs [%]
Additional indicators	**EE14 – Expenses on R & D**	Total costs * 100 / value added [%]
	EE15 – Number of employees	Average number of employees in the year (persons) [number]

assessing the sustainability of crop production systems for the conditions of the Czech Republic [15]. The rationale behind these choices is that we want to test both simpler (and generic) reporting systems such as the one listed in Table 1 and complex reporting systems of indicators such as SAFA or [15] and [18].

3 Reporting Information System

Our goal is to provide an architectural proposal generic enough to contain all of the above-described methodologies. To implement such a system successfully, several abstractions are required. These need to be carefully constructed to ensure that they are hidden before the end-use and that they do not add any more complexity to the reporting itself [20].

3.1 System Requirements and Goals

In our vision, we foresee two main goals of the reporting information system. A company may use such a system to create various reports and share data with both its stakeholders and in the future, with state institutions. The second goal is company performance assessment performed by evaluating key performance indicators by means of one of the proposed methodologies. This way, the company can share its performance with the public, or check performance development.

The reporting system should include a flexible layout design, rich visualization, business requirements and data logic definition. In addition, there is a need for translation or of publishing requirements, the central deployment and customization of interactive reports. These primary functions represent good a reporting system with strong delivery capabilities of relevant reports.

For all the use-cases of the generic actors – *reporter* and *company performance evaluator* – we consider two scenarios. The system may be used either on a regular basis (*registered company scenario*) to generate scheduled company reports (annual, quarterly ...) of performance, and to check trailing company performance and its development. A second scenario (*anonymous company*) represents the case when a company wants to generate an ad-hoc report. This can occur, e.g., when a company is applying for a subsidy or grant or when a company is evaluating changes in company operation (e.g., restructuring) where a report would be generated for the period before the change and for the period after the change. The second scenario can also be seen as an entry point into the reporting system for companies that do not use it regularly.

3.2 Indicator Abstraction

To encompass the different methodologies described in first chapter we propose the following abstraction of performance indicators (applicable universally to both key performance indicators and additional performance indicators). For the prototype application, we have used a relational database as a storage system. Hence, the following description uses relational database technology. The basic indicator abstraction entities are: *indicator group, indicator, indicator item, indicator item aspect.*

The entities described above define the core of the data model of the information system. The following scheme (Fig. 2) shows further details with associated entities. Here we defined a report type, which represents an instance of a given methodology e.g. SAFA version 1.2. We also introduce a report instance that represents a concrete filled report for a specific company for a specified time period. The indicator value, which is an actual value entered by the end user, may be bound to either an indicator item or indicator item aspect but not both. Note that indicator value is not actually related to the report instance. This is intentional, so that a single indicator value may be reused in several report instances. As long as the period is the same, the value should be reused. This is crucial to maintain a consistency of data across different reports and it is a highly important feature to make reporting as easy and quick as possible.

There is also an indicator report group entity, which defines grouping of indicators in the defined report types. The linkage is provided by the indicator report types association table which assigns indicators to specific groups in specified report types. Note that it is possible that an indicator is used in more than one report definition, which is desired behavior.

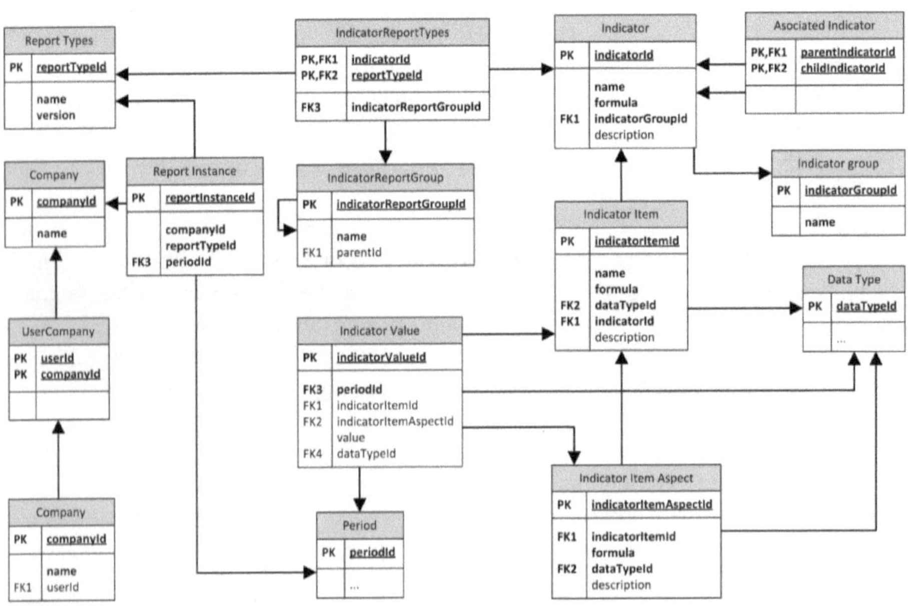

Fig. 2. Core data model of the reporting system

Closely linked to indicator value computation is the *Associated Indicator* table. This is used to track dependencies between indicators in a case when one indicator is used for the computing of another indicator. A concrete example can be seen in the methodology enterprises [14] where there is an indicator *EE14 – Expenses on R&D*, which is computed as *Total costs of R&D* divided by *Value added* and expressed in per cent. The *value added* is another indicator, although it is not directly expressed in the report. There is only *EE01 – value added per employee* and *EE02 –value added to personnel costs*. Both of which are computed as *value added*, divided by the *average number of employees* for the former and by *the payroll costs* for the latter.

4 Conclusions

In the article we describe the data model for an information system for sustainable reporting. We have implemented a prototype of the described specialized information system. There are two primary goals of the prototype implementation: to verify that the proposed indicator abstraction can be implemented successfully, and to verify the usability of the system for end-users. This is in accordance with the SAFA framework testing. As we have shown in our previous research, the critical issue in data-processing and data-analysis tasks is to get the right information quickly, nearly in real-time, in a targeted way, and effectively, and the future will be extended with a more flexible report-generation and other features adding to the end user experience.

Acknowledgments. This paper is supported by the Czech Science Foundation. Name of the Project: Construction of Methods for Multifactor Assessment of Company Complex Performance in Selected Sectors. Reg. No P403/11/2085.

References

1. Eurostat, European Commision: Analytical framework (2010), http://epp.eurostat.ec.europa.eu/portal/page/portal/agri_environmental_indicators/introduction/analytical_framework
2. Kubásek, M., Hřebíček, J., Kalina, J., Dušek, L., Holoubek, I.: Semantics Annotations of Ontology for Scenario: Anthropogenic Impact and Climate Change Issues. In: Hřebíček, J., Schimak, G., Denzer, R. (eds.) ISESS 2011. IFIP AICT, vol. 359, pp. 398–406. Springer, Heidelberg (2011)
3. SAFA Sustainability Assessment of Food and Agriculture systems. Draft Guidelines, Version 2.0 (2012), http://www.fao.org/fileadmin/templates/nr/sustainability_pathways/docs/SAFA_Guidelines_final_draft.pdf
4. Food and Agriculture Organization of United Nations: Sustainability Pathways: SAFA Pilot studies (2013), http://www.fao.org/nr/sustainability/sustainability-assessments-safa/safa-pilot-studies/en/
5. Food and Agriculture Organization of United Nations: SAFA Practitioners and Partners' Workshop Summary Report. March 18–19 2013, Italy (2013), http://www.fao.org/fileadmin/templates/nr/sustainability_pathways/docs/SAFA_Workshop_Report_final.pdf
6. Globalreporting: Guidelines on Globalreporting (2011), https://www.globalreporting.org/reporting/latest-guidelines/g3-1-guidelines/Pages/default.aspx
7. Hodinka, M., Štencl, M.: Reporting Tools for Business Rules. Informační technologie pro praxi. Ostrava (2011)
8. Hodinka, M., Štencl, M., Hřebíček, J., Trenz, O.: Business intelligence in Environmental reporting powered by XBRL. Acta univ. Agric. et silvic. Mendel. Brun. (in print, 2013)

9. Hřebíček, J., Hodinka, M., Motyčka, A., Popelka, O., Trenz, O.: Environmental and Sustainability Indicators: Case Study for Agriculture and Food Processing Sector. In: 26th International Conference on Informatics for Environmental Protection (EnviroInfo 2012), pp. 83–93. Shaker Verlag, Aachen (2012) ISBN 978-3-8440-1248-4
10. Hřebíček, J., Popelka, O., Štencl, M., Trenz, O.: Corporate performance indicators for agriculture and food processing sector. Acta Univ. Agric. et silvic. Mendel. Brun. LX(4), 121–132 (2012)
11. Hřebíček, J., et al.: Sustainability indicators: development and application for the agriculture sector. In: Erechtchoukova, M.G., Khaiter, P. A., Golinska, P. (eds.)Sustainability Appraisal: Quantitative Methods and Mathematical Techniques for Environmental Performance Evaluation. Springer, Heidelberg (2013)
12. ISEAL Alliance: Assessing the impacts of environmental and social standards systems. ISEAL Code of Good Practice, version 1.0. ISEAL Alliance, London (2010), `http://www.isealalliance.org/resources/p041-impactscode-of-good-practice`
13. Kocmanová, A., Dočekalová, M.: Construction of the economic indicators of performance in relation to environmental, social and corporate governance (ESG) factors. Acta Univ. Agric. et Silvic. Mendel. Brun. LX(4) (2012)
14. Kocmanová, A., et al.: Sustainability: Environmental, Social and Corporate Governance Performance in Czech SMEs. In: The 15th World Multi-Conference on Systemics, Cybernetics and Informatics (IFSR), pp. 94–99. WMSCI Orlando, USA (2011)
15. Křen, J.: Methodology for assessing the sustainability of crop production systems for the conditions of the Czech Republic, 48 p. Mendel University in Brno, Brno (2011)
16. OECD: Environmental Indicators for Agriculture Volume 1: Concepts and framework. OECD, Paris (1999), `http://www.oecd.org/agriculture/sustainable agriculture/40680795.pdf`
17. Soukopová, J., Bakoš, E.: Assessing the efficiency of municipal expenditures regarding environmental protection. In: Environmental Economics and Investment Assessment. III. ed., pp. 107–119. WIT Press, Cyprus (2010)
18. Valtýniová, S., Křen, J.: Indicators used for assessment of the ecological dimension of sustainable arable farming – review. Acta Universitatis Agriculturae et Silviculturae Mendelianae Brunensis LIX(3), 247–256 (2011)
19. G4 Guidelines website on Global reporting initiative (2013), `https://www.globalreporting.org/reporting/g4/Pages/default.aspx`
20. Chalupová, N., Motyčka, A.: Situation and trends in trade-supporting information technologies. Acta univ. Agric. et silvic. Mendel. Brun. LVI(6), 25–36 (2008)

The Model of Territory Unit Evaluation for Allocation of Resources on Flood Protection

Jana Soukopová[1], Eduard Bakoš[2], Jan Šelešovský[1], and Lenka Furová[1]

[1] Masaryk University, Faculty of Economics and Administration, Department of Public Economics, Brno, Czech Republic
{soukopova,jansel}@econ.muni.cz, lenka.furova@gmail.com
[2] University of Defence, Faculty of Economics and Management, Civil Protection Department
eduard.bakos@unob.cz

Abstract. Within implementation of Directive of the European Parliament and of the Council 2007/06/ES of 23 October 2007 on the assessment and management of flood risks the classification was recommended to Czech Republic which sorties the areas affected by floods according to the degree of threat and economic potential of these areas. For this purpose was developed the model of territory unit evaluation. This model is formulated as a combination of two models including model of risk matrix and model of the area value expressed by available statistical data. These data reflect the different levels of areas development and their possible future development through weighted multi-criteria decision making. The aim of our paper is to present the model and its use in terms of increasing the efficiency of spending resources in area in relation to flood protection. The model may serve as an inspiration in terms of use for such a large territorial units in European countries applying the Directive.

Keywords: Flood protection, territory unit, economic potential, damage.

1 Introduction

In economic sciences different categories which represent economic level, economic power, economic dimension, economic performance and economic potential of the territory unit are used. Usually are derived from the gross domestic product of the state, or from the gross domestic product per capita [5]. In a more sophisticated form are structured and decomposed into sub-indicators reflecting diversified sectorial and cross-cutting components. In a dynamic concept of the theory this indicators reflect performance factors, the country's competitiveness, development and growth potential based on multi-criteria basis. For example, categories of competitiveness of the national state have been in the U.S. in the eighties explicated not only theoretically, but also operationalized in government documents. However, it is debated to these days [4]. Similarly, the European Union puts the date in both the Lisbon strategy and the Europe 2020 strategy. In this context, as well as regular charts of the countries are compiled by some countries or international institutions.

J. Hřebíček et al. (Eds.): ISESS 2013, IFIP AICT 413, pp. 641–650, 2013.

In a similar way, the economic power of territorial units within individual states [1] is defined. At the regional level are most commonly used categories of gross domestic product, but also other indicators describing the sub-segments of the economy of the monitored territory [3] can be used too. When determining the potential of regions is mostly based on the so-called "driving" factors, resulting in the identification of settlement patterns, economy and industry, demography and employment and social sphere [2] [6] [8 - 11]. Individual areas are associated with a set of relevant indicators. For multi-dimensional assessment of regional development could be used significance „trees“ of the factors, which is usually named „Ishikawa diagram“[7]. With this diagram can be described development potential of the area. Unlike this method there are other methods which could be used of assessment as is the method of paired comparisons, and other available methods. Partial indicators are then usually assigned appropriate weight and according to certain algorithms are aggregated.

Unlike the national view, where there is sufficient data base, for analysis, comparison and evaluation of lower level of territorial and administrative units are not always complex indicators and indicators. In the present paper, we offer one of the possible variants of creation the index of economic potential of the region, taking into account the available data base.

However, it is clear that in the regional domestic and foreign literature there are different approaches determining the economic potential of the region and are proposed different criteria and indicators for their synergistic expression. Compared to the macroeconomic level, there is much less consensus of experts.

The occurrences of many catastrophic floods in recent years led to significant increase interest of policies in improve flood protection. The main subject of discussion is the impact of measures to protect again negative effect of floods in individual regions and the influence of these measures on both economic growth and other economic fundamentals. This is directly connected with the problem of allocation of public spending in these areas.

The aim of our paper is to present the model of territory unit evaluation (hereinafter referred to as “the MTUE”) and its use in terms of increasing the efficiency of spending resources in area in relation to flood protection. This model is formulated as a combination of two models including model of risk matrix and model of the area value. In developing Model of the Czech Republic, we came out the models available in literature.

2 The Model of Territory Unit Evaluation

The basic concept of the model is the assumption of its application for comparison of individual territorial units and for evaluation of efficiency of spending resources in area in relation to flood protection. The total amount of aid should be based on character of the affected area and risk of threats. The MTUE is developed for the evaluation territory unit. In the MTUE we have used the concept of risk matrix and the concept of economic potential of the area.

The MTUE consist of the combination of two sub-models: the risk matrix model and the territory unit model.

2.1 Risk Matrix Model

The risk matrix model takes into account the risk of property loss and risk of human life loss in the territory unit.

The Property Loss. The coefficient of the property loss is based on potential loss of property during the floods. It is determined as follows:

$$Em = w_1 Em_1 + w_2 Em_2 \tag{1}$$

where Em_1 is the property loss,
Em_2 is the property loss in area with the property,
$w_{1,2}$ is the value for the coefficient.
The values were set 0.5 for both coefficient based on an expert assessment.

Risk of Human Life Loss. The coefficient of risk of human life loss is designed as weighted sum, see the following formula:

$$Eo = w_1 Eo_1 + w_2 Eo_2 \tag{2}$$

where Eo_1 is the coefficient of human life loss,
Eo_2 is the coefficient of human life loss per population of the territory (district),
$w_{1,2}$ is the value for the coefficient.
The values were set 0.5 for both coefficient based on an expert assessment.

2.2 Territory Unit Model

Territory unit model comprises:

- **basic indicators** defining the territory (number of inhabitants and area, number inhabitants per ha);
- **economic indicators** considering the riches of the region from the perspective of tangible assets (property values and structures), investment and business (Foreign Direct Investment and number of business entities);
- **income indicators** (tax revenue per capita, average wage and employment).

The economic potential is then a function of the above variables:

$$Ep_j = f(x_1, \dots, x_n) \tag{3}$$

where x_i are variables affecting the value of the economic potential

For the construction of the economic potential we used concept multi-criteria evaluation and weight point method. Thus

$$Ep_j = \sum_{i=1}^{7} ep_{ij} w_i \qquad (4)$$

where Ep_j is the value of economic potential of j-territory unit, $j=1,\ldots,n$,
ep_{1j} is the coefficient of population per hectare of the territory unit (TU),
ep_{2j} is the coefficient values and structures of the property of the TU
ep_{3j} is the coefficient of the tax revenue of the TU,
ep_{4j} is a coefficient of the average salary of a TU,
ep_{5j} is the coefficient of the employment of the TU,
ep_{6j} is the coefficient representing the amount of foreign direct investment per capita of the TU,
ep_{7j} is a coefficient representing the number of enterprises per inhabitant of the TU and $ep_j \in [0,1]$.

Subsequently, the expert review determined weights for individual coefficients, where:

$$w_1 = 0.05,\ w_2 = 0.05,\ w_3 = 0.4,\ w_4 = 0.2,\ w_5 = 0.05,\ w_6 = 0.15,\ w_7 = 0.1.$$

Resulting evaluation of territory unit is then weighted coefficient of the above three coefficients:

$$TUE_j = w_1 Em_j + w_2 Eo_j + w_3 Ep_j \in \langle 0,1 \rangle \qquad (5)$$

where TUE_j is j-TU evaluation, $j=1,\ldots,n$,
Em_j is the coefficient of risk of property loss of j-TU, $j=1,\ldots,n$,
Eo_j is the coefficient of risk of human life loss of j-TU, $j=1,\ldots,n$,
Ep_j is economic potential of j-TU, $j=1,\ldots,n$,
w_i is standardise value for the coefficient.

The expert review determined weights for individual coefficients, where:

$$w_1 = 0.3, w_2 = 0.3, w_3 = 0.4.$$

3 Case Study in the Czech Republic

The developed Model is constructed on the basis of available data in the Czech Republic. As the basic territory unit we choose district.

All coefficients used in Model are tracked and available and were determined on the basis of an expert assessment and the possibility of freely available data based on a database of the Czech Statistical Office and is available on www.czso.cz. The average wage is also based on company data, and server Profesia City for business operating on site http://www.mestoprobyznys.cz. Only in the case of tax revenues per capita, we use data from the Ministry of Finance. For the case study we choose the year 2010 with the exception of data which we obtained from the Ministry of Finance.

3.1 The Property Loss of District

Coefficient Em_1 was calculated according to loss of fixed assets of districts. Based on the distribution of the data the reference value of 3 billion CZK per year was determined by expert way. The districts that exceeded the reference values have been assigned a value of 1. Data of other districts were recalculated as a proportion of potentially affected assets by flood in million CZK per year and the reference value.

Coefficient Em_2 was calculated as the proportion of potentially affected property by floods in urban area of the region (district). Based on the distribution of the data the reference value for the calculation of coefficient was set out 1.5 million CZK. The districts that exceeded the reference value have been assigned a value of 1. Data of other district were recalculated according to the following formula:

$$Em_{2j} = \frac{M_j}{ZU_j} \tag{6}$$

where M_j is value of potentially affected property by floods in million CZK per year,
ZU_j is the urban area of district j in ha, where $j = 1, \ldots, n$ $(n = 76)$.

3.2 Human Life Loss of District

Coefficient Eo_1 was calculated according human life loss of district. Based on the distribution of the data the reference value of 1000 inhabitants per year was determined by expert way. The districts that exceeded the reference value have been assigned a value of 1. Data of other district were recalculated as a proportion of human life loss of the district per year and reference value.

Coefficient Eo_2 was calculated as a proportion of human life loss per population of territory (district). When calculating the risk of human life loss we have used data on the number of inhabitants of the territory in 2010. Based on the distribution of data the reference value for the calculation of the coefficient was set out 0.01. Data of other districts was recalculated as the proportion of the human life loss per population of territory in year and reference value.

3.3 The Economic Potential of the District

Coefficient of population per hectare. For the purpose of the creation this coefficient was used indicator on the total area of the region (district). As a reference value was determined by expert assessment value of 10 inhabitants per hectare. This value is based on the distribution of the data. While districts that exceed this value were assigned a value of 1 and other districts of the coefficient of population per hectare of the district recalculated using the following formula:

$$ep_{1j} = \frac{Po_j}{10} \tag{7}$$

where Po_j is the number of inhabitants per hectare.

Coefficient values and structures of the property. The value and structure of assets is based on the conversion of built up area to the total area of the region. For this indicator was using the expert assessment of the minimum reference value of 10%. This is based on data distribution. Districts with the reference value higher 10% assigned value 1.In other districts the coefficient values and structures of assets is calculated using the following formula:

$$ep_{2j} = \frac{SM_j}{0{,}1} \tag{8}$$

where SM_j is the proportion of built-up areas and courtyards in the total area.

Coefficient of the tax revenue. In the case of tax revenues per capita, we use data from the Ministry of Finance. The reference value was determined according to the distribution function of 50 thousand CZK. This is based on data distribution. Districts with tax revenues per capita higher than reference value assigned value 1. Other values were recalculated using the following formula:

$$ep_{3j} = \frac{D}{50{,}000} \tag{9}$$

where D_j is *j*-district tax revenue per capita.

*Coefficient of the average salary.*For the conversion coefficient of the average wage was, according to the distribution of the dataset reference minimum average wage of 26 thousand CZK. This value exceeds only districts: Prague, Mladá Boleslav, Brno city and Pilsen-city. Average wages were recalculated using reference value. When districts exceeded this value they were assigned a value of 1. In other districts the coefficient values and structures of assets is calculated using the following formula:

$$ep_{4j} = \frac{m_j}{26{,}000} \tag{10}$$

where m_j is the average wage in *j*-district.

Coefficient of employment based on the unemployment rate in the districts. The reference value was determined natural rate of unemployment at 4%. Taking districts, who had lower unemployment than 4%, has been assigned a value of 1. Other districts were assigned a value according to the following formula:

$$ep_{5j} = \frac{0{,}04}{u_j} \tag{11}$$

where u_j is the unemployment rate of the district

Coefficient of foreign direct investment was based on the foreign direct investment. For this indicator was using the expert assessment of the minimum reference value 400 thousand CZK per capita. This is based on data distribution. Districts that the reference value exceeded 400 thousand CZK per capita has been assigned the value of 1. In other districts the average coefficient of FDI is calculated according to following formula:

$$ep_{6j} = \frac{PZI_j}{400{,}000} \tag{12}$$

where PZI_j is the average foreign direct investment flowing into the *j*-district.

Coefficient of enterprises. Coefficient of enterprises is constructed by dividing the number of business entities in the population. For the construction was used data from year 2010. As reference value was used 0.32. This is based on data distribution. Counties that exceeded the reference value (only districts Prague and Brno-city) was assigned a value of 1. In other districts the coefficient values and structures of assets is calculated using the following formula:

$$ep_{7j} = \frac{ES_j}{0.32} \tag{13}$$

Subsequently, it was determined coefficient of the economic potential of the district according to the formula (4). A coefficient of the economic potential is shown in next figure.

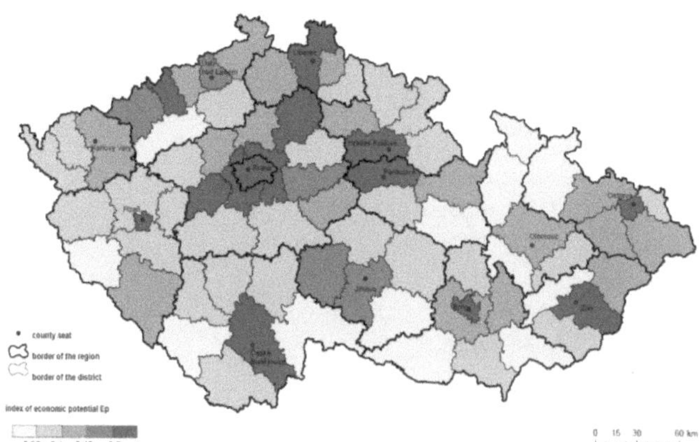

Fig. 1. Map of economic potential for the districts of the Czech Republic in 2010 (Source: Authors)

4 Results and Discussion

The coefficient of the territory value evaluation of the district according to the formula (5) is shown in next table.

Table 1. Coefficient of the district value in the Czech Republic in 2010 (Source: Authors)

District	*Em*	*Eo*	*Ep*	*TUE*	*District*	*Em*	*Eo*	*Ep*	*TUE*
Praha	**1,00**	0,41	**1,00**	0,82	Liberec	0,71	0,73	0,93	0,81
Benešov	0,13	0,06	0,40	0,22	Semily	0,41	0,41	0,80	0,56
Beroun	0,24	0,17	0,63	0,38	Hradec Králové	0,40	0,26	0,80	0,52
Kladno	0,22	0,08	0,46	0,27	Jičín	0,14	0,25	0,74	0,41
Kolín	0,25	0,23	0,49	0,34	Náchod	0,65	**1,00**	0,71	0,78
KutnáHora	0,17	0,17	0,45	0,28	RychnovnadKněžnou	0,15	0,39	0,68	0,43
Mělník	0,92	0,49	0,47	0,61	Trutnov	0,45	0,94	0,76	0,72
MladáBoleslav	**1,00**	0,50	0,88	0,80	Chrudim	0,13	0,25	0,38	0,27
Nymburk	0,30	0,28	0,38	0,33	Pardubice	0,20	0,08	0,66	0,35
Praha-východ	0,22	0,08	0,72	0,38	Svitavy	0,17	0,72	0,76	0,57
Praha-západ	0,28	0,10	0,61	0,36	Ústí nad Orlicí	0,53	**1,00**	0,59	0,70
Příbram	0,10	0,07	0,39	0,21	Havlíčkův Brod	0,09	0,04	0,38	0,19
Rakovník	0,17	0,04	0,42	0,23	Jihlava	0,29	0,12	0,52	0,33
České Budějovice	0,66	0,33	0,69	0,57	Pelhřimov	0,06	0,06	0,50	0,24
Český Krumlov	0,20	0,13	0,39	0,26	Třebíč	0,15	0,12	0,34	0,21
Jindřichův Hradec	0,14	0,14	0,35	0,22	ŽďárnadSázavou	0,10	0,08	0,38	0,20
Písek	0,10	0,25	0,41	0,27	Blansko	0,20	0,11	0,37	0,24
Prachatice	0,18	0,18	0,33	0,24	Brno-město	0,62	0,23	0,86	0,60
Strakonice	0,19	0,24	0,37	0,28	Brno-venkov	0,30	0,29	0,45	0,36
Tábor	0,31	0,19	0,41	0,32	Břeclav	0,16	0,11	0,40	0,24
Domažlice	0,05	0,17	0,36	0,21	Hodonín	0,05	0,03	0,36	0,17
Klatovy	0,29	0,43	0,44	0,39	Vyškov	0,10	0,31	0,43	0,29
Plzeň-město	0,71	0,05	0,86	0,57	Znojmo	0,02	0,01	0,36	0,15
Plzeň-jih	0,06	0,07	0,39	0,19	Jeseník	0,08	0,14	0,33	0,20
Plzeň-sever	0,13	0,21	0,39	0,26	Olomouc	0,79	**1,00**	0,45	0,72
Rokycany	0,11	0,29	0,42	0,29	Prostějov	0,03	0,01	0,37	0,16
Tachov	0,22	0,10	0,40	0,25	Přerov	0,74	0,98	0,39	0,67
Cheb	0,14	0,08	0,39	0,22	Šumperk	0,54	**1,00**	0,36	0,61
Karlovy Vary	0,45	0,07	0,43	0,33	Kroměříž	0,31	0,73	0,34	0,45
Sokolov	0,41	0,59	0,38	0,45	Uherské Hradiště	0,23	0,46	0,38	0,36
Děčín	0,51	0,31	0,46	0,43	Vsetín	0,43	0,66	0,42	0,50
Chomutov	0,27	0,11	0,49	0,31	Zlín	0,70	0,83	0,59	0,70
Litoměřice	0,99	0,44	0,41	0,59	Bruntál	0,21	0,40	0,32	0,31
Louny	0,35	0,19	0,35	0,30	Frýdek-Místek	0,14	0,34	0,46	0,33
Most	0,13	0,02	0,56	0,27	Karviná	0,22	0,21	0,37	0,28
Teplice	0,09	0,06	0,46	0,23	NovýJičín	0,11	0,17	0,42	0,25
Ústí nad Labem	0,64	0,21	0,51	0,46	Opava	0,13	0,15	0,41	0,25
Česká Lípa	0,15	0,27	0,46	0,31	Ostrava-město	0,22	0,05	0,92	0,45
Jablonec nad Nisou	0,33	0,37	0,45	0,39					

For the expression of economic potential of the region there are not always sufficient sources of data. This is probably one of the reasons why experts in this research area design different patterns and indicators which are usually based on the available information. Diversification in view of economic (social-economic) aspects of the described area is also conditioned their professional orientation (e.g. regionalists, economists, experts in labour market and social policy, etc.). In this sense, there is no uniform methodology for the evaluation of territorial or administrative units of state, or this approach has own purposeful nature (e.g. comparison of cohesion regions NUTS 2 within regional policy of EU). If the data exist (e.g. gross domestic product), then could be this data apply generally to the middle level of territorial units or public administration of the country. For lower levels the synthetic economic characteristics are not usually stated and are published only partial indicators such as population, unemployment, etc.

Therefore, we recommended using for assessment of regional development for named lower level of territorial units (on the example of district in the Czech Republic) the appropriate indicators and their immanent coefficients (population per hectare of the district, the value and structure of the property in district, the tax revenue of the district, the average wage of the district, employment the district, the amount of foreign direct investment per capita in district and the number of enterprises of the district).

Number of indicators is not random. It is not only based on their traceability but also from expert panel. Using only one or two indicators could possibly lead to distortion in description of real economic power of examined territorial units. Distortion is to a certain extent reduced by using multi-criteria evaluation. It is obvious that into account could be taken other indicators (e.g. purchasing power of population). However, on the one hand with the rise in number of indicators the complexity of the calculation is increased, on the other hand their marginal value added decreases, as many indicators correlate or have some substitutes. For example, the purchasing power of the population is usually conditioned the average wage of reporting district, that is not a border tourist destination. It is therefore a selection of indicators that are available, complementary and balance, and while providing relatively easy operationalize.

5 Conclusion

The proposed structure of economic potential of region may have a wide application, e.g. in developing strategic, conceptual and program documents of public administration on the state or regional level, especially in regional politics, when deciding how to allocate resources in the area and establish the priorities for disparately regions. In our case this procedure, calculation respectively, has been use in the context of the project which examined the flood risk in flood affected districts in the Czech Republic in terms of flood protection. In addition to the categorization of floodplains according to frequency of recurring floods and flood of threat (Q20, Q50, Q100, etc.), the classification of districts according to their economic potential was performed. Based on their interaction there was recommended model for determination of order for investment in flood protection. This case study shows practical applicability of assessment of regional development within the investment in the region. Much more could be done to improve this model. That could be subject of further research.

Acknowledgements. This article has been elaborated as one of the findings of specific research project MUNI/A/0786/2012 "Quality evaluation of public policies in the context of restrictive constraints of public finances".

References

1. Armstrong, M., De Kervenoael, R.J., Li, X., Read, R.: A comparison of the economic performance of different micro-states, and between micro-states and larger countries. World Dev. 26, 639–656 (1998)
2. Agarwal, S., Rahman, S., Errington, A.: Measuring the determinants of relative economic performance of rural areas. J. Rural Stud. 25, 309–321 (2009)
3. Bai, C., Ma, H., Pan, W.: Spatial spillover and regional economic growth in China. China Econ. Rev. 23, 982–990 (2012)
4. Delgado, M., Ketels, C., Porter, M.E., Stern, S.: The Determinants of National Competitiveness, NBER Working Paper No. 18249 (2012)
5. Chlad, M.: Diferencovaný ekonomický rozvoj v regionech ČR – potřeba makroekonomických indikátorů pro řízení v regionech. Urbanismus a územní rozvoj 5, 40–43 (2010)
6. Llner, M., Kostelecký, T., Patočková, V.: Jak fungují kraje – příspěvek k hodnocení výkonu krajských vlád. Sociol cas. 43, 967–992 (2007)
7. Ishikawa, K.: Guide to Quality Control. JUSE, Tokyo (1968)
8. Kostelecký, T., Patočková, V., Vobecká, J.: Kraje v České republice – existujísouvislosti mezi ekonomickým rozvojem, sociálním kapitálem a výkonem krajských vlád? Sociol Cas. 43, 911–944 (2007)
9. Malý, I.: Veřejné statky a veřejně poskytované statky. Polit ekon 46, 861–868 (1998)
10. Porter, M.E.: The economic performance of regions. Reg. Stud. 37, 549–578 (2003)
11. Terluin, I.J.: Differences in economic development in rural regions of advanced countries an overview and critical analysis of theories. J. Rural Stud. 19, 327–344 (2009)

Decision Support of Waste Management Expenditures Efficiency Assessment

Jana Soukopová[1], Ivan Malý[1], Jiří Hřebíček[2], and Michal Struk[1]

[1] Masaryk University, Faculty of Economics and Administration, Department of Public Economics, Brno, Czech Republic
{soukopova,ivan}@econ.muni.cz, struk@mail.muni.cz
[2] Masaryk University, Institute of Biostatistics and Analyses, Brno, Czech Republic
hrebicek@iba.muni.cz

Abstract. This paper is devoted to the development of methodology and information and communication technology tools for decision support in the public sector. It analyses appropriate metrics for a municipal solid waste management expenditure (MSWE) efficiency assessment using cost-effectiveness Analysis (CEA). In addition to many other methodological issues, finding a proper output (performance, outcome) measurement is important. From the point of view of municipalities, such a measurement ought to be as clear and simple to use as possible. We analyse three possible criteria – total generated municipal solid waste, population, and municipality area – for evaluating MSWE efficiency in order to examine their appropriateness for municipal administration. The analysis covers three years, from 2009 to 2011, and municipalities from the South Moravian Region of the Czech Republic. We focus on a sample of 21 municipalities with specific administrative status. Expenditures were estimated using open public data from the Czech Ministry of Finance municipal accounts database. Correlation analysis showed a very strong relationship between the three chosen criteria. Public administration can certainly use all of the criteria for an efficiency assessment of MSWE to aid in decision making. However, the most suitable criterion to be population, since efficiency analysis results showed a strong correlation between population and both CEA for waste amount and CEA for municipality area. Moreover, population has a stronger relationship with MSWE than either of the other two criteria.

Keywords: cost-effectiveness analysis, municipal expenditures, efficiency, waste management, decision support, ICT, municipal solid waste.

1 Introduction

Defining and measuring the efficiency of environmental protection expenditures in the process of using public resources and their transformation into outputs and outcomes is an important issue of decision support in the public sector [22-23], [26]. Efficiency evaluation and its methodology and use of information and communication technology (ICT) tools have greatly improved and advanced in recent decades.

J. Hřebíček et al. (Eds.): ISESS 2013, IFIP AICT 413, pp. 651–660, 2013.

However, efficiency evaluation still remains a conceptual challenge in relation to public expenditures. Although the amount of expenditures spent in a specific area is important, another key characteristic for decision support in public administration is (or should be) the efficiency of the spending. The efficiency affects the extent of a utility and the rate of satisfying people's needs. Efficiency can be perceived as a rational criterion for the actions of involved subjects and a key category of the economic approach for analysing and evaluating social processes [20], [34]. This economic rationality comes from the idea that rational action consists of the efficient use of limited resources for the purpose of maximizing goals or achieving desired use [34]. This is complicated by the fact that public sector outcomes (especially in environmental protection) are often off-market, lacking relevant data, and thus difficult to quantify, as stated by a collective group at the European Commission [21]. In such cases, cost-effective Analysis (CEA) appears to be the most useful tool for efficiency assessment [5], [19]. Quite a few studies dealing with this topic have been published in economic journals. A CEA of environmental protection expenditures was examined by [1], [9-11], [24], [28], [32-33]. A CEA for the centralized waste treatment industry was discussed in [9], [11] and [32]. Anderson [1] conducted a CEA of proposed effluent limitations and standards for industrial waste combustors [10] focused on remediation. Papageorgiou, Barton and Karagiannidis [24] focused on the impact of technologies used for energy recovery from municipal waste. Market structure and refuse collection of municipal waste was studied in [28]. Willan and Briggs [33] compared the results of relevant CEA studies.

Studies [7], [22-23], [27], [29-30] dealing with issue of efficiency in the field of waste management in the Czech Republic and Slovakia have been focused on contractual issues and on the impact of competitiveness on efficiency [27]. Data envelopment analysis for evaluating MSWE was used in [7] and [23]. We developed a simple model for calculating an ideal amount of municipal solid waste expenditures [29] based on the characteristics of the municipality.

Municipal solid waste management costs (and consequently the expenditures (MSWE) that are at the centre of our concern) are determined by numerous variables. There is extensive literature dealing with this issue, including [1], [11-12], [17-18], [24], [27]. The following variables have the largest impact:

- total generated solid municipal waste [t];
- population [number of inhabitants];
- area of municipality [km^2];
- distance from the municipality to the waste facility site [km];
- competitive environment;
- ownership type of collecting company (private, public);
- size of household;
- impact of income;
- other socioeconomic variables.

Given this number of variables, it could be quite costly to obtain the information needed for a serious efficiency assessment for public administration bodies such as municipalities. It is necessary to decide how much and what type of information is worth gathering [26]. Even a simple comparison in order to obtain a benchmark could be complicated. However, municipality representatives cannot conduct a complicated and sophisticated analysis on a daily basis in order to understand if they are doing well with their budgets and receiving good value for their money. A simple and robust efficiency assessment tool is needed to support their decision making.

There is a wide range of indicators for measuring *output* in waste management and in any other infrastructure service, as well as for measuring *outcomes* (defined as the impact of a service on its recipients [26] – for example, the impact of solid waste collection on environment protection). Esfahani [26] offers more than ten "performance measures" for sanitation infrastructure; some of them can be easily adapted to waste management.

We analyse a) *amount of solid waste*; b) *population*; and c) *municipal area* as criteria for "effectiveness measures" [26] in a CEA evaluating MSWE efficiency, in order to examine the use of these criteria for municipalities. The analysis covers three years, from 2009 to 2011, and the selected analysed sample consists of all the municipalities from the South Moravian Region of the Czech Republic. We focus on 21 municipalities with specific administrative status. Expenditures were estimated using open government data from the Czech Ministry of Finance municipal accounts database [2], [31]. The aim of this paper is to create a methodology and ICT tools for decision support of municipality decision makers enabling benchmarking with other municipalities.

2 Methods and Data for Decision Support

We chose cost-effectiveness analysis (CEA) as a simple metric for evaluating MSWE efficiency in public sector decision support. We intended to find the simplest method for considering and evaluating select criteria c_{ij}, $i = 1,..,k$ for j-municipality of a set of n municipalities of a given region.

The basic algorithm for decision support in MSWE efficiency consists of the following steps:

1. collecting the set of MSWE data $\boldsymbol{e} = \{e_1, e_2..., e_n\}$ from linked open data, where n is the number of considered municipalities;
2. setting the matrix C of selected criteria $\{c_{ij}\}$, $i = 1...k$, $j = 1...n$ for the evaluation of cost-effectiveness of MSWE in n municipalities;
3. collecting data of criteria c_{ij}, $i = 1...k$, $j = 1...n$ from linked open data;
4. calculating the ratios $\text{CEA}_{ij} = e_j/c_{ij}$, $i = 1...k$, $j = 1...n$ for each individual criterion c_{ij} and each municipality expenditure e_j;
5. trimming values (reducing extreme values within samples) by chosen statistical tests;
6. comparing CEA_{ij} results in analysed samples of n municipalities;
7. choosing individual criteria c_{mj} in CEA m from $\{1,2,...,k\}$, for $j = 1...n$ and their utilization.

The above algorithm for decision support in evaluating MSWE efficiency was implemented in Maple [8], [13] connected with MS Excel. We analyse current MSWE e_j, $j = 1\ldots n$. The idea behind CEA is either minimizing CEA_{ij} i.e. e_j/c_{ij} ratio or maximizing $1/\text{CEA}_{ij}$, i.e. c_{ij}/e_j ratio, for $i=1,..,k$, $j=1,\ldots,n$.

The most efficient municipality administration of the given region is then considered as the administration that attains either the lowest e_j/c_{ij} ratio or the highest c_{ij}/e_j ratio [19].

When we use the e_j/c_{ij} ratio, the cost-effectiveness of a given expenditure can be expressed as follows:

$$\min \{\text{CEA}_{ij}\} \tag{1}$$

There are several sources of linked open data in the above algorithm: *municipal area* and *population* were taken from the Czech Statistical Office (CZSO); *amount of generated municipal solid waste* was taken from the information system of the waste management database [16] provided by CENIA (Czech Environmental Information Agency): and MSWE were downloaded from the Czech Ministry of Finance's ARIS [2] and ÚFIS [31] databases. CZSO publishes a MSWE database as well, however, these linked open data are published only for larger administrative municipality units and are not exactly those we need. Such slight differences among data provided by CZSO and the Ministry of Finance in terms of expenditures have been noted by [3] and are most likely results of either incorrect reporting or differences in methodology.

Our sample from all of the municipalities of the Czech Republic consists of 672 municipalities[1] from the South Moravian Region (SMR) of the Czech Republic. We analysed municipalities in the whole SMR, and then conducted a more detailed analyses on a selected sample of 21 municipalities with specific administrative status usually abbreviated as ORPs. These municipalities are former county administrative capitals and represent over 50% of total population of the SMR.

3 Outcomes and Discussion

We chose the following three criteria for the CEA:

- c_{1j} – total generated solid municipal waste [t] for the first CEA_1,
- c_{2j} – population for the second CEA_2 [number of inhabitants]

c_{3j} – area of municipality [km^2] for the third CEA_3We next calculated the individual CEA_{ij} for a selected sample of municipalities. We used data trimming [16] to reduce the original sample to an appropriate size. The following table contains the calculated ratios for municipalities of the SMR.

[1] The South Moravian Region officially has 673 municipalities. One of them, Březina, is a former military area that still operates under a special regime and has only a few permanent residents. We did not include it in our analysis. Some municipalities did not report certain characteristics, and are thus not included in sample for certain analyses.

Table 1. Results of CEA for all municipalities from the South Moravian Region

	CEA_1			CEA_2			CEA_3		
Year	2009	2010	2011	2009	2010	2011	2009	2010	2011
Sample siz n	595	602	542	672	672	672	672	672	672
mean	3 126.6	2 779.4	93 376.1	565.9	578.5	577.2	59 044.2	61 071.5	60 986.2
σ	10 993.5	3 282.0	1 737 375	258.0	257.4	258.9	103 130.8	102 731.2	100 421.6
γ_2	515.3	460.6	517.8	70.7	55.2	58.9	151.5	96.9	92.6
γ_1	22.2	20.3	22.6	6.5	5.7	5.9	10.5	8.6	8.4
$mean_{(0.05)}$	2 526.6	2 587.2	2 740.5	546.5	559.7	557.8	49 115.4	50 576.2	50 665.6
$\sigma_{(0.05)}$	749.8	678.8	951.0	132.5	140.1	139.1	40 479.2	41 475.4	41 737.6
$\gamma_{2(0.05)}$	0.9	0.0	9.4	0.2	0.5	0.3	5.8	5.7	6.0
$\gamma_{1(0.05)}$	0.8	0.5	2.3	0.7	0.8	0.8	2.2	2.2	2.3

Source: Authors *σ - standard deviation, γ_2 – kurtosis, γ_1 – skewness, (0.05) – 5% trimmed sample

When considering c_{1j} criterion, we can see that CEA_1 values show significant differences, especially in 2011[2]. We used data trimming methods [33] to provide results for a 5% trimmed sample, using Maple [13] for the whole SMR. The following table contains calculated ratios CEA_i , i=1,2,3 for 21 ORPs of the SMR.

Table 2. Results of CEA for 21 ORPs from the South Moravian Region

	CEA_1			CEA_2			CEA_3		
Year	2009	2010	2011	2009	2010	2011	2009	2010	2011
Sample size n	21	21	21	21	21	21	21	21	21
mean	2 836.3	3 006.3	3 109.9	673.1	683.4	681.9	318 123.2	315 086.3	308 977.9
σ	1 081.0	1 222.4	1 106.1	209.0	219.7	197.5	353 421.2	294 602.2	270 815.1
γ_2	2.4	3.7	1.5	4.8	4.6	5.7	16.6	11.1	11.9
γ_1	0.1	0.7	-0.5	-1.1	-0.5	-1.6	3.9	3.1	3.1
$mean_{(0.1)}^*$	2 841.6	2 971.9	3 168.7	684.6	690.0	700.9	255 775.4	271 115.9	268 670.5
$\sigma_{(0.1)}^*$	689.2	700.4	803.4	115.8	116.1	116.7	257 434.8	247 240.9	258 790.1
$\gamma_{2(0.1)}^*$	0.2	0.1	-0.8	-0.2	0.3	0.0	0.4	5.2	1.1
$\gamma_{1(0.1)}^*$	0.6	1.0	0.7	0.4	0.9	0.9	0.4	1.8	1.0

Source: Authors *σ - standard deviation, γ_2 – kurtosis, γ_1 – skewness, (0,1) – 10% trimmed sample

Table 2 shows that the presented data do not have a high value of standard deviation. We calculated results for a 5% trimmed sample of ORPs as in Table 1; however, due to the low number of municipalities the results were not different from the original sample. We also analysed values for a 10% trimmed sample. Even this adjusted sample did not show significant differences in results. Therefore we used the original sample of 21 ORPs for further analyses. We compared the calculated values

[2] The extreme values of mean and σ in 2011 (CEA_1) are mainly due to the municipalities Zbýšov and Telnice, both of which have a population of around 600 and relatively high MSWE, reporting just 60 and 150 kg of collected municipal solid waste. This resulted in extreme CEA_{ij} ratios that affect the mean of the whole sample.

among selected municipalities. We assume that CEA$_i$ results for criteria c_{1j}, c_{2j} and c_{3j} j=1,..,n will not differ significantly among n municipalities, and thus they can be used for MSWE efficiency evaluation. The strength of the relationship between calculated criteria $\mathbf{c}_i = (c_{i1},\ldots, c_{in})$ is shown in the following table, with $\boldsymbol{c}_1$ and $\boldsymbol{c}_2$ having strongest relationship.

Table 3. Correlation coefficients between analysed criteria for different municipality samples

	c_1 & c_2			c_1 & c_3			c_2 & c_3		
Year	2009	2010	2011	2009	2010	2011	2009	2010	2011
SMR	0.9979	0.9972	0.9990	0.8096	0.8061	0.8026	0.7781	0.7767	0.7767
ORPs	0.9987	0.9981	0.9997	0.9475	0.9498	0.9417	0.9384	0.9379	0.9378

Source: Authors

The following table contains comparison between CEA$_i$ results for all three variables in 2009-2011 for a 5% trimmed sample of municipalities from the SMR and then for all 21 ORPs of the SMR.

Table 4. Correlation coefficients between CEA$_i$ for different municipality samples

	CEA$_1$ & CEA$_2$			CEA$_1$ & CEA$_3$			CEA$_2$ & CEA$_3$		
Year	2009	2010	2011	2009	2010	2011	2009	2010	2011
SMR	0.4224	0.4379	0.4473	0.0603	0.0102	0.0852	0.2102	0.1868	0.1647
ORPs	0.8185	0.8764	0.9140	0.7025	0.6205	0.5144	0.6055	0.5124	0.4112

Source: Authors

Table 4 shows that CEA$_i$ results for the variables of solid waste amount and population are significantly similar, especially in the ORPs. Moreover, in the ORPs all three criteria can be used, as correlations between CEA$_i$ acquire significant values. For this reason, we tested CEA$_i$ rankings for individual i-criteria among ORPs to see whether there were significant differences among them. The average rankings and standard deviations of rankings between 2009 and 2011 are presented in Table 5.

Table 5 shows that municipalities acquire relatively stable positions if comparing efficiency in individual categories among the sample. To verify whether municipalities acquire similar positions in individual efficiency categories, we calculated standard deviations of differences among the positions that municipalities acquired in different categories.The majority of municipalities acquired positions in individual years that vary only by one or two positions. Larger standard deviations were rather rare. Municipalities acquired very close positions especially in CEA$_3$, where the largest standard deviation was 1.7 and the sum of standard deviations for 21 ORPs was only 15.3.

Table 5. CEA average ranking among ORPs and standard deviations of rankings in years

Municipality	CEA_1		CEA_2		CEA_3	
Brno	20	0.8	19	1.6	21	0.0
Znojmo	20	0.8	20	1.2	19	0.5
Hodonín	11	1.7	15	2.2	15	0.8
Břeclav	**7**	**4.1**	**16**	**0.5**	10	0.5
Vyškov	9	4.2	5	2.5	10	1.7
Blansko	17	0.5	14	0.9	17	0.5
Veselí n. M	8	0.5	10	0.5	7	0.5
Kyjov	2	0.5	4	2.4	8	1.2
Boskovice	12	1.2	10	1.7	12	0.9
Kuřim	5	1.4	4	1.9	17	0.8
Ivančice	15	1.7	17	1.2	5	0.8
Tišnov	6	1.9	10	3.6	16	1.2
Mikulov	17	0.9	20	0.5	5	0.8
Šlapanice	5	0.8	6	0.5	14	0.8
Bučovice	10	0.9	12	2.1	3	0.0
Slavkov u Brna	**17**	**1.7**	**4**	**0.8**	10	1.2
Moravský Krumlov	1	0.0	1	0.0	1	0.0
Hustopeče	11	5.4	12	2.6	5	1.2
Rosice	10	2.1	6	3.3	12	1.2
Pohořelice	8	3.3	8	0.8	2	0.0
Židlochovice	20	1.2	18	2.9	20	0.5

Source: Authors

The observation that the results are closest between CEA_1 and CEA_2 was further verified by correlation analysis (see Table 5). Average rankings for the majority of municipalities did not differ by more than 3 positions. In the ORPs, Hodonín, Tišnov, and Rosice municipalities differed by 4, resulting in a standard deviation value of around 2. We can see significant differences in rankings only in the ORPs Břeclav (9 positions) and Slavkov u Brna (13 positions)[3]. These two municipalities acquired more similar results between CEA_2 and CEA_3. For this reason, we examined the relationships of individual criteria and MSWE.

Table 6. Correlation coefficients between analysed criteria and MSWE for different municipality samples

	$e \& c_1$			$e \& c_2$			$e \& c_3$		
Year	2009	2010	2010	2009	2010	2011	2009	2010	2011
SMR	0.9128	0.8900	0.9237	0.9232	0.8775	0.9583	0.5077	0.5062	0.5050
ORPs	0.9128	0.8959	0.9731	0.9626	0.9087	0.9767	0.7957	0.7980	0.7773

Source: Authors

[3] Bold values in Table 5.

Correlation analysis shows that municipalities can use all three examined criteria for MSWE efficiency evaluation. Nevertheless, based on Tables 4, 5, and 6 we suggest that the most suitable criterion for CEA evaluation is $\boldsymbol{c}_2$ – *population*. Population has strongest correlation with MSWE. The results of CEA_2 are in strong correlations with CEA_1, but compared to results of CEA_1 the results of CEA_2 have stronger correlation with CEA_3, if taking into account the sample of all municipalities from the SMR.

The amount of solid municipal waste criterion is a generally recommended indicator for measuring efficiency of MSWE, see for instance [1], [9-11], [27], [29] and [32]. This makes the results of our analysis and the optimization of the decision making process of municipal administration using CEA even more interesting, as the most suitable criterion seems to actually be population. This criterion can be recommended for two more reasons – it has a very significant relationship with the amount of solid municipal waste verified by many analyses (e.g. [4], [14-15], [27]), and the criterion is easy to acquire from open linked data [2], [25], [31] databases and generally available for all municipalities and public.

4 Conclusion

Managing the information resources in decision making processes is one of the most prominent challenges of the public sector. Better methodology and internal information management create opportunities for innovation in reducing bureaucracy and diminishing administrative costs. Local governments (municipalities) face the complexities of uncertain and evolving environment protection, including external constraints such as the changing legal framework or the evaluation of public expenditure, especially in municipal waste management. Using linked open government data in current economic theory (CEA) together with advanced ICT brings new ideas in decision support for public administration about MSWE efficiency that allow decision makers to create benchmarks for municipalities in a given region.

The completed results of the paper:

- designed an appropriate methodology for rating the cost-effectiveness of MSWE, which was implemented with the use of ICT tools;
- concluded that in similar communities (ORPs) it is possible to use any of the selected three criteria and the evaluation results (order) are similar (very distinctive colinearity);
- from the perspective of different municipalities and in terms of benchmarks, the best metric to use is *per capita expenditure*.

Acknowledgements. This paper elaborates one of the findings of the specific research project MUNI/A/0786/2012 "Quality evaluation of public policies in the context of restrictive constraints of public finances". The authors express their gratitude to CENIA for providing data on waste generation.

References

1. Anderson, W.: Economic analysis and cost-effectiveness analysis of proposed effluent limitations guidelines and standards for industrial waste combustors. US Environmental Protection Agency, Washington, DC (1998)
2. ARIS- Automatized budget information system, http://wwwinfo.mfcr.cz/aris/
3. Bakoš, E., Soukopová, J., Kaplanová, B.: Porovnání vykazovaných environmentálních dat se zaměřením na úroveň municipalit. In: Účetnictví a reporting udržitelného rozvoje na mikroekonomické a makroekonomické úrovni, pp. 79–85. Linde nakladatelstv dí Praha, Praha (2009)
4. Berger, C., Savard, G., Wisere, A.: EUGENE: An Optimisation Model for Integrated Regional Solid Waste Management Planning. Int. J. Environ. Pollut. 12, 280–307 (1999)
5. Boardman, A.E.: Cost-benefit analysis: concepts and practice. Prentice Hall, Upper Saddle River (2001)
6. Drummond, M.F., O´Brian, B., Stoddart, G., Torrance, G.W.: Methods for the Economics Evaluation of Health Care Programmes. Oxford University Press, Oxford (2001)
7. Fiala, P., Šauer, P.: Household Waste Management: Municipality Behavior Analysis by DEA Method. In: Aktan, C., Savanas, F. (eds.) Selected Proceedings of the First International Conference on Social Sciences, pp. 149–165. Izmir, Birlesik Matbaacilik (2008)
8. Gander, W., Hřebíček, J.: Solving problems in scientific computing using Maple and MATLAB. Springer, Heidelberg (2004)
9. Goulder, L.H., Parry, I.W., Williams, R.C., Burtraw, D.: The cost-effectiveness of alternative instruments for environmental protection in a second-best setting. J. Public Econ. 72, 329–360 (1999)
10. Hamilton, J.T., Viscusi, W.K.: How costly is "clean"? An analysis of the benefits and costs of Superfund site remediations. J. Policy Anal. Manag. 18, 2–27 (1999)
11. Hirsch, W.: Cost functions of an urban government service: refuse collection. Rev. Econ. Stat. 47(1), 87–92 (1965)
12. Hodinka, M., Štencl, M., Hřebíček, J., Trenz, O.: Current trends of corporate performance reporting tools and methodology design of multifactor measurement of company overall performance. Acta Universitatis Agriculturae et Silviculturae Mendelianae Brunensis 60, 85–90 (2012)
13. Hřebíček, J.: Mathematical Modeling of Economic Phenomena with Maple. In: Ramík, J., Stavárek, D. (eds.) Proceedings of 30th International Conference Mathematical Methods in Economics, pp. 326–331. Silesian University, Karviná (2012)
14. Hřebíček, J., Kalina, J., Soukopová, J.: Integrated economic model of waste management: Case study for South Moravia region. Acta Universitatis Agriculturae et Silviculturae Mendelianae Brunensis 61, 913–918 (2013)
15. Chen, H.W., Chang, N.: Prediction Analysis of solid waste generation based on grey fuzzy dynamic modelling, Resources. Conserv Recycling 29, 1–18 (2000)
16. Information system of waste management, http://isoh.cenia.cz/groupisoh/
17. Jenkins, R.R.: The Economics of Solid Waste Reduction: The Impact of User Fees. Edward Elgar Press, Brookfield (1993)
18. Kinnaman, T.C.: The economics of residential solid waste management. Ashgate, Aldershot (2003)
19. Levin, H.M., McEwan, P.J.: Cost-effectiveness analysis: Methods and applications. Sage Publications, Inc., Thousand Oaks (2000)

20. Malý, I.: Problém optimální alokace zdrojů ve zdravotnictví. Masarykova Univerzita, Brno (1993)
21. Mandl, U., et al.: The effectiveness and efficiency of public spending (No. 301). European Commission: Directorate General Economic and Monetary Affairs (2008)
22. Nemec, J., Merickova, B., Vitek, L.: Contracting-out at local government level: Theory and selected evidence from the Czech and Slovak Republics. Pub. Manag. Rev. 7, 637–647 (2005)
23. Nemec, J., Merickova, B., Ochrana, F.: Introducing benchmarking in the Czech Republic and Slovakia. Pub. Manag. Rev. 10, 673–684 (2008)
24. Papageorgiou, A., Barton, J.R., Karagiannidis, A.: Assessment of the greenhouse effect impact of technologies used for energy recovery from municipal waste: a case for England. J. Environ. Manage. 90, 2999–3012 (2009)
25. RIS - Regional Information Service, `http://www.risy.cz/en/`
26. Shah, A. (ed.): Public Services Delivery. The World Bank, Washington (2005)
27. Soukopová, J., Malý, I.: Competitive environment in waste management and its impact on municipal expenditures. Acta Universitatis Agriculturae et Silviculturae Mendelianae Brunensis 4, 173–183 (2013)
28. Stevens, B.: Scale, market structure and cost of refuse collection. Rev. Econ. Stat. 60, 438–448 (1978)
29. Struk, M., Soukopová, J.: Efficiency of the current municipal waste expenditure - methodology approach and its application. Acta Universitatis Agriculturae et Silviculturae Mendelianae Brunensis 59, 379–386 (2011)
30. Šauer, P., Dvořák, A., Lisa, A., Fiala, P.: A procedure for negotiating pollution reduction under information asymmetry. Surface water quality case. Environmental and Resource Economics 24, 103–119 (2003)
31. UFIS, `http://wwwinfo.mfcr.cz/ufis/`
32. Wheeler, W.: Cost-effectiveness analysis of effluent limitation guidelines and standards for the centralized waste treatment industry. US Environmental Protection Agency Washington, DC (1998)
33. Willan, A., Brigs, A.: Statistical analysis of cost-effectiveness data. John Wiley & Sons, New York (2006)
34. Winkler, J.: Filozoficko-metodologická východiska. In: Malý, I., et al. (eds.) Vliv veřejného sektoru na efektivnost rozvoje regionů. Masarykova univerzita, Brno (2001)

GeneDBase – Genetic Database of Selected Species of Mammals in the Czech Republic

Luděk Skočovský[1], Tomáš Šikula[2], Martin Ernst[3], and David Kouřil[2]

[1] Freelance Programme Writer, Brno, Czech Republic
ludek@skocovsky.cz
[2] HBH Projekt spol. s r.o., Brno, Czech Republic
{t.sikula,d.kouril}@hbh.cz
[3] Mendel University, Brno, Czech Repulic
martin.ernst@mendelu.cz

Abstract. GeneDBase is a part of a research project titled "Creating a genetic database of selected species of mammals in the Czech Republic to be used for sustainable transport development". The aim of the research project is to assess the impact of roads on the fragmentation of animal populations and their genetic variability. Further, the objective is to create a genetic database of mammals comprising the data from collected samples of selected wild animals which could be used widely, not only for road planning, but also in other fields (agriculture, environmental protection, land-use planning).

Keywords: GeneDBase, migration, genetic variability, landscape fragmentation, genetic isolation, genetic database, database application, sustainable road development.

1 Introduction

Wildlife migration preserves balance among populations, enables exchange of genetic information and the use of food sources and maintains genetic variability. Due to migration, populations are able to adapt to changes in their environment. The development of transport infrastructure leads to landscape fragmentation and decreases the migration permeability. A grant project was established to assess the impact of landscape fragmentation onto genetic variability of populations. A part of the grant project aims to create a database application called GeneDBase.

2 About the Project

GeneDBase is a name for a database application, which is one of the outcomes of a grant project named: "Creating a genetic database of selected species of mammals in the Czech Republic to be used for sustainable transport development". The grant project is supported by the Technology Agency of the Czech Republic (within Alfa Programme). The contractor is a private road design company, HBH Projekt spol. s r. o., together with a project participant, Mendel University in Brno, Department of Forest Protection and Wildlife Management.

J. Hřebíček et al. (Eds.): ISESS 2013, IFIP AICT 413, pp. 661–668, 2013.

The aim of the grant project is to create and evaluate a genetic database, which will contribute to a more objective and more complex assessment of the barrier effect of the transport infrastructure on wildlife populations, especially on their genetic variability. To reach the aim, the following steps were defined:

- establishing a DNA-bank with genetic samples (especially samples of tissue, hair bulb) of selected mammals;
- creation of a database with records of collected samples (not only results of genetic analyses, but also other entries will be stored);
- creation of a database application to view and edit the database using the internet.

Expected outcomes of the grant projects are:

- Evaluation of a landscape fragmentation impact on genetic variability of populations in selected areas, also concerning an isolation level of populations.
- Universal database application which can be continuously edited and developed.
- Certified methodology to maintain and develop the database.
- Certified methodology for the landscape fragmentation assessment in relation to genetic factors, in order to supplement existing methodologies used to assess decreasing landscape fragmentation level.

The grant project was launched in January 2012 and will be completed in December 2015. The completion of subsequent issues is designed/planned as follows: The following works in designated years are expected:

2012 – analysis of the current status, literature research;
2013 – sample collection and processing, complementing research, database design;
2014 – data analysis, database application;
2015 – testing of the designed genetic database and completion of the methodology.

At the beginning of the grant project, areas for collecting samples were chosen. The key factors in the selection process included the presence of distinctive migration barriers (roads) where the migration routes disappeared. The selection of the area was also based on the previous studies, and on the knowledge of the area. The localities for the collection of the samples are defined by hunting districts. The sample collection is performed in cooperation with local game keepers. The sampling will be concentrated on following species: Eurasian elk (*Alces alces*), red deer (*Cervus elaphus*), sika deer (*Cervus nippon*), roe deer (*Capreolus capreolus*), wild boar (*Sus scrofa*), European hare (*Lepus europaeus*), European badger (*Meles meles*) and red fox (*Vulpes vulpes*). Biological samples (tissue, skin, hair bulb, droppings) are stored in a DNA-bank deep frozen within Mendel University in Brno. An analysis of selected microsatellites will be performed and genetic indicators will be calculated, based on the results of the analysis.

The indicators include: inbreeding coefficient (indicates the degree of genetic relatedness within the population, the higher the value is, the greater risk of population destabilization may be), genetic variability, heterozygosity, polymorfic information content (the higher the values are, the greater is the presumption of stability, adaptability or reproduction of the population). The above mentioned genetic indicators refer to certain condition of the population. Isolation of populations of the same species can increase the inbreeding coefficient, decrease genetic variability and gradually deprive the gene pool together with vanishing of certain alleles. If the reliability of individual identification using the microsatellite analysis is 99,99 %, the results can be used for migration studies (droppings or hair bulb need to be used for sampling).

3 Software Solution

A database application called GeneDBase is being prepared for storing and handling all the data gathered during the project. The author of the software is a freelance program writer Luděk Skočovský. The database application should meet the following requirements:

- the database will record selected genetic properties of selected species of mammals
- the database will be based on DNA-bank, i.e. collected genetic samples of individuals and their genetic analysis, i.e. the record of microsatellites and the calculated indicators of genetic variability
- accompanying visual material will be part of the database (not just individuals themselves, but also elements related to the registration of the individual and its properties that are important for the evaluation)
- the database will enable export/import of data to/from different formats for their processing by third-party programs (e.g. for the calculation of genetic variability), or for their use in mapping servers to visualize the movement of registered individuals, or for share data with other databases of similar nature
- the database must be generally usable, so that the user will be able to add another species and attributes associated with them
- the content of the database must be professionally guaranteed and at the same time accessible to the public

Based on set requirements, it is clear that a suitable database should mainly provide professionally guaranteed storing of described genetic sample records and their analyses.

Further, as stated above, another important parameter of the database is the flexibility and connectivity to external data sources. The database should enable to edit e.g. new species (flexibility in terms of data structures), but also to add new user interfaces defined on the basis of newly formulated requirements on the data viewing or handling).

With respect to the status and availability of current technologies, the suitable and functional database should be implemented so that it is visible on the Internet, while visibility means availability of all its functions including editing and data management. Using the Internet it should be possible to export and import the data or to share data using inter-database protocols.

3.1 Data Model

The proposal of a data model is based on the current needs which have been defined in the previous chapter. It is the evidence of:

- selected species;
- individuals of selected species, and their relationships;
- collected samples for genetic analyses;
- results of genetic analyses (the evidence of microsatellites and the calculated genetic variability indicators);
- photographs and related documentation;
- database users and their access rights;
- editing history.

3.2 Conceptual Schema

A conceptual schema of the data model is given in Fig. 1.

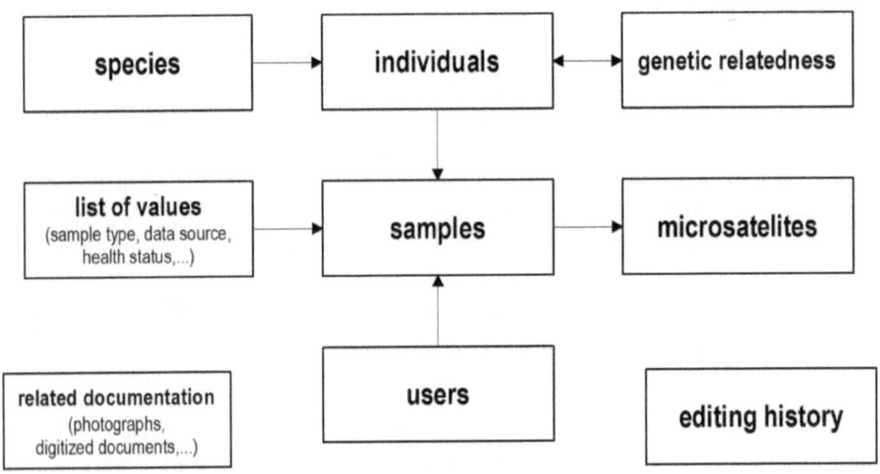

Fig. 1. GeneDBase – conceptual schema

The basic platform of the database is **the table of samples**, which is shown in the center of the diagram. Each sample belongs to a particular individual (see Fig. 1). The table of samples and **the table of individuals** are separated and therefore it is possible to register several samples from one individual without redundancy of data. Separate tables are used to register the species and relatedness of each individual.

Samples, other than direct numerical or text data, also include given options, i.e. references to so called lists of values. For example, the list of value of a sample type includes the following options: (1. hair bulb, 2. tissue, 3. inner tissue, 4. bone, 5.dropping, 6. other)

The table of samples records all important data regarding collected sample so that the sample can be related to the specific individual that was at the time of sample collection present at a particular location and was in a particular status. Therefore, the table includes:

- the method of sample collection (reference to the list of values);
- sample type (reference to the list of values);
- location of sample collection (GPS coordinates);
- health status of the individual;
- weight of the individual;
- age of the individual;
- morphological and colour variations of the individual.

The table of samples is connected with the table of microsatelites so that each sample can be assigned with more microsatellites.

The table of microsatellites contains genetic information on the uniqueness of the individual and also the genetic variability indicators.

The table of individuals, the table of species, the table of samples and the table of microsatellites can be connected with the table of photographs and digitized documents in the way, that each record in these tables can be connected to any number of binary files.

The table of users is used for the purposes of access mechanisms to the database in a protected mode. It keeps a record of users that access the database in a protected mode using the login and password on the entrance portal of GeneDBase. According to user type, there are set levels with different access rights. User types are set as follows:

- ordinary editor;
- professional editor;
- administrator;
- database administrator;
- ordinary user.

An **ordinary editor** is a user with the permission to fill in the data or to edit the data that has been filled in. The data include information from the table of samples and individuals, or photographs and other digital documents.

A **professional editor** can not only fill in and edit own data, but also act as a competent authority confirming the data filled in by other users before these data become a part of the published database.

An **administrator** has additional rights to work with the table of microsatellites. Administrator can export the data and process them using external programs to calculate indicators of genetic variability, which can be then uploaded back to the database.

A **database administrator** can insert new records into the table of species and add new values to the lists of values. Database administrator has access to all data and can process the data using controlled algorithms according to the needs of the development of the database, can also use the history record of the development of the database tables to get the insight into the data and, if it is necessary, to restore some of the original data.

An **ordinary user** has a permission to read data or export them to the selected format.

Access to the database is also possible in a public mode. In this mode, users do not have to log in and the access portal offers the display of the contents of the database through commonly available websites as it is common using the public Internet. Users entering a public mode can only read data and the data cannot be exported.

3.3 Implementation

GeneDBase will be implemented in open source technologies of current Internet. MySQL will be used as the database. Access algorithms will be programmed using PHP application server.

GeneDBase will be implemented as a typical network application of the client-server type, where the client (user) would use an Internet browser (Mozilla Firefox, Internet Explorer, Google Chrome etc.).

The output of the application server on the database side will be HTML and CSS code, the JavaScript client programming will be used too.

The running platform of the implementation is given in Fig. 2.

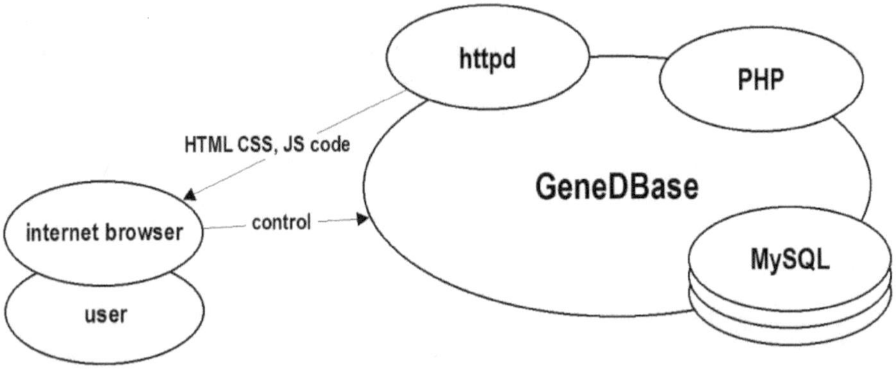

Fig. 2. The running platform of GeneDBase

No special hardware is needed for the implementation of GeneDBase, the commonly offered hosting of current Internet providers is sufficient in the initial phase. The domain name genedbase.eu (genedbase.cz) is already registered and used. Only in the case of a significant increase in the number of concurrent database accesses (app. 100 and above), special hardware should be considered. An increase in other parameters (number of records, total volume of the database etc.) will not require special hardware.

GeneDBase will be developed in strict compliance with all standard recommendations (especially ISO) and all technologies used (especially the database based on general SQL, application language based on Java, and transmission and communication protocols based on XML). Therefore, any necessary transformation to another (commercially supported) platform will be possible without technological problems in the future.

3.4 Functionalities

The functionalities of the application, based on the prerequisites, are following:

- Data Collection Module – this is the basic functionality of GeneDBase. Data collection can be done by both ordinary and proffesional editor. Data can be loaded manually or by a batch file. Import batch should be in XML format.
- Verification Module, Microsatellite editing – all the loaded data must be verified by proffesional editor, who has access rights to all levels of records and is abble to edit the data. Editing history is also recorded. The proffesional editor also have the rights to edit the table of microsatellites.
- Data Viewing Module – stored data can be listed and sorted by selected properties (date, individual…). As viewing is also considered sharing the database with other participants (mapservers, statistical applications)
- Data Sharing Module, Export-import – it is possible to share data with other users and with other partner projects. Data could be gained or received in different formats. Output formats are XLS, TXT, XML. Other possibilities are continuously checked and implemented into the database application.
- Data Administration Module – the module is specified for database administrator to run and maintain the application. New users can be added and their rights specified, the tables can be edited and lists of values can be added. Also to return changes is possible. Further development is expected.

3.5 List of Species (Taxalist)

A structure of Taxalist was designed by Ludek Skocovsky during the last decade. The structure has been implemented in several government information systems (ARROW, ENVIVoda, Salamander) and is guaranteed by the Ministry of Environment and the Ministry of Agriculture of the Czech Republic, it is also operated by the Czech Hydrometeorological Institute and basin management state enterprises. The same Taxalist is used for GeneDBase project in case of data interoperability.

4 GeneDBase Application

Taking advantages and combining knowledge from different scientific branches contributes to finding connections and bringing new views into already known principles. Usage and future development of GeneDBase by different users can make the data accessible for other subjects. The database is accessible through www.genedbase.eu.

Acknowledgement. The grant project is supported by the Technology Agency of the Czech Republic, identification code TA02031259.

References

1. Ernst, M., Kuciel, J., Urban, T.: Analysis of genetic variation of eight candidate genes in two wild boar subspecies. Czech Journal of Animal Science 48, 533–539 (2003)
2. Feuereisel, J., Ernst, M.: Genetische Untersuchungen zum Einfluss langzeitiger Isolation einer Population weißer Hirsche (Cervus elaphus L.) in der Tschechischen Republik. Fragmentierung der Landschaft und andere anthropogene Einflüsse auf Wildtierpopulationen, Gesellschaft für Wildtier- und Jagdforschung 32, 327–337 (2007)
3. Ernst, M., Kliment, J., Levý, E., Kourková, L., Stejskal, M.: Populace bílých jelenů – Využití mikrosatelitních analýz při šlechtění populace bílých jelenů u LČR, s.p. Edice Grantové služby LČR, Hradec Králové (2008)
4. Ernst, M., Matoušková, J., Křížek, L.: Využití mikrosatelitních analýz při monitoringu populace a přeshraniční migrace losa evropského v oblasti LS Vyšší Brod. Závěrečná zpráva projektu Grantové služby Lesy ČR, s.p (2011)
5. Ernst, M., Putnová, L., Pokoradi, J., Matoušková, J., Židek, R., Maršálková, L., Lamka, J.: Microsatellite Analyses DNA for Controlled Breeding of Cervidae and Genetic Breeding of White Colored Cervus elaphus in Czech Republic. Iranian Journal of Applied Animal Science 2, 233–237 (2012)
6. Skočovský, L.: Informační systém SALAMANDER monitoringu ZVHS: Sběr biotických dat: Implementace. ZVHS Brno (2002)
7. Skočovský, L.: ARROW Supported Database: Sběrná podpůrná databáze informačního systému ARROW MŽP OOV pro zavedení potřebných opatření k implementaci Rámcové směrnice v oblasti vodní politiky. Ministerstvo životního prostředí ČR, Praha (2006)

Vattenwebb: A Transparent Service to Support Decision Makers in Achieving Improved Water Status

Lena Strömbäck, Niclas Hjerdt, Lena Eriksson Bram, and Per Lewau

Swedish Meteorological and Hydrological Institute,
Folkborgsvägen 1, 601 76 Norrköping, Sweden
lena.stromback@smhi.se

Abstract. The *vattenwebb.smhi.se* service provides simulated as well as measured data on water flow and water quality for Sweden. The available data is used by the water authorities and decision makers in Sweden. The development of the site has been driven by the needs of the users and resulting in a site that is very appreciated by its users. An important aim in the development has been to make the data transparent for the end users, i.e. to explain the model assumptions and data quality in a way that is easy accessible. Therefore the site contains explanations about the model setup, how this data has been computed and information about the performance of the model. In this paper we will describe the service and its features with an emphasis on features used for achieving transparency.

Keywords: Hydrology, HYPE, environmental modeling, water management.

1 Introduction

Water management in Sweden is focused on characterization of water bodies and establishing action plans to achieve a good ecological status according to the framework directives stated by the European Union. To support this work SMHI has been commissioned to make databases of hydrography, statistics, water flows and scenario models freely available on the web. As a result of this we have created the open website *vattenweb.smhi.se* making collected and simulated information regarding discharge and nutrient load available for the end users. However, important when users are assessing this data is that they understand the viability, assumptions and uncertainties within the modeled results. Therefore, the development of the site has been done in close cooperation with the end users. This has resulted in a transparent web site that exposes information on the model setup and result evaluations for the end user to enable them to assess the validity of the presented values.

The aim of the website [1] is to provide historical data about water quantity and quality in Sweden. These data will provide an important basis for assessing water quality and establishing action plans in Sweden. On the site simulated data on fresh water quantity and quality are provided by the S-HYPE model [2], [3], a Swedish setup of the Open Source HYPE model [4], [5]. The model provides historical simulations of daily values for discharge and monthly values for transport of nitrogen

J. Hřebíček et al. (Eds.): ISESS 2013, IFIP AICT 413, pp. 669–678, 2013.

and phosphorous for the sub basins defined in the Swedish Water ARchive (SVAR [6]). Currently, the site provides observed and modeled data for fresh water and coastal areas as time series and statistics. In addition to this, the site also exposes data on wetlands and results from climate scenario simulations.

There are many challenges with developing and making this kind of dataset available to decision makers and to the general public. One such challenge is, of course, the development of the S-HYPE simulation model to provide an as good dataset as possible. Another, as important, challenge is to make this data available in a format that makes it useful for the decision makers. This includes usability issues, but also to make the simulation model transparent, enabling decision makers to understand assumptions and uncertainties within the model and use the data in a correct way.

In the paper we will give a general overview of the design and features of the web site and the data it provides. We will start with giving a general overview of the provided functionality on the web site, including a short introduction to the hydrological simulation model S-Hype. We will then describe the design process and user involvement in the development of the design, and describe the main design principles this has led to. After this we focus on the issue of transparency and describe how this has been achieved by making more information about the S-hype model available for the end users on the site.

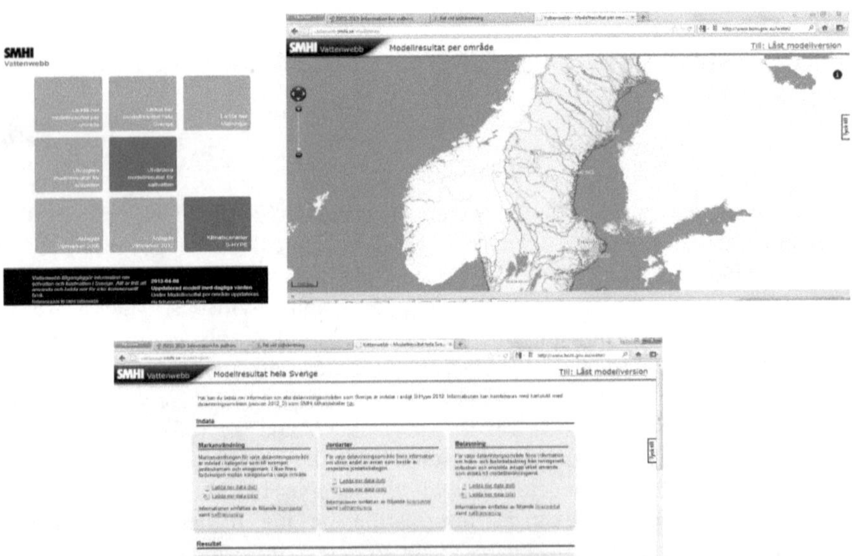

Fig. 1. This figure gives an overview of vattenwebb.smhi.se. The start page where each blue square represents one application is shown in the top left part of the figure. To the right we show the map view used for selecting an area of interest. The bottom right part of the figure shows the view used for downloading input data and statistics for the whole Sweden.

2 An Overview of the Service

As explained above the vattenweb.smhi.se service provides measured and simulated data for rivers and the coast of Sweden. Figure 1 shows the front page of vattenweb.smhi.se service where the different services are available as applications or apps for the end users. Each of the apps is shown as a blue box on the front page. The top line of applications is used to download data. The leftmost application gives the user the ability to select a sub basin of his interest and then download simulated data for this area. The data available includes historical time series for discharge and water quality, statistics and model input for the selected basin. The middle app gives similar information represented as summaries for the whole Sweden. The rightmost application gives the ability for the user to download observed values. The second line of application provides interactive evaluation tools that shows how well the models perform compared to the observed values. The applications of the bottom line concerns wetlands and climate predictions and will not be further discussed in this paper. In addition to data the web site contains a reference guide explaining details about the model setup.

In this paper we focus on fresh water and water quality modeling applications of the site. The simulated data for fresh water and water quality is provided by hydrological simulation model HYPE (Hydrological Predictions for the Environment) [4], [5] for modeling the hydrological processes in an area. The HYPE model simulates water flow and substances on their way from precipitation through soil, river and lakes to the river outlet. It is developed for being easy to set up for large scale applications and to make use of existing data about geography. The geographic area to be modeled consists of one or several catchment areas each catchment is divided into subbasins which are the basic unit for describing the hydrological processes. The HYPE model is developed for being easy to set up for large scale applications and to make use of existing data about geography where such are available.

For S-HYPE 2012, which is the HYPE model for Sweden currently used in vattenweb.smhi.se, the model setup consist of around 37000 subbasins derived from the SVAR [6] database. For each subbasin different properties relevant for water and water quality is described. This includes how the land is used i.e. whether it is covered by e.g. forest, lake, or open land. For agricultural land we describe which crops are most common and how these are fertilized. Elevation is used to describe temperature variations within a subbasin to influence the snow conditions. To be able to model water quality it is also important to describe emissions from different sources and their content of phosphorus and nitrogen. To describe nutrient transformation the soil is modeled as several layers different for each land use. As an example agricultural land classes commonly use three soil layers. Table 1 gives an overview of data sources for the S-HYPE 2012 model setup, while more details is available in our reference guide [1]. The final model setup is calibrated and evaluated against observations for a high number of gauging stations, for most of these the observed data is available on the website.

Table 1. Facts about the S-HYPE 2012 model setup.

Modeled geographic area:	Sweden
Total area (km^2):	525 000
No. of sub-basins:	~37000
No. of discharge stations:	303
Topography:	SVAR (SMHI)
Land Use:	CORINE updated with information from SVAR for lakes and SJV (Board of agriculture) for agricultural land
Soil:	SGU (Geological Survey of Sweden)
Discharge observations:	SMHI
Precipitation:	PTHBV (SMHI)
Temperature:	PTHBV (SMHI)
Lakes and reservoirs:	SVAR (SMHI)
Point sources:	FUT environmental monitoring (SMED)
Crop types:	SCB (Statistics Sweden)

3 User Involvement and Main Design Principles

The main users of the site are the Swedish water authorities and municipalities that need this data as background for decisions and reports. Therefore, the development of the site has been done in close cooperation with the end users at the water authorities to meet their needs and requirements. The users have been active in the design of the site and are involved in all steps of the development of new features. In practice, users are welcome to suggest new data and new features. These features are discussed and prioritized together with the existing development plan in a selected group consisting of SMHI officials and representatives for the users in the water authorities. Moreover, users take part in the development phase of the new features as participants in reference and expert groups.

This process has been very important for the design of the site. As expected, large parts of the discussions with end users focus on exactly which data to publish, what it represents and how it is computed. However, during this process it became obvious that an as important issue for the end users is to understand the S-Hype simulation model and its underlying encoded assumptions in order to assess how to use the results in relation to a specific task. There were also many questions regarding quality of the results. This process has resulted in a web site where we have adopted four main design principles for the site. These principles are further described below:

3.1 Data Is Available in the Way the User Needs Them

One of the most important properties of the site is that data is available as desired, in a way that makes it easy to select interesting data. Therefore the basic design is focused around the first page presenting a selection of available data as different applications. As one example, for download of data we have two options where the user can decide whether he is interested in data for a particular geographic area or general information for the whole country. For each of these choices it is easy to select data by clicking in the map or choosing dataset for Sweden. Figure 1 gives an overview of these selections on the site, while figure 2 shows one area selected for download of data.

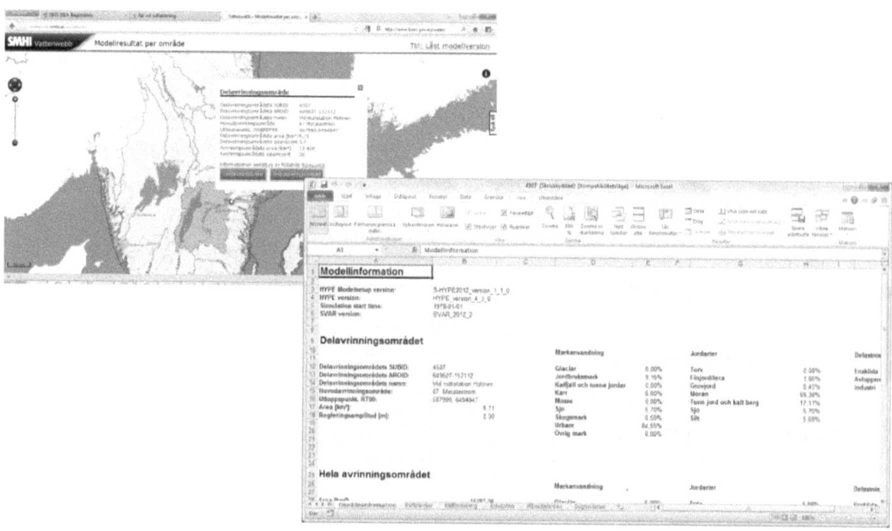

Fig. 2. At the top of this figure we show the map view with one specific area selected. To the bottom we show the Excel file resulting in downloading data for this area. The view show parts of the provided input data for the model, however, the file also contains statistics and time series for discharge and nutrients, source appointment, natural discharge and information about background and anthropogenic nutrient load.

3.2 Data Easily Imported into Other Tools

A strong desire of the users at the water authorities was to be able to easily use the tools they use in their daily work for further analysis of data. Most important here was Excel and various GIS-tools. Therefore, an important principle for the design was that available data can easily be imported into other tools for further use and analysis. This has led to the decision to offer data about a specific geographic area in Excel format. The data is organized in sheets so that it is easy to read and understand as shown in the example in figure 2. However, it can also easily be further analyzed in Excel or by export to other tools by the end user. Data about the whole country is available as Excel sheets and as text files. These files can easily be imported in GIS tools together with the shape file for the polygons which is also provided on the site.

3.3 Transparency of Results

During the meetings with the users we had many questions about the quality of the results and how they were computed. In particular this is important for a user which have access to more detailed information about a specific local area than is encoded in the input data for S-Hype. Therefore, it is important for the users to be able to assess the validity of the simulated data. On the site we have implemented this by making the model transparent in several ways: by explaining the S-Hype model; showing input data for the model and by a tool assessing the validity of results. All these features will be further explained in the following section.

3.4 Agile Development That Allows Fast Changes

The design has enabled easy access to the desired data by the end users. However, an as important issue is that it is also favorable for redesign and adding new features. From a technical perspective this has been provided in two ways. Firstly, the general design with a number of applications where each application has few dependencies to other parts of the site. This makes it easy to add a new application to the site without affecting any of the other applications. Secondly, the data representation where we use text files with data organized close to the model results achieved from S-Hype. This makes it very fast to publish a new release of the model and to add new features and results to the end users.

4 Data Validity and Transparency

During the discussions with end users an important issue was quality and uncertainty of the data provided by the model. The users are interested in comparing simulated results with older simulations and with local knowledge for their specific area. Whenever there are differences, it is important for the users to understand more about the simulated values to analyze the cause of these differences. To support them we decided that it was important to make the simulated data more transparent, i.e. to explain the underlying assumptions and setup of the model. To achieve transparency we have added three different kinds of features; a thorough reference guide explaining the S-HYPE model and how it was setup; input data for the model exposed on the site and finally an evaluation application which makes it possible to investigate the validity of the model from different perspectives.

The data used for model setup includes data on, for instance, land use, soils and point sources. As with result data from the simulations, the input data is provided for easy download on the site, for each subbasin as well as for the whole country. The data is collected directly from the model setup, but summarized in a way that makes it easier to understand for the end user that is not an expert on the model. Based on this data the users can easily asses the validity of the model set up compared to their local knowledge. The model setup builds on global databases (see the reference guide [1] for more information) and in some cases these do not match recent local changes. For the large scale, this has minor effect on the result of the simulation, but it might make a difference for a local area. Figure 2 shows an example of the provided input data for one subbasin. As shown in the figure, the data is provided as Excel, which makes it easy to read and further process. In the example we see input data on land use, soil and point sources.

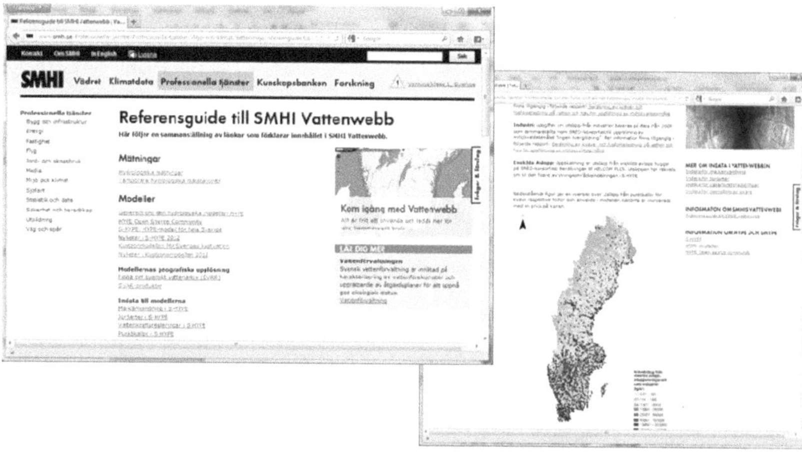

Fig. 3. Examples from the vattenweb.smhi.se reference guide. To the left we show the index page which is the entry point to the information. To the right we show one example describing information on point sources.

Publishing the model data made the users interested in where these data originate from and how they were used by the model. This raised a need for explaining how the model data was derived and we decided to provide a reference guide where origin of data as well as the processing or assimilation performed for the model setup is explained. This guide was provided as a set of web pages. Figure 3 shows the introduction page together with parts of the information available for soils. This allows users to understand the model assumptions and explain reasons for differences to more detailed information on local conditions they might have access to.

The features discussed above give the user ways to understand the model validity based on the model setup. However, there was also a need to present the quality of the results. To allow the end users to further evaluate quality, the site provides an interactive tool to compare the simulated results with measured values. Figure 4 shows a sample view of the tool. The tool presents model uncertainty in a map which gives an easy to understand overview of the model performance. The user can then select basins of his or her interest based on catchment size or geographic area, or make a detailed study of the validity of results for a particular catchment or group of catchments. The tool allows selection of which variable and size of catchment that are of interest, but also to zoom in to a specific geographic area. For each subbasin where measured data exist the user can inspect the measured time series and compare them to the simulated values. This gives valuable information about model performance for different geographic locations.

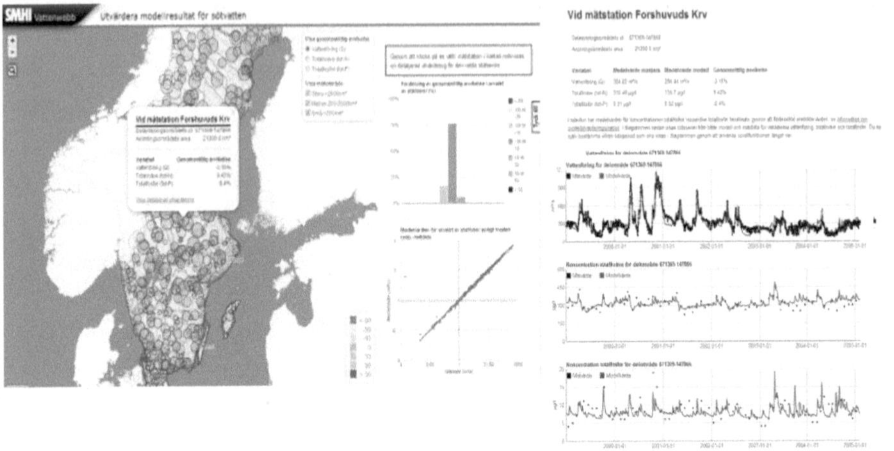

Fig. 4. This figure shows a sample view of the interactive evaluation tool. The map to the left gives an evaluation of the model performance compared to measured data. The user can select whether he is interested in discharge, nitrogen or phosphorous, it is also possible to focus on a specific size of basin or a geographical area by zooming in the map. To the left we show the more detailed information available for each station, where the user can compare details of modeled and measured time series.

5 Related Work

There are many available sites that present water related information. Several countries, such as Germany [7], United Kingdom [8], Belgium [9], Australia [10] and the United States [11] have publicly available water information services. These sites publish information on measurements and statistics. The Australian site is of particular interest since it publishes measurements and warnings together with data on water and water quality as yearly means. The American site [11] and the CUASHI service [12] are designed as portals where time series data is available for download. At the CUASHI site, time series is available in WaterML, together with a client program that can be downloaded by the user for further analysis of the data.

None of the above sites combines measured and modeled data in the same way as *vattenweb.smhi.se* site does. This means that they do not have to address the issues of transparency and validity of model data and how to make this information transparent for the end user. Such questions have been addresses by researchers working with water related environmental actions and how to involve stakeholders in the process. In this area there are a lot of interesting work in how to present water related information for stake holders and how to build tools that actively involve users in the decision process, e.g. [13], [14], [15]. These tools concentrate on a particular kind of problem and based on this problem try methods for decision support for this question.

The tools and methods used by these researchers are very interesting for us when developing *vattenweb.smhi.se*. However, as our site provide measurements and model data for a relatively general purpose, our focus is a bit different. For many problems addressed today within stakeholder research the focus is more on supporting the decision making process for a specific problem than providing simulated data and explanations for a more general use.

6 Conclusion and Ongoing Work

The new features for making the data more transparent have been appreciated by the end users and contributed to their understanding on how to use and interpret model results. The site is constantly getting new users and the user satisfaction is shown by many positive comments on the site. The users are very active in contributing with feedback for further development and new suggested features. During spring 2013 we have been working on further development of the website to allow interactive analysis of the impact of actions to improve water status. The interactive scenario tool is designed to support decisions on measures to reduce emissions from, e.g., industries, treatment plants and sewage systems. This is an important step towards the aim of using the site as a basis for planning actions to achieve an improved water status in Sweden. With current methods these kinds of assessments are done by changing input data to the model and then make a new simulation, which require high processing resources and result in long turnaround times for the end users. This old approach has made it hard to evaluate and compare a large number of combinations with possible actions, why a more interactive tool is very valuable for the end users.

Acknowledgments. The development of the S-Hype model and the web site has been funded by the Swedish water authorities. The development is also dependent on funding from Swedish Research Council (Vetenskapsrådet), The Swedish Environmental Protection Agency, The Swedish Research Council Formas and the European Commission, for model development and Scientific aspects. We are also grateful to our web team, in particular, Jens Gustafsson, Tomas Petersson and Anders Årstrand that has contributed largely to the features of the site.

References

1. SMHI Vattenwebb, `http://www.vattenwebb.se`
2. Arheimer, B., Dahné, J., Lindström, G., Marklund, L., Strömqvist, J.: Multi-variable evaluation of an integrated model system covering Sweden (S-HYPE). IAHS Publ. 345, 145–150 (2011)
3. Arheimer, B., Lindström, G., Pers, C., Rosberg, J., Strömqvist, J.: Development and test of a new Swedish water quality model for small-scale and large-scale applications. XXV Nordic Hydrological Conference, Reykjavik, August 11-13, 2008. NHP Report No. 50, pp. 483-492 (2008)
4. Lindström, G., Pers, C.P., Rosberg, R., Strömqvist, J., Arheimer, B.: Development and test of the HYPE (Hydrological Predictions for the Environment) model – A water quality model for different spatial scales. Hydrology Research 41(3-4), 295–319 (2010)

5. Strömqvist, J., Arheimer, B., Dahné, J., Donnelly, C., Lindström, G.: Water and nutrient predictions in ungauged basins – Set-up and evaluation of a model at the national scale. Hydrological Sciences Journal 57(2), 229–247 (2012)
6. Swedish water archive (SVAR), http://www.smhi.se/klimatdata/hydrologi/sjoar-och-vattendrag/ladda-ner-data-fran-svenskt-vattenarkiv-1.20127
7. Hochwasserportal Länder- Übergreifendes, http://www.hochwasserzentralen.de/
8. National river flood archive, Centre for Ecology & hydrology, http://www.ceh.ac.uk/data/nrfa/index.html
9. WEISS Planning support system, http://weiss.vmm.be/
10. Australian Government, Bureau of Meteorology, http://www.bom.gov.au/water/
11. USGS Water Resources for the United States, http://water.usgs.gov/
12. CUASHI HIS, http://his.cuahsi.org/
13. Brugnach, M., Dewulf, A., Henricksen, H.J., van der Keur, P.: More is not always better: Coping with ambiguity in natural resources management. Journal of Environmental Management 92(1), 78–84 (2011)
14. Carr, G., Loucks, D.P., Bloeschl, G.: An Analysis of Public Participation in the Lake Ontario – St. Lawrence River Study. In: Gover, V.I., Kratzenberg, G. (eds.) Water Co-mangement, ch. 4, CRC Press, Taylor and Francis Group, Boca Raton, USA (2013)
15. Jonsson, A.C., Andersson, L., Alkan Olsson, J., Jonsson, M.: Defining goals in participatory water management: merging local visions and expert judgements. Journal of Environmental Planning and Management 54(6) (2011), doi:10.1080/09640568.2010.541738

Pedestrian Localization in Closed Environments
Android Prototype

Jonáš Ševčík

Masaryk University, Faculty of Informatics, Brno, Czech Republic
jonas.sevcik@mail.muni.cz

Abstract. This research presents techniques suitable for pedestrian localization in closed environments using mobile devices without the need of GPS technology. The objective of this research is to design and implement a pedestrian localization system, which can be used directly without investments into building a support infrastructure and acquiring expensive devices. The research problem is that GPS signal is weak or absent in closed spaces, thus cannot be used to identify location. Several technologies, which are using mobile sensors, are used as part of the experimental methodology to implement the system. These include tracking of wireless networks, dead reckoning, step detection, and barcode scanning. These technologies were combined and coded in the Java programming language to form the localization system. Beside the technologies mentioned above, crowdsourcing is used for gathering environment data needed for calculation of location estimates. Currently, the implementation has been done for the Android platform, but it is designed to be universal, and can be expanded to other mobile platforms. Preliminary results of the prototype application report a positioning error (standard deviation) of roughly 2 meters.

Keywords: Indoor localization, Android, prototype, Wi-Fi tracking, step detection, dead reckoning, SMC filtering.

1 Introduction

Navigation throughout vast complex buildings e.g. hospitals or school buildings can be difficult even if the building is densely equipped with navigation signs. For acquiring location coordinates, the GPS can be used. This technique is commonly used in transportation and nowadays also in mobile phones to calculate coordinates on a virtual map. Even though, GPS tracking can be useful outdoors, it cannot be used indoors, because there is no direct visibility to GPS satellites [5].

The indoor environment positioning in comparison to outdoor positioning is problematic also for the reason that for meaningful localization it requires more accurate measurements. Where standard deviation of 10 meters may be sufficient outdoors, it may be insufficient indoors, due to the complexity of inner structure of indoor spaces.

In contrast to outdoor spaces, indoor environments offer systems other than GPS. Public buildings tend to have dense installations of WLAN infrastructure, which can be used for positioning.

J. Hřebíček et al. (Eds.): ISESS 2013, IFIP AICT 413, pp. 679–687, 2013.

Additionally, current mobile phones are not limited to perform localization only via GPS. They are equipped with various sensors such as magnetometer, linear accelerometer, gyroscope, and wireless network adapters. Even a low-end smartphone contains at least a magnetometer and an accelerometer. Therefore, they can be used as a sensor and computational devices for a localization system.

In order to achieve truly independent solution, we believe it must be as universal as possible. Therefore it is necessary to use only already present infrastructure (i.e. WLAN) and devices used for measurements being standard Android powered mobile phones. Even though we used Android devices, all the presented techniques are generally applicable to any mobile platform.

In this paper we will present an approach that combines an existing WLAN infrastructure and mobile device, which will be used as a pedometer, compass and computational device. We address, that using a mobile device in this way causes increased battery drain, and therefore there must be performed code optimizations for sensor handling, which are not the case of this study. The goal is to obtain and process all the sensor data locally. Main localization technique used is inertial navigation system. Mobile sensor fusion is used to provide position estimate, which is combined with WLAN positioning [1] and particle filtering [3] used to maintain a long term accuracy of this system. We enhance techniques presented in [4], where authors suggest a solution based on accelerometers mounted on shoe, which is only partially applicable to our approach, where we assume the user is holding the device in hand, thus we have to apply algorithms for filtering noise and false data generated by hand movement.

This work presents WLAN positioning and principles of gathering required data, sensor fusion and its application in dead reckoning, and particle filtering. Finally these techniques are combined into an Android pedestrian localization prototype. Subsequently, the prototype was tested for the positioning accuracy against ground locations. Achieved results were as accurate as those in [4], except our solution does not require additional hardware and is generally usable. The resulting system was shown to provide sufficient accuracy to locate user within a building. The paper concludes with recommendations for extensions and directions of further work.

2 Wi-Fi Localization

The first technique is indoor tracking based on wireless networks. We decided not to use standard triangulation based on access point (AP) locations and signal regression, as instead, we use Received Signal Strength (RSS) fingerprinting, as presented in [1] to create a database of unified AP identifiers and their received signal strength, which is mapped to location coordinates. RSS measurements at an unknown location are then compared to the signal strength maps to estimate the receiver's location. This method proves to be useful, when the information about location of APs is not known.

RSS fingerprinting consist of two phases. Offline phase consists of fingerprint gathering. During this phase the database of RSS fingerprints is created. Fingerprint is symbolized as a pair (x, y), where x are a coordinates in the real environment; y is a unique access point identifier and the strength of received signal measured in dBm.

During the online phase, the measured signal strength along with AP identifier are compared against the database of gathered fingerprints. The best matching candidate is chosen as an estimated location provider.

The phase of gathering fingerprints can be simplified as shown in [2]. This method consists of gathering a smaller amount of fingerprints and interpolating the gathered data to obtain the full map image. Measured map is triangulated using Delaunay algorithm.

The triangulation algorithm splits the area into non-overlapping triangle mesh, where each location in the plane is as-signed three unique vertices of the triangle. Full linear interpolation of the map can be recreated from the Barycentric coordinates of the location in the triangle.

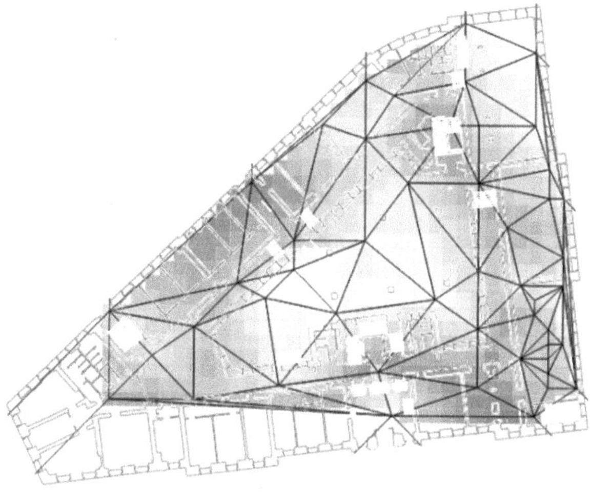

Fig. 1. Interpolated RSS map with Delaunay triangulation

Fingerprint gathering in offline phase is time consuming activity, because it must be done accurately and time of taking one fingerprint should be, from our testing, at least 20 seconds. This amount of time is optimal for collecting enough data. To shorten total sampling time, we implemented Wi-Fi fingerprinting as a crowdsourcing activity. Users are encouraged to perform measurements and upload them to our online service, from where they can be distributed into other client applications. Users are for their work awarded with more precise positioning capabilities.

3 Dead Reckoning

The second used technique is dead reckoning – a process of calculating current position by using a previously determined coordinates, which are advanced by known speed and course. We use dead reckoning in inertial navigation system, where initial location is determined via WLAN fingerprinting. Alternatively, coordinates can be encoded into a barcode, which when placed in to the position of those coordinates, can be scanned by a camera and then decoded. Position tracking is done by calculating number of steps. Stride length and its deviation are calculated by letting user walk fixed distance 5 times as presented in [2]. Course of steps is determined by a compass.

Sensor data from mobile device cannot be used as they are to provide reliable measurements. We use magnetometer to obtain physical position of a device; gyroscope to measure rotational forces along device's three axes; and accelerometer to measure acceleration of the device. Unfortunately, all those sensors have their flaws. Magnetometer can be easily affected by external electro-magnetic and magnetic fields. If the magnetometer is not parallel to the Earth's surface, the z component (pointing downward) of the Earth's magnetic vector is projected into the x and y axes of the mobile device's magnetic sensor causing the compass to rotate when the device is tilted. Gyroscope measurements tend to cumulate errors as the device moves, and accelerometer measurements are influenced by Earth's gravity. However, all those flaws can be suppressed by fusion of all above mentioned sensors as described in [5]. Accelerometer-gyroscope fusion provides gravity and linear acceleration. Compass-gravity fusion provides compass tilt compensation. The gravity is already a derived measurement coming from the accelerometer-gyroscope fusion. Compass-gyroscope fusion provides orientation and external magnetic field.

Even though linear acceleration data do not contain gravity, they still need to be filtered to be used for further calculations. For filtering we use moving averaging. We use 2 pass moving average filer with window size of 0.2. This low-pass filtering smooths the signal and subsequently we calculate moving average with window size of 0.5. When it intersects with rising edge of the signal, we state the user made a step forward (we assume user does not walk backwards, and strafe left and right).

Fig. 2 shows signal filtering and step detection. Red line (bottom) corresponds to raw linear acceleration data measurement. Green line corresponds to filtered signal. Blue line represents moving average. Green dots (first 10) are positive steps. Red dots are false positive steps. To filter white noise and slight movements of hand we use threshold approach to discard any steps which might have been detected. We calculate accumulative signal power (ASP) and only those values are marked as positive, which ASP is higher than threshold value. Also, while continuously walking, elapsed time between 2 following steps cannot exceed 2 seconds.

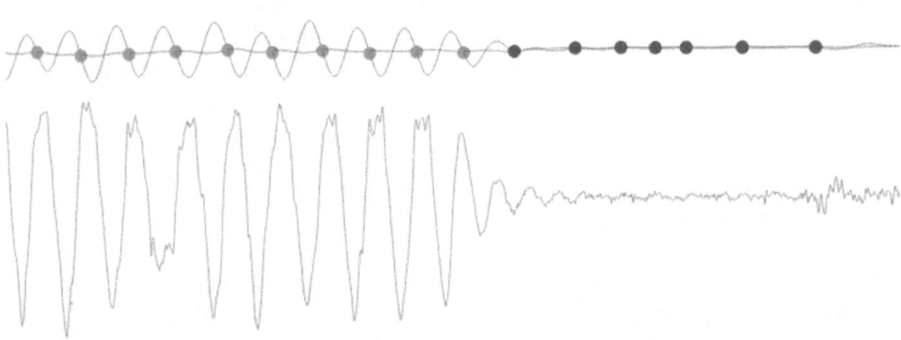

Fig. 2. Signal filtering

4 Particle Filtering

The third technique is Sequential Monte Carlo filtering (SMC) [3]. It is modeling the state of a dynamic system by approximating the posterior density function by a set of random samples of the state vector while sequences of noisy measurements are made on the system. The computation of the SMC requires two models [3]:

1. System model – model describing the evolution of the state and time;
2. Measurement model – model relating the noisy measurements to the state.

Filter processes events from the step detector (step event, length estimate) and a probability density function from the Wi-Fi AP scans, which uses particles evenly spread in the probable location determined by RSS fingerprint database. These particles are set to motion with events generated by step detection. Consequently, those particles, which hypothetical motion leads through impassable obstacles, e.g. walls, are eliminated (Fig. 3). This results in improvement of location estimation.

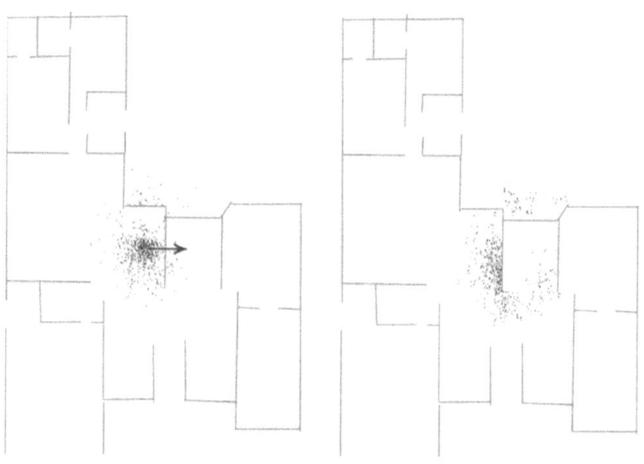

Fig. 3. Particle elimination

5 Implementation and Evaluation

5.1 Test Environment and Device

We chose hallway of university building, which was covered with a total of 6 APs used for public internet access. There were also present several devices transmitting Wi-Fi signal, but these were discarded from the measured sample due to the possible loss of accuracy. Prior to the test, we marked 11 locations such that distance between 2 adjacent points was 2.5 meters. The exact coordinates of marked points were placed into the system. Initial location was set programmatically (top-right circle in Fig. 4). We performed a circular walk taking 4 laps around selected markers (movement was

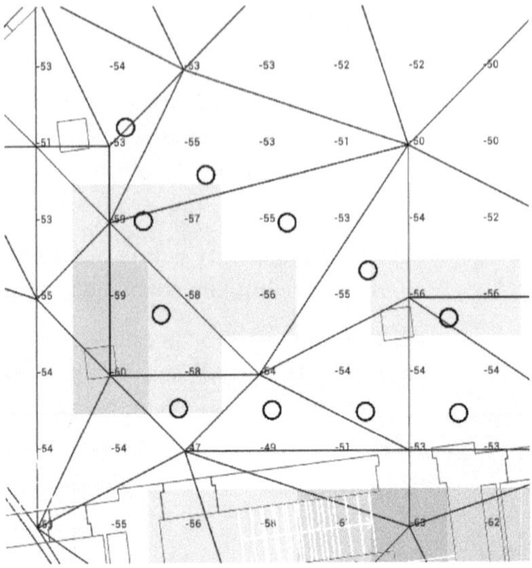

Fig. 4. Test route

done counterclockwise). After reaching a marked physical point, we took a sample, which coordinates were compared to the real coordinates of the marked point and distance between these points was calculated. As a testing device we used Asus Nexus 7 tablet running Android 4.2.2 JellyBean. During testing the device was held in constant height.

5.2 Implementation

The implementation of the positioning system was realized in three separate components: RSS position estimation provider, pedometer, and particle filter. These are encapsulated in an application service.

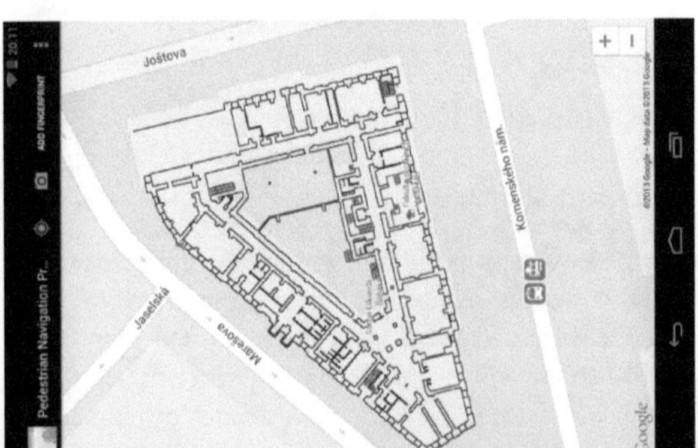

Fig. 5. Building overlay over Google Maps

This service merges all components together to supply location data for UI component. All the code was implemented in Java using Android API v8 and compiled using API v17 to assure the best possible performance. Google Maps API with custom overlay was used for the visual presentation as seen in Fig. 5.

5.3 Evaluation

Evaluation aim was to determine accuracy of the standalone WLAN location estimation, determine the accuracy of inertial navigation, and finally to compare both mentioned approaches.

To evaluate the gathered data, we compared user's position obtained from the fusion of WLAN fingerprinting and dead reckoning to location estimate obtained only via WLAN fingerprinting. All the data samples for both approaches were taken simultaneously at each marked location. Fig. 6 shows each of 44 recorded measurements. WLAN estimate data are presented by the red line (top), fused location data are presented by blue line (bottom). Black vertical lines show start of the new lap.

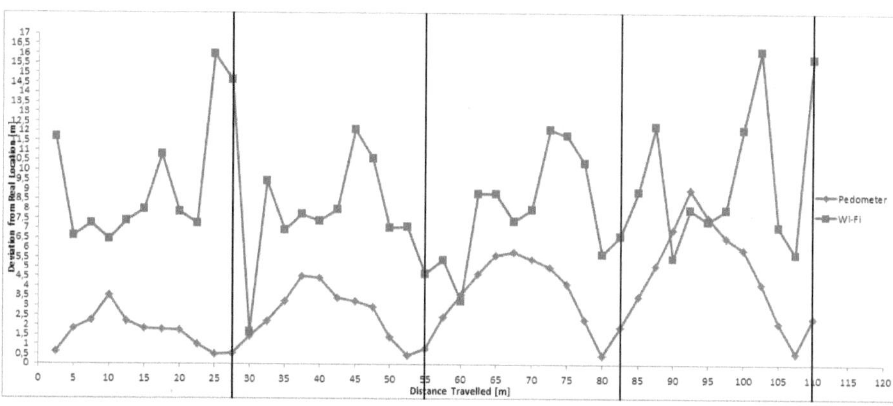

Fig. 6. Position error over walked distance

Arithmetic mean for WLAN fingerprinting is 8.6557 m, for fused location is 3.2023 m. Standard deviations are 3.1770 m, and 2.0911 m respectively. As seen from the graph, fused solution results into more stable and less erroneous location estimation than WLAN fingerprinting. Measured data show regular tendency in respect to lap location. Nevertheless, it is visible, that the error cumulates with travelled distance.

6 Conclusion

In this paper we have presented indoor localization system for Android devices combining WLAN fingerprinting with pedometer made as a fusion of gyroscope, accelerometer, and magnetometer. Presented solution is platform independent and requires no additional changes to existing Wi-Fi infrastructure. The only a priori requirements are map materials and WLAN fingerprint gathering. The accuracy of the combined system was quantitatively evaluated in a real building and shows that it is much higher than that of the fingerprinting alone.

Designed solution proved to be efficient enough to be used in crisis management. In case of a fire emergency it is crucial to provide escape strategies for effective emergency management. Map data can provide user with accessible information about water hydrants, fire extinguishers, and possible escape routes/emergency exits. The system can be easily extended to provide visual location of particular users, therefore it may be possible to show also positions of people connected to the system. This functionality can be used in 2 scenarios.

The first possible scenario presents a situation when a group of people is trapped inside a building due to the fire, which has closed all possible escape routes. Trapped people are unable to escape. Therefore a rescue team must strike in for the rescue. With indoor localization system it would be possible to make crisis managers aware of trapped people's positions inside the building, thus keeping count of endangered victims and rescuing them easily.

Second possible scenario represents a fire brigade fighting a fire in a vast building. It may be difficult for the rescue team to track positions of its members. Every individual can monitor other team members' position and proceed efficiently. Of course this functionality can be combined with all above mentioned features.

6.1 Future Work

Further work should focus on estimating stride length using a neural network. Constant stride length results into cumulative error increase. Also, current solution filters false positive steps, caused by hand movement, by threshold approach. This should be substituted by focusing on the frequency of steps and acceleration growth. Sustained walking can be recognized by periodical frequency and even pitches in acceleration growth. Recognizing left leg from the right may also provide interesting results, when treated separately. Last but not least, future work should focus on extending recognizing changes altitude.

Acknowledgments. The coding of the base of the Android prototype was made by Michal Holčík. Further enhancements and map convertor were implemented by Adam Berthóty. All map data and their partial format conversion was kindly provided by Masaryk University Department of Passportisation.

References

1. Chen, A., Harko, C., Lambert, D., Whiting, P.: An Algorithm for Fast, Model-Free Tracking Indoors. ACM SIGMOBILE Mobile Computing and Communications Review 11(3), 48–58 (2007)
2. Cho, S.Y., Park, C.G.: MEMS Based Pedestrian Navigation System. Journal of Navigation 59, 135–153 (2006)

3. Arulampalam, M.S., Maskell, S., Gordon, N.: A tutorial on particle filters for online nonlinear/non-Gaussian Bayesian tracking. IEEE Transactions on Signal Processing 50(2), 174–188 (2002)
4. Frank, K., Krach, B., Catterall, N., Robertson, P.: Development and Evaluation of a Combined WLAN & Inertial Indoor Pedestrian Positioning System. In: 4th International Symposium on Location and Context Awareness, pp. 538–546 (2001)
5. Feliz, R., Zalama, E., Bermejo, J.G.G.: Pedestrian tracking using inertial sensors. Journal of Physical Agents 3(1), 35–43 (2009)
6. Woodman, O., Harle, R.: Pedestrian localisation for indoor environments. In: Proceedings of the 10th International Conference on Ubiquitous Computing, pp. 114–123. ACM (2008)

Environmental Infrastructures and Platforms with Citizens Observatories and Linked Open Data

Arne J. Berre[1], Sven Schade[2], and Dumitru Roman[1]

[1] SINTEF, Norway
{Arne.J.Berre,Dumitru.Roman}@sintef.no
[2] European Environment Agency, Denmark
Sven.Schade@eea.europa.eu

Abstract. A number of past and current research and development projects aim to improve the sharing and use of environmental information. In 2010, the Environmental Infrastructures and Platforms (ENVIP) initiative was introduced as a means to identify the European potentials and specify common building blocks ('services and enablers') related to these projects. Work began with a set of projects supporting the Shared Environmental Information System (SEIS). This paper briefly summarizes the ongoing result collection and introduces the next wave of activities, which will follow two newly emerging trends: Citizens Observatories and Linked Open Data. We invite interested parties and project consortia to provide their contributions for future analysis and synergies through the CEN/TC287 TR 15449-2 Best practices registry and the ENVIP initiative.

Keywords: Environmental, Infrastructure, Platform, Architecture.

1 Introduction

In the context of environmental systems, Information and Communication Technologies (ICT) have an essential role to play as they provide the necessary support in terms of tools, systems and protocols to establish a dynamic environmental space of collaboration in a more and more sophisticated digital world. Core challenges are not only related to providing seamless environmental data access to public authorities, businesses and the public at large, but also to allowing for interoperable environmental services based on Web technologies, and stimulating innovation and growth.

The Shared Environmental Information System (SEIS) [1] is one of three major initiatives along with the INSPIRE Directive [2] and the Global Monitoring for Environment and Security (Copernicus) [3] undertaken by Europe to collect and share environmental information for the benefit of the global society. Different efforts emerged towards the creation of infrastructures and platforms for Environmental Information Systems and Services – including a rich set of European research and development projects. Spatial data infrastructure (SDI) is a general term for the computerised environment for handling data that relates to a position on or near the surface of the earth. It may be defined in a range of ways, in different circumstances, from the local up to the global level.

J. Hřebíček et al. (Eds.): ISESS 2013, IFIP AICT 413, pp. 688–696, 2013.

In this context, the Environmental Infrastructures and Platforms (ENVIP) initiative [4-7] was introduced as a means to identify the European potentials, specify common services and building blocks ('enablers') and their relations, and provide a framework to sustain the results of research and development projects that are usually funded for up to four years.

This position paper summarizes initial findings (Section 2), shows the recent structure of CEN/TC287 TR 15449-2 (Section 3), and points to upcoming areas with a new set of projects (Section 4). We conclude with an open invitation for further contributions (Section 5).

2 ENVIP Analysis of Past Research and Development Projects

A number of SEIS contributing projects have recently been completed. Some of the SEIS related projects under the environment topic have been investigated in previous ENVIP events, in particular at ENVIP'2010 [5-6] and at the ENVIP session during ISESS'2011 [7]. Table 1 shows the results from ENVIP'2010.

Table 1. ENVIP reference architecture contributions from ENVIP'2010

Project	Ontologies/domain models	(Service) discovery	(Service) composition	Sensors access / streams	(Web) visualisation	Multilingualism	Transformation/mapping	Standards contributions	Security
TATOO	X	X			X	X			
PESCADO		X	X		X				
UncertWeb		X	X					X	
SUDPLAN			X		X				
HYDROSYS				X	X				
NETMAR	X	X	X		X	X	X		
OEPI	X	X	X				X		
GENESI-DEC	X	X	X		X				X
LARKC				X				X	
ENVISION	X	X	X	X	X	X	X	X	
DIADEM	X		X		X		X		

The ENVIP'2010 workshop analysed a number of projects that were presented (Tatoo [8], PESCADO [9], UncertWeb [10], SUDPLAN [11], HYDROSYS [12], NETMAR [13], OEPI [14], GENESI-DEC [15], LARKC [16], ENVISION [17] and DIADEM [18], with respect to the their use and contributions of technologies to the areas of ontology models, service discovery and composition, sensor usage, visualisation, multi language support, transformation/mapping technologies, planned standards contributions and security.

Table 2. ENVIP reference architecture contributions from ISESS'2011

Project	Reference Architectures	Semantic Annotation, Ontologies & Publication	Resource Discovery	(Data, Service and Model) Composition	Sensors and Observation Web	(Web) visualization and Portals	Context and Multilingualism	Security, Trust, Privacy
TATOO	X	X	X			X		
PESCADO	X		X	X		X		
SUDPLAN	X			X		X		
data.reglee. Info	X	X	X					
SA-DEM	X							
GENESIS	X	X	X	X		X		X
GS-SOIL	X	X					X	
ENVISION	X	X	X	X	X	X	X	
DIADEM	X	X		X		X		
CryoLand	X			X				
RBIS	X			X				
ENVIROFI	X		X	X	X	X	X	X

The ENVIP session at ISESS'2011 [7] adopted some of the analysis structure from ENVIP'2010 by starting the creation of a reference architecture, extending to broader resource discovery, and composition. At ISESS'2011 some of the projects from ENVIP'2010 presented their further progress, and in addition a number of new projects were presented, in particular data.reglee [19], SA-DEM [20], GENESIS [21], GS-SOIL [22], CryoLand [23], RBIS [24] and ENVIROFI [25-26].

The ENVIROFI project has also extended work on the reference model and on the further identification of specific environmental enablers and generic enablers as reported in the paper [27] at ISESS'2013.

A next step is now the more detailed analysis of the actual results from the various projects, where the CEN/TC287 TR 154449-2 Best practices foundation for this is presented in the next section.

3 A Structure for SDI Best Practices – CEN/TC287 TR 15449-2

A recent activity in CEN/TC287 [33] has aimed at creating a foundation for the development of SDIs, including a collection of best practices and results from recent SDI related projects and activities, through the creation of the technical report TR 15449 Spatial Data Infrastructures, with five different parts.

The intended readership of this Technical Report are those people who are responsible for creating frameworks for SDIs, experts contributing to INSPIRE, experts in information and communication technologies and e-government that need to familiarize themselves with geographic information and SDI concepts, and standards developers and writers.

The CEN/TR 15449 parts on Geographic Information – Spatial data infrastructures consists of five parts: part 1: Reference model – providing a foundation for SDI and the other parts, part 2: Best practices – collecting results from recent European projects, part 3: Data centric view – providing guidelines on data specifications for an SDI, part 4: Service centric view – providing guidelines on service specifications for an SDI, and part 5: Validation and Testing – supporting validation and testing for the data and services of an SDI.

Part 2 "Best practices" is aimed at collection of best practices results from recent and ongoing projects in particular for projects in the frame of the European Union funding programmes. Experiences from ENVIP were used as input to the creation of the TR 15449-1 Reference model, and the corresponding best practices structure in TR 15449-2. This part of the technical report provides best practices regarding SDIs, referencing to the outcomes of the projects in the frame of the European Union funding programmes. It summarizes the deliverables of projects, structured according to the reference model defined in part 1, to be made available in an on-line repository where the relevant outcomes are collected and classified in order to provide a structured sets of recommendations for implementing SDIs at the European, national and sub-national level. The outcomes delivered by these relevant practices are collected into a document registry available through the CEN/TC287 web site [33].

The following table shows the structure for capturing various elements of project results related to various dimensions, such as the provisioning of SDI reference model components, the support for the various phases of an SDI, and the services from the architectural service reference model, (derived from the ISO/TC211 ISO 19119 service taxonomy).

Table 3. TR 15449-2 template for recording best practices project results

Project	Name	
Relevant classification	SDI reference model components	☐ Data ☐ Register ☐ Discovery ☐ View ☐ Invoke ☐ Download ☐ GeoRM ☐ Orchestration and Composition
	Phases of an SDI	☐ Concept and design ☐ Implementation ☐ Validation
	Architectural reference model services	☐ Human Interaction Services ☐ Model Management Services ☐ Workflow/Task Services ☐ System Management Services ☐ Processing Services ☐ Communication Services
	Project document type	☐ Standards ☐ Specifications ☐ Technical reports ☐ Guidelines
	Free Keywords	
Relevant outcomes		

Finalised and ongoing projects and activities are now encouraged to make their relevant deliverables and results available through the CEN/TC287 Best practices project registry. Information about how to do this can be found on the CEN/TC287 website [33].

4 ENVIP Evolution

A number of the initial projects and activities that contributed to ENVIP have now finished, while new projects are emerging, and the next step is to analyse which of the results that should be captured as best practices to be used by other projects and activities in the future, and potentially also fed into future standardisation activities. We are now encouraging these projects to provide their best practices input to the CEN/TC287 best practices collection activity and to support further elaboration of the usability of results through the ENVIP community.

There are two major trends that also will influence the future evolution of ENVIP. Firstly, user contributed content and citizen science with crowd sourcing and volunteered geographic information (VGI) has gained major attention over the last few years. Secondly, the request for 'raw data now' and for the combination of data from various sources, has resulted in an immense interest and activity in Linked Open Data.

The remainder of this section describes these areas and identifies some on-going projects of interest. It is the intention to update and complement the services and enablers that now are being collected from the last four years of SEIS related projects, in order to complete the picture of emerging environmental infrastructures and platforms and to make diverse research results better accessible and comparable.

4.1 The Raise of Citizens' Observatories

The European Commission recently funded a total of five Citizens' Observatories Projects under the topic "Developing community-based environmental monitoring and information systems, using innovative and novel earth observation applications". All are tasked to develop highly innovative technologies, while each project addresses a particular domain and selected locations in and beyond Europe and connections with international standardization organizations, such as CEN/TC287, ISO/TC211, OGC, INSPIRE and GEOSS are foreseen.

The following five projects received funding by the European FP7 for Research, under the Environment theme of the Cooperation Specific Programme: CITI-SENSE Development of Sensor-based Citizens' Observatory Community for Improving Quality of Life in Cities [34], WeSenseIt Citizen Observatory of Water [35], COBWEB Citizen Observatory Web [36], Citclops Citizens' Observatory for Coast and Ocean Optical Monitoring [37], OMNISCIENTIS Odour Monitoring and Information System based on Citizen and Technology Innovative Sensors [38]. These activities are complemented by two COST actions. “Mapping and the citizen sensor” [39]: Which focuses on the rise of citizen sensors e.g. volunteers contributing information from remote devices, and its potential to radically change mapping, and “ENERGIC” [40] a European Network Exploring Research into Geospatial Information Crowdsourcing: software and methodologies for harnessing geographic information from the crowd.

The five projects of the environment topic, together with the two COST actions, will design, develop and test a new environmental information system founded on the concept of the citizens’ observatories that will help to empower citizens allowing them to influence in the environmental governance processes, as well provide models for decision-makers that facilitate connections between environmental governance, global policy objectives and citizens’ needs.

It is assumed that these projects can take advantage of previous developed environmental infrastructures and platforms, as well as provide further input to these, in particular related to the areas of mobile computing, offline support and user interaction.

4.2 The Use of Linked Open Data

The European Commission recently funded a number of projects related to the use of Linked Open data (LOD) in combination with geospatial and environmental information, and also a set of upcoming projects focused on publication and consumption of open data. LOD is becoming a source of unprecedented visibility for environmental data that is expected to enable the generation of new businesses as well as a significant advance for research in the environmental area.

The following projects has received funding by the European FP7 for Research, under the Environment theme of the Cooperation Specific Programme: GeoKnow - Geospatial Data and the Semantic Web [41], Linked Geo Data [42] and SmartOpenData [43]. Such projects will provide advances in publication and consumption of existing environmental data, with a focus on how LOD technologies can be applied generally to spatial data resource and specifically to public open data portals, GEOSS Data-CORE, GMES, INSPIRE and voluntary data (OpenStreetMap, GEP-WIKI, etc.), and how it can impact on the economic and sustainability progress in environment research.

Addressing environmental geospatial data management based on LOD will require harmonization of geospatial metadata (ISO19115/19119 based) with principles of Semantic Web, provisioning of spatial data fusion introducing principles of Open Linked Data, improvement of spatial data visualisation of Geospatial Open Linked Data and publication of the resulting information according to user requirements and Linked Open Data principles to provide new opportunities for research and business.

Wide applicability of LOD in the environmental geospatial domain will have to be complemented by adoption of new standards covering alignment and compatibility of semantic technologies, geospatial data interoperability, new RDF-based vocabularies for integrated location description and Open Data Quality, Certification and Reputation. This work will require semantic harmonisation of different datasets of INSPIRE -INSPIRE compliance doesn't guarantee harmonisation-, GEOSS and GMES and should be consistent with the ISO/TC 211 standards.

5 Conclusions and Outlook

A broader initiative has been started for the further identification of services and architectural components through the ENVIP community [6] and the CEN/TC287 TR 15449-2 Best practices [12].

The objective is that recently finished and ongoing projects and initiatives around SDI can contribute with their results and best practices to the various architectural elements of an SDI, including the experiences from the use of various geospatial and ICT based standards. With an expansion in new projects for citizen observatories and Linked Open Data it is seen that these projects can benefit from the best practices of the recent activities, as well as contribute with new potentially shared services. This will also serve as a foundation for change requests for existing standards and for potentially new standard through the collaboration with CEN/TC287, OGC, ISO/TC211, GEOSS and INSPIRE.

For an optimal evolution from past to present projects we encourage project consortia for these projects, as well as other organizations working on related topics, to contribute their relevant results to the CEN/TC287 TR 15449-2 Best practices registry and to join the ENVIP initiative at www.envip.eu for the further analysis and synergies of the results. The ENVIP'2013 workshop session during ISESS'2013 aims a further refinement of the ENVIP reference architecture and also continued the reporting and analysis of emerging project results, including reports from emerging projects in the areas of Citizens observatories and Linked Open Environmental data.

Acknowledgments. We thank the initial contributors to the ENVIP community for their foundation for the work, and for the partners in the supporting projects ENVISION (FP7- 249120), ENVIROFI (FP7 – 284898) and CITI-SENSE (FP7-308524) for their further input. This position paper is based on the ENVIP community activities, extended from previous work in various European research projects and also from standardisation work in the context of ISO/TC211, CEN/TC287, OGC and OMG.

References

1. Commission of the European Communities: Communication COM(2008)0046 - Towards a Shared Environmental Information System (SEIS). 2008/0046 (2008)
2. The European Parliament and of the Council: Directive 2007/2/EC of the European Parliament and of the Council of 14 March 2007 establishing an Infrastructure for Spatial Information in the European Community (INSPIRE). Official Journal on the European Parliament and of the Council (2007)
3. Commission of the European Communities: Communication COM, – Copernicus, was GMES (2011), http://www.copernicus.eu/
4. ENVIP, http://www.envip.eu
5. ENVIP 2010, Proceedings of the Workshop "Environmental Information Systems and Services - Infrastructures and Platforms" at EnviroInfo (2010), http://ceur-ws.org/Vol-679
6. ENVIP 2010 online presentations, http://ifgi.uni-muenster.de/~pajoma/persistent/ENVIP10/
7. ENVIP at ISESS 2011, http://www.isess2011.org
8. TaToo, Tagging Tool based on a Semantic Discovery Framework, http://www.tatoo-fp7.eu/tatooweb
9. PESCADO, Personalized Environmental Service Configuration and Delivery Orchestration, http://www.pescado-project.eu
10. UncertWeb, The Uncertainty Enabled Model Web, http://www.uncertweb.org/
11. SUDPLAN, Sustainable Urban Development Planner for Climate Change Adaptation, http://www.smhi.se/sudplan
12. HYDROSYS, http://www.hydrosysonline.eu/
13. NETMAR, Open service network for marine environmental data, http://netmar.nersc.no/
14. OEPI, Exploring and Monitoring Any Organisation's Environmental Performance Indicators, http://www.oepi-project.eu/
15. GENESI-DEC, http://www.genesi-dec.eu
16. LARKC, The Large Knowledge Collider, http://www.larkc.eu/
17. ENVISION, ENVIronmental Services Infrastructure with Ontologies, http://envision-project.eu
18. DIADEM, Distributed information acquisition and decision-making for environmental management, http://www.ist-diadem.eu/

19. data.reglee, http://data.reegle.info/
20. SA-DEM – from ISESS 2011, http://www.isess2011.org/
21. GENESIS, GENeric European Sustainable Information Space for environment, http://www.genesis-fp7.eu
22. GS-SOIL, Geodata-services for European Soil Data, http://www.gssoil-portal.eu
23. CryoLand, The GMES Service Snow and Land Ice, http://www.cryoland.eu/
24. RBIS - An Environmental Information System for Integrated Landscape Management, http://www.geoinf.uni-jena.de/
25. ENVIROFI - The Environmental Observation Web and its Service Applications within the Future Internet, http://www.envirofi.eu/
26. Schade, S., Fogarty, B., Kobernus, M., Schleidt, K., Gaughan, P., Mazzetti, P., Berre, A.-J.: Environmental Information Systems on the Internet: A Need for Change. In: Hřebíček, J., Schimak, G., Denzer, R. (eds.) ISESS 2011. IFIP AICT, vol. 359, pp. 144–153. Springer, Heidelberg (2011)
27. Usländer, T., Berre, A.J., Granell, C., Havlik, D., Lorenzo, J., Sabeur, Z., Modafferi, S.: The future internet enablement of the environment information space. In: Hřebíček, J., Schimak, G., Kubásek, M., Rizzoli, A. (eds.) ISESS 2013. IFIP AICT, vol. 413, pp. 111–122. Springer, Heidelberg (2013)
28. European Committee for Standardization (CEN): TR15449 Geographic information — Spatial data infrastructures — Part 1: Reference model (2011)
29. European Committee for Standardization (CEN): TR15449 Geographic information — Spatial data infrastructures Part 2: Best practices (2012)
30. European Committee for Standardization (CEN): TR15449 Geographic information — Spatial data infrastructures —Part 3: Data centric view (2012)
31. European Committee for Standardization (CEN): TR15449 Geographic information — Spatial data infrastructures —Part 4: Service centric view (2012)
32. European Committee for Standardization (CEN): TR15449 Geographic information — Spatial data infrastructures —Part 5: Validation and Testing
33. CEN/TC287 website, http://www.centc287.eu/
34. CITI-SENSE - Development of Sensor-based Citizens' Observatory Community for Improving Quality of Life in Cities, http://www.citisense.eu/
35. WeSenseIt - Citizen Observatory of Water, http://www.wesenseit.eu
36. COBWEB - Citizen Observatory Web, http://www.cobwebproject.eu
37. Citclops - Citizens' Observatory for Coast and Ocean Optical Monitoring, http://www.citclops.eu
38. OMNISCIENTIS - Odour Monitoring and Information System based on Citizen and Technology Innovative Sensors, http://www.omniscientis.eu
39. Mapping and the citizen sensor, http://www.cost.eu/domains_actions/ict/Actions/TD1202
40. ENERGIC, http://www.cost.eu/domains_actions/ict/Actions/IC1203
41. GeoKnow - Geospatial Data and the Semantic Web, http://geoknow.eu/
42. Linked Geo Data, http://linkedgeodata.org/
43. SmartOpenData, http://www.smartopendata.eu/

Author Index

Abodeely, Jared M. 121
Aitkenhead, Matthew 129
Anastasiadou, Despina 143
Athanasiadis, Ioannis N. 143

Báčová, Radka 285
Bakoš, Eduard 560, 641
Baranizadeh, Elham 98
Barredo, José I. 271
Barta, Jiří 294, 302
Bayer, Nina 196
Berre, Arne J. 109, 688
Bezáková, Zuzana 579
Bisaro, Alexander 489
Black, Helaina 129
Bonatsos, Antonios 311
Borůvková, Jana 222, 230, 253
Bosco, Claudio 11, 321, 367
Bram, Lena Eriksson 669
Bruen, Michael 262
Bryden, Kenneth M. 121
Burritt, Roger L. 500
Busetto, Lorenzo 271

Camia, Andrea 11
Caudullo, Giovanni 271
Cellina, Francesca 154
Čerba, Otakar 1
Charvát, Karel 1, 588
Charvát Jr., Karel 588
Chvátalová, Zuzana 31, 451
Cizek, Markus 164
Coull, Malcolm 129

Denzer, Ralf 403, 464
de Rigo, Daniele 11, 271, 321, 367
de Rooij, Erik 54
Dijkstra, Tom 321
Di Leo, Margherita 11, 367
Dočekalová, Marie 44
Donnelly, David 129
Düpmeier, Clemens 196
Dušek, Ladislav 204, 222, 230, 240, 253, 597
Dvořák, Antonín 391

Egly, Maria 164
Eibenová, Patrícia 525
Ernst, Martin 661

Falgenhauer, Markus 62, 164
Fiala, Petr 391
Flachberger, Christian 345
Foitik, Gerry 345
Förster, Anna 154
Furová, Lenka 641

Giacomelli, Andrea 85
Gidhagen, Lars 336, 464
Granell, Carlos 109
Gregor, Jakub 204, 222, 230, 253
Güttler, Reiner 403

Hájek, Miroslav 508
Hamed, Amar 98
Havlik, Denis 85, 109, 164
Hell, Thorsten 403, 464
Herber, Vladimír 285
Herman, Lukáš 414
Hjerdt, Niclas 669
Hodinka, Michal 630
Holoubek, Ivan 204, 230, 253
Hřebíček, Jiří 204, 230, 253, 425, 543, 597, 630, 651
Huber, Hermann 164
Hůlek, Richard 204, 222, 230, 240, 253
Huttunen, Markku 381
Hyršlová, Jaroslava 508

Jager, Bettina 345
Jakubínský, Jiří 285
Jarkovský, Jiří 204, 222, 230, 240, 253
Ježek, Jan 588

Kafka, Štěpán 1
Kalina, Jiří 222, 240, 597
Kaufmann, Alexander 23
Kiourtsis, Fotis 143
Klánová, Jana 204, 222, 230, 253
Kloyber, Christian 345

Kobernus, Mike 62
Koch, Joshua B. 121
Koch, Miloš 31, 451
Kocmanová, Alena 44
Kohlhas, Eckhard 403
Kolehmainen, Mikko 98, 357, 381, 533
Komenda, Martin 618
Komppula, Mika 98
Kotovirta, Ville 381, 533
Koulinas, Kostas 143
Kouřil, David 661
Kropp, Jürgen P. 489
Kubásek, Miroslav 177, 230, 253, 597
Kubíček, Petr 285
Kučera, Adam 608
Kutschera, Peter 164, 336

Laaksonen, Ari 98
Laurenne, Nina 85
Lewau, Per 669
Lobo, Tomas Pariente 71
Lorenzo, José 109
Luňáček, Jiří 44

Malý, Ivan 651
Melas, Panos 311
Middleton, Lee 311
Mildorf, Tomáš 1, 588
Ministr, Jan 571
Mockler, Eva 262
Modafferi, Stefano 109
Muth, David J. 121
Mužík, Jan 597

Neubauer, Georg 345
Nowak, Andrea 345

Obrová, Vladěna 517
Olsson, Jonas 336

Packham, Ian 262
Pampuri, Luca 154
Pekárková, Lucie 525
Peters-Anders, Jan 23
Petroliagkis, Thomas 11
Petronzio, Luca 23, 71
Pickl, Stefan 357
Pielorz, Jasmin 62
Pisaniello, John D. 500
Pitner, Tomáš 525, 608
Popelka, Ondřej 630
Portin, Harri 98
Prášek, Jan 425
Preda, Alexander 357
Procházková, Dana 472

Rajdlová, Gabriela 508
Reusser, Dominik 489
Řezník, Tomáš 414
Ritschelová, Iva 543
Rivola, Davide 154
Rizzoli, Andrea Emilio 154
Rodriguez-Aseretto, Dario 11, 367
Roman, Dumitru 688
Rönkkö, Mauno 98, 357, 381, 533
Rood, Ente 188
Rudel, Roman 154

Sabeur, Zoheir 109, 311
Sadovská, Veronika 302
Sander, Graham 321
San-Miguel-Ayanz, Jesús 11, 271, 367
Šauer, Petr 391
Schade, Sven 688
Schillinger, Wolfgang 196
Schimak, Gerald 71, 345
Schlachter, Thorsten 196
Schleidt, Katharina 85, 439
Schlobinski, Sascha 403, 464
Schwarz, Daniel 222, 618
Šebková, Kateřina 222, 230, 253
Šelešovský, Jan 641
Ševčík, Jonáš 679
Sidorov, Egor 543
Šikula, Tomáš 661
Skočovský, Luděk 661
Smolíková, Lenka 517
Soukopová, Jana 560, 641, 651
Srník, Albert 302
Stocker, Markus 98, 381
Strömbäck, Lena 669
Struk, Michal 651
Svobodová, Eva 285

Tingey-Holyoak, Joanne L. 500
Trenz, Oldřich 630

Urbánek, Jiří F. 302
Usländer, Thomas 109

Valta, Jiří 425
van der Schaaf, Hylke 62, 188
van Heeringen, Klaas-Jan 54
Virtanen, Annele 98
Vohnout, Přemysl 1

Wasowski, Janusz 321
Watson, Kym 62, 188
Weidemann, Rainer 196
Wrobel, Markus 489

Yurtsever, Sinan 23

FSC
www.fsc.org
MIX
Papier | Fördert
gute Waldnutzung
FSC® C083411

Kolibri 360 GmbH
Ferdinand-Jühlke-Straße 7
99095 Erfurt, Deutschland
produktsicherheit@kolibri360.de